Y0-BTA-960

1992 LECTURES IN COMPLEX SYSTEMS

1992 LECTURES IN COMPLEX SYSTEMS

Editors

Lynn Nadel
Department of Psychology
University of Arizona

Daniel Stein
Department of Physics
University of Arizona

Lecture Volume V

Santa Fe Institute
Studies in the Sciences of Complexity

Addison-Wesley Publishing Company
The Advanced Book Program
Reading, Massachusetts Menlo Park, California New York
Don Mills, Ontario Wokingham, England Amsterdam Bonn
Sydney Singapore Tokyo Madrid San Juan
Paris Seoul Milan Mexico City Taipei

Publisher: *David Goehring*
Editor in Chief: *Jack Repcheck*
Production Manager: *Michael Cirone*
Production Supervisor: *Lynne Reed*

Director of Publications, Santa Fe Institute: *Ronda K. Butler-Villa*
Publications Assistant, Santa Fe Institute: *Della L. Ulibarri*

This volume was typeset using TEXtures on a Macintosh II computer. Camera-ready **output from a Hewlett-Packard LaserJet 4M Printer.**

1 2 3 4 5 6 7 8 9 10 – MA – 96959493
First printing, November 1993

About the Santa Fe Institute

The *Santa Fe Institute* (SFI) is a multidisciplinary graduate research and teaching institution formed to nurture research on complex systems and their simpler elements. A private, independent institution, SFI was founded in 1984. Its primary concern is to focus the tools of traditional scientific disciplines and emerging new computer resources on the problems and opportunities that are involved in the multidisciplinary study of complex systems—those fundamental processes that shape almost every aspect of human life. Understanding complex systems is critical to realizing the full potential of science, and may be expected to yield enormous intellectual and practical benefits.

All titles from the *Santa Fe Institute Studies in the Sciences of Complexity* series will carry this imprint which is based on a Mimbres pottery design (circa A.D. 950–1150), drawn by Betsy Jones. The design was selected because the radiating feathers are evocative of the outreach of the Santa Fe Institute Program to many disciplines and institutions.

Contributors to This Volume

Robert H. Austin, Princeton University
Cathleen Barczys, University of California, Berkeley
Subbiah Baskaran, University of Vienna
Laura Bloom, University of California, San Diego
E. Bonabeau, CNET Lannion B-OCM/TEP
T. David Burns, George Mason University
Joshua M. Epstein, Princeton University
Igor Fedchenia, University of Umeå
Barry Feldman, State University of New York at Stony Brook
Raymond E. Goldstein, Princeton University
Charles M. Gray, The Salk Institute for Biological Studies
Tad Hogg, Xerox Palo Alto Research Center
V. Holden, The University, England
Bernardo A. Huberman, Xerox Palo Alto Research Center
A. Atlee Jackson, Beckman Institute, University of Illinois
Leslie Kay, University of California, Berkeley
Brian L. Keeley, University of California at San Deigo
Robert S. Maier, University of Arizona
Gottfried Mayer-Kress, University of Illinois
Melanie Mitchell, Santa Fe Institute
Kai Nagel, Universität zu Köln
David Noever, George C. Marshall Space Flight Center
Garry D. Peterson, University of Florida
Stefan Schaal, MIT
U. R. Smith, Yale University
Dagmar Sternad, University of Connecticut
Villy Sundstrom, University of Umeå
A. T. Winfree, University of Arizona
David H. Wolpert, Santa Fe Institute
G. Yagil, The Weizmann Institute of Science
Jonathan S. Yedidia, Harvard University
Kay-Pong Yip, University of Southern California
Henggui Zhang, The University, England

Santa Fe Institute
Studies in the Sciences of Complexity

Lectures Volumes

Vol.	Editor	Title
I	D. L. Stein	Lectures in the Sciences of Complexity, 1989
II	E. Jen	1989 Lectures in Complex Systems, 1990
III	L. Nadel & D. L. Stein	1990 Lectures in Complex Systems, 1991
IV	L. Nadel & D. L. Stein	1991 Lectures in Complex Systems, 1992
V	L. Nadel & D. L. Stein	1992 Lectures in Complex Systems, 1993

Lecture Notes Volumes

Vol.	Author	Title
I	J. Hertz, A. Krogh, & R. Palmer	Introduction to the Theory of Neural Computation, 1990
II	G. Weisbuch	Complex Systems Dynamics, 1990
III	W. D. Stein & F. J. Varela	Thinking About Biology, 1993

Reference Volumes

Vol.	Author	Title
I	A. Wuensche & M. Lesser	The Global Dynamics of Cellular Automata: Attraction Fields of One-Dimensional Cellular Automata, 1992

Proceedings Volumes

Vol.	Editor	Title
I	D. Pines	Emerging Syntheses in Science, 1987
II	A. S. Perelson	Theoretical Immunology, Part One, 1988
III	A. S. Perelson	Theoretical Immunology, Part Two, 1988
IV	G. D. Doolen et al.	Lattice Gas Methods for Partial Differential Equations, 1989
V	P. W. Anderson, K. Arrow, D. Pines	The Economy as an Evolving Complex System, 1988
VI	C. G. Langton	Artificial Life: Proceedings of an Interdisciplinary Workshop on the Synthesis and Simulation of Living Systems, 1988
VII	G. I. Bell & T. G. Marr	Computers and DNA, 1989
VIII	W. H. Zurek	Complexity, Entropy, and the Physics of Information, 1990
IX	A. S. Perelson & S. A. Kauffman	Molecular Evolution on Rugged Landscapes: Proteins, RNA and the Immune System, 1990
X	C. G. Langton et al.	Artificial Life II, 1991
XI	J. A. Hawkins & M. Gell-Mann	The Evolution of Human Languages, 1992
XII	M. Casdagli & S. Eubank	Nonlinear Modeling and Forecasting, 1992
XIII	J. E. Mittenthal & A. B. Baskin	Principles of Organization in Organisms, 1992
XIV	D. Friedman & J. Rust	The Double Auction Market: Institutions, Theories, and Evidence, 1993
XV	A. S. Weigend & N. A. Gershenfeld	Time Series Prediction: Forecasting the Future and Understanding the Past

Contents

Preface

The 1992 Complex Systems Summer School once again provided an exciting atmosphere for research, learning, and discussion in a wide variety of fields and topics. As in previous volumes, the contents of this book reflect the topics discussed in the 1992 summer school, although a few do not appear within. We make special note of the fact that one of our lecturers, Joshua Epstein of the Brookings Institution, has used his Summer School lectures as the basis for a book, *Nonlinear Dynamics, Mathematical Biology, and Social Science*, to be published as a separate volume within the SFI complexity series. We are also pleased to include a number of contributions from the participants themselves. These are the result of research by individuals or working groups set up during the school. The results are quite impressive. Special thanks to Brian Keeley for his efforts on this part of the volume, and to Cathleen Barczys for arranging the student seminar series during the Summer School itself.

We are also pleased to note that Una Smith, a participant in the School, has prepared a biologist's guide to Internet resources. This contains an overview and lists of free Internet resources such as scientific discussion groups and mailing lists; research newsletters, directories, and bibliographies; huge data and software archives; and tools for finding and retrieving information. The guide is a formal Usenet FAQ; that is, it is posted in various *.answers newsgroups in Usenet and archived in the FAQ repository on rtfm.mit.edu, where the most current version can be found via FTP or e-mail. For more detailed information, please refer to Una Smith's contribution in the text.

ACKNOWLEDGMENTS

Many people contributed to the success of the summer school. The planning for the school, its day-by-day functioning, and the follow-up after the school finishes are all a reflection of the efforts of a number of people at the Santa Fe Institute. Ed Knapp and Mike Simmons gave much of their time and effort to the Summer School; Ginger Richardson and Andi Sutherland were indispensable from start to finish, as usual; Ronda Butler-Villa and Della Ulibarri played a major role in getting this volume together; Marcella Austin handled the rather complex financial side; and Brent McClure got the computational laboratory up and running, and kept it that way. Stuart Kauffman of the Santa Fe Institute committed much of his time to the students, and could often be seen in heated discussion with groups of them in the courtyard of the Institute. Most critical of all, Peter Hraber performed myriad tasks which kept the school running smoothly.

We thank also our advisory board, and several institutions that provided computers and associated peripherals. We also thank the University of Arizona, and its Center for the Study of Complex Systems, for permitting the two of us to spend time on this rewarding but time-consuming enterprise. Finally, we must thank those agencies that contributed the funds needed to make the school a reality: financial support was provided by the National Science Foundation, the Department of Energy, Office of Naval Research, the National Institute of Mental Health, Sandia National Laboratories, the Center for Nonlinear Study at Los Alamos National Laboratory, Institutional Collaborative Research Program of the University of California (INCOR), the Los Alamos Graduate Center of the University of New Mexico, the University of Arizona, and Professor Marcus Feldman; student support was provided by the University of Florida, University of Michigan, and Stanford University; and computer equipment was provided by the University of New Mexico, Digital, Sun Microsystems, Silicon Graphics, and Computerland of Santa Fe.

Lynn Nadel
University of Arizona
Tucson, ZA 85721

Daniel L. Stein
University of Arizona
Tucson, AZ 85721

May 31, 1993

Lectures

Melanie Mitchell
Santa Fe Institute, 1660 Old Pecos Trail, Suite A, Santa Fe, NM 87501

Genetic Algorithms

CONTENTS

1992 Lectures in Complex Systems, Eds. L. Nadel & D. Stein, SFI Studies in the Sciences of Complexity, Lect. Vol. V, Addison-Wesley, 1993

1. INTRODUCTION

The advent of electronic computers has brought about a revolution in all areas of science and engineering and has opened up the possibility for scientific investigations and technological accomplishments of a wholly new kind. Computers have permitted the in-depth study and modeling of systems of great complexity, such as stellar and galactic dynamics, atmospheric processes, biological cells, brains, the human immune system, natural ecologies, and economies. The importance of understanding such systems is enormous: many of the most serious challenges facing humanity—e.g., environmental sustainability, economic stability, or the control of disease—as well as many of the hardest scientific questions—e.g., the nature of intelligence or the origin of life—will require a deep understanding of complex systems. Computers have also provided the ability to address previously intractable practical problems such as large-scale combinatorial optimization, the automatic analysis of complex data, and the creation of autonomous learning systems, all of which will have tremendous significance for science and technology.

As research in such areas has progressed, the need for increasingly powerful and sophisticated computational systems has become critical. The recent development of massively parallel computers holds much potential promise for addressing these problems. However, powerful hardware is almost never enough for making significant progress on the types of problems listed above; at present the main bottleneck lies in the creation of new computational methods—algorithms, interfaces, and analysis tools—that are more sophisticated and that fit these problems more naturally than do traditional computational and mathematical methods.

What is required are methods that naturally take advantage of parallel processing, methods in which appropriate complex behavior emerges from the interaction of simple parts rather than being laboriously (and most often inadequately) preprogrammed, and methods that can efficiently search through large spaces, that have sophisticated pattern-recognition abilities, and that are "adaptive"—i.e., able to automatically improve their performance (according to some measure) over time in response to what has been encountered previously.

Such features have been the basis of some novel approaches to computation that have been developed in recent years, many of them inspired by natural adaptive systems. In particular, almost since the advent of the computer age, a small number of computer scientists have been inspired by the process of biological evolution. They have attempted to develop "evolutionary" approaches to computational problems and, in turn, to use computers to model evolutionary processes. The field of genetic algorithms springs from one such approach.

1.1 THE APPEAL OF EVOLUTION

Why use evolution as an inspiration for solving computational problems? Natural evolution addresses many of the requirements discussed above, in the context of biology. Evolution can be thought of as a massively parallel search through a huge space of possible solutions to a problem, where the "problem" is to create an organism that can survive and flourish in a given environment and the "solutions" are the genetic blueprints for different organisms. Evolution results in emergent complexity from simple rules—the rules of evolution under natural selection are simple to state, and yet their effects are hard to predict and their repeated action has given rise to extremely complex structures. One example is the human nervous system, which is, among other things, the paramount existing system for sophisticated pattern recognition.

1.2 ELEMENTS OF GENETIC ALGORITHMS

Genetic algorithms (GAs), computational methods inspired by ideas from evolution under natural selection, were invented in the 1960s by John Holland of the University of Michigan and were first described at length in his book *Adaptation in Natural and Artificial Systems.*[38] A GA searches through a space of "chromosomes," each of which represents a potential solution to a given problem (in some cases, a solution consists of a set of chromosomes). These chromosomes often take the form of bit strings; each bit position ("locus") in the chromosome has two possible values ("alleles"), 0 and 1. (These biological terms are used in the spirit of analogy with real biology, though the entities they refer to are, of course, much simpler than the real biological entities.) Each chromosome can be thought of as a point in the search space of potential solutions. The search takes place by processing populations of chromosomes, moving from one population to another. This is different from most search methods, which move between single points in the search space. The GA most often requires a "fitness function" that assigns a score (fitness) to each chromosome in the current population. The fitness of the chromosome depends on how good a solution that chromosome is to the problem at hand.

For example, a common application of GAs is in function optimization, where the goal is to find a set of parameter values that optimize a complex multiparameter function. As a simple example, one might want to maximize the one-dimensional function[75]

$$f(x) = x + |\sin(32x)|, x\epsilon[0 \ldots \pi].$$

Here the potential solutions are values of x; these might be encoded as bit strings. The fitness function would translate a given bit string into a real number and then apply the function at that value. The fitness of a string would be the function value at that point.

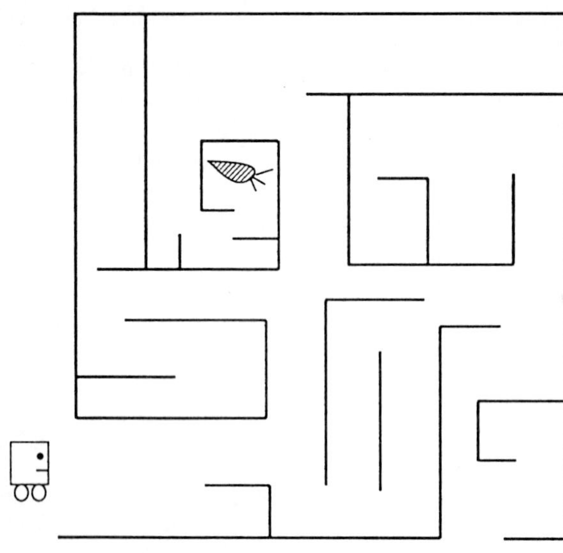

0 0 = Backward; 0 1 = Left; 1 0 = Forward; 1 1 = Right

1 0 1 0 0 1 1 0 1 1 0 0 1 0 . . .

F F L F R B F

FIGURE 1 A robotic navigation problem, in which one potential solution (a series of "forward," "backward," "left, and "right" moves) is encoded as a bit string.

As another example, the GA might be applied to the robotics problem illustrated in Figure 1. Here the problem is to find a good strategy for traversing the maze efficiently to reach the goal, where a strategy is a sequence of FORWARD, BACKWARD, LEFT, and RIGHT moves. As shown in the figure, each of these possibilities can be represented by two bits. A bit string encoding a potential solution is also shown. The fitness of a strategy might be calculated by letting the robot follow the strategy and measuring the Euclidean distance between its final position and the goal.

The preceding two examples show two different contexts in which potential solutions to a problem are encoded as abstract bit-string "chromosomes" and fitness functions are defined on the resulting space of bit strings. A search space together with a fitness function is known as a "fitness landscape," analogous to the notion of a fitness landscape in biology.[90] The GA is a method for searching a fitness landscape for high-fitness strings. The GA, in its simplest form, involves three "genetic" operators:

- **Reproduction:** This operator makes identical copies of some of the chromosomes in the population (the fitter the chromosome, the more copies are likely to be made).

- **Crossover:** This operator exchanges subparts of two chromosomes. For example, the strings

  10000100

  and

  11111111

  could be crossed over after the third bit to produce two offspring:

  10011111

  and

  11100100.

  This operator roughly mimics sexual recombination between two single-chromosome organisms.
- **Mutation:** This operator randomly flips some bits in the chromosome. For example, the string 00000100 might be mutated in its second position to yield 01000100. Mutation can occur at each bit position in a string with some probability, usually very small.

1.3 A SIMPLE GA

With these three genetic operators, we can now give a simple genetic algorithm:

1. Start with a randomly generated population of chromosomes (potential solutions to a problem).
2. Calculate the fitness of each chromosome in the population.
3. Apply genetic operators (reproduction, crossover, and mutation) to the population to create a new population.
4. Go to step 2.

This process is iterated over a number of time steps ("generations"). After several generations, the result is often one or more highly fit solutions in the population.

This simple procedure is the basis for most applications of the GA. There are a number of details to fill in, such as the size of the population or how exactly to apply the genetic operators, and often the success of the algorithm depends very much on these decisions. There are also much more complex versions of the GA (e.g., GAs that work on representations other than bit strings or GAs that have different types of crossover and mutation operators). Some examples will be given later in this chapter.

As a more detailed example of the simple GA, suppose that the fitness of a bit string is equal to the number of 1's in the string and suppose that the population contains four strings.

The initial (randomly generated) population might look like:

	Fitness
00000110	2
11101110	6
00100000	1
00110100	3

A common way to perform reproduction in GAs is known as *fitness-proportionate reproduction*: each individual in the population is assigned an *expected number of copies* value equal to its fitness divided by the average fitness of the population. These values are given below:

	Fitness	ExpectedCopies
00000110	2	2/3
11101110	6	6/3
00100000	1	1/3
00110100	3	3/3

Since an individual cannot have a fractional number of copies, some kind of sampling procedure must be used to assign an integral number of copies to each individual, based on the expectation values. Usually this is done so that the total actual number of copies equals the population size. The distribution above might, for example, yield zero copies for the first string, two copies for the second, and one each for the third and fourth under a random sampling procedure biased by expectation value. (The zero copies for the first string is just the luck of the draw here. If the selection procedure were repeated several times, the average results would be closer to the expected values.) Pairs of strings are then randomly chosen from the copies, and each pair crosses over to produce two offspring. The simplest form of crossover is *single-point crossover*, in which a single crossover point is chosen randomly with uniform probability over the entire string, and substrings of the parents before and after the crossover point are exchanged. The two offspring from the cross then undergo mutation with some probability—for example, the probability might be fixed at 0.01 per bit, which, in the case of strings of length 8, would mean an 0.08 chance that a given string would be mutated in one position. The resulting strings then are placed in the new population, and the entire procedure is repeated for the next generation.

The GA described above is very simple; many other, more complicated versions have been developed, some of which are described in this chapter. This chapter does not include a general discussion of the issues involved in implementing a GA; such discussions can be found in Goldberg,[27] Davis,[16] and Michalewicz.[59]

1.4 OVERVIEW OF SOME APPLICATIONS OF GENETIC ALGORITHMS

The algorithm described above is very simple, but variations on this basic theme have been used in a large number of scientific and engineering problems and models, including the following:

- **Optimization**: GAs have been used in a wide variety of optimization tasks, including numerical optimization (e.g., De Jong[18]) as well as combinatorial optimization problems such as circuit design (e.g., Shahookar and Mazumder[82]) and job shop scheduling (e.g., Nakano[65]).
- **Automatic programming**: GAs have been used as a means to evolve computer programs to perform various tasks[33,43,44]; one such project will be discussed in detail later in this chapter.
- **Machine and robot learning**: GAs have been used in some machine learning tasks such as classification and prediction (e.g., prediction of dynamical systems,[58] weather prediction,[76] and prediction of protein structure[81]), the design of neural networks (e.g., Belew et al.,[8] Chalmers,[12] Harp and Samad,[31] Miller et al.,[60] and Montana and Davis[63]), and the evolution of behavioral rules for a cognitive system.[35]
- **Economic models**: GAs have been used by economists to model processes of innovation, to model the development of bidding strategies, and to model the emergence of economic markets.[2,37]
- **Immune system models**: GAs have been used to model the evolution of immunological antibodies in a changing environment of antigens.[23]
- **Ecological models**: GAs have been used in models of ecological phenomena such as biological arms races, host-parasite coevolution, symbiosis, and resource flow in ecologies.[38,39,50,73,74]

This list is by no means exhaustive, but it gives the flavor of the kinds of things GAs have been used for, both in problem-solving and scientific contexts. Because of the GA's success in these and other areas, interest has been growing rapidly in the last several years among researchers in many disciplines and the field of GAs is becoming its own subdiscipline of computer science, with its own conferences, journals, and scientific society.

A BRIEF EXAMPLE: USING GENETIC ALGORITHMS TO EVOLVE STRATEGIES TO THE "PRISONER'S DILEMMA"

As a warm-up to more extensive discussions of GA applications, I will describe an application of the GA to evolve strategies for the Prisoner's Dilemma.[4]

The Prisoner's Dilemma is a simple two-person game that has been studied extensively in game theory, economics, and political science because it can be seen as an idealized model for real-world phenomena such as arms races.[3] On a given turn,

		Player B	
		Cooperate	*Defect*
Player A	Cooperate	**3, 3**	**0, 5**
	Defect	**5, 0**	**1, 1**

FIGURE 2 The payoff matrix for the Prisoner's Dilemma (adapted from Axelrod[4]). The numbers given in each box are the respective payoffs for players A and B in that situation.

each player independently decides whether to "cooperate" or "defect." The game is summarized in the payoff matrix shown in Figure 2. If both players cooperate, they each get three points. If player A defects and player B cooperates, then player A gets five points and player B gets zero points; vice versa if the situation is reversed. Finally, if both players defect, they each get one point. What is the best strategy to take? If there is only one turn to be played, then clearly the best strategy is to defect: the worst consequence for a defector is to get one point and the best is to get five points, which are better than the worst score and the best score, respectively, for a cooperator. The dilemma is that if the game is *iterated*, that is, if two players play several turns in a row, the strategy of always defecting will lead to a much lower total payoff than the players would get if they both cooperated. How can reciprocal cooperation be induced? This question takes on special significance when the notions of "cooperating" and "defecting" correspond to actions in, say, a real-world arms race.

Robert Axelrod of the University of Michigan has studied the Prisoner's Dilemma and related games extensively. His interest in what makes for a good strategy led him to organize two Prisoner's Dilemma tournaments (described in Axelrod[3]). He solicited strategies from researchers in a number of disciplines. Each participant submitted a computer program that implemented his or her strategy, and the various programs played iterated games with each other. During each iterated game, each program remembered what move its opponent made on the three previous turns, and its strategy was based on this memory. The programs were paired in a round-robin tournament, where each played with many or all of the other programs over a number of turns. The first tournament contained 14 different programs and the second tournament contained 62 programs. Some of the submitted strategies were rather complicated, using techniques such as Markov processes and Bayesian inference to model other players and to determine the best move. However, in both tournaments the winner (the strategy with the highest average score) was the simplest of the submitted strategies: TIT FOR TAT. TIT FOR TAT cooperates on the first move and then, on subsequent moves, does whatever the other player did last. That is, it offers cooperation and then reciprocates it, but if the other player defects, TIT FOR TAT will punish that with a defection.

After the two tournaments, Axelrod decided to see if the GA could *evolve* strategies to play this game successfully.[4] The first problem was figuring out how to best encode a strategy as a bit string. The encoding used by Axelrod follows. Suppose the memory of each strategy is one previous move. There are four possibilities for the previous move:

CC (case 1)
CD (case 2)
DC (case 3)
DD (case 4)

Case 1 is when both players cooperated on the previous move, case two is when player A cooperated and player B defected, and so on. A strategy is simply a rule that specifies an action in each case. For example, TIT FOR TAT is the following strategy:

If CC (case 1), then C.
If CD (case 2), then D.
If DC (case 3), then C.
If DD (case 4), then D.

This strategy can be encoded by a string of length 4 which says what to do in each of the four cases:

C D C D.

To use the strategy, the player determines the case corresponding to the previous move and uses the letter corresponding to that case in the string (e.g., in case 1, the player uses the letter in the first position in the string, here C).

Axelrod's tournaments involved strategies that used three previous moves. There are 64 possibilities for the previous three moves:

CC CC CC (case 1)
CC CC CD (case 2)
CC CC DC (case 3)
etc.

Thus a strategy can be encoded by a 64-bit string, e.g.,

C D C C C D D C C C D D . . . .

Axelrod actually used a 70-bit string where the six extra bits (C's or D's) were not part of the strategy but encoded three hypothetical "previous moves" used by the strategy to decide what to do on the very first move of the game. The number of possible strategies is thus 2^{70}; the search space is thus far too big to search exhaustively.

In Axelrod's first experiment, the GA had a population of 20 strategies. The fitnesses of strategies in the population were determined as follows. Axelrod had found earlier that eight of the human-generated strategies from the second tournament were representative, in the sense that a given strategy's score playing with these eight was a good predictor of the strategy's score playing with all 62 entries. This set of eight strategies (which did not include TIT FOR TAT) served as the "environment" for the evolving strategies in the population. Each strategy S in the population played iterated games with each of the eight fixed strategies, and S's fitness was its average score over all the games it played.

The GA was run for 50 generations, with fitness-proportionate reproduction, crossover, and mutation being applied at each generation. Forty replications were made of the GA run, with different random number seeds used for each replication. Most of the strategies that evolved were similar to TIT FOR TAT, having many of the properties that make TIT FOR TAT successful. However, the GA often found strategies that scored substantially higher than TIT FOR TAT. This is a striking result, especially in view of the fact that in a given run the GA is testing only $20 * 50 = 1000$ individuals, out of a huge search space of 2^{70} individuals.

It is not correct to conclude that the GA evolved strategies that are "better" than any human-designed strategy. The performance of a strategy depends very much on its environment—that is, the other strategies that it is playing with. Here the environment was fixed—it consisted of eight human-designed strategies that did not change over the course of a run. The highest-scoring strategies produced by the GA were ones that "learned" how to exploit specific weaknesses of the eight fixed strategies. It is not necessarily true that these high-scoring strategies would also score highly in some other environment. TIT FOR TAT is a generalist, whereas the highest-scoring evolved strategies were more specialized to their given environment. Axelrod concluded that the GA is good at doing what evolution often does: developing highly specialized adaptations to specific characteristics of the environment.

To see the effects of a *changing* (as opposed to fixed) environment, Axelrod carried out another experiment in which the fitness of a strategy was determined by allowing the strategies in the population to play with *each other* rather than with the fixed set of eight strategies. The environment changes from generation to generation because the strategies themselves are evolving. At each generation, each strategy played iterated games with each of the nineteen other members of the population, and its fitness was again its average score over all games.

In this second set of experiments, Axelrod observed the following phenomena. The GA initially evolves uncooperative strategies, because strategies that tend to cooperate early on do not find reciprocation among their fellow population members and thus tend to die out. But after about 10 to 20 generations, the trend starts to reverse: the GA discovers strategies that reciprocate cooperation and that punish defection (i.e., variants of TIT FOR TAT). These strategies do well with each other and are not completely defeated by other strategies, as were the initial cooperative strategies. The reciprocators score better than average, so they spread in the population, resulting in more and more cooperation and higher and higher fitness.

This example illustrates how one might use a GA both to evolve solutions to a complex problem and to model evolution and coevolution in an idealized way. One can think of many additional possible experiments, such as running the GA without crossover and seeing the effect this has on the evolution of strategies (this experiment was done by Axelrod[4]) or allowing a more open-ended kind of evolution where the amount of memory available to a given strategy is allowed to increase with evolution (such an experiment was performed by Lindgren[50]).

1.6 HOW AND WHY DO GENETIC ALGORITHMS WORK?

GAs are simple to describe and program, but their behavior can be complex and many open questions exist about how and why they work and what they are good for. Much work has been done on the foundations of GAs (see, for example, Holland,[38] Goldberg,[27] Rawlins,[72] and Whitley[89]). The last section of this chapter describes some approaches toward answering these questions. Here I give a brief overview of some fundamental concepts related to the theory of GAs.

At a very general level of description, it is believed that GAs work by discovering, emphasizing, and recombining high-quality *building blocks* of solutions in a highly parallel way. The idea here is that good solutions tend to be made up of good building blocks—combinations of bit values that often confer higher fitness to the string in which they are present.

Most studies of the theory of GAs start with the notion of *schemas* (or "schemata"),[38] which formalizes the informal notion of "building blocks." A schema is a set of bit strings that can be described by a template made up of 1's, 0's, and *'s, where the *'s represent wild cards (or "don't cares"). For example, the schema s = 1****1 represents the set of all 6-bit strings that begin and end with 1. The strings that fit this template (e.g., 100111 or 110011) are said to be *instances* of s. The schema s is said to have two *defined* bits (the number of non-*'s) or, equivalently, to be of *order* 2. Its *defining length* (the distance between its outermost defined bits) is 5.

Note that not every possible subset of the search space of bit strings can be described as a schema; in fact, the huge majority cannot. In a search space of bit strings of length l, there are 2^l possible strings and thus 2^{2^l} possible subsets of strings, but only 3^l possible schemas. However, a central tenet in GA theory is that schemas are the building blocks that the GA processes effectively under the operators of reproduction, mutation, and single-point crossover.

How does the GA process schemas? Any given bit string of length l is an instance of 2^l different schemas. For example, the string 11 is an instance of ** (the schema that contains the entire search space), *1, 1*, and 11 (the schema that contains only one string, 11). Thus any given population of N strings contains between 2^l and $N \times 2^l$ different schemas (if all the strings are identical, then there are exactly 2^l different schemas; otherwise, the number is less than $N \times 2^l$). This means that at a given generation, while the GA is explicitly evaluating the fitnesses of the N strings in the population, it is actually *implicitly* estimating the average fitness of a much larger number of schemas. For example, in a randomly generated population of N strings, on average half the strings will be instances of 1*** ... * and half will be instances of 0*** ... *. The evaluations of the approximately $N/2$ strings that are instances of 1*** ... * give an estimate of the average fitness of that schema. (The average fitness of a schema is defined to be the average fitness of all strings in the search space that are instances of that schema.) Similarly, in evaluating a population of N strings, the GA is implicitly estimating the average fitnesses of all schemas that are present in the population. This simultaneous evaluation of large numbers of schemas in a population of N strings is known as *implicit parallelism*.[38] The effect of reproduction is to gradually bias the sampling procedure toward schemas whose fitness is estimated to be above average. Over time, the estimate of a schema s's average fitness should in principle become more and more accurate since the GA is sampling more and more instances of s (some possible problems with this assumption are discussed in the last section in this chapter).

We can calculate the dynamics of this sample biasing as follows. Let s be a schema present in the population at time t (i.e., there is at least one instance of s at time t). Let $N(s,t)$ be the number of instances of s at time t, and $\hat{u}(s,t)$ be the observed average fitness of s at time t (i.e., the average fitness of instances of s in the population at time t). We want to calculate $N(s,t+1)$, the expected number of instances of s at time $t+1$. Assume that reproduction is carried out as described earlier: the expected number of copies of a string x is equal to $F(x)/\overline{F}(t)$, where $F(x)$ is the fitness of string x in the population and $\overline{F}(t)$ is the average fitness of the population at time t. (For now, we will ignore the effects of crossover and mutation.) Then,

$$\begin{aligned} N(s,t+1) &= \sum_{x \epsilon s} \frac{F(x)}{\overline{F}(t)} \\ &= \frac{[\sum_{x \epsilon s} F(x)]}{\overline{F}(t)} \\ &= \frac{\hat{u}(s,t)}{\overline{F}(t)} N(s,t), \end{aligned} \tag{1.1}$$

by definition, since $\hat{u}(s,t) = \sum_{x \epsilon s} F(x)/N(s,t)$ for x in the population at time t.

This is known as the Schema Theorem[38] (see also Goldberg[27]). It says that schemas whose observed average fitness stays above the population average fitness will receive exponentially increasing numbers of samples over time.

Crossover and mutation can both destroy and create instances of s, so the right side of Eq. (1.1) can be thought of as a lower bound on $N(s, t+1)$ if we include the effects of crossover and mutation. First, let us consider the disruptive effects of crossover. Let p_c be the probability that single-point crossover will be applied to a string. Then we can state a lower bound on the probability $S_c(s)$ that a schema s will survive under crossover:

$$S_c(s) \geq 1 - p_c\left(\frac{d(s)}{l-1}\right),$$

where $d(s)$ is the defining length of s and l is the length of bit strings in the search space. That is, crossovers occurring within the defining length of the schema can destroy the schema, so we multiply the fraction of the string that the schema occupies by the crossover probability to obtain an upper bound on the probability that it will be destroyed. (The value is an upper bound because some crossovers inside a schema will not destroy it, e.g., if two identical strings cross with each other.) In short, the probability of survival under crossover is higher for shorter schemas.

The disruptive effects of mutation can be quantified as follows. Let p_m be the probability of any bit being mutated. Then $S_m(s)$, the probability that schema s will survive under mutation, is the following:

$$S_m(s) = (1 - p_m)^{o(s)},$$

where $o(s)$ is the order of s (i.e., the number of defined bits in s). That is, for each bit, the probability that it will not be mutated is $1 - p_m$, so the probability that no bits of schema s will be mutated is this quantity multiplied by itself $o(s)$ times. In short, the probability of survival under mutation is higher for lower-order schemas.

These disruptive effects can be used to amend Eq. (1.1):

$$N(s, t+1) \geq \frac{\hat{u}(s,t)}{\overline{F}(t)} N(s,t) \left[1 - p_c \frac{d(s)}{l-1}\right] \left[(1 - p_m)^{o(s)}\right]. \tag{1.2}$$

The conclusion is that short, low-order schemas whose average fitness remains above the mean will receive exponentially increasing numbers of samples over time.

The Schema Theorem as stated in Eq. (1.2) is incomplete in that it only deals with the destructive effects of crossover and mutation. However, crossover is believed to be a major source of the GA's search power, taking the high-fitness schemas that are emphasized in the population and recombining them to form even fitter higher-order schemas that are themselves then emphasized via reproduction. The supposition that this is the process by which GAs work is known as the "Building-Block Hypothesis"[38,27]: it proposes that the GA produces fitter and fitter strings by combining building blocks.

The Schema Theorem and the Building-Block Hypothesis deal with the roles of reproduction and crossover in GAs. What is the role of mutation? Holland[38] proposes that mutation is what prevents loss of diversity at a given bit position. For example, without mutation, all the strings in the population might come to have a 1 at the first bit position, and there would be no way to obtain a string beginning with a zero. Mutation provides a kind of "insurance policy" against such fixation.

The reader may have noticed that the Schema Theorem given in Eq. (1.1) applies not only to schemas but to any subset of strings in the search space. The reason for specifically focusing on schemas is that they (in particular, short, high-fitness schemas) are a good description of the types of building blocks that are combined effectively by single-point crossover. Thus, a belief underlying this formulation of the GA is that schemas will be a good description of the relevant building blocks of a good solution. GA researchers have defined other types of crossover operators that deal with different types of building blocks and have analyzed the generalized "schemas" that a given crossover operator effectively manipulates (e.g., Radcliffe[71] and Vose[85]).

2. GENETIC ALGORITHMS IN PROBLEM SOLVING

In the previous section some applications of GAs were listed, including applications both for solving practical problems and for modeling natural evolutionary systems. This section describes three projects in which the GA is used in problem solving. These three projects each include different GA representations and techniques and give a good flavor for the diversity of possible uses for GAs.

2.1 AUTOMATIC PROGRAMMING

Automatic programming—having computer programs automatically write computer programs—has a long history in the field of artificial intelligence, but automatic programming methods have not had much success in producing the complex and robust programs needed for real applications. John Koza of Stanford University has used a form of the GA to evolve computer programs to perform various tasks[43,44] and claims that his method—"Genetic Programming" (GP)—has the potential to produce programs of the necessary complexity and robustness. The programs are expressed in the programming language Lisp. Programs in Lisp can easily be expressed in the form of a "parse tree," the object the GA will work on.

For example, consider a program to compute the area of a circle. In a programming language such as FORTRAN, such a program might be:

```
PROGRAM AREA-OF-CIRCLE
     R = 45
     PI = 3.1415
     AREA = PI * (R * R)
     PRINT AREA
END AREA-OF-CIRCLE
```

In Lisp, this program could be written as

```
(DEFUN AREA-OF-CIRCLE ()
     (SETF R 45)
     (SETF PI 3.1415)
     (* PI (* R R)))
```

(In Lisp, the value of the last expression in the program is automatically printed.) Assuming we know PI and R, the important statement here is (* PI (* R R)), which can be expressed as the parse tree shown in Figure 3. In Koza's GP algorithm, the population does not consist of bit strings but of such trees, and new genetic operators are defined to work on them.

Each tree consists of *functions* and *terminals.* In the tree shown in Figure 3, the multiplication operator * is a function that takes two arguments. PI and R are terminals. Notice that the argument to a function can be the result of another function, as in the expression above where one of the arguments to the top-level * is (* R R).

Koza's algorithm is:

1. Choose a set of possible functions and terminals for the program. The idea behind GP is, of course, to *evolve* programs that are difficult to write and, in general, one does not know ahead of time precisely which functions and terminals will be used in a successful program. So the user of GP has to make an intelligent guess as to a reasonable set of functions and terminals for the problem at hand. For example, for the area-of-circle problem, the function set might be $\{+, -, *, /, \sqrt{}\}$ and the terminal set might be {PI, R, C, D} (where C is the circle's circumference and D is its diameter).
2. Generate an initial population of random trees (programs) using the set of possible functions and terminals. These random trees must be syntactically correct programs—that is, the number of branches extending from each function node must equal the number of arguments taken by that function. Three programs from a possible randomly generated initial population are displayed in Figure 4. Notice that the randomly generated programs can be different sizes (different numbers of nodes and levels in the trees). In principle a randomly generated tree can be any size, but in practice Koza restricts the size of the initially generated trees.
3. Calculate the fitness of each program in the population. The fitness of a program is calculated by running it on a set of "training cases"—a set of inputs for which

the correct output is known. For the area-of-circle example, the training cases might be a set of experimental observations in which the areas of a set of circles were measured experimentally (assuming that the user did not know the formula ahead of time). Another application might be to evolve a robot program to navigate a maze; there the training cases might be a number of mazes, and the desired output is a set of moves that takes the robot to the goal of the maze. The fitness of a program is a function of the number of training cases on which it performs correctly (some fitness functions might give partial credit to a program for getting close to the correct output).

The randomly generated programs in the initial population are not likely to perform well but, with a large enough population, some of them will perform better than others by chance. This initial fitness differential provides a basis for "natural selection" to get off the ground.

4. Apply reproduction, crossover, and mutation to the population. Reproduction in GP is the same as in the simple GA described above, with the expected number of copies of a program being its fitness divided by the average fitness of the population. Crossover is performed by selecting pairs of parents from the set of copies made under reproduction and allowing them to exchange subtrees. Figure 5 displays one possible crossover. Here, a random point is chosen in each parent (the top two trees) and the subtrees beneath those two points are exchanged, to produce two offspring (the bottom two trees). Notice that crossover allows the size of a program to increase or decrease.

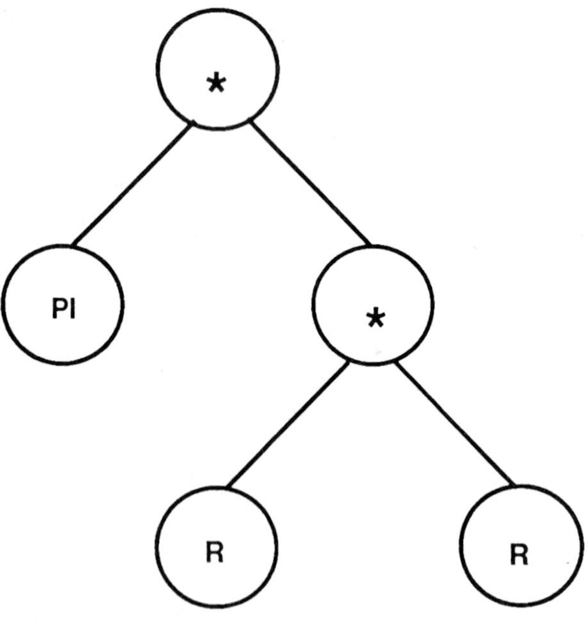

FIGURE 3 Parse tree for the Lisp expression for PI*R^2.

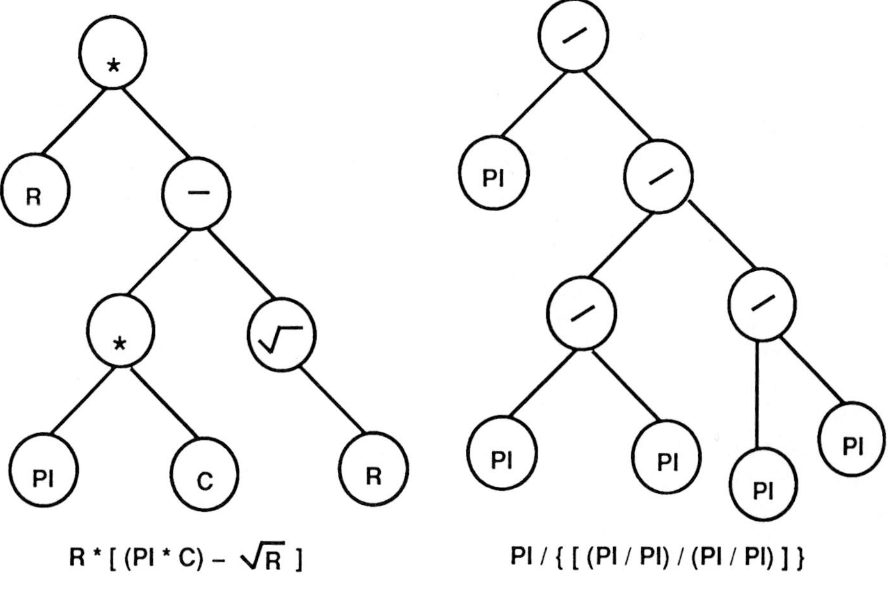

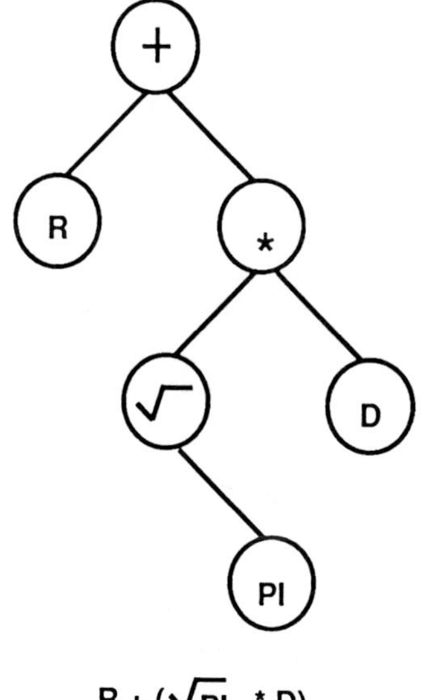

FIGURE 4 Three programs from a possible randomly generated initial population for the area-of-circle task. The expression represented by each tree is printed beneath the tree.

Mutation might performed by choosing a random point in a tree and replacing the subtree beneath that point by a randomly generated subtree. Koza generally does not use a mutation operator in his applications.[44]

Steps 3 and 4 are repeated for some number of generations.

It may seem difficult to believe that this procedure would ever result in a program that would perform the desired task; the famous example of a monkey randomly hitting the keys on a typewriter and producing the works of Shakespeare comes to mind. But surprising as it might seem, the GP technique has succeeded in evolving correct programs to solve a large number of specific problems in various domains, including optimal control, planning, sequence induction, symbolic regression, image compression, robotics, and many others. One example (described in detail in Koza[44]) is the problem of block stacking. This is a common "microworld" used to develop and test planning methods in artificial intelligence. The specific problem to which GP was applied is illustrated in Figure 6. Here, the problem is to find an algorithm that takes any initial configuration of blocks—some on the table, some in a stack—and places them in the stack in the correct order. Here the correct order spells out the word "universal."

For the terminals and functions for this problem, Koza used the set defined by Nilsson.[66] The terminals consisted of three sensors (available to a hypothetical robot to be controlled by the resulting program):

- **CS** ("current stack") returns the name of the top block of the stack. If the stack is empty, this sensor returns NIL.
- **TB** ("top correct block") returns the name of the topmost block on the stack such that it and all blocks below it are in the correct order.
- **NN** ("next needed") returns the block needed immediately above **TB** in the goal "universal." If no more blocks are needed, this sensor returns NIL.

In addition to these terminals, there were five functions available to GP:

- **MS**(x) ("move to stack") moves block x to the top of the stack if x is on the table.
- **MT**(x) ("move to table") moves the top of the stack to the table if block x is anywhere in the stack.
- **DU**(*expression1*, *expression2*) ("do until") executes *expression1* until *expression2* (a predicate) becomes satisfied (i.e., returns TRUE).
- **NOT**(*expression1*) returns TRUE if *expression1* is NIL and returns NIL otherwise.
- **EQ**(*expression1*, *expression2*) returns TRUE if *expression1* and *expression2* are equal (i.e., return the same value).

The programs in the population are generated from these two sets. The fitness of a given program is the number of sample environmental cases (initial configurations of blocks) for which the stack is correct (i.e., spells "universal") after the program is run. Koza used 166 different environmental cases, carefully constructed to cover the various classes of possible initial configurations (see Koza[44] for details).

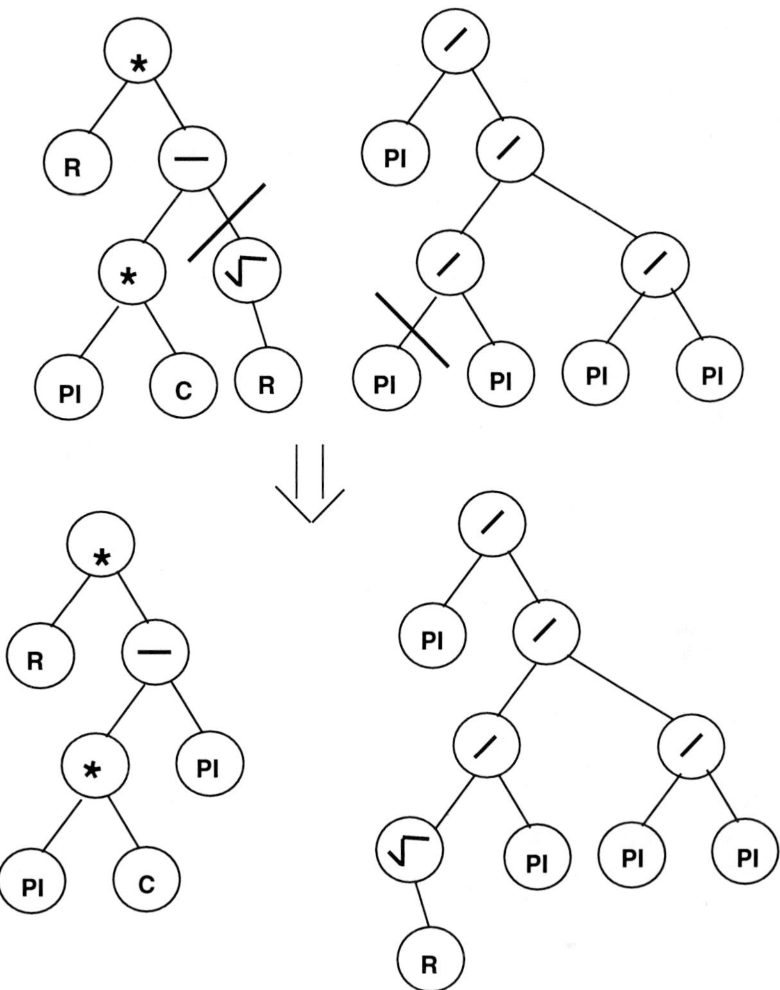

FIGURE 5 An example of crossover in the Genetic Programming algorithm. The two parents are shown at the top of the figure and the two offspring are shown below. The crossover points are indicated by slashes in the parent trees.

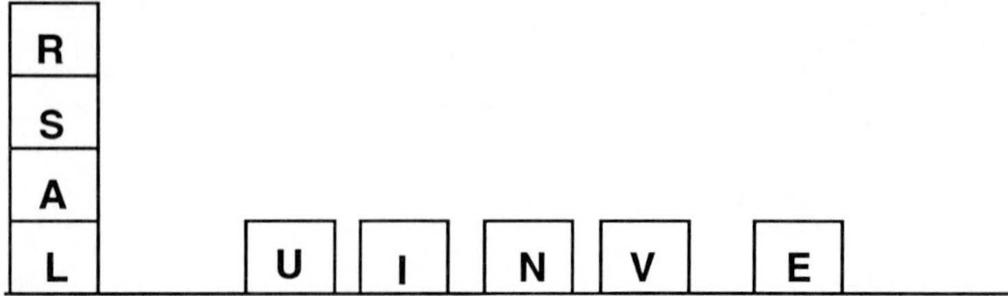

FIGURE 6 One initial state for the block-stacking problem. The goal is to find a plan that will stack the blocks correctly (spelling "universal") from any initial state. (Adapted from Koza.[43])

The initial population contained 300 randomly generated programs. Some examples (written in Lisp style rather than tree style) follow:

- (EQ (MT CS) NN)
  "Move the current top of stack to the table, and see if it is equal to the next needed." This clearly does not make any progress in sorting the blocks, and the program's fitness is 0.
- (MS TB)
  "Move the top block on the stack to the stack." This program effectively does nothing, but doing nothing allows it to get one environmental case correct: the case where all the blocks are already in the stack in the correct order. Thus this program's fitness is 1.
- (EQ (MS NN) (EQ (MS NN) (MS NN)))
  "Move the next needed block to the stack three times." This program makes some progress and gets four environmental cases right, giving it fitness 4. (Here EQ serves merely as a control structure. Lisp executes the first expression, then executes the second expression, and then compares their value. EQ thus performs the desired task of executing the two expressions in the proper order—we don't actually care whether their values are equal.)

By generation 5, the population contains some much more successful programs. The best one is:

- (DU (MS NN) (NOT NN))
  "Move the next needed block to the stack until no more blocks are needed." Here we have the basics of a reasonable plan. This works in all cases in which the blocks on the stack are already in the correct order; the program moves the remaining blocks on the table onto the stack in the correct order. There were ten such cases in the total set of 166, so this program's fitness is 10. Notice that this program uses a building block—(MS NN)—that was discovered in the first generation and found to be useful there.

In generation 10 a completely correct program (fitness 166) was discovered:

- (EQ (DU (MT CS) (NOT CS)) (DU (MS NN) (NOT NN)))
  This is an extension of the best program of generation 5. The program empties the stack onto the table, and then moves the next needed block to the stack until no more blocks are needed. GP thus discovered a plan that works in all cases, although it is not very efficient. Koza[44] discusses how to amend the fitness function to produce a more efficient program to do this task.

Koza's GP technique has produced some interesting and impressive results, but there are some open questions about its capabilities. Does it work well because the space of Lisp expressions is in some sense "dense" with correct programs for the tasks Koza has tried? This was given as one reason for the success of the artificial intelligence programs AM and Eurisko,[45] which used Lisp expressions to discover "interesting" conjectures in mathematics, such as the Goldbach conjecture (every even number is the sum of two primes). Koza refutes this hypothesis about GP by demonstrating how difficult it is to randomly generate a successful program to perform some of the tasks for which GP evolves successful programs. However, one could speculate that the space of Lisp expressions (with a given function and terminal set) is dense with useful intermediate-size building blocks for the tasks on which GP has been successful. The fact that GP is often extremely quick to find solutions (e.g., within 10 generations using a population of 300) lends credence to this hypothesis.

Some other questions are:

- Will the technique scale up to more complex problems for which larger programs are needed?
- Will the technique work if the function and terminal sets are large?
- How many environmental cases are typically needed? In many of Koza's examples, the evolving programs are tested on all possible environmental cases. In most real-world problems, such exhaustive (or even near-exhaustive) testing is infeasible. It is important to know the extent to which GP produces programs that generalize well after seeing only a small fraction of possible environmental cases.
- To what extent can programs be optimized for correctness, size, and efficiency at the same time?

The success of GP over a wide range of problems is encouraging and makes it well worth the effort to address these questions in future research.

2.2 COMPLEX DATA ANALYSIS AND PREDICTION

One major impediment to scientific progress in many fields is the inability to make sense of huge amounts of data that have been collected via experiment or computer simulation. There has been much work on developing automatic methods for finding significant and interesting patterns in complex data, or for forecasting the future from such data, but in general this remains an open problem.

Norman Packard of the Prediction Company has developed a form of the GA to address this problem[68] and has applied his method to a number of data analysis and prediction problems, including work with Thomas Meyer on forecasting a particular chaotic dynamical system.[58]

The general problem can be stated as follows. A series of observations from some process (e.g., a physical system or a formal dynamical system) take the form of a set of pairs:

$$\{(\vec{x}^1, y^1), \ldots, (\vec{x}^N, y^N)\},$$

where $\vec{x} = (x_1, \ldots, x_n)$ are independent variables and y is a dependent variable. For example, in a visual pattern-recognition task, the independent variables might be some set of features of the visual image (e.g., number of edges, number of vertices, curvature of lines, etc.) and the dependent variable might be the category of the visual image (e.g., "the letter 'A'"). Or in a time-series prediction task, the independent variables might be $\vec{x} = (x(t_1), x(t_2), \ldots, x(t_n))$, representing the values of a state variable at successive time steps, and the dependent variable might be $y = x(t_{n+k})$, representing the value of the state variable at some time in the future. (In these examples there is only one dependent variable y, but a more general form of the problem would allow any number of dependent variables.)

Packard used the GA to search through the space of *sets of conditions* on the independent variables for sets of conditions that give good predictions for the dependent variable. An individual in the GA population is a set of conditions such as:

$$C = \{(x_1 = c_1 \vee c_2 \vee c_3) \wedge (x_5 = d_1 \vee d_2) \wedge (x_6 = e_1 \vee e_2)\}.$$

This individual represents all the situations in which x_1 is equal to one of the values c_1, c_2, or c_3, and x_5 is equal to either d_1 or d_2, and x_6 is equal to either e_1 or e_2. Here c_1, c_2, c_3, etc. are the values of some feature; they do not have to be numerical. No requirements are made on the values of any of the other independent variables.

If the independent variables are numerical, a set of conditions might be

$$C = \{(c \leq x_1 \leq c') \wedge (d \leq x_5 \leq d') \wedge \ldots\}.$$

Such a condition set C specifies a particular subset of the data points. Packard's goal is to use a GA to search for condition sets that are good predictors of *something*—in other words, to search for condition sets that specify subsets of data points whose dependent-variable values are close to being uniform. In the character-recognition example, if the GA found a condition set such that all the characters satisfying that set are instances of the letter 'A,' then we might be confident to predict that some new character satisfying those conditions is also an 'A.'

The fitness of each individual C is calculated by running all the data points $(\vec{x}^i, y^i)$ in the training set through C and, for each $\vec{x}^i$ that satisfies C, collecting the corresponding y^i. After this has been done, a measurement is made of the uniformity of the resulting set of y's. For numerical y's, Meyer and Packard used the following fitness function:

$$F(C) = -\log\frac{\sigma}{\sigma_0} - \frac{\alpha}{N_C}.$$

Here σ is the standard deviation of the set of y's for data points satisfying C, σ_0 is the standard deviation of the distribution of y's over the entire data set, N_C is the number of data points satisfying condition C, and α is a constant. The first term of the fitness function measures the amount of information in the distribution of y's for points satisfying C, and the second term is a penalty term for poor statistics—if there is a small number of points satisfying C, then the first term is less reliable, so C should have lower fitness. The constant α can be adjusted for each particular application.

Meyer and Packard use the following version of the GA:

1. Initialize the population with a random set of C's.
2. Calculate the fitness of each C by running all the data points through it (note that this can be very computationally expensive!).
3. Rank the population by fitness.
4. Discard some fraction of the lower fitness individuals and replace them by new C's obtained by applying crossover and mutation to the remaining C's.
5. Go to step 2.

Meyer and Packard use a form of crossover known in the GA literature as "uniform crossover."[83] This operator takes two C's, and exchanges approximately half the "genes" (conditions). That is, at each gene position in parent A and parent B, a random decision is made whether that gene should go into offspring A or offspring B. An example follows:

$$\begin{aligned}
\text{Parent A}: &\quad \{(3.2 \le x_6 \le 5.5) \wedge (0.2 \le x_8 \le 4.8) \wedge (3.4 \le x_9 \le 9.9)\} \\
\text{Parent B}: &\quad \{(\mathbf{6.5} \le \mathbf{x_2} \le \mathbf{6.8}) \wedge (\mathbf{1.4} \le \mathbf{x_4} \le \mathbf{4.8}) \wedge (\mathbf{1.2} \le \mathbf{x_9} \le \mathbf{1.7}) \\
&\quad \wedge (\mathbf{4.8} \le \mathbf{x_{16}} \le \mathbf{5.1})\} \\
\text{Offspring A}: &\quad \{(\mathbf{6.5} \le \mathbf{x_2} \le \mathbf{6.8}) \wedge (\mathbf{1.4} \le \mathbf{x_4} \le \mathbf{4.8}) \wedge (3.4 \le x_9 \le 9.9)\} \\
\text{Offspring B}: &\quad \{(3.2 \le x_6 \le 5.5) \wedge (0.2 \le x_8 \le 4.8) \wedge (\mathbf{1.2} \le \mathbf{x_9} \le \mathbf{1.7}) \\
&\quad \wedge (\mathbf{4.8} \le \mathbf{x_{16}} \le \mathbf{5.1})\}
\end{aligned}$$

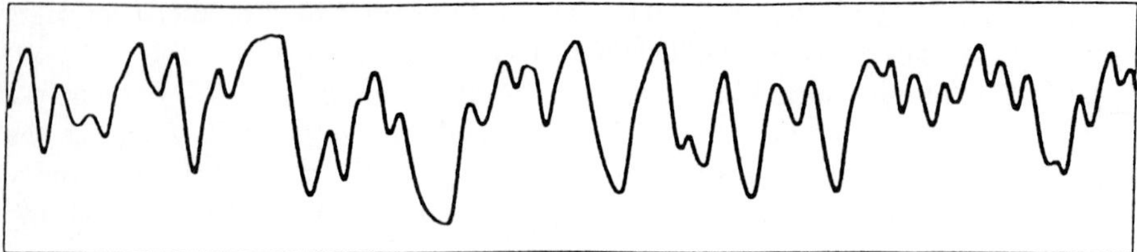

FIGURE 7 Plot of time series from Mackey-Glass system with $\tau = 150$. Time is plotted on the horizontal axis and $x(t)$ is plotted on the vertical axis. (Reprinted from Meyer and Packard.[58] Copyright © 1992 by T. P. Meyer.)

Offspring A has one gene from Parent A and two genes from Parent B. Offspring B has two genes from Parent A and two genes from Parent B.

In addition to crossover, four different mutation operators are used:

- Add a new condition:
  $\{(3.2 \le x_6 \le 5.5) \wedge (0.2 \le x_8 \le 4.8)\} \longrightarrow \{(3.2 \le x_6 \le 5.5) \wedge (0.2 \le x_8 \le 4.8) \wedge (\mathbf{3.4 \le x_9 \le 9.9})\}$
- Delete a condition:
  $\{(3.2 \le x_6 \le 5.5) \wedge (\mathbf{0.2 \le x_8 \le 4.8}) \wedge (3.4 \le x_9 \le 9.9)\} \longrightarrow \{(3.2 \le x_6 \le 5.5) \wedge (3.4 \le x_9 \le 9.9)\}$
- Broaden or shrink a range:
  $\{(\mathbf{3.2} \le x_6 \le \mathbf{5.5}) \wedge (0.2 \le x_8 \le 4.8)\} \longrightarrow \{(\mathbf{3.9} \le x_6 \le \mathbf{4.8}) \wedge (0.2 \le x_8 \le 4.8)\}$
- Shift a range up or down:
  $\{(3.2 \le x_6 \le 5.5) \wedge (\mathbf{0.2} \le x_8 \le \mathbf{4.8})\} \longrightarrow \{(3.2 \le x_6 \le 5.5) \wedge (\mathbf{1.2} \le x_8 \le \mathbf{5.8})\}$

Meyer and Packard applied this technique to the problem of finding "regions of predictability" in time series generated by a chaotic dynamical system. The particular system they used is defined by the Mackey-Glass equation[52]:

$$\frac{dx}{dt} = \frac{ax(t-\tau)}{1+[x(t-\tau)]^c} - bx(t).$$

Here $x(t)$ is the state variable, and a, b, c, and τ are constants. A plot of the time series from this system (with τ set to 150) is given in Figure 7.

To form the data set, Meyer and Packard did the following: for each data point i, the independent variables $\vec{x}^i$ are 50 consecutive values of $x(t)$ (one per second):

$$\vec{x}^i = [x_1^i, x_2^i, \ldots, x_{50}^i].$$

The dependent variable for data point i, y^i, is the state variable at a given time in the future: $y^i = x_{50+t'}^i$. Each data point $(\vec{x}^i, y^i)$ is formed by iterating the Mackey-Glass equation with a different initial condition, where an initial condition consists of values for $\{x_{1-\tau}, \ldots, x_0\}$.

$$C_a = \left\{ \begin{array}{lllll} (x_{20} > 1.122) & \wedge & (x_{25} < 1.330) & \wedge & (x_{26} > 1.168) \quad \wedge \\ (x_{35} < 1.342) & \wedge & (x_{41} > 1.304) & \wedge & (x_{49} > 1.262) \end{array} \right\} \rightarrow y = 0.18 \pm 0.014$$

$$C_b = \left\{ \begin{array}{lllll} (x_{25} < 1.330) & \wedge & (x_{26} > 1.177) & \wedge & (x_{31} > 1.127) \quad \wedge \\ (x_{38} < 1.156) & \wedge & (x_{40} < 1.256) & \wedge & (x_{46} > 1.194) \quad \wedge \\ (x_{47} < 1.311) & \wedge & (x_{49} > 1.070) & & \end{array} \right\} \rightarrow y = 0.27 \pm 0.019$$

$$C_c = \left\{ \begin{array}{lllll} (x_{24} > 0.992) & \wedge & (x_{29} < 1.150) & \wedge & (x_{30} > 1.020) \quad \wedge \\ (x_{34} < 1.090) & \wedge & (x_{40} < 0.951) & \wedge & (x_{42} > 0.599) \quad \wedge \\ (x_{45} > 0.591) & \wedge & (x_{49} < 0.763) & \wedge & (x_{50} > 0.576) \end{array} \right\} \rightarrow y = 1.22 \pm 0.024$$

$$C_d = \left\{ \begin{array}{lllll} (x_{19} < 0.967) & \wedge & (x_{22} < 1.049) & \wedge & (x_{26} > 0.487) \quad \wedge \\ (x_{29} < 1.066) & \wedge & (x_{33} > 0.416) & \wedge & (x_{34} < 1.008) \quad \wedge \\ (x_{37} < 1.331) & \wedge & (x_{40} < 0.941) & \wedge & (x_{41} > 0.654) \quad \wedge \\ (x_{42} > 0.262) & \wedge & (x_{48} > 0.639) & \wedge & (x_{49} < 0.814) \end{array} \right\} \rightarrow y = 1.34 \pm 0.034$$

FIGURE 8 The four best condition sets found by the GA for the Mackey-Glass system with $\tau = 150$. (Adapted from Meyer and Packard.[58])

The results of running the GA using this data from the $\tau = 150$ time series with $t' = 150$ are illustrated in Figures 8 and 9. Figure 8 gives the four best condition sets found by the GA, and Figure 9 shows the four results of those condition sets. Each of the four plots in Figure 9 shows the trajectories corresponding to data points $(\vec{x}^i, y^i)$ that satisfied the condition set. The leftmost white region is the initial 50 time steps during which the data was taken. The vertical lines in that region represent the various conditions on $\vec{x}$ given in the condition set. For example, in plot a, the leftmost vertical line represents a condition on x_{20} (this set of trajectories is plotted starting at this point), and the rightmost vertical line in that region represents a condition on x_{49}. The shaded region represents the period of time not covered by $\vec{x}^i$, and the rightmost vertical line represents the point at which the y^i observation was made. Notice that in all of these plots, the values of the y^i's fall into a very narrow range, which means that the GA was successful in finding subsets of the data for which it is possible to make highly accurate predictions. (Other results along the same lines are reported in Meyer.[56])

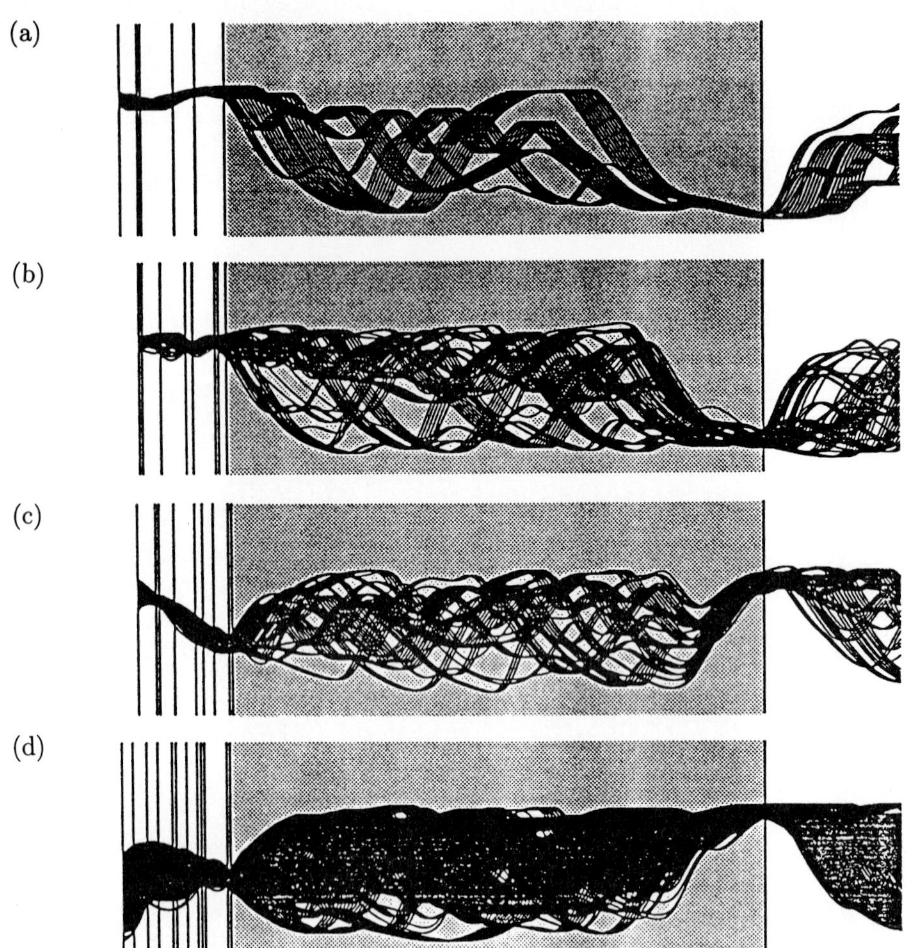

FIGURE 9 Results of the four best condition sets from Figure 8. Each plot shows trajectories of data points that satisfied that condition set. The leftmost white region is the initial 50 time steps during which data was taken. The vertical lines in that region represent the various conditions on $\vec{x}$ given in the condition set. The vertical line at the right-hand side represents the time at which the prediction is to be made. Note how the trajectories narrow at that region, indicating that the GA has found conditions for good predictability. (Reprinted from Meyer and Packard.[58] Copyright © 1992 by T. P. Meyer.)

These results are very striking, but some questions immediately arise. First and most important, are the results significant? That is, do the discovered conditions yield correct predictions for data points outside the training set (i.e., the set of data points used to calculate fitness) or do they merely describe chance statistical fluctuations in the data that were learned by the GA? Meyer and Packard performed a number of "out of sample" tests with data points outside the training set that satisfied the evolved condition sets and found that the results were robust—the y^i values for these data points also tended to be in the narrow range.

Another question is: how exactly is the GA solving the problem? What are the schemas that are being processed? What is the role of crossover in finding a good solution? Uniform crossover of the type used here has very different properties than single-point crossover, and its use makes it harder to figure out what schemas are being recombined. Meyer found that turning crossover off and relying solely on the four mutation operators did not have a large effect on the solution time or solution quality[57]; this brings up the question as to whether or not the GA is the best method for this task. An interesting extension of this work would be to perform control experiments in which the performance of the GA is compared with other search methods such as hill climbing.

Yet another question is: to what extent are the results restricted by the fact that only certain conditions are allowed (i.e., ones that are conjunctions of ranges on independent variables)? Packard[68] proposes a more general form for conditions that also allows disjunctions ($\vee$'s); an example might be:

$$\{[(3.2 \leq x_6 \leq 5.5) \vee (1.1 \leq x_6 \leq 2.5)] \wedge [0.2 \leq x_8 \leq 4.8]\}.$$

Here we are given two non-overlapping choices for the conditions on x_6. A further generalization proposed by Packard would be to allow disjunctions between sets of conditions.

A final question is: to what extent will this method succeed on other types of prediction tasks? Packard[68] proposes applying this method to tasks such as weather prediction, financial-market prediction, speech recognition, and visual pattern recognition. It is an open question to what extent this method will succeed on such "real-world" prediction tasks.

2.3 NEURAL NETWORKS

Neural networks are becoming an increasingly popular approach to machine learning, and recently some efforts have been made to combine GAs and neural networks, by using GAs to evolve aspects of neural networks.

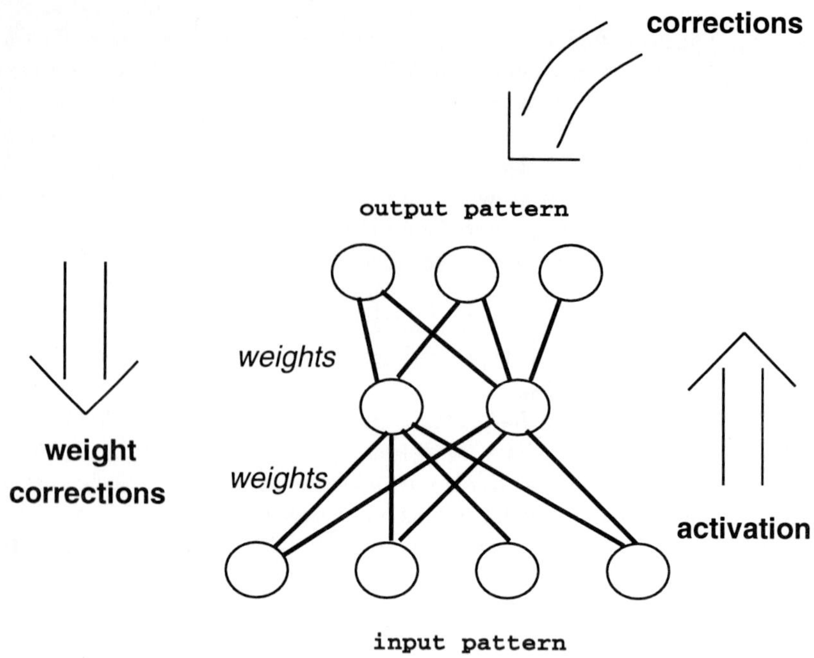

FIGURE 10 A schematic diagram of a simple feedforward neural network and the learning process by which weight values are adjusted.

In its simplest form (illustrated in Figure 10), a neural network is a collection of connected units in which the connections are weighted, usually with real-valued weights. The network is presented with an input pattern on its input units (e.g., a set of numbers representing features of an image to be classified). Activation spreads over the weighted connections according to some predefined method and winds up as an activation pattern over the output units that encodes the network's "answer" to the input (e.g., a classification of the input pattern). In many applications, the network learns a correct mapping between input and output patterns via a learning algorithm in which a set of inputs are presented to the network. After each input has propagated through the network and an output has been produced, the weights in the network are adjusted in order to reduce the difference between the network's output and the correct desired output. The most common weight adjustment method for feedforward networks is known as *back-propagation*.[77] (For an overview of neural networks and their applications, the reader can consult Rumelhart and McClelland[55,78] and Hertz, Krogh, and Palmer.[32])

There are many possible ways to apply GAs to neural networks. Some possible aspects that can be evolved are:

- the weights (and thresholds) in a fixed network;
- the network architecture; and
- the learning rule used by the network.

In this subsection, I will describe three different projects, each using a GA to evolve one of these aspects.

EVOLVING WEIGHTS IN A FIXED NETWORK. David Montana and Lawrence Davis[63] took the first approach—evolving the weights and thresholds in a fixed network. That is, Montana and Davis were using the GA *instead* of back-propagation as a way of finding a good set of weights and thresholds. Several problems associated with the back-propagation algorithm (including the tendency to get stuck at local optima in weight space) often make it desirable to find alternative weight-training schemes.

Montana and Davis were interested in using neural networks to classify underwater sonic "lofargrams" (similar to spectrograms) into two classes: "interesting" and "not interesting." The overall goal is to "detect and reason about interesting signals in the midst of the wide variety of acoustic noise and interference which exist in the ocean." The networks were to be trained from a database containing lofargrams and classifications made by experts as to whether or not a given lofargram is "interesting." Each network had four input units representing four parameters used by an expert system that performs the same classification. Each network had one output unit and two layers of hidden units with seven and ten units respectively. The networks were fully connected feedforward, meaning that each unit was connected to every unit in the next higher layer. Thus there were a total of 108 weights on connections between units, and an additional 18 weights on connections between the non-input units and a bias (threshold) unit—a total of 126 weights to evolve.

The GA was used as follows. Each chromosome is a list (or "vector") of 126 weights. Figure 11 shows (for a much smaller network) how the encoding was done: the weights are read off the network in a fixed order and placed in a list. Notice that, here, each "gene" in the chromosome is a real number rather than a bit. To calculate the fitness of a given chromosome, the weights in the chromosome are assigned to the links in the corresponding network, the network is run on the training set (here 236 examples from the database of lofargrams), and the sum of the squares of the errors (i.e., differences between the desired output and the actual output) is returned. Here, low error means high fitness.

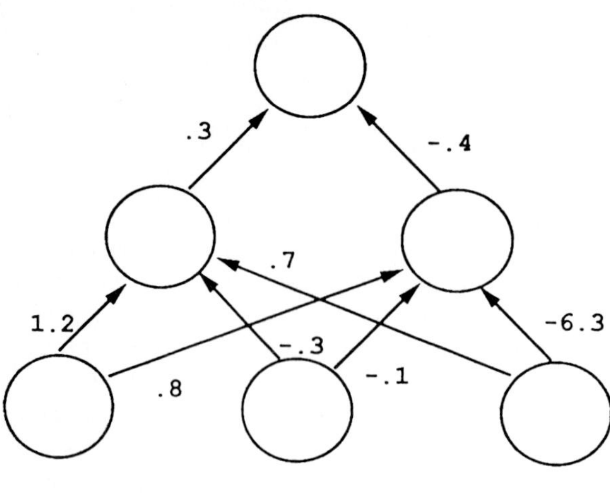

(.3 –.4 1.2 .8 –.3 –.1 .7 –6.3)

FIGURE 11 Illustration of Montana and Davis' encoding of network weights into a list that serves as a "chromosome" for the GA.

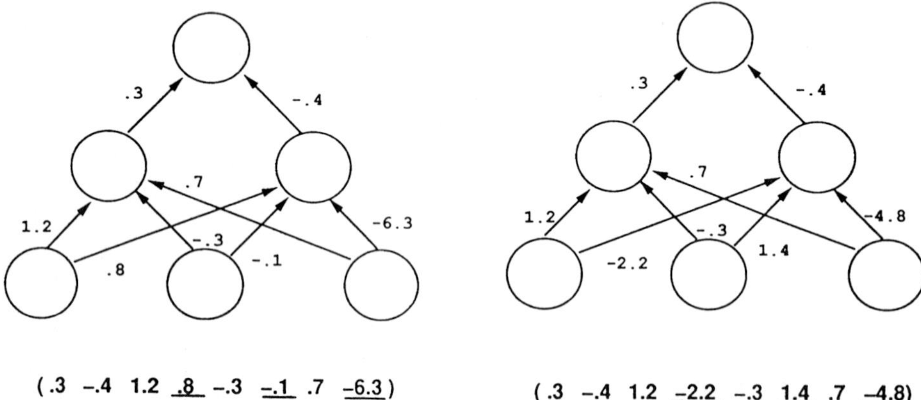

FIGURE 12 Illustration of Montana and Davis' mutation method. The weights on incoming links to the right-hand node in the middle layer (underlined weights) are mutated.

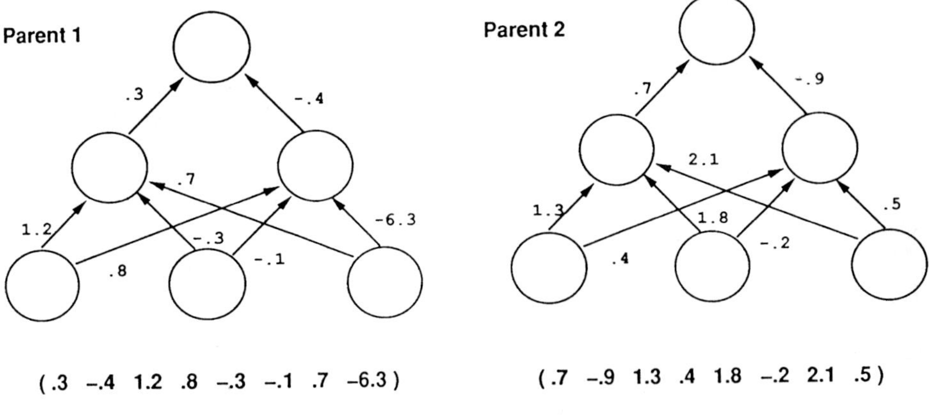

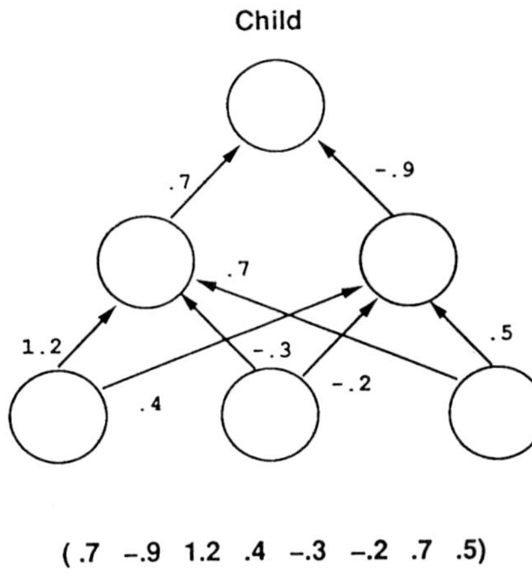

FIGURE 13 Illustration of Montana and Davis' crossover method. The offspring is created as follows: for each non-input node, a parent is chosen at random and the weights on the incoming links to that node are copied from the chosen parent.

An initial population of 50 weight vectors was chosen randomly, according to the probability distribution given by $e^{-|x|}$. (Each weight is between –1.0 and +1.0.) Montana and Davis tried a number of different genetic operators in various experiments. The mutation and crossover operators they used for their comparison of the GA with back-propagation are illustrated in Figures 12 and 13. The mutation operator selects n non-input units and, for each incoming link, adds a random value to the weight on the link. In their experiments, $n = 2$ and the random values were selected between –1.0 and +1.0 from the $e^{-|x|}$ distribution. The crossover operator

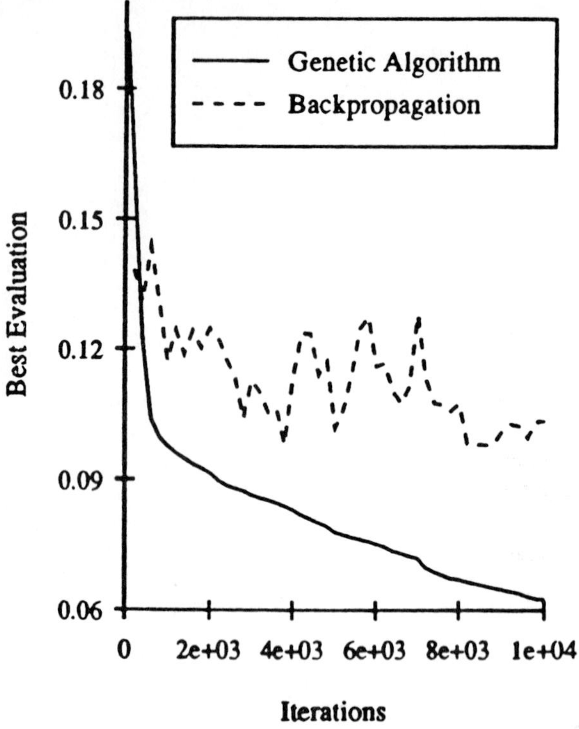

FIGURE 14 Montana and Davis' results comparing the performance of the GA with back-propagation. The figure plots the best evaluation (lower is better) at a given iteration. (Reprinted from Montana and Davis[63] by permission of the authors. Copyright ©International Joint Conference on Artificial Intelligence, *Proceedings of the International Joint Conference on Artificial Intelligence*, Morgan Kaufmann, 1989. Copies of this and other IJCAI proceedings are available from Morgan Kaufmann Publishers, Inc., 2929 Campus Drive, San Mateo, CA 94403.)

takes two parent weight vectors and, for each non-input unit in the offspring vector, selects one of the parents at random and copies the weights on the incoming links from that parent to the offspring.

The performance of a GA using these operators was compared with the performance of a back-propagation algorithm. The GA had a population of 50 weight vectors, and a selection method was used in which the population was ranked by fitness and the rankings (rather than absolute fitness) determined the probability of allowing a given weight vector to reproduce (either via direct copying, crossover with another weight vector, or mutation). The GA was allowed to run for 200 generations (or 10,000 network evaluations). The back-propagation algorithm was allowed to run for 5,000 iterations, where one iteration is a complete cycle through the training data. Montana and Davis reasoned that two network evaluations under the GA are equivalent to one back-propagation iteration, since back-propagation on a given training example consists of two parts—the forward propagation of activation (and the calculation of errors at the output units) and the backward error propagation (and adjusting of the weights). The GA performs only the first part, and since the second part requires more computation, one GA evaluation takes less than half the computation of a single back-propagation iteration.

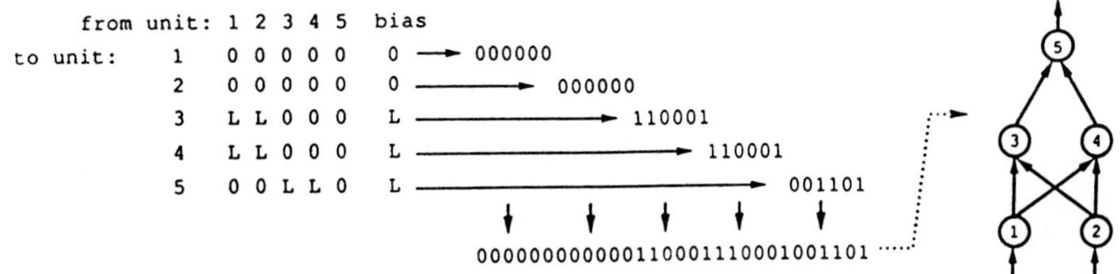

FIGURE 15 An illustration of Miller, Todd, and Hegde's representation scheme. Each entry in the matrix represents the type of connection on the link between the *from* unit (column) and the *to* unit (row). The rows of the matrix are strung together to make the bit-string encoding of the network, given at the bottom of the figure. The resulting network is shown at the right. (Reprinted from Miller et al.[60] by permission of the authors.)

The results of the comparison are displayed in Figure 14. Here one back-propagation iteration is plotted for every two GA evaluations. The x-axis gives the number of iterations, and the y-axis gives the best evaluation (lowest sum of squares of errors) found by that time. It can be seen that the GA significantly outperforms back-propagation on this task, obtaining better weight vectors more quickly. (Montana and Davis also presented other experimental results not discussed here. See Montana and Davis[63] for details.)

This experiment shows that in some situations, the GA is a better training method for networks than is simple back-propagation. This does not mean that the GA will outperform back-propagation in all cases. It is also possible that enhancements of back-propagation might help it overcome some of the problems that prevented it from performing as well as the GA in this experiment. More research needs to be done to characterize the problems for which the GA will outperform back-propagation for discovering weight vectors.

EVOLVING NETWORK ARCHITECTURES. The second approach to applying GAs to neural networks—evolving the network architecture—is illustrated in work done by Geoffrey Miller, Peter Todd, and Shailesh Hegde (MT&H).[60] Here, "network architecture" refers to structural aspects of the network: the number of units in the network and their topological arrangement in terms of interconnections. The neural network community has produced many heuristics for designing network architecture (e.g., "more hidden units are required for more difficult problems"), but there is no sure recipe to follow in designing the best architecture for a given problem. MT&H propose that the GA is a promising method to automate the design procedure.

Of course, the first problem is to decide on a scheme for representing network architectures as chromosomes. MT&H restricted their initial project to feedforward networks with a fixed number of units, for which the GA will evolve the connection topology. MT&H used the representation scheme illustrated in Figure 15. The connection topology is represented by a 2×2 matrix, in which each entry represents the type of connection from the "from unit" to the "to unit." Here there are only two possible elements: "0," meaning no connection, and "L," meaning a "learnable" connection—i.e., one for which the weight can be changed through learning. Figure 15 also shows how the connection matrix is transformed into a bit-string chromosome for the GA ("0" corresponds to 0 and "L" to 1) and how the bit string is decoded into a network (the connections from a bias unit to units 3, 4, and 5 are not shown). Connections that were specified to be learnable are initialized with small random weights. Since MT&H currently restrict these networks to be feed-forward, any connections to input units or feedback connections specified in the chromosome are ignored.

MT&H used a simple fitness-proportionate reproduction method and the usual mutation operator (bits in the string were flipped with some low probability). Their crossover operator chose the bits corresponding to a random row in the matrix, and those bits (representing the entire row) were swapped between the two parents to produce the two offspring. The intuition behind that operator is similar to that behind Montana and Davis' crossover operator—each row represents all the incoming connections to a single unit, and this set is thought to be a functional building block of the network.

The fitness of a chromosome is calculated in the same way as in Montana and Davis' project: for a given problem, the network is trained on a training set for a certain number of epochs (one "epoch" is one pass through the training set), using back-propagation to modify the weights. The fitness of the chromosome is a function of the sum of the squares of the errors on the training set at the last epoch. Again, low error translates to high fitness.

MT&H tried their GA on three tasks (XOR, a "four quadrant" problem, and pattern copying) that are relatively easy for neural networks to learn. The networks had different numbers of units for different tasks (ranging from five units for the XOR task to 20 units for the pattern-copying task); the goal was to see if the GA could discover a good connection topology for each task. For each run the population size was 50, the crossover rate was 0.6 (probability for a given pair of parents to cross over), and the bitwise mutation rate was 0.005. In all three tasks, the GA was easily able to find networks that readily learned to map inputs to outputs over the training set with little error. However, the three tasks were too easy to be a rigorous test of this method—it remains to be seen if this method can scale up to more complex tasks that require much larger networks with many more interconnections.

EVOLVING A LEARNING RULE. David Chalmers[12] took the idea of applying GAs to neural networks one step further and applied GAs to the task of evolving a good learning rule for neural networks. Chalmers limited his initial study to single-layer, fully connected feedforward networks. A learning rule is used during the training procedure for modifying network weights in response to the network's performance on the training data. At each training cycle, one training pair is given to the network, which then produces an output. Assuming a single-layer, fully connected feedforward network, a learning rule might use the following local information for a given training cycle to modify the weight on the link from input unit i to output unit j:

- a_i: the activation of input unit i;
- o_j: the activation of output unit j;
- t_j: the training signal (i.e., correct activation) on output unit j; and
- w_{ij}: the current weight on the link from i to j.

The amount to modify the weight w_{ij} is a function of these values:

$$\Delta w_{ij} = F(a_i, o_j, t_j, w_{ij}).$$

The chromosomes in the GA population encode such functions.

Chalmers made the assumption that the learning rule should be a linear function of these variables and all their pairwise products. That is, the general form of the learning rule is:

$$\Delta w_{ij} = k_0(k_1 w_{ij} + k_2 a_i + k_3 o_j + k_4 t_j + k_5 w_{ij} a_i + k_6 w_{ij} o_j + k_7 w_{ij} t_j + k_8 a_i o_j + k_9 a_i t_j + k_{10} o_j t_j).$$

(Here, k_0 is a scale parameter, which affects the learning rate of a network.) The assumption about the form of the learning rule came in part from the fact that a known good learning rule for such networks—the "Widrow-Hoff" or "delta" rule—has this form: $\Delta w_{ij} = \eta(t_j o_j - a_i o_j)$.[78] (Here, η is a constant representing the learning rate.) One goal of this work was to see if the GA could evolve a rule that is as good as the delta rule.

The task of the GA is to evolve values for the k_i's. The chromosome encoding for the set of k_i's is illustrated in Figure 16. The scale parameter k_0 is encoded as five bits, with the zeroth bit encoding the sign (1 encoding + and 0 encoding −), and the first through fourth bits encoding an integer n: $k_0 = 0$ if $n = 0$; otherwise, $|k_0| = 2^{n-9}$. Thus k_0 can take on the values 0, $\pm 1/256$, $\pm 1/128$, ..., ± 32, ± 64. The other coefficients k_i are encoded by three bits each, with the zeroth bit encoding the sign and the first and second bits encoding an integer n. For $i = 1 \ldots 10$, $k_i = 0$ if $n = 0$; otherwise, $|k_i| = 2^{n-1}$.

It is known that single-layer networks can learn only input-output mappings that are linearly separable.[78] As an "environment" for the evolving learning rules,

Genome encoding:

k_0 k_1 k_2 k_3

1 0 0 1 0   0 0 1   0 0 0   1 1 0   ...

k_0 encoded by 5 bits:

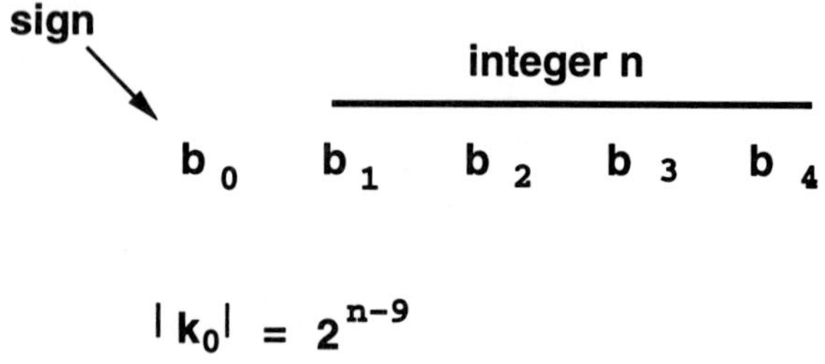

Other k's encoded by 3 bits each:

sign

integer n

b_0 b_1 b_2

$|k_i| = 2^{n-1}$

FIGURE 16 Illustration of the method for encoding the k's in Chalmers' system.

Chalmers used 30 different linearly separable mappings to be learned via the learning rules. The mappings always had a single output unit and between two and seven input units.

The fitness of each chromosome (learning rule) was determined as follows. A subset of 20 mappings was selected from the full set of 30 mappings and, for each mapping, 12 training examples were selected. For each of these mappings, a network was created with the appropriate number of input units for the given mapping (each network had one output unit). The network's weights were initialized randomly. The network was run on the training set for some number of epochs (typically 10), using the learning rule specified by the chromosome. The performance of the learning rule on a given mapping was a function of the network's error on the training set, with low error meaning high performance. The overall fitness of the learning rule was a function of the average error of the 20 networks over the chosen subset of 20 mappings—low average error translated to high fitness. This fitness was then transformed to be a percentage, where a high percentage means high fitness.

Using this fitness measure, the GA was run on a population of 40 learning rules, with two-point crossover (crossover was performed at two points along the chromosome rather than at one point) and standard mutation. The crossover rate (probability of two parents crossing over) was 0.8 and the bitwise mutation rate was 0.01.

The results of a run of the GA were that, over 1000 generations, the fitness of the best learning rules in the population rose from between 40%–60% in the initial generation (indicating no significant learning ability) to between 80% and 98%, with a mean (over several runs) of about 92%. The fitness of the delta rule is around 98% and, on one out of ten runs, the GA discovered a successful form of this rule with 98% fitness. (On three other runs, it discovered slight variations of this rule with lower fitness.)

These results show that, given a somewhat constrained representation, the GA is able to evolve a successful learning rule for simple single-layer networks. To what extent this method will be successful in finding learning rules for more complex networks (including networks with hidden units) remains an open question, but these results are a first step in that direction. Chalmers points out that it is unlikely that evolutionary methods will discover learning methods that are more powerful than back-propagation, but speculates that the GA might be a powerful method for discovering learning rules for unsupervised learning paradigms (e.g., reinforcement learning) or for new classes of network architectures (e.g., recurrent networks).

Chalmers performed a second interesting study in which he asked the question: How much diversity in learning tasks is needed to produce a general learning rule? That is, to what extent were the learning rules that evolved effective only on the specific environment of the given 20 mappings and to what extent were they more general? This is similar to the issues brought up in Axelrod's Prisoner's Dilemma study described earlier, in which the initial experiment yielded rules that were specifically adapted to a fixed environment of strategies and in which more generally successful strategies evolved only in a more diverse environment (the environment made up of the other evolving strategies).

To study this issue, Chalmers first measured the generality of the best rules evolved in the set of ten runs by testing each one on the ten mappings that had not been used in the fitness calculation for that rule. The mean fitness of the best

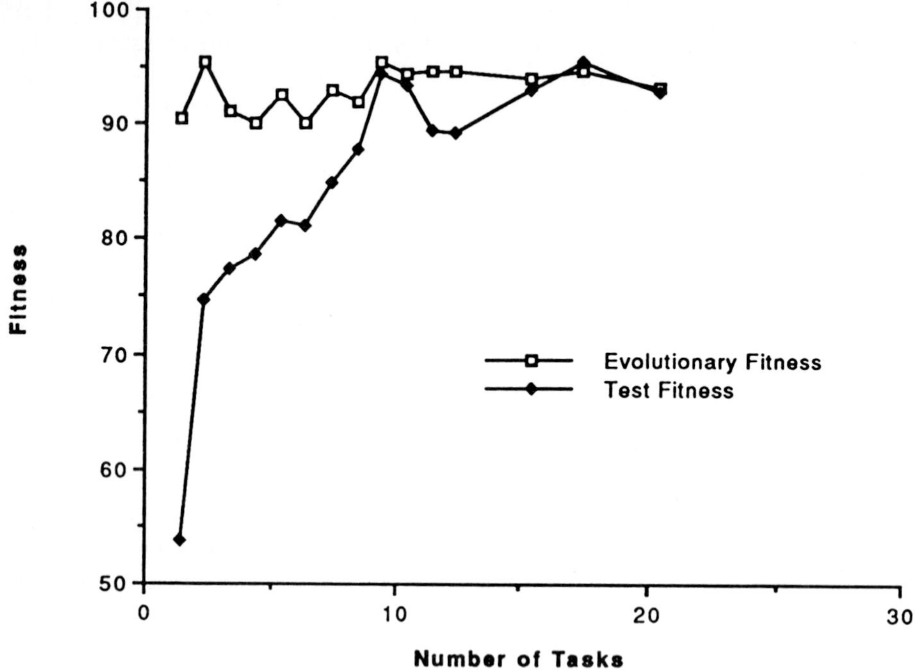

FIGURE 17 Results of Chalmers' experiments testing the effect of diversity of environment on generalization ability. The plot gives the evolutionary fitness and test fitness as a function of the number of tasks in the environment. (Reprinted from Chalmers[12] by permission of the author.)

rules on the original mappings was 92%, and Chalmers found that the mean fitness of these rules on the test set was 91.9%, indicating that the environment of 20 mappings was sufficiently diverse for the GA to evolve general rules.

Chalmers then looked at the question of how diverse the environment has to be to produce general rules. He repeated the original experiment, varying the number of mappings in each original environment between 1 and 20. A rule's *evolutionary fitness* is the fitness obtained by testing a rule on its original environment. A rule's *test fitness* is the fitness obtained by testing a rule on ten additional tasks not in the original environment. Chalmers then measured these two quantities as a function of the number of tasks in the original environment. The results are shown in Figure 17. The two curves are the mean evolutionary fitness and the mean test fitness for rules that were tested in an environment with the given number of tasks. This plot shows that while the evolutionary fitness stays roughly constant for different numbers of environmental tasks, the test fitness increases sharply with the number of tasks, leveling off somewhere between 10 and 20 tasks. The conclusion is that the evolution of a general learning rule requires a sufficiently diverse environment of tasks although, in this case of simple single-layer networks, the necessary degree of diversity is fairly small.

3. GENETIC ALGORITHMS IN SCIENTIFIC MODELS

In this section I describe two modeling projects, one project on modeling the interaction between evolution and learning, and a related project in which a simple model of culture is added to the original model.

3.1 MODELING THE INTERACTION BETWEEN LEARNING AND EVOLUTION

Many people have drawn analogies between learning and evolution as two adaptive processes—one taking place during the lifetime of an organism, and the other taking place over the evolutionary history of life on Earth. A major question in evolutionary theory and in psychology is: to what extent do these processes interact? In particular, can learning that occurs over the course of an individual's lifetime guide the evolution of that individual's species to any extent? The famous (or infamous) Lamarckian hypothesis states that traits acquired during the lifetime of an organism can be transmitted genetically to the organism's offspring. Lamarck's hypothesis is generally interpreted to refer to acquired physical traits (such as physical defects due to environmental toxins), but something learned during an organism's lifetime also can be thought of as a type of acquired trait. Thus, according to Lamarck, learning might guide evolution directly. However, because of overwhelming evidence against it, the Lamarckian hypothesis has been rejected by virtually all biologists; in addition, it is very hard to imagine a direct mechanism for "reverse transcription" of acquired traits into a genetic code.

Does this mean that learning can have no effect on evolution? In spite of the rejection of Lamarckianism, the (perhaps surprising) answer seems to be that learning can indeed have significant effects on evolution, though in less direct ways than Lamarck proposed. One proposal of a mechanism by which learning affects evolution is due to J. M. Baldwin, and is known as the "Baldwin Effect."[6] (Similar mechanisms were proposed by Lloyd Morgan[51] and Waddington.[86]) Baldwin pointed out that if learning helps survival, then organisms best able to learn will have the most offspring, thus increasing the frequency of the genes responsible for learning. And if the environment remains relatively fixed so that the best things to learn remain constant, then this can lead, via selection, to a genetic encoding of a trait that originally had to be learned.[54] For example, an organism that has the capacity to learn that a particular plant is poisonous will be more likely to survive (by learning not to eat the plant) than organisms that are unable to learn this information, and thus will be more likely to produce offspring that also have this learning capacity. Evolutionary variation will have a chance to work on this line of offspring, allowing for the possibility that the trait—avoiding the poisonous plant—will be discovered genetically rather than learned anew each generation. Having the desired behavior encoded genetically would give an organism a selective advantage over organisms that were merely able to learn the desired behavior during their lifetimes, because learning a behavior is generally a less robust process than developing a genetically encoded behavior. Too many unexpected things could get in the way of learning during an organism's lifetime. In short, the capacity to acquire a certain desired trait allows the learning organism to survive preferentially, thus giving genetic variation the possibility to independently discover the desired trait. Without such learning, the likelihood of survival—and thus the opportunity for genetic discovery—decreases. In this indirect way, learning *can* guide evolution, even if what is learned cannot be transmitted genetically.

Some computer scientists and computational biologists have constructed computer models that explore issues related to the interaction between learning and evolution (e.g., Hinton and Nowlan,[34] Belew,[7] Nolfi et al.,[67] Fontanari and Meir,[20] Ackley and Littman,[1] and Parisi et al.[70]). In this section I describe one model constructed by Geoffrey Hinton and Steven Nowlan and an extension constructed by Richard Belew.

Hinton and Nowlan used the GA to construct a computer model of the Baldwin effect.[34] Their goal was to empirically demonstrate this effect and to measure its magnitude, using the simplest possible model. An extremely simple neural-network learning algorithm modeled learning, and the GA played the role of evolution, evolving a population of neural networks with varying learning capabilities. In Hinton and Nowlan's model, each individual is a neural network with 20 potential connections. Each connection can have one of three values: "present," "absent," and "learnable." These are specified by "1," "0," and "?," respectively, where each "?" connection can be set during learning to either 1 or 0. There is only one correct setting for the connections (i.e., only one correct set of 1's and 0's). The problem to be solved (the learning goal) is for each network to find this single correct set of connections. This will not be possible for those networks that have incorrect fixed connections (e.g., a 1 where there should be a 0), but those networks that have correct settings in all places except where there are ?'s have the capacity to learn the correct settings.

Hinton and Nowlan used the simplest possible "learning" method: random guessing. On each learning trial, a network simply guesses a 1 or 0 at random for each of its learnable connections.

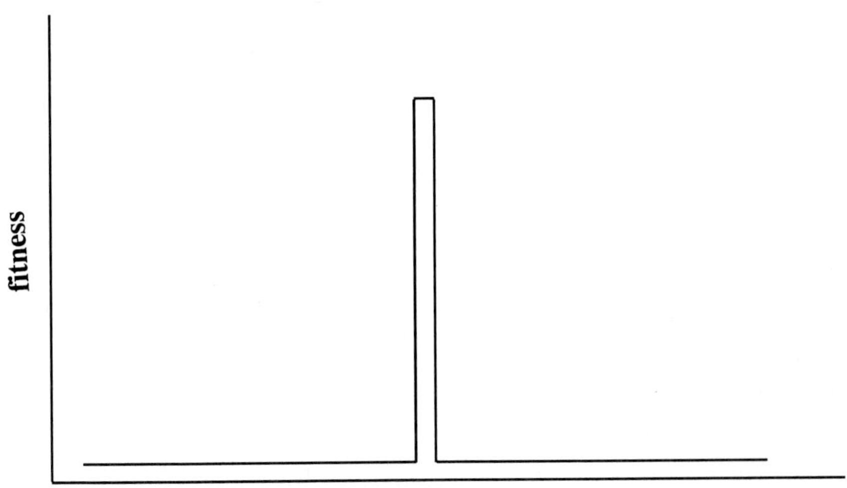

FIGURE 18 Illustration of the fitness landscape for Hinton and Nowlan's search problem. (Adapted from Hinton and Nowlan[34]; copyright © 1987 by *Complex Systems* and reprinted by permission.)

This is, of course, a "needle in a haystack" search problem, since there is only one correct setting in a space of 2^{20} possibilities.[1] The fitness landscape for this problem is illustrated in Figure 18—the single spike represents the single correct connection setting. Introducing the ability to learn changes the shape of this landscape, as shown in Figure 19. Here the spike is smoothed out into a "zone of increased fitness," within which it is possible to learn the correct connections.

For the GA, each network is represented by a string of length 20 consisting of the 1's, 0's, and ?'s making up the settings on the network's connections. The initial population consists of 1,000 individuals, generated at random but with each individual having on average 25% 0's, 25% 1's, and 50% ?'s. The fitness of an individual is calculated as follows. Each individual is given 1,000 learning trials—on each learning trial, the individual tries a random combination of settings for

[1]The problem as stated has little to do with the usual notions of neural-network learning; Hinton and Nowlan presented this problem in terms of neural networks so as to keep in mind the possibility of extending the example to more standard learning tasks and methods.

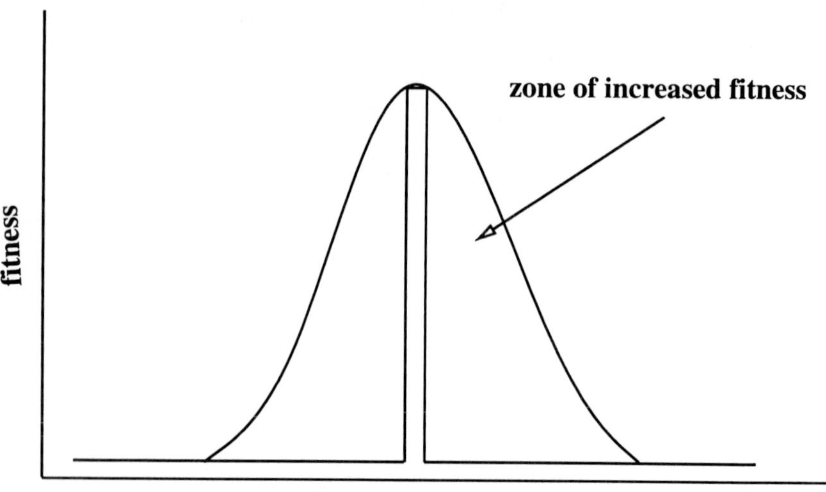

FIGURE 19 With the possibility of learning, the fitness landscape for Hinton and Nowlan's search problem is smoother, with a zone of increased fitness. (Adapted from Hinton and Nowlan[34]; copyright © 1987 by *Complex Systems* and reprinted by permission.)

the ?'s. The fitness is a function of the number of trials needed to find the correct solution:

$$\text{Fitness} = 1 + \frac{19n}{1000},$$

where n is the number of trials (out of 1000) left after the correct solution has been found. Thus an individual that already has all its connections set correctly would have fitness 20, and an individual that never finds the correct solution would have fitness 1. Hence, a tradeoff exists between efficiency and flexibility: having many ?'s means that, on average, many guesses are needed to arrive at the correct answer, but the more connections that are fixed, the more likely it is that one or more of them will be fixed incorrectly, meaning that there is no possibility to find the correct answer.

Hinton and Nowlan's GA is similar to the simple GA described in subsection 1.3. An individual is selected to be a parent with probability proportional to its fitness, and can be selected more than once. The next generation is created by 1,000 single-point crossovers between pairs of parents. No mutation occurs. An individual's chromosome is, of course, not affected by the learning that takes place during its lifetime—parents pass on their original alleles to their offspring.

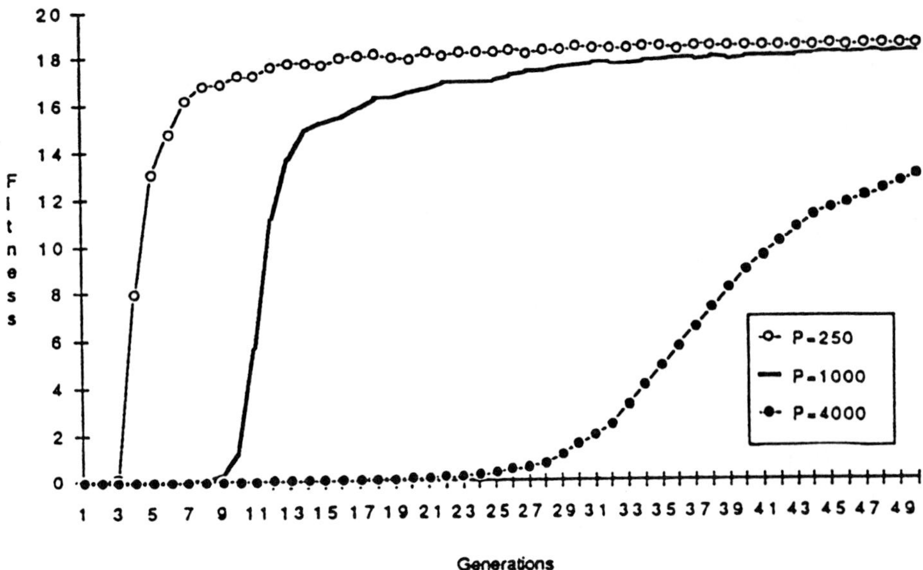

FIGURE 20 Mean fitness versus generations for one run of the GA on each of three population sizes. The solid line gives the results for population size 1000, the size used in Hinton and Nowlan's experiments. These plots are from a replication by Belew.[7] (Reprinted from Belew[7]; copyright © 1990 by *Complex Systems* and reprinted by permission.)

Hinton and Nowlan ran the GA for 50 generations. A plot of the mean fitness of the population versus generation for one run on each of three population sizes is given in Figure 20. (This plot is from a replication of Hinton and Nowlan's experiments performed by Belew.[7]) The solid curve gives the results for population size 1000, the size used in Hinton and Nowlan's experiments. This plot shows that learning during an individual's "lifetime" indeed guides evolution by allowing the population fitness to increase. This increase in fitness is due to a Baldwin-like effect: those individuals that are able to efficiently learn the task tend to be selected to reproduce, and crossovers among these individuals tend to increase the number of correctly fixed alleles, increasing the learning efficiency of the offspring. With learning, evolution could discover individuals with all their connections fixed correctly (and such individuals were discovered in these experiments). Without learning, the evolutionary search never discovered such an individual.

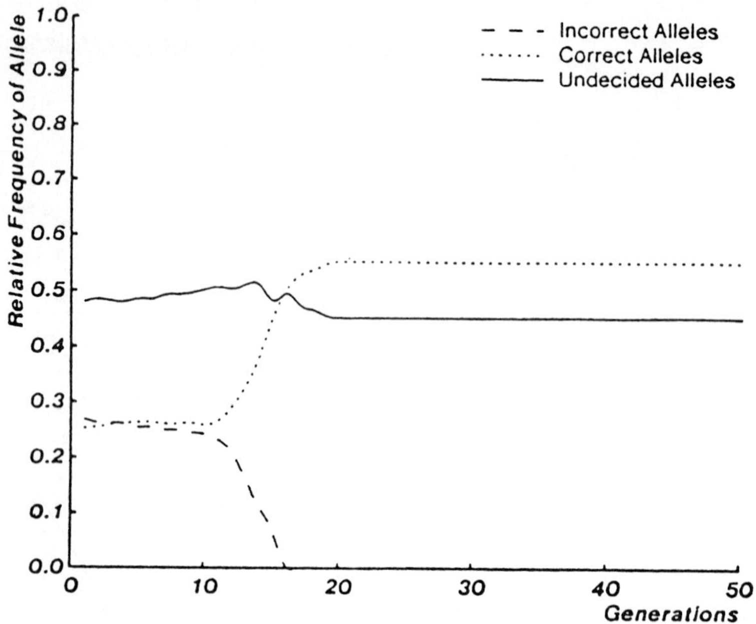

FIGURE 21 Relative frequencies of correct, incorrect, and undecided (i.e., "?") alleles in the population plotted over 50 generations. (Reprinted from Hinton and Nowlan[34]; copyright © 1990 by *Complex Systems* and reprinted by permission.)

Figure 21 shows the relative frequencies of the correct, incorrect, and undecided (i.e., "?") alleles in the population plotted over 50 generations. As can be seen, over time the frequency of fixed correct connections increases and the frequency of fixed incorrect connections decreases.

On inspection of Figure 21, one question immediately comes up: why does the frequency of undecided alleles stay so high? Hinton and Nowlan answer that there is not much selective pressure to fix all the undecided alleles, since individuals with a small number of ?'s can learn the correct answer in a small number of learning trials. If the selection pressure were increased, then the Baldwin effect would be stronger. Figure 22 shows these same results over a much-extended run (these results come from Belew's replication and extension of Hinton and Nowlan's original experiments[7]). This plot shows that the frequency of ?'s goes down to about 30%. It would go down to zero given enough time but, under this selection regime, the convergence is extremely slow.

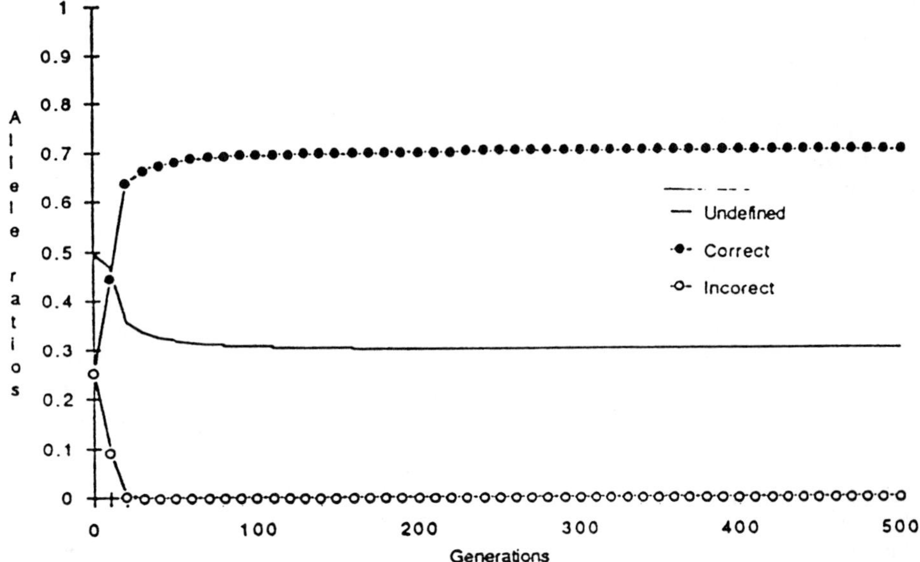

FIGURE 22 Relative frequencies of correct, incorrect, and undecided (i.e., "?") alleles in the population plotted over 500 generations, from Belew's replication of Hinton and Nowlan's experiments. (Reprinted from Belew[7]; copyright © 1990 by *Complex Systems* and reprinted by permission.)

In short, learning allows genetically coded partial solutions to get partial credit, rather than the all-or-nothing reward that an organism would get without learning. A common argument for the benefits of learning is that learning makes it possible to deal with genetically unpredictable aspects of the environment (e.g., aspects that change too quickly for evolution to keep up). While this is clearly one benefit of learning, the Baldwin effect is different: the Baldwin effect says that learning helps organisms adapt to genetically *predictable* (but difficult) aspects of the environment, and that learning allows these adaptations to eventually become genetically encoded. Thus the Baldwin effect is important only on fitness landscapes that are hard to search by evolution alone, such as the extreme example given by Hinton and Nowlan.

The "learning" mechanism used in Hinton and Nowlan's experiments—random guessing—is, of course, completely unrealistic as a model of learning. However, their goal was to keep the model as simple as possible, without compromising the model's generality too much. Hinton and Nowlan point out that "a more sophisticated learning procedure only strengthens the argument for the importance of the Baldwin effect" (see Hinton and Nowlan,[34] p. 500). This is true insofar as a more sophisticated learning procedure would further broaden the "zone of increased fitness" shown in Figure 19. However, if the learning procedure were *too* sophisticated—that is, if learning the necessary trait were too easy—then there would be little selection pressure for evolution to move from the ability to learn a trait to a genetic hardwiring of that trait. Such tradeoffs occur in evolution and can be seen even in Hinton and Nowlan's simple model. Computer simulations such as theirs can help us to understand and to measure the details of such tradeoffs, as well as other details of subtle processes such as the Baldwin effect.

3.2 ADDING CULTURE

Belew[7] carried out a careful replication and analysis of Hinton and Nowlan's model of learning and evolution and also extended the model to incorporate a third factor, culture. He defined culture as an "adaptive system that allows the hard-won knowledge learned by an individual to improve the evolutionary fitness of other conspecifics (i.e., members of the same species) via nongenetic informational pathways." Belew's models of culture were, like Hinton and Nowlan's model of learning, extremely simple; these models are not meant to be realistic in their details but rather to be first steps in assessing the effects of the interaction among learning, evolution, and culture.

In his first experiment, Belew made one addition to Hinton and Nowlan's model: an offspring from a successful parent (one who has found the correct solution within the allotted number of learning trials) is given a "cultural advantage" (CA). This cultural advantage gives the offspring a better than 0.5 chance of guessing the correct values of learnable connections. In most of Belew's simulations, $CA = 0.1$, which means that the offspring of a successful parent has a 0.6 chance of guessing the correct value for any learnable connection.

The results of adding this form of "culture" to the model are shown in Figure 23, where plots (for a single, typical run) of the mean population fitness versus generation with and without CA are shown. The population fitness with $CA = 0.1$ rises much more quickly than that without CA, but does not get as high. This is because a culturally advantaged individual with many learnable connections (?'s) has a much better chance of being successful (guessing the right answer)—the more effective learning procedure (guessing plus CA) serves to broaden the "zone of increased fitness" shown in Figure 19. Culturally advantaged individuals thus tend to reproduce more quickly and, via crossover, spread ?'s in the population, leading to worse average performance. In the model with CA, the population tends to converge to a suboptimal solution (an individual with a relatively large number of ?'s).

To counter this convergence, Belew introduced a small probability of mutation into the GA (recall that Hinton and Nowlan's GA had no mutation). The results are shown in Figure 24. The four curves give (in the order given in the plot's key) the original results with no mutation and no CA, the results with no mutation but

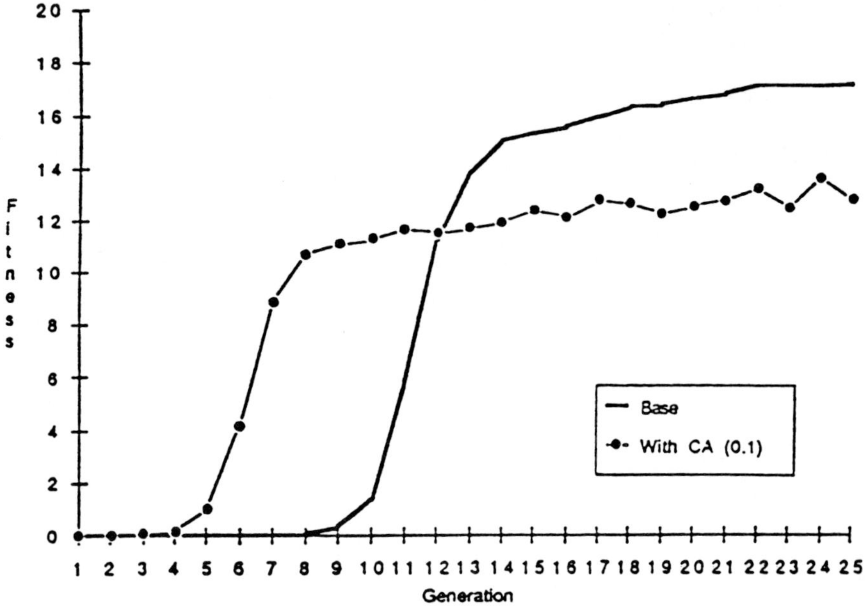

FIGURE 23 Mean population fitness versus generation with and without "cultural advantage" (*CA*). Each plot is for a single, typical run from Belew's experiments. (Reprinted from Belew[7]; copyright 1990 by *Complex Systems* and reprinted by permission.)

with $CA = 0.1$, the results with no *CA* but with a mutation probability of 0.001 per bit, and the results with both *CA* and mutation. It can be seen that the model with both *CA* and mutation retains the advantages of *CA* alone and avoids the disadvantages.

Belew carried out a third experiment, in which he studied how culture helps a system to deal with a changing environment. The environment here is the specific correct connection settings. Belew changed the environment by varying several of the bits in that correct solution every 25 generations. He found that the original model (with no mutation or *CA*) was not able to adapt to these changes in the environment; the population converged on certain alleles that were correct during the first 25 generations and was not able to modify these when the environment changed. Introducing a small probability of mutation into the original model allowed the population to adapt to changes, but the adaptation process was relatively slow. On the other hand, introducing cultural advantage with mutation allowed the population to adapt very quickly to changes. Cultural advantage allowed the population to stay flexible (i.e., retain more learnable connections) so it could adapt to environmental changes, while mutation prevented a too-fast convergence to many ?'s and thus kept the mean fitness of the population high.

Belew also experimented with a different model of culture—a "broadcast" model—which will not be described here.

Belew's model of cultural advantage is, of course, much too simple to be realistic but, even in its simplicity, it captures something about how culture interacts with evolution and learning. These simulations are a first step in using abstract computer models to help us to understand these complex interactions. Belew's results show two advantages for culture: it allows faster convergence on a solution and it allows robustness in the face of a changing environment.

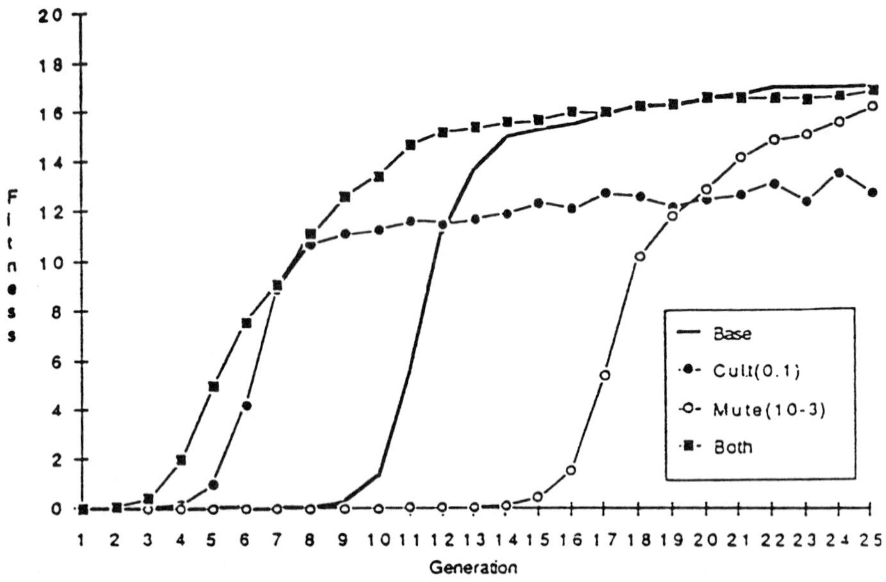

FIGURE 24 Mean population fitness versus generation with and without *CA*, and with and without *CA* with a small probability of mutation added. Each plot is for a single, typical run from Belew's experiments. (Reprinted from Belew[7]; copyright © 1990 by *Complex Systems* and reprinted by permission.)

4. THEORETICAL FOUNDATIONS OF GENETIC ALGORITHMS

As GAs become more and more widely used for practical problem solving and for scientific modeling, increasing emphasis has been put on understanding the theoretical foundations of this class of algorithms. Some major questions in this area are:

- What are the laws describing the behavior of schemas in GAs?
- How can we characterize the types of fitness landscapes on which the GA is likely to perform well?
- What does it mean for a GA to "perform well"? That is, what is the GA good at doing?
- How can we characterize the types of fitness landscapes on which the GA outperforms other search methods, e.g., hill climbing?

The first question above is answered in part by the Schema Theorem. As was described in subsection 1.6, the Schema Theorem states that those schemas whose average fitness remains above the population mean will receive exponentially increasing numbers of samples over time. This idea is related to the solution of the two-armed bandit problem discussed in Holland.[38] The two-armed bandit problem asks: Given a slot machine with two arms, each with an unknown average payoff rate, what strategy of dividing one's play between the two arms is optimal for making a profit? The solution states that the optimal strategy is to be *willing* at all times to sample either arm, but with probabilities whose ratio diverges increasingly fast as time progresses. In particular, as more and more information is gained through sampling, the optimal strategy is to exponentially increase the probability of sampling the better-seeming arm relative to the probability of sampling the worse-seeming arm. (One never knows with absolute certainty which of the two actually *is* the better arm, since all information gained is merely statistical evidence.) The possible schemas in a search space can be likened to the arms on a multiarmed bandit, and the evaluation of a given string in a population is like sampling a number of arms at once—the arms corresponding to the schemas of which the string is an instance. The Schema Theorem shows how the GA implicitly obtains statistical averages (without explicit calculations) for the various schemas (arms), and then implicitly allocates exponentially increasing numbers of samples to those schemas (arms) that are observed to be above-average.

There are some limitations, though, to this analogy between the behavior of GAs and the solution to the multiarmed bandit problem. The analogy assumes that the observed average fitness of a schema is close to its actual average fitness. However, this can fail for several reasons,[30] including large variation within a schema, small population size, and biased sampling due to premature convergence (e.g., the population might converge to a set of strings that are mostly instances of 1111** ... *; this would result in a biased estimate of the average fitness of, say, schema 11**** ... *).

In addition, the Schema Theorem addresses only the negative aspects of crossover—i.e., to what extent it disrupts schemas. It does not address the question of how crossover works to recombine highly fit schemas. What is needed is a more detailed description of the dynamics of building-block processing and combination.

There have been other approaches to understanding GAs and characterizing the landscapes on which they will perform well. In this section I will describe three of these approaches: Walsh analysis and GA deception, characterizing the effects of the statistical structure of fitness landscapes, and studying schema processing in detail on specially designed fitness landscapes. Several other approaches are described in the proceedings of the Foundations of Genetic Algorithms workshops.[72,89]

4.1 WALSH ANALYSIS AND GA DECEPTION

(This subsection is adapted from Forrest and Mitchell[22] by permission of the authors; copyright © by Morgan Kaufmann.)

As mentioned above, two of the goals for a theory of GAs are (1) to describe in detail how schemas are processed and (2) to predict the degree to which a given problem will be easy or difficult for a GA. Albert Bethke addressed these issues by applying Walsh functions[87] to the study of schema processing in GAs.[10] In particular, Bethke developed the *Walsh-schema* transform, in which discrete versions of Walsh functions are used to efficiently calculate the average fitnesses of schemas. He then used this transform to characterize functions as easy or hard for the GA to optimize. Bethke's work was further developed and explicated by Goldberg.[25,26] In this subsection I introduce Walsh functions, describe how the Walsh schema transform can be used to understand GAs, and sketch Bethke's use of this transform for characterizing different functions. This discussion is similar to that given by Goldberg.[25]

SCHEMAS AND PARTITIONS

Before introducing Walsh functions, it is necessary to explain the notion of a *partition* of the search space. Schemas can be viewed as defining hyperplanes in the search space $\{0,1\}^l$, as shown in Figure 25. Figure 25 shows four hyperplanes (corresponding to the schemas 0****, 1****, *0***, and *1***). Any point in the space is simultaneously an instance of two of these schemas. For example, the point in the figure is a member of both 1**** and *0*** (and also of 10***). The hyperplanes defined by schemas induce a *partitioning* of the search space.[36] For example, as seen in Figure 25, the partition d**** (where "d" means "defined bit") divides the search space into two halves, corresponding to the schemas 1**** and 0****. That is,

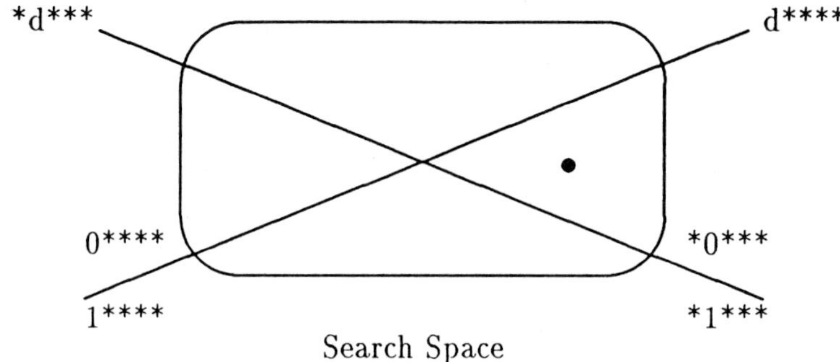

FIGURE 25 Schemas define hyperplanes in the search space.

the notation d**** represents the partitioning that divides the space into two halves consisting of schemas with a single defined bit in the leftmost position. Similarly, the partition *d*** divides the search space into two different halves, corresponding to the schemas *1*** and *0***. The partition dd*** represents a division of the space into four quarters, each of which corresponds to a schema with the leftmost two bits defined. Any partitioning of the search space can be written as a string in $\{d, *\}^l$, where the *order* of the partition is the number of defined bits (number of d's). Each partitioning of n defined bits contains 2^n *partition elements* and each partition element corresponds to a schema. Each different partitioning of the search space can be indexed by a unique bit string in which 1's correspond to the partition's defined bits and 0's correspond to the nondefined bits. For example, under this enumeration, the partition d***...* has index $j = 1000\ldots0$, and the partition dd***...* has index $j = 11000\ldots0$.

WALSH FUNCTIONS AND WALSH DECOMPOSITIONS

Walsh functions are a complete orthogonal set of basis functions that induce transforms similar to Fourier transforms. However, Walsh functions differ from other bases (e.g., trigonometric functions or complex exponentials) in that they have only two values, $+1$ and -1. Bethke demonstrated how to use these basis functions to construct functions with varying degrees of difficulty for the GA. In order to do this, Bethke used a discrete version of Walsh's original continuous functions. These functions form an orthogonal basis for real-valued functions defined on $\{0, 1\}^l$.

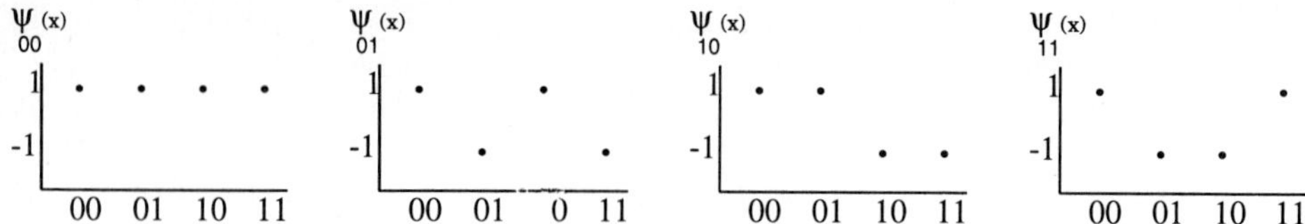

FIGURE 26 Plots of the four Walsh functions defined on two bits.

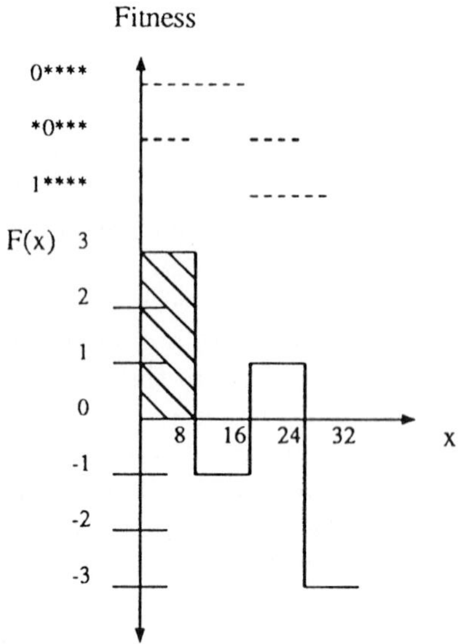

FIGURE 27 Example Function. The solid line indicates the function and the dashed lines indicate some schemas. The shaded region is the fittest region of the space.

The discrete Walsh functions map bit strings x into $\{1, -1\}$. Each Walsh function is associated with a particular partitioning of the search space. The Walsh function corresponding to the jth partition (where, as was described above, the index j is a bit string) is defined as follows[10,84]:

$$\psi_j(x) = \begin{cases} 1, & \text{if } x \wedge j \text{ has even parity (i.e., an even number of 1's);} \\ -1, & \text{otherwise.} \end{cases}$$

Here, $\wedge$ stands for bitwise AND. For example, $\psi_{0001}(1001) = 0$ since $1001 \wedge 0001 = 0001$ which has an odd number of 1's. Notice that $\psi_j(x)$ has the property that the only bits in x that contribute to its value are those that correspond to 1's in j. This is an important property, as will be seen below.

Plots of the four Walsh functions defined on two bits are given in Figure 26.

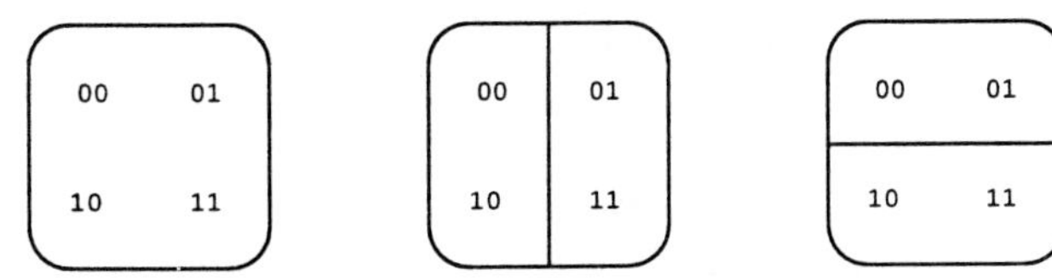

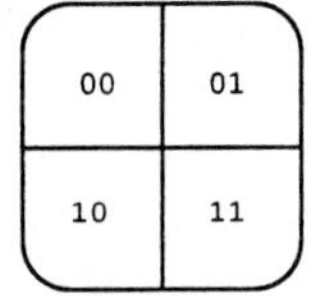

A. Partition = **, j = 00 B. Partition = *d, j = 01 C. Partition = d*, j = 10 D. Partition = dd, j = 11

FIGURE 28 Four different partitionings of the space of 2-bit strings.

Since the Walsh functions form a basis set, any function $F(x)$ defined on $\{0,1\}^l$ can be written as a linear combination of Walsh functions:

$$F(x) = \sum_{j=0}^{2^l-1} \omega_j \psi_j(x),$$

where x is a bit string, l is its length, and each ω_j is a real-valued coefficient called a *Walsh coefficient*. For example, the function shown in Figure 27 can be written as

$$F(x) = 2\psi_{01000}(x) + \psi_{10000}(x).$$

The Walsh coefficients ω_j of a given function F can be obtained via the *Walsh transform*, which is similar to a Fourier transform. Below I will explain how the Walsh transform works and discuss the close relationship between Walsh analysis and schemas.

As a simple example of the Walsh transform, consider the function $F(x) = x^2$, where x is a two-bit string. The space of two-bit strings can be partitioned into sets of schemas in four different ways, as illustrated in Figure 28.

The Walsh transform works by transforming $F(x)$ into the summed series of Walsh terms $F(x) = \sum_{j=0}^{2^l-1} \omega_j \psi_j(x)$, in which increasingly longer partial sums provide progressively better estimates of the value of $F(x)$. The terms in the sum are obtained from the average values of F in progressively smaller partition elements.

In this example Walsh analysis will be used to get better and better estimates for $F(11)$ $(= 9)$.

Consider first the average value of F on the entire space, which is the same as the average fitness $u(**)$ of the schema ** in the partition $j = 00$ (part A of Figure 28):

$$u(**) = \overline{F} = (F(00) + F(01) + F(10) + F(11))/4 = 14/4.$$

Let $\omega_{00} = u(**) = \overline{F}$. This could be said to be a "zeroth order" estimate of $F(11)$ (or of $F(x)$ for any x).

Now to get a better estimate for $F(11)$, some corrections need to be made to the zeroth-order estimate. One way to do this is to look at the average value of F in a smaller partition element containing $F(11)$—say, *1 (the right-hand element shown in part (B) of Figure 28). The average value of the schema *1 is

$$u(^*1) = \omega_{00} - \textit{deviation}_{*1};$$

that is, it is equal to the average of the entire space minus the deviation of $u(^*1)$ from the global average. Likewise, the average value of the complement schema *0 is

$$u(^*0) = \omega_{00} + \textit{deviation}_{*1},$$

since $u(^*1) + u(^*0) = 2u(^{**}) = 2\omega_{00}$. (The assignment of $+$ or $-$ to *deviation*$_{*1}$ here is arbitrary; it could have been reversed.) The magnitude of the deviation is the same for both schemas (*1 and *0) in partition *d. Call this magnitude ω_{01}. A better estimate for F(11) is then $\omega_{00} - \omega_{01}$.

The same thing can be done for the other order 1 schema containing 11, namely 1*. Let ω_{10} be the deviation of the average value in d* from the global average. Then,

$$u(1^*) = \omega_{00} - \omega_{10}.$$

An even better estimate for $F(11)$ is $\omega_{00} - \omega_{01} - \omega_{10}$. This is a first-order estimate (based on 1-bit schemas). The two deviation terms are independent of each other, since they correct for differences in average values of schemas defined on different bits, so we subtract them both. If the function were linear, this would give the exact value of $F(11)$. (In some sense, this is what it *means* for a function defined on bit strings to be linear.)

However, since $F(x) = x^2$ is nonlinear, one additional correction needs to be made to account for the difference between this estimate and the average of the order 2 schema (i.e., the string 11 itself):

$$F(11) = \omega_{00} - \omega_{01} - \omega_{10} + \textit{correction}_{11}.$$

The magnitude of the order 2 correction term is the same for each $F(x)$. This can be shown as follows. We know that

$$F(11) = \omega_{00} - \omega_{01} - \omega_{10} + \textit{correction}_{11},$$

and, by a similar analysis,

$$F(10) = \omega_{00} + \omega_{01} - \omega_{10} + \textit{correction}_{10}.$$

Adding both sides of these two equations, we get

$$F(11) + F(10) = 2\omega_{00} - 2\omega_{10} + \textit{correction}_{11} + \textit{correction}_{10}.$$

But $F(11) + F(10) = 2u(1^*)$ (by definition of $u(1*)$), so we have

$$F(11) + F(10) = 2u(1*) = 2\omega_{00} - 2\omega_{10}$$

since, as was discussed above, $u(1^*) = \omega_{00} - \omega_{10}$. Thus, $correction_{11} = -correction_{10}$.

Similarly,

$$F(01) = \omega_{00} - \omega_{01} + \omega_{10} + correction_{01},$$

so

$$F(11) + F(01) = 2\omega_{00} - 2\omega_{01} + correction_{11} + correction_{01}$$

and, since

$$F(11) + F(01) = 2u(^*1) = 2\omega_{00} - 2\omega_{01},$$

we have $correction_{11} = -correction_{01}$.

Finally,

$$F(00) = \omega_{00} + \omega_{01} + \omega_{10} + correction_{00},$$

so

$$F(00) + F(01) = 2\omega_{00} + 2\omega_{10} + correction_{11} + correction_{01},$$

and since

$$F(00) + F(01) = 2u(0^*) = 2\omega_{00} + 2\omega_{10},$$

we have $correction_{00} = -correction_{01}$. Thus the magnitudes of the second-order correction terms are all equal. Call this common magnitude ω_{11}.

This discussion shows that, for this simple function, each partition j' has a single $\omega_{j'}$ associated with it, representing the deviation of the real average fitness of each schema in partition j' from the estimates given by the combinations of lower-order ω_j's. The magnitude of this deviation is the same for all schemas in partition j'. This was easy to see for the first-order partitions and, as shown, it is also true for the second-order partitions (which are the highest-order partitions in this simple example). In general, for any partition j, the average fitnesses of schemas are mutually constrained in ways similar to those shown above, and the uniqueness of ω_j can be similarly demonstrated for j's of any order.

Table 1 gives the exact Walsh decomposition for each $F(x)$.

TABLE 1 Expressions for $F(x)$ for each $x\epsilon\{0,1\}^2$.

$F(00) = \omega_{00} + \omega_{01} + \omega_{10} + \omega_{11}$.
$F(01) = \omega_{00} - \omega_{01} + \omega_{10} - \omega_{11}$.
$F(10) = \omega_{00} + \omega_{01} - \omega_{10} - \omega_{11}$.
$F(11) = \omega_{00} - \omega_{01} - \omega_{10} + \omega_{11}$.

It has now been shown how function values can be calculated in terms of Walsh coefficients, which represent progressively finer correction terms to lower-order estimates in terms of schema averages. A converse analysis demonstrates how the ω_j's are calculated:

$$\begin{aligned}
\omega_{00} &= u(^{**}) \\
&= \frac{0+1+4+9}{4} \\
&= \frac{14}{4}. \\
\omega_{01} &= \omega_{00} - u(^*1) \\
&= \frac{0+1+4+9}{4} - \frac{1+9}{2} \\
&= \frac{0-1+4-9}{4} \\
&= -\frac{6}{4}. \\
\omega_{10} &= \omega_{00} - u(1^*) \\
&= \frac{0+1+4+9}{4} - \frac{4+9}{2} \\
&= \frac{0+1-4-9}{4} \\
&= -\frac{12}{4}. \\
\omega_{11} &= F(11) - \text{first-order estimate} \\
&= F(11) - (\omega_{00} - \omega_{01} - \omega_{10}) \\
&= 9 - \left(\frac{14}{4} + \frac{6}{4} + \frac{12}{4}\right) \\
&= \frac{4}{4}.
\end{aligned}$$

And to check:

$$F(11) = \omega_{00} - \omega_{01} - \omega_{10} + \omega_{11} = 14/4 + 6/4 + 12/4 + 4/4 = 9.$$

In general,

$$\omega_j = \frac{1}{2^l} \sum_{x=0}^{2^l-1} F(x)\psi_j(x).$$

This is the *Walsh transform* (it is derived more formally in Goldberg[25]). Once the ω_j's have been determined, F can be calculated as

$$F(x) = \sum_{j=0}^{2^l-1} \omega_j \psi_j(x).$$

How does one decide whether or not a deviation term ω_j is added or subtracted in this expression? The answer to this question depends on some conventions: e.g., whether $u(^*1)$ is said to be $\omega_{00} - \omega_{01}$ or $\omega_{00} + \omega_{01}$. Once these conventions are decided, they impose constraints on whether higher-order Walsh coefficients will be added or subtracted in the expression for $F(x)$. If x happens to be a member of a schema s whose average deviates in a positive way from the lower-order estimate, then the positive value of the ω_j corresponding to s's partition goes into the sum. All that is needed is a consistent way of assigning these signs, depending on the partition j and what element of j a given bit string x is in. The purpose of the Walsh functions $\psi_j(x)$ is to provide such a consistent way of assigning signs to ω_j's, via bitwise AND and parity. This is not the only possible method; a slightly different method is given by Holland for his *hyperplane transform.*[36]

THE WALSH-SCHEMA TRANSFORM

There is a close connection between the Walsh transform and schemas. It was shown above that, using Walsh coefficients, a function's value on a given argument x can be calculated using the average fitnesses of schemas of which that x is an instance. An analogous method, proposed by Bethke,[10] can be used to calculate the average fitness $u(s)$ of a given schema s. Bethke called this method the *Walsh-schema transform.* This transform gives some insight into how schema processing is thought to occur in the GA. It also allowed Bethke to state some conditions under which a function will be easy for the GA to optimize, and allowed him to construct functions that can be difficult for the GA because low-order schemas lead the search in the wrong direction.

Formal derivations of the Walsh-schema transform are given by Bethke,[10] Goldberg,[25] and Tanese.[84] Here the the transform is presented informally.

Using the same example as before, the average fitness of the schema *1 is $u(^*1) = \omega_{00} - \omega_{01}$; this comes from the definition of ω_{01}. The value of $u(*1)$ does not depend on, say, ω_{10}; it depends only on Walsh coefficients of partitions that either contain *1 or contain a superset of *1 (e.g., ** $\supset$ *1). In general, a partition j is said to *subsume* a schema s if it contains as an element some schema s' such that $s' \supseteq s$. For example, the 3-bit schema 10* is subsumed by four partitions: dd*, d**, *d*, and ***, which correspond respectively to the j values 110, 100, 010, and 000. Notice that j subsumes s if and only if each defined bit in j (i.e., each 1) corresponds to a defined bit in s (i.e., a 0 or a 1, not a *).

The Walsh-schema transform expresses the average fitness of a schema s as a sum of progressively higher-order Walsh coefficients ω_j, analogous to the expression of $F(x)$ as a sum of progressively higher-order ω_j's. Just as each ω_j in the expression for $F(x)$ is a correction term for the average fitness of some schema in partition j containing x, each ω_j in the expression for $u(s)$ is a correction term, correcting the estimate given by some lower-order schema that contains s. The difference is that, for $F(x)$, all 2^l partition coefficients must be summed (although some of them may be zero). But to calculate $u(s)$, only coefficients of the subsuming partitions ("subsuming coefficients") need to be summed.

The 2-bit function example given above is too simple to illustrate these ideas, but an extension to three bits suffices. Let $F(x) = x^2$ as before, but let x be defined over three bits instead of two. The average fitness of the schema *01 is a sum of the coefficients of partitions that contain the schemas ***, **1, *0*, and *01. An easy way to determine the sign of a subsuming coefficient ω_j is to take any instance of s and to compute $\psi_j(x)$. This value will be the same for all $x \in s$, as long as j is a subsuming partition, since all the ones in j are matched with the same bits in any instance of s. For example, the partition **d ($j = 001$) subsumes the schema *11, and $\psi_{001}(x) = -1$ for any $x \in$ *11. Using a similar method to obtain the signs of the other coefficients, we get

$$u(^*11) = \omega_{000} - \omega_{001} - \omega_{010} + \omega_{011}.$$

In general,

$$u(s) = \sum_{j:j \text{ subsumes } s} \omega_j \Psi_j(s)$$

where $\Psi_j(s)$ is the value of $\psi_j(x)$ ($= +1$ or -1) for any $x \in s$.

The sum

$$u(^*11) = \omega_{000} - \omega_{001} - \omega_{010} + \omega_{011}$$

gives the flavor of how the GA actually goes about estimating $u(^*11)$. To review, a population of strings in a GA can be thought of as a number of samples of various schemas, and the GA works by using the fitness of the strings in the population to estimate the fitness of schemas. It exploits fit schemas via reproduction by allocating more samples to them, and it explores new schemas via crossover by combining fit low-order schemas to sample higher-order schemas that will hopefully also be fit. In general, there are many more instances of low-order schemas in a given population than high-order schemas (e.g., in a randomly generated population, about half of the strings will be instances of 1** . . . *, but very few, if any, will be instances of 111 . . . 1). Thus, accurate fitness estimates will be obtained much earlier for low-order schemas than for high-order schemas. The GA's estimate of a given schema s can be thought of as a process of gradual refinement, where the algorithm initially bases its estimate on information about the low-order schemas containing s and gradually refines this estimate from information about higher and higher order schemas containing s. Likewise, the terms in the sum above represent increasing refinements to the estimate of how good the schema *11 is. The term ω_{000} gives the population average (corresponding to the average fitness of the schema ***) and the increasingly higher-order ω_j's in the sum represent higher-order refinements of the estimate of *11's fitness, where the refinements are obtained by summing ω_j's corresponding to higher and higher order partitions j containing *11.

Thus, one way of describing the GA's operation on a fitness function F is that it makes progressively deeper estimates of what F's Walsh coefficients are, and biases the search towards partitions j with high-magnitude ω_j's, and to the partition elements (schemas) for which these correction terms are positive.

THE WALSH-SCHEMA TRANSFORM AND GA-DECEPTIVE FUNCTIONS

Bethke[10] used Walsh analysis to partially characterize functions that will be easy for the GA to optimize. This characterization comes from two facts about the average fitness of a schema s. First, since $u(s)$ depends only on ω_j's for which j subsumes s, then if the order of j (i.e., the number of 1's in j) exceeds the order of s (i.e., the number of defined bits in s), then ω_j does not affect $u(s)$. For example, ω_{111} does not affect $u(*11)$: *11's two instances 011 and 111 receive opposite-sign contributions from ω_{111}. Second, if the defining length of j (i.e., the distance between the leftmost and rightmost 1's in j) is greater than the defining length of s (i.e., the distance between the leftmost and rightmost defined bits in s), then $u(s)$ does not depend on ω_j. For example, ω_{101} does not affect $u(*11)$, since $u(*11)$'s two instances again receive opposite-sign contributions from ω_{101}.

Bethke suggested that if the magnitude of the Walsh coefficients of a function decrease rapidly with increasing order and the defining length of the j's—that is, the most important coefficients are associated with short, low-order partitions—then the function will be easy for the GA to optimize. In such cases, the location of the global optimum can be determined from the estimated average fitness of low-order, low-defining-length schemas. As was described above, such schemas receive many more samples than higher-order, longer-defining-length schemas: "low order" means that they define larger subsets of the search space and "short defining length" means that they tend to be kept intact under crossover. Thus the GA can estimate their average fitnesses more quickly than those of higher-order, longer-defining-length schemas.

Thus, all else being equal, a function whose Walsh decomposition involves high-order j's with high-magnitude coefficients should be harder for the GA to optimize than a function with only lower-order j's, since the GA will have a harder time constructing good estimates of the higher-order schemas belonging to the higher-order partitions.

Bethke's analysis was not intended as a practical tool for use in deciding whether a given problem will be hard or easy for the GA. A Walsh transform of F requires evaluating F at every point in its argument space (this is also true for the "Fast Walsh Transform,"[25]) and is thus an infeasible operation for most fitness functions of interest. It is much more efficient to run the GA on a given function and to measure its performance directly than to decompose the function into Walsh coefficients and then to determine from those coefficients the likelihood of GA success. However, Walsh analysis can be used as a theoretical tool for understanding the types of properties that can make a problem hard for the GA. For example, Bethke used the Walsh-schema transform to construct functions that mislead the GA, by directly assigning the values of Walsh coefficients in such a way that the average values of low-order schemas give misleading information about the average values of higher-order refinements of those schemas. Specifically, Bethke chose coefficients so that some short, low-order schemas had relatively low average fitness, and then chose other coefficients so as to make these low-fitness schemas actually contain the global optimum. Such functions were later termed "deceptive" by Goldberg[24,26,28] who carried out a number of theoretical studies of such functions. Deception has since been a central focus of theoretical work on GAs.[13,17,26,48,49,88] Walsh analysis can be used to construct problems with different degrees and types of deception, and the GA's performance on these problems can be studied empirically. The goal of such research is to learn how deception affects GA performance (and thus why the GA might fail in certain cases) and to learn how to modify the GA or the problem's representation in order to improve performance.

Intuitively, it seems that a deceptive problem will be difficult for the GA. The GA works by accumulating information about schemas and using this information to bias its future samples. If some schemas give the GA the wrong information about the location of the global optimum, then the GA should have difficulty finding the global optimum. However, the GA is often able to find the optimum fairly readily even on functions with a large number of deceptive schemas.[27] There is not yet any rigorous understanding of the relation between different types of deception and the performance of the GA. Some critical discussions of the notion of deception and its role in understanding GAs are given in Forrest and Mitchell[22] and Grefenstette.[29]

4.2 STATISTICAL STRUCTURE OF FITNESS LANDSCAPES

Stuart Kauffman of the Santa Fe Institute has studied in detail how certain statistics of a particular class of fitness landscapes affect the process of evolution over those landscapes[41]. In particular, Kauffman has defined a class of parameterizable fitness landscapes called "*NK* landscapes."[2] The purpose is to define the simplest class of landscapes whose "ruggedness" can be varied; one can then create landscapes with various degrees of ruggedness and study the effects of the degree of ruggedness on evolution over these landscapes.

A *NK* landscape is defined over a space of bit strings. To create an *NK* landscape, one chooses a value for N and K, where N is the number of bits in the string and K is the degree of "epistasis"—the number of other bits that each bit's fitness contribution depends on. One then chooses, for each locus i, the K other

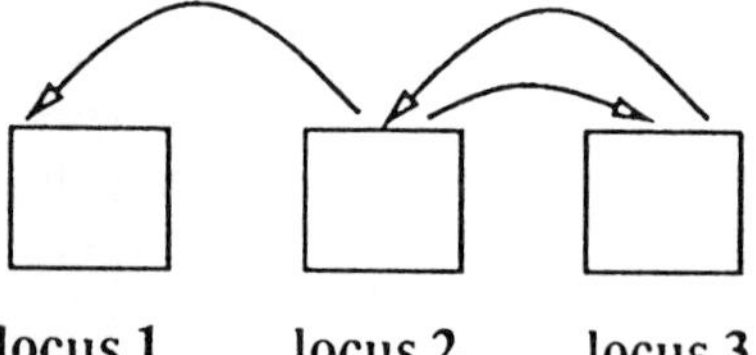

FIGURE 29 The network of dependencies for the example *NK* landscape.

[2] Kauffman's *NK* landscapes are a slightly different formulation of his "random Boolean networks."[40,42]

loci that affect i's fitness contribution. That is, the fitness contribution of the allele at i depends on itself and the alleles at these K other loci. There are 2^{K+1} possible configurations of these $K+1$ alleles, and one assigns to each such configuration a randomly chosen fitness contribution between 0 and 1. The fitness of the entire string is defined to be the average of the contributions of each locus.

For example, consider a simple *NK* landscape with $N = 3$ and $K = 1$. Suppose we have decided that the fitness contribution of locus 1 (the leftmost locus) depends on locus 2, the contribution of locus 2 depends on locus 3, and the contribution of locus 3 depends on locus 2. This network of dependencies is illustrated in Figure 29. Now we randomly assign fitness contributions for each possible configuration (the asterisks denote the loci that are not taken into account in determining the fitness contribution for a given locus):

Fitness contribution of locus 1:

$$\begin{aligned} 00* &: \text{ contribution} = 0.6 \\ 01* &: \text{ contribution} = 0.2 \\ 10* &: \text{ contribution} = 0.3 \\ 11* &: \text{ contribution} = 0.8 \end{aligned}$$

Fitness contribution of locus 2:

$$\begin{aligned} *00 &: \text{ contribution} = 0.2 \\ *01 &: \text{ contribution} = 0.4 \\ *10 &: \text{ contribution} = 0.1 \\ *11 &: \text{ contribution} = 0.3 \end{aligned}$$

Fitness contribution of locus 3:

$$\begin{aligned} *00 &: \text{ contribution} = 0.1 \\ *01 &: \text{ contribution} = 0.9 \\ *10 &: \text{ contribution} = 0.2 \\ *11 &: \text{ contribution} = 0.1 \end{aligned}$$

To calculate the fitness of a given string, we determine the contributions from each locus in that string and average them. For example, $F(000) = (0.6 + 0.2 + 0.1)/3$. Now that the fitness of each possible chromosome has been determined, the entire fitness landscape has been defined.

NK landscapes, though highly simplified, are meant to capture some important aspects of fitness landscapes in nature. In particular, the N sites can roughly be thought of as representing N traits in an organism, with the fitness contribution of each trait depending on the value of K other traits. With a fixed N, when K is tuned from 0 to $N - 1$, the resulting landscape goes from being very smooth (few local optima) to very rugged (many local optima). *NK* landscapes are closely related to spin glass models in physics.[41]

Kauffman has looked in detail at how the statistical structure of *NK* landscapes varies for different values of N and K.[41] The statistical structure includes, among other properties, the number of local fitness optima in the landscape, the average length of an "adaptive walk" from a given point to a fitness optimum, and the average number of alternative optima that can be reached via an adaptive walk from a given point. Kauffman has used his results on the statistical structure of *NK* landscapes to predict some aspects of the dynamics of evolution on these landscapes and to hypothesize about what features of landscapes would allow organisms of significant complexity to evolve.

The notion of the statistical structure of a fitness landscape assumes a metric over the space of genotypes (here, bit strings). For example, in defining the "average length of an adaptive walk to a fitness optimum," one needs to specify a metric in terms of which length will be measured. Kauffman's metric (like that for most studies of fitness landscapes) is in terms of single mutations, or *Hamming distance.* The Hamming distance H between two bit strings is the number of bit positions in which they differ (e.g., $H(1001, 1000) = 1$ and $H(1001, 0110) = 4$). Kauffman's statistics on *NK* landscapes are all in terms of this metric. For example, the length of a "walk" from string A to string B is the number of single mutations needed to transform A into B (an "adaptive walk" is a walk in which each step leads to an improvement in fitness).

If mutation were the only genetic operator used in a GA, one could straightforwardly apply Kauffman's results to predict some aspects of the performance of GAs on *NK* landscapes. However, the main source of variation in GAs is crossover, which defines a different metric on the space of genotypes. What characteristics must a fitness landscape have in order for crossover to be a useful operator? Kauffman offers some intuitive answers in the context of *NK* landscapes.[41] If $K = 0$, then the contribution of each of the N loci is independent, and each locus can be thought of as an independent building block. In this case, crossover may help speed up the search for the optimum since, in a single step, it can combine different substrings of optimized alleles from different strings. That is, if string A and string B have high fitness, then their offspring are likely to have even higher fitness. Likewise, if K is small and the epistatic interactions are restricted to be among near neighbors, then crossover may be useful since it can combine distant regions of strings that are functionally independent. Kauffman also hypothesizes that crossover will be useful when the fitness landscape contains pairs of fitness peaks that together contain mutual information about good regions of the space. When strings located at such peaks cross over, there is a good possibility that they will produce fit offspring.

These are all intuitive arguments similar to the Building-Block Hypothesis mentioned earlier, and Kauffman has given some experimental evidence to support these hypotheses.

Bernard Manderick, Mark de Weger, and Piet Spiessens of the Free University of Brussels have extended Kauffman's work and have applied their results to the problem of predicting the GA's performance on a given landscape.[53] In their work they looked at two types of correlation measures:

- *Operator-specific correlation coefficients:*

$$\frac{\text{covariance}(F_{\text{parents}}, F_{\text{offspring}})}{\sigma(F_{\text{parents}})\sigma(F_{\text{offspring}})}$$

  Given a certain operator that creates offspring from parents, these correlation coefficients measure how well the offsprings' fitness is correlated with the parents' fitness.
- *Correlation length of landscape:*
  This measure gives the length (in Hamming distance) at which fitnesses are no longer correlated (see Manderick et al.[53] for details).

Manderick et al. measured the correlation coefficient of crossover as a function of the Hamming distance between the two parents for four different *NK* landscapes, in which *N* was fixed at 10 and *K* was varied. Their results are displayed in Figure 30.

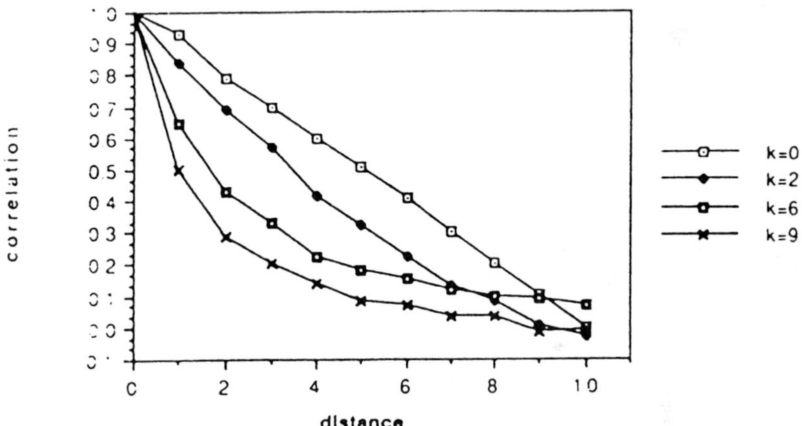

FIGURE 30 Correlation coefficient of crossover as a function of the Hamming distance between the two parents for four different *NK* landscapes, in which *N* was fixed at 10 and *K* was varied. (Reprinted from Manderick et al[53] by permission of the authors; copyright © 1991 by Morgan Kaufmann Publishing.)

TABLE 2 Relation between correlation length τ of an NK landscape and GA performance $Imp.$ (number of improvements after 2,048 generations). These results (each averaged over five runs) are given for several different values of K. N is fixed at 96. (Adapted from Manderick et al.[53] by permission of the authors.)

K	0	1	2	4	8	16	32	48	95
τ	29.96	24.37	19.51	14.15	7.06	3.90	1.72	1.00	0.52
$Imp.$	19.80	16.00	15.20	11.60	8.60	6.20	3.80	5.40	5.20

Each curve corresponds to the results on a landscape with the given K value, and each point on a curve was obtained by randomly selecting 100 pairs of parents (each separated by the given Hamming distance), crossing over each pair at a random locus, and computing the correlation coefficient between the fitnesses of the parents and the fitnesses of their offspring. As can be seen, the correlation goes down as Hamming distance increases and, the greater the K, the faster the decrease in correlation. Assuming that crossover is useful only when there is a correlation between the fitness of parents and the fitness of offspring, Manderick et al. point out that their results help to quantify a qualitative heuristic used by some GA practitioners: crossover performs best when it is restricted to similar parents.

Manderick et al. also measured the relation between the correlation length of a landscape and GA performance. These results are given in Table 2. Here N was fixed at 96 and K was varied. For each K, the correlation length was calculated by sampling a number of random walks on the landscape. The correlation length τ is roughly the average number of single mutation steps from a given string needed until the fitness of the resulting string and the fitness of the original string are no longer correlated (see Manderick et al.[53] for details). The performance of the GA was measured in terms of the number of improvements in a fixed number M of generations—that is, the number of strings in generation M whose fitnesses are higher than that of any string in the initial population (here $M = 2048$ and the results are averaged over five runs). Table 2 shows that both correlation length and GA performance decrease with K. Manderick et al.'s conclusion is that the GA performs well on an NK landscape when the correlation length of the landscape is high.

The results of Kauffman and of Manderick et al. are initial steps in characterizing, in terms of statistical properties, the types of fitness landscapes on which GAs are likely to perform well. However, the NK landscapes are a limited class that may not capture the landscape features most relevant to the performance of GAs. The next subsection describes an attempt to define a parametrizable class of landscapes that more directly captures such features and that yields some surprising results related to the Building-Block Hypothesis.

4.3 ROYAL ROAD LANDSCAPES

(This subsection is adapted from Forrest and Mitchell[21] by permission of the authors; copyright © 1993 by Morgan Kaufmann.)

Stephanie Forrest of the University of New Mexico and I are currently carrying out research to answer the four questions posed at the beginning of this section.[61,21] Our strategy for answering these questions consists of the following general approach. We begin by identifying *features* of fitness landscapes that are particularly relevant to the GA's performance. A number of such features have been discussed in the GA literature, including local hills, "deserts," deception, hierarchically structured building blocks, noise, and high fitness-variance within schemas. We then design simplified landscapes containing different configurations of such features, for example, varying the distribution, frequency, and size of different features in the landscape. We then study in detail the effects of these features on the GA's behavior. A longer-term goal of this research is to develop statistical methods of classifying any given landscape in terms of our spectrum of hand-designed landscapes, thus being able to predict some aspects of the GA's performance on the given landscape.

It should be noted that by stating this problem in terms of the GA's performance on fitness landscapes, we are sidestepping the question of how a particular problem can best be represented to the GA. The success of the GA on a particular function is certainly related to how the function is encoded (e.g., using Gray codes for numerical parameters can greatly enhance the performance of the GA on some problems[11]) but, since we are interested in biases that pertain directly to the GA, we will simply consider the landscape that the GA "sees."

In this subsection I describe some initial results from this long-term research program. A starting point for our research is the Building-Block Hypothesis, which states that the GA works well when short, low-order, highly-fit schemas ("building blocks") recombine to form even more highly fit, higher-order schemas. In Goldberg's words, "...we construct better and better strings from the best partial solutions of past samplings" (see Goldberg,[26] p. 41). As has been emphasized in earlier sections, the ability to produce fitter and fitter partial solutions by combining building blocks is believed to be the primary source of the GA's search power. However, in spite of the presumed central role of building blocks and recombination, the GA research community lacks precise and quantitative descriptions of how schemas interact and combine during the typical evolution of a GA search. Thus, we are interested in isolating landscape features implied by the Building-Block Hypothesis, and studying in detail the GA's behavior—the way in which schemas are processed and building blocks are combined—on simple landscapes containing those features.

One major component of this endeavor is to define the simplest class of landscapes on which the GA performs "as expected," thus confirming the broad claims of the Building-Block Hypothesis. However, the task of designing such landscapes has turned out to be substantially more difficult and more subtle than we originally anticipated. Our initial choices of simple landscapes have revealed some surprising and unanticipated phenomena. The story of how small variations of a basic landscape can make GA search much less effective reveals a great deal about the complexity of GAs and points out the need for a deeper theory of how low-order building blocks are discovered and combined into higher-order solutions.

Below I introduce the *Royal Road* functions, a class of nondeceptive functions in which the building blocks are explicitly defined. I then present experimental results that demonstrate how simple variants of these functions can have quite different effects on the performance of the GA and discuss the reasons for these differences.

STEPPING STONES IN THE CROSSOVER LANDSCAPE

The Building-Block Hypothesis suggests two landscape features that are particularly relevant for the GA: (1) the presence of short, low-order, highly fit schemas and (2) the presence of intermediate "stepping stones"—intermediate-order higher-fitness schemas that result from combinations of the lower-order schemas and that, in turn, can combine to create even higher-fitness schemas. Two basic questions about stepping stones are: How much higher in fitness do the intermediate stepping stones have to be for the GA to work well? And how must these stepping stones be configured? To investigate these questions, we first defined the Royal Road functions which contain these features explicitly.

s_1 = 11111111**; $c_1 = 8$
s_2 = ********11111111**; $c_2 = 8$
s_3 = ****************11111111**; $c_3 = 8$
s_4 = ************************11111111********************************; $c_4 = 8$
s_5 = ********************************11111111************************; $c_5 = 8$
s_6 = **11111111****************; $c_6 = 8$
s_7 = **11111111********; $c_7 = 8$
s_8 = **11111111; $c_8 = 8$
s_{opt}=1111111111111111111111111111111111111111111111111111111111111111

FIGURE 31 An optimal string broken up into eight building blocks. The function $R1(x)$ (where x is a bit string) is computed by summing the coefficients c_s corresponding to each of the given schemas of which x is an instance. For example, $R1(1111111100\ldots0) = 8$, and $R1(1111111100\ldots011111111) = 16$. Here $c_s = order(s)$.

To construct a Royal Road function, we select an optimum string and break it up into a number of small building blocks, as illustrated in Figure 31. We then assign values to each low-order schema and each possible intermediate combination of low-order schemas, and use those values to compute the fitness of a bit string x in terms of the schemas of which it is an instance.

As illustrated in Figure 31, the function $R1$ is computed very simply: a bit string x gets 8 points added to its fitness for each of the given order 8 schemas of which it is an instance. For example, if x contains exactly two of the order 8 building blocks, $R1(x) = 16$. Likewise, $R1(111\ldots1) = 64$. Stated more generally, the value $R1(x)$ is the sum of the coefficients c_s corresponding to each given schema of which x is an instance. Here c_s is equal to $order(s)$. The fitness contribution from an intermediate stepping stone (such as the combination of s_1 and s_3 in Figure 31) is thus a linear combination of the fitness contribution of the lower-level components. $R1$ is similar to the "plateau" problem described by Schaffer and Eshelman.[79]

According to the Building-Block Hypothesis, $R1$'s building-block and stepping-stone structure should lay out a "royal road" for the GA to follow to the global optimum. In contrast, an algorithm such as simple steepest-ascent hill climbing, which systematically tries out single-bit mutations and only moves in an uphill direction, cannot easily find high values in such a function, since a large number of single bit positions must be optimized simultaneously in order to move from an instance of a lower-order schema (e.g., 11111111**...*) to an instance of a higher-order intermediate schema (e.g., 11111111********11111111**...*). While some initial random search may be involved in finding the lowest-level building blocks (depending on the size of the initial population and the size of the lowest-level blocks), the interesting aspect of $R1$ is studying how lower-level blocks are combined into higher-level ones, and this is the aspect with which we are most concerned. Part of our purpose in designing the Royal Road functions is to construct a class of fitness landscapes that distinguishes the GA from other search methods such as hill climbing. This actually turned out to be more difficult than we anticipated, as will be discussed below.

This class of functions provides an ideal laboratory for studying the GA's behavior:

- The landscape can be varied in a number of ways. For example, the "height" of various intermediate stepping stones can be increased or decreased. Also, the size of the lowest-order building blocks can be varied, as can the degree to which they cover the optimum. Finally, different degrees of deception can be introduced by allowing the lower-order schemas to differ in some bits from the higher-order stepping stones, effectively creating low-order schemas that lead the GA away from the good higher-order schemas. The effects of these variations on the GA's behavior then can be studied in detail.
- Since the global optimum and, in fact, all possible fitness values are known in advance, it is easy to compare the GA's performance on different variations of Royal Road functions.

- All of the desired schemas are known in advance, since they are explicitly built into the function. Therefore, the dynamics of the search process can be studied in detail by tracing the histories of individual schemas.

We are using the Royal Road functions to study some questions about the effects of crossover on various landscapes, including the following: For a given landscape, to what extent does crossover help the GA find highly fit schemas? What is the effect of crossover on the waiting times for desirable schemas to be discovered? What are the bottlenecks in the discovery process? How does the configuration of stepping stones and size of steps defined by stepping stones affect the GA's performance? Answering these questions in the context of the idealized Royal Road functions is a first step toward answering them for more general cases.

We first investigated the effect of the step size of the intermediate stepping stones on the GA's performance. To do this, we compared the performance of the GA on $R1$ with its performance on a second function $R2$, where the fitness contributions of certain intermediate stepping stones are much higher. $R2$ is illustrated in Figure 32. $R2$ is computed in the same way as $R1$: the fitness of a bit string x is the sum of the coefficients corresponding to each schema (s_1–s_{14}) of which it is an instance. For example, $R2(1111111100\ldots011111111) = 16$, since the string is an instance of both s_1 and s_8, but $R2(1111111111111111100\ldots0) = 32$ since the string is an instance of s_1, s_2, and s_9. Thus, a string's fitness depends not only on the number of 8-bit schemas to which the string belongs, but also on their positions in the string. The optimum string $11111111\ldots1$ has fitness 192, since the string is an instance of each schema in the list.

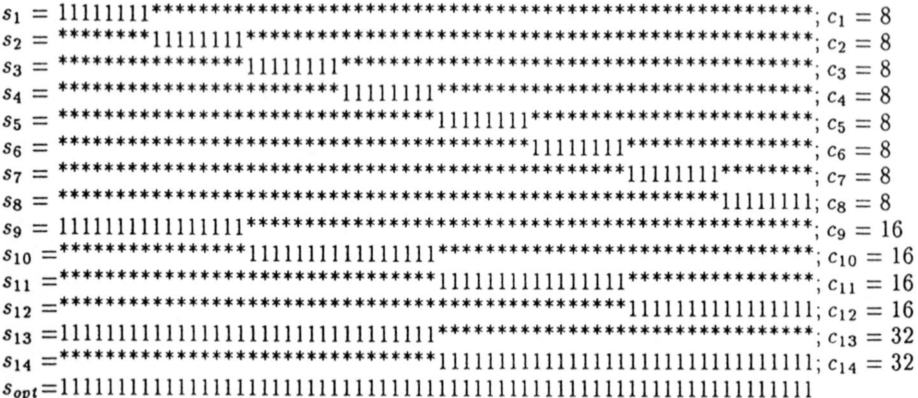

FIGURE 32 Royal Road Function $R2$. $R2(x)$ is computed in the same way as $R1$: by summing the coefficients c_s corresponding to each of the given schemas of which x is an instance. For example, $R2(1111111100\ldots011111111) = 16$, but $R2(1111111111111111100\ldots0) = 32$. $R2(11111111\ldots1) = 192$.

ROYAL ROAD EXPERIMENTS

For our initial experiments, we used functions defined over strings of length 64. The GA population size was 128, with the initial population generated at random. In each run the GA was allowed to continue until the optimum string was discovered, and the total number of function evaluations performed was recorded. We used a standard GA with single-point crossover and sigma scaling, an alternative scheme for assigning the expected number of copies to each individual. Under sigma scaling, an individual i's expected number of offspring is $1 + (F_i - \overline{F})/2\sigma$, where F_i is i's fitness, $\overline{F}$ is the mean fitness of the population, and σ is the standard deviation. The maximum expected offspring of any string was 1.5—if the above formula gave a higher value, the value was reset to 1.5. This is a strict cutoff, since it implies that most individuals will reproduce only 0, 1, or 2 times. The effect of this selection scheme is to slow down convergence by restricting the effect that a single individual can have on the population, regardless of how much more fit it is than the rest of the population. Even with this precaution, we observe some interesting premature convergence effects, described below. The crossover probability was 0.7 per pair of parents and the mutation probability was 0.005 per bit.

EXPERIMENTS ON $R1$ AND $R2$

We expected the GA to perform better—that is, find the optimum more quickly—on $R2$ than on $R1$. In $R2$ there is a very clear path via crossover from pairs of the eight initial order 8 schemas (s_1–s_8) to the four order 16 schemas (s_9–s_{12}), and from there to the two order 32 schemas (s_{13} and s_{14}), and finally to the optimum (s_{opt}). We believed that the presence of this stronger path would speed up the GA's discovery of the optimum, but our experiments showed the opposite: the GA performed significantly better on $R1$ than on $R2$. Statistics summarizing the results of 500 runs on each function are given in Table 3. This table gives the mean and median number of function evaluations taken to find the optimum over 500 runs each on $R1$ and $R2$. Each run on a given function uses identical parameters but starts with a different random-number seed.

If we hope to understand the GA's performance in general, we need to understand in detail what are the potential bottlenecks for discovering desirable schemas. This has been studied extensively in the deception literature, but $R2$ is a nondeceptive function that nonetheless contains some features that keep the GA from discovering desirable schemas as quickly as in $R1$. What slows down the GA in the case of $R2$? To investigate this, we took a typical run of the GA on $R2$ and graphically traced the evolution of each schema shown in Figure 32. Figure 33 gives this trace for three sets of schemas: s_1, s_2, and s_9; s_3, s_4, and s_{10}; and s_5, s_6, and s_{11} (see Figure 32). In each plot, the density (% of population) of each schema is plotted against time (generations). The density is sampled every 10 generations.

TABLE 3 Summary of results of running the GA on $R1$ and $R2$. The table gives the mean and median number of function evaluations taken to find the optimum over 500 runs on each function. The numbers in parentheses are the standard errors.

500 runs	Function Evaluations to Optimum $R1$	$R2$
Mean	62099 (std err: 1390)	73563 (std err: 1794)
Median	56576	66304

These plots show a striking phenomenon. In the top plot in Figure 33, s_1 and s_2 appear early and instances of them quickly combine to form s_9. Once each schema is discovered, its density in the population rises quite quickly to over 90% of the population by generation 60 or so. Around generation 220 there is a distinct dip in the densities of these three schemas.

The middle plot shows a very different evolution for s_3, s_4, and s_{10}. The schemas s_3 and s_4 are both present in the initial (randomly generated) population (though s_3's presence at generation 0 is not visible on the plot) but, while s_4 rises quickly, s_3 dies out by generation 10, is fleetingly rediscovered (along with s_{10}) at generation 120 (see blip on the x-axis), and does not return until the very end of the run, at which point a mutation brings it (along with s_{10}) back (see blip on the x-axis). This same mutation is responsible for creating s_{opt} at generation 535, when the run ends. After a quick initial rise, the schema s_4 enters a pronounced dip at the same time the milder dip can be seen in the top plot of Figure 33, around generation 220.

What is the cause of these dips, and what prevents s_3 from persisting in the population? A likely answer can be inferred from the bottom plot. Schema s_6 appears around generation 30, rises fairly quickly, and takes a sharp upturn around generation 220, rising to about 95% of the population. Schema s_5 appears briefly around generation 20 (dot close to the x-axis) and dies out, but appears again at generation 220. The instance of it in the population is also an instance of s_{11}, and instances of s_{11} rise very quickly. This rise exactly coincides with the minor dip in s_1, s_2, and s_9 and the major dip in s_4. What appears to be happening is: in the first few instances of s_{11}, along with the sixteen 1's in the fifth and sixth blocks are several 0's in the first through fourth blocks. An instance of s_{11} has fitness $8+8+16=32$, whereas an instance of an order 8 schema such as s_4 has fitness 8. This difference causes s_{11} to rise very quickly compared to s_4, and instances of s_{11} with some 0's in the fourth block tend to push out many of the previously existing instances of s_4 in the population, and prevent the rediscovery of s_3. This phenomenon has been called

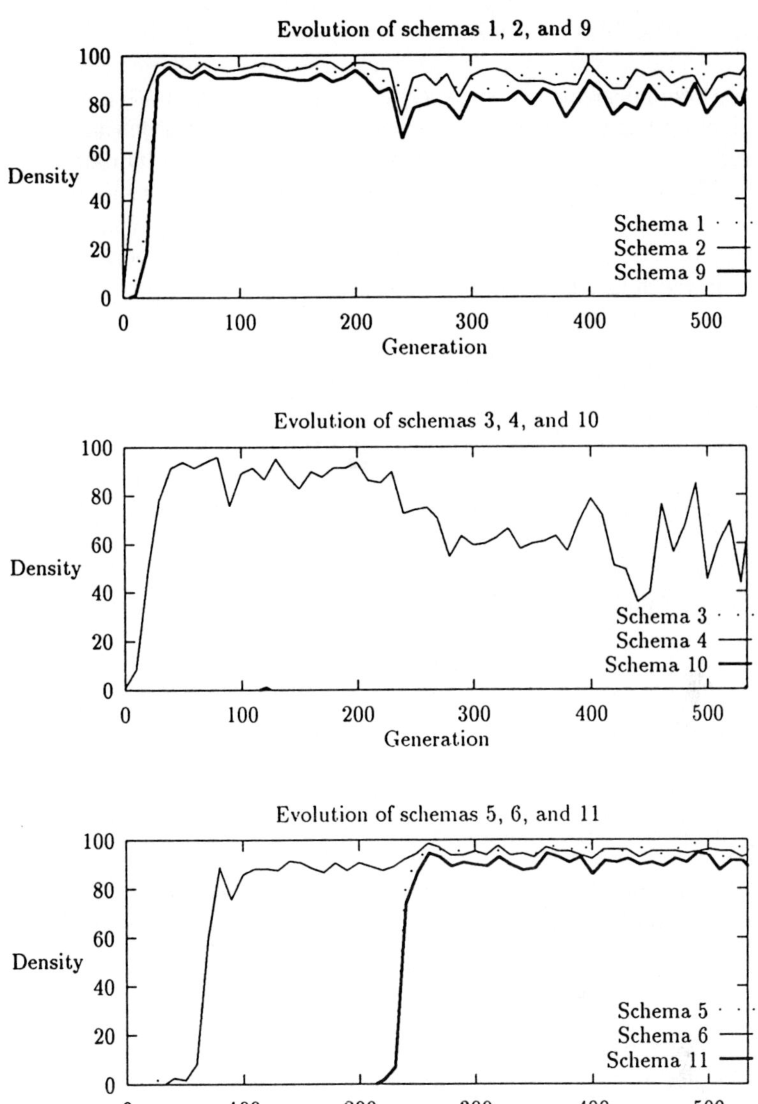

FIGURE 33 Evolution of three sets of schemas in a typical run of the GA on $R2$. (See Figure 32 for schema numbers.) In each plot, the density of each schema (% of population) is plotted against the generation. Note that in the middle plot, schemas 3 and 10 are visible only as tiny bumps on the x-axis at generations 120 and 535.

"hitchhiking," where 0's in other positions in the string hitchhike along with the highly fit s_{11}. The most likely positions for hitchhikers are those close to the highly fit schema's defined positions, since they are less likely to be separated from the schema's defined positions under crossover. Such effects are seen in real population genetics and have been discussed in the context of GAs by Schraudolph and Belew,[80] and Das and Whitley,[13] among others. Note that this effect is pronounced even with the relatively weak form of selection used in our GA. (We also compared the GA's performance on $R1$ and $R2$ using a linear rank-scaling method[5] instead of the sigma-scaling method described above, and obtained results similar to those given in Table 3.)

The plots given in Figure 33 come from a single run, but this run was typical; the same type of phenomenon was observed on many of the other runs on $R2$ as well. Our hypothesis is that this hitchhiking effect is what causes the relatively slower times (on average) for the GA to find the optimum on $R2$. The power of crossover to combine lower-level building blocks was hampered, since some of the necessary building blocks were either partially or totally suppressed by the quick rise of disjoint building blocks. This suggests that there is more to characterizing a GA landscape than the absolute direction of the search gradients. In these functions, it is the actual differences in relative fitnesses for the different schemas that are relevant.

In $R1$, which lacks the extra fitness given to some intermediate-level schemas, the hitchhiking problem does not occur to such a devastating degree. The fitness of an instance of, say, s_{11} in $R1$ is only 16, so its discovery does not have such a dramatic effect on the discovery and persistence of other order 8 schemas in the function. Contrary to our initial intuitions, it appears that the extra reinforcement from some intermediate-level stepping stones actually harms the GA in these functions.

These results point to a pervasive and important issue in the performance of GAs: the problem of premature convergence, in which the GA population converges on some suboptimal set of alleles at some set of loci. The fact that we observe a form of premature convergence even in this very simple setting suggests that it can be a factor in any GA search in which the population is simultaneously searching for two or more nonoverlapping high-fitness schemas (e.g., s_4 and s_{11}), which is often the case. The fact that the population loses useful schemas once one of the disjoint good schemas is found suggests one reason that the rate of effective implicit parallelism of the GA[38,27] may need to be reconsidered. (For another discussion of implicit parallelism in GAs, see Grefenstette and Baker.[30])

DO INTRONS SUPPRESS HITCHHIKERS?

In order to understand the hitchhiking behavior more precisely, we performed an experiment that we believed would eliminate it to some degree. Our hypothesis was that hitchhiking occurred in the loci that were spatially adjacent to the high-fitness schemas (e.g., s_{11} above). In order to reduce this effect, we constructed a new function, $R2_{\text{introns}}$, by introducing blocks of 8 "introns"—8 additional *'s—in between each of the 8-bit blocks of 1's. Thus in $R2_{\text{introns}}$, strings were of length 128 instead of 64. For example, in $R2_{\text{introns}}$, s_1 is 11111111**********...*, s_2 is ****************11111111**...*, and their combination, s_9, is

11111111********11111111**...*.

The optimum is the string containing each block of eight 1's, where the blocks are each separated by eight loci that can contain either 0's or 1's. The idea here was that a potentially damaging hitchhiker would be at least 8 bits away from the schema on which it was hitchhiking, and would thus be likely to be lost under crossover. (Levenick[46] found that inserting introns into individuals improved the performance of the GA in one particular set of environments.)

As shown in column 1 of Table 4, the GA on $R2_{\text{introns}}$ was not faster than the GA on $R2$. This was contrary to our expectations, and the reasons for this result are not clear, but one hypothesis is that once an instance of a higher-order schema (e.g., s_{11}) is discovered, convergence is so fast that hitchhikers are possible even in loci that are relatively distant from the schema's defined positions.

VARYING THE COEFFICIENTS IN $R2$

It is clear that some intermediate-level reinforcement is necessary for the GA to work. Consider $R1'$, a variant of $R1$, where $R1'(x) = 8$ if x is an instance of at least one of the 8-bit schemas, and $R1'(x) = 64$ if x is an instance of *all* the 8-bit schemas. Here the GA would have no reason to prefer a string with block of sixteen 1's over a string with a block of eight 1's, and thus there would be no pressure to increase the number of 1's. Intermediate schemas in $R1$ provide *additive* reinforcement, since

TABLE 4 Summary of results of 200 runs of the GA on two variants of $R2$.

	Function Evaluations to Optimum	
200 runs	$R2_{\text{introns}}$	$R2_{\text{flat}}$
Mean	75599 (std err: 2697)	62692 (std err: 2391)
Median	70400	56448

the fitness of an instance of an intermediate-order schema is always the sum of the fitnesses of instances of the component order 8 schemas. Some schemas in $R2$ provide additional reinforcement: the fitness of an instance of, say, s_9 is much higher than the sum of the fitnesses of instances of the component order 8 schemas s_1 and s_2. Our results indicate that the extra reinforcement given by some schemas is too high—it hurts rather than helps the GA's performance.

Does such additional reinforcement ever help the GA rather than hinder it? To study this we constructed a new function, $R2_{\text{flat}}$, with a much weaker reinforcement scheme: for this function, c_1–c_{14} are each set to the flat value 1. Here there is still additional reinforcement (an instance of s_9 will have fitness $1 + 1 + 1$, which is greater than the sum of the two components), but the amount of reinforcement is reduced considerably.

The results of running the GA on $R2_{\text{flat}}$ is given in the second column of Table 4. The average time to optimum for this function is approximately the same as for $R1$. Thus the smaller fitness reinforcement in $R2_{\text{flat}}$ does not seem to hurt performance, although it does not result in *improved* performance over that on $R1$.

DISCUSSION

The results described above show that the GA's ability to process building blocks effectively depends not only on their presence but also on their relative fitness. If some intermediate stepping stones are too much fitter than the primitive components, then premature convergence slows down the discovery of some necessary schemas. Simple introns do not seem to alleviate the premature convergence and hitchhiking problems.

Our results point out the importance of making the Building-Block Hypothesis a more precise and useful description of building-block processing. While the disruptive effects that we observed (hitchhiking, premature convergence, etc.) are already known in the GA literature, as yet no theorem exists associating them with the building-block structure of a given problem.

In our experiments we have observed that the role of crossover varies considerably throughout the course of the GA search. In particular, three stages of the search can be identified: (1) the time it takes for the GA to discover the lowest-order schemas, (2) the time it takes for crossover to combine lower-order schemas into a higher-order schema, and (3) the time it takes for the higher-order schema to take over the population. In multilevel functions, such as the Royal Road functions, these phases of the search overlap considerably, and it is essential to understand the role of crossover and the details of schema processing at each stage (this issue has also been investigated by Davis[14] and by Schaffer and Eshelman,[79] among others). In previous work,[61] we have discussed the complexities of measuring the relative times for these different phases.

EXPERIMENTS WITH HILL CLIMBING

As was mentioned earlier, part of our purpose in designing the Royal Road functions is to construct the simplest class of fitness landscapes on which the GA will not only perform well, but on which it will outperform other search methods such as hill climbing. In addition to our experiments comparing the GA's performance on $R1$ and $R2$, we compared the GA's performance with that of three commonly used iterated hill-climbing schemes: steepest-ascent hill climbing, next-ascent hill climbing,[64] and a scheme we call "random-mutation hill climbing," that was suggested by Richard Palmer.[69] Our implementation of these various hill-climbing schemes follows:

- **Steepest-ascent hill climbing (SAHC):**

  1. Choose a string at random. Call this string *current-hilltop.*
  2. Systematically mutate each bit in the string from left to right, recording the fitnesses of the resulting strings.
  3. If any of the resulting strings give a fitness increase, then set *current-hilltop* to the resulting string giving the highest fitness increase. (Ties are decided at random.)
  4. If there is no fitness increase, then save *current-hilltop* and go to step 1. Otherwise, go to step 2 with the new *current-hilltop.*
  5. When a set number of function evaluations has been performed, return the highest hilltop that was found.

- **Next-ascent hill climbing (NAHC):**

  1. Choose a string at random. Call this string *current-hilltop.*
  2. Mutate single bits in the string from left to right, recording the fitnesses of the resulting strings. If any increase in fitness is found, then set *current-hilltop* to that increased-fitness string without evaluating any more single-bit mutations of the original string. Go to step 2 with the new *current-hilltop*, but continue mutating the new string starting after the bit position at which the previous fitness increase was found.
  3. If no increases in fitness were found, save *current-hilltop* and go to step 1.
  4. When a set number of function evaluations has been performed, return the highest hilltop that was found.

  This method is similar to Davis' "bit-climbing" scheme.[15] In his scheme, the bits are mutated in a random order, and *current-hilltop* is reset to any string having *equal* or better fitness than the previous best evaluation.

TABLE 5 Summary of results of 200 runs of various hill-climbing algorithms on $R2$.

	Function Evaluations to Optimum		
200 runs	SAHC	NAHC	RMHC
Mean	> 256,000 (std err: 0)	> 256,000 (std err: 0)	6551 (std err: 212)
Median	> 256,000	> 256,000	5925

- **Random-mutation hill climbing (RMHC):**
  1. Choose a string at random. Call this string *best-evaluated.*
  2. Choose a locus at random to mutate. If the mutation leads to an equal or higher fitness, then set *best-evaluated* to the resulting string.
  3. Go to step 2.
  4. When a set number of function evaluations has been performed, return the current value of *best-evaluated.*

Table 5 gives results from running these three hill-climbing schemes on $R2$. In each run the hill-climbing algorithm was allowed to continue either until the optimum string was discovered or until 256,000 function evaluations had taken place, and the total number of function evaluations performed was recorded. The entries in the table are each means over 200 runs. As can be seen, steepest-ascent and next-ascent hill climbing never found the optimum during the allotted time, but random-mutation hill climbing found the optimum on average more than ten times faster than the GA. Note that random-mutation hill climbing as we have described it differs from the bit-climbing method used by Davis[15] in that it does not systematically mutate bits and it never gives up and starts from a new random string; rather it continues to wander around on plateaus indefinitely. Eshelman[19] has pointed out that the random-mutation hill-climber is ideal for the Royal Road functions—in fact, much better than Davis' bit-climber—but will have trouble with any function with local minima. (Eshelman found that Davis' bit-climber does very poorly on $R1$, never finding the optimum in 50 runs of 50,000 function evaluations each.)

These results are a striking demonstration that, when comparing the GA with hill climbing on a particular problem or test-suite, it matters *which* type of hill-climbing algorithm is used. Davis[15] has also made this point.

CONCLUSION

The research on Royal Road landscapes is an initial step in understanding more precisely how schemas are processed under crossover. By studying the GA's behavior on simple landscapes in which the desirable building blocks are explicitly defined, we have discovered some unanticipated phenomena related to the GA's ability to process schemas efficiently, even in nondeceptive functions. The Royal Road functions capture, in an idealized and clear way, some landscape features that are particularly relevant for the GA, and we believe that a thorough understanding of the GA's behavior on these simple landscapes will be very useful in developing more detailed and useful theorems about GA behavior.

The results described in this subsection represent work in progress, and there are several directions for future investigation.

In the short term, we plan to study more carefully the bottlenecks in the discovery of desirable schemas and to quantify more precisely the relationship between the fitness values of the various building blocks and the degree to which these bottlenecks will occur. Hitchhiking is evidently one bottleneck, and we need to understand better the way in which it occurs and under what circumstances. Once we have described the phenomena in more detail, we can begin developing a mathematical model of the schema competitions we observe (illustrated in Figure 33) and how they are affected by different building-block fitness schemes. The hill-climbing results need to be further analyzed and explained. An important part of this research is to construct versions of "royal road" landscapes that will fulfill our goal of finding simple functions that distinguish GAs from hill climbing. Some steps in that direction are reported in Mitchell and Holland.[62]

The Royal Road functions explore only one type of landscape feature that is relevant to GAs: the presence and relative fitnesses of intermediate-order building blocks. Our longer-term plans include extending the class of fitness landscapes under investigation to include other types of relevant features; some such features were described in Mitchell et al.[61] We are also interested in developing statistical measures that could determine the presence or absence of the features of interest. These might be related to work on determining the statistical structure of fitness landscapes described in subsection 4.2. If such measures could be developed, they could be used to help predict the likelihood of successful GA performance on a given landscape.

ACKNOWLEDGMENTS

A number of people have contributed to this chapter in direct and and indirect ways. The work described in subsections 4.1 and 4.3 was carried out in collaboration with Stephanie Forrest, and much of the text there is excerpted from two papers co-authored with her. Emily Dickinson and Julie Rehmeyer helped produce the figures given in subsection 4.1. Thanks to Robert Axelrod, Richard Belew, Stuart Kauffman, and Thomas Meyer for answering questions and for helpful discussions about the research projects described here. Thanks to Robert Axelrod, Aviv Bergman, Lashon Booker, Arthur Burks, Michael Cohen, Marcus Feldman, Stephanie Forrest, Rick Riolo, and Carl Simon for many helpful discussions on issues related to genetic algorithms, and special thanks to John Holland for continuing to produce and share his insights about genetic algorithms and adaptive complex systems. Finally, thanks to Daniel Stein for inviting me to lecture at the 1992 Santa Fe Institute Summer School, and to the Summer School students for incisive questions and comments on my lectures.

REFERENCES

1. Ackley, D., and M. Littman. "Interactions Between Learning and Evolution." In *Artificial Life II*, edited by C. G. Langton et al. Santa Fe Institute Studies in the Sciences of Complexity, Proc. Vol. X, 487–507. Reading, MA: Addison-Wesley, 1991.
2. Andreoni, J., and J. H. Miller. "Auctions with Adaptive Artificial Agents." Working Paper 91-01-004, Santa Fe Institute, Santa Fe, New Mexico, 1991.
3. Axelrod, R. *The Evolution of Cooperation.* New York: Basic Books, 1984.
4. Axelrod, R. "The Evolution of Strategies in the Iterated Prisoner's Dilemma." In *Genetic Algorithms and Simulated Annealing*, edited by L. D. Davis, chapter 3. Research Notes in Artificial Intelligence. Los Altos, CA: Morgan Kaufmann, 1987.
5. Baker, J. E. "Adaptive Selection Methods for Genetic Algorithms." In *Proceedings of the First International Conference on Genetic Algorithms and Their Applications*, edited by J. J. Grefenstette. Hillsdale, NJ: Lawrence Erlbaum, 1985.
6. Baldwin, J. M. "A New Factor in Evolution." *Amer. Natur.* **30** (1896): 441–451.
7. Belew, R. K. "Evolution, Learning, and Culture: Computational Metaphors for Adaptive Algorithms." *Complex Systems* **4** (1990): 11–49.
8. Belew, R. K., J. McInerney, and N. N. Schraudolph. "Evolving Networks: Using the Genetic Algorithm with Connectionist Learning." In *Artificial Life II*, edited by C. G. Langton et al. Santa Fe Institute Studies in the Sciences of Complexity, Proc. Vol. X, 511–547. Reading, MA: Addison-Wesley, 1991.
9. Bergman, A., and M. W. Feldman. "More on Selection For and Against Recombination." *Theor. Pop. Biol.* **38(1)** (1990): 68–92.
10. Bethke, A. D. *Genetic Algorithms as Function Optimizers.* Ph.D. Thesis, University of Michigan, Ann Arbor, MI. Dissertation Abstracts International, 41(9), 3503B, No. 8106101. Ann Arbor, MI: University Microfilms, 1980.
11. Caruana, R. A., and J. D. Schaffer. "Representation and Hidden Bias: Gray vs. Binary Coding for Genetic Algorithms." In *Proceedings of the Fifth International Conference on Machine Learning.* San Mateo, CA: Morgan Kaufmann, 1988.
12. Chalmers, D. J. "The Evolution of Learning: An Experiment in Genetic Connectionism." In *Proceedings of the 1990 Connectionist Models Summer School*, edited by D. S. Touretzky et al. San Mateo, CA: Morgan Kaufmann, 1990.
13. Das, R., and L. D. Whitley. "The Only Challenging Problems are Deceptive: Global Search by Solving Order 1 Hyperplanes." In *Proceedings of the Fourth International Conference on Genetic Algorithms*, edited by R. K. Belew and L. B. Booker. San Mateo, CA: Morgan Kaufmann, 1991.

14. Davis, L. D. "Adapting Operator Probabilities in Genetic Algorithms." In *Proceedings of the Third International Conference on Genetic Algorithms*, edited by J. D. Schaffer. San Mateo, CA: Morgan Kaufmann, 1989.
15. Davis, L. "Bit-Climbing, Representational Bias, and Test Suite Design." In *Proceedings of the Fourth International Conference on Genetic Algorithms*, edited by R. K. Belew and L. B. Booker. San Mateo, CA: Morgan Kaufmann, 1991.
16. Davis, L. D., ed. *Handbook of Genetic Algorithms.* New York: Van Nostrand Reinhold, 1991.
17. Deb, K., and D. E. Goldberg. "Sufficient Conditions for Deceptive and Easy Binary Functions." IlliGAL Report No. 92001, Illinois Genetic Algorithms Laboratory, Department of General Engineering, University of Illinois at Urbana-Champaign, 1992.
18. De Jong, K. A. "An Analysis of the Behavior of a Class of Genetic Adaptive Systems." Ph.D. Thesis, University of Michigan, Ann Arbor, MI, 1975.
19. Eshelman, L. Private communication, 1992.
20. Fontanari, J. F., and R. Meir. "The Effect of Learning on the Evolution of Asexual Populations." *Complex Systems* **4** (1990): 401–414.
21. Forrest, S., and M. Mitchell. "Relative Building Block Fitness and the Building Block Hypothesis." In *Foundations of Genetic Algorithms 2*, edited by L. D. Whitley. Los Altos, CA: Morgan Kaufmann, 1993.
22. Forrest, S., and M. Mitchell. "What Makes a Problem Hard For a Genetic Algorithm? Some Anomalous Results and Their Explanation." *Mach. Learn.* (1993): to appear.
23. Forrest, S., and A. S. Perelson. "Genetic Algorithms and the Immune System." In *Parallel Problem Solving from Nature*, edited by H. Schwefel and R. Maenner. Lecture Notes in Computer Science. Berlin: Springer-Verlag, 1990.
24. Goldberg, D. E. "Simple Genetic Algorithms and the Minimal Deceptive Problem." In *Genetic Algorithms and Simulated Annealing*, edited by L. D. Davis, chapter 6. Research Notes in Artificial Intelligence. Los Altos, CA: Morgan Kaufmann, 1987.
25. Goldberg, D. E. "Genetic Algorithms and Walsh Functions: Part I, A Gentle Introduction." *Complex Systems* **3** (1989): 129–152.
26. Goldberg, D. E. "Genetic Algorithms and Walsh Functions: Part II, Deception and Its Analysis." *Complex Systems* **3** (1989): 153–171.
27. Goldberg, D. E. *Genetic Algorithms in Search, Optimization, and Machine Learning.* Reading, MA: Addison Wesley, 1989.
28. Goldberg, D. E. "Construction of High-Order Deceptive Functions Using Low-Order Walsh Coefficients." Technical Report 90002, Illinois Genetic Algorithms Laboratory, Department of General Engineering, University of Illinois, Urbana, IL, 1990.

29. Grefenstette, J. J. "Deception Considered Harmful." In *Foundations of Genetic Algorithms 2*, edited by L. D. Whitley. Los Altos, CA: Morgan Kaufmann, 1993.
30. Grefenstette, J. J., and J. E. Baker. "How Genetic Algorithms Work: A Critical Look at Implicit Parallelism." In *Proceedings of the Third International Conference on Genetic Algorithms*, edited by J. D. Schaffer. San Mateo, CA: Morgan Kaufmann, 1989.
31. Harp, S. A., and T. Samad. "Genetic Synthesis of Neural Network Architecture." In *Handbook of Genetic Algorithms*, edited by L. D. Davis, 202–221. New York: Van Nostrand Reinhold, 1991.
32. Hertz, J., A. Krogh, and Richard G. Palmer. *Introduction to the Theory of Neural Computation.* Santa Fe Institute Studies in the Sciences of Complexity, Lect. Notes Vol. I. Reading, MA: Addison-Wesley, 1991.
33. Hillis, W. D. "Co-Evolving Parasites Improve Simulated Evolution as an Optimization Procedure." Special issue edited by S. Forrest. *Physica D* **42** (1990): 228–234.
34. Hinton, G. E., and S. J. Nowlan. "How Learning Can Guide Evolution." *Complex Systems* **1** (1987): 495–502.
35. Holland, J. H. "Escaping Brittleness: The Possibilities of General-Purpose Learning Algorithms Applied to Parallel Rule-Based Systems." In *Machine Learning II*, edited by R. S. Michalski, J. G. Carbonell, and T. M. Mitchell, 593–623. San Mateo, CA: Morgan Kaufmann, 1986.
36. Holland, J. H. "The Dynamics of Searches Directed by Genetic Algorithms." In *Evolution, Learning, and Cognition*, edited by Y. C. Lee, 111–128. Teaneck, NJ: World Scientific, 1988.
37. Holland, J. H., and J. H. Miller. "Artificial Adaptive Agents in Economic Theory." Working Paper 91-05-025, Santa Fe Institute, Santa Fe, New Mexico, 1991.
38. Holland, J. H. *Adaptation in Natural and Artificial Systems*, 2nd Edition. Cambridge, MA: MIT Press, 1992. (First edition, 1975.)
39. Holland, J. H. "Echoing Emergence: Objectives, Rough Definitions, and Speculations for Echo-Class Models." Working Paper 93-04-023, Santa Fe Institute, Santa Fe, New Mexico, 1993.
40. Kauffman, S. A. "Emergent Properties in Random Complex Automata." *Physica D* **10** (1984): 145–156.
41. Kauffman, S. A. "Adaptation on Rugged Fitness Landscapes." In *Lectures in the Sciences of Complexity*, edited by D. Stein. Santa Fe Institute Studies in the Sciences of Complexity, Lect. Vol. I, 527–618. Reading, MA: Addison-Wesley, 1989.
42. Kauffman, S. A. "Requirements for Evolvability in Complex Systems: Orderly Dynamics and Frozen Components." *Physica D* **42** (1990): 135–152.

43. Koza, J. R. "Genetic Programming: A Paradigm for Genetically Breeding Populations of Computer Programs to Solve Problems." STAN-CS-90-1314, Department of Computer Science, Stanford University, Stanford, CA, 1990.
44. Koza, J. R. *Genetic Programming: On the Programming of Computers by Means of Natural Selection.* Cambridge, MA: MIT Press, 1993.
45. Lenet, D. B., and J. S. Brown. "Why AM and Eurisko Appear to Work." *Art. Intel.* **23** (1984): 260–294.
46. Levenick, J. R. "Inserting Introns Improves Genetic Algorithm Success Rate: Taking a Cue from Biology." In *Proceedings of the Fourth International Conference on Genetic Algorithms*, edited by R. K. Belew and L. B. Booker, 123–127. San Mateo, CA: Morgan Kaufmann, 1991.
47. Liberman, U., and M. W. Feldman. "A General Reduction Principle for Genetic Modifiers of Recombination." *Theor. Pop. Biol.* **30(3)** (1986): 341–371.
48. Liepins, G. E., and M. D. Vose. "Representational Issues in Genetic Optimization." *J. Exper. & Theor. Art. Intel.* **2** (1990): 101–115.
49. Liepins, G. E., and M. D. Vose. "Deceptiveness and Genetic Algorithm Dynamics." In *Foundations of Genetic Algorithms*, edited by G. Rawlins. San Mateo, CA: Morgan Kaufmann, 1991.
50. Lindgren, K. "Evolutionary Phenomena in Simple Dynamics." In *Artificial Life II*, edited by C. G. Langton et al. Santa Fe Institute Studies in the Sciences of Complexity, Proc. Vol. X, 295–312. Reading, MA: Addison-Wesley, 1991.
51. Lloyd Morgan, C. "On Modification and Variation." *Science* **4** (1896): 733–740.
52. Mackey, M. C., and L. Glass. *Science* **197** (1977): 297.
53. Manderick, B., M. de Weger, and P. Spiessens. "The Genetic Algorithm and the Structure of the Fitness Landscape." In *Proceedings of the Fourth International Conference on Genetic Algorithms*, edited by R. K. Belew and L. B. Booker. San Mateo, CA: Morgan Kaufmann, 1991.
54. Maynard Smith, J. "When Learning Guides Evolution." *Nature* **329** (1987).
55. McClelland, J. L., and D. E. Rumelhart. *Parallel Distributed Processing*, Vol. 2. Cambridge, MA: MIT Press, 1986.
56. Meyer, T. P. "Long-Range Predictability of High-Dimensional Chaotic Dynamics." Ph.D. Thesis, University of Illinois at Urbana-Champaign, Urbana, IL, 1992.
57. Meyer, T. P. Personal communication, 1992.
58. Meyer, T. P., and N. H. Packard. "Local Forecasting of High-Dimensional Chaotic Dynamics." In *Nonlinear Modeling and Forecasting*, edited by M. Casdagli and S. Eubank. Santa Fe Institute Studies in the Sciences of Complexity, Proc. Vol. XII, 249–264. Reading, MA: Addison-Wesley, 1992.

59. Michalewicz, Z. *Genetic Algorithms + Data Structures = Evolution Programs.* Artificial Intelligence Series. Berlin: Springer-Verlag, 1992.
60. Miller, G. F., P. M. Todd, and S. U. Hegde. "Designing Neural Networks Using Genetic Algorithms." In *Proceedings of the Third International Conference on Genetic Algorithms*, edited by J. D. Schaffer, 379–384. San Mateo, CA: Morgan Kaufmann, 1989.
61. Mitchell, M., S. Forrest, and J. H. Holland. "The Royal Road for Genetic Algorithms: Fitness Landscapes and GA Performance." In *Proceedings of the First European Conference on Artificial Life.* Cambridge, MA: MIT Press/ Bradford Books, 1992.
62. Mitchell, M., and J. H. Holland. "When Will a Genetic Algorithm Outperform Hill-Climbing?" In *Advances in Neural Information Processing Systems 6*, edited by J. D. Cowan, G. Tesauro, and J. Alspector. San Mateo, CA: Morgan Kaufmann, to appear.
63. Montana, D. J., and L. D. Davis. "Training Feedforward Networks Using Genetic Algorithms." In *Proceedings of the International Joint Conference on Artificial Intelligence.* San Mateo, CA: Morgan Kaufmann, 1989.
64. Mühlenbein, H. "Evolution in Time and Space—The Parallel Genetic Algorithm." In *Foundations of Genetic Algorithms*, edited by G. Rawlins, 316–337. San Mateo, CA: Morgan Kaufmann, 1991.
65. Nakano, R. "Conventional Genetic Algorithm for Job Shop Problems." In *Proceedings of the Fourth International Conference on Genetic Algorithms*, edited by R. K. Belew and L. B. Booker, 474–479. San Mateo, CA: Morgan Kaufmann, 1991.
66. Nilsson, N. J. "Action Networks." In *Proceedings from the Rochester Planning Workshop: From Formal Systems to Practical Systems*, edited by J. Tenenberg et al. Technical Report 284, Computer Science Department, University of Rochester, Rochester, New York, 1989.
67. Nolfi, S., J. L. Elman, and D. Parisi. "Learning and Evolution in Neural Networks." Technical Paper CRL 9019, Center for Research in Language, University of California, San Diego, CA, 1990.
68. Packard, N. H. "A Genetic Learning Algorithm for the Analysis of Complex Data." *Complex Systems* **4(5)** (1990).
69. Palmer, R. G. Personal communication, 1992.
70. Parisi, D., S. Nolfi, and F. Cecconi. "Learning, Behavior, and Evolution." In *Proceedings of the First European Conference on Artificial Life.* Cambridge, MA: MIT Press/Bradford Books, 1992.
71. Radcliffe, N. J. "Equivalence Class Analysis of Genetic Algorithms." *Complex Systems* **5(2)** (1991): 183–205.
72. Rawlins, G. *Foundations of Genetic Algorithms.* San Mateo, CA: Morgan Kaufmann, 1991.

73. Ray, T. S. "An Approach to the Synthesis of Life." In *Artificial Life II*, edited by C. G. Langton et al. Santa Fe Institute Studies in the Sciences of Complexity, Proc. Vol. X, 371–408. Reading, MA: Addison Wesley, 1991.
74. Ray, T. S. "Is It Alive, or Is It GA?" In *Proceedings of the Fourth International Conference on Genetic Algorithms*, edited by R. K. Belew and L. B. Booker, 527–534. San Mateo, CA: Morgan Kaufmann, 1991.
75. Riolo, R. L. "Survival of the Fittest Bits." *Sci. Am.* **July** (1992): 114–116.
76. Rogers, D. "Weather Prediction Using a Genetic Memory." Technical Paper 90.6, Research Institute for Advanced Computer Science, NASA Ames Research Center, Moffett Field, CA, 1990.
77. Rumelhart, D. E., G. E. Hinton, and R. J. Williams. "Learning Internal Representations by Error Propagation." In *Parallel Distributed Processing*, edited by D. E. Rumelhart and J. L. McClelland. Cambridge, MA: MIT Press, 1986.
78. Rumelhart, D. E., and J. L. McClelland. *Parallel Distributed Processing*, Vol. 1. Cambridge, MA: MIT Press, 1986.
79. Schaffer, J. D., and L. J. Eshelman. "On Crossover as an Evolutionarily Viable Strategy." In *Proceedings of the Fourth International Conference on Genetic Algorithms*, edited by R. K. Belew and L. B. Booker, 61–68. San Mateo, CA: Morgan Kaufmann, 1991.
80. Schraudolph, N. N., and R. K. Belew. "Dynamic Parameter Encoding for Genetic Algorithms." Technical Report CS 90-175, Computer Science and Engineering Department, University of California, San Diego, CA, 1990.
81. Schulze-Kremer, S. "Genetic Algorithms for Protein Tertiary Structure Prediction." In *Parallel Problem Solving from Nature 2*, edited by R. Manner and B. Manderick, 391–400. Amsterdam: North Holland, 1992.
82. Shahookar, K., and P. Mazumder. "A Genetic Approach to Standard Cell Placement Using Meta-Genetic Parameter Optimization." *IEEE Trans. Comp. Design* **9(5)** (1990): 500–511.
83. Syswerda, G. "Uniform Crossover in Genetic Algorithms." In *Proceedings of the Third International Conference on Genetic Algorithms*, edited by J. D. Schaffer, 2–9. San Mateo, CA: Morgan Kaufmann, 1989.
84. Tanese, R. "Distributed Genetic Algorithms for Function Optimization." Ph.D. Thesis, Computer Science Department, University of Michigan, Ann Arbor, MI, 1989.
85. Vose, M. D. "Generalizing the Notion of Schema in Genetic Algorithms." *Art. Intel.* **50** (1991): 385–396
86. Waddington, C. H. "Canalization of Development and the Inheritance of Acquired Characters." *Nature* **150** (1942): 563–565.
87. Walsh, J. L. "A Closed Set of Orthogonal Functions." *Am. J. Math.* **55** (1923): 5–24.
88. Whitley, L. D. "Fundamental Principles of Deception in Genetic Search." In *Foundations of Genetic Algorithms*, edited by G. Rawlins. San Mateo, CA: Morgan Kaufmann, 1991.

89. Whitley, L. D. *Foundations of Genetic Algorithms 2.* San Mateo, CA: Morgan Kaufmann, 1993.
90. Wright, S. "Evolution in Mendelian Populations." *Genetics* **16** (1931): 97–159.

Charles M. Gray
The Salk Institute for Biological Studies, La Jolla, CA 92186–5800;
Present Address: Center for Neuroscience, University of California–Davis, Davis, CA 95616

Rhythmic Activity in Neuronal Systems: Insights Into Integrative Function

INTRODUCTION

Most functions carried out by the nervous system are integrative in nature. Our ability to see depends on the integration of color, motion, form, and depth signals from many different locations in the visual field. Our sense of hearing requires the temporal integration of many different frequencies and intensities of sound occurring at different locations at the same time. When we move we most often do so in a coordinated way involving the concerted action of many muscles. These functions rarely take place in isolation but rather occur in parallel as integrative sensorimotor functions involving aspects of cognition and memory.

From a physiological point of view we know very little about the neural mechanisms underlying such integrative actions of the nervous system. A simple example illustrates this point. Consider the act of picking up a glass of water and taking a drink. Such a task engages vast numbers of cells in many different parts of the brain. The visual system must detect, localize, and recognize the glass. The motor system must initiate and control the movement of the eyes, the hand, the arm, and the shoulder. The somatosensory system must tactilely sense the object and provide information about the strength of grip required to hold it. And all of this must take place in a continuous fashion requiring the parallel and coordinated involvement of each system. This entire process, one of the simplest examples of normal sensorimotor behavior, requires the coordinated, dynamic interaction between a host of systems in the brain, each of which must carry out parallel computations involving millions of interconnected neurons. How does the nervous system control the dynamic coordination of multiple systems operating in parallel?

It seems reasonable to assume that a mechanism or a class of mechanisms has evolved to enable the temporal coordination of activity within and between subsystems of the central nervous system. For several reasons, rhythmic neuronal activity has long been thought to play an important role in such coordination. Since the discovery of the electroencephalogram (EEG) over 60 years ago, it has been known that neural systems in the mammalian brain often engage in rhythmic activities. These patterned neuronal oscillations take many forms. They occur over a broad range of frequencies and are present in a multitude of different systems in the brain during a variety of different behavioral states. They are often the most salient aspect of observable electrical activity in the brain and typically encompass widespread regions of cerebral tissue. Our understanding of such processes, however, is only beginning to emerge.

With the advent of sophisticated new techniques in multielectrode recording and neural imaging, studies of the behavior and function of neuronal populations have entered a new era. It is now within the realm of possibility to record from 100 single neurons simultaneously,[203] to optically measure the activity in an entire cortical area,[29,190] or to noninvasively image the pattern of electrical current flow in an alert human being performing a task.[153] In a number of instances the application of these new techniques has revealed that spatially and temporally organized activity in a distributed population of cells often takes the form of synchronous rhythms. When taken together with cellular neurophysiological and anatomical studies, these findings provide new insights into the behavior and mechanisms controlling the coordination of activity in neuronal populations.

In this chapter I will review recent advances in six areas of systems neurophysiology where synchronous rhythmic activity has been observed and investigated. I have made no attempt to exhaustively review the literature on the occurrence of neuronal oscillations in the nervous system. Rather I have chosen to focus on particular areas where it appears that coordinated activity in a neural system may play a functional role. These areas include the olfactory bulb, thalamocortical spindle rhythms, and the visual cortex, each reviewed in some detail, and the olivocerebellar system, hippocampal rhythms, and the somatomotor cortices, which I discuss more briefly. In each case I have attempted, when possible, to combine a discussion of the macroscopic behavior of the system and its relation to behavior with the cellular mechanisms thought to control the rhythmic activity and its synchronization.

THE OLFACTORY SYSTEM: THE INDUCED WAVE AND THE SPATIAL CODING HYPOTHESIS

THE INDUCED WAVE

Adrian, a pioneer in the investigation of rhythmic neuronal activities, discovered a unique form of oscillatory activity in the olfactory system that he later termed "the induced wave."[2,3,4,5] This was a rhythmic fluctuation of voltage evoked in the EEG of the olfactory bulb of rabbits and hedgehogs by the stimulation of the olfactory receptor sheet. It ranged in frequency from 40–80 Hz and was associated with increased firing in the output neuron population within the bulb (Figure 1).

Since then, subsequent studies by Adrian and others have revealed that the induced wave is a general property of the olfactory bulbs of a number of vertebrate species. Induced waves have been observed in the olfactory bulbs of amphibia and fish,[113,185] a variety of mammalian species,[35,66] and humans.[96] In each instance the occurrence of the rhythmic activity was found to depend on input from the sensory receptor sheet, itself showing no evidence of oscillatory activity.[66,152] In later studies, most notably by Freeman and his colleagues, it was discovered that olfactory structures receiving input from the bulb, such as the anterior olfactory nucleus, the prepiriform, and piriform and entorhinal cortices also displayed an induced wave dependent on olfactory receptor input[31,62,66] (Figure 2). These rhythmic activities often had a broader and somewhat lower range of frequencies. If bulbar output

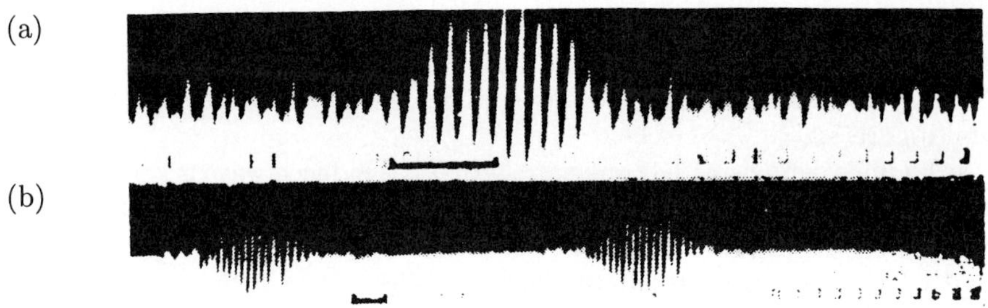

FIGURE 1 "Induced waves" due to olfactory stimulation (amyl acetate). The records are both from rabbits, anesthetized deeply with urethane. This electrode was placed in the surface layer of olfactory bulb. Time mark (black line) gives 0.1/sec. The frequency of the waves is (a) 60/sec and (b) 55/sec. (See Figure 1 of Adrian,[3] "The Electrical Activity of the Mammalian Olfactory Bulb," *Electroenceph. Clin. Neurophysiol.* **2** (1950): 377–388.) Reprinted by permission.

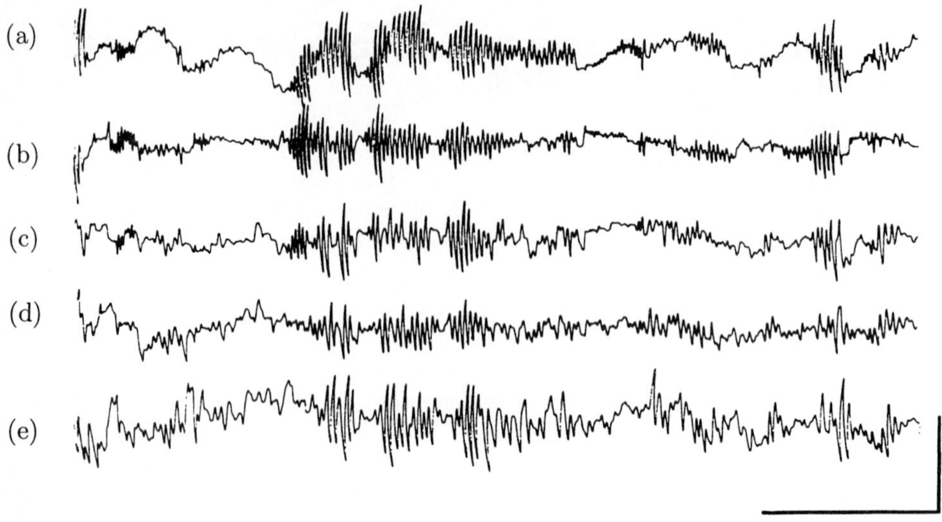

FIGURE 2 (a) Bursts of EEG activity from olfactory bulb and (b)–(e) prepyriform cortex in waking cat. (See Figure 6.28 of Freeman,[66] *Mass Action in the Nervous System*, New York: Academic Press, 1975.) Reprinted by permission.

to these structures was blocked by severing the lateral olfactory tract—the main output pathway of the bulb—the induced waves could no longer be evoked in the piriform cortex.[19] However, even when the bulbar stalk was completely transected the induced wave in the bulb persisted. This effect was also revealed in a later study by Gray and Skinner[84] using reversible cryogenic inactivation of the olfactory peduncle (Figure 3).

These findings demonstrated conclusively that the induced wave is generated by mechanisms intrinsic to the olfactory bulb. Gradually a number of studies initiated in the 1960s began to unravel these basic neural mechanisms.[66] Anatomical studies revealed that the principal circuit in the bulb consisted of a negative feedback loop between the mitral cells, the output neuron of the bulb, and the granule cells, an inhibitory interneuron population[44,172] (Figure 4). An essential feature of this circuit was the dendrodendritic reciprocal synapse.[161] Mitral and granule cells were found to make bidirectional synaptic contact at their apical dendrites in the external plexiform layer of the bulb, thus providing a structural basis for the recurrent inhibition of mitral cells. These synapses were found to be located exclusively in the external plexiform layer directly above the mitral-cell body layer of the bulb.[172]

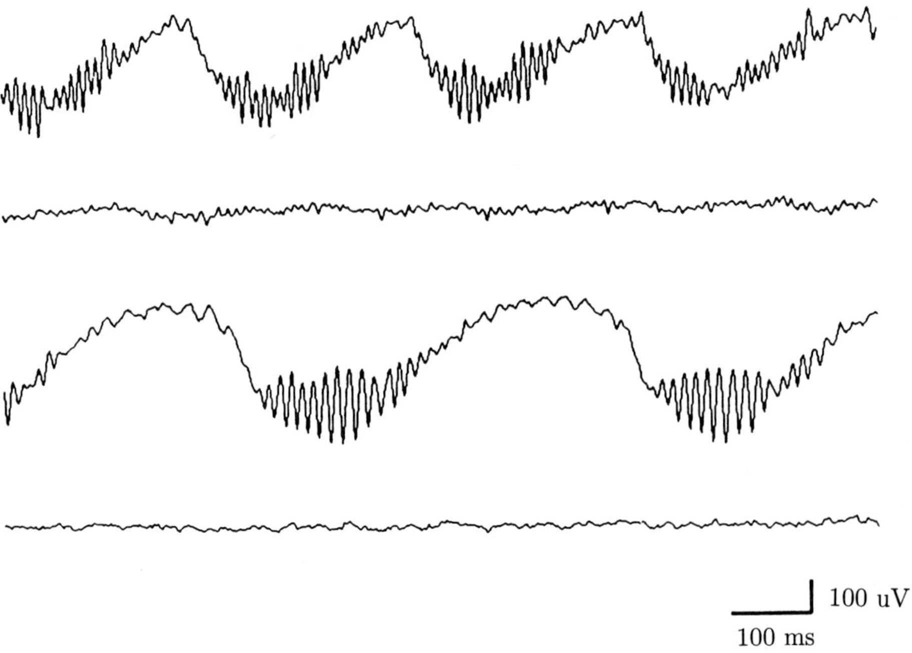

FIGURE 3 The combined effects of nostril closure and cryogenic blockade on the bulbar EEG recorded from a depth electrode in the olfactory bulb in an alert rabbit. Each trace consists of a one-second epoch of EEG activity recorded at 1–300 Hz bandpass. Four separate conditions are displayed from top to bottom: control, nostrils lightly pinched shut, cryogenic blockade of the olfactory peduncle, and cryogenic blockade combined with nostril closure. (See Figure 3 in Gray and Skinner,[84] "Centrifugal Regulation of Neuronal Activity in the Olfactory Bulb of the Waking Rabbit as Revealed by Reversible Cryogenic Blockade," *Exp. Brain Res.* **69** (1988): 378–386.) Reprinted by permission.

This unique synaptic relay combined with the particular morphological structure of the mitral and granule cells led Rall and Shepherd[162] to the prediction that field potentials in the bulb, observable as the induced wave or evoked potential, were generated by synaptic current flow in the granule cell population. Their argument was based primarily on the morphology of the dendrites of mitral and granule cells (Figure 4). The former have radially organized dendrites[151] and were thought to generate a closed dipole field of potential that did not sum with other surrounding mitral-cell dendritic potentials. In contrast the granule cells, which vastly outnumber the mitral cells, have well-aligned longitudinally oriented dendrites.[140] Excitatory synaptic current in this population of cells was thought to give rise to

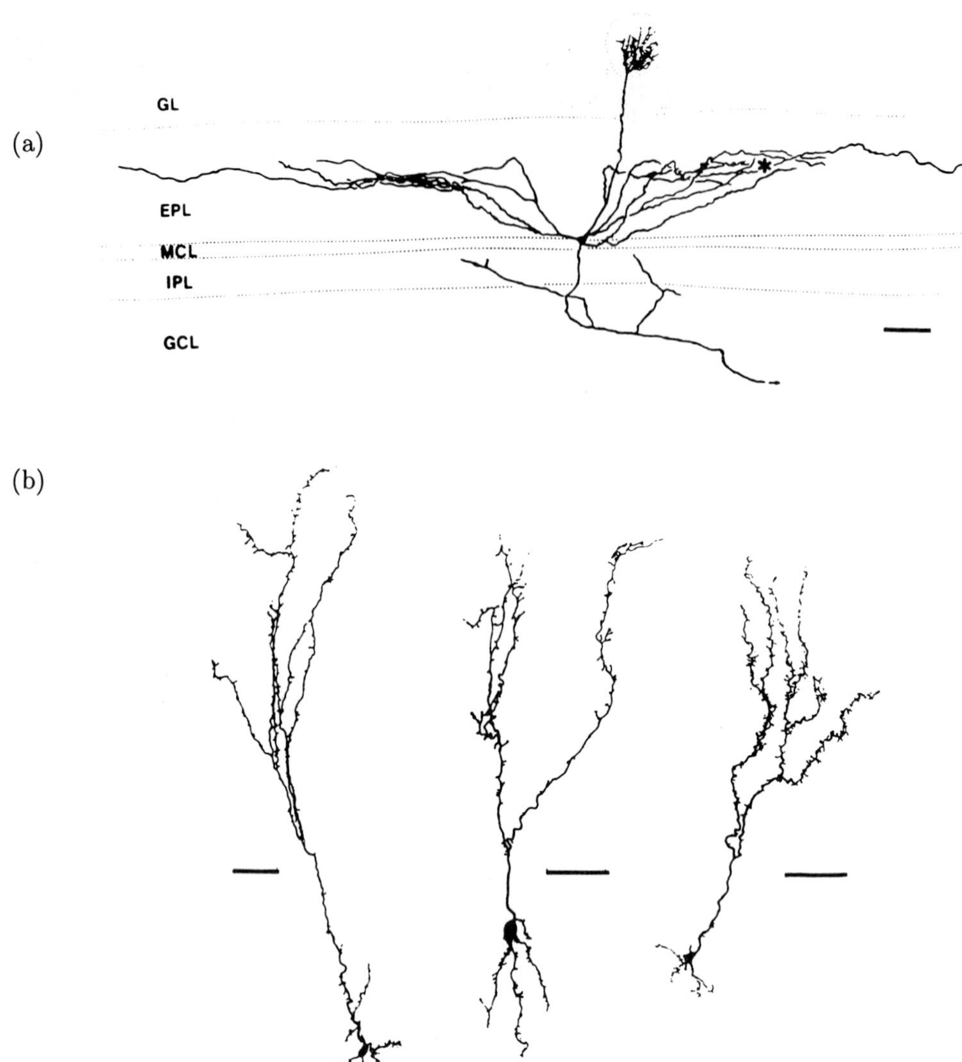

FIGURE 4 (a) Reconstruction of an intracellularly stained type II mitral cell in the olfactory bulb of a rat. GL, EPL, MCL, IPL, and GCL refer to the glomerular layer, the external plexiform layer, mitral cell layer, internal plexiform layer, and the granule cell layer, respectively. (b) Reconstruction of intracellularly stained granule cells in the olfactory bulb of a rabbit. Each scale bar is 50 μm. (See Figure 3(a) of Orona, Rainer, and Scott,[151] "Dendritic and Axonal Organization of Mitral and Tufted Cells in the Rat Olfactory Bulb," *J. Comp. Neurol.* **226** (1984): 346–356; and Figure 2 of Mori and Kishi,[140] "The Morphology and Physiology of the Granule Cells in the Rabbit Olfactory Bulb Revealed by Intracellular Recording and HRP Injection," *Brain Res.* **247** (1982): 129–133.) Reprinted by permission.

an open dipole field easily detected at the surface of the bulb. The resulting model suggested that the field potential was the direct result of excitatory synaptic activity in the granule cell population. It predicted both the laminar distribution of the evoked potential and its triphasic waveform.[162]

This finding was particularly significant because it suggested that the induced wave resulted from the combined activity of the granule cell population. Subsequent experiments by Freeman and his colleagues confirmed this prediction. They showed that the spike discharge of mitral cells rarely exhibited any evidence of rhythmicity as expected. Rather they found that mitral cells most often fire at low rates around 10 spikes/sec and show interspike interval and autocorrelation histograms having properties consistent with a Poisson process.[66] This seeming lack of rhythmicity in the mitral-cell firing patterns proved to be misleading. Subsequent investigations demonstrated that the timing of mitral-cell spikes was often related to the phase of the induced wave. In particular it was found that the mitral cells showed their highest firing probability shortly after the negative peak of the induced wave when the latter was measured at the surface of the bulb.[66] This delay in firing was approximately 1/4 of the period length of the induced wave itself. Thus, even though the spike trains of mitral cells showed little evidence of periodicity, their firing nonetheless was closely related to the oscillatory pattern in the induced wave.

These data confirmed two predictions. First, they demonstrated that mitral -cell activity did not give rise to the extracellular voltage measured in the induced wave. Second, the 1/4-cycle phase lead of mitral-cell activity relative to the induced wave was consistent with the prediction that bulbar rhythmicity was generated by a recurrent inhibitory interaction between the mitral and granule cell populations.[66,162]

Subsequent studies confirmed the basic tenets of the Rall/Shepherd model. Using a two-dimensional multielectrode array, Freeman[64] directly measured the spatial distribution of potential throughout the depth of the bulb during different phases of the evoked response following electrical stimulation of the lateral olfactory tract (LOT). He found the potential field to be concentrically organized during the peak of mitral-cell activity consistent with a closed field. In contrast during the surface negative phase of the evoked response, when the granule cells are maximally excited, he observed an open dipole field of potential that reversed polarity at the mitral-cell body layer. These findings gained further support in a later study by Mori and Kishi.[140] Utilizing intracellular recording, they demonstrated that suprathreshold EPSPs in granule cells occurred during the surface negative phase of the LOT evoked response.

The combination of these anatomical and physiological studies eventually led to the following scheme thought to underlie the rhythmicity in the induced wave.[65] Afferent input on the olfactory nerve axons excites the apical dendrites of the mitral cells. This depolarization propagates along the mitral-cell dendritic membrane resulting in both spike discharge at the soma and the synaptic excitation of the granule cells at the dendrodendritic synapses. Activity in the granule cells leads to a recurrent inhibition of the mitral-cell population, the decreased activity of which results in a disexcitation of the granule cell population. This latter effect, in turn, results in a disinhibition of the mitral-cell population. If the afferent input coming from the olfactory nerve is sustained, the mitral cells will be reexcited and the cycle will repeat.[65,75]

Induced rhythms in the piriform cortex (Figure 2) have been shown to depend on the presence of the induced wave in the bulb.[19,66] If the output of the bulb, the LOT, is severed, rhythmic activity is abolished and electrical stimulation of the central stump produces a biphasic evoked potential having little or no oscillatory component. Such results have led to the view that rhythmic activity in the piriform cortex is tuned to and dependent on the rhythmic afferent drive from the bulb for its expression.

SPATIAL DISTRIBUTION OF ACTIVITY

On the basis of his studies in the 1940s and early 1950s, Adrian[3,4] proposed what has come to be known as the spatial coding hypothesis. He reasoned that, for each discriminable odor, there should exist a unique spatial pattern of activity representing a given odor quality that persists transiently throughout the olfactory system. Adrian's ability to test his own hypothesis unfortunately was hampered by technical limitations. He was limited to recording field potentials or unit activity from one or a few electrodes at a time. In subsequent years, however, Freeman and his colleagues began a concerted and sustained effort to test Adrian's basic hypothesis. Their approach was founded on two specific assumptions: First, the representation of an odor in the olfactory bulb and cortex should exist during the inspiratory phase of respiration when the induced wave is present. Second, the close correlation between the phase of the induced wave and the firing patterns of cells in both the bulb and cortex should enable one to indirectly measure the spatial pattern of activity of large populations of cells in the bulb and cortex by recording the induced wave at many locations simultaneously.

In order to test the hypothesis, it is necessary to obtain an adequate spatial sample of the induced wave in either the bulb or the cortex in a behaving animal. On the basis of a detailed analysis of the spatial frequency of the induced wave over the surface of the olfactory bulb,[68] Freeman[67] developed a technique for measuring the induced wave over the lateral surface of the bulb or the cortex at 64 locations simultaneously. Each recording site was separated by 0.5 mm. This provided a sample of activity from roughly 20% of the bulbar surface and 30–50% of the piriform cortical surface at an optimal spatial resolution. Using this technique the first glimpse of the spatial distribution of the induced wave yielded a remarkable picture[67]: the oscillatory activity was synchronous over broad expanses of the bulb and cortex (Figure 5). In fact, it appeared as though the entire structures of both the bulb and cortex were often oscillating at a common phase and frequency. Phase differences were present but showed no clear dependence on odor. The frequencies in the signals changed over time within a burst but were found to covary across space. Fluctuations in the signal at one site were accompanied by similar changes across the structure. The most notable inhomogeneities were in the amplitude of the signals. The induced wave was often found to have one or more foci or hot spots where the amplitude of the signals over a region were significantly higher than in surrounding regions. These spatial patterns varied slightly from burst to burst but often took the form of a characteristic signature for each animal. The patterns were never the same for any two animals.[67]

These remarkable patterns raised a number of interesting questions. How can such synchrony be established among hundreds of millions of cells within and between two separate cortical structures? What relationship do the patterns show to odor stimuli and behavioral state? What is the significance of the oscillation? Does it carry information in a temporal code or simply provide a mechanism for controlling synchrony? Freeman and his colleagues addressed these questions through a combined approach of physiological experiment and theoretical modeling.

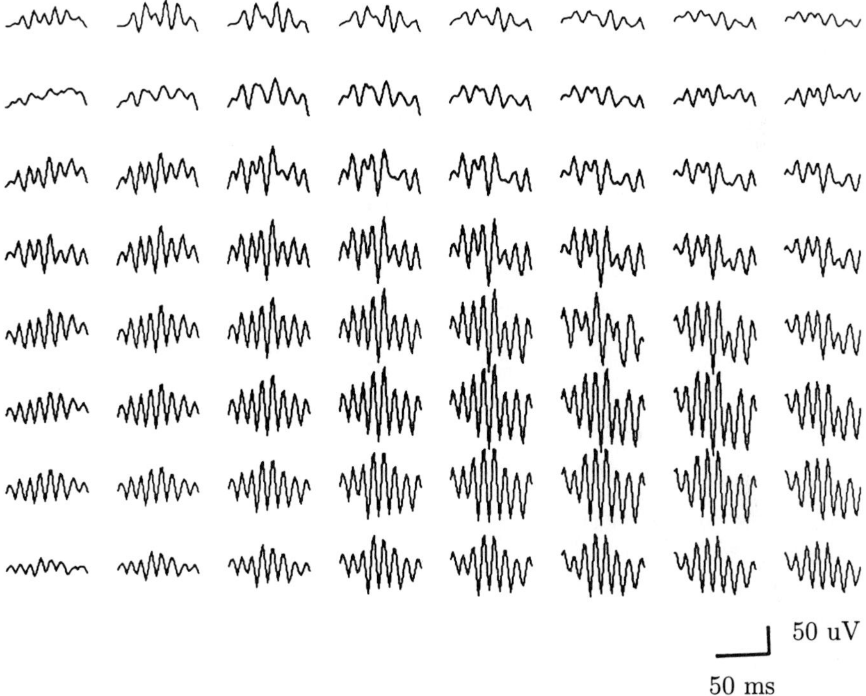

FIGURE 5 Induced wave recorded at 64 locations over the lateral surface of the olfactory bulb in an alert rabbit breathing purified air. The interelectrode spacing is 0.5 mm.

Subsequent studies of the influence of odor stimuli on the spatial patterns revealed another remarkable result: there was essentially no effect. Freeman measured the spatial pattern of phase, frequency, and root mean squared amplitude of the induced wave separately in both the bulb and the cortex of awake rabbits during the passive presentation of several different odors. None of the measures revealed a significant change in the pattern related to the odor stimuli. Spatial phase changed unpredictably in the bulb showing no consistent relation to the odors. In the cortex the principal anteroposterior phase gradient of the induced wave was found to be the result of conduction delays in the propagation of activity in the LOT. Outside the trajectory of the LOT the phase patterns bore no relation to the odors presented. The frequency content of the signals was equally uninformative. The induced wave consisted of multiple frequencies in both the bulb and cortex and no relation could be found between frequency content and odor presentation. Similarly for amplitude the rms power of the induced wave surprisingly showed no consistent relationship to the odor.[67]

These studies prompted a number of conclusions. First, the synchrony of activity was robust and could be established within the bulb and cortex very rapidly, on the order of 10–20 msec. The spatiotemporal properties of the induced wave, its frequency and its phase, appeared to show no relation to odor input. This suggested that the oscillations were not carriers of information but rather a basic neural mechanism for establishing synchronous interactions among widely distributed populations of cells. The anatomical substrate for establishing synchrony in the piriform cortex was clear. The long-range excitatory collaterals of the pyramidal cells could easily provide the necessary drive for synchronizing the activity in a distributed population of cells.[92,93,94] In the bulb the answer to the question was a bit more perplexing. Freeman and his colleagues had suggested that a system of mutually excitatory collaterals of mitral cells, as in the piriform cortex, would provide the ideal substrate. Evidence for such connections has remained rather sparse,[200] but a system of axon collaterals of the superficial tufted cells has been discovered that also could serve such a function.[103,170] Alternatively it is conceivable that the extensively branching dendritic arbors of the mitral cells could provide the substrate for synchronization (Figure 4). Propagation of activity along the dendritic tree would act to excite large groups of inhibitory granule cells. These cells, in turn, would deliver inhibitory feedback onto groups of mitral cells, thereby controlling the timing of their ouput. In addition to this mechanism it is possible that the dendritic release of excitatory transmitter from mitral/tufted cells could act to excite the same or adjacent mitral/tufted cells, thereby providing another form of positive feedback.[44,148,149]

The absence of changes in the amplitude patterns of the induced wave in response to different odor inputs remained a mystery. Either the spatial measure of the activity was too crude to reveal the associated changes or the behavioral conditions of the experiment precluded the observation of specific changes in relation to the odor stimuli. To address these issues the experiments were repeated with two significant changes in the paradigm. First, the animals were engaged in a classical discriminative conditioning protocol in order to present the odors within a behaviorally significant context. Second, a new quantitative statistical test was devised to measure amplitude pattern differences with much greater sensitivity.

These improvements in technique led to some astonishing results.[71,193] When odors were presented within a behavioral context (i.e., either positively or negatively reinforced), the animals reliably discriminated between them by sniffing and/or licking. This insured that the animals were actually detecting and classifying the stimuli. Under these conditions the statistical measure revealed significant differences between the control amplitude pattern of the induced wave and the pattern evoked during the response to the odor. This difference was transient and was most pronounced during the first three inspirations following odor presentation (Figure 6). If odors were presented without reinforcement, the animals rapidly habituated their behavioral responses[84] and no reliable spatial pattern differences could be detected.[67,71] These findings clearly suggested that the spatial coding of odor information was a learning-dependent process.[72] Thus, in a behaviorally meaningless context it was thought that the patterns evoked by unreinforced odors rapidly habituated and, as a result of the low sensitivity of the method, were not detectable.

A second equally surprising result was revealed by these experiments. In the course of evaluating pattern differences in the induced wave evoked by odor stimuli, it was necessary to measure the pattern present in the control state when the animals were breathing purified air. As stated above, these measurements revealed a rather stable spatial amplitude pattern of the induced wave that was unique for each animal. This control pattern, however, was observed to change during the course of behavioral training (Figure 7). For example, after acquiring a particular discrimination behavior to a pair of odors, the foci in the control patterns were seen to shift contiguously in position. Such spatial shifts in the control pattern continued for as long as the animal learned new behavioral discriminations. If an initial discrimination set was extinguished and then later retrained after an intervening sequence of different discriminations, the control pattern did not revert back to its original form. This result revealed clear, long-term, learning-dependent changes in the organization of the olfactory bulb. Moreover, it suggested that the control pattern reflected a sum over all learning and not some simple spatial representation of the currently trained odors.[71,72,193]

In spite of these remarkable findings it was still not possible to confirm or reject Adrain's hypothesis on the basis of the available data. A difference in the pattern of activity produced by a reinforced and an unreinforced odor could not be distinguished. Freeman conjectured that this failure was still the result of inadequate resolution of the measurement techniques. To resolve the uncertainty, he applied a number of data-processing techniques designed to remove sources of noise

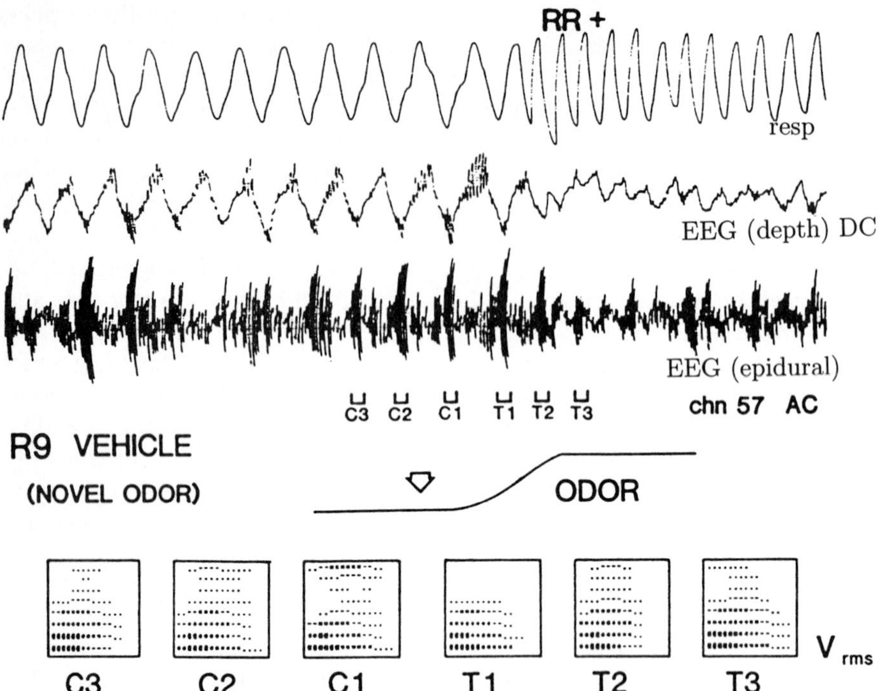

FIGURE 6 Illustration of event-related changes in bulbar activity that accompany the presentation of a novel odor during a single six-second trial. The three upper traces from top to bottom are respiration, the bulbar EEG recorded from a depth electrode, and the bulbar EEG recorded from one of the electrodes in the array. Three bursts were sampled for spatial pattern analysis prior to (C1,C2,C3) and immediately following (T1,T2,T3), the odor presentation. The root mean square voltage was calculated for each of these bursts and displayed as an amplitude density plot. The number symbols represent the highest amplitudes. (See Figure 2 of Gray, Freeman, and Skinner,[82] "Chemical Dependencies of Learning in the Rabbit Olfactory Bulb: Acquisition of the Transient Spatial Pattern Change Depends on Norepinephrine," *Behav. Neurosci.* **100(4)** (1986): 585–596.) Reprinted by permission

from the signals, both behavioral and neural.[72,74] Analysis was restricted to those signals recorded on sessions after the acquisition of discrimination behavior was complete, and only to those trials in which a correct behavioral response was given by the animals. Within each trial, only the first three inspiratory bursts (80 msec in duration) were evaluated (Figure 6). Here it was reasoned that the relevant odor-specific patterns should be present during the burst at the beginning of a sniffing response.[193]

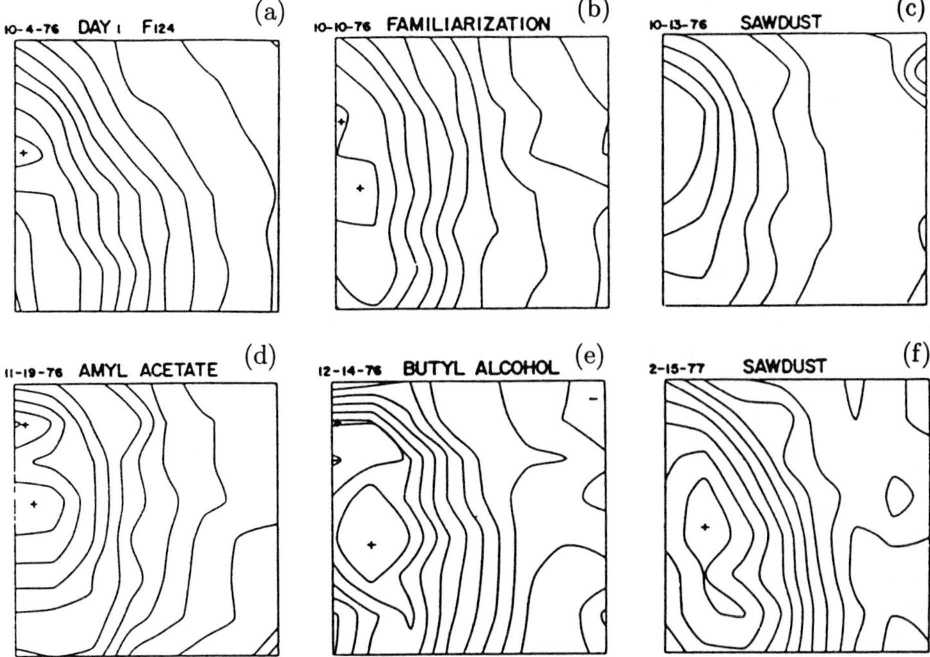

FIGURE 7 Contour plots of the mean EEG rms amplitude during the induced wave from a trained rabbit over a period of four months. Under aversive or appetitive conditioning the patterns changed with each new set of stimulus response contingencies. Shown are examples of five such changes. In the last stage of conditioning the presentation of the odorant "sawdust" did not result in the return of the EEG pattern of the first stage. (See Figure 3 of Freeman and Skarda,[72] "Spatial EEG Patterns, Nonlinear Dynamics and Perception: The Neo-Sherringtonian View," *Brain Res. Rev.* **10** (1985): 147–175.) Reprinted by permission.

A number of powerful signal-processing techniques were applied to the data. First, the principal oscillatory component of the induced waves was extracted using a curve-fitting method[72] and the peak frequency in the signals was computed using spectral analysis. This served to remove some of the biological noise from the signals. Two different spatial-filtering techniques then were applied to the resulting signals, a spatial filter[68] and a compensation for volume conduction of the field potential.[69] These techniques served to remove additional sources of biological noise. Initial parameters were chosen for these calculations and the resulting processed signals for three stimulus conditions (control, unreinforced odor, and reinforced odor) were then submitted to a nonlinear mapping algorithm. An additional constraint was applied that only those signals exceeding a specified frequency were included in the analysis. A statistical test was applied to determine if the centroids of the distributions of the three sets of data were significantly different. This entire process was iteractively repeated with each repetition utilizing a different set of values for the spatial filter, the frequency cutoff, and the volume conduction compensation.[72]

Without the use of these data-processing techniques, it was not possible to clearly resolve differences in the spatial patterns of the induced wave evoked by air and the two odors.[193] The inclusion of the additional data-selection criteria and imaging methods now made it possible, however, to resolve significant differences in the spatial patterns evoked by the two odors.[72] The best resolution of the pattern differences was obtained when (1) spatial frequencies above 0.5 cycles/mm were filtered out, (2) volume conduction compensation was focused at the external plexiform layer of the bulb (i.e., a focal depth of 0.5 mm), and (3) burst frequencies below 55 Hz were excluded from the analysis. The latter result proved to be particularly important. Oscillatory bursts of frequencies lower than 55 Hz had a higher degree of frequency variability and were spatially more disordered. In effect these bursts appeared to be spatially nonspecific, suggesting that they were not signaling the presence of the odor but rather a failure of classification.

These studies revealed that different odors indeed do evoke different spatial patterns of activity in the olfactory bulb. The conditions under which this process occurs however, appear to be quite different from what Adrian had hypothesized. Rather than a hard-wired, labeled-line type of process, the spatially specific coding of odor information in the bulb appears to be learning dependent. Spatial coding relies on experience in a behaviorally meaningful context. If odors are presented without reinforcement, a process of habituation ensues and it appears that no spatially specific patterns are evoked. It is likely, although not yet confirmed, that these processes depend on synaptic plasticity. Moreover, the structural changes that take place in the bulb are expressed mainly when the odor evokes a response in the nasal epithelium. If the mechanisms controlling convergence of the activity to a particular pattern fail, the process can be repeated by continued sniffing. This may explain why many of the lower frequency bursts were spatially disorganized.

Finally, another, even more remarkable result was revealed by these studies. The spatial representation of odor quality is not spatially specific; it is distributed, not localized. This conclusion was reached by a modification of the data analysis protocol described above. Freeman simply repeated the data analysis after deleting channels from the data sample at random. He found that the classification of pattern differences between the two odors was degraded in proportion to the number of channels removed. The spatial location of the channels within the array that were excluded from the analysis had no significant effect. In essence the information representing the odor was distributed equally throughout the bulb, in some pseudo-holographic manner. The obvious spatial variation of the amplitude of the induced wave over the surface of the bulb meant that sites with low amplitude were just as important as sites of high amplitude. The long-term changes in the location of the hot spots indicate that, during learning, the high- and low-amplitude regions tend to migrate. The significance of this effect remains unclear, however.

So what is the role of neuronal oscillations in these exceedingly complicated olfactory processes? That is a difficult question to answer. We can say that it is highly unlikely that the oscillations per se play any direct role in the actual coding of information, such as a temporal code. They are simply too variable and unspecific. The most obvious attributes of the rhythmic activity are that it is synchronous and that it is amplitude-modulated in space. This suggests that the oscillations provide a local mechanism to enable the establishment of synchrony among a much larger population of cells. In effect the oscillation acts as an AM carrier wave. Once a group of neurons become synchronized, can they then act as an assembly to code, represent, and process information. This is a difficult hypothesis to test. But synchronous rhythmic activity is a common property of neural structures and it would seem unlikely that the nervous system would not take advantage of it for some functions.

THALAMOCORTICAL SYNCHRONY AND THE SLEEP SPINDLE

Soon after the discovery of the electroencephalogram, it was realized that the properties of the EEG often show a clear relationship to behavioral state. One of the most striking changes in the EEG was observed during the transition from waking to sleep. It was found that, as a person begins to doze off, the EEG typically changes from a disorganized pattern of low-amplitude fluctuations containing many different frequency components, often referred to as "desynchronized," to a more coherent pattern of fluctuations having lower frequencies and higher voltage, or "synchronized." During this period a distinct pattern of 7- to 12-Hz regular oscillations often appears for brief periods encompassing 5–10 cycles. This rhythmic activity often takes the form of a waxing and waning of amplitude resembling the shape of a bundle of thread wound onto a weaver's spindle. The name "sleep spindle" stuck (Figure 8).

Over the course of the next several decades, and particularly in the 1940s and 1950s, the neural mechanisms responsible for controlling the sleep-waking cycle became the focus of an intensive research effort.[128] In the process, it became apparent that the sleep spindle is a general property of mammalian brains. Spindle activity was found to be present over large areas of the cerebral cortex and thalamus. Simultaneous measurements at multiple sites revealed that spindle oscillations are global events involving an interaction between thalamus and cortex (Figure 9(a)). These findings sparked widespread interest in the mechanisms underlying the generation and synchronization of rhythmic spindle activity. In fact, the research spawned during this period led to the study of the sleep spindle as a model system of rhythmogenesis and thalamocortical synchrony.[12]

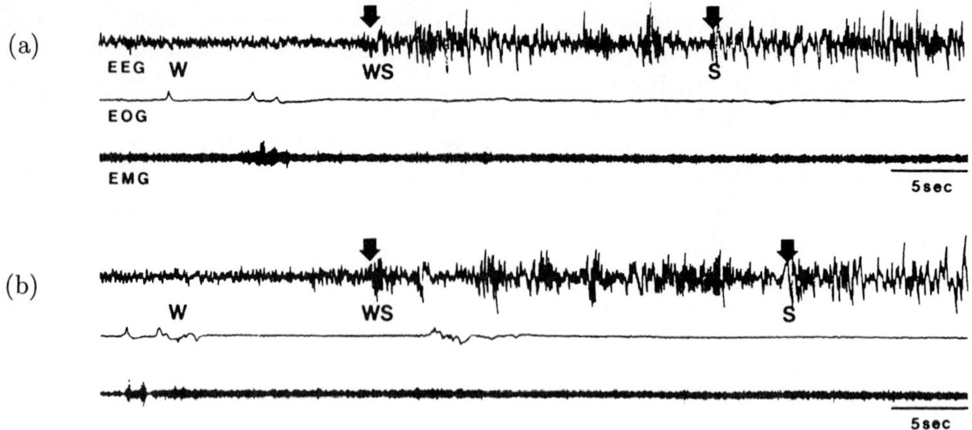

FIGURE 8 Electrographic criteria of waking (W) and slow-wave sleep (S) states in a behaving cat. WS and SW are transitional states between W and S, and S and W. (a)–(b) EEG waves from the surface of the precruciate gyrus, ocular movements (EOG), and electromyogram of neck muscles (EMG). Note the appearence of spindle sequences during WS and repeated desynchronizing reactions; S begins when slow waves appear and EEG desynchronizations no longer occur. (See Figure 5.5 of Steriade, Jones, andLlinas,[182] *Thalamic Oscillations and Signaling*, New York: John Wiley, 1990.) Reprinted by permission.

Although the functional significance of sleep spindles is unclear the behavioral dependence of these synchronous events in the EEG is well established. As a result the vast majority of studies on spindle oscillations have focused primarily on the mechanisms of their generation. Much of this work has been reviewed in a recent book.[182] Here, I will focus my discussion on some of the key mechanisms controlling spindle oscillations and not attempt to exhaustively review the literature. In addition, much has recently been learned of the cellular basis of spindle rhythms and I will discuss these findings where relevant.

Two discoveries facilitated the study of spindle oscillations. Early on it was found that transection of the midbrain at the intercollicular level[34] produced a dramatic enhancement of spindling. It was found that a similar state could be produced by the administration of barbiturate anesthetics. Both techniques essentially produced a state similar to permanent sleep in which there is a high degree of thalamocortical synchrony in the frequency range of 7–12 Hz. Subsequent research efforts relied heavily on these effects and, as a result, many of the basic neural mechanisms controlling spindle activity have been investigated in barbiturate-anesthetized or "cerveau isolé" animals.

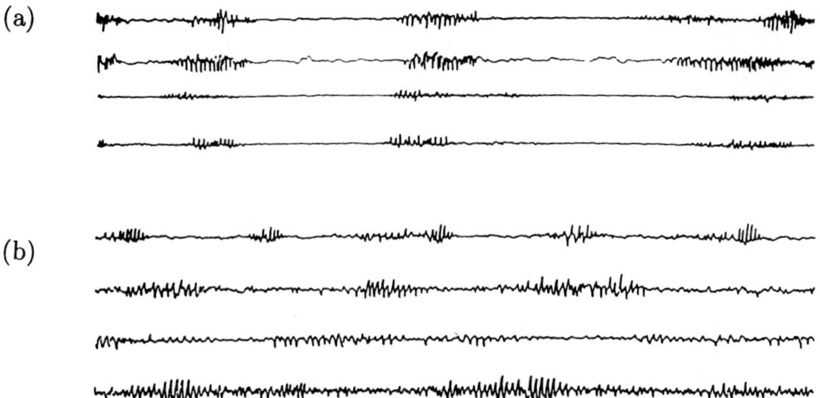

FIGURE 9 Thalamic origin of spindle rhythmicity. (a) Synchrony of spindle activity in the cat thalamus and various parts of the neocortex under nembutal anesthesia. The four traces depict activities recorded from (top to bottom): the medial part of the intralaminar thalamic nuclei, the anterior sigmoid gyrus, the arm sensory area, and the middle suprasylvian gyrus. Calibration time: 5 mm/sec. (b) Spindle sequences recorded from the thalamic intralaminar region in a bilaterally decorticated cat, with both optic nerves divided and brainstem transection at the intercollicular level. The four traces are taken at (top to bottom) 2 hours, 8 hours 45 min, 19 hours, and 72 hours after operation. (See Figure 1.1 of Steriade, Jones, and Llinas,[182] *Thalamic Oscillations and Signaling*, New York: John Wiley, 1990.) Reprinted by permission.

Where do spindle rhythms originate? This question dominated research on spindle activity for many years. Although, early on, it was an exceedingly difficult question to answer, we now know that the thalamus, and specific circuits within the thalamus, are responsible (reviewed in Steriade et al.[182]). Morison and Bassett[141] determined that spindle activity could be recorded in the intralaminar thalamic nuclei after complete removal of the neocortex and sectioning of the optic nerves in a cerveau isolé preparation (Figure 9(b)). Villablanca[194] found that spindle rhythms could be recorded in the thalamus for many hours and for days following the complete removal of the neocortex, striatum, and rhinencephalon. Moreover, he never found cortical spindle rhythms in animals that had had their thalamus completely removed by suctioning. These data along with many earlier and subsequent reports firmly established the thalamic nuclei as the structures essential for generating spindle activity in thalamocortical networks.[182]

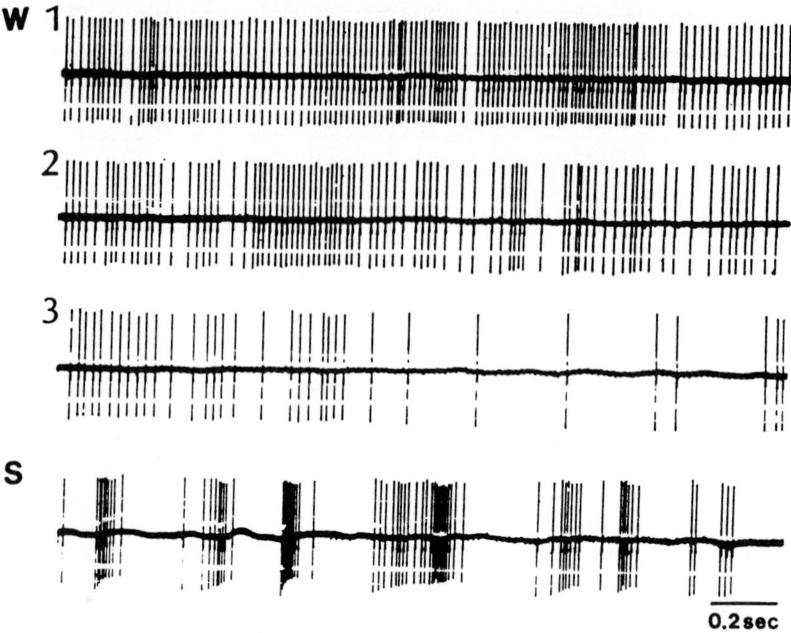

FIGURE 10 Discharge characteristics of cat reticular nucleus neurons during states of waking (W) and slow-wave sleep (S). Three epochs are illustrated during W: (1) from the onset of arousal from S, (2) in the middle part of W (15 sec after 1), and (3) at the end of W, just before the transitional epoch to S (11 sec after 2). (See Figure 6.12 of Steriade, Jones, and Llinas,[182] *Thalamic Oscillations and Signaling*, New York: John Wiley, 1990.) Reprinted by permission.

Of course, this raised the further question of which structures in the thalamus engage in spindle rhythms. The answer to this question, which evolved over a period of 20–30 years, turned out to be rather simplistic. Spindle discharges, recorded in both the EEG and single-unit activities, were found to be a ubiquitous property of thalamic nuclei. During sleep onset or under the influence of barbiturates the vast majority of thalamocortical relay cells were found to shift from a tonic firing mode to one of intermittent burst firing associated with spindling. This activity was evident in most thalamic nuclei but was particularly pronounced in the lateral structures[12,13] (Figure 10). Subsequent lesion studies revealed that removal of the midline and intralaminar thalamic nuclei produced no obvious changes in spindle activity while lesions of the lateral nuclear groups had the dramatic effect of completely abolishing thalamic spindling.[13] In addition, the thalamic reticular nucleus, a thin sheath of cells surrounding the outer surface of the thalamus,[168] was found to exhibit robust spindle rhythms associated with burst firing.

These and other studies led to a focus on the thalamic circuitry thought to underlie spindle rhythmogenesis. The lateral thalamic structures, in simple terms, consist of three basic circuit elements: (1) thalamocortical relay cells, the principal cell type which receives afferent input from sensory structures such as the optic tract; (2) local interneurons that receive excitatory input from the collaterals of relays cells and deliver recurrent inhibition onto the relay cells; and (3) the thalamic reticular nucleus, a thin sheath of inhibitory cells that covers the entire thalamic complex. This latter structure receives excitatory input from the axon collaterals of thalamic relay cells as well as the corticothalamic cells projecting into the thalamus from the cortex. In turn, it sends axons into the thalamic nuclei which make inhibitory synaptic connections onto the relays cells. There are thus two well-defined forms of synaptic inhibition acting on relay cells which long have been thought to provide the basic mechanism for generating rhythmic patterns of discharge.[12,182]

However, electrophysiological and lesion studies over the years, have provided several competing views of the mechanisms underlying the genesis of spindle rhythms. In one view, a purely circuit-based model emerged as a likely explanation for spindling.[12] In this scenario, afferent input acts to excite a population of relay cells that in turn deliver synaptic excitation onto the local intranuclear interneurons and the cells of the thalamic reticular nucleus. Activity in these cell groups then feeds back a powerful synaptic inhibition onto the relay cells to suppress their ouput. The hyperpolarization of the relay cells leads to a post-inhibitory rebound excitation[127] that, when combined with sustained afferent input, leads to reexcitation and a repetition of the cycle. Because of the interactions within the thalamic network the spindle rhythm is a highly synchronous event that effectively drives the cortex into a similar rhythm. The cortex in turn delivers excitatory feedback input to the thalamic relay cells as well as the reticular nucleus neurons and acts to facilitate the entire sequence, thereby giving rise to a thalamocortical spindle oscillation.

In conjunction with the view that spindle rhythms emerge as a property of an interconnected network, evidence began to accumulate suggesting that single cells are tuned to oscillate at particular frequencies by virtue of their intrinsic membrane properties. A key turning point for this concept came after the discovery that many thalamic cells possess a low-threshold calcium conductance[100,101,118,119] responsible for the rebound bursts described by Maekawa and Purpura.[127] This current was found to play a key role in the generation of spindle rhythms in thalamus largely because of its voltage dependence on activation and inactivation. The conductance was found to be activated at membrane potentials more positive than -65 mV and deinactivated at membrane potentials negative to -65 mV.[100,101,134] This relationship results in a profound effect on the response of relay cells to input. A cell receiving a depolarizing input at a resting level of -55 mV produces a train of single spikes as output. If the resting level of the cell is hyperpolarized to -70 mV, its response to input consists of a calcium spike upon which a burst of fast sodium and potassium action potentials is superimposed (Figure 11). The interplay between the low-threshold calcium conductance and the inhibitory influence of the nRT provide the most likely mechanism for the generation of spindle rhythms in relay cells.[182]

More recently, an additional cation current has been discovered that, when acting in concert with the low-threshold calcium conductance, confers upon the thalamic relay cells an intrinsic tendency to oscillate at 1–4 Hz.[133,134] When activated by hyperpolarization, depolarizes the cell resulting in the activation of the low-threshold calcium conductance and the generation of a calcium spike. The resulting depolarization during the burst leads to the deactivation of the cation current and the inactivation of the low-threshold calcium current. These events cause a net hyperpolarization of the cell and the consequent reactivation of the cation current combined with the removal of inactivation of the low-threshold calcium current. Under appropriate conditions the time course of these conductances leads to an oscillation of the membrane potential at 1–4 Hz.[133,134]

In combination with these studies on relay cells, it was found that cells of the thalamic reticular nucleus (nRT) also possess a set of conductances giving them an intrinsic tendency to oscillate[18,182] (Figure 12). In these cells the low-threshold calcium current and a calcium-dependent potassium current are critical elements. The latter current is both calcium and voltage dependent and is responsible for the powerful after-hyperpolarization (AHP) that is seen to follow the calcium spike.[182] A depolarizing input delivered during a period of relative hyperpolarization of the cell

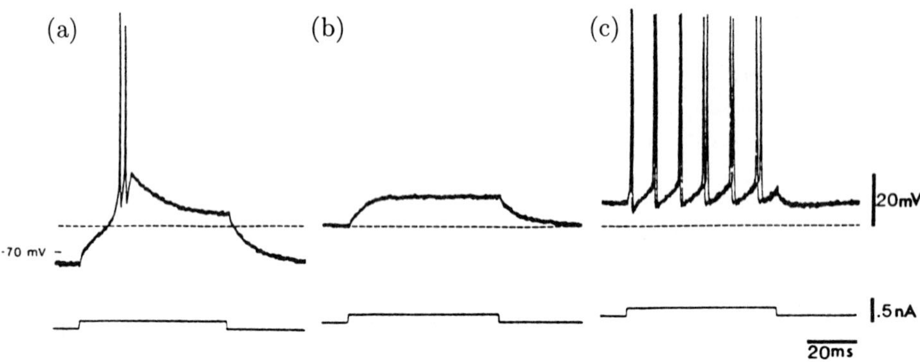

FIGURE 11 Electrophysiological properties of thalamic neurons. A subthreshold depolarization of the cell at resting level (broken line) is illustrated in (b). After a DC depolarization (c), repetitive firing of the cell is observed during a current pulse having the same amplitude as the one in (b). Following DC hyperpolarization, current pulses similar to those in (b) produce a single high-frequency burst (a). (See Figure 4.3 of Steriade, Jones, and Llinas,[182] *Thalamic Oscillations and Signaling*, New York: John Wiley 1990.) Reprinted by permission.

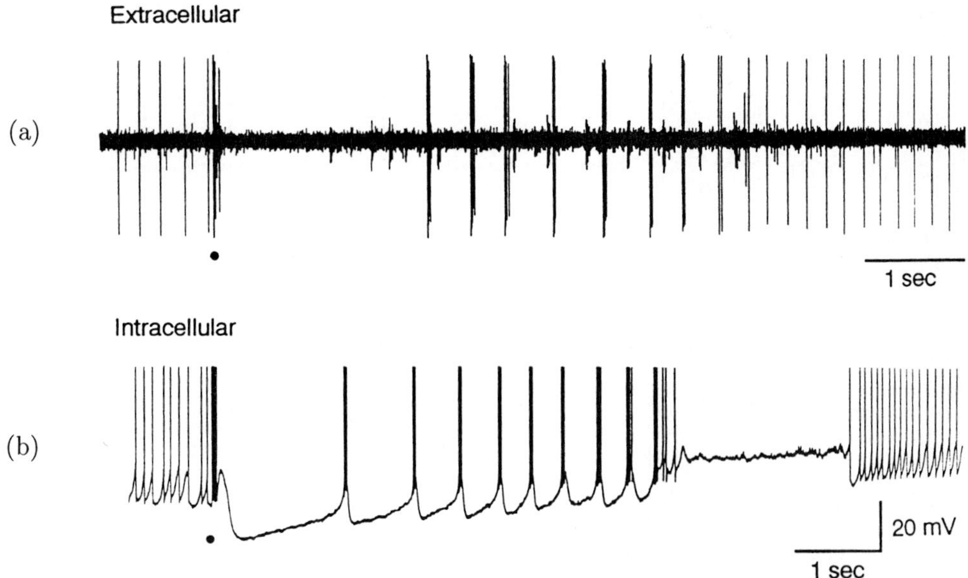

FIGURE 12 Rhythmic response of nucleus reticularis neurons to orthodromic activation. (a) Electrical stimulation of corticothalamic inputs results in a burst of action potentials followed by an inhibitory pause, followed by rhythmic burst firing in a guinea pig nRT cell recorded extracellularly. (b) Intracellular recording of this activity revealed that the burst of action potentials is associated with a large excitatory postsynaptic potential followed by a hyperpolarization of the membrane potential and the rhythmic appearance of low-threshold calcium spike-mediated bursts of action potentials and burst after hyperpolarizations. (See Figure 1 of Bal and McCormick,[18] "Ionic Mechanisms of Rhythmic Burst Firing and Tonic Activity in the Nucleus Reticularis Thalami: A Mammalian Pacemaker," *J. Physiol.* (1992): Submitted.) Reprinted by Permission.

leads to a low-threshold calcium spike and a burst of action potentials followed by an AHP. The resulting hyperpolarization reinitiates the sequence by deinactivating the low-threshold calcium conductance. The time course of these two currents enables the nRT cells to generate 7- to 12-Hz burst discharges intrinsically.[18] An additional calcium-activated cation current has recently been discovered in nRT cells which may contribute to the termination of a spindle oscillation by gradually depolarizing the cells and counteracting the hyperpolarizing influence of the calcium-dependent potassium current.[18]

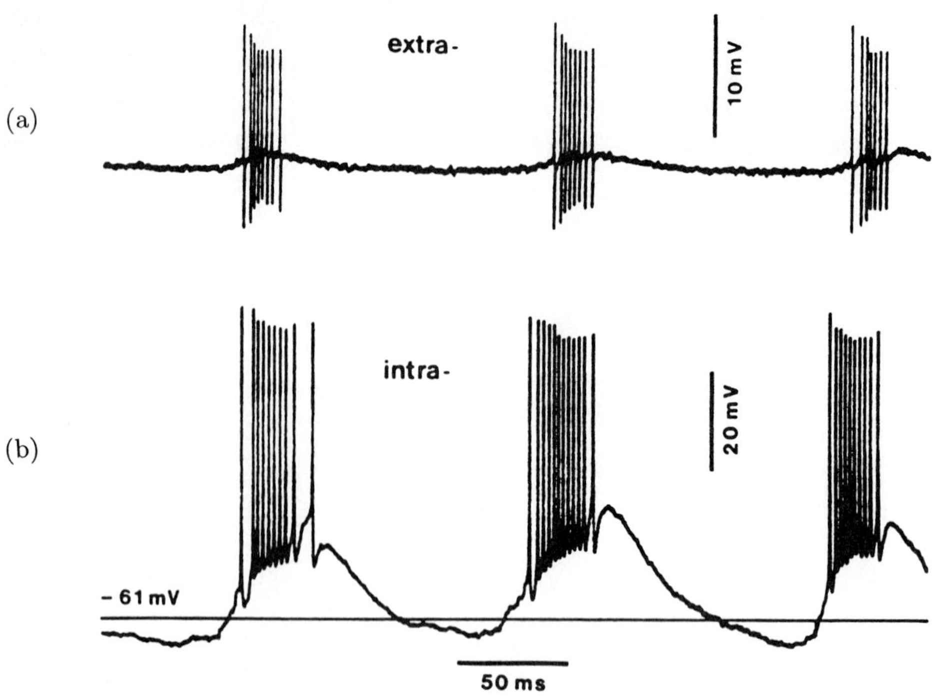

FIGURE 13 Episodes of rhythmic burst discharges recorded (a) extracellularly and then (b) intracellularly from the same reticular nucleus neuron of a cat under barbiturate anesthesia. (See Figure 6.16 of Steriade, Jones, and Llinas,[182] *Thalamic Oscillations and Signaling*, New York: John Wiley, 1990.) Reprinted by permission

The unique anatomical organization of the nRT and its powerful tendency to exhibit intrinsic spindle-like oscillations (Figure 13) led naturally to the notion of this structure as a pacemaker for spindle bursts.[182] Three important findings provided support for this hypothesis. Several anterior thalamic nuclei known to be naturally devoid of input from the nRT do not exhibit spindle oscillations.[142,154,179,182] Relay nuclei deprived of their input from the nRT by chemical or surgical lesions fail to exhibit the normal pattern of spindle rhythms[180,182] (Figure 14). And spindle oscillations are present in the nRT after surgical isolation *in vivo*.[181,182] Thus the view emerged that thalamocortical spindle rhythms are largely, if not exclusively, dependent on the nRT for their generation.

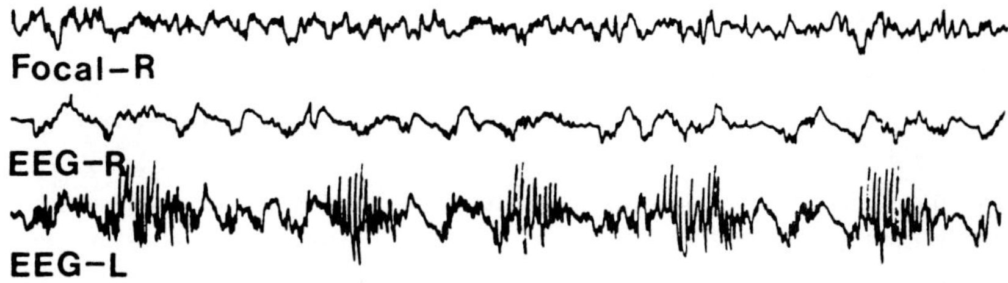

FIGURE 14 Abolition of spindle waves in thalamic nuclei disconnected from the cat reticular nucleus. (a) Acute experiment with right thalamic transection. Absence of focal spindling in the right centrolateral nucleus and in the ipsilateral EEG contrasts to the spindle sequences in the contralateral (left) EEG. (See Figure 6.23a of Steriade, Jones, and Llinas,[182] *Thalamic Oscillations and Signaling*, New York: John Wiley, 1990.) Reprinted by permission.

Nonetheless, it is clear that the intrinsic properties of relay cells and the anatomical organization of the thalamic network play a facilitatory, if not synergistic, role in the generation of thalamocortical spindles.[18,41] A typical spindle sequence is likely to include the following events. Synaptic input to nRT under appropriate conditions leads to burst discharge in a population of cells due to the activation of the low-threshold calcium current. This response of the nRT cells to input has two effects: the cells of the nRT are hyperpolarized by the calcium-activated potassium current, and a group of thalamic relay cells is synaptically inhibited by input from the nRT. The relay cells exhibit a rebound excitation due to the low-threshold calcium current. The nRT cells in turn receive two depolarizing inputs, (1) an intrinsic drive from the calcium-activated cation current and the low-threshold calcium current and (2) a synaptic excitation coming from the relay cells. The nRT cells are driven to burst and the cycle repeats. These mutually reinforcing influences act to recruit large numbers of cells in the thalamic network and thereby deliver a synchronized rhythmic input to the cortex. Thalamically projecting cells in layer 6 of the cortex in turn deliver excitatory input onto cells of the nRT and relay nuclei further facilitating the oscillatory sequence. Within one to two cycles of the oscillation, large networks in thalamus and cortex become engaged in a spindle burst.

During this synchronized oscillatory mode of firing, the thalamus becomes less sensitive to afferent input. Herein lies the possible function of the spindle rhythm and its association with the onset and occurrence of sleep. The thalamic network generates a state of internal activation during which powerful intrinsic and synaptic influences dominate its activity. The normally powerful influence of afferent inputs become ineffective in overriding the synchronous activity and sensory information is no longer able to reach the cerebral cortex.

THE VISUAL CORTEX AND FEATURE INTEGRATION

The process of visual pattern recognition presents a formidable task to the mammalian central nervous system. Within a period that may be as brief as 200 milliseconds, complex combinations of visual features making up a retinal image must be unambiguously grouped together giving visual scenes a segmented appearance as a collection of independent entities or objects. The number of possible interrelationships that features and objects can have within visual scenes is nearly infinite. And yet the visual system appears to have adapted an extremely effective strategy to enable the grouping of features pertaining to objects while avoiding false conjunctions with nearby features belonging to other objects.

The obvious ease with which mammalian nervous systems deal with complex visual environments is all the more surprising when one considers the underlying organization of the visual system. The bulk of the mammalian visual system is subdivided into a collection of maps of the visual field.[171] The majority of these maps reside in the neocortex and each map or area is laid out as a complete or partial representation of the visual field. This collection of areas exhibits a parallel hierarchical organization in which there exist many interconnections within and between different levels of the hierarchy.[58] At low levels in the hierarchy, cortical areas are retinotopically organized and this point-to-point mapping of the visual field gradually becomes less precise until at the highest levels of the system, cells in the cortex respond to visual stimuli placed in nearly any location within the visual field. The representation becomes nonretinotopic.

Superimposed on this organization is the additional property of feature specificity. In primary visual cortex and its satellite areas, cells respond most effectively to particular categories or combinations of visual features, such as motion, form, and color.[115] Cells having similar response properties are grouped together and this compartmentalization is preserved throughout the system yielding, at higher stages of the hierarchy, areas that are devoted to the analysis of particular aspects of visual images.[48] At the highest levels in the primate, this segregation of function is thought to yield two separate functional streams, one for object recognition and discrimination located in the ventrolateral regions of the temporal lobe and another for spatial localization and motion processing located in the parietal lobe.[191]

Taken at face value it is apparent that, when one views a visual image consisting of numerous identifiable objects, vast numbers of neurons in many different cortical areas become active simultaneously. The representation of an object thus is not likely to take place at any given fixed location but rather involves a large population of cells distributed over the hierarchy of visual cortical areas. This view of visual pattern recognition as a distributed population-based code presents several formidable challenges to our understanding. How are the neuronal elements comprising the representation of a particular object bound together? How does the binding process achieve its combinatorial flexibility to enable the recognition of a nearly infinite variety of objects? How can multiple object representations coexist at the same time within the same set of cortical areas without being confounded? How do these representations change with time as the visual image changes or as the observer moves its eyes?

In the mid to late 1970s it was recognized that population or ensemble coding as a mechanism for representing sensory information is not without problems.[195,137] Foremost among these is the problem associated with the superposition of objects in an image.[195,196,174] If ensembles encoding properties of a visual image are solely distinguished by the amplitude of the constituent neuronal responses, some ambiguities remain unresolved. Problems arise when an image contains more than one coherent figure or when the background itself has some coherent properties, such as a regular texture. If ensembles were to form for each coherent object, it would become impossible to identify which active neurons belong to the ensembles coding for each particular figure and the background. Individual figures would no longer be distinguishable. It was proposed that an additional signal is needed to label the relevant neurons as belonging to the representation of a coherent object and not to part of another object or the background. von der Malsburg,[195] drawing on earlier theoretical ideas,[95,137] proposed that temporal correlation among neuronal populations could provide the requisite label for the selective and transient formation of ensembles of neurons representing specific objects. Such a mechanism, it was proposed, would allow multiple independent ensembles to be formed simultaneously within the same cortical network and thereby avoid the confusion associated with the superposition of images.

SYNCHRONOUS NEURONAL RESPONSES IN CAT VISUAL CORTEX

The first evidence for the prediction that selective temporal correlations among neuronal populations might play a role in cortical information processing came from work on the olfactory system. As described in detail earlier, Freeman and his colleagues demonstrated that spatially correlated oscillatory neuronal activity in the olfactory bulb and cortex contributes to olfactory discrimination (for a review see Freeman[66]). These findings led to the prediction that similar synchronous oscillatory responses should also exist in neocortical structures.[70,72] Prompted by such predictions Gray and Singer observed, several years later, that neurons in area 17 of the cat visual cortex often exhibit synchronous rhythmic responses to optimal visual stimuli in a frequency range of 40–60 Hz (Figure 15).[83,85] This temporal pattern was observed both in the responses of single neurons and in the local field potential (LFP).[51,83,85] In those recordings where the spike train was periodic the action potentials typically occurred during the peak negativity of the LFP oscillation. Analysis of the autocorrelation histograms of the spike trains and the

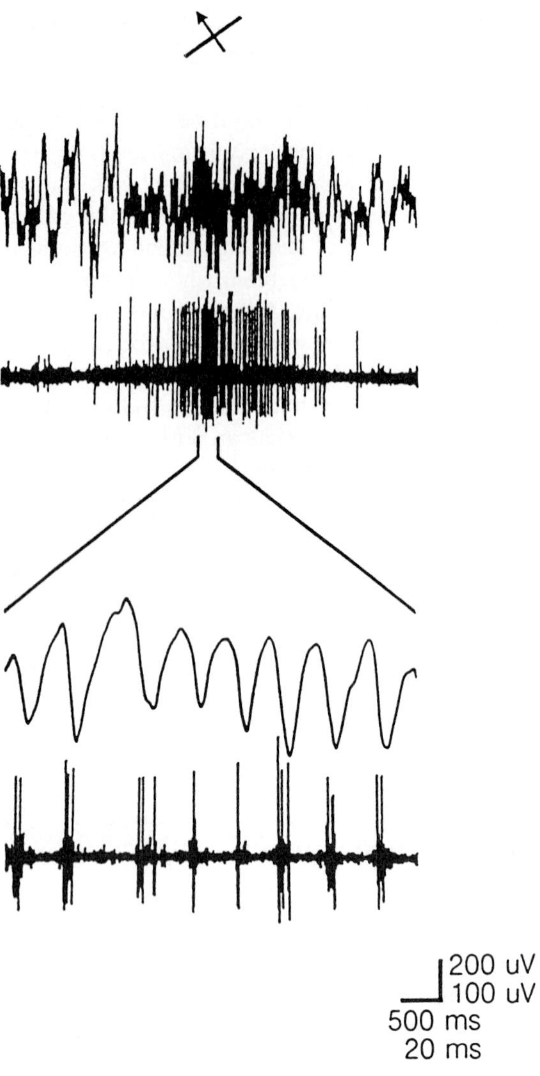

FIGURE 15 Multiunit activity and local field potential responses recorded from area 17 in an adult cat to the presentation of an optimally oriented light bar moving across the receptive field of the recorded cells. (a) Oscilloscope records of a single trial showing the response to the preferred direction of movement. In the upper two traces, at a slow time scale, the onset of the response is associated with an increase in high-frequency activity in the local field potential. The lower two traces display the activity at an expanded time scale. Note the presence of rhythmic oscillations in the local field potential and the multiunit activity that are correlated in phase. (See Figure 1(a) of Gray and Singer,[85] "Stimulus-Specific Neuronal Oscillations in Orientation Columns of Cat Visual Cortex," *Proc. Nat. Acad. Sci.* **86** (1989): 1698–1702.) Reprinted by permission.

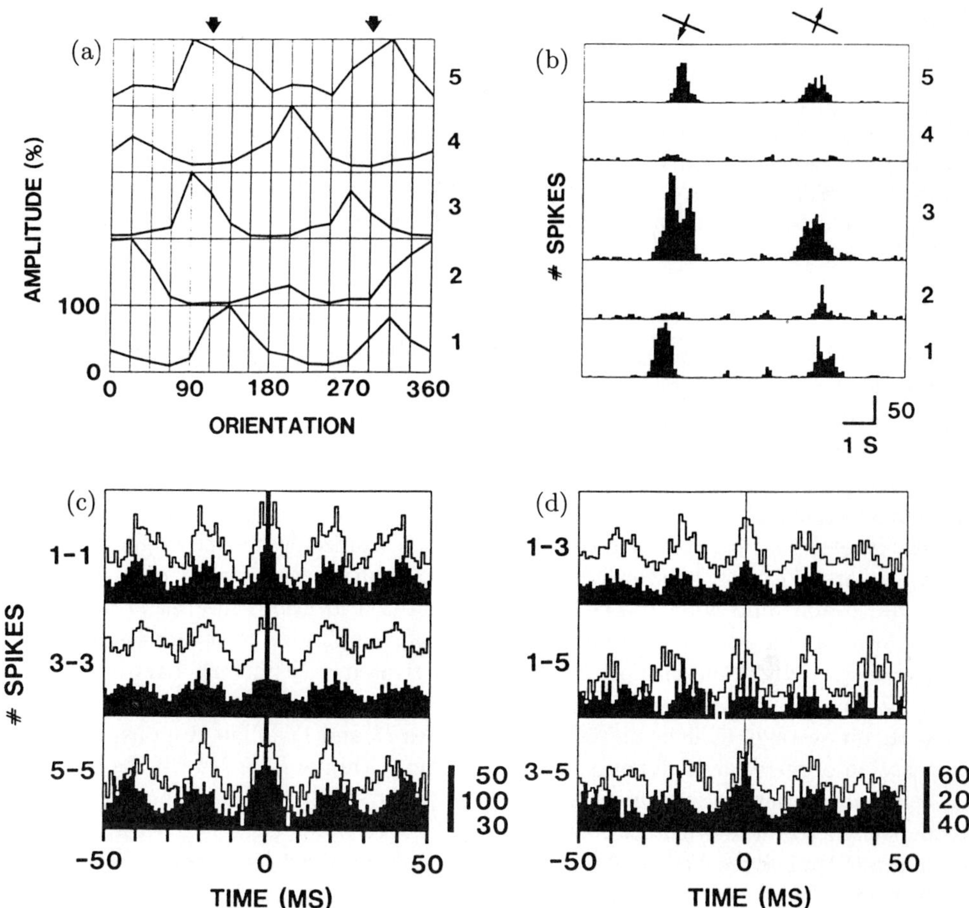

FIGURE 16 Orientation-specific intercolumnar synchronization of oscillatory neuronal responses in area 17 of an adult cat. (a) Normalized orientation tuning curves of the neuronal responses recorded from five electrodes spaced 400 um apart and centered on the representation of the area centralis. Response amplitudes (ordinate) to stimuli of different orientations (absscissa) are expressed as a percentage of the maximum response for each electrode. The arrows indicate the stimulus orientation (112 degrees) at which the responses were recorded in (b), (c), and (d). (b) Post-stimulus-time histograms recorded simultaneously from the same five electrodes at an orientation of 112 degrees. Note the small difference in the latencies of the responses.

FIGURE 16 (cont'd.) indicating overlapping but slight offset receptive field locations. (c) Autocorrelograms of the responses recorded at sites 1 (1–1), 3 (3–3), and 5 (5–5). (d) Cross correlograms computed for the three possible combinations (1–3, 1–5, 3–5) between responses recorded on electrodes 1, 3, and 5. Correlograms computed for the first direction of stimulus movement are displayed with unfilled bars with the exception of comparison 1-5 in (d). (See Figure 1 of Gray, Koenig, Engel, and Singer,[85] "Stimulus-Specific Neuronal Oscillations in Cat Visual Cortex Exhibit Inter-Columnar Synchronization Which Reflects Global Stimulus Properties," *Nature* **338** (1989): 334–337.) Reprinted by permission.

frequency spectra of the LFPs revealed that the two signals were, on the average, of the same frequency. And spike-triggered averaging of the LFP demonstrated that the two signals were closely correlated in time.

Taken together, the results of these experiments demonstrated that units close enough to be recorded with a single electrode, if responsive to the same stimulus, show a synchronization of their respective oscillatory activity.[85] Subsequent studies revealed that oscillatory responses often synchronize over much larger distances within area 17.[51,53,86] Recordings were made of multiunit activity from both closely (0.5–2.0 mm) and widely (4.0–10.0 mm) separated locations in area 17 such that the corresponding cells had either spatially overlapping or nonoverlapping receptive fields. When the groups of cells were activated with moving bars of optimal direction and orientation the oscillatory responses were often found to synchronize with, on average, little or no phase lag (Figures 16 and 17). This response synchronization occurred over distances of up to 7 mm within area 17.[86,53] If the recorded cells had overlapping receptive fields, the response synchronization showed no clear dependence on the orientation preferences of the recorded cells.[53,86] If the cells had nonoverlapping receptive fields, synchronization occurred less frequently and was primarily found between cells with similar orientation preference.[53,86]

These studies lent support to the notion that synchronous interactions among cells within a cortical area could contribute to the integration of information across the visual field. The integration of disparate visual features, however, is likely to involve the interaction of cells in different cortical areas. It thus became important to determine if response synchronization could be measured between cells in different cortical areas. The first experiments addressing this question were conducted by Eckhorn et al.[51] They demonstrated that oscillatory field-potential activity recorded in area 18 of the cat was often synchronous with unit activity recorded simultaneously in area 17. The observed correlations, as those measured in area 17, were found to be stimulus dependent but not locked to the visual stimulus. No consistent phase relation could be observed between the stimulus onset and the oscillatory responses.[51,83,85] Subsequently, Engel et al.[55] demonstrated synchronized neuronal responses among cells recorded simultaneously in areas 17 and PMLS

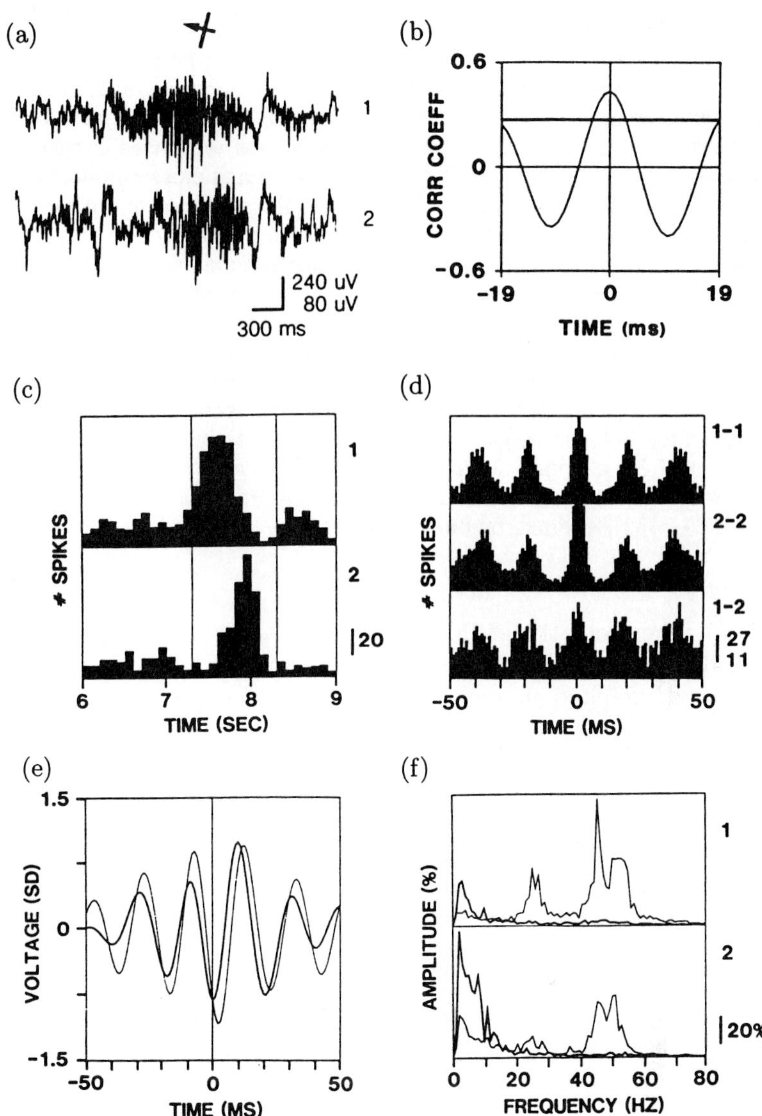

FIGURE 17 The local field potential and multiunit activity recorded at two sites in area 17 separated by 7 mm show similar temporal properties and correlated interactions. (a) Plot of the LFP responses (1- to 100-Hz bandpass) recorded on a single trial to the presentation of two optimally oriented light bars passing over the receptive fields of the recorded neurons at each site. The peak of the responses overlap in time but are not in precise register. (b) The average cross correlogram computed between the (cont'd)

FIGURE 17 (cont'd.) two LFP signals (20–100 Hz bandpass) at a latency corresponding to the peak of the oscillatory responses. The thick horizontal line represents the 95% confidence limit for significant deviation from random correlation. (c) Peri-stimulus-time histograms (PSTH) of the multiunit activity recorded over ten trials at the same two cortical sites as shown in (a). Again the responses overlap but are not in precise register. (d) Autocorrelograms (1–1,2–2) and cross correlograms (1–2) of the multiunit activity recorded at each site. Note the presence of a clear periodicity in each correlogram indicating that the responses are oscillatory and that they show a consistent phase relationship. (e) Plots of the spike-triggered averages of the LFP signals at each site computed over all 10 trials. The thick and thin lines correspond to electrodes 1 and 2 respectively. Note that the peak negativity of the waveform is correlated with the occurrence of neuronal spikes at 0-ms latency. (f) Normalized average power spectrum of the LFP signals computed from periods of spontaneous (thick line) and stimulus-evoked (thin line) activity. The frequency of the activity is similar in both the autocorrelograms of the MUA and the power spectra of the LFPs. (See Figure 1 of Gray, Engel, Koenig, and Singer,[88] "Synchronization of Oscillatory Neuronal Responses in Cat Striate Cortex: Temporal Properties," *Visual Neurosci.* 8 (1992): 337–347.) Reprinted by permission.

(Figure 18). The properties of this synchronous activity were very similar to that observed within area 17 and between areas 17 and 18. The synchrony was stimulus dependent and exhibited an oscillatory temporal structure.[55] These studies clearly demonstrated that temporal correlations in neuronal activity, in fact, do occur over large distances in visual cortex spanning different cortical maps. Thus, in principle the integration of visual information across feature domains could be achieved by such a mechanism.[55]

If the mechanisms controlling response synchronization are general, then it is reasonable to assume that such processes should not be limited to one hemisphere. In a further set of experiments Engel et al.[56] investigated the occurrence of temporally correlated neuronal activity recorded in area 17 of the two hemispheres. They found that cells having overlapping receptive fields near the vertical midline showed a high incidence of response synchronization to optimal visual stimulation (Figure 19). These interactions were found to be absent in animals that had previously had their corpus callosum severed.[56] These results have been confirmed recently and extended to include synchronization of activity between cells in area PMLS of the two hemispheres.[37] Although these investigators did not explicitly observe oscillatory responses, the implications of the results were similar. Synchronized neuronal activity can be established over large distances spanning the two cerebral hemispheres.

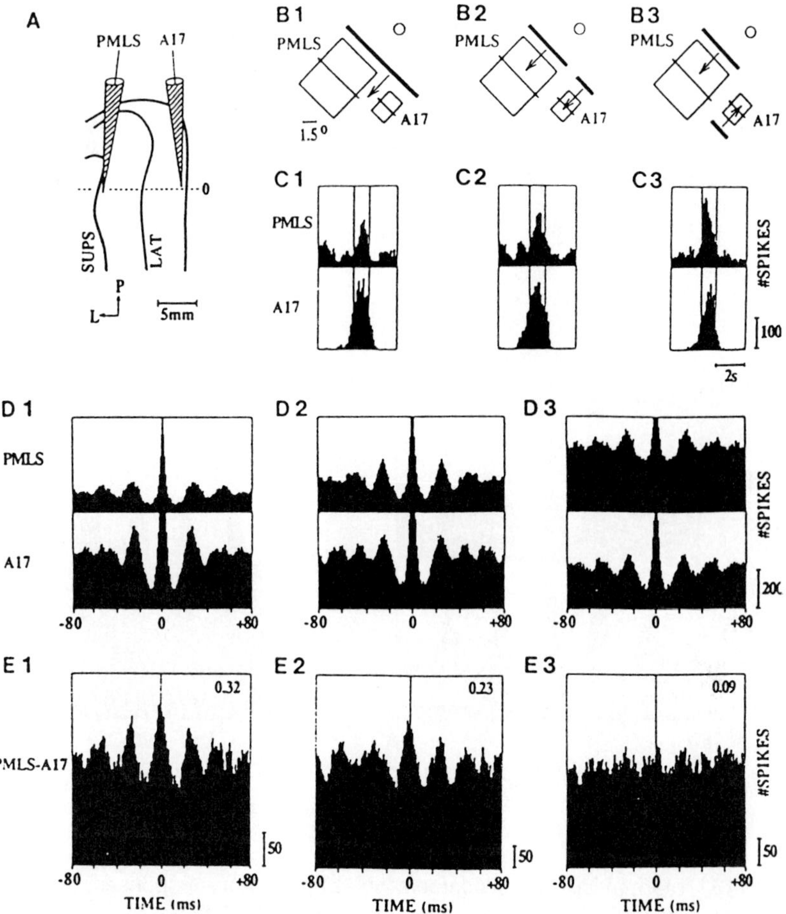

FIGURE 18 Interareal synchronization is sensitive to global stimulus features. (a) Position of the recording electrodes. A17, area 17; LAT, lateral sulcus; SUPS, suprasylvian sulcus; P, posterior; L, lateral. (b)–(d) Plots of the receptive fields of the PMLS and area 17 recording. The diagrams depict the three stimulus conditions tested. The circle indicates the visual field center. (e)–(g) Peristimulus-time histograms for the three stimulus conditions. The vertical lines indicate 1-sec windows for which autocorrelograms and cross correlograms were computed. (h)–(j) Comparison of the autocorrelograms computed for the three stimulus paradigms. (k)–(m) Cross correlograms computed for the three stimulus conditions. (See Figure 3 of Engel, Kreiter, Koenig, and Singer,[55] "Synchronization of Oscillatory Neuronal Responses Between Striate and Extrastriate Visual Cortical Areas of the Cat," *Proc. Natl. Acad. Sci.* 88 (1991): 6048–6052.) Reprinted by permission.

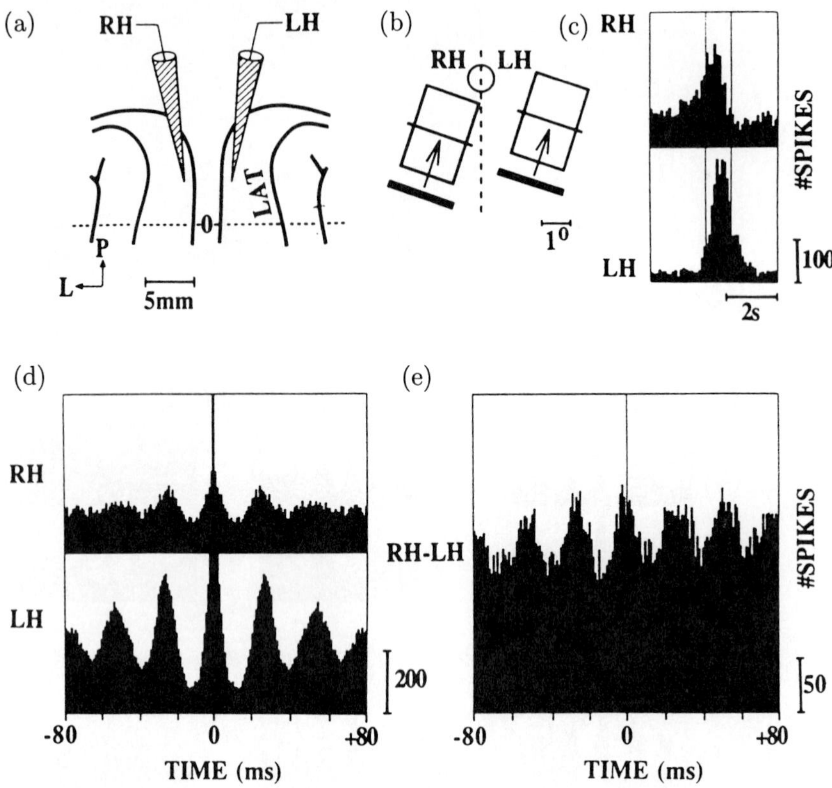

FIGURE 19 Interhemispheric synchronization of oscillatory responses. (a) The electrodes were located in area 17 of the right and left hemisphere close to the representation of the vertical meridian. (b) The receptive fields of the two multiunit recordings. The cells had the same orientation preferences, and the fields were located in the respective contralateral hemifields within 4 degrees of the vertical meridian (dashed line). Circle, center of the visual field. (c) Histograms of the responses evoked simultaneously with two light bars of optimal orientation. (d) Autocorrelograms computed for the two responses in a 1-sec window centered on the peak of the response. (e) The cross correlogram of the two responses computed within the same time window. (See Figure 3 of Engel, Kreiter, Koenig, and Singer,[56] "Interhemispheric Sychronization of Oscillatory Responses in Cat Visual Cortex," *Science* **252** (1991): 1177–1179.) Reprinted by permission.

SYNCHRONOUS NEURONAL RESPONSES IN MONKEY VISUAL CORTEX

The occurrence of high-frequency oscillatory activity in the visual cortex of monkeys is a controversial subject. Reports of the phenomenon have been both positive and negative. In one report, Young et al.[206] found little or no evidence for oscillatory activity in areas V1, MT, and IT in macaque monkeys. This study was conducted under conditions very similar to those used in the earlier cat studies.[53,85,87] Recordings in areas V1 and MT were performed while the animals were under Halothane anesthesia and paralysis, while those in area IT were performed on alert animals trained to fixate a visual target.[206] Multiple-unit activity and field potentials were recorded from single electrodes and the signals were subjected to autocorrelation and spectral analysis, respectively. Very little evidence for rhythmic firing was obtained even though the methods of data analysis used were very similar to those employed in the cat experiments.[53] Similar, though not identical, studies by Tovee and Rolls[187] and Bair et al.[17] found little evidence for high-frequency oscillatory activity in areas IT and MT of alert macaque monkeys, respectively.

Contrary to these results, Livingstone[114] found that many cells in area V1 of squirrel monkeys exhibit a pattern of rhythmic burst firing in response to optimal visual stimulation. These cells were found to be located in layers 2, 3, and 4 in both blob and interblob regions but not in layers 5 or 6.[116] Moreover, she performed multielectrode recordings and found that cells showing this rhythmic burst firing were often synchronized, as in the cat experiments.[114] This synchrony was found to extend up to 5 mm. Also contrary to the above results, Kreiter and Singer[105] found evidence for rhythmic burst firing in cells recorded in area MT of the alert macaque monkey. This activity was found to be synchronous among nearby cells recorded with closely spaced electrodes.[105,106] Finally, a recently published report has demonstrated robust oscillatory activity in the striate cortex of alert Macaque monkeys (Eckhorn et al.[52]). These signals exhibit most if not all of the characteristics of oscillatory responses recorded in cat striate cortex, and were reported to be more robust than in the cat.

The data on oscillatory activity in the primate visual cortex are thus not consistant across laboratories. Differences in results could arise from different methods of visual stimulation, behavioral training, and data analysis or be due to a sampling bias from different layers or cells types. In the reports citing positive evidence for oscillatory activity in the monkey striate cortex the principal difference with the cat is a higher frequency range. Otherwise, the phenomena appear to be very similar.

STIMULUS DEPENDENCE OF RESPONSE SYNCHRONIZATION

If the synchronization of neuronal responses contributes to the binding of distributed features in the visual field, then it is conceivable that cross-columnar interactions should not only depend on receptive field properties of the recorded cells, but should change dynamically with variations of the visual stimulus.[195,196] This appears from a theoretical point of view to be important since any given cell in the visual cortex is likely to participate in the representation of large numbers of visual stimuli. Therefore a mechanism enabling cells to synchronize their activity with different groups of cells under differing conditions of stimulation could potentially provide a robust combinatorial flexibility in the coding of visual or other sensory information.

To date, a number of examples of stimulus-dependent changes in the magnitude of synchrony among neurons in visual cortex have been observed. These data, collected from a number of different experiments, must still be considered somewhat preliminary but, nonetheless, provide provocative food for thought. One example of stimulus-dependent changes in correlated firing comes from the work of T'so et al.[189] They recorded the activity from two cells in area 17 of the cat having overlapping receptive fields of different orientation preference. The arrangement of the fields were such that the cells could either be activated by two optimal bars moving across each receptive field or by a single bar moving across both fields having an orientation capable of activating both cells. They found the correlation of firing to be greater in the latter condition than in the former, even though the cells generated vigorous responses under both stimulus conditions[189] (Figure 20).

A similar result was obtained by Gray et al.[86] when observing oscillatory neuronal responses in cat area 17. They recorded multiunit activity from two locations separated by 7 mm. The receptive fields of the cells were nonoverlapping, had nearly identical orientation preferences, and were spatially displaced along the axis of preferred orientation. This enabled stimulation of the cells with bars of the same orientation under three different conditions: two bars moving in opposite directions, two bars moving in the same direction, and one long bar moving across both fields coherently. Under these conditions no significant correlation was found when the cells were stimulated by oppositely moving bars. A weak correlation was present for the coherently moving bars. But the long-bar stimulus resulted in a robust synchronization of the activity at the two sites (Figure 21). This effect occurred in spite of the fact that the overall number of spikes produced by the two cells was similar in the three conditions.[86]

In a nearly identical experiment Engel et al.[55] demonstrated that the synchronization of activity between cells in areas 17 and PMLS of the cat also depends on the properties of the visual stimulus. They recorded from cells having nonoverlapping receptive fields with similar orientation preference that were aligned co-linearly.

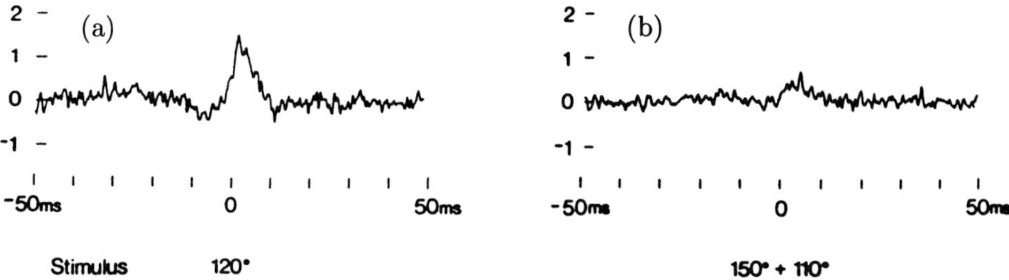

FIGURE 20 Correlograms from a cell pair with visual stimulation oriented differently. One cell of the pair had an orientation preference of 100 degrees, and the second cell had an orientation preference of 130 degrees. The two cells had matched directionality and ocular dominance and were separated by 350 um. The pair was first stimulated with a single light slit oriented at 120 degrees (a), then with two slits simultaneously at 100 degrees and 150 degrees (b). (See Figure 11 of T'so, Gilbert, and Wiesel,[189] "Relationships Between Horizontal Interactions and Functional Architecture in Cat Striate Cortex as Revealed by Cross-correlation Analysis," *J. Neurosci.* **6(4)** (1986): 1160–1170.) Reprinted by permission.

This enabled them to conduct the same test for the effects of coherent motion on response synchronization. They found little or no correlation when the cells were activated by oppositely moving contours and a robust synchronization of activity when the cells were stimulated by a single long bar moving over both fields[55] (Figure 18). These findings, combined with the earlier results, indicate that the global properties of visual stimuli can influence the magnitude of synchronization between widely separated cells located within and between different cortical areas.

A more detailed analysis of the influence of visual stimuli on response synchronization was conducted by Engel et al.[57] In this study, multiunit activity was recorded from up to 4 electrodes simultaneously having a spacing of approximately 0.5 mm. The proximity of the electrodes yielded recordings in which all the cells had overlapping receptive fields and a range of orientation preferences. This configuration often led to the opportunity of comparing the correlation of activity among cells coactivated by either one or two moving bars. In a number of cases, cells having different orientation preferences were found to fire asynchronously when activated by two independent bars of differing orientation. The same cells, however, fired synchronously when coactivated by a single bar of intermediate orientation,[57] much in the same manner as demonstrated previously by T'so et al.[189] These results provide further evidence that the synchronization of activity, and hence the formation of neuronal assemblies in visual cortex, is a dynamic process under the influence of visual stimuli.

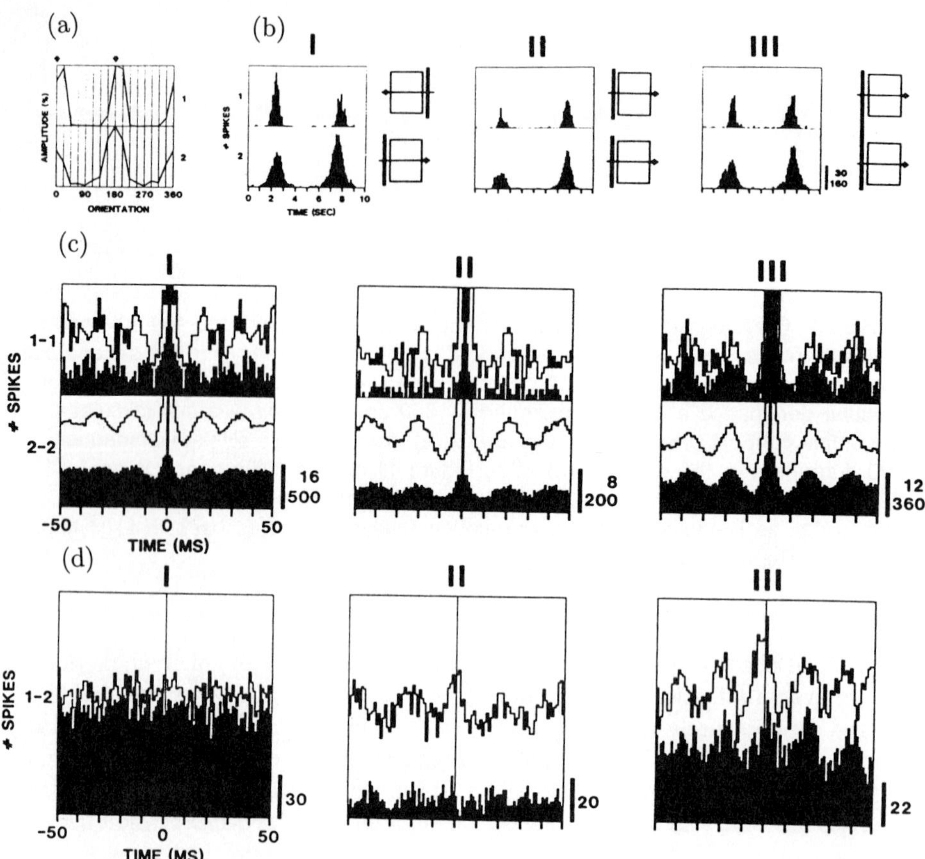

FIGURE 21 Long-range synchronization in area 17 reflects global stimulus properties. (a) Orientation tuning curves of multiunit responses recorded at two sites separated by 7 mm. At both sites, the responses were tuned to vertical orientations (arrows). (b) Peri-stimulus-time histograms of the responses recorded at each site for three different stimulus conditions: (I) two light bars moved in opposite directions; (II) two light bars moved in the same direction; and (III) one long light bar moved across both receptive fields. A schematic plot of the receptive fields and the stimulus configuration is displayed to the right of each peri-stimulus-time histogram. (c) Autocorrelation (1–1,2–2) histograms computed for the responses recorded at both sites for each of the three stimulus conditions (I–III). (d) Cross-correlation histograms computed for the same responses. Note that the strongest response synchronization is observed with the continuous long light bar. For each pair of correlograms except the two displayed in (c) (I,1–1) and D (I), the second direction of stimulus movement is shown with unfilled bars. (See Figure 3 of Gray, Koenig, Engel, and Singer,[85] "Stimulus-Specific Neuronal Oscillations in Cat Visual Cortex Exhibit Inter-Columnar Synchronization Which Reflects Global Stimulus Properties," *Nature* **338** (1989): 334–337.) Reprinted by permission.

That this process is a general property of visual cortex and not confined to the visual cortex of anesthetized cats was recently demonstrated by Kreiter et al.[106] Recordings of multiunit activity were made from two electrodes in area MT of a macaque monkey. The electrode separation was less than 0.5 mm yielding cells with nearly completely overlapping receptive fields but often differing direction preferences.[8] This enabled them to repeat the earlier experiment conducted in the cat. When possible the cells were coactivated by first two moving bars and then one. Under these conditions little or no correlation in activity was observed when the cells were activated by two independently moving bars. The firing of the cells synchronized, however, when responses were evoked by a single bar moving over both fields in a direction capable of activating both cells together.[106] Repeated measures of the responses from the same cells under identical conditions revealed the effect to be stable.

These numerous examples demonstrate that under appropriate conditions the synchronization of activity of two or more groups of neurons in cat and monkey visual cortex can be influenced by the properties of visual stimuli. The results further suggest that response sychronization occurs preferentially when the cells are activated by stimuli having coherent properties. These data thus support the hypothesis that synchronization of activity can act as a mechanism for establishing relations between spatially distributed features in a visual image. However, it must be pointed out that, for every positive example cited above, there are also instances in which the correlation of neuronal responses is not influenced by changes in visual stimulation.[169] This question has not been extensively investigated and much research still needs to be done to rigorously determine the stimulus dependence of response synchronization in visual cortex.

TEMPORAL PROPERTIES OF RESPONSE SYNCHRONIZATION

One of the more impressive aspects of mammalian visual pattern recognition is its rapidity. Psychophysical and behavioral studies have demonstrated that various forms of pattern recognition can be performed within 100–300 ms.[21,23,97,159,160] These findings provide a key temporal constraint on the neuronal mechanisms underlying pattern recognition. They suggest that if response synchronization, oscillatory or otherwise, is utilized for the processing of visual information, the interactions must be rapid. In fact, oscillatory responses of 40–60 Hz should couple within 1–5 cycles. Thus, a measure of the temporal properties of the synchronized oscillations becomes an important step in determining their functional significance.

This question was addressed in a study by Gray et al.[88] Recordings of field potential and unit activity were performed at two sites in cat visual cortex having

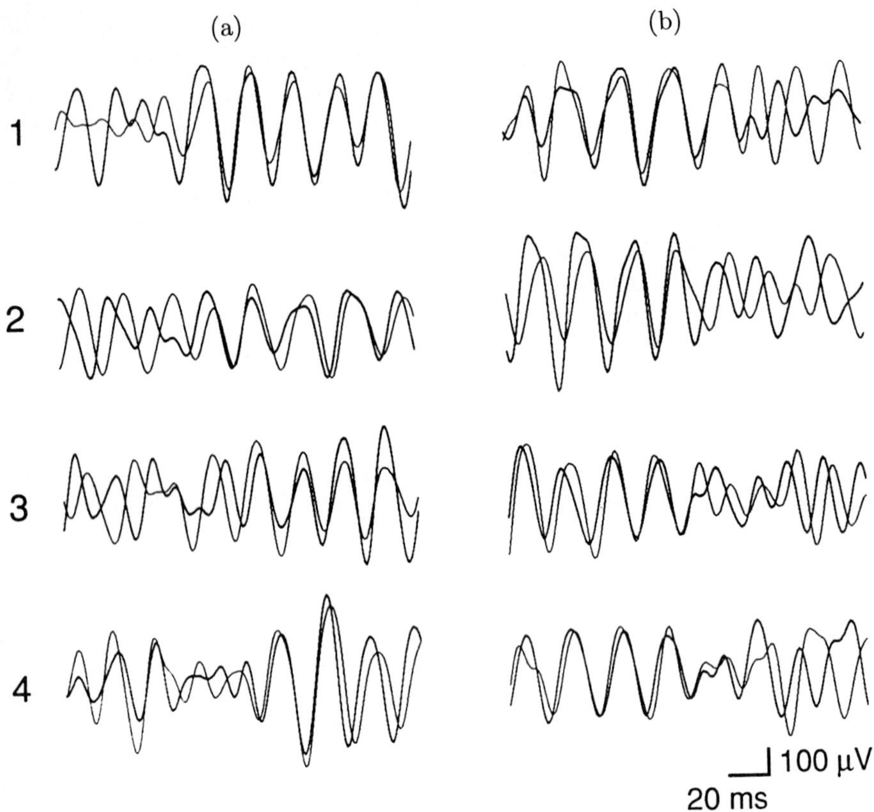

FIGURE 22 The onset and offset of response synchronization is a rapid process. Shown here are four examples each of the onset ((a),1–4) and offset ((b),1–4) of synchronized oscillations at near 0-ms phase-lag. The thick and thin lines represent the two LFP signals recorded. The examples were chosen for their clear and rapid transitions and therefore reflect the most robust cases recorded. The epochs were sampled from the data recorded during the peak of the visual response in each case, and no care was taken to select these examples with respect to the latency of onset of the neuronal response. Note in each case that transition to or from synchrony appears to occur within 20–40 ms. In the example shown in A4, there is a transition from synchrony to asynchrony and back to synchrony again within a period of approximately 100 ms. (See Figure 7 of Gray, Engel, Koenig, and Singer,[88] "Synchronization of Oscillatory Neuronal Responses in Cat Striate Cortex: Temporal Properties," *Visual Neuroscience* 8 (1992): 337–347.) Reprinted by permission.

a separation of at least 4 mm. This insured that the receptive fields of the recorded cells would not overlap and that there would be little contamination by volume conduction in the field potential signals from one recording site to the other.[53] Field potential responses were chosen for analysis in which the signals displayed a particularly robust oscillation, a close correlation to the simultaneously recorded unit activity, and a statistically significant, average cross correlation. Under these conditions it became possible to determine (1) the onset latency of the synchrony between the two recordings; (2) the time-dependent changes in the phase, frequency, and duration of the synchrony within individual trials; and (3) the intertrial variation in each of these parameters.

Combined with previous observations,[53] the results demonstrated that correlated oscillatory responses in cat visual cortex exhibit a high degree of dynamic variability. The amplitude, frequency, and phase of the synchronous oscillations fluctuate over time. The onset of the synchrony is variable and bears no fixed relation to the stimulus. Multiple epochs of synchrony can occur on individual trials and the duration of these events also fluctuates from one stimulus presentation to the next. Most importantly, the results demonstrated that response synchronization can be established within 50–100 ms (Figure 22), a time scale consistent with behavioral performance on visual discrimination tasks.[88]

MECHANISMS UNDERLYING THE GENERATION OF OSCILLATORY RESPONSES IN VISUAL CORTEX

As discussed earlier, for the olfactory system and thalamocortical spindle rhythms, there are three basic mechanisms likely to underlie the generation of oscillatory neuronal activity in the visual cortex. The first and most obvious mechanism to consider is oscillatory afferent input. It is conceivable that the periodic temporal structure of cortical responses could be simply due to an oscillatory input from the lateral geniculate nucleus in much the same way as input from the olfactory bulb drives the piriform cortex into an oscillatory pattern. Support for this view has come from a number of studies demonstrating both in the retina and the LGN the existence of robust oscillatory activity in the frequency range of 40–60-Hz.[14,16,24,50,76,78,110,143] In both structures the oscillatory activity is present in a subpopulation of roughly 20% of the cells and appears to be largely spontaneous in nature.[110,163] Diffuse changes in illumination also evoke oscillatory activity[110] but often the oscillations are suppressed or uninfluenced by visual stimuli.[78]

Such input to the cortex could provide a catalyst or even a driving influence for oscillatory responses. However, the data on retinogeniculate oscillations do not appear to account for several aspects of the cortical oscillatory activity. Cortical oscillations are often synchronous across multiple hypercolumns within an area,[53,86] between cells in different areas,[51,55] and between cells in area 17 of the two cerebral hemispheres.[56] An explanation of these phenomena in terms of afferent input from the LGN would require that the synchronously firing cells in cortex be driven by a common source. Such common input could occur through a divergence of connectivity from LGN to cortex or by synchronization of the activity within the LGN.

It is well established that afferents to the cortex from the LGN can have terminal arbors that span up to 3–4 mm in cortex.[60] However, few thalamic afferent connections to area 17 can account for the synchronization of activity that has been observed across 6–7 mm of cortex[53,86] (or between two cortical areas[51,55]). It is likely that such interactions result largely from the influence of intracortical connections.[79,80,81,131,147,166] Moreover, LGN afferents to the cortex project only to the ipsilateral hemisphere and therefore cannot provide the necessary common drive to synchronize activity in the two cerebral hemispheres.[56] In fact, several lines of evidence indicate that the common input underlying interhemispheric synchronization passes through the corpus callosum.[37,56]

Synchronization of activity at the thalamic level in principal could provide the structural basis for synchrony in the cortex. This is, of course, what happens during a thalamocortical spindle.[182] Although synchrony among thalamic cells is pronounced during low frequency spindling, correlated 40- to 60-Hz oscillatory activity in the LGN has only been observed for closely spaced cells.[16] To date no correlations of 40–60 Hz oscillations have been observed between LGN cells having nonoverlapping receptive fields.

Nevertheless, it is possible that rhythmic input from the LGN could make a significant contribution to the generation of oscillatory responses and their local synchronization in cortex.[78] One important experiment to test this hypothesis would be to simultaneously measure oscillatory activity in LGN and cortex in retinotopic correspondence. The occurrence of significant synchronization in the activity of these two structures would support a role for the LGN in generating cortical high-frequency oscillations.

The other mechanisms likely to underlie the generation of cortical oscillatory responses rely on purely intracortical processes. In one scenario, rhythmic 40- to 60-Hz firing could arise soley from intracortical network interactions. A possible substrate for this could be inhibitory feedback in much the same manner as that described for the olfactory bulb and cortex.[66] Although direct evidence that recurrent inhibition underlies high-frequency cortical oscillations is very limited,[89] such a prediction appears plausible since local circuit inhibitory interneurons exist in abundance in the neocortex,[125,130] and artificial neuronal networks containing recurrent inhibitory connections readily exhibit oscillatory activity.[38,66,104,201,202] Demonstration of such interactions is complicated, however, by the difficulties associated with recording from identified inhibitory cells. Thus, one may be largely limited to indirect sources of evidence.

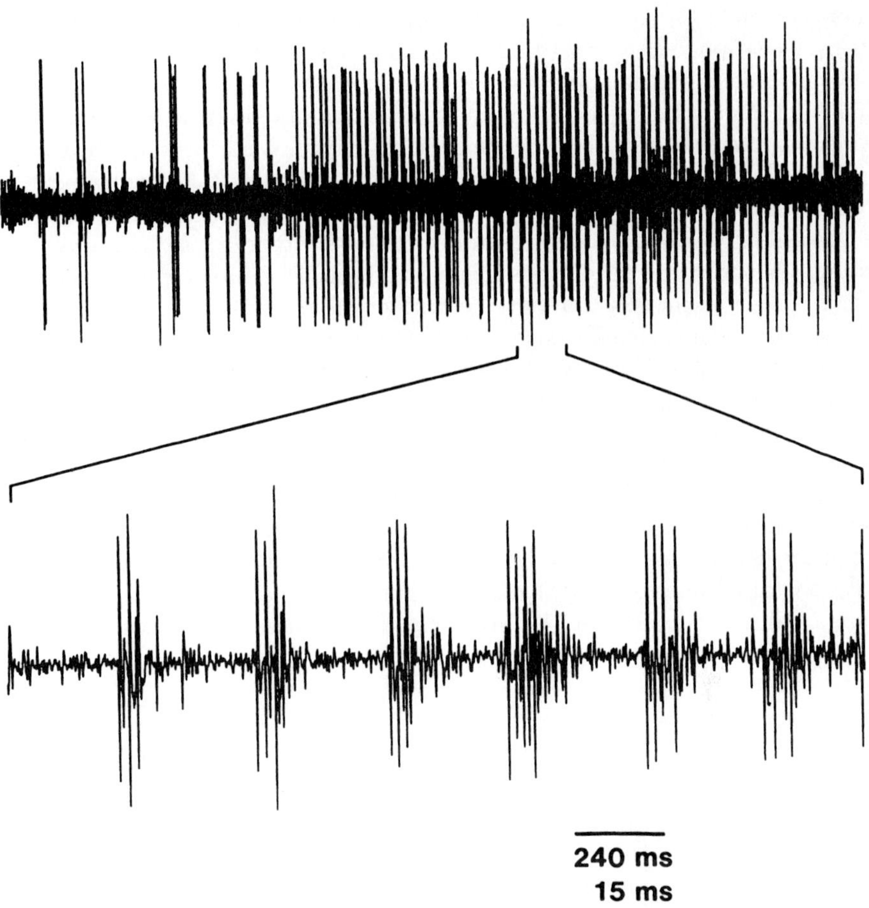

FIGURE 23 Repetitive burst-firing cell recorded extracellularly from area 17 of an awake behaving cat. Upper trace shows the response of the cell to the presentation of an optimally oriented light bar passed over the cell's receptive field. The lower trace shows part of the response at an expanded time scale. Note the presence of repetitive burst discharges at a frequency near 40 Hz. (See Gray.[90])

Alternatively, oscillatory activity in cortex may arise as a consequence of the intrinsic membrane properties of cortical neurons,[122] much as it does in thalamocortical relay neurons[133] and cells of the thalamic reticular nucleus.[18,182] Support for this conjecture has recently been obtained from *in vitro* intracellular recordings in rat frontal[124] and sensorimotor cortex.[173] In the former study, intrinsic 10- to 50-Hz oscillations in membrane potential were observed in inhibitory interneurons of layer 4. These fluctuations were produced by membrane depolarization and were dependent on both sodium and potassium conductances.[124] In the latter study, intrinsic low-frequency 5- to 10-Hz oscillations produced by membrane depolarization were found in a subpopulation of layer 5 cells.[173] These cells are known to be capable of intrinsically generating bursts of action potentials in response to depolarizing inputs.[45,108,132] This phenomenon of intrinsic bursting may prove relevant since cortical 40- to 60-Hz oscillatory activity is often associated with burst firing in single cells[87,90] (Figure 23).

Recent intracellular recordings from cat striate cortex *in vivo* have attempted to distinguish among the various possible mechanisms. High-frequency 30–60-Hz oscillations of membrane potential have been observed in response to visual stimuli[59,99] (Figure 24). These signals show a remarkable similarity to oscillatory activity recorded extracellularly.[85] The oscillations in membrane potential are stimulus dependent and orientation specific and occur less often in simple cells receiving monosynaptic input from the LGN.[99] Action potentials occur on the depolarizing phase of the oscillation mirroring the negativity observed extracellularly.[85] Two further lines of evidence indicate that the oscillations result from extrinsic as opposed to intrinsic mechanisms.[99] Oscillations in membrane potential increase in amplitude in response to visual input when the cells are hyperpolarized, suggesting that they arise from excitatory synaptic input rather than voltage-dependent membrane conductances (Figure 24). Intracellular depolarization reveals no consistent

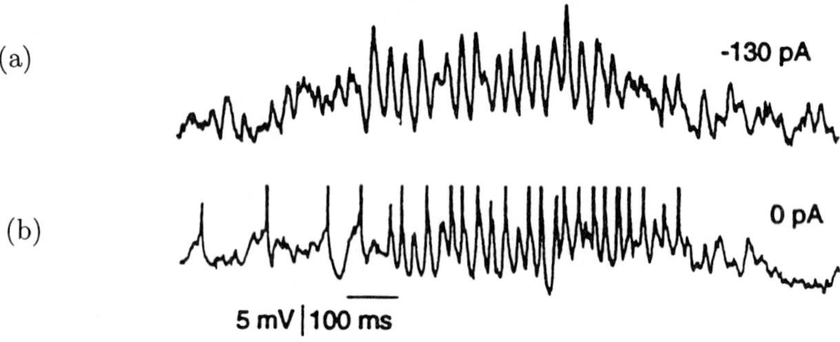

FIGURE 24 Intracellular recording of the response of a complex cell in area 17 of an anesthetized cat to the presentation of an optimally oriented light bar passed over the cell's receptive field. In (a) the cell is being hyperpolarized by the injection of −130 pA of current. The action potentials in A have been truncated. (See Figure 1(a),(b) of Jagadeesh, Gray, and Ferster,[99] "Visually-Evoked Oscillations of Membrane Potential in Neurons of Cat Striate Cortex Studied with In Vivo Whole Cell Patch Recording," *Science* **257** (1992): 552–554.) Reprinted by permission.

tendency of the cells to fire at frequencies in the range of 30–60 Hz. These data therefore suggest that oscillatory responses in visual cortex reflect a property of the cortical network.

Because of the small sample size (28 cells), it is likely that cells exhibiting an intrinsic capacity to oscillate could have easily been missed. In fact, no burst-firing cells were observed in the study of Jagadeesh at el.[99] Thus one cannot rule out the possibility that a subpopulation of cells in visual cortex have an intrinsic tendency to fire in repetitive bursts at a frequency of 30–60 Hz.[90]

HIPPOCAMPAL RHYTHMS: CARRIER WAVES FOR SYNAPTIC PLASTICITY?

The hippocampus is an archicortical structure in the limbic system known for the similarity of its shape to a seahorse. During a number of different behavioral states, it is known to exhibit some of the most robust forms of synchronous rhythmic activity to be observed in the central nervous system. Foremost among these is the theta rhythm, a sinusoidal-like oscillation of neuronal activity at 4–10 Hz that occurs during particular behavioral states (see below). In addition, two other neuronal rhythms have been discovered, one having a frequency in the range of 30–100 Hz that occurs during both anesthesia- and theta-related behavioral states[39,112] and another more recently discovered signal having a frequency around 200 Hz associated with alert immobility and the presence of sharp waves in the hippocampal EEG.[43]

THETA RHYTHM

The theta rhythm or rhythmic slow activity (RSA) was originally discovered in 1938 by Jung and Kornmuller and later investigated more extensively by Green and Arduini[91] among others. It occurs primarily in nonprimate mammals and is broadly distributed throughout the hippocampus during two different behavioral states.[28] The most prominent of these two is type I or movement related theta.[192] During exploratory behaviors such as walking, shifts in posture, or the manipulation of objects with the forelimbs, type I theta is prominent throughout the hippocampus. A second form of RSA termed type II theta occurs during periods of behavioral immobility.[28] This form of theta can also be elicited by sensory stimulation under anesthesia, making it quite amenable to detailed analysis.

In both behaving and anesthetized animals, theta activity is localized largely to the hippocampus,[25] the surrounding entorhinal cortex,[1,9,138] and the medial septal nucleus[157] (Figure 25). The generation of hippocampal RSA has long been thought to depend largely on input from the medial septal nucleus a structure thought to act as a pacemaker.[9,28,39,157] Both the medial septal nucleus and the entorhinal cortex appear to be capable of intrinsically generating RSA either through network interactions[28,186] or by intrinsic cellular membrane properties.[11] Stellate cells in layer 2 of the entorhinal cortex, for example, are intrinsically oscillatory by virtue of a combination of sodium and potassium conductances.[11] These neurons project via the perforant path to area CA1 and the dentate gyrus of the hippocampus where they produce an additional 4–10 Hz drive on the system.[39] Although it is clear that both the medial septum and entorhinal cortex do provide a rhythmic afferent drive to hippocampus, recent intracellular recordings indicate

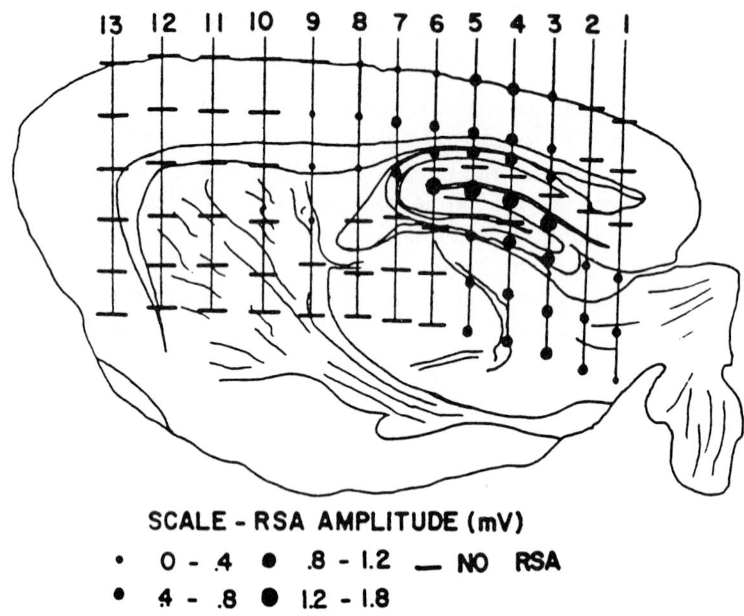

FIGURE 25 Topography of hippocampal rhythmic slow activity. A diagrammatic representation of a saggital section of the brain showing the reconstruction of microelectrode tracks made during a mapping experiment. During the mapping, a reference microelectrode was fixed in the stratum moleculare. Tracks 1–8 were each 0.75 mm apart and tracks 8–13 were each 1 mm apart. Each individual track went to a depth of 5 mm from the surface of the brain. The filled circles represent the location and approximate amplitude (mV) of RSA in each track and the dashes indicate locations where no RSA was recorded. (See Figure 9 of Bland and Wishaw,[25] "Generators and Topography of Hippocampal Theta (RSA) in the Anesthetized and Freely Moving Rat," *Brain Res.* **118** (1976): 259–280.) Reprinted by permission.

that hippocampal pyramidal neurons are also capable of generating intrinsic oscillations of membrane potential in the 4- to 10-Hz range.[150,111] Thus there exists a redundant set of mechanisms in these structures capable of generating theta activity.

The intrinsic membrane properties and the local network interactions among cells in the hippocampus, entorhinal cortex, and medial septal nucleus lead to a wide range of rhythmic synchronous activities displayed in the discharges of single cells in these structures. Some cells show a rather precise pattern of repetitive burst firing synchronized to the phase of the RSA observed in the field potential.

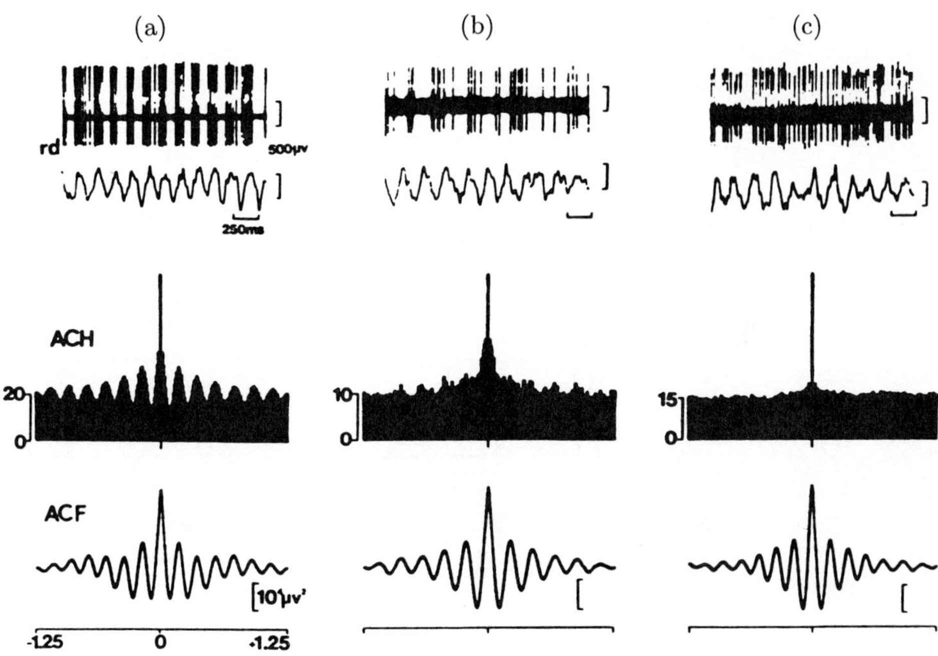

FIGURE 26 Types of entorhinal cortex neuronal discharges. rd: short samples of raw data, unit activity (upper) and reference theta rhythm (lower). Autocorrelation functions (ACF) indicate the presence of a stationary theta rhythm during the epochs processed. (a) Rhythmic cell at the theta frequency as demonstrated by the autocorrelation histogram (ACH). (b) Intermediate cell with almost no periodic ACH. (c) Nonrhythmic cell as demonstrated by the aperiodic ACH. Vertical calibration in all figures for ACH: impulses/sec. (See Figure 1(a),(b),(c) of Alonso and Garcia-Austt,[10] "Neuronal Sources of Theta Rhythm in the Entorhinal Cortex of the Rat II. Phase Relations Between Unit Discharges and Theta Field Potentials," *Exp. Brain Res.* **67** (1987): 502–509.) Reprinted by permission.

Autocorrelation and interval histograms of the spike trains of these cells reveal a clear 4- to 10-Hz periodicity. In the hippocampus proper, these neurons are termed theta cells and are largely thought to be inhibitory interneurons.[28,61] In the entorhinal cortex such cells have been classed simply as rhythmic[10] whereas in the medial septum they are termed "B" cells.[156] In each of these structures, there are significant numbers of cells that display little or no rhythmicity in their discharge patterns. These cells on first glance appear not to participate in theta activity. Further studies, however, have shown that the vast majority of these nonrhythmically firing cells show a clear relation with the phase of the theta rhythm observed in the field potential[10,27,156] (Figure 26), a result similar to that observed for mitral cells and pyramidal neurons in the olfactory bulb and piriform cortex in relation to the induced wave (see above, Freeman[66]). Thus, although often not readily apparent in the activity of single cells, the hippocampal theta rhythm represents the coordinated activation of very large populations of synchronously active cells.

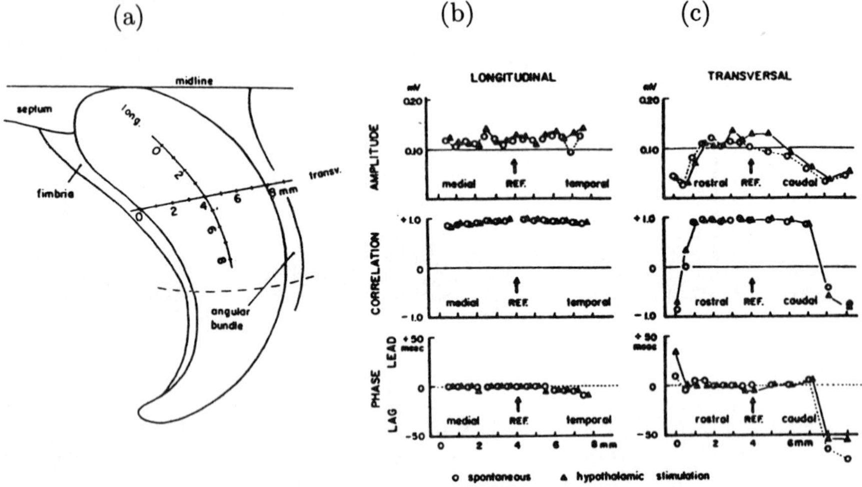

FIGURE 27 Spatial distribution of theta activity. (a) Diagrammatic representation of the hippocampal formation seen from above with the temporal part deflected laterally. A reference electrode was situated in the cross between the lines labelled long and trans. The moving surface electrode was shifted along the two lines in steps of 0.5 mm. (b) The upper graph gives computer-averaged amplitude. The middle and lower graphs give the correlation coefficients and phase shift between the reference and the moving electrode records, plotted as a function of the longitudinal position of the latter. (c) Same as (b), but with the moving electrode shifted along the transverse axis (transverse) in (a). Data obtained during spontaneous theta activity are indicated by open circles and data recorded during hypothalamic stimulation by filled triangles. (See Figure 2 of Bland, Andersen, and Ganes,[26] "Two Generators of Hippocampal Theta Activity in Rabbits," *Brain Res.* **94** (1975): 199–218.) Reprinted by permission.

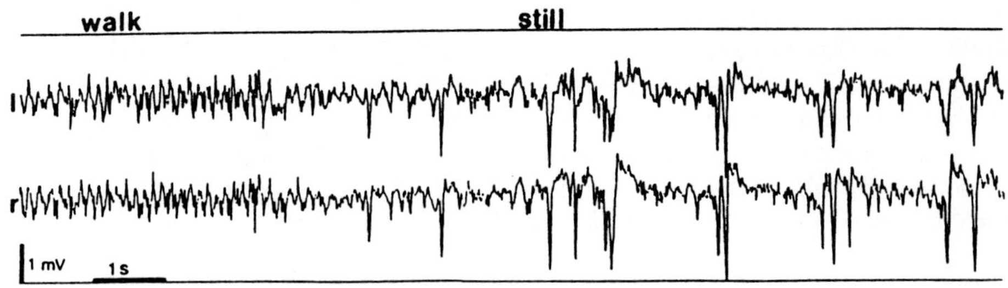

FIGURE 28 EEG recorded from the stratum radiatum of the left (l) and right (r) CA1 region of the hippocampus during walk-immobility (still) transition. Note regular theta waves during walking and large monophasic sharp waves during immobility. Note also the bilaterally synchronous nature of sharp waves. (See Figure 1 of Buzsaki.[42] "Two-Stage Model of Memory Trace Formation: A Role for "Noisy" Brain States." *Neuroscience* **31(3)** (1989): 551–570.)

This synchronization process was observed within CA3 by Kuperstein et al.[107] at a much finer scale of resolution using multiple microelectrode recording techniques. By recording single cell activity at up to 24 sites simultaneously, they were able to avoid the potential problem of volume conduction of the field potentials generated within the hippocampus. Their results revealed a complex but synchronous pattern of activity during the theta rhythm. Different cells tended to fire at different phases of the theta rhythm, but these relations were consistent and tended to yield a coordinated pattern of activity from the sample as a whole.[107]

HIPPOCAMPAL FAST ACTIVITY

Another form of rhythmic activity in the hippocampus about which much less is known is the so-called fast activity.[112] These rhythmic field potential signals have a frequency ranging from 30–100-Hz occur during theta-related behaviors and at higher amplitudes and lower frequencies (30–50 Hz) under anesthesia. The signals have been recorded in CA1 and in the dentate but are most prominent in the hilus of the dentate gyrus. Fast activity has been observed to be synchronous between the two hippocampi. And the activity of interneurons and granule cells is often in phase with the fast oscillations observed in the field potential signals.[39] Little else is known regarding the functional significance of this form of activity in the hippocampus.[112] These fast rhythms however, are indicative of the propensity of the hippocampus to engage in large-scale cooperative patterns of synchronous rhythmic activity.

of the dentate gyrus. Fast activity has been observed to be synchronous between the two hippocampi. And the activity of interneurons and granule cells is often in phase with the fast oscillations observed in the field potential signals.[39] Little else is known regarding the functional significance of this form of activity in the hippocampus.[112] These fast rhythms however, are indicative of the propensity of the hippocampus to engage in large-scale cooperative patterns of synchronous rhythmic activity.

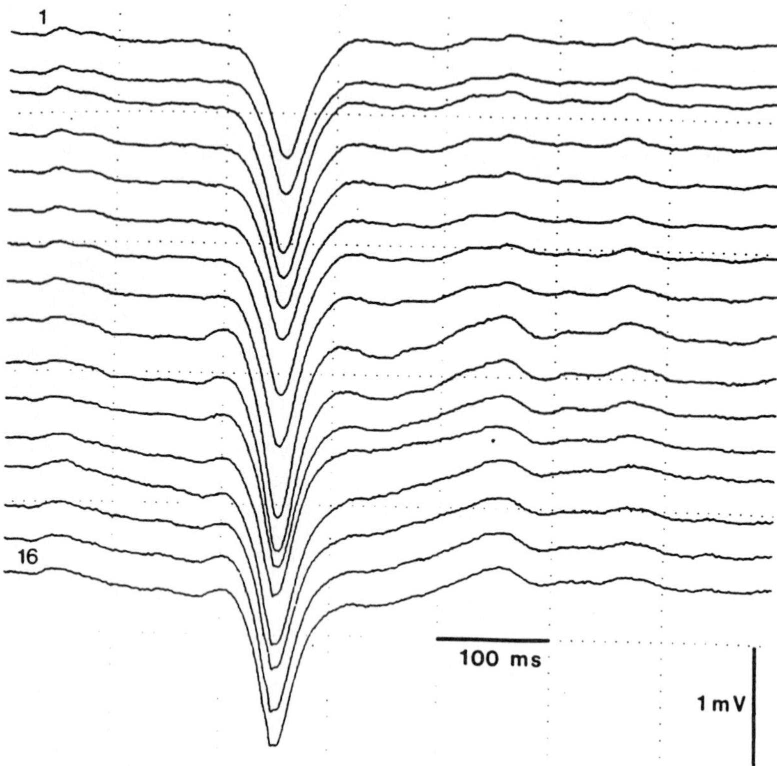

FIGURE 29 Averaged sharp wave ($n = 50$) recorded simultaneously from 16 microelectrodes along the longitudinal axis of the CA1 region of the hippocampus. Spacing between electrodes was 200 um. All tips were positioned in the stratum radiatum. Note the simultaneous occurrence of sharp waves over a distance of 3.2 mm. (See Figure 2 of Buzsaki,[42] "Two-Stage Model of Memory Trace Formation: A Role for "Noisy" Brain States," *Neuroscience* **31(3)** (1989): 551–570.) Reprinted by permission.

HIPPOCAMPAL SHARP WAVES

During the transition from exploratory behavior to alert immobility, associated with acts such as eating and grooming, hippocampal activity in the rat shows another remarkable change. Highly synchronous theta activity vanishes and is largely replaced by a pattern of irregular activity intermingled with large amplitude (of 1–3 mV) sharp waves of voltage lasting 40–100 ms.[39,40,42] These events are synchronous in the two hippocampi (Figure 28) and show a high degree of coherence across the transverse and longitudinal axes of the hippocampus[42] (Figure 29). Moreover, each sharp wave is associated with a synchronous burst discharge of action potentials in a population of pyramidal cells.[42]

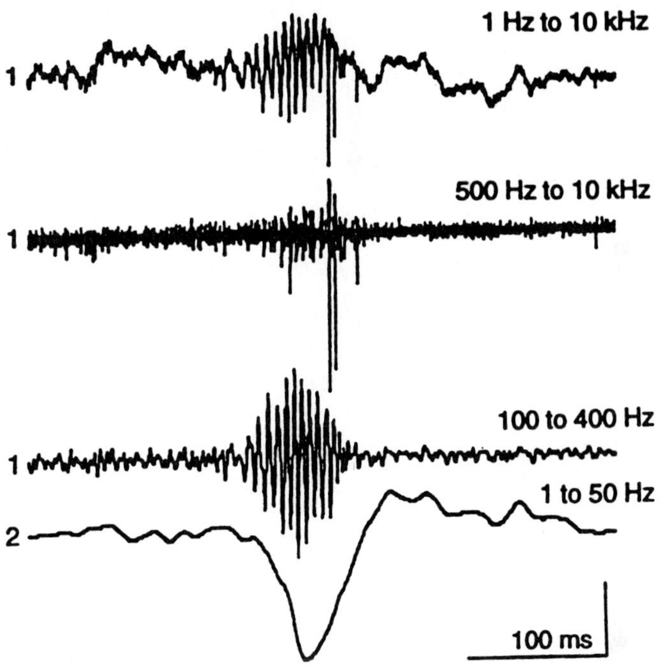

FIGURE 30 Fast field oscillation in the CA1 region of the dorsal hippocampus. Simultaneous recordings from the CA1 pyramidal cell layer (1) and stratum radiatum (2). Note the simultaneous occurrence of fast field oscillations, unit discharges, and sharp wave. Calibrations: 0.5 mV (trace 1), 0.25 mV (traces 2 and 3), and 1.0 mV (trace 4). (See Figure 1 of Buzsaki, Horvath, Urioste, Hetke, and Wise,[43] "High-Frequency Network Oscillation in the Hippocampus," *Science* **256** (1992): 1025–1027.) Reprinted by permission.

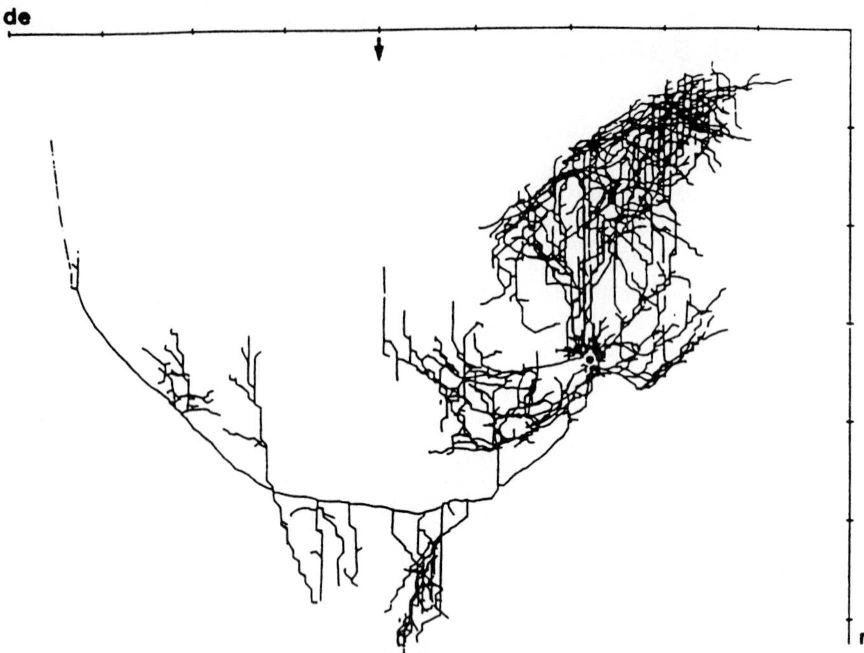

FIGURE 31 Dorsal projection of a CA2 pyramidal neuron in the hippocampus. Arrow indicates the midline of the brain and the closed circle indicates the position of the soma. Graduations on the axes are at 1-mm intervals. (See Figure 3(b) of Tamamaki, Abe, and Nojyo,[184] "Three-Dimensional Analysis of the Whole Axonal Arbors Originating from Single CA2 Pyramidal Neurons in the Rat Hippocampus with the Aid of a Computer Graphic Technique," *Brain Res.* **452** (1988): 255–272. Reprinted by permission.

Recently these signals have been examined at a higher degree of spatial and temporal resolution using new multielectrode recording techniques.[43] These studies have revealed that the sharp wave actually consists of a synchronous oscillation of activity in a population of cells at a frequency near 200 Hz (Figure 30). Individual cells were found to fire at low rates, but their coordinated activation yielded an emergent pattern of activity having an oscillatory character. Single cells recorded in isolation in the pyramidal cell layer showed no indication of rhythmicity but were often found to fire in phase with the 200-Hz oscillation recorded in the field potential. In a few recordings the activity of identified interneurons was also measured. These cells fired consistently at phases near 180 degrees different from that of the pyramidal cell population, suggesting a role for the interneuron population in the generation of the oscillations. When observed at spatially separate sites extending up to 1.2 mm, the activities of single cells were found to be correlated with each other and the oscillatory field potential. It remains to be determined if similar synchronous interactions extend over larger distances.

FUNCTIONAL SIGNIFICANCE OF SYNCHRONOUS HIPPOCAMPAL RHYTHMS

The preceeding discussions clearly demonstrate that the hippocampus can exhibit a wide range of well-organized coherent states of rhythmic activity. These states are often closely linked to particular behavioral states or acts. And the synchronous activity patterns can often encompass both hippocampi and surrounding limbic structures. In fact, as is the case for hippocampal theta, synchronous states of activation can be coupled to motor acts such as exploratory sniffing behavior.[126]

Although the functional significance of these synchronous rhythms is largely unknown, such stereotyped and global patterns of activity are suggestive of an important function. Multiple, redundant mechanisms have evolved to enable the generation of various forms of rhythmic activity[39,188] and the anatomical organization of hippocampus clearly lends itself to the establishment of macroscopic patterned states. Hippocampal pyramidal cells have widely arborizing axonal collaterals which extend over many millimeters including the contralateral hippocampus[184] (Figure 31). These long-range connections in combination with the intrinsic membrane properties of cells and local network organization yield a structure capable of some of the most robust and organized states of synchronous activity to be observed in the central nervous system. The question is, of course, what are the functions of these coherent states of activity.

Despite extensive studies into the mechanisms of generation, the behavioral dependence, and the spatiotemporal distribution of hippocampal rhythmic activity, we still do not know the answer to this question. A number of attractive proposals have been put forth, however. Several of these relate closely to the well-established role of the hippocampus in learning and memory[136] and the propensity of hippocampal neurons to exhibit synaptic plasticity such as long-term potentiation and depression.[30,178]

Studies investigating the latter phenomena have established that correlation of activity among synaptically connected neurons is a critical factor in the induction of changes in synaptic strength. Correlated activity is, of course, the essence of hippocampal rhythms. Thus these coherent rhythmic states may provide the substrate for synaptic modifications and the storage of information. For example, it has recently been observed that tetanic stimulation of excitatory pathways in the hippocampus in phase with the theta rhythm[155] or delivered *in vitro* at the frequency of the theta rhythm,[109] leads to a marked long-term increase in synaptic strength. If stimulation is given at frequencies significantly above or below the theta range, the

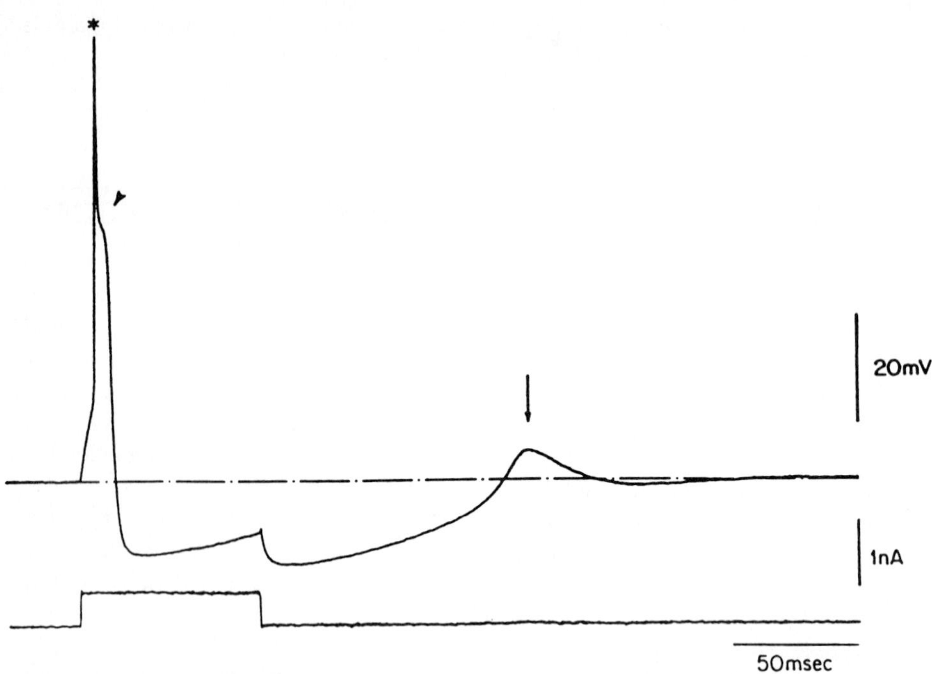

FIGURE 32 Intracellular recording from an olivary neuron in a slice preparation of a guinea pig brain stem. The response was elicited by a 70-ms, 0.5-nA positive current pulse. Asterisk, the sodium-dependent component of the response; arrowhead, the prolonged depolarization generated by the high-threshold calcium current; and arrow, the low-threshold calcium spike. Dashed line indicates the resting potential. (See Figure 1 of Yarom,[205] "Oscillatory Behavior of Olivary Neurons," in *The Olivocerebellar System in Motor Control*, edited by Piergiorgio Strata, Berlin: Springer-Verlag, 1989.) Reprinted by permission.

induction of synaptic gain changes is less pronounced or absent. Moreover, if 5-Hz tetanic stimulation is given to two pathways together but 180 degrees out of phase, synaptic connections can actually be significantly weakened or depressed.[178] These data suggest an important role for hippocampal theta in the control of mnemonic functions.

Another important factor regulating synaptic plasticity in the hippocampus is the degree of cooperactivity of the convergent input onto any given neuron.[135] Consistent with this it is well known that high-frequency tetanic stimulation provides an effective stimulus for the induction of LTP in hippocampal neurons. The

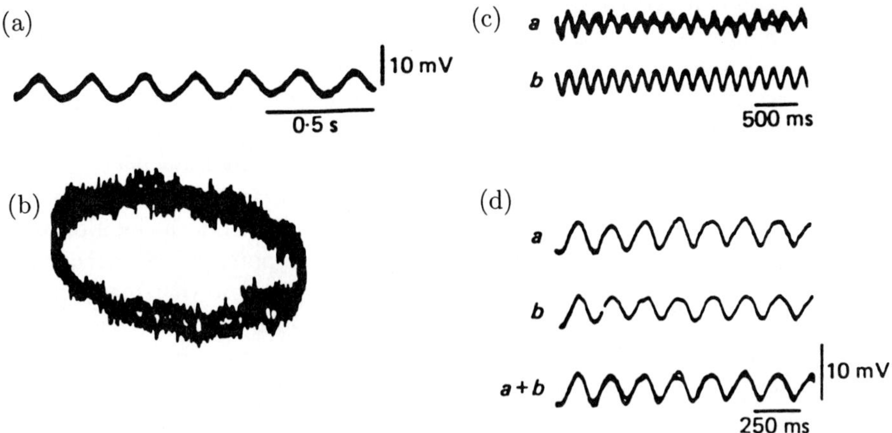

FIGURE 33 Spontaneous oscillation of the membrane potential and its synchronicity in two inferior olive neurons simultaneously recorded intracellularly. (a) Spontaneous oscillatory property of the membrane potential. (b) Lissajous figure to illustrate the regularity of the spontaneous oscillation. The x-axis of the oscilloscope was derived by a sinusoidal waveform of 4 Hz. (c) Superimposed traces of spontaneous membrane potential oscillations recorded simultaneously from two olivary neurons (a,b). (d) An average record of 6 traces recorded during the same time interval from the two cells in B and superimposed in (a) and (b). (See Figure 4 of Llinas and Yarom,[121] "Oscillatory Properties of Guinea-Pig Inferior Olivary Neurons and Their Pharmacological Modulation: An *In Vitro* Study," *J. Physiol.* **376** (1986): 163–182.) Reprinted by permission.

hippocampal sharp wave represents perhaps the closest physiological correlate to a tetanic electrical stimulus, brief high-frequency synchronous bursts of activity in a population of cells. These waves, however, have the added dimension of spatial organization, making it likely that a large number of coherent states of sharp wave activity could exist at different times. Such spatial variation could provide the hippocampus with a powerful combinatorial capacity for information storage.[42]

THE OLIVOCEREBELLAR SYSTEM AND PHYSIOLOGICAL TREMOR

One of the most robust and regular oscillatory systems in the brain resides in the inferior olivary nucleus. This subcortical motor structure provides the climbing fiber input to the cerebellum and exerts a powerful control on cerebellar output generated by purkinje cells. Single units in the inferior olive of the intact animal demonstrate a pronounced rhythmicity in their discharge patterns.[15,98] This rhythmicity can be readily observed *in vitro*[121] and is known to arise intrinsically through a combination of specific membrane conductances.[118,119,205] As in the generation of thalamocortical spindles, the interplay between a low-threshold calcium conductance and a calcium-dependent potassium conductance is integral to the oscillatory behavior[205] (Figure 32). These conductances endow the inferior olive neurons with an intrinsic capacity to oscillate at 4–10 Hz upon receiving input.

In a small percentage of *in vitro* slice preparations, neurons in the inferior olive exhibit a sustained subthreshold oscillation of membrane potential at a frequency of 4–10 Hz. When this behavior is observed, it is often the case that many, if not all, of the cells in the slice show continuous oscillations at the same frequency[121,205] (Figure 33). This behavior is thought to depend on electrotonic coupling between cells in the nucleus.[117,118] Because of this coupling, single cells continue to oscillate when they are depolarized or hyperpolarized away from their resting potential by intracellular current injection.[121] Thus disruption of the oscillation in a single cell does not disrupt the pattern present in the rest of the network. External stimulation

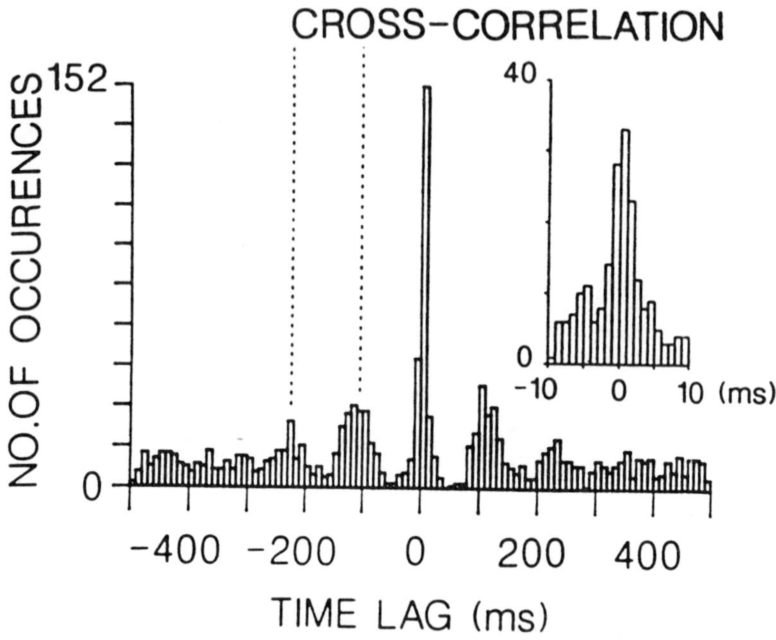

FIGURE 34 Cross correlogram of the activity of two simultaneously recorded purkinje cells in the cerebellar cortex of an anesthetized rat. (See Figure 7(b) of Sasaki, Bower, and Llinas,[167] "Multiple Purkinje Cell Recording in Rodent Cerebellar Cortex," *Europena J. Neurosci.* **1** (1989): 572–586.) Reprinted by permission.

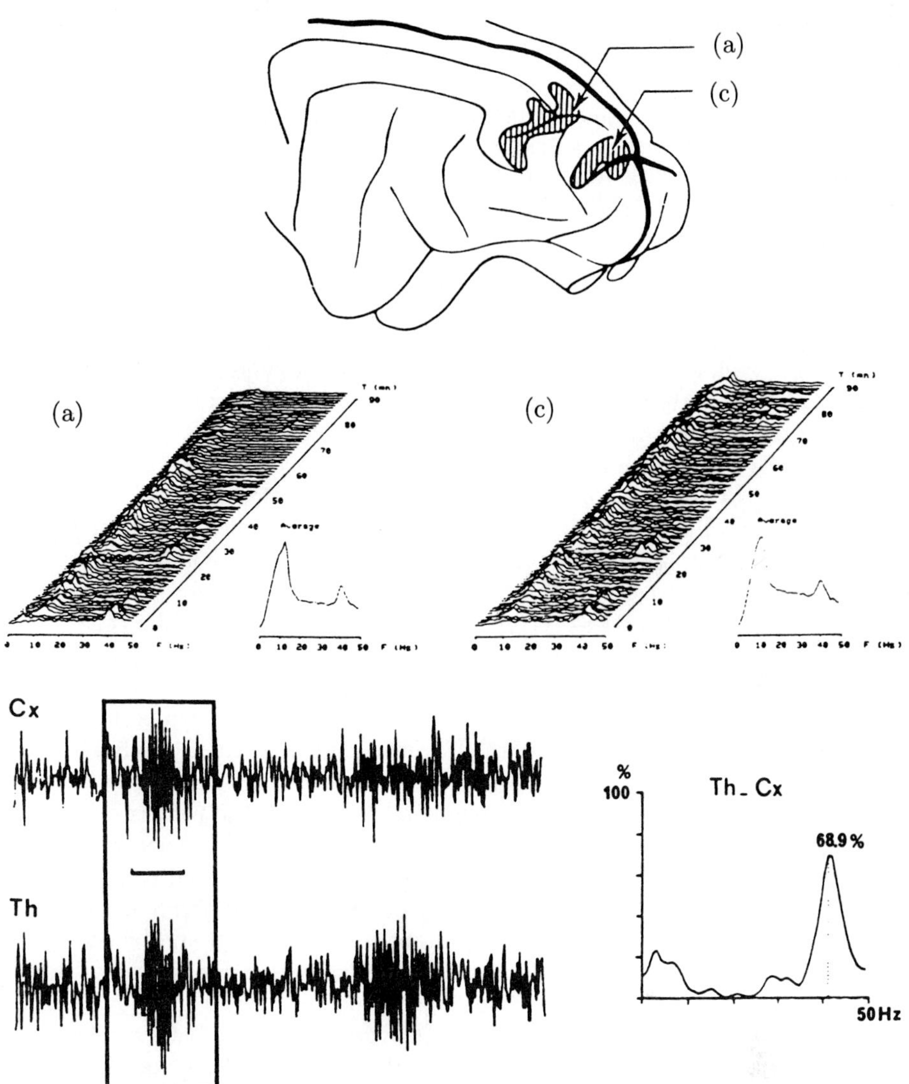

FIGURE 35 Distribution of fast (35–45 Hz) rhythms over the cat anterior cortex. The hatched zones represent data pooled from 20 cats, encompassing the extreme limits of two foci found to display this activity. Below are presented two compressed spectral arrays taken at 1-min intervals of the EEG recorded at A and C, with a total recording time of 90 min. Episodes of fast rhythms (35–45 Hz) that accompany high vigilance and immobility can be seen. They alternate with rhythmic activities in the (cont'd)

FIGURE 35 (cont'd.) lower frequency ranges that correspond to quiet vigilance, drowsiness, and then sleep. Added to each set of spectral arrays is the average spectrum computed over the 90-min recording time. Lower left: Fast hypervigilance rhythms simultaneously recorded over the parietal cortex (Cx) and the posterior thalamic group (Th). Time scale = 0.5 sec. Lower right: Coherence spectrum of the activity between the thalamus and cortex. (See Figure 5.3 of Steriade, Jones and Llinas,[182] *Thalamic Oscillations and Signaling*, New York: John Wiley, 1990. Reprinted by permission.

delivered to a group of cells can transiently disrupt the rhythm, however.[121] These findings suggest that the sustained nature of the oscillations in the inferior olive is an emergent property of the coupled network. The oscillatory behavior critically depends on the intrinsic membrane properties of the cells but is expressed when a network of cells are electronically coupled.[205]

This emergent property of the inferior olive has a profound impact on the activity of the cerebellar cortex. In effect the sustained and synchronous oscillations in the climbing fiber input result in the synchronous and rhythmic discharge of populations of purkinje cells.[20,167] Using a sophisticated system for monitoring the activity at up to 32 sites Sasaki et al.[167] demonstrated that purkinje cells within separate cerebellar folia show a high degree of synchronous rhythmic discharge at 6–10 Hz (Figure 34). The synchrony extends over distances of 2 mm in the rostrocaudal direction but falls off sharply within 250 um in the mediolateral direction. This effect results from a variation in the frequency of the rhythmic discharge of the purkinje cells along this axis.[167] However, the spatial organization of the synchrony observed in the cerebellar cortex is thought to depend on the pattern of electrotonic coupling of cells in the inferior olive. Llinas and Sasaki[123] demonstrated that blockade of GABAergic transmission in the inferior olive leads to a disruption of the specificity of the spatial pattern of synchrony observed among purkinje cells. Thus the inferior olive appears to act as an important control structure in the generation and regulation of synchronous rhythmic patterns of activity in the cerebellar cortex.

As in other systems the functional significance of these synchronous olivocerebellar rhythms is not fully understood. However, an attractive hypothesis has been put forward by Llinas and his colleagues.[120,123] In essence the rhythms of 6–10 Hz are thought to provide the underlying mechanism controlling physiological tremor.[129] This tremor may provide a time frame, or carrier frequency, for the generation of complex coordinated movements requiring distributed populations of cells to be coactive. Such a mechanism may explain why both voluntary and involuntary movements occur in phase with physiological tremor[129] and why the upper limit of the rate of repetitive movements is at or near 10 Hz.[167]

THE SOMATOMOTOR CORTEX AND ATTENTIVE BEHAVIOR

Another, less well understood, but equally impressive form of synchronous rhythmic activity occurs in the somatosensory and motor cortices and thalamus of cats and monkeys. This activity has been observed primarily in intracortical field potential recordings. It ranges in frequency from 15–45 Hz and is most prevalent and highest in amplitude under conditions of alert attentive behavior. Early studies of this phenomenon revealed that, during periods of attentive immobility, the EEG recorded from the somatosensory cortex often shifted from a disorganized pattern of low frequencies to a clearly oscillatory pattern having a dominant frequency around 15 Hz in the monkey[164] and 35–45 Hz in the cat.[32,33] These rhythmic patterns of activity persisted for as long as the animal maintained a state of attentive behavior.

This effect was found to be particularly striking in the cat when the animal was able to view, but not gain access to, a mouse housed in another cage. Under these circumstances the cat displayed a sterotypical pattern of attentive behavior in which it remained completely motionless, its ears pointed forward and its gaze directed towards the mouse.[32] During these periods the EEG in both the somatosensory cortex and ventrobasal thalamus showed a pronounced spectral peak at 35–40 Hz that was coherent between the two structures (Figure 35). These data provided one of the first clear demonstrations of thalamocortical synchrony at a frequency range above that observed in spindle activity and in the alpha rhythm. Subsequently these high-frequency rhythmic activities in the cat have been localized to several areas of the somatosensory cortex[33] and have been found to depend on dopaminergic input from the ventral midbrain for their occurrence.[139]

A similar form of synchronous high-frequency rhythmic activity has recently been observed in the motor cortex of alert macaque monkeys. In studies by two groups, field potentials[49,77,144,145] and multiple unit activity[144,145] were recorded from a number of electrodes simultaneously in different regions of the motor cortical map. In one study, activity was also recorded from the adjacent somatosensory cortex.[144] The animals were trained to perform a stereotyped motor task as well as to perform fine voluntary movements of the hands in the absence of visual guidance.[144] During the periods prior to execution of the trained task, activity of 25–35 Hz was readily apparent in both the field potential and less so in the unit activity. Correlation measurements revealed that the field potentials were synchronous over distances spanning up to 10 mm in motor cortex.[77,144] These signals were found to rapidly abate during the onset and execution of the movement. If, however, the animals were allowed to extract a raisin from a Kluver board without the benefit of visual guidance, a task requiring significant attention, the 25- to 35-Hz activity increased in amplitude, and synchrony could be observed within the motor cortex as well as between the motor and somatosensory cortex (Figure 36), a distance approaching 20 mm in the macaque.[144,145]

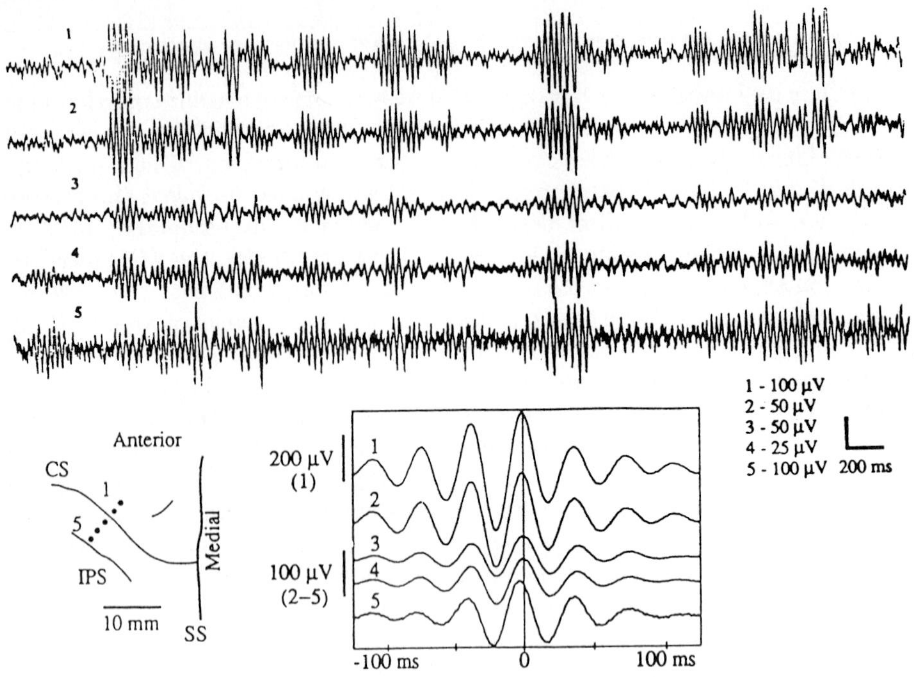

FIGURE 36 Local field potentials recorded simultaneously in five anterior-posterior tracks separated by 2 mm each. Electrode sites, marked on the sketch of the cortical surface (lower left), straddled the central sulcus. Averages of the local field potentials aligned on triggers from oscillatory cycles in trace 1. The monkey was reaching for a raisin offered to the side of its head by the experimenter. (See Figure 4 of Murthy and Fetz,[144] "Coherent 25–35 Hz Oscillations in the Sensorimotor Cortex of the Awake Behaving Monkey," *Proc. Natl. Acad. Sci.* **89** (1992): 5670–5674.) Reprinted by permission.

These data suggest a possible role of synchronous activation of motor and somatosensory cortex during the preparation of planned movement. The clear abatement of the rhythmic activity during the execution of a trained motor act suggests that such activity plays little or no role in learned or overtrained movements. However, during novel tasks requiring a degree of attention, the rhythmic activity is highly synchronous over large regions of somatomotor cortex. Thus, among a number of possible interpretations, it is conceivable that in novel situations the somatomotor cortical map requires a higher degree of temporal coordination in order to execute previously unlearned movements.

CONCLUSIONS

It is clear from the foregoing discussions that synchronized rhythmic activity is a general property of neuronal systems in the mammalian brain. Such activity occurs in a number of different systems, over a range of spatial and temporal scales, during different behavioral states and can be generated by a variety of different mechanisms. At present, the functional significance of such coherent macroscopic states of activity is largely a matter of speculation. And the degree to which these states of activity can be related to particular functions depends on the system under study.

It does appear, however, that the generation of rhythmic activity patterns in groups of neurons can often arise from what otherwise seem to be mechanisms for maintaining stability. For instance, excitatory neurons in the nervous system are almost invariably connected to inhibitory cells. This is a simple and obvious requirement for maintaining stability and preventing runaway excitation. The result of such an anatomical arrangement is often the appearance of rhythmic discharge patterns, a result well known to neuronal network modelers.[66,104,176,201,204] The time course of transmission through the negative feedback circuit determines the frequency of oscillation.[66]

At the cellular level, there is a similar requirement for stability. Excitatory conductances such as fast sodium, persistent sodium, and low- and high-threshold calcium are often accompanied by inhibitory conductances such as delayed rectifier, A current, and calcium-dependent potassium currents. The interplay of these conductances constitute negative feedback circuits at the membrane level that often give rise to intrinsic oscillatory behaviors.[11,18,121,124,173,182]

It comes as no surprise then that the nervous system should exhibit a wide range of oscillatory phenomena expressed at the level of single cells and local networks of cells. The question then arises as to whether such oscillatory activities are simply epiphenomena reflecting the physiological characteristics of neural structures, alternatively, has the nervous system taken advantage of a class of ubiquitous neural mechanisms for specific functional purposes? The fact that oscillatory activities are often associated with or give rise to global macroscopic states of synchronous activity related to specific behaviors lends support to the latter alternative. Perhaps then evolution has taken advantage of the ubiquitous nature of rhythmic activity in the nervous system and has utilized it as a means for forming large-scale coherent patterns of activity. Such patterned states of activity, once they are formed, may have emergent computational capacities that vastly exceed their component neuronal elements. Most functional tasks performed by the nervous system appear to require a means for the integration of information distributed in both space and time. As suggested by anatomy,[58] this requires in turn that large populations of neurons act in concert. Neuronal rhythms may simply be a useful mechanism for dynamically grouping populations of cells into organized assemblies.

ACKNOWLEDGMENTS

This work was supported by grants from the Office of Naval Research, the National Science Foundation, the National Eye Institute, and fellowships from the Klingenstein Foundation and the Sloan Foundation. I thank Lynn Nadel, David McCormick, Paul Rhodes, György Buzsáki, Ted Bullock, and Walter Freeman for their valuable comments on an earlier version of this manuscript.

REFERENCES

1. Adey, W. R., C. W. Dunlop, and C. E. Hendrix. "Hippocampal Slow Waves: Distribution and Phase Relations in the Course of Approach Learning." *Arch. Neurol.* **3** (1960): 74–90.
2. Adrian, E. D. "Olfactory Reactions in the Brain of the Hedgehog." *J. Physiol.* **100** (1942): 459–473.
3. Adrian, E. D. "The Electrical Activity of the Mammalian Olfactory Bulb." *Electroenceph. Clin. Neurophysiol.* **2** (1950): 377–388.
4. Adrian, E. D. "Sensory Discrimination: With Some Recent Evidence from the Olfactory Organ." *Brit. Med. Bull.* **6(4)** (1950): 330–333.
5. Adrian, E. D. "The Basis of Sensation: Some Recent Studies of Olfaction." *Brit. Med. J.* Feb. 6th (1954).
6. Agmon, A., and B. W. Connors. "Repetitive Burst-Firing Neurons in the Deep Layers of Mouse Somatosensory Cortex." *Neurosci. Lett.* **99** (1989): 137–141.
7. Ahissar, E., and E. Vaadia. "Oscillatory Activity of Singel Units in a Somatosensory Cortex of an Awake Monkey and Their Possible Role in Texture Analysis." *Proc. Natl. Acad. Sci.* **87** (1990): 8935–8939.
8. Albright, T. D. "Direction and Orientation Selectivity of Neurons in Visual Area MT of the Macaque." *J. Neurophysiol.* **52(6)** (1984): 1106–1130.
9. Alonso, A., and E. Garcia-Austt. "Neuronal Sources of Theta Rhythm in the Entorhinal Cortex of the Rat. I. Laminar Distribution of Theta Field Potentials." *Exp. Brain Res.* **67** (1987): 493–501.
10. Alonso, A., and E. Garcia-Austt. "Neuronal Sources of Theta Rhythm in the Entorhinal Cortex of the Rat. II. Phase Relations Between Unit Discharges and Theta Field Potentials." *Exp. Brain Res.* **67** (1987): 502–509.
11. Alonso, A., and R. R. Llinas. "Subthreshold Na^+-Dependent Theta-Like Rhythmicity in Stellate Cells of Entorhinal Cortex Layer II." *Nature* **342** (1989): 175–177.
12. Andersen, P., and S. A. Andersson. *Physiolgical Basis of the Alpha Rhythm.* New York: Appelton-Century-Crofts, 1968.
13. Andersen, P., L. Olsen, K. Skrede, and O. Sveen. "Mechanism of the Thalamocortical Rhythmic Activity with Special Reference to the Motor System."

In *Third Symposium on Parkinson's Disease*, edited by F. G. Gillingham and I. M. L. Donaldson, 112–118. Edinburgh: Livingstone, 1969.

14. Ariel, M., N. W. Daw, and R. K. Rader. "Rhythmicity in Rabbit Retinal Ganglion Cell Responses." *Vision Res.* **23(12)** (1983): 1485–1493.
15. Armstrong, D. M., J. C. Eccles, R. J. Harvey, and P. B. S. Matthews. "Responses in the Dorsal Accessory Olive to Stimulation of the Hind Limb Afferents." *J. Physiol.* **194** (1968): 125–145.
16. Arnett, D. W. "Correlation Analysis of Units Recorded in the Cat Dorsal Lateral Geniculate Nucleus." *Exp. Brain Res.* **24** (1975): 111–130.
17. Bair, W., C. Koch, W. Newsome, K. Britten, and E. Niebur. "Power Spectrum Analysis of MT Neurons from Awake Monkey." *Soc. Neurosci. Abs.* **18** (1992): 11–12.
18. Bal, T., and D. McCormick. "Ionic Mechanisms of Rhythmic Burst Firing and Tonic Activity in the Nucleus Reticularis Thalami: A Mammalian Pacemaker." *J. Physiol.* (1992): In press.
19. Becker, C. J., and W. J. Freeman. "Prepyriform Electrical Activity After Loss of Peripheral or Central Input, or Both." *Physiol. & Behav.* **3** (1968): 597–599.
20. Bell, C. C., and T. Kawasaki. "Relations Between Climbing Fiber Responses of Nearby Purkinje Cells." *J. Neurophysiol.* **35** (1972): 155–169.
21. Bergen, J. R., and B. Julesz. "Parallel Versus Serial Processing in Rapid Pattern Discrimination." *Nature* **303** (1983): 696–698.
22. Berger. "Ueber das Elektroenkephalogramm des Menschen." *Arch. Psychiatr. Nervenkr.* **87** (1929): 527–570.
23. Biederman, I. "On the Semantics of a Glance at a Scene." In *Perceptual Organization*, edited by M. Kubovy and J. R. Pomerantz. Hillsdale, NJ: Erlbaum, 1981.
24. Bishop, P. O., W. R. Levick, and W. O. Williams. "Statitical Analyses of the Dark Discharge of Lateral Geniculate Neurons." *J. Physiol.* **170** (1964): 598–612.
25. Bland, B. H., and I. Q. Wishaw. "Generators and Topography of Hippocampal Theta (RSA) in the Anesthetized and Freely Moving Rat." *Brain Res.* **118** (1976): 259–280.
26. Bland, B. H., P. Andersen, and T. Ganes. "Two Generators of Hippocampal Theta Activity in Rabbits." *Brain Res.* **94** (1975): 199–218.
27. Bland, B. H., P. Andersen, T. Ganes, and O. Sveen. "Automated Analysis of Rhythmicity of Physiologically Identified Hippocampal Formation Neurons." *Exp. Brain Res.* **38** (1980): 205–219.
28. Bland, B. H. "The Physiology and Pharmacology of Hippocampal Formation Theta Rhythms." *Prog. in Neurobiol.* **26** (1986): 1–54.
29. Blasdel, G. G., and G. Salama. "Voltage-Sensitive Dyes Reveal a Modular Organization in Monkey Striate Cortex." *Nature* **321** (1986): 579–585.
30. Bliss, T. V. P., and T. Lomo. "Long-Lasting Potentiation of Synaptic Transmission in the Dentate Area of the Anesthetized Rabbit Following Stimulation of the Perforant Path." *J. Physiol (Lond.)* **232** (1973): 331–356.

31. Boeijinga, P. H., and F. H. Lopes da Silva. "Modulations of EEG Activity in the Entorhinal Cortex and Forebrain Olfactory Areas During Odour Sampling." *Brain Res.* **478** (1989): 257–268.
32. Bouyer, J. J., M. F. Montaron, and A. Rougeul. "Fast Fronto-Parietal Rhythms During Combined Focused Attentive Behavior and Immobility in Cat: Cortical and Thalamic Localizations." *Electroenceph. & Clin. Neurophysiol.* **51** (1981): 244–252.
33. Bouyer, J. J., M. F. Montaron, J. M. Vahnee, M. P. Albert, and A. Rougeul. "Anatomical Localization of Cortical Beta Rhythms in Cat." *Neuroscience* **22(3)** (1987): 863–869
34. Bremer, F. "Cerveau Isole et Physiologie du Sommeil." *C. R. Soc. Biol. (Paris)* **118** (1935): 1235–1241.
35. Bressler, S. L., and W. J. Freeman. "Frequency Analysis of Olfactory System EEG in Cat, Rabbit and Rat." *Electroenceph. and Clinical Neurophysiol.* **50** (1980): 19–24.
36. Bressler, S. L. "The Gamma Wave: A Cortical Information Carrier?" *Trends in Neurosci.* **13(5)** (1990): 161–162.
37. Bullier, J., M. H. J. Munk, and L. G. Nowak. "Synchronization of Neuronal Firing in Areas V1 and V2 of the Monkey." *Soc. Neurosci. Abstr.* **18** (1992): 11.7.
38. Bush, P. C., and R. J. Douglas. "Synchronization of Bursting Action Potential Discharge in a Model Network of Neocortical Neurons." *Neural Comp.* **3** (1991): 19–30.
39. Buzsáki, G., L. S. Leung, and C. F. Vanderwolf. "Cellular Bases of Hippocampal EEG in the Behaving Rat." *Brain Res. Rev.* **6** (1983): 139–171.
40. Buzsáki, G. "Hippocampal Sharp Waves: Their Origin and Significance." *Brain Res.* **398** (1986): 242–252.
41. Buzsáki, G. "The Thalamic Clock: Emergent Network Properties." *Neuroscience* **4** (1991): 351–364.
42. Buzsáki, G. "Two-Stage Model of Memory Trace Formation: A Role for 'Noisy' Brain States." *Neurosci.* **31(3)** (1989): 551–570.
43. Buzsáki, G., Z. Horvath, R. Urioste, J. Hetke, and K. Wise. "High-Frequency Network Oscillation in the Hippocampus." *Science* **256** (1992): 1025–1027.
44. Cajal, S. R. *Studies on the Cerebral Cortex (limbic structures)*, translated by L. M. Kraft. London: Lloyd-Luke, 1955.
45. Connors, B. W., M. J. Gutnick, and D. A. Prince. "Electrophysiological Properties of Neocortical Neurons *in Vitro*." *J. Neurophysiol.* **48(6)** (1982): 1302–1320.
46. Connors, B. W. "Initiation of Synchronized Neuronal Bursting in Neocortex." *Nature* **310** (1984): 685–687.
47. Crick, F., and C. Koch. "Towards a Neurobiological Theory of Consciousness." *Sem. in Neurosci.* **2(4)** (1990): 263–275.

48. Desimone, R., and L. G. Ungerleider. "Neural Mechanisms of Visual Processing in Monkeys." In *Handbook of Neuropsychology*, edited by F. Boller and J. Grafman, Vol. 2, Ch. 14. Amsterdam: Elsevier Science Publishers, 1989.
49. Donoghue, J. P., and J. N. Sanes. "Dynamic Modulation of Primate Motor Cortex Output During Movement." *Neurosci. Soc. Abstr.* **17** (1991): 407.5.
50. Doty, R. W., and D. S. Kimura. "Oscillatory Potentials in the Visual System of Cats and Monkeys." *J. Physiol.* **168** (1963): 205–218.
51. Eckhorn, R., R. Bauer, W. Jordan, M. Brosch, W. Kruse, M. Munk, and H. J. Reitboeck. "Coherent Oscillations: A Mechanism of Feature Linking in the Visual Cortex?" *Biol. Cybern.* **60** (1988): 121–130.
52. Eckhorn, R., A. Frien, R. Bauer, T. Woelbern, and H. Kehr. "High-Frequency (60–90 Hz) Oscillations in Primary Visual Cortex of Awake Monkey." *Neuroreport* **4** (1993): 243–246.
53. Engel, A. K., P. Koenig, C. M. Gray, and W. Singer. "Stimulus-Dependent Neuronal Oscillations in Cat Visual Cortex: Inter-Columnar Interaction as Determined by Cross-Correlation Analysis." *Eur. J. Neurosci.* **2** (1990): 588–606.
54. Engel, A. K., P. Koenig, A. K. Kreiter, C. M. Gray, and W. Singer. "Temporal Coding by Coherent Oscillations as a Potential Solution to the Binding Problem: Physiological Evidence." In *Nonlinear Dynamics and Neural Networks*, edited by H. Schuster, 3–25. Germany: VCH Verlag, 1991.
55. Engel, A. K., A. K. Kreiter, P. Koenig, and W. Singer. "Synchronization of Oscillatory Neuronal Responses Between Striate and Extrastriate Visual Cortical Areas of the Cat." *Proc. Natl. Acad. Sci.* **88** (1991): 6048–6052.
56. Engel, A. K., P. Koenig, A. K. Kreiter, and W. Singer. "Interhemispheric Sychronization of Oscillatory Responses in Cat Visual Cortex." *Science* **252** (1991): 1177–1179.
57. Engel, A. K., P. Koenig, and W. Singer. "Direct Physiological Evidence for Scene Segmentation by Temporal Coding." *Proc. Natl. Acad. Sci.* **88** (1991): 9136–9140.
58. Felleman, D. J., and D. C. Van Essen. "Distributed Hierarchical Processing in the Primate Cerebral Cortex." *Cerebral Cortex* **1(1)** (1991): 1–47.
59. Ferster, D. "Orientation Selectivity of Synaptic Potentials in Neurons of Cat Primary Visual Cortex." *J. Neurosci.* **6(5)** (1986): 1284–1301.
60. Ferster, D., and S. LeVay. "The Axonal Arborizations of Lateral Geniculate Neurons in the Striate Cortex of the Cat." *J. Comp. Neurol.* **182** (1978): 923–944.
61. Fox, S. E., and J. B. Ranck. "Electrophysiological Characterisitics of Hippocampal Complex-Spike Cells and Theta Cells." *Exp. Brain Res.* **41** (1981): 399–410.
62. Freeman, W. J. "Distribution in Space and Time of Prepyriform Electrical Activity." *J. Neurophysiol.* **22** (1959): 644–666.
63. Freeman, W. J. "Relations Between Unit Activity and Evoked Potentials in Prepyriform Cortex of Cats." *J. Neurophysiol.* **31** (1968): 337–348.

64. Freeman, W. J. "Depth Recording of Averaged Evoked Potential of Olfactory Bulb." *J. Neurophysiol.* **35** (1972): 780–796.
65. Freeman, W. J. "Average Transmission Distance from Mitral-Tufted to Granule Cells in Olfactory Bulb." *Electroenceph. & Clin. Neurophysiol.* **36** (1974): 609–618.
66. Freeman, W. J. *Mass Action in the Nervous System.* New York: Academic Press, 1975.
67. Freeman, W. J. "Spatial Properties of an EEG Event in the Olfactory Bulb and Cortex." *Electroenceph. & Clin. Neurophysiol.* **44** (1978): 586–605.
68. Freeman, W. J. "Spatial Frequency Analysis of an EEG Event in the Olfactory Bulb." In *Multidisciplinary Perspectives in Event-Related Brain Potential Research,* edited by D. A. Otto, 531–546. U.S. Government Printing Office, Washington, DC, EPA-600/9-77-043, 1978.
69. Freeman, W. J. "Use of Spatial Deconvolution to Compensate for Distortion of EEG by Volume Conduction." *IEEE Trans. Biomed. Engr.* **BME-27(8)** (1980): 421–429.
70. Freeman, W. J. "A Physiological Hypothesis of Perception." In *Perspectives in Biology and Medicine,* 561–592. Chicago, IL: University of Chicago Press, 1981.
71. Freeman, W. J., and W. Schneider. "Changes in Spatial Patterns of Rabbit Olfactory EEG with Conditioning to Odors." *Psychophysiology* **19(1)** (1982): 44–56.
72. Freeman, W. J. "Analytic Techniques Used in the Search for the Physiological Basis for the EEG." In *Handbook of Electroencephalography and Clinical Neurophysiology,* edited by A. Gevins, and A. Remond, Vol 3A, Part 2, Ch. 18. Amsterdam: Elsevier, 1985.
73. Freeman, W. J., and C. A. Skarda. "Spatial EEG Patterns, Nonlinear Dynamics and Perception: The Neo-Sherringtonian View." *Brain Res. Rev.* **10** (1985): 147–175.
74. Freeman, W. J., and G. Viana Di Prisco. "EEG Spatial Pattern Differences with Discriminated Odors Manifest Chaotic and Limit Cycle Attractors in Olfactory Bulb of Rabbits." In *Brain Theory,* edited by G. Palm and A. Aertsen, 97–119. Berlin: Springer-Verlag, 1986.
75. Freeman, W. J. "The Physiology of Perception." *Sci. Am.* **264(2)** (1991): 78–85.
76. Fuster, J. M., A. Herz, and O. D. Creutzfeldt. "Interval Analysis of Cell Discharge in Spontaneous and Optically Modulated Activity in the Visual System." *Arch. Ital. Biol.* **103** (1965): 159–177.
77. Gaal, G., J. N. Sanes, and J. P. Donoghue. "Motor Cortex Oscillatory Neural Activity During Voluntary Movement in Macaca Fascicularis." *Soc. Neurosci. Abs.* **18** (1992): 355.14.
78. Ghose, G. M., and R. D. Freeman. "Oscillatory Discharge in the Visual System: Does it Have a Functional Role?" *J. Neurophysiol.* **68(5)** (1992): 1558–1574.

79. Gilbert, C. D., and T. N. Wiesel. "Morphology and Intracortical Projections of Functionally Characterized Neurons in the Cat Visual Cortex." *Nature* **280** (1979): 120–125.
80. Gilbert, C. D., and T. N. Wiesel. "Clustered Intrinsic Connections in Cat Visual Cortex." *J. Neurosci.* **3(5)** (1983): 1116–1133.
81. Gilbert, C. D., and T. N. Wiesel. "Columnar Specificity of Intrinsic Horizontal and Corticocortical Connections in Cat Visual Cortex." *J. Neurosci.* **9(7)** (1989): 2432–2442.
82. Gray, C. M., W. J. Freeman, and J. E. Skinner. "Chemical Dependencies of Learning in the Rabbit Olfactory Bulb: Acquisition of the Transient Spatial Pattern Change Depends on Norepinephrine." *Behav. Neurosci.* **100(4)** (1986): 585–596.
83. Gray, C., and W. Singer. "Stimulus-Specific Neuronal Oscillations in the Cat Visual Cortex: A Cortical Functional Unit." *Soc. Neurosci. Abs.* **13** (1987): 404.3
84. Gray, C. M., and J. E. Skinner. "Centrifugal Regulation of Neuronal Activity in the Olfactory Bulb of the Waking Rabbit as Revealed by Reversible Cryogenic Blockade." *Exp. Brain Res.* **69** (1988): 378–386.
85. Gray, C. M., and W. Singer. "Stimulus-Specific Neuronal Oscillations in Orientation Columns of Cat Visual Cortex." *Proc. Nat. Acad. Sci.* **86** (1989): 1698–1702.
86. Gray, C. M., P. Koenig, A. K. Engel, and W. Singer. "Stimulus-Specific Neuronal Oscillations in Cat Visual Cortex Exhibit Inter-Columnar Synchronization Which Reflects Global Stimulus Properties." *Nature* **338** (1989): 334–337.
87. Gray, C., A. K. Engel, P. Koenig, and W. Singer. "Stimulus-Dependent Neuronal Oscillations in Cat Visual Cortex: Receptive Field Properties and Feature Dependence." *Eur. J. Neurosci.* **2** (1990): 607–619.
88. Gray, C. M., A. K. Engel, P. Koenig, and W. Singer. "Synchronization of Oscillatory Neuronal Responses in Cat Striate Cortex: Temporal Properties." *Visual Neurosci.* **8** (1992): 337–347.
89. Gray, C. M., A. K. Engel, P. Koenig, and W. Singer. "Mechanisms Underlying the Generation of Neuronal Oscillations in Cat Visual Cortex." In *Induced Rhythmicities in the Brain*, edited by T. Bullock and E. Basar. 1992.
90. Gray, C. M. "Bursting Cells in Visual Cortex Exhibit Properties Characteristic of Intrinsically Bursting Cells in Sensorimotor Cortex." *Soc. Neurosci. Abs.* **18** (1992): 131.2.
91. Green, J. D., and A. Arduini. "Hippocampal Electrical Activity in Arousal." *J. Neurophysiol.* **17** (1954): 533–557.
92. Haberly, L. B., and J. L. Price. "Association and Commissural Fiber Systems of the Olfactory Cortex of the Rat: Systems Originating in the Piriform Cortex and Adjacent Areas." *J. Comp. Neurol.* **178** (1978): 711–740.

93. Haberly, L. B., and J. M. Bower. "Analysis of Association Fiber System in Piriform Cortex with Intracellular Recording and Staining Techniques." *J. Neurophysiol.* **51(1)** (1984): 90–112.
94. Haberly, L. B., and J. M. Bower. "Olfactory Cortex: Model Circuit for Study of Associative Memory?" *TINS* **12(7)** (1989): 258–264.
95. Hebb, D. O. *The Organization of Behavior: A Neuropsychological Theory.* New York: Wiley, 1949
96. Hughes, J. R., D. E. Hendrix, N. S. Wetzel, and J. J. Johnston. "Correlations Between Electrophysiological Activity from the Human Olfactory Bulb and the Subjective Response to Odoriferous Stimuli." In *Olfaction and Taste III*, edited by C. Pfaffman. New York: Rockefeller, 1969.
97. Intraub, H. "Identification and Naming of Briefly Glimpsed Visual Scenes." In *Eye Movements: Cognition and Visual Perception*, edited by D. F. Fisher, R. A. Monty and J. W. Senders. Hillsdale, NJ: L. Erlbaum, 1981.
98. Ito, M. *The Cerebellum and Neuronal Control*, 100–102. New York: Raven, 1984.
99. Jagadeesh, B., C. M. Gray, and D. Ferster. "Visually-Evoked Oscillations of Membrane Potential in Neurons of Cat Striate Cortex Studied with *In Vivo* Whole Cell Patch Recording." *Science* **257** (1992): 552–554.
100. Jahnsen, H., and R. Llinas. "Electrophysiological Properties of Guinea-Pig Thalamic Neurons: An *In Vitro* Study." *J. Physiol.* **349** (1984): 205–226.
101. Jahnsen, H., and R. Llinas. "Ionic Basis for the Electroresponsiveness and Oscillatory Properties of Guinea-Pig Thalamic Neurons *In Vitro*." *J. Physiol.* **349** (1984): 227–247.
102. Jung, R., and A. Kornmuller. "Eine Methodik der Abteilung Lokalisierter Potential Schwankingen aus Subcorticalen Hirnyebieten." *Arch. Psychiat. Neruenkr.* **109** (1938): 1–30.
103. Kishi, K., K. Mori, and H. Ojima. "Distribution of Local Axon Collaterals of Mitral, Displaced Mitral, and Tufted Cells in the Rabbit Olfactory Bulb." *J. Comp. Neurol.* **225** (1984): 511–526.
104. Koenig, P., and T. B. Schillen. "Stimulus Dependent Assembly Formation of Oscillatory Responses: I. Synchronization." *Neural Comp.* **3** (1991): 155–166.
105. Kreiter, A. K., and W. Singer. "Oscillatory Neuronal Responses in the Visual Cortex of the Awake Macaque Monkey." *Eur. J. Neurosci.* **4** (1992): 369–375.
106. Kreiter, A. K., A. K. Engel, and W. Singer. "Stimulus Dependent Synchronization in the Caudal Superior Temporal Sulcus of Macaque Monkeys." *Soc. Neurosci. Abs.* **18** (1992): 11.11.
107. Kuperstein, M., H. Eichenbaum, and T. VanDeMark. "Neural Group Properties in the Rat Hippocampus During the Theta Rhythm." *Exp. Brain Res.* **61** (1986): 438–442.
108. Larkman, A., and A. Mason. "Correlations Between Morphology and Electrophysiology of Pyramidal Neurons in Slices of Rat Visual Cortex. I. Establishment of Cell Classes." *J. Neurosci.* **10(5)** (1990): 1407–1414.

109. Larson, J., D. Wong, and G. Lynch. "Patterned Stimulation at the Theta Frequency is Optimal for the Induction of Hippocampal Long Term Potentiation." *Brain Res.* **368** (1986): 347–350.
110. Laufer, M., and M. Verzeano. "Periodic Activity in the Visual System of the Cat." *Vision Res.* **7** (1967): 215–229.
111. Leung, L. S. and C. C. Yim. "Intrinsic Membrane Potential Oscillations in Hippocampal Neurons *in vitro*." *Brain Resh.* **553** (1991): 261–274.
112. Leung, L. S. "Fast (Beta) Rhythms in the Hippocampus: A Review." *Hiipocampus* **2(2)** (1992): 93–98.
113. Libet, B., and R. W. Gerard. "Control of the Potential Rhythm of the Isolated Frog Brain." *J. Neurophysiol.* **2** (1939): 153–169.
114. Livingstone, M. S. "Visually Evoked Oscillations in Monkey Striate Cortex." *Soc. Neurosci. Abs.* **17** (1991): 73.3.
115. Livingstone, M. S., and D. H. Hubel. "Segregation of Form, Color, Movement, and Depth: Anatomy, Physiology, and Perception." *Science* **240** (1988): 740–749.
116. Livingstone, M. S. Personal communication.
117. Llinas, R., R. Baker, and C. Sotelo. "Electrotonic Coupling Between Neurons in Cat Inferior Olive." *J. Neurophysiol.* **37** (1974): 560–571.
118. Llinas, R., and Y. Yarom. "Electrophysiology of Mammalian Inferior Olivary Neurons *In Vitro*. Different Types of Voltage Dependent Ionic Conductances." *J. Physiol.* **315** (1981): 549–567.
119. Llinas, R., and Y. Yarom. "Properties and Distribution of Ionic Conductances Generating Electroresponsiveness of Mammalian Inferior Olivary Neurons *In Vitro*." *J. Physiol.* **315** (1981): 569–584.
120. Llinas, R. "Rebound Excitation as the Physiological Basis for Tremor: A Biophysical Study of the Oscillatory Properties of Mammalian Central Neurons *in vitro*." In *Movement Disorders: Tremor*, edited by L. J. Findley and R. Capildeo, 165–181. McMillan, 1984.
121. Llinas, R., and Y. Yarom. "Oscillatory Properties of Guinea-Pig Inferior Olivary Neurons and Their Pharmacological Modulation: An *In Vitro* Study." *J. Physiol.* **376** (1986): 163–182.
122. Llinas, R. R. "The Intrinsic Electrophysiological Properties of Mammalian Neurons: Insights into Central Nervous System Function." *Science* **242** (1988): 1654–1664.
123. Llinas, R., and K. Sasaki. "The Functional Organization of the Olivo-Cerebellar System as Examined by Multiple Purkinje Cell Recordings." *Eur. J. Neurosci.* **1** (1989): 587–602
124. Llinas, R. R., A. A. Grace, and Y. Yarom. "*In Vitro* Neurons in Mammalian Cortical Layer 4 Exhibit Intrinsic Oscillatory Activity in the 10- to 50-Hz Frequency Range." *Proc. Natl. Acad. Sci.* **88** (1991): 897–901.

125. Lund, J. S., G. H. Henry, C. L. MacQueen, and A. R. Harvey. "Anatomical Organization of the Primary Visual Cortex (Area 17) of the Cat: A Comparison with Area 17 of the Macaque Monkey." *J. Comp. Neurol.* **184** (1979): 599–618.
126. Macrides, F., H. B. Eichenbaum, and W. B. Forbes. "Temporal Relationship Between Sniffing and the Limbic Theta Rhythm During Odor Discrimination Reversal Learning." *J. Neurosci.* **2(12)** (1982): 1705–1717.
127. Maekawa, K., and D. P. Purpura. "Intracellular Study of Lemniscal and Nonspecific Synaptic Interactions in Thalamic Ventrobasal Neurons." *Brain Res.* **4** (1967): 308–323.
128. Magoun, H. W. *The Waking Brain.* Springfield IL: Charles C. Thomas Pub., 1958.
129. Marshall, J., and E. G. Walsh. "Physiological Tremor." *J. Neurol. Neurosurg. Psychiatry* **19** (1956): 260.
130. Martin, K. A. C. "Neuronal Circuits in Cat Striate Cortex." In *Cerebral Cortex: Functional Properties of Cortical Cells*, edited by E. G. Jones and A. Peters, Vol. 2. New York: Plenum Press, 1984.
131. Martin, K. A. C., and D. Whitteridge. "Form, Function and Intracortical Projections of Spiny Neurons in the Striate Visual Cortex of the Cat." *J. Physiol.* **353** (1984): 463–504.
132. McCormick, D. A., B. W. Connors, J. W. Lighthall, and D. A. Prince. "Comparative Electrophysiology of Pyramidal and Sparsely Spiny Stellate Neurons of the Neocortex." *J. Neurophysiol.* **54(4)** (1985): 782–806.
133. McCormick, D. A., and H. Pape. "Properties of a Hyperpolarization-Activated Cation Current and Its Role in Rhythmic Oscillation in Thalamic Relay Neurones." *J. Physiol.* **431** (1990): 291–318
134. McCormick. D. A., J. Huguenard, and B. W. Strowbridge. "Determination of State-Dependent Processing in Thalamus by Single Neuron Properties and Neuromodulators." In *Single Neuron Computation. Neural Nets: Foundations to Applications.* New York: Academic Press, 1992.
135. McNaughton, B. L., R. M. Douglas, and G. V. Goddard. "Synaptic Enhancement in Fascia Dentata: Cooperativity Among Coactive Afferents." *Brain Res.* **157** (1978): 277–293.
136. Milner, B. "Amnesia Following Operation on the Temporal Lobes." In *Amnesia*, edited by C. W. M. Whitty and O. L. Zangwill, 109–133. London: Butterworths, 1966.
137. Milner, P. "A Model for Visual Shape Recognition." *Psychol. Rev.* **81(6)** (1974): 521–535.
138. Mitchell, S. J., and J. B. Ranck. "Generation of Theta Rhythm in Medial Entorhinal Cortex of Freely Moving Rats." *Brain Res.* **178** (1980): 49–66.
139. Montaron, M., J. Bouyer, A. Rougeul, and P. Buser. "Ventral Mesencephalic Tegmentum (VMT) Controls Electrocortical Beta Rhythms and Associated Attentive Behaviour in the Cat." *Behav. Brain Res.* **6** (1982): 129–145

140. Mori, K., and K. Kishi. "The Morphology and Physiology of the Granule Cells in the Rabbit Olfactory Bulb Revealed by Intracellular Recording and HRP Injection." *Brain Res.* **247** (1982): 129–133.
141. Morison, R. S., and D. L. Bassett. "Electrical Activity of the Thalamus and Basal Ganglia in Decorticate Cats." *J. Neurophysiol.* **8** (1945): 309–314.
142. Mulle, C., M. Steriade, and M. Deschenes. "Ansence of Spindle Oscillations in the Cat Anterior Thalamic Nuclei." *Brain Res.* **334** (1985): 169–171.
143. Munemori, J., K. Hara, M. Kimura, and R. Sato. "Statistical Features of Impulse Trains in Cat's Lateral Geniculate Neurons." *Biol. Cybern.* **50** (1984): 167–172.
144. Murthy, V. N., and E. E. Fetz. "Coherent 25–35 Hz Oscillations in the Sensorimotor Cortex of the Awake Behaving Monkey." *Proc. Natl. Acad. Sci.* **89** (1992): 5670–5674
145. Murthy, V. N., D. F. Chen, and E. E. Fetz. "Spatial Extent and Behavioral Dependence of Coherence of 25–35 Hz Oscillations in Primate Sensorimotor Cortex." *Soc. Neurosci. Abs.* **18** (1992): 355.12.
146. Nakashima, M., K. Mori, and S. Takagi. "Centrifugal Influence on Olfactory Bulb Activity in the Rabbit." *Brain Res.* **154** (1978): 301–316.
147. Nelson, J. I., L. G. Nowak, G. Chouvet, M. H. J. Munk, and J. Bullier. "Synchronization Between Cortical Neurons Depends on Activity in Remote Areas." *Soc. Neurosci. Abs.* **18** (1992): 11.8.
148. Nicoll, R. A. "Recurrent Excitation of Secondary Olfactory Neurons: A Possible Mechanism for Signal Amplification." *Science* **171** (1971): 824–826.
149. Nicoll, R. A., and C. E. Jahr. "Self Excitation of Olfactory Bulb Neurons." *Nature* **296** (1982): 441–444.
150. Nunez, A., E. Garcia-Austt, and W. Buno. "Intracellular Theta-Rhythm Generation in Identified Hippocampal Pyramids." *Brain Res.* **416** (1987): 289–300.
151. Orona, E., E. C. Rainer, and J. W. Scott. "Dendritic and Axonal Organization of Mitral and Tufted Cells in the Rat Olfactory Bulb." *J. Comp. Neurol.* **226** (1984): 346–356.
152. Ottoson, D. "Studies on Slow Potentials in the Rabbit's Olfactory Bulb and Nasal Mucosa." *Acta. Physiol. Scand.* **47** (1959): 136–148.
153. Pantev, C., S. Makeig, M. Hoke, R. Galambos, S. Hampson, and C. Galen. "Human Auditory Evoked Gamma-Band Magnetic Fields." *Proc. Natl. Acad. Sci.* **88** (1991): 8996–9000.
154. Pare, D., M. Steriade, M. Deschenes, and G. Oakson. "Physiological Properties of Anterior Thalamic Nuclei, a Group Devoid of Inputs from the Reticular Thalamic Nucleus." *J. Neurophysiol.* **57** (1987): 1669–1685.
155. Pavlides, C., Y. J. Greenstein, M. Grudman, and J. Winson. "Long-Term Potentiation in the Dentate Gyrus is Induced Preferentially on the Positive Phase of Theta Rhythm." *Brain Res.* **439** (1988): 383–387.

156. Petsche, H., G. Stumpf, and G. Gogolak. "The Significance of the Rabbit's Septum as a Relay Station Between the Midbrain and the Hippocampus." *Electroenceph. Clin. Neurophysiol.* **19** (1962): 25–33.
157. Petsche, H., G. Gogolak, and P. A. Van Zwieten. "Rhythmicity of Septal Cell Discharges at Various Levels of Reticular Excitation. Electroenceph." *Clin. Neurophysiol.* **19** (1965): 25–33.
158. Pinault, D., and M. Deschenes. "Voltage-Dependent 40–Hz Oscillations in Rat Reticular Thalamic Neurons." *Neurosci.* **51(2)** (1992): 245–258.
159. Potter, M. C. "Meaning in Visual Search." *Science* **187** (1975): 965–966.
160. Potter, M. C. "Short-Term Conceptual Memory for Pictures." Special Issue: Human Learning and Memory. *J. Exp. Psychol.* **2(5)** (1976): 509–522.
161. Rall, W., G. M. Shepherd, T. S. Reese, and M. W. Brightman. "Dendrodendritic Synaptic Pathway for Inhibition in the Olfactory Bulb." *Exp. Neurol.* **14** (1966): 44–56.
162. Rall, W., and G. M. Shepherd. "Theoretical Reconstruction of Field Potentials and Dendrodendritic Synaptic Interactions in Olfactory Bulb." *J. Neurophysiol.* **31** (1968): 884–915.
163. Robson, J. G., and J. B. Troy. "Nature of the Maintained Discharge of Q, X, and Y Retinal Ganglion Cells of the Cat." *J. Opt. Soc. Am. A.* **4(12)** (1987): 2301–2307.
164. Rougeul, A., J. J. Bouyer, L. Dedet. and O. Debray. "Fast Somato-Parietal Rhythms During Combined Focal Attention and Immobility in Baboon and Squirrel Monkey." *Electro. Clin. Neurophysiol.* **46** (1979): 310–319.
165. Sakai, H. M., and K. Naka. "Dissection of the Neuron Network in the Catfish Inner Retina V. Interactions Between NA and NB Amacrine Cells." *J. Neurophysiol.* **63(1)** (1990): 120–130.
166. Salin, P. A., J. Bullier, and H. Kennedy. "Convergence and Divergence in the Afferent Projections to Cat Area 17." *J. Comp. Neurol.* **283** (1988): 486–512.
167. Sasaki, K., J. M. Bower, and R. Llinas. "Multiple Purkinje Cell Recording in Rodent Cerebellar Cortex." *Eur. J. Neurosci.* **1** (1989): 572–586
168. Scheibel, M. E., and A. B. Scheibel. "The Organization of the Nucleus Reticularis Thalami: A Golgi Study." *Brain Res.* **1** (1966): 43–62.
169. Schwarz, C., and J. Bolz. "Functional Specificity of a Long-Range Horizontal Connection in Cat Visual Cortex: A Cross-Correlation Study." *J. Neurosci.* **11(10)** (1991): 2995–3007
170. Schoenfeld,T. A., J. E. Marchand, and F. Macrides. "Topographic Organization of Tufted Cell Axonal Projections in the Hamster Main Olfactory Bulb: An Intrabulbar Associational System." *J. Comp. Neurol.* **235** (1985): 503–518.
171. Sereno, M. I., and J. M. Allman. "Cortical Visual Areas in Mammals." In *The Neural Basis of Visual Function*, edited by A. Leventhal, 160–172. MacMillan, 1990.
172. Shepherd, G. M. "Synaptic Organization of the Mammalian Olfactory Bulb." *Physiol. Rev.* **52** (1972): 864–917.

173. Silva, L. R., Y. Amitai, and B. W. Connors. "Intrinsic Oscillations of Neocortex Generated by Layer 5 Pyramidal Neurons." *Science* **251** (1990): 432–435.
174. Singer, W. "Search for Coherence." *Concepts in Neurosci.* **1(1)** (1990).
175. Sompolinsky, H., D. Golomb, and D. Kleinfeld. "Global Processing of Visual Stimuli in a Neural Network of Coupled Oscillators." *Proc. Natl. Acad. Sci.* **87** (1990): 7200–7204.
176. Sporns, O., J. A. Gally, G. N. Reeke, and G. M. Edelman. "Reentrant Signaling Among Simulated Neuronal Groups Leads to Coherency in Their Oscillatory Activity." *Proc. Natl. Acad. Sci.* **86** (1989): 7265–7269.
177. Sporns, O., G. Tononi, and G. M. Edelman. "Modeling Perceptual Grouping and Figure-Ground Segregation by Means of Active Reentrant Connections." *Proc. Natl. Acad. Sci.* **88** (1991): 129–133.
178. Stanton, P. K., and T. J. Sejnowski. "Associative Long-Term Depression in the Hippocampus Induced by Hebbian Covariance." *Nature* **339** (1989): 215–218.
179. Steriade, M., A. Parent, and J. Hada. "Thalamic Projections of Nucleus Reticularis Thalami of Cat: A Study Using Retrograde Transport of Horseradish Peroxidase and Double Flourescent Tracers." *J. Comp. Neurol.* **229** (1984): 531–547.
180. Steriade, M., M. Deschenes, L. Domich, and C. Mulle. "Abolition of Spindle Oscillations in Thalamic Neurons Disconnected from Nucleus Reticularis Thalami." *J. Neurophysiol.* **54** (1985): 1473–1497.
181. Steriade, M., L. Domich, G. Oakson, and M. Deschenes. "The Deafferented Reticularis Thalami Nucleus Generates Spindle Rhythmicity." *J. Neurophysiol.* **57** (1987): 260–273
182. Steriade, M., E. G. Jones, and R. R. Llinas. *Thalamic Oscillations and Signaling.* New York: John Wiley, 1990.
183. Steriade, M., R. Curro Dossi, D. Pare, and G. Oakson. "Fast Oscillations (20-40 Hz) in Thalamocortical Systems and Their Potentiation by Mesopontine Cholinergic Nuclei in the Cat." *Proc. Natl. Acad. Sci.* **88** (1991): 4396–4400.
184. Tamamaki, N., K. Abe, and Y. Nojyo. "Three-Dimensional Analysis of the Whole Axonal Arbors Originating from Single CA2 Pyramidal Neurons in the Rat Hippocampus with the Aid of a Computer Graphic Technique." *Brain Res.* **452** (1988): 255–272.
185. Thommesen, G. "The Spatial Distribution of Odor-Induced Potentials in the Olfactory Bulb of Char and Trout (Salmonidae)." *Acta Physiol. Scand.* **102** (1978): 205–217.
186. Tombol, T., and H. Petsche. "The Histological Organization of the Pacemaker for the Hippocampal Theta Rhythm in the Rabbit." *Brain Res.* **12** (1969): 414–426.
187. Tovee, M. J., and E. T. Rolls. "Oscillatory Activity is Not Evident in the Primate Temporal Visual Cortex with Static Stimuli." *Neuroreport* **3** (1992): 369–372.

188. Traub, R. D., R. Miles, and R. K. S. Wong. "Model of the Origin of Rhythmic Population Oscillations in the Hippocampal Slice." *Science* **243** (1989): 1319–1325.
189. T'so, D. Y., C. D. Gilbert, and T. N. Wiesel. "Relationships Between Horizontal Interactions and Functional Architecture in Cat Striate Cortex as Revealed by Cross-Correlation Analysis." *J. Neurosci.* **6(4)** (1986): 1160–1170.
190. T'so, D. Y., R. D. Frostig, E. E. Lieke, and A. Grinvald. "Functional Organization of Primate Visual Cortex Revealed by High Resolution Optical Imaging." *Science* **249** (1990): 417–420.
191. Ungerleider, L. G., and M. Mishkin. "Two Cortical Visual Systems." In *Analysis of Visual Behavior*, edited by D. J. Ingle, M. A. Goodale, and R. J. W. Mansfield, 549–586. Cambridge, MA: MIT Press, 1982.
192. Vanderwolf, C. H. "Hippocampal Electrical Activity and Voluntary Movement in the Rat." *Electroenceph. Clin. Neurophysiol.* **26** (1969): 407–418.
193. Viana Di Prisco, G., and W. J. Freeman. "Odor-Related Bulbar EEG Spatial Pattern Analysis During Appetitive Conditioning in Rabbits." *Behav. Neurosci.* **99(5)** (1985): 964–978.
194. Villablanca, J. "Role of the Thalamus in Sleep Control: Sleep Wakefulness Studies in Chronic Diencephalic and Athalamic Cats." In *Basic Sleep Mechanisms*, edited by O. Petre-Quadens and J. Schlag, 51–81. New York: Academic, 1974.
195. von der Malsburg, C. "The Correlation Theory of Brain Function." Internal Report, Max-Planck-Institute for Biophysical Chemistry, Gottingen, West Germany, 1981.
196. von der Malsburg, C. "Nervous Structures with Dynamical Links." *Ber. Bunsenges. Phys. Chem.* **89** (1985): 703–710.
197. von der Malsburg, C. "Am I Thinking Assemblies?" In *Brain Theory*, edited by G. Palm and A. Aertsen, 161–176. Berlin: Springer Verlag, 1986.
198. von der Malsburg, C., and W. Schneider. "A Neural Cocktail-Party Processor." *Biol. Cybern.* **54** (1986): 29–40.
199. von Krosigk, M., T. Bal, and D. A. McCormick. "Cellular Mechanisms of a Synchronized Oscillation in the Thalamus." *Science*: submitted.
200. Willey, T. J. "The Ultrastructure of the Cat Olfactory Bulb." *J. Comp. Neurol.* **152** (1973): 211–232.
201. Wilson, H. R., and J. D. Cowan. "Excitatory and Inhibitory Interactions in Localized Populations of Model Neurons." *Biophys. J.* **12** (1972): 1–24.
202. Wilson, M., and J. M. Bower. "Cortical Oscillations and Temporal Interactions in a Computer Simulation of Piriform Cortex." *J. Neurophysiol.* **67(4)** (1992): 981–995
203. Wilson, M., B. L. McNaughton, and K. Stengel. "Large-Scale Parallel Recording of Multiple Single Unit Activity in the Hippocampus and Parietal Cortex of the Behaving Rat." *Soc. Neurosci. Abs.* **18** (1992): 508.8.
204. Wilson, M. A., and J. M. Bower. "Computer Simulations of OScillatory Behavior in Visual Cerebral Cortex." *Neural Comp.* **3** (1989): 415–419.

205. Yarom, Y. "Oscillatory Behavior of Olivary Neurons." In *The Olivocerebellar System in Motor Control*, edited by Piergiorgio Strata. Berlin: Springer-Verlag, 1989.

206. Young, M. P., K. Tanaka, and S. Yamane. "On Oscillating Neuronal Responses in the Visual Cortex of the Monkey." *J. Neurophysiol.* **67(6)** (1992): 1464–1474.

Tad Hogg and Bernardo A. Huberman
Xerox Palo Alto Research Center, Palo Alto, California 94304 USA

Better Than the Best: The Power of Cooperation

We show that when agents cooperate in a distributed search problem, they can solve it faster than any agent working in isolation. This is accomplished by having agents exchange hints within a computational ecosystem. We present a quantitative assessment of the value of cooperation for solving constraint satisfaction problems through a series of experiments. Our results suggest an alternative method to existing techniques for solving these problems in computer science and distributed artificial intelligence.

1. INTRODUCTION

The development of parallelism in computation allows a number of agents to share the work required to solve computational problems. The potential speedup offered by this approach has led to a large effort devoted to the design of parallel algorithms and architectures. In spite of its obvious advantages however, the effective use of concurrency is fraught with difficulties. Most of these difficulties stem from the fact that the experience gained from programming single processor machines cannot be simply extrapolated to large number of computational agents, because parallel computing involves a number of new issues: how tasks can be usefully divided among many agents, how one program can exploit the knowledge generated by another, and how the agents can communicate efficiently with each other. These issues are particularly important for large-scale distributed processing in which individual agents operate largely without central controls. If the task is easily decomposed into fairly independent subtasks, requiring little communication, a parallel implementation is relatively easy. However, this is not always possible and eliminates the possibility of using communication to significantly help with individual subtasks.

Some insights into how these issues can be effectively addressed for more complex cases can be gained from studying the way human societies solve problems of collective interest. Although the individuals differ from these computational agents in many important aspects, they nevertheless face the same general problems of coordination and communication described above. In human societies, the benefit of cooperation underlies the existence of firms, scientific and professional communities, and the use of committees charged with solving particular problems. Often groups of people can solve a problem more effectively than any single individual acting alone. This suggests implementing, in a computational context, the mechanisms that seem to work among humans.

The existence of computational ecologies[10] provides a natural framework for these methods because they share a number of key features in common with human societies. These include asynchronous independent agents that solve problems from their local perspective involving uncertain and delayed information that they can retrieve from the system. A number of attempts at collective problem solving from this perspective have been made. These include work by several authors who have pointed out the beneficial effects of cooperation on hard problems by constructing models in which agents cooperate to accomplish a task.[2,5,8,14,19]

In a computational context, cooperation involves a collection of agents that interact by communicating information, or *hints*, to each other while solving a problem. The most natural way to think of cooperation is as a collection of independent processes, possibly running on separate processors. However, it is always possible to have a single computational process that, in effect, multiplexes among the procedures followed by this diverse set of agents. In this way, a single agent could also obtain the benefit of cooperation discussed here. This ability of one computational process to emulate a collection of other processes is quite distinct from other cases of cooperation, e.g., human societies, where individuals have different skills that are not easily transferred to others. Most important to increasing in performance is the diversity of approaches available by having many agent processes.

The information exchanged between agents may be incorrect at times and should alter the behavior of the agents receiving it. An example of cooperative problem solving is the use of a genetic algorithm[6] to find states of high value in some space of possibilities. In a genetic algorithm, members of a population of states exchange pieces of themselves or mutate to create a new population, often containing states of high value. In a neural network, the output of one neuron affects the behavior of the neuron receiving it.

We will concentrate on a particular type of computational task—searching, an important general task arises that when no algorithmic method is known for directly constructing a solution. Instead, one must examine a large number of alternative candidate states to identify a satisfactory solution. Typically, the number of considered states grows exponentially as larger problems are considered, making these problems considerably more difficult than many numerical operations such as linear algebra or solution of differential equations whose computational cost generally grows as a low-degree polynomial as problems scale up. Because of the huge number of considered states in a search, many heuristic methods have been developed to guide the selection of states. While not always correct, they can considerably reduce the time required to find a solution, by guiding the search toward states that are more likely to lead to solutions. Most heuristics are meant to improve individual searches. By contrast, the cases that we will discuss highlight the potential of cooperative methods that can be thought of as heuristics in which information obtained by one agent is used to guide the search of another. We also present a number of more practical issues that arise in applying cooperation to problems in computer science and distributed artificial intelligence.[5]

As a concrete illustration of the value of cooperation for the search, we solve discrete constraint satisfaction problems in which values from a finite set must be assigned to a finite set of variables such that a number of conditions (the constraints) are satisfied. Constraint satisfaction problems lie at the heart of human and computer problem solving.[13,16,18] Examples are scheduling, navigating through a maze, and crossword puzzles, to name a few. A *complete state* in the search is a set of assignments for all the variables and a *partial state* has only some of the variables assigned.

To evaluate the usefulness of cooperation in computational problems, we examine its behavior for two specific problems. At one extreme, a cryptarithmetic problem with a simple individual search method shows how even very simple methods can benefit from an exchange of information. By contrast, our second example, graph coloring, is a computationally hard problem that illustrates how simple hints can be used in conjunction with an effective heuristic search method.

2. COOPERATIVE SEARCHES

The success of cooperation may be explained by observing that hints change the way different agents find the solution by combining these hints with their own current state. Although not always successful, those cases in which hints do combine well allow the agent to proceed to a solution by searching in a reduced space of possibilities. Even if many of the hints are not successful, this results in a larger variation of performance and hence can still improve the performance of the group when measured by the time it takes for the first agent to finish.

The speed at which an agent can solve the problem depends on the initial conditions and the particular sequence of actions it chooses as it moves through a search space. This sequence relies on the knowledge, or heuristics, that an agent has about which state should be examined next. The better the agent uses the heuristics, the quicker it will solve the problem. When many agents work on the same problem, this knowledge can include hints from other agents suggesting where solutions are likely to be.

Cooperative search methods are based on modifying individual search methods. A useful distinction is whether a method is *complete* or *incomplete*. Complete methods systematically examine states and are guaranteed to either eventually find a solution or terminate when no solution exists. By contrast, incomplete methods explore more opportunistically and may miss some states in the search space; hence, they can never guarantee that a solution does not exist. For parallel searches, a further issue is whether to split the search space among the agents. In the simplest case, each agent examines the entire search space. However, this can mean a single state is examined by more than one agent during the search. This can be avoided by partitioning the search space into disjoint parts and assigning one to each agent. In this partitioned search, agents only examine states in their assigned part of the space, thus avoiding unnecessary duplicate examination of the states. Restricting each agent to examine a state at most once and partitioning the search space so that a state is not examined by more than one agent improve performance somewhat, but far less than the enhancement due to cooperation.[3]

2.1 THE USE OF HINTS

There are a number of search methods an individual agent can use to solve a problem, as well as a variety of methods for combining the partial information obtained from other agents. These choices determine if hints build on each other and, if so, how the search improves.

2.1.1 SEARCHING WITH COMPLETE STATES. The most straightforward search method is "generate and test." In this case, at each step an agent generates a complete state and tests whether it is a solution. This generation can be done in a simple pre-specified order or new states can be generated randomly. In random generation, states can be selected completely at random (which we refer to as random selection with replacement) or the selection can be made only from states that have not yet been examined. The latter case avoids some unnecessary searching and guarantees the search will terminate after all search states are examined, but does introduce an additional requirement of storing previously examined states and the cost of checking that they are not subsequently generated.

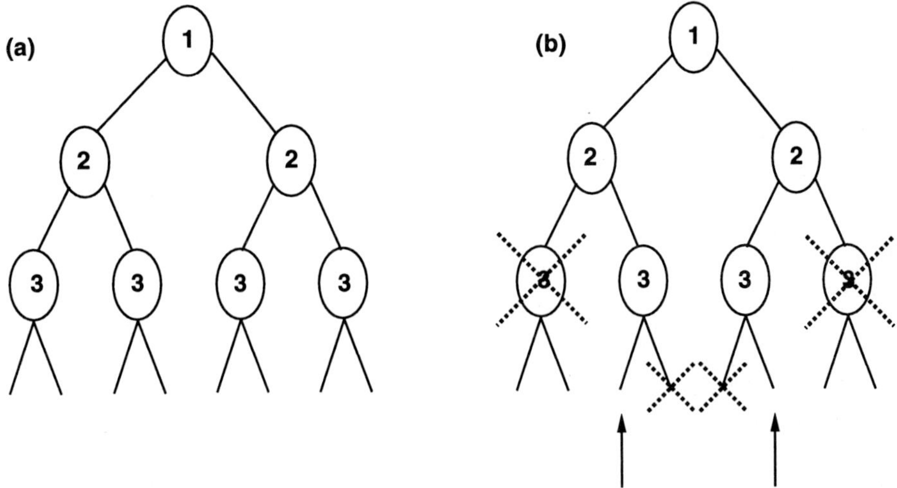

FIGURE 1 (a) Illustration of the tree-structured search space resulting from three variables, v_1, v_2, and v_3, (corresponding to the nodes in the tree) each with two possible values (corresponding to the branches). Searches generally start at the top of the tree and examine successive branches until a complete state (corresponding to a leaf, at the bottom of the tree) that satisfies the constraints is found. (b) The pruning of the search tree for the constraint problem $\{v_1 \neq v_2, v_2 \neq v_4\}$. The crosses indicate those states that violate one or more of the constraints. The arrows point to the leaves corresponding to solutions, i.e., $\{v_1 = 1, v_2 = 2, v_3 = 1\}$ and $\{v_1 = 2, v_2 = 1, v_3 = 2\}$.

Other restrictions on the generation of new states are possible as well. For instance, the assignments to all the variables can be replaced in one step (which we refer to as "jumping" around the search space) or some assignments can remain unchanged, with the extreme case being a change to only a single assignment ("walking"). Walking rather than jumping through the space preserves the property that an agent near or far from a solution is still fairly near or far after one step.

There are more sophisticated methods that share the same basic strategy, i.e., start from some randomly selected initial state and attempt to make a series of zero or more small adjustments to the state, attempting to satisfy all the constraints. If these adjustments do not produce a solution, a new initial state is selected. Examples of this strategy include simulated annealing,[12] heuristic repair,[17] and simple hill climbing. By contrast, generate and test makes no adjustments and simply tests the initial state itself.

With this search strategy, if hints are only used to guide the selection of the initial state and each new hint completely overwrites the old state, there will be no build up of progress from one hint to another. Alternatively, if the new hint modifies just part of the state, then successive hints could correspond to a kind of random walk in the state space in which there is (at least for the lucky agents) an overall bias to move successive initial states closer to a solution that is eventually found by the local adjustments.

2.1.2 CONSTRUCTING SOLUTIONS FROM PARTIAL STATES. Other search methods rely on a more systematic exploration of the space, attempting to construct a complete solution by incrementally extending partial solutions. Combined with some backtrack scheme when further progress is impossible, such a hierarchical construction of a solution, allows for pruning regions of the search space that would be unproductive. With this depth-first search method, some ordering of the variables is selected (e.g., either fixed in advance or chosen randomly) and partial states are constructed using this ordering until a full solution is found or enough assignments are made to violate one of the constraints, indicating that no solution corresponds to this partial state. Where these constraint violations occur well before all assignments have been made, backtracking avoids a considerable amount of unnecessary search.

A simple illustration of the resulting tree-structured search is shown in Figure 1(a). Specifically this is for a constraint problem with three variables, v_1, v_2, and v_3, each of which can be value 1 or 2. The nodes in the tree represent the variables, and the links from a node represent the two choices for the values to assign to that node's variable (corresponding to the value 1 for the left branch and 2 for the right branch). The leaves of the tree correspond to complete search states in which each variable has a value. For example, the leftmost leaf corresponds to the assignments $\{v_1 = 1, v_2 = 1, v_3 = 1\}$. Partial states, in which some variables are not assigned, are found higher in the tree (in the ordering illustrated, v_1 is assigned first, v_2 next, and v_3 last). Adding consideration of these partial states means that these search methods could potentially examine more states than the those that use only complete states. However, this increase in total states is usually more than offset by the ability to prune, high in the tree, many states at one time. This pruning is illustrated in Figure 1(b) for the constraints $\{v_1 \neq v_2, v_2 \neq v_3\}$; i.e., the values for the first two variables, and the last two, are required to be different. For example, the leftmost pruned node is due to the partial state $\{v_1 = 1, v_2 = 1\}$ that already violates the first constraint so there is no need to consider possible values for the third variable.

These basic methods can be improved with the use of heuristics to guide the selection of states. An important class of heuristics uses information obtained in prior steps of the search. Such heuristics are readily modified for cooperative search and allow us to directly evaluate the effect of cooperation. Specifically, in a noncooperative search, an agent using such a method could only use information that it had found previously, while cooperative search also allows the agent to use information found by others.

Hints can naturally be used to guide the ordering of backtrack choices, which can be viewed as moving in a tree structure. When a hint gives the correct choice for an agent, the remaining choices, in effect, are pruned. More generally, these hints can give large partial solutions from other regions of the search space. This is the case, for example, when putting together a puzzle by working on different regions and then combining them. Genetic algorithms are another instance of this general strategy.

2.1.3 DIVERSITY. A more interesting possibility is to have a group of agents use different search methods. Such diverse communities are particularly well suited for the use of cooperation since a particular agent may not be able to utilize all the information it generates, whereas another agent using a different strategy can. For example, a systematic backtrack search method rapidly may find promising regions of the search space but take a long time to finally reach a solution when some changes to choices made early in the backtracking are required. This could be quickly fixed by other methods that make adjustments opportunistically with no prespecified ordering. Thus the exchange of information among methods can improve performance beyond that possible without cooperation.

The effectiveness of these hints will depend on the search choices made by the agents. For example, as the search progresses, agents may find better partial solutions so that hint quality increases over time. Conversely, as agents get near the solution, hints become less important since they will tend to duplicate partial solutions already found or, in fact, incorrect hints may even become more detrimental.

2.2 IMPLEMENTATION ISSUES

From this general discussion of using hints with various search methods, we now turn to a number of implementation issues and how they were resolved in our experiments. From the many ways to address these issues, we made fairly simple choices. We can expect further improvements from more sophisticated use of hints, but the choices made here illustrate the potential of this method and have many direct correspondences with a wide range of constraint satisfaction problems. As a note of caution in developing more sophisticated strategies, the choices made should tend to promote high diversity among the agents[7,9] so there will be many opportunities to try hints in different promising contexts. This means that when viewed from the perspective of a single agent, some choices that appear reasonable could result in lowered performance for the group as a whole.

2.2.1 COOPERATIVE SEARCH. There are two basic steps in implementing a cooperative search based on individual algorithms. First, the algorithms themselves must be modified to enable them to produce and incorporate information from other agents, i.e., read and write hints. We should note that the first step, in itself, may change the performance of the initial algorithm or its characteristics (e.g., changing a complete search method into an incomplete one). Since this may change the absolute performance of the individual algorithm, a proper evaluation of the benefit of cooperation should compare the behavior of multiple agents, exchanging hints, to that of a single one running the same, modified algorithm but unable to communicate with other agents. In that way, the effect of cooperation, due to obtaining hints from other agents, will be highlighted.

In the second step, decisions as to exactly what information to use as hints, when to read them, etc. must be made. The hints consist of any useful information concerning regions of the search space to avoid or that are likely to contain solutions. A simple choice for constraint satisfaction problems is to use partial solutions, i.e., partial states whose assignments do not violate any constraint. We must also specify the organizational structure, i.e., which agents communicate with each other. In our experiments, all hints were written to a central blackboard, so each agent could access the results of any other agent. Hierarchical organizations more suitable to larger populations have also been studied.[3]

The next major question is: during its search when should an agent produce a hint. Generally, agents should tend to write hints that are likely to be useful in other parts of the search space. Possible methods include only writing the largest partial solutions an agent finds (i.e., at the point it is forced to backtrack) or only the hints comparable in size to those already on the blackboard.

Another set of complementary questions concerns when an agent decides to read a hint from the blackboard, which one it should choose, and how it should use the information for its subsequent search. Once again, a number of reasonable choices have different benefits in avoiding search and costs in their evaluation, as well as more global consequences for the diversity of the agent population. For instance, agents could select hints whenever a sufficiently good hint is available, whenever the agent is about to make a random choice in its search method (i.e., use the hint to break ties), or whenever the agent is in some sense stuck, e.g., needing to backtrack, or at a local optimum of a hill-climbing search method. For deciding which available hint to use, methods range from random selection[2] to picking one that is a good match, in some sense, to the agent's current state.

A final issue concerns the memory requirements for the hints. To avoid the potential of an unbounded growth in the size of the blackboard, one can limit the number of hints it could store. Once this limit is reached, some hints have to be discarded. For our experiments, the oldest (i.e., added to the blackboard before any others) of the smallest (i.e., involving the fewest assignments) hints were overwritten with new hints. We found that relatively small blackboards were sufficient to obtain significantly better performance than the independent searches.

2.2.2 PERFORMANCE MEASURES. Before turning to our experimental comparison of cooperating and noncooperating agents, we must specify how the performance of a group of agents is to be measured. The appropriate performance measure depends on the nature of the problem.[3] In many cases, one is interested in finding a single solution to the problem and each agent is individually capable of finding a complete solution. This means that the search is complete as soon as one agent finds a solution. The appropriate overall performance measure then is just the time required for some agent in the group to find a solution.

As a simple performance criterion we use the number of search steps required for the first agent to find a solution. However, we should note that this ignores the additional overhead involved in selecting and incorporating hints. Including such costs in simple cases doesn't change the qualitative observation of cooperative improvement.[3] Whether this remains true for the more sophisticated search methods remains open and is ultimately best addressed by comparing execution times of careful implementations of the algorithms. Moreover, an actual parallel implementation would also face possible communication bottlenecks at the central blackboard though this is unlikely to be a major problem with the small blackboards considered here due to the relatively low reading rate and the possibility of caching multiple copies of the blackboard which are only slowly updated with new hints. Nevertheless, the improvement in the number of search steps reported below, and comparisons of the execution time of our unoptimized code, suggest the cooperative methods are likely to be beneficial for large, hard problems.

3. CRYPTARITHMETIC

For our first example, we consider a simple search method, used in the familiar problem of solving cryptarithmetic codes. These problems require finding a unique digit assignment to each of the letters of a word addition so that the numbers represented by the words add up correctly. An example is this sum: $DONALD + GERALD = ROBERT$, which has one solution given by $A = 4$, $B = 3$, $D = 5$, $E = 9$, $G = 1$, $L = 8$, $N = 6$, $O = 2$, $R = 7$, and $T = 0$. In general, n letters yields 10^n possible states. However, the requirement of a unique digit for each letter means that there are $\binom{10}{n}$ ways to choose the values and $n!$ ways to assign them to the letters, which reduces the total number of search states to $n!\binom{10}{n} = 10!/(10-n)!$. Thus the above example, which has 10 letters, has 10! states in its search space.

Solving a cryptarithmetic problem involves performing a search. Although clever heuristics can be used to rapidly solve the particular case of cryptarithmetic,[15] our purpose is to address the general issue of cooperation in parallel search using cryptarithmetic as a simple example. Thus we focus on simple search methods, without clever heuristics that can lead to quick solutions by a single agent. This is precisely the situation faced with more complex constraint problems where searches remain long even with the best available heuristics.

The basic search paradigm we have used in the cryptarithmetic problem is "random generate and test with replacement." We used hints consisting of letter-digit assignments in columns that add correctly. These hints were posted to a blackboard. Agents used the available hints to select their next state. In a noncooperative search, an agent using this method could use only hints that it had previously found so each agent had a separate blackboard. Cooperative search allowed the agent also to use hints found by others, using a single blackboard.

For each search step, an agent chooses a hint randomly from the blackboard and replaces assignments in its current state with those specified by the hint. If there are no hints, it chooses a random letter-digit assignment using random generate and test. Once the agent obtains the new state, it generates and posts all possible hints from its state, if any. Thus, assignments that work for more than one column are posted as several different hints. When random states are generated by jumping, rather than single-letter replacements, there is a greater possibility of generating more hints faster but at the expense of frequently overwriting partially correct states.

As an example of this search method, consider an agent solving the problem $AB+AC = DE$. This problem has $10!/5! = 30,240$ possible states and 144 solutions (determined by exhaustive search). In the first step, each agent selects a random set of letter-digit assignments such that no digit is assigned to more than one letter. Suppose the letter-digit assignments, or state, of the first agent are $A = 4$, $B = 2$, $C = 7$, $D = 3$, and $E = 9$. In this case the assignments do not correspond to a solution since $42 + 47$ does not equal 39. However, the rightmost column, $B+C = E(2+7 = 9)$, does add up correctly so that the agent's state is *partially* (or *locally*) correct. Partial correctness includes cases where a carry has been brought over from the previous column or may be sent to the next column. Note that although a particular column may be locally correct, it may not lead to a solution. In this example, the agent has one column correct (three letters: B, C, and E). If these letter assignments do lead to a solution, then there are only two letters that need to be assigned from seven possible choices. Thus the agent went from a search space of size 30,240 to one of $7!/5! = 42$ states, a reduction by a factor of nearly 1,000. In a cooperative search, this reduction could also be used by other agents, perhaps in other regions of the search space where this hint is more successfully used.

3.2 RESULTS

As a specific case, we examine the effect of cooperation for groups of 100 agents solving the problem $WOW + HOT = TEA$. This problem, with six distinct letters, has 151,200 search states and 82 different solutions. The comparative performance of cooperation is shown in Table 1.

It also worthwhile to note the effect of cooperation as the problems become more difficult. One way of measuring the difficulty of problems is by the ratio of the number of states in the search space, T, to the number of solutions, S. Table 2 shows the relative speed for the first finisher of 100 agents for four problems of vastly different complexities. The data for the cooperative case came from experiments while the behavior of the noncooperative case was obtained theoretically[3] by noting that each random-generate-and-test step has probability S/T to find a solution.

TABLE 1 Average performance of 10 trials of 100 agents solving $WOW + HOT = TEA$ for different search methods. The relative time is the average time required for the first agent of the group to find a solution, divided by the average time required for the cooperative case. The relative deviation is the standard deviation in the time to first solution divided by the average time for each method. The benefit of cooperation, i.e., sharing hints among the agents, is shown by the comparison between the cooperative case and the case where the agents used the same method, i.e., had memory, but did not share it. For comparison the last row shows the theoretical performance of the unmodified random-generate-and-test method.

search method	relative time	relative deviation
cooperative	1	.87
independent, with memory	7.5	.49
independent, no memory	23.9	1

TABLE 2 Scaling of cooperative performance for cryptarithmetic problems of increasing difficulty for 100 agents. The second column comparing the ratio of speeds, the cooperative search to independent agents with no memory, represents an average over about 100 trials for the first three cases and a few trials for the last case. (Note that the entry for the second problem is 45 compared to 23.9 of Table 1; this was from a separate run of the experiment and indicates the degree of statistical fluctuation in the cooperative search.) The fourth column shows the range in the fraction of hints on the final blackboard that were subsets of some solution for some of these cooperative searches. Typically these were added just before the end of the search by the agent that found the first solution.

Problem	ratio of speeds	T/S	Fraction of hints that are subsets of solutions
AB + AC = DE	7	210	0.9 – 1.0
WOW + HOT = TEA	45	1844	0.5 – 0.6
CLEAR + WATER = SCOTT	145	181440	0.1 – 0.2
DONALD + GERALD = ROBERT	315	3628800	0.004

Note that as the problem becomes more difficult the importance of cooperation and use of memory in speedup is increased. The relative increase becomes even more startling when one considers that the fraction of hints posted on the blackboard that are subsets of *any* of the solutions (not necessarily the one found first) decreases as the problems become more complex. Thus the high performance is due to some agents finding combinations of hints that lead to solutions even though the full hints are rarely part of a solution.

Another way of studying the effect of cooperation vs. problem complexity is to vary the effectiveness of the search performed by the agent itself, i.e., the *self-work*, without utilizing the hints from the other agents. For example, suppose that, when the agents are not using hints, they perform a depth-first backtrack search, each using a randomly selected ordering of the variables. During the depth-first search the agents have the opportunity to prune partial states that do not lead to any solution. For example, if some columns do not add up correctly, there is no point in considering assignments to uninstantiated letters for this state. Whenever a hint

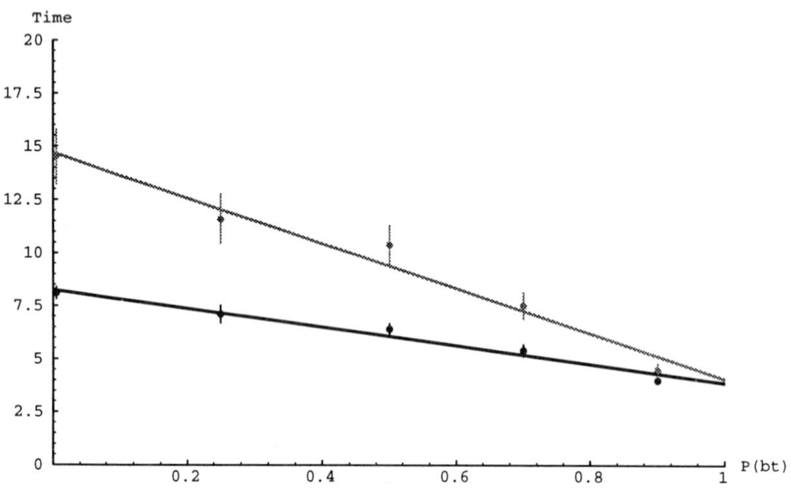

FIGURE 2 Cooperation works best for harder problems. The plot shows the average time to the first solution for 100 agents solving the $AB + AC = DE$ as a function of the probability of pruning, $P(bt)$, a state that is known not to lead to a solution. The left side of the plot corresponds to "hard" problems where pruning of the search space is very poor and the right side of the plot corresponds to "easy" problems where pruning is very effective. The light line, for the case of noncooperating agents, is a depth-first search. The dark line is for the case where the agents spend 80% of their time doing depth-first self-work and 20% cooperating, i.e., using hints from the blackboard. The lines show the best linear fits to the data. The data points correspond to the average solution time from 50–100 runs. The error bars are the error of the mean.

is used, it overwrites the current partial state in the same manner as for the agents using simple generate and test, so that there may be large jumps through the search space and the resulting search is no longer complete. We can simulate the effect of this pruning by probabilistically pruning partially assigned states that are known not to lead to a solution. (We can do this with cryptarithmetic by generating all the solutions ahead of time.) When the probability of pruning is small, this corresponds to difficult problems because the agents must instantiate nearly all the letters of an incorrect assignment before pruning. The results of this study, shown in Figure 2, show the greater relative importance of cooperation for harder problems.

In summary, these results show the value of cooperation in solving a relatively easy constraint satisfaction problem using simple search methods. One question remains: how can this method be used in solving harder problems.

4. GRAPH COLORING

The distinction between easy and hard problems is important in determining the feasibility of computations, and a great deal of research has been devoted to it.[4] An important distinction among problems is based on how rapidly the number of elementary operations required to solve them increases as the problems scale up to larger instances, particularly whether the scaling is dominated by polynomial or exponential growth. An elementary operation could typically be an arithmetic operation for a numerical problem or an examination of a single state in a search problem.

A surprising result is that sometimes the difference between these classes of problems is extremely subtle. For instance, consider two given nodes of a graph, which consists of a number of nodes and links between them. The problem of deciding whether there is a path between them, i.e., a series of distinct linked nodes that connect the two given nodes, whose total length is less than a given bound M can be solved in polynomial time with respect to the number of nodes in the graph. On the other hand, the similar problem of whether there is a path with length greater than M has no known solution in polynomial time. However, if one is given a path whose length is claimed to be larger than M so that such a path exists, an algorithm will quickly verify that the answer is correct; it counts the links in the path and checks that the length is indeed larger than M. This procedure operates in linear time in the length of the path which in turn is no more than the total number of nodes in the graph. This is an example of a simple yes-or-no problem in which an affirmative answer can be verified in polynomial time, even though there may be no way to actually construct the answer readily.

Such problems are said to belong to the class NP (for nondeterministic polynomial). Conceptually, these problems can be rapidly solved by a *nondeterministic* algorithm, i.e., one that can somehow guess the correct answer, and then rapidly verify it. Actual implementations, however, are deterministic and appear to be unable to solve the problem in polynomial time. Note that NP includes all problems in *P*, the class of problems for which there is a deterministic polynomial time algorithm. Whether NP is in fact the same as *P* remains an open question.

Although the class NP is based on the ability to easily verify solutions, it also can be shown to include many optimization problems whose solutions would seem more difficult to check. For instance, corresponding to the path problems mentioned above are the optimization problems of determining the *shortest* and *longest* paths between the vertices, respectively. The shortest path can be found in polynomial time, but there is no known rapid solution (i.e., short of checking all possible paths) for determining the longest one. In the latter case, being given a path that is claimed to be the longest is difficult to directly verify since not only its length must be determined—it must also be compared to all other possible paths. However, this latter problem does in fact belong to NP because it can be transformed into a series of verifiable problems involving specified bounds on the lengths such that the total time to verify all the subproblems is still polynomial. Another example of such a problem is the travelling salesman problem, in which a collection of cities

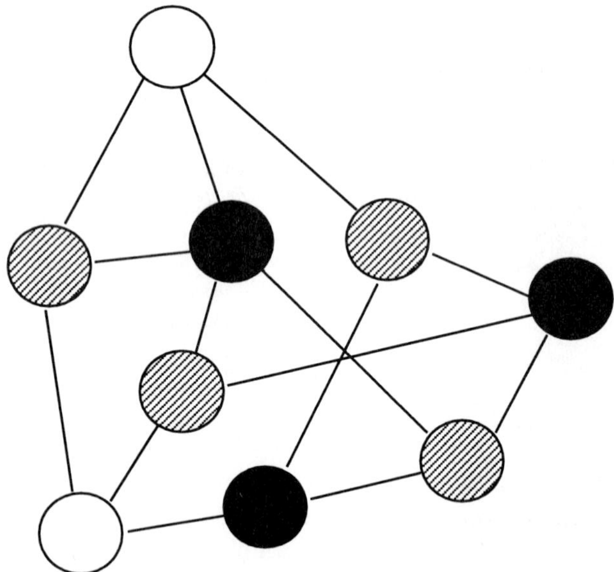

FIGURE 3 A graph with nine nodes of three colors (black, gray, and white) such that no two adjacent nodes have the same color.

and distances between them is given, and the task is to find the shortest path that visits each city. Among the problems in the class NP, some are known to be at least as hard, up to a polynomial factor, as any other problem in the class. In this sense, these so-called NP-complete problems constitute the most difficult problems in NP. As far as currently known, the solution cost grows exponentially in the worst case as the size of the problem increases.

As our second example of cooperative search, we consider the NP-complete problem of graph coloring. The problem consists of coloring the nodes in a graph, from a limited set of colors, such that no two adjacent nodes (i.e., nodes linked by an edge in the graph) have the same color. An example of a colored graph is shown in Figure 3. Graph coloring has received considerable attention and a number of search methods have been developed.[11] Paradoxically, although some graphs are very hard to color, among graphs of a given size, there is considerable variation in the difficulty of finding a solution, and most of them can be colored (or determined to have no coloring) quite rapidly with existing heuristic methods.

For this problem, the average degree of the graph γ (i.e., the average number of edges coming from a node in the graph) distinguishes relatively easy from harder problems, on average. For the case of three-coloring, (i.e., when three different colors are available) which we focus on in this paper, the region of hardest problems is empirically observed to occur near $\gamma = 5$.[1] We used the Brelaz search heuristic[11] which effectively finds colorings by assigning the most constrained nodes first (i.e., those with the most distinct colored neighbors) and breaking ties by choosing nodes with the most uncolored neighbors (ties that remain after applying this criterion are broken randomly). For the chosen node, the smallest available color is examined first, with successive colors considered when the search is forced to backtrack. For the graph shown in Figure 3, this heuristic would first color the central node with four neighbors, then randomly select one of those neighbors to color (since after the first node is colored, each of its neighbors will have the same number of uncolored neighbors), etc.; this continues until a complete coloring is found or the search is forced to backtrack because no consistent coloring is possible for the next node selected. By focusing attention on the most constrained nodes first, this will generally rapidly determine if a proposed partial coloring is inconsistent, thus pruning unproductive searches high in the tree and avoiding substantial wasted effort. This heuristic is considerably more efficient than simple generate-and-test or backtracking with a random ordering of the nodes.

To generate a collection of hard problems, we examined a large number of random graphs. Trivial cases with underconstrained nodes were removed by ensuring each node had at least three edges. Notice that nodes with fewer edges are underconstrained in that they can always be colored differently from the nodes they are linked to when there are three available colors. The resulting graphs were searched repeatedly with the Brelaz heuristic, and only those with high search cost were retained. Moreover, to correspond with the cooperative methods used for the cryptarithmetic example and to simplify the use of hints, we considered only graphs that had solutions. In addition to high average cost for solution with the Brelaz heuristic, these graphs also had a large variance in the cost of repeated searches due to different choices made at tie points. This variance gives rise to improved performance by multiple independent searches in parallel, stopping when the first one finishes. The experiments reported here show the additional benefit from exchanging hints.

At any point in a backtracking search, the current partial state is a consistent coloring of some subset of the graph's nodes. When writing a hint to the blackboard, the Brelaz agents simply wrote their current state. Specifically, each agent independently wrote its current state at each step with a fixed probability q.

Each time the agent was about to expand a node in its backtrack search, instead it would attempt, with probability p, to read a compatible hint from the blackboard, i.e., a hint whose assignments: (1) were consistent with those of the agent (up to a permutation of the colors)[1] and (2) specified at least one node not already assigned in the agent's current state. Frequently, there was no such compatible hint (especially when the agent was deep in the tree and hence had already made assignments to many of the nodes), in which case the agent continued with its own search.

When a compatible hint was found, its overlap with the agent's current state was used to determine a permutation of the hint's colors that made it consistent with the state. This permutation was applied to the remaining colorings of the hint and then used to extend the agent's current state as far as possible (ordering the new nodes as determined by the Brelaz heuristic), thus retaining necessary backtrack points so that the overall search remained complete. In effect, this hint simply replaced decisions that the Brelaz heuristic would have made regarding the initial colors for a number of nodes. Thus, this amounts to a fairly conservative use of hints compared to the backtrack search for cryptarithmetic in Figure 2, where hints overwrote the agent's state without retaining backtrack points.

4.2 RESULTS

The experimental results show the benefit of cooperation for graph coloring using a variety of search methods.[8] In Figure 4, we compare the performance of a group of ten independent and ten cooperative agents, all using the same Brelaz search algorithm described above. We generated a set of graphs whose search cost was one to three orders of magnitude more than the minimum possible. To highlight the benefit of cooperation beyond that achieved with multiple runs of independent agents, we compare the cooperative case with the same number of agents running

[1]We thus used the fact that, for graph coloring, any permutation of the color assignments for a consistent set of assignments is also consistent.

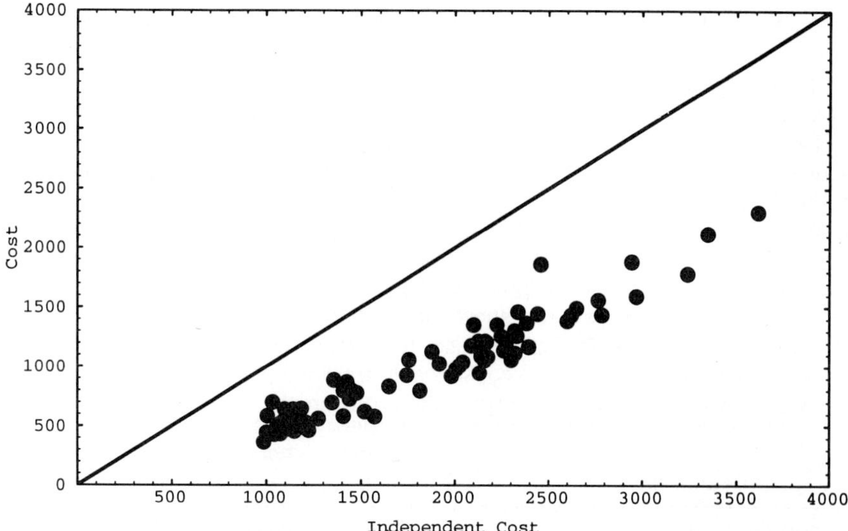

FIGURE 4 Performance of groups of ten cooperating agents using the Brelaz search method on a range of graphs vs. the performance of a group of ten independent agents using the same method. The performance values used for each graph are the median over ten trials of the search steps required for the first agent in the group to find a solution. For comparison, the line shows the performance of the independent agents. In these experiments, the blackboard was limited to hold 100 hints, and we used $p = 0.4$, $q = 0.1$, and graphs with 100 nodes.

TABLE 3 Extreme cases from Figure 4. Note that the search space for this problem has $3^{100} \approx 5 \times 10^{47}$ states, giving much larger values of T/S than for the cryptarithmetic problems. The number of solutions was found by exhaustive search. The fourth column shows the average size (i.e., number of colored nodes) of the hints on the blackboard at the time the solution was found. The fifth shows the fraction that are subsets of a solution.

example	independent search cost	T/S	avg. hint size	fraction of hints that are subsets of solutions
A	3614	9×10^{42}	64.2	0.02
B	985	7×10^{41}	42.3	0.05

TABLE 4 Performance for the examples, A and B, given in Table 3 for different search methods. The relative time is the median time required for the first agent of the group to find a solution, divided by the median for the cooperative case. The relative deviation is the standard deviation in the time to first solution divided by the median time for each method. The benefit of cooperation, i.e., sharing hints among the agents, is shown by the comparison between the cooperative case and the case where the agents used the same method, i.e., had memory but did not share it. For comparison, the last row shows, the performance of the unmodified backtrack using the Brelaz heuristic. Note that the deviations for this backtrack search method are considerably smaller than for the generate-and-test search used for the cryptarithmetic example.

	relative time		relative deviation	
search method	A	B	A	B
cooperative	1	1	0.03	0.14
independent, with memory	1.6	2.7	0.03	0.05
independent, no memory	1.8	3.1	0.06	0.11

independently. Note that in both cases, cooperation gives better performance than simply taking the best of ten independent agents. Moreover, cooperation appears to be more beneficial as problem hardness (measured by the performance of a group of independent agents) increases. We obtained a few graphs of significantly greater hardness than those shown here confirming this trend.

As with cryptarithmetic, most hints on the blackboard are not subsets of solutions. As an example, for two of the cases shown in Figure 4, Table 3 shows the number of hints on the final blackboard (for a single run) that are subsets of solutions. Note that unlike the cryptarithmetic case, here the blackboard is limited to 100 hints. Finally, Table 4 shows the speedup obtained for some of the graph-coloring cases. These are considerably less than those obtained from the simple generate-and-test search with cryptarithmetic, but are comparable to the speedup obtained with the backtrack search shown in Figure 2.

Similar cooperative improvements are obtained for other search methods,[8] including heuristic repair,[17] in which changes are made to complete colorings that minimize the number of violated constraints, and a mixed group of agents in which some use the Brelaz heuristic with backtracking, as described above, while others use heuristic repair.

5. DISCUSSION

We have shown how cooperating agents working toward the solution of a constraint satisfaction problem can lead to a marked increase in the speed with which they solve it compared to their working in isolation. A summary of the cases studied is shown in Table 5.

In our implementation we defined hints in terms of information that moved the agents toward a region of the space that could have a solution. Another possibility is for hints to contain information that tends to move them away from regions that can have no solutions. More generally, any search algorithm that agents may use will have parameters that will affect the benefit of cooperation. Another consideration is when are the hints most useful for problem solving. At the beginning of a problem the hints provide crucial information for starting the agents off on a plausible course, but usually they will be fairly nonspecific. Near the end of the problem however, there are likely to be many detailed hints but of less relevance to the agents since they may have already discovered that information themselves. This suggests that typical cooperative searches will both start and end with agents primarily working on their own and that the main benefit of exchanging hints will occur in the middle of the search.

This work suggests an alternative to the current mode of constructing task-specific computer programs that deal with constraint satisfaction problems. Rather than developing a monolithic program or perfect heuristic, it may be better to have a set of relatively simple cooperating processes work concurrently on the problem while communicating their partial results. This would imply the use of "hint engineers" for coupling previously disjoint programs into interacting systems that are able to use each others' (imperfect) knowledge.

This new method may be particularly useful in areas of artificial intelligence such as design, qualitative reasoning, truth maintenance systems, and machine learning. Researchers in these areas are just starting to consider the benefits brought about by massive parallelism and concurrency, and our work suggests the additional benefits that could be obtained from cooperation.

In closing, we have seen how computational ecosystems can be used to solve complex problems by exploiting the benefit of cooperation in a distributed context. We believe this is just the beginning; one can envision systems where the demands of a particular task will dynamically spawn new processes to work on promising avenues while deleting those agents that are not making much progress. This will require new programming methods for resource allocation in these systems. The spread of these ecosystems will make it easier to program them in order to use cooperative methods for the solution of even harder problems.

TABLE 5 Comparison of cooperative search methods used for cryptarithmetic and graph coloring, except that the cryptarithmetic results shown in Figure 2 use simple backtrack and only use hints on some of the search steps.

search problem	cryptarithmetic	graph coloring
individual method	random generate and test	backtracking using Brelaz heuristic
blackboard size	unlimited	100 hints, old ones overwritten with new ones
hints were partial solutions	digits for some letters that added correctly	consistent colors for some nodes
when to write a hint	whenever some columns added correctly	randomly with probability $q = 0.1$ at each step
when to read a hint	every step when hint was available	randomly with probability $p = 0.5$ at each step a compatible hint was available
how to use a hint	overwrite current state	extend current state

ACKNOWLEDGMENTS

We thank S. Clearwater and C. Williams for helpful discussions. This work was partially supported by the Air Force Office of Scientific Research Contract No. F49620–90–C-0086.

REFERENCES

1. Cheeseman, P., B. Kanefsky, and W. M. Taylor. "Where the Really Hard Problems Are." In *Proceedings of IJCAI 91*, edited by J. Mylopoulos and R. Reiter, 331–337. San Mateo, CA: Morgan Kaufmann, 1991.
2. Clearwater, S. H., B. A. Huberman, and T. Hogg. "Cooperative Solution of Constraint Satisfaction Problems." *Science* **254** (1991): 1181–1183.

3. Clearwater, S. H., B. A. Huberman, and T. Hogg. "Cooperative Problem Solving." In *Computation: The Micro and the Macro View*, edited by B. Huberman, 33–70. Singapore: World Scientific, 1992.
4. Cormen, T. H., C. E. Leiserson, and R. L. Rivest. *Introduction to Algorithms.* Cambridge, MA: MIT Press, 1990.
5. Gasser, L., and M. N. Huhns, eds. *Distributed Artificial Intelligence*, vol. 2. Menlo Park, CA: Morgan Kaufmann, 1989.
6. Goldberg, D. E. *Genetic Algorithms in Search, Optimization and Machine Learning.* Reading, MA: Addison-Wesley, 1989.
7. Hogg, T. "The Dynamics of Complex Computational Systems." In *Complexity, Entropy, and the Physics of Information*, edited by W. Zurek. Santa Fe Institute Studies in the Sciences of Complexity, Proc. Vol. VIII, 207–222. Reading, MA: Addison-Wesley, 1990.
8. Hogg, T., and C. P. Williams. "Solving the Really Hard Problems with Cooperative Search." Technical Report, Xerox PARC, Palo Alto, CA, October 1992.
9. Huberman, B. A. "The Performance of Cooperative Processes." *Physica D* **42** (1990): 38–47.
10. Huberman, B. A., and T. Hogg. "The Behavior of Computational Ecologies." In *The Ecology of Computation*, edited by B. A. Huberman, 77–115. Amsterdam: North-Holland, 1988.
11. Johnson, D. S., C. R. Aragon, L. A. McGeoch, and C. Schevon. "Optimization by Simulated Annealing: An Experimental Evaluation; Part II, Graph Coloring and Number Partitioning." *Oper. Res.* **39(3)** (1991): 378–406.
12. Kirkpatrick, S., C. D. Gelatt, and M. P. Vecchi. "Optimization by Simulated Annealing." *Science* **220** (1983): 671–680.
13. Kornfeld, W. A., and C. E. Hewitt. "The Scientific Community Metaphor." *IEEE Trans. Sys., Man & Cyber.* **SMC-11** (1981): 24–33.
14. Kornfeld, W. A. "The Use of Parallelism to Implement Heuristic Search." Technical Report, A.I. Memo No. 627 Artificial Intelligence Laboratory, Massachusetts Institute of Technology, 1981.
15. Lauriere, J.-L. "A Language and a Program for Stating and Solving Combinatorial Problems." *Artificial Intelligence* **10** (1978): 29–127.
16. Mackworth, A. K. "Constraint Satisfaction." In *Encyclopedia of Artificial Intelligence*, edited by S. Shapiro and D. Eckroth, 205–211. New York: John Wiley, 1987.
17. Minton, S., M. D. Johnston, A. B. Philips, and P. Laird. "Solving Large-Scale Constraint Satisfaction and Scheduling Problems Using a Heursitic Repair Method." In *Proc. AAAI-90*, 17–24, Menlo Park, CA: AAAI Press, 1990.
18. Newell, A., and H. A. Simon. *Human Problem Solving.* New Jersey: Prentice Hall, 1972.

19. Nii, H. P., N. Aiello, and J. Rice. "Experiments on Cage and Poligon: Measuring the Performance of Parallel Blackboard Systems." In *Distributed Artificial Intelligence*, edited by Les Gasser and Michael N. Huhns, vol. 2, 319–383. San Mateo, CA: Morgan Kaufmann, 1989.

Bernardo A. Huberman and Tad Hogg
Xerox Palo Alto Research Center, Palo Alto, California 94304 USA

The Emergence of Computational Ecologies

We describe a form of distributed computation in which agents have incomplete knowledge and imperfect information on the state of the system, and an instantiation of such systems based on market mechanisms. When agents can choose among several resources, the dynamics of the system can be oscillatory and even chaotic. A mechanism is described for achieving global stability through local controls.

1. INTRODUCTION

Propelled by advances in software design and increasing connectivity of computer networks, distributed computational systems are starting to spread throughout offices, laboratories, countries, and continents. In these systems, computational processes consisting of the active execution of programs can spawn new ones in other machines as they make use of printers, file servers, and other machines of the network as the need arises. In the most complex applications, various processes can collaborate to solve problems, while competing for the available computational resources, and may also directly interact with the physical world. This contrasts with the more familiar stand-alone computers, with traditional methods of centralized scheduling for resource allocation and programming methods based on serial processing.

The effective use of distributed computation is a challenging task, since the processes must obtain resources in a dynamically changing environment and must be designed to collaborate despite a variety of asynchronous and unpredictable changes. For instance, the lack of global perspectives for determining resource allocation requires a very different approach to system-level programming and the creation of suitable languages. Even implementing reliable methods whereby processes can compute in machines with diverse characteristics is difficult.

As these distributed systems grow, they become a community of concurrent processes, or a *computational ecosystem*,[5] which, in their interactions, strategies, and lack of perfect knowledge, are analogous to biological ecosystems and human economies. Since all of these systems consist of a large number of independent actors competing for resources, this analogy can suggest new ways to design and understand the behavior of these emerging computational systems. In particular, these existing systems have methods to deal successfully with coordinating asynchronous operations in the face of imperfect knowledge. These methods allow the system as a whole to adapt to changes in the environment or disturbances to individual members, in marked contrast to the brittle nature of most current computer programs which often fail completely if there is even a small change in their inputs or an error in the program itself. To improve the reliability and usefulness of distributed computation, it is therefore interesting to examine the extent to which this analogy can be exploited.

Based on the law of large numbers statistical mechanics has taught us that many universal and generic features of large systems can be quantitatively understood as approximations to the average behavior of infinite systems. Although such infinite models can be difficult to solve in detail, their overall qualitative features can be determined with a surprising degree of accuracy. Since these features are universal in character and depend only on a few general properties of the system, they can be expected to apply to a wide range of actual configurations. This is the case when the number of relevant degrees of freedom in the system, as well as the number of interesting parameters, is small. In this situation, it becomes useful to treat the unspecified internal degrees of freedom as if they are given by a probability distribution. This implies assuming a lack of correlations between the unspecified and specified degrees of freedom. This assumption has been extremely successful in statistical mechanics. It implies that although degrees of freedom may change according to purely deterministic algorithms, because they are unspecified, they appear to be effectively random to an outside observer.

Consider, for instance, massively parallel systems which are desired to be robust and adaptable. They should work in the presence of unexpected errors and with changes in the environment in which they are embedded (i.e., fail soft). This implies that many of the system's internal degrees of freedom will be adjustable by taking on a range of possible configurations. Furthermore, their large size will necessarily enforce a perspective that concentrates on a few relevant variables. Although these considerations suggest that the assumptions necessary for a statistical description hold for these systems, experiments will be necessary for deciding their applicability.

While computational and biological ecosystems share a number of features, we should also note there are a number of important differences. For instance, in contrast to biological individuals, computational agents are programmed to complete their tasks as soon as possible, which in turn implies a desirability for their earliest death. This task completion may also involve terminating other processes spawned to work on different aspects of the same problem, as in parallel search, where the first process to find a solution terminates the others. This rapid turnover of agents can be expected to lead to dynamics at much shorter time scales than seen in biological or economic counterparts.

Another interesting difference between biological and computational ecologies lies: in the fact that for the latter the local rules (or programs for the processes) can be arbitrarily defined, whereas in biology those rules are quite fixed. Moreover, in distributed computational systems the interactions are not constrained by a Euclidean metric, so that processes separated by large physical distances can strongly affect each other by passing messages of arbitrary complexity between them. And last but not least, in computational ecologies the rationality assumption of game theory can be explicitly imposed on their agents, thereby making these systems amenable to game dynamic analyses suitably adjusted for their intrinsic characteristics. On the other hand, computational agents are considerably less sophisticated in their decision-making capacity than people, which could prevent expectations based on observed human performance from being realized.

By now there are a number of distributed computational systems that exhibit many of the above characteristics and offer increased performance when compared with traditional operating systems. *Enterprise*[8] is a marketlike scheduler where independent processes or agents are allocated at run time among remote idle workstations through a bidding mechanism. A more evolved system, *Spawn*,[12] is organized as a market economy composed of interacting buyers and sellers. The commodities in this economy are computer-processing resources, specifically, slices of CPU time on various types of computers in a distributed computational environment. The system has been shown to provide substantial improvements over more conventional systems, while providing dynamic response to changes and resource sharing.

From a scientific point of view, the analogy between distributed computation and natural ecologies brings to mind the spontaneous appearance of organized behavior in biological and social systems, where agents can engage in cooperating strategies while working on the solution of particular problems. In some cases, the strategy mix used by these agents evolves towards an asymptotic ratio that is constant in time and stable against perturbations. This phenomenon sometimes goes under the name of evolutionarily stable strategy (ESS). Recently, it has been shown that spontaneous organization can also exist in open computational systems when agents can choose among many possible strategies while collaborating in the solution of computational tasks. In this case, however, imperfect knowledge and delays in information introduce asymptotic oscillatory and chaotic states that exclude the existence of simple ESS's. This is an important finding in light of studies which resort to notions of evolutionarily stable strategies in the design and prediction of an open system's performance.

In what follows we will describe a market-based computational ecosystem and a theory of distributed computation. The theory describes the collective dynamics of computational agents, while incorporating many of the features endemic to such systems, including distributed control, asynchrony, resource contention, and extensive communication among agents. When processes can choose among many possible strategies while collaborating in the solution of computational tasks, the dynamics leads to asymptotic regimes characterized by complex attractors. Detailed experiments have confirmed many of the theoretical predictions while uncovering new phenomena, such as chaos induced by overly clever decision-making procedures.

Next, we deal with the problem of controlling chaos in such systems, for we have discovered ways of achieving global stability through local controls inspired by fitness mechanisms found in nature. Furthermore, we show how diversity enters into the picture, along with the minimal amount of such diversity that is required to achieve stable behavior in a distributed computational system.

2. COMPUTATIONAL MARKETS FOR RESOURCE ALLOCATION

Allocating resources to competing tasks is one of the key issues for making effective use of computer networks. Examples include deciding whether to run a task in parallel on many machines or serially on one, and whether to save intermediate results or recompute them as needed. The similarity of this problem to resource allocation in market economies has prompted considerable interest in using analogous techniques to schedule tasks in a network environment. In effect, a coordinated solution to the allocation problem is obtained using Adam Smith's "invisible hand."[10] Although unlikely to produce the same optimal allocation made by an omniscient controller with unlimited computational capability, it can perform well compared to other feasible alternatives.[1,7] As in economics,[3] the use of prices provides a flexible mechanism for allocating resources, with relatively low information requirements: a single price summarizes the current demand for each resource, whether processor time, memory, communication bandwidth, use of a database, or control of a particular sensor. This flexibility is especially desirable when resource preferences and performance measures differ among tasks. For instance, an intensive numerical simulation's need for fast floating-point hardware is quite different from an interactive text editor's requirement for rapid response to user commands or a database's search requirement for rapid access to the data and fast query matching.

As a conceptual example of how this could work in a computational setting, suppose that a number of database search tasks are using networked computers to find items of interest to various users. Furthermore, suppose that some of the machines have fast floating-point hardware but are otherwise identical. Assuming the search tasks make little use of floating-point operations, their performance will not depend on whether they run on a machine with fast floating-point hardware. In a market based system, these programs will tend to value each machine based on how many other tasks it is running, leading to a uniform load on the machines. Now suppose some floating-point intensive tasks arrive in the system. These will definitely prefer the specialized machines and consequently bid up the price of those particular resources. Observing that the price for some machines has gone up, the databse tasks will then tend to migrate toward those machines without the fast floating-point hardware. Importantly, because of the high cost of modifying large existing programs, the database tasks will not need to be rewritten to adjust for the presence of the new tasks. Similarly, there is no need to reprogram the scheduling method of a traditional central controller, which is often very time consuming.

This example illustrates how a reasonable allocation of resources could be brought about by simply having the tasks be sensitive to current resource price. Moreover, adjustments can take place continually as new uses are found for particular network resources (which could include specialized databases or proprietary algorithms as well as the more obvious hardware resources) that do not require all users to agree on, or even know about, these new uses thus encouraging an incremental and experimental approach to resource allocation.

While this example motivates the use of market-based resource allocation, a study of actual implementations is required to see how large the system must be for its benefits to appear and whether any of the differences between simple computer programs and human agents pose additional problems. In particular, a successful use of markets requires a number of changes to traditional computer systems. First, the system must provide an easily accessible, reliable market so that buyers and sellers can quickly find each other. Second, individual programs must be price sensitive so they will respond to changes in relative prices among resources. This implies that the programs must, in some sense at least, be able to make choices among various resources based on how well suited they are for the task at hand.

A number of marketlike systems have been implemented over the years.[8,11,12] Most instances focus on finding an appropriate machine for running a single task. While this is important, further flexibility is provided by systems that use market mechanisms to also manage a collection of parallel processes contributing to the solution of a single task. In this latter case, prices give a flexible method for allocating resources among multiple competing heuristics for the same problem based on their perceived progress. Thus it greatly simplifies the development of programs that adjust to unpredictable changes in resource demand or availability. So we have a second reason to consider markets: not only may they be useful for flexible allocation of computational resources among competing tasks, but the simplicity of the price mechanism could provide help with designing cooperative parallel programs.

One such system is Spawn,[12] in which each task, starting with a certain amount of money corresponding to its relative priority, bids for the use of machines on the network. In this way, each task can allocate its budget toward those resources most important for it. In addition, when prices are low enough, some tasks can split into several parts which run in parallel, as shown in Figure 1, thereby adjusting the number of machines devoted to each task based on the demand from other users. From a user's point of view, starting a task with the Spawn system amounts to giving a command to execute it and the necessary funding for it to buy resources. The Spawn system manages auctions on each of the participating machines and the use of resources by each participating task, and provides communication paths

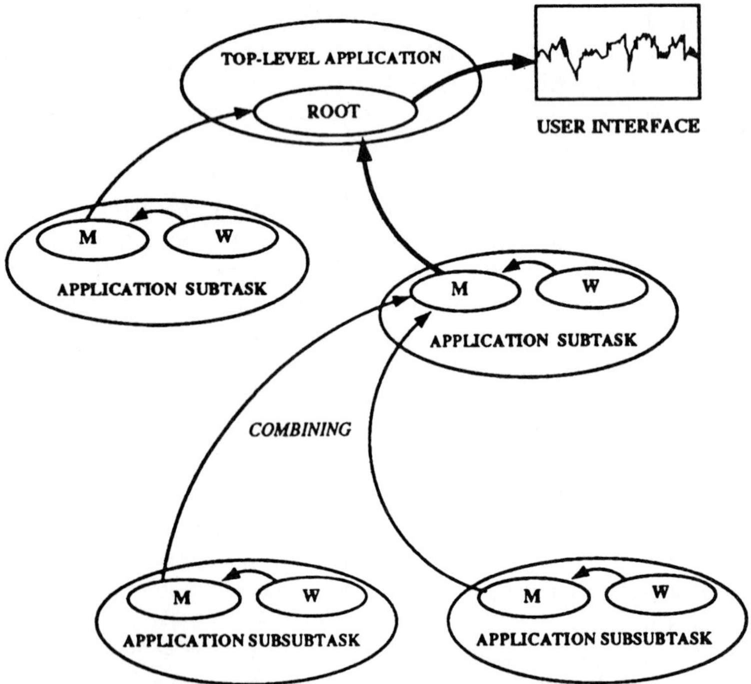

FIGURE 1 Managing parallel execution of subtasks in Spawn. Worker processes (W) report progress to their local managers (M) who in turn make reports to the next higher level of management. Upper management combines data into aggregate reports. Finally, the root manager presents results to the user. Managers also bid for the use of additional machines and, if successful, spawn additional subtasks on them.

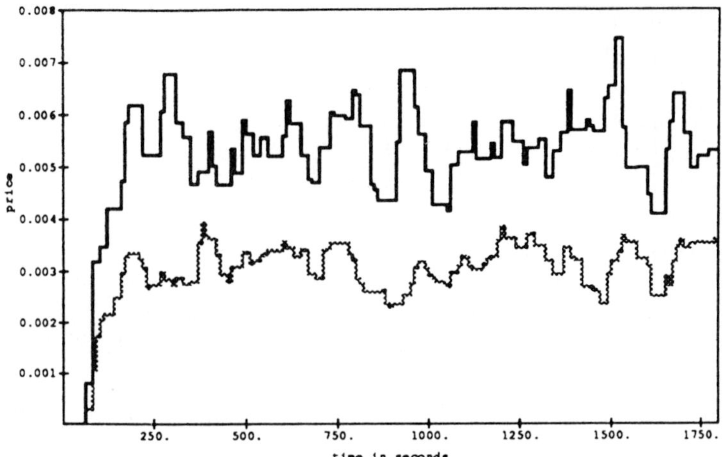

FIGURE 2 Price as a function of time (in seconds) in an inhomogeneous Spawn network consisting of three Sun 4/260's and six Sun 4/110's running four independent tasks. The average price of the 260's is in black, the less powerful 110's in gray.

among the spawned processes. It remains for the programmer to determine the specific algorithms to be used and the meaningful subtasks into which to partition the problem. That is, the Spawn system provides the price information and a market, but the individual programs must be written to make their own price decisions to effectively participate in the market. To allow existing, nonprice-sensitive programs to run within the Spawn system without modification, we provided a simple default manager that simply attempted to buy time on a single machine for that task. Users could then gradually modify this manager for their particular task, if desired, to spawn subtasks or to use market strategies more appropriate for the particular task.

Studies with this system show that an equilibrium price can be meaningfully defined with even a few machines participating. A specific instance is shown in Figure 2. Despite the continuing fluctuations, this small network reaches a rough price equilibrium. Moreover, the ratio of prices between the two machines closely matches their relative speeds, which was the only important difference between the two types of machine for these tasks. An additional experiment studied a network with some lengthy, low-priority tasks to which was added a short, high-priority task. The new task rapidly expands throughout the network by outbidding the existing tasks and driving the price of CPU time up, as shown in Figure 3. It is able therefore to utilize briefly a large number of networked machines and to illustrate the inherent

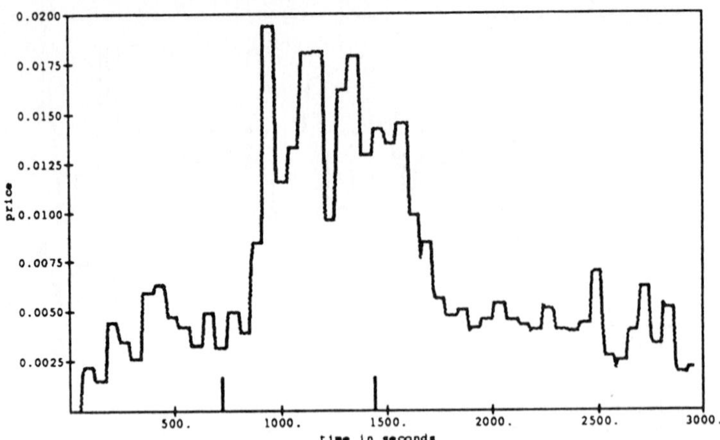

FIGURE 3 Price as a function of time (in seconds) when a high-priority task is introduced into a Spawn network running low-priority jobs. The first vertical line segment on the time axis marks the introduction of the high-priority task, and the second one the termination of its funding.

flexibility of market-based resource allocation. Although the very small networks used in these experiments could be adequately managed centrally, these results do show that expected market behavior can emerge even in small cases.

Computer market systems can be used to experimentally address a number of additional issues. For instance, they can help in understanding what happens when more sophisticated programs begin to use the network, e.g., processes that attempt to anticipate future loads so as to maximize their own resource usage. Such behavior can destabilize the overall system. Another area of interest is the emergence of diversity or specialization from a group of initially similar machines. For example, a machine might cache some of the routines or data commonly used by its processes, giving it a comparative advantage in bids for similar tasks in the future. Ultimately this could result in complex organizational structures embedded within a larger market framework.[9] Within these groups, some machines could keep track of the kinds of problems for which others perform best and use this information to guide new tasks to appropriate machines. In this way the system could gradually learn to perform common tasks more effectively.

These experiments also highlighted a number of more immediate practical issues. In setting up Spawn, it was necessary to find individuals willing to allow their machines to be part of the market. While it would seem simple enough to do so, in practice a number of incentives were needed to overcome the natural reluctance of people to have other tasks running on their machines. This reluctance is partly based on perceived limitations on the security of the network and the individual operating systems, for it was possible that a remote procedure could crash an individual machine or consume more resources than anticipated. In particular, users with little need for computer-intensive tasks saw little benefit from participating since they had no use for the money collected by their machines. This indicates the need to use real money in such situations so that these users could use their revenues for their own needs. This in turn brings the issue of computer security to the forefront so users will feel confident that no counterfeiting of money takes place and tasks in fact will be limited to use only resources they have paid for.

Similarly, for those users participating in the system as buyers, they need to have some idea of what amount of money is appropriate to give a task. In a fully developed market, there could easily be tools to monitor the results of various auctions and, hence, give a current market price for resources. However, when using a newly created market with only a few users, tools are not always available to give easy access to prices and, even if they are, the prices have large fluctuations. Effective use of such a system also requires users to have some idea of what resources are required for their programs or, better yet, to encode that information in the program itself so it will be able to respond to available resources—e.g., by spawning subtasks—more rapidly than the users can. Conversely, there must be a mechanism whereby sellers can make available information about the characteristics of their resources (e.g., clock speed, available disk space, or special hardware). This can eventually allow for more complex market mechanisms, such as auctions that attempt to sell simultaneous use of different resources (e.g., CPU time and fast memory) or future use of currently unavailable resources to give tasks a more predictable use of resources. Developing and evaluating a variety of auction and price mechanisms that are particularly well suited to these computational tasks is an interesting open problem.

Finally, these experimental systems help clarify the differences between human and computer markets. For instance, computational processes can respond to events much more rapidly than people, but they are far less sophisticated. Moreover, unlike the situation with people, particular incentive structures, rationality assumptions, etc. can be explicitly built into computational processes, allowing for the possibility of designing particular market mechanisms. This could lead to the ironic situation in which economic theory has greater predictability for the behavior of computational markets than for that of the larger, and more complex, human economy.

3. CHAOS IN COMPUTATIONAL ECOSYSTEMS

The systems we have been discussing are basically made up of simple agents with fast response times, compared to human agents in more complex and slower economic settings. This implies that an understanding of the behavior of computational ecosystems requires focusing on the dynamics of collections of agents capable of a set of simple decisions.

Since decisions in a computational ecosystem are not centrally controlled, agents independently and asynchronously select among the available choices based on their perceived payoff. These payoffs are actual computational measures of performance, such as the time required to complete a task, accuracy of the solution, amount of memory required, etc. In general, the payoff G_r for using resource r depends on the number of agents already using it. In a purely competitive environment, the payoff for using a particular resource tends to decrease as more agents make use of it. Alternatively, the agents using a resource could assist one another in their computations, as might be the case if the overall task could be decomposed into a number of subtasks. If these subtasks communicate extensively to share partial results, the agents will be better off using the same computer rather than running more rapidly on separate machines and then being limited by slow communications. As another example, agents using a particular database could leave index links that are useful to others. In such cooperative situations, the payoff of a resource then would increase as more agents use it, until it became sufficiently crowded.

Imperfect information about the state of the system causes each agent's perceived payoff to differ from the actual value, with the difference increasing when there is more uncertainty in the information available to the agents. This type of uncertainty concisely captures the effect of many sources of errors such as some program bugs, heuristics incorrectly evaluating choices, errors in communicating the load on various machines, and mistakes in interpreting sensory data. Specifically, the perceived payoffs are taken to be normally distributed, with standard deviation σ, around their correct values. In addition, information delays cause each agent's knowledge of the state of the system to be somewhat out of date. Although for simplicity we will consider the case in which all agents have the same effective delay, uncertainty, and preferences for resource use, we should mention that the same range of behaviors is also found in more general situations.[4]

As a specific illustration of this approach, we consider the case of two resources, so the system can be described by the fraction f of agents which are using resource 1 at any given time. Its dynamics is then governed by[5]

$$\frac{df}{dt} = \alpha(\rho - f) \tag{1}$$

where α is the rate at which agents reevaluate their resource choice and ρ is the probability that an agent will prefer resource 1 over 2 when it makes a choice. Generally, ρ is a function of f through the density-dependent payoffs. In terms of the payoffs and uncertainty, we have

$$\rho = \frac{1}{2}\left(1 + \operatorname{erf}\left(\frac{G_1(f) - G_2(f)}{2\sigma}\right)\right) \tag{2}$$

where σ quantifies the uncertainty. Notice that this definition captures the simple requirement that an agent is more likely to prefer a resource when its payoff is relatively large. Finally, delays in information are modeled by supposing that the payoffs that enter into ρ at time t are the values they had at a delayed time $t - \tau$.

For a typical system of many agents with a mixture of cooperative and competitive payoffs, the kinds of dynamical behaviors exhibited by the model are shown in

(a)

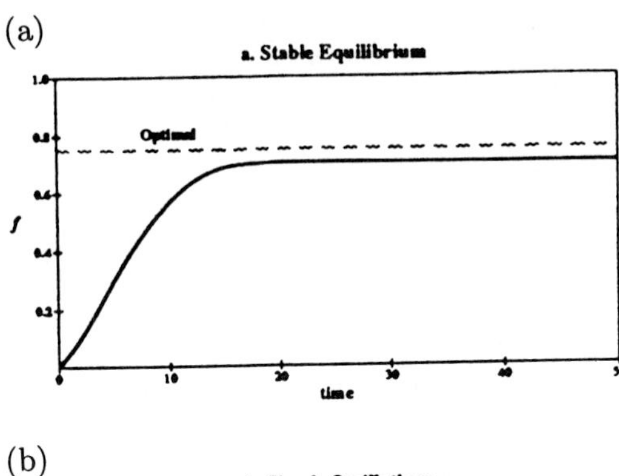

(b)

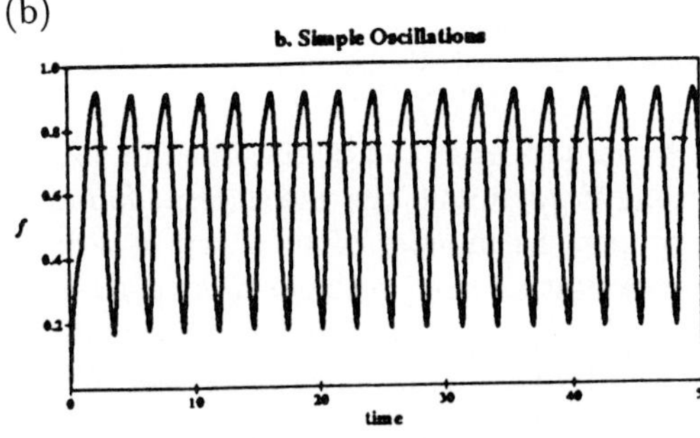

(c)

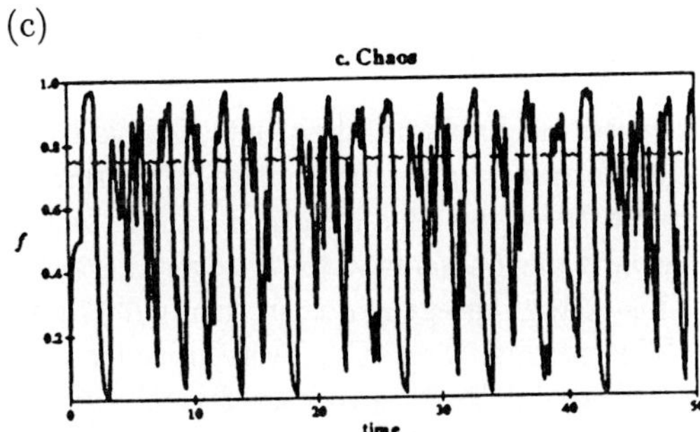

FIGURE 4 Typical behaviors for the fraction f of agents using resource 1 as a function of time for successively longer delays: (a) relaxation toward stable equilibrium, (b) simple persistent oscillations, and (c) chaotic oscillations. The payoffs are $G_1 = 4 + 7f - 5.333f^2$ for resource 1 and $G_2 = 4 + 3f$ for resource 2. The time scale is in units of the delay time τ, $\sigma = 1/4$, and the dashed line shows the optimal allocation for these payoffs.

Figure 4. When the delays and uncertainty are fairly small, the system converges to an equilibrium point close to the optimal obtainable by an omniscient, central controller. As the information available to the agents becomes more corrupted, the equilibrium point moves further from the optimal value. With increasing delays, the equilibrium eventually becomes unstable, leading to the oscillatory and chaotic behavior shown in the figure. In these cases, the number of agents using particular resources continues to vary so that the system spends relatively little time near the optimal value, with a consequent drop in its overall performance. This can be due to the fact that chaotic systems are unpredictable, hence making it difficult for individual agents to automatically select the best resources at any given time.

4. THE USES OF FITNESS

We will now describe an effective procedure for controlling chaos in distributed systems.[4] It is based on a mechanism that rewards agents according to their actual performance. As we shall see, such an algorithm leads to the emergence of a diverse community of agents out of an essentially homogenous one. This diversity in turn eliminates chaotic behavior through a series of dynamical bifurcations which render chaos a transient phenomenon.

The actual performance of computational processes can be rewarded in a number of ways. A particularly appealing one is to mimic the mechanism found in biological evolution, where fitness determines the number of survivors of a given species in a changing environment. In computation this mechanism is called a *genetic algorithm*.[2] Another example is provided by computational systems modeled on ideal economic markets,[9,12] which reward good performance in terms of profits. In this case, agents pay for the use of resources, and they in turn are paid for completing their tasks. Those making the best choices collect the most currency and are able to outbid others for the use of resources. Consequently they come to dominate the system.

While there is a range of possible reward mechanisms, their net effect is to increase the proportion of agents that are performing successfully, thereby decreasing the number of those who are less successful. It is with this insight in mind that we developed a general theory of effective reward mechanisms without resorting to the details of their implementations. Since this change in agent mix in turn will change the choices made by every agent and their payoffs, those that were initially most successful need not be so in the future. This leads to an evolving diversity whose eventual stability is by no means obvious.

Before proceeding with the theory, we point out that the resource payoffs that we will consider are instantaneous ones (i.e., shorter than the delays in the system), e.g., work actually done by a machine, currency actually received, etc. Other reward mechanisms, such as those based on averaged past performance, could lead to very different behavior from the one exhibited in this paper.

In order to investigate the effects of rewarding actual performance, we generalize the previous model of computational ecosystems by allowing agents to be different types, a fact which gives them different performance characteristics. Recall that the agents need to estimate the current state of the system based on imperfect and delayed information in order to make good choices. This can be done in a number of ways, ranging from extremely simple extrapolations from previous data to complex forecasting techniques. The different types of agents then correspond to the various ways in which they can make these extrapolations.

Within this context, a computational ecosystem can be described by specifying the fraction of agents, f_{rs} of a given type s using a given resource r at a particular time. We will also define the total fraction of agents using a resource of a particular type as

$$\begin{aligned} f_r^{\text{res}} &= \sum_s f_{rs} \\ f_s^{\text{type}} &= \sum_r f_{rs} \end{aligned} \tag{3}$$

respectively.

As mentioned previously, the net effect of rewarding performance is to increase the fraction of highly performing agents. If γ is the rate at which performance is rewarded, then Eq. 1 is enhanced with an extra term which corresponds to this reward mechanism. This gives

$$\frac{df_{rs}}{dt} = \alpha(f_s^{\text{type}}\rho_{rs} - f_{rs}) + \gamma(f_r^{\text{res}}\eta_s - f_{rs}) \tag{4}$$

where the first term is analogous to that of the previous theory and the second term incorporates the effect of rewards on the population. In this equation, ρ_{rs} is the probability that an agent of type s will prefer resource r when it makes a choice and η_s is the probability that new agents will be of type s, which we take to be proportional to the actual payoff associated with agents of type s. As before, α denotes the rate at which agents make resource choices and the detailed interpretation of γ depends on the particular reward mechanism involved. For example, if they are replaced on the basis of their fitness, it is the rate at which this happens. On the other hand, in a market system, γ corresponds to the rate at which agents are paid. Notice that in this case, the fraction of each type is proportional to the wealth of agents of that type.

Since the total fraction of agents of all types must be one, a simple form of the normalization condition can be obtained if one considers the relative payoff, which is given by

$$\eta_s = \frac{\sum_r f_{rs} G_r}{\sum_r f_r^{\text{res}} G_r} \, . \tag{5}$$

Note that the numerator is the actual payoff received by agents of type s given their current resource usage and the denominator is the total payoff for all agents in the system, both normalized to the total number of agents in the system. This form assumes positive payoffs: e.g., they could be growth rates. If the payoffs can be negative (e.g., they are currency changes in an economic system), one can use instead the difference between the actual payoffs and their minimum value m. Since the η_s must sum to 1, this will give

$$\eta_s = \frac{\sum_r f_{rs} G_r - m}{\sum_r f_r^{\text{res}} G_r - Sm} \tag{6}$$

where S is the number of agent types, and which reduces to the previous case when $m = 0$.

Summing Eq. 4 over all resources and types gives

$$\begin{aligned} \frac{df_r^{\text{res}}}{dt} &= \alpha \left(\sum_s f_s^{\text{type}} \rho_{rs} - f_r^{\text{res}} \right) \\ \frac{df_s^{\text{type}}}{dt} &= \gamma \left(\eta_s - f_s^{\text{type}} \right) \end{aligned} \tag{7}$$

which describe the dynamics of overall resource use and the distribution of agent types, respectively. Note that this implies that those agent types which receive greater than average payoff (i.e., types for which $\eta_s > f_s^{\text{type}}$ will increase in the system at the expense of the low performing types).

Note that the actual payoffs can only reward existing types of agents. Thus, in order to introduce new variations into the population, an additional mechanism is needed (e.g., corresponding to mutation in genetic algorithms or learning).

5. RESULTS

In order to illustrate the effectiveness of rewarding actual payoffs in controlling chaos, we examine the dynamics generated by Eq. 4 for the case in which agents choose among two resources with cooperative payoffs, a case which, as we have shown, generates chaotic behavior in the absence of rewards.[5,6] As in the particular example of Figure 4(c), we use $\tau = 10$; $G_1 = 4 + 7f_1 - 5.333f_1^2$; $G_2 = 7 - 3f_2$; $\sigma = 1/4$; and an initial condition in which all agents start by using resource 2.

One kind of diversity among agents is motivated by the simple case in which the system oscillates with a fixed period. In this case, those agents that are able to discover the period of the oscillation can then use this knowledge to reliably estimate the current system state in spite of delays in information. Notice that this

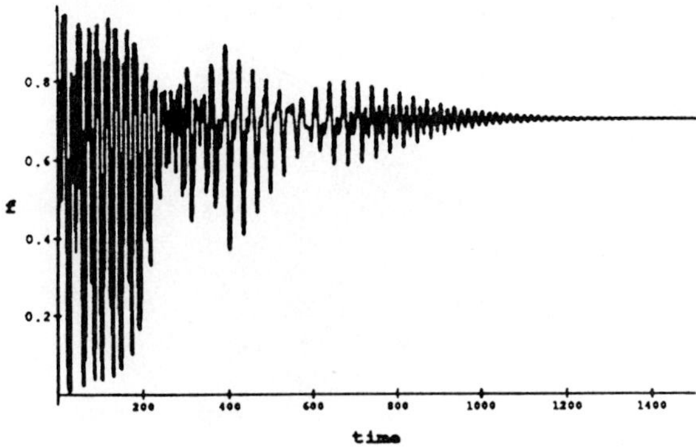

FIGURE 5 Fraction of agents using resource 1 as a function of time with adjustment based on actual payoff. These parameters correspond to Figure 4(c), so without the adjustment, the system would remain chaotic.

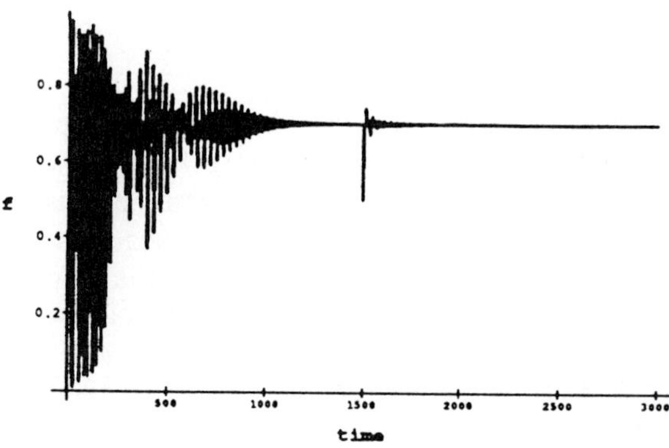

FIGURE 6 Behavior of the system shown in Figure 5 with a perturbation introduced at time 1500.

estimate does not necessarily guarantee that they will keep performing well in the future, for their choice can change the basic frequency of oscillation of the system.

In what follows, he diversity of agent types corresponds to the different past horizons, or extra delays, that they use to extrapolate to the current state of the system. These differences in estimation could be due to the variety of procedures for analyzing the system's behavior. Specifically, we identify different agent types with the different assumed periods that range over a given interval. Thus, agents of type s use an effective delay of $\tau + s$ while evaluating their choices.

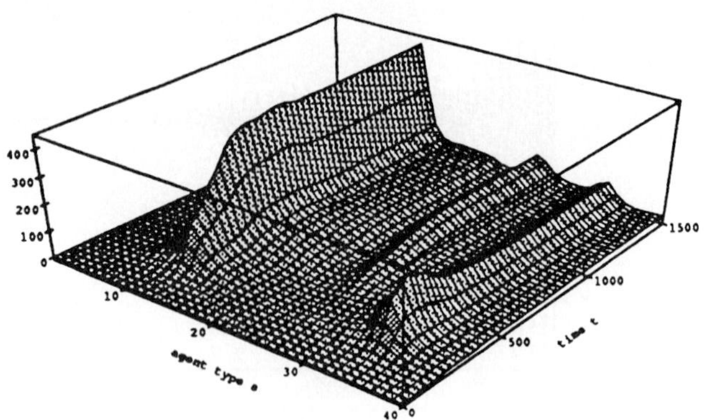

FIGURE 7 Ratio $f_s^{\text{type}}(t)/f_s^{\text{type}}(0)$ of the fraction of agents of each type, normalized to their initial values, as a function of time. Note there are several peaks, which correspond to agents with extra delays of 12, 26, and 34 time units. Since $\tau = 10$, these match periods of length 22, 36, and 44 respectively.

The resulting behavior, shown in Figure 5, should be contrasted with Figure 4(c). We used an interval of extra delays ranging from 0 to 40. As shown, the introduction of actual payoffs induces a chaotic transient that, after a series of dynamical bifurcations, settles into a fixed point that signals stable behavior. Furthermore, this fixed point is exactly that obtained in the case of no delays. This equilibrium is stable against perturbations because, if the system were perturbed again (as shown in Figure 6), it rapidly returns to its previous value. In additional experiments, with a smaller range of delays, we found that the system continued to oscillate without achieving the fixed point.

This transient chaos and its eventual stability can be understood from the distribution of agents with extra delays as a function of time. As can be seen in Figure 7, actual payoffs lead to a highly heterogeneous system characterized by a diverse population of agents of different types. It also shows that the fraction of agents with certain extra delays increases greatly. These delays correspond to the major periodicities in the system.

6. STABILITY AND MINIMAL DIVERSITY

As we showed in the previous section, rewarding the performance of large collections of agents engaging in resource choices leads to a highly diverse mix of agents that stabilize the system. This suggests that the real cause of stability in a distributed system is sufficient diversity, and the reward mechanism is an efficient way of automatically finding a good mix. This raises the interesting question of the minimal amount of diversity needed in order to have a stable system.

The stability of a system is determined by the behavior of a perturbation around equilibrium, which can be found from the linearized version of Eq. 4. In our case, the diversity is related to the range of different delays that agents can have. For a continuous distribution of extra delays, the characteristic equation is obtained by assuming a solution of the type $e^{\lambda t}$ in the linearized equation, giving

$$\lambda + \alpha - \alpha\rho' \int ds f(s) e^{-\lambda(s+\tau)} = 0 \,. \tag{8}$$

Stability requires that all the values of λ have negative real parts, so that perturbations will relax back to equilibrium. As an example, suppose agent types are uniformly distributed in $(0, S)$. Then $f(s) = 1/S$, and the characteristic equation becomes

$$\lambda + \alpha - \alpha\rho' \frac{1 - e^{-\lambda S}}{\lambda S} e^{-\lambda\tau} = 0 \,. \tag{9}$$

Defining a normalized measure of the diversity of the system for this case by $\eta \equiv S/\tau$, introducing the new variable $z \equiv \lambda\tau(1-\eta)$, and multiplying Eq. 9 by $\tau(1+\eta)ze^{z}$ introduces an extra root at $z = 0$ and gives

$$(z^2 + az)e^{z} - b + be^{rz} = 0 \tag{10}$$

where

$$\begin{aligned} a &= \alpha\tau(1+\eta) > 0 \,, \\ b &= -\rho' \frac{\alpha\tau(1+\eta)^2}{\eta} > 0 \,, \\ r &= \frac{\eta}{1+\eta} \in (0,1) \,. \end{aligned} \tag{11}$$

The stability of the system with a uniform distribution of agents with extra delays thus reduces to finding the condition under which all roots of Eq. 10, other than $z = 0$, have negative real parts. This equation is a particular instance of an *exponential polynomial*, having terms that consist of powers multiplied by exponentials. Unlike regular polynomials, these objects generally have an infinite number of roots and are important in the study of the stability properties of differential-delay equations. Established methods can be used then to determine when they have roots with positive real parts. This in turn defines the stability boundary of the equation. The result for the particular case in which $\rho' = -3.41044$, corresponding to the parameters used in Section 5, is shown in Figure 8(a).

(a)

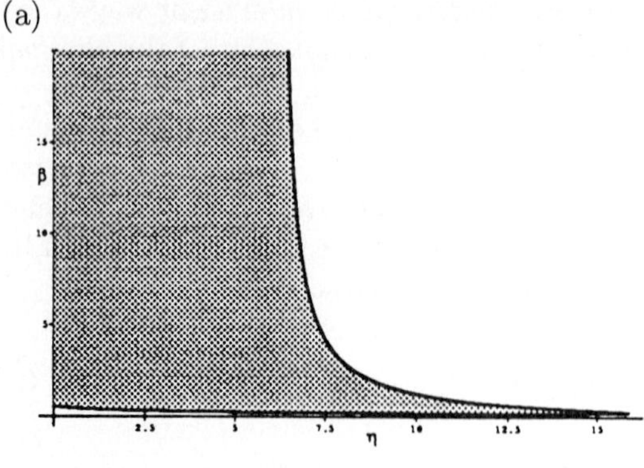

(b)

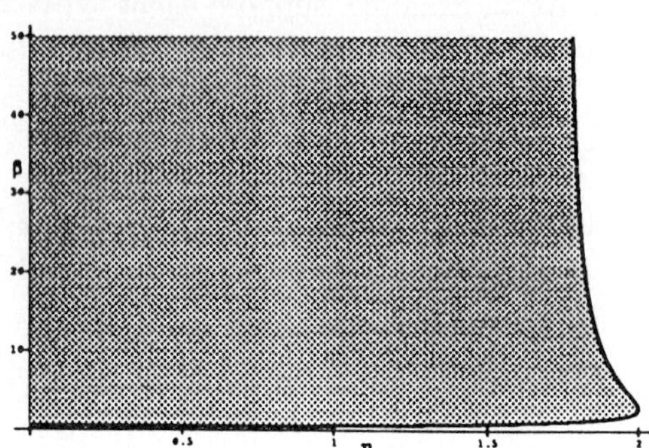

FIGURE 8 Stability as a function of $\beta = \alpha\tau$ and $\eta = S/\tau$ for two possible distributions of agent types: (a) $f(s) = 1/S$ in $(0, S)$ and (b) $f(s) = 1/Se^{-s/S}$. The system is unstable in the shaded regions and stable to the right and below the curves.

Similarly, if we choose an exponential distribution of delays, i.e., $f(s) = 1/Se^{-s/S}$ with positive S, the characteristic equation acquires the form

$$(z^2 + pz + q)e^z + r = 0 \tag{12}$$

where

$$\begin{aligned} p &= \alpha\tau + \frac{1}{\eta} > 0\,, \\ q &= \frac{\alpha\tau}{\eta} > 0\,, \\ r &= \frac{-\alpha\tau\rho'}{\eta} > 0\,, \end{aligned} \tag{13}$$

and $z \equiv \lambda\tau$. An analysis similar to that for the uniform distribution case leads to the stability diagram shown in Figure 8(b).

Although the actual distributions of agent types can differ from these two cases, the similarity between the stability diagrams suggests that, regardless of the magnitude of β, one can always find an appropriate mix that will make the system stable. This property follows from the vertical asymptote of the stability boundary. It also illustrates the need for a minimum diversity in the system to stablize it when the delays are not too small.

Having established the right mix that produces stability one may wonder whether a static assignment of agent types at an initial time would not constitute a simpler and more direct procedure to stabilize the system without resorting to a dynamic reward mechanism. While this is indeed the case in a nonfluctuating environment, such a static mechanism cannot cope with changes in both the nature of the system (e.g., machines crashing) and the arrival of new tasks or fluctuating loads. A dynamic procedure is needed precisely to avoid this vulnerability by keeping the system adaptive.

Having seen how sufficient diversity stabilizes a distributed system, we now turn to the mechanisms that can generate such heterogeneity as well as the time that it takes for the system to stabilize. In particular, the details of the reward procedures determine whether the system can even find a stable mix of agents. In the cases describe above, reward was proportional to actual performance, as measured by the payoffs associated with the resources used. One might also wonder whether stability would be achieved more rapidly by giving greater (than their fair share) increases to the top performers.

We have examined two such cases: (a) rewards proportional to the square of their actual performance and (b) one giving all the rewards to top performers (e.g., those performing at the 90th percentile or better in the population). In the former case we observed stability with a shorter transient whereas, in the latter case, the mix of agents continued to change through time, thus preventing stable behavior. This can be understood in terms of our earlier observation that, whereas a small percentage agents can identify oscillation periods and thereby reduce their amplitude, a large number of them no longer can perform well.

Note that the time to reach equilibrium is determined by two parameters of the system. The first is the time that it takes to find a stable mix of agent types, which is governed by γ, and the second is the rate at which perturbations relax, given the stable mix. The latter is determined by the largest real part of any of the roots, λ, of the characteristic equation.

7. DISCUSSION

In this paper we have presented a case for treating distributed computation as an ecosystem, an analogy that turns out to be quite fruitful in the analysis, design, and control of such systems. Resource contention, complex dynamics, and reward mechanisms seem to be ubiquitous in distributed computation, making it also a tool for the study of natural ecosystems in spite of the many differences between computational processes and organisms,.

Since chaotic behavior seems to be the natural result of interacting processes with imperfect and delayed information, the problem of controlling such systems is of paramount importance. We discovered that rewards based on the actual performance of agents in a distributed computational system can stabilize an otherwise chaotic or oscillatory system. This leads in turn to greatly improved system performance.

In all these cases, stability is achieved by making chaos a transient phenomena. In the case of distributed systems, the addition of the reward mechanism has the effect of dynamically changing the control parameters of the resource allocation dynamics in such a way that a global fixed point of the system is achieved. This brings the issue of the length of the chaotic transient as compared to the time needed for most agents to complete their tasks. Even when the transients are long, the results of this study show that the range gradually decreases, thereby improving performance even before the fixed point is achieved.

A particularly relevant question for distributed systems is the extent to which these results generalize beyond the mechanism that we studied. We only considered the specific situation of a collection of agents with different delays in their appraisal of the system evolution.—Hence it remains an open question whether using rewards to increase diversity works more generally than in the case of extra delays.

Since we only considered agents choosing between two resources, it is important to understand what happens when the agents have many resources to choose from. One may argue that since diversity is the key to stability, a plurality of resources provides enough channels to develop the necessary heterogeneity, which is what we observed in situations with three resources. Another note of caution: While we have shown that sufficient diversity can, on average, stabilize the system, in practice a fluctuation could wipe out those agent types that otherwise would be successful in stabilizing the system. Thus, we need either a large number of each kind of agent or a mechanism, such as mutation, to create new kinds of agents.

A final issue concerns the time scales over which rewards are assigned to agents. In our treatment, we assumed the rewards were always based on the performance at the time they were given. Since in many cases this procedure is delayed, there remains the question of the extent to which rewards based on past performance are also able to stabilize chaotic distributed systems.

The fact that these simple resource allocation mechanisms work and produce a stable environment provides a basis for developing more complex software systems that can be used for a wide range of computational problems.

REFERENCES

1. Ferguson, D., Y. Yemini, and C. Nikolaou. "Microeconomic Algorithms for Load Balancing in Distributed Computer Systems." In *International Conference on Distributed Computer Systems*, 491–499. Washington, DC: IEEE, 1988.
2. Goldberg, D. E. *Genetic Algorithms in Search, Optimization and Machine Learning.* Reading, MA: Addison-Wesley, 1989.
3. Hayek, F. A. "Competition as a Discovery Procedure." In *New Studies in Philosophy, Politics, Economics and the History of Ideas*, 179–190. Chicago: University of Chicago Press, 1978.
4. Hogg, T., and B. A. Huberman. "Controlling Chaos in Distributed Systems." *IEEE Trans. Sys., Man & Cyber.* **21(6)** (1991): 1325–1332.
5. Huberman, B. A., and T. Hogg. "The Behavior of Computational Ecologies." In *The Ecology of Computation*, edited by B. A. Huberman, 77–115. Amsterdam: North-Holland, 1988.
6. Kephart, J. O., T. Hogg, and B. A. Huberman. "Dynamics of Computational Ecosystems." *Phys. Rev. A* **40** (1989): 404–421.
7. Kurose, J. F., and R. Simha. "A Microeconomic Approach to Optimal Resource Allocation in Distributed Computer Systems." *IEEE Trans. Comp.* **38(5)** (1989): 705–717.
8. Malone, T. W., R. E. Fikes, K. R. Grant, and M. T. Howard. "Enterprise: A Market-Like Task Scheduler for Distributed Computing Environments." In *The Ecology of Computation*, edited by B. A. Huberman, 177–205. Amsterdam: North-Holland, 1988.
9. Miller, M. S., and K. E. Drexler. "Markets and Computation: Agoric Open Systems." In *The Ecology of Computation*, edited by B. A. Huberman, 133–176. Amsterdam: North-Holland, 1988.
10. Smith, A. *An Inquiry into the Nature and Causes of the Wealth of Nations.* Chicago: University of Chicago Press, 1976. Reprint of the 1776 edition.
11. Sutherland, I. E. "A Futures Market in Computer Time." *Comm. ACM* **11(6)** (1968): 449–451.
12. Waldspurger, C. A., T. Hogg, B. A. Huberman, J. O. Kephart, and W. S. Stornetta. "Spawn: A Distributed Computational Economy." *IEEE Trans. Software Engr.* **18(2)** (1992): 103–117.

A. T. Winfree
326 BSW, University of Arizona, Tucson, AZ 85721; e-mail: 73257.2443@compuserve.com

The Geometry of Excitability

This chapter presents a series of lectures given from this manuscript (with many digressions and less coherently) at St. John's College in Santa Fe. There was quite a lot of illustrative material, including reprints, laboratory demonstrations, computer programs, 135 color slides, and 4 videos. Two dozen printed pictures stand in for it here. I am keeping the first three of these five lectures unpedantic and devoid of scholarly apparatus. There are few citations in the first three, and I am leaving out all physiological detail and mathematics, to try instead to convey the context in which they may have interest to you when you are ready. The last two lectures are deliberately redundant. They review and supplement by traversing much of the same material from a somewhat different direction. Every step is keyed to a bibliography of about 247 publications. These should suffice for whatever follow-up you choose.

1992 Lectures in Complex Systems, Eds. L. Nadel & D. Stein, SFI Studies in the Sciences of Complexity, Lect. Vol. V, Addison-Wesley, 1993

FIRST LECTURE

Many were startled by a tragic incident in 1990, when basketball player Hank Gathers unexpectedly collapsed on the court and died of heart failure on national television. Sudden cardiac death was a new experience for most witnesses. Though it seldom happens to people in youthful vigor, it does occur about 1,000 times a day even in the United States alone, so perhaps 20,000 times a day in the world at large, or about once every 4 seconds. In fact, "the majority of deaths in the developed countries of the world are caused by coronary artery disease, with the majority of these deaths occurring suddenly due to. . .ventricular fibrillation."[238]

What the problem boils down to is that the muscle of the heart has basically two modes of activity. It has a nice orderly mode of synchronous squeezing about once a second, or up to about twice or three times that fast when pumping more blood to your brain and muscles during vigorous exercise or making love or watching a horror show like CNN. *And* it has a second mode in which contraction occurs locally but not in a globally synchronous way, about 5–10 times per second. This second mode looks uncoordinated and does *not* manage to pump blood to your brain and muscles. So you faint and in a couple more minutes your brain fails irreversibly, then respiration stops, then metabolism slows drastically throughout the body. That is called death; in this case, specifically, "sudden cardiac death." This *second* mode of perfectly healthy heart muscle is not at all understood, but like most things that are not well understood, it has an imposing name: it is a cardiac arrhythmia called ventricular fibrillation. The word "cardiac" means "having to do with the heart." "Ventricular" means "more specifically, having to do with the ventricular chambers of the heart," the heavy musculature that you are using in mode 1 right now. The word "arrhythmia" *sounds like* it means "not rhythmic," but as used by cardiologists it actually means *any departure, whether rhythmic or not, from the usual rhythm.* The word "fibrillation" sounds like it means "falling apart into fibrils," as though each tiny part of the muscle were twitching independently of the others. That is what people *thought* until 50 years ago, when the name had already taken root.

The cardiac arrhythmia called ventricular fibrillation is probably the least well understood of all arrhythmias. The name has been around for close to a century and much study within the usual context of medicine and physiology has been invested in figuring out what it means. I am going to tell you about an approach to the problem from the perspective of dynamical systems theory.

The center of attention will be solutions of partial differential equations purporting to represent the electrical activity of heart muscle, which precedes and determines its mechanical contractions. These equations support stable particle-like pulsating/rotating modes of activity in two- or three-dimensional continua called "excitable media."

EXCITABILITY

Before saying what an excitable medium is, I should say what *excitability* is. First of all, it is an adjective. You find this adjective stuck onto materials, and people, and organs that are content to just sit there and not do anything until stimulated in the right way, and then still not to do anything dramatic unless stimulated beyond a certain threshold, but *then* to make a big fuss and go *in*excitable before gradually returning to the excitable state. Excitable systems usually have a threshold. To excite promptly and dramatically, they need more than that much stimulation. Less won't set off the dramatic response that gives excitable systems their name. They also have a so-called refractory state just after excitation, while they are recovering and cannot yet be excited again. When stated verbally like this, these several properties sound like separate ad hoc gimmicks concatenated arbitrarily, but, when you look at excitability in terms of vector fields, it is evident that they are just three aspects of one thing.

For an example of excitability, consider yawning. Yawning can be provoked by someone else yawning sufficiently nearby. Just *thinking* of someone else yawning is usually not a sufficient stimulus, but *seeing* your neighbor yawn widely often makes it irresistible, exceeding your threshold.

One simple idealization of excitability in nerve membranes was provided during the middle of this century by K. F. Bonhoeffer in Germany, Richard FitzHugh in the United States, and Jin-Ichi Nagumo in Japan:

$$\begin{aligned} \frac{\partial u}{\partial t} &= \frac{u - \frac{u^3}{3} - v}{\varepsilon}, \\ \frac{\partial v}{\partial t} &= \varepsilon(u + \beta - \gamma v). \end{aligned} \tag{1}$$

These equations represent the following situation. At each point in a spatial continuum, two local state variables, denoted u and v, change continuously in time. Quantity u represents an electric potential difference across the cell membrane and v represents one of the ionic currents responsible for maintaining that gradient. The local kinetics are described by two first-order ordinary differential equations. The kinetics adopted are quite simple: but for one term, rates of change are linear in two variables. The diffusing quantity u, called the "excitation variable," "activator" or "propagator," is degraded at a rate proportional to v, which is called the "recovery variable," "inhibitor" or "controller." Quantity u is also autocatalytically regenerated in proportion to u. Its "nullcline" is the locus on the (u, v) plane of Figure 1 where u's rate of change is 0. This would be the horizontal line $u = 0$ were there no more to the equation. But with the additional cubic degradation rate, the first equation's nullcline becomes a Z-shaped zigzag. Meanwhile, the recovery variable, v, grows at a rate proportional to u plus a constant, while also undergoing first-order decay. Its nullcline is a straight line. At the intersection between the

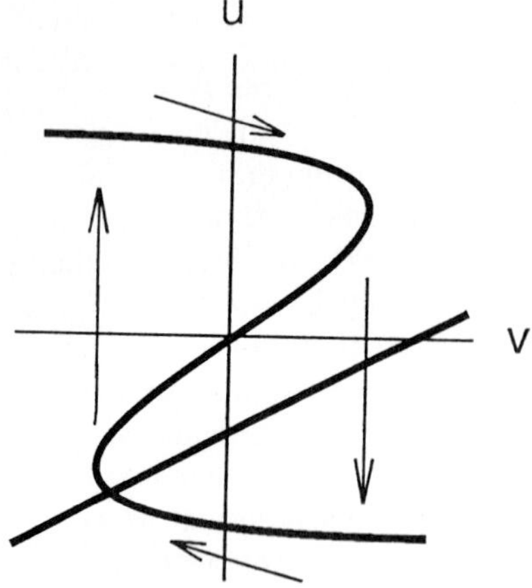

FIGURE 1 A sketch of the phase portrait of the FitzHugh-Nagumo model of excitability, per Eq. (1), supposing spatial uniformity or $D = 0$. The heavy curve is the "nullcline" locus $du/dt = 0$; to its left $du/dt > 0$. The heavy line is the "nullcline" locus $dv/dt = 0$, supposing β and γ both about 1/2; to its left $dv/dt > 0$. Both rates are 0 where nullclines cross in a unique (attracting, in this instance) equilibrium.

linear and cubic nullclines, there is a globally attracting equilibrium. If the local state is perturbed from equilibrium by displacing u across the middle branch of the cubic nullcline (thus crossing a threshold), then it departs farther before returning clockwise toward equilibrium.

Now consider the constants (parameters). Parameter ε governs the rate of this excitation relative to the recovery process. If parameter ε is small, then the $\partial u/\partial t$ is much quicker than $\partial v/\partial t$: u increases until it crosses the upper branch of the cubic, then glides along it until jumping to the lower branch, which escorts it back to the original equilibrium. The offset parameter β and slope parameter γ locate the flat nullcline, so they locate equilibrium along the Z-nullcline and so determine the threshold for excitation from that resting state. This is a generic picture of the local dynamics of one very common kind of continuous excitable kinetics.

EXCITABLE MEDIA

The general idea of excitability, in this zero-dimensional or spatially uniform sense, is thus that a small but not too small stimulus, applied to such a dynamical system while it is in a certain range of states called "excitable" states, may provoke it to execute a relatively large and rapid excursion. In an excitable *medium*, the local dynamics are excitable *and* neighboring regions are so coupled that excitation in one region can provide its neighbors with the kind of stimulus required to provoke excitation *there*. An excitable *medium* is an array of *many* excitable units, each coupled to neighbors so that one can trigger the next, like those lines of dominoes you sometimes see in TV commercials. One domino falls, kicking off the next, which shortly after kicks off the next, and a wave of falling propagates. A solitary pulse of asymptotically stable shape and speed then propagates by this mechanism.

It is also not hard to imagine a wave of yawns propagating across a lecture theater. Excitable media play a fundamental role in phenomena that interest physical chemists and physiologists, just as hydrodynamic media and the electromagnetic continuum and the complex-valued field of quantum mechanics play fundamental roles in phenomena that interest physicists. All are described by partial differential equations describing the rates of change of locally defined quantities in terms of all those quantities and their spatial derivatives.

One way to make an excitable medium from such local dynamics is by spatially coupling a continuum of such sites through the second spatial derivative of the propagator variable, u. The local time rate of change of u now consists of local excitable kinetics plus this new "Laplacian" term representing diffusion of u from adjacent sites. In these lectures I represent the idea of excitability in such equations, and numerically study their properties alongside parallel laboratory investigations of two representative excitable media, one physiological and one chemical. The self-organizing stable objects that I want to show you were discovered while trying to understand the kind of turbulence called "fibrillation" in the excitable medium called "heart muscle." They also occur in a chemically excitable medium called the "Belousov-Zhabotinsky reagent." They are probably endemic to all sorts of excitable media. Such excitable media are commonly written as partial differential equations in two (or more) local scalar variables, both of which diffuse:

$$\begin{aligned} \frac{\partial u}{\partial t} &= f(u,v) + D_u \nabla^2 u, \\ \frac{\partial v}{\partial t} &= g(u,v) + D_v \nabla^2 v; \end{aligned} \tag{2}$$

for example,

$$\frac{\partial u}{\partial t} = \frac{u - \frac{u^3}{3} - v}{\varepsilon} + D\nabla^2 u, \tag{3a}$$

$$\frac{\partial v}{\partial t} = \varepsilon(u + \beta - \gamma v) + \delta D \nabla^2 v. \tag{3b}$$

In *electrophysiological* applications, propagator variable u represents local electric potential across the cell membrane, a quantity which diffuses easily by field-guided transport of ions. Controller variable v represents the ionic conductivity of relatively immovable proteins embedded in the cell membrane which act as voltage-sensitive ionic channels. In this application the recovery variable (the second equation, governing v) is strictly local: $\delta = 0$. These two equations (3) taken together are then called "the cable equation" (because the essential physics was borrowed from Lord Kelvin's analysis of a proposed trans-Atlantic undersea telegraph cable).

In context of *chemical* excitability, both u and v are the local concentrations of reacting molecular species. The Laplacian introduces Fick's law of molecular diffusion, added to the local kinetics of synthesis and degradation—to all rate equations, not just the propagator's, if the chemical species are all comparably mobile.

Without the cubic degradation term, we have a linear partial differential equation. With parameters in the range that will interest us, its analytical solution relaxes boringly from any initial distribution of u and v promptly toward uniform quiescence. But with the nonlinear term added, so that the local dynamic is excitable, we encounter the possibility of interesting behavior: a self-sustaining impulse may propagate like a shock front. This is supposed to represent a chemical wave or an electrophysiological impulse. In these lectures I will draw your attention to the geometrical forms of this shock front in two- and three-dimensional media, and the roles they play in cardiac physiology and in physical chemistry.

HEART MUSCLE ("MYOCARDIUM")

The study of excitable media leads to one of the central problems of cardiology, the problem of ventricular fibrillation, and I think to a *quantitative* understanding of certain aspects of it. Let's start with basic facts about the electrical mechanisms of the normal heartbeat. The human heart is, first of all, a mechanical pump. Like any other mechanical pump, it has lots of parts. We focus on one part, the thick ventricular muscle that provides power for the pump. The key to normally coordinated contraction is an electrical signal called an action potential. The action potential is usually depicted in textbooks as an impulse propagating one-dimensionally along a nerve fiber. It starts, propagates, and is snuffed out when there is no more membrane to activate. Similarly in the thick muscle of the human ventricle, it propagates two-dimensionally or even three-dimensionally, normally from inside to outside. The additional dimensions introduce qualitatively new possibilities for more complex modes of activity. But now for starters consider just the normal spread of an action potential in the human ventricular wall. It is a one-way trip from inside to outside and it takes only 1/14 second: activation is essentially synchronous. This normally synchronous activation shows as an electrical spike once per second on an electrocardiogram. (The electrocardiogram integrates into a spot reading of voltage the sum over all boundaries of the excitable medium of the component in the direction of the electrode of the activation front's propagation vector.) If the ventricular muscle becomes uncoordinated in its normally synchronous contractions, then the ventricular chambers cease to rhythmically contract in unison, blood ceases to surge once a second into the brain, and the deprived brain within minutes changes irreversibly, as mentioned above.

In this tragic episode the electrocardiogram's discrete spikes turn into 5-hertz smooth wiggles (which more commonly accelerate to about 10 hertz). What's going on at this stage is called "reentry." Five times per second the action potential is somehow reentering muscle that it already passed through, without ever snuffing out: the same action potential circulates endlessly. How do we know this? Well, when people are having open-chest surgery in hopes of correcting a life-threatening cardiac arrhythmia, you can fit an array of electrodes around the whole heart so as to monitor the times when action potentials arrive at hundreds of sites around the surface. Then the heart is stimulated electrically to try to provoke the specific cardiac arrhythmia that this particular heart easily falls into. A computer gathers these hundreds of voltage measurements once every thousandth of a second, and paints a picture on a TV screen showing how the action potential is moving across the heart surface. This is called an "epicardial map." On such a screen you see the activation front circulating several times each second. This high-frequency reactivation apparently makes the muscle vulnerable to fibrillation: the 5-hertz regularity begins to break up and degenerates into a ragged wiggle. Meanwhile, the blood is not circulating, the brain dies, and respiration stops.

This kind of trace provides an important clue to the initial stage of *vulnerability* to fibrillation and sudden cardiac death. But giving it a name—reentry—doesn't quite answer the question: what is going on here? To find out, it is instructive to digress for a close look at such waves in *other* excitable media that are, at the moment, experimentally more tractable. This digression is intended to bring you to back to cardiac arrhythmias from a perspective that may have biomedical engineering applications which remain to be exploited.

Several excitable media share certain basic features with mammalian heart muscle. For example, they have a threshold, and a refractory state between excitation and recovery. For some excitable media, the degree of correspondence is pretty amazing. Even the *equations* of their *mechanism* have the same form as the *electrophysiologist's* equation for electrical conduction in heart muscle. Of course, they lack a lot of the distinguishing idiosyncrasies of heart muscle; e.g., they are not made of cells. But the point of doing science is *supposed to be* to simplify things, to find out what are the *essentials*, and to tease them apart from the incidentals. No one knows beforehand which are the essentials and which are the incidentals, so we make models and do experiments and eventually find out. In the present case the striking result is that, in quite a broad range of analogous excitable media, we find behaviors strikingly similar to reentry and the early stages of fibrillation. These behaviors include a vulnerable phase, a special moment when a stimulus can evoke this alternative mode of spontaneous activity.

THE BELOUSOV-ZHABOTINSKY MEDIUM

To understand the geometry of action potential propagation in excitable media, it would be convenient to study an excitable medium in which everything happens slower—in seconds or minutes rather than in fractions of a second. And it would be nice if the propagating disturbance were not an invisible electrical disturbance, for example, if it were visibly indicated by color changes. Exactly such a *chemical* excitable medium was discovered in the Soviet Union by Boris Belousov in 1950. Belousov was a biochemist interested in the basic metabolic cycle of oxidative metabolism in all cells called the Krebs' Cycle. He was trying to simplify the Krebs' Cycle in a cell-free extract so that its essential functions could be studied in glassware. He overdid it, reducing the Krebs' cycle to just one substrate (citric acid), replacing the enzymes by a metallic catalyst, the cerium ion, and replacing the mitochondrial electron-transport system by an inorganic oxidizer, the bromate ion.

In this drastic oversimplification he found a reaction that oscillates spontaneously while changing color. His report was imagined at the time to violate sacred dogmas of theoretical thermodynamics, so it was persistently refused publication in Russian journals from 1950 until Belousov aged, withdrew, and died embittered in 1970. In the late 1970s, several years too late, I was privileged to be party to a letter-writing campaign that secured the 1980 Leninski Primia (Lenin Prize) for his work. That is a considerable honor, but Belousov never knew about it.

This Lenin Prize was shared by *five* people. Among the four still living were A. Zhabotinsky and A. Zaikin. They had discovered 20 years after Belousov's first unpublished report that this oscillating reaction also *propagates* an impulse of oxidation/reduction activity. Until then this reaction had been studied mainly as an instance of limit-cycle kinetics in homogeneous solution, so modeling started by adding molecular diffusion to a limit-cycle kinetics like the Brusselator or a version of the Ginzburg-Landau equation (Eq. (5) below). Yet one might ask whether the phenomena of propagation necessarily have anything to do with the limit cycle, or might persist unaltered in its absence. I posed this question experimentally and found that the BZ medium (as it came to be called[211]) can be altered to *not* oscillate, but instead just be an ordinary excitable medium with the main properties that enable heart muscle to propagate an impulse three-dimensionally. To extricate it from the overwhelming context of limit-cycle modeling, I called the nonoscillating variant "the malonic acid reagent," but the name didn't stick: instead the original name generalized. In these lectures I use "BZ" for all versions.

Figure 2 shows a thin, essentially *two*-dimensional layer of this chemical solution in a plastic dish such as I was using at the time for experiments about pattern formation in fungi. A few sites of spontaneous periodic activity radiate pulses of oxidative activity at regular intervals. Notice that each pacemaker has its own period. The mechanism of propagation is that bromous acid is created in the blue zone and diffuses into the orange region just ahead until it too gets excited beyond a threshold and turns blue itself, emitting bromous acid molecules to topple the next domino, so to speak. This pulse propagates at a fixed speed in the order of millimeters per *minute* rather than half a millimeter per *millisecond* as in heart muscle. But just as in heart muscle and just as in grass fires, each pulse has a refractory wake. You can see the consequence wherever two fronts collide: neither continues through the ashlike wake of the other. Behind the tapering wake, the medium recovers and again grows excitable.

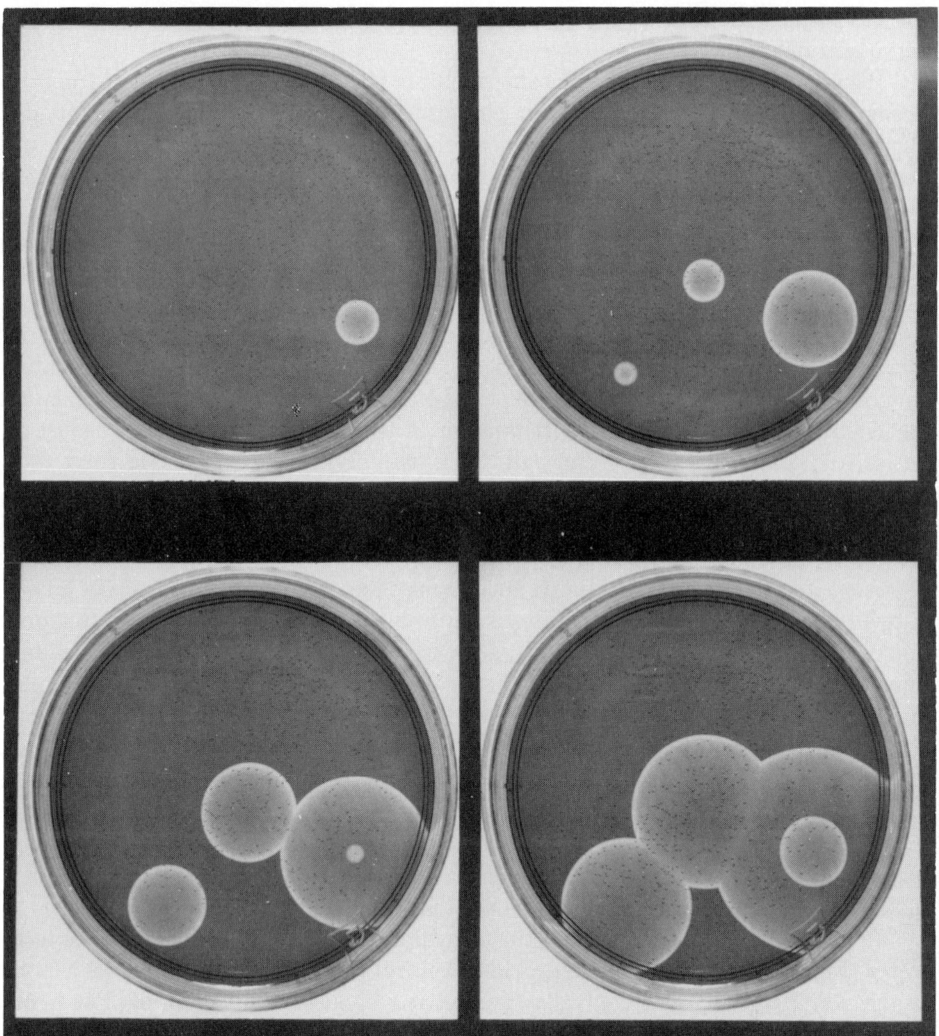

FIGURE 2 Circular waves radiating from pacemaker points in a 1.6-mm layer of the Belousov-Zhabotinsky excitable medium. The dish is 47 mm in diameter. Snapshots are 30 sec apart.

The mechanism of this reaction and its three-dimensional coupling through molecular diffusion are now understood in enough detail so that they can be written down as fairly exact equations for numerical solution in a computer. The remarkable feature that makes it an instructive analogy to heart muscle is that these equations turn out to be identical in their essential features to the one written above and to the electrophysiologist's cable equation for action potential propagation in living nerve and muscle.

It was in this chemically excitable medium that something *strange* turned up in my lab (and in the same year, in Zhabotinsky's near Moscow). This happened due to an experiment that I designed to check one of the implications of my understanding of the mechanism at that time. This was a topological inference that *rotating* waves could not exist in any such medium. Since I already believed it, I was eventually able to *prove* it (not rigorously) on paper. Another essential part of any mathematical proof taken as a metaphor for physical/chemical phenomena remained to be carried through: illustration in the laboratory to check whether the assumptions and approximations behind the mathematics were valid in this particular instance. If this "theorem" were right and pertinent, then if we stir up random complicated patterns of excitation in the chemical medium, we should see them all resolve into wave fronts of the sort familiar in two-dimensional physics, acoustics, optics, radio and TV engineering, and so forth, viz., closed concentric rings radiating away from sources here and there. You would never see, for example, wave fronts that just end without closing, leaving a dangling stump.

Or would you? Try looking at it this way:

Any excitable medium supports such a pulselike plane wave *or* a *train* of pulses. A train of equispaced traveling pulses is equivalent to a single pulse circulating on a one-dimensional ring. There is a one-parameter *family* of such solutions, depending on wave spacing or, equivalently, on the ring's perimeter. One might expect similar circulation on a thin enough planar annulus or within a thin enough solid torus. This circulating solution persists if the annulus or torus is fattened. It persists *even* when the hole in the annulus or torus is finally closed. What happens when the hole closes? Does the entire wave front vanish, perhaps by swift erosion from the unsupported tip? In passive media (elastic continua, water surface, the electromagnetic continuum) and in some excitable media, it does. Let me leave it there for a moment, and direct your attention to another way of posing the question.

Imagine doing *this* experiment in any familiar wave-supporting medium, e.g., water (Figure 3, top). When you provide a central perturbation by throwing a stone in the pond, you expect a centrally symmetric response, a collection of closed concentric wave fronts. This is indeed what you get in solving any familiar so-called "wave equation" for passive media, whether for water waves, sound waves, radio waves, or light waves. It is also what you get in any excitable medium, initially at rest then excited from a central disk. Now try throwing a second stone in the water. You expect to see a second, interpenetrating set of circles concentric to the new splash site. But excitable media and such wave equations as shown above turn out to respond in a topologically different way (Figure 3, second panel). The second excitation fails to propagate, wherever its wave fronts overlap the refractory wake of the first waves. For this to happen, the stimulus must be given at a place and moment after the trailing edge of the refractory wake has passed it, and its wave front must originate at some radius from the splash site so that it can intersect the

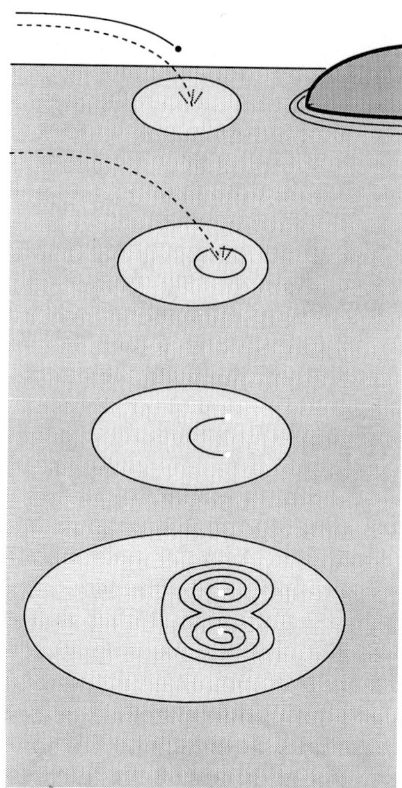

FIGURE 3 The bottom panel is not what one would expect of familiar wave equations (e.g., for water ripples in a pond), but it is what occurs in excitable media.

circle of the trailing edge of wake. If all that were satisfied, then you might expect the surviving arc of a circle to propagate normally, just disappearing wherever it overlaps that moving wake (Figure 3, third panel). The endpoints of the arc would thus move apart along the hyperbola on which the two moving circles can intersect.

SPIRAL WAVES RADIATED FROM ROTORS

But surprisingly, contradicting Wiener and Rosenblueth's[200] Figure 19(a)(b), for example, in real chemical and physiological excitable media, the endpoints of a broken wave front turn out not to propagate but to stay where they were first created. The wave front coils around them into two counter-rotating spirals radiating out of a tiny vortex core (Figure 3, bottom), each like a rotating garden sprinkler spinning out a wave front of excitation. The wave front is a spiral with equal distances between successive turns, so along every radius there is an outgoing periodic wave train at the period of source rotation. Because these wave trains are progressively phase-shifted from northward to eastward to southward to westward, the wave front ends not far from the pivot around which the spiral turns. Except in media that are only marginally excitable, the vortex core, alias "rotor," is about 1/3 wavelength in diameter: 140 microns in the chemical medium, about 10 mm in myocardium. Its wavelength squared and divided by its period seldom differs more than four-fold from $8\pi^2$ times the diffusion coefficient. Unlike any pacemaker point, the core's interior is *not* going through the same cycle of excitation and recovery experienced by every other point. This source, paradoxically, is quiescent unless "meandering" starts (see below), and if the spiral meanders, then the center excites at intervals distinct from the rotation period of the spiral. At least, it seems paradoxical if you imagine that the waves acquire their energy from the source, and carry it away, as in passive media like a vacuum or air. In excitable media, however, the energy is produced locally and the passing wave merely milks the local medium, much as light waves do in a laser cavity.

Such behavior is never seen in autonomous solutions to "the wave equation(s)" of passive media in familiar physics. You don't see things like this in two-dimensional optics, acoustics, and so on (though they can occur in active media, e.g., Coullet et al.,[36] Brambilla et al.,[26] and in cross sections transverse to the propagation direction, in three-dimensional wave fields, e.g., Wright and Berry[240]). Solutions of the familiar "wave equation" in empty space *cannot* support a dangling endpoint or radiate continually from a pivot in a uniform continuum because that would constitute an endless source of energy. Of course, one *might* provide an energy source at the center and let it rotate: a rotating neutron star does emit an electromagnetic wave in the shape of a equispaced spiral. A motorboat racing in a tight circle creates a similar scene in water. And in an active medium such as the inverted population of gases in a laser, every point in the medium is an energy source: rather than Maxwell equations, one uses Maxwell-Bloch equations, and spirals become possible. In such cases, and in the chemical and electrophysiological cases we will attend to in these lectures, the contours of fixed level of any local state variable (water depth, magnetic field intensity, chemical concentration, voltage) are indeed closed rings (very elongated and rolled-up) as they must be for topological reasons unless the medium has a boundary near enough to matter. But each contour runs through *two* mirror-image spirals, changing from wave *front* to wave back as it passes near the center of each. Only by ignoring the wave *back* part of one contour can we get the misleading appearance of a dangling end. The feature that makes excitable media different from the familiar media of physics is not the dangling wave fronts, but the fact that the medium *does* contain an energy source, in fact at every point.

What we see outside the rotor in any given medium is not another kind of wave, but just the same kind of wave, radiating from a different kind of source. Instead of an ephemeral source, a single stimulus that sends out an impulse in all directions, we have a self-sustaining structure, a tiny disk of rotating concentration gradients that excites the surrounding medium. The mystery is not so much about the wave as about the tiny disk, the rotor: how is it created, why does it persist, what are its dynamical properties, how can it perish?

Since it continues to rotate, it continues to excite, and provides a perpetual source. Without changing the parameters of the medium, its period can be made longer or shorter. This can be done by carving out of it a hole of the right size. In media with parameters in a certain range, it can also be done by tickling the rotor to switch to an alternative stable anatomy that rotates at a different period. Then the spiral wave is radiated at that new period and wavelength. The propagating wave front is an interesting structure, but it differs in no important way from one emitted from a radially symmetric source. The interesting difference here is in the source, a reaction-diffusion structure that forms near the endpoint of a broken front. The rotor looks like an energy source, like the rotating neutron star mentioned above, but please remember that in excitable media "energy" is not a useful concept and waves do *not* attenuate as they spread out because every point, not just the rotor, is an energy source. The rotor is only an organizing center, a signal source at the origin of the reentrant propagation pattern.

THE FUNDAMENTAL PROBLEM OF ELECTROPHYSIOLOGY

The experience of discovering the rotor while preparing an experiment to illustrate its nonexistence illustrates instead the utility of experiment in pointing to faulty assumptions. In retrospect, the assumption that seemed to imply that such vortices could not arise in a uniformly excitable continuum turned out to be the assumption that the dynamics of the medium could be realistically approximated by attention to a single internal state variable (like u-like membrane potential in heart cells, or the v-like colored catalyst in the BZ reaction). The actual existence and stability of these vortices derives instead from the crucial role of *additional* less-visible variables of state involved in processes of refractoriness and recovery similar to those in myocardium.

From one point of view, this is the Fundamental Problem of Electrophysiology. The field is limited to observing electrical potentials at each point in space, but they constitute just one component of a multicomponent state vector at each point, and the other components determine the rate of change of the potential. Without observing the dynamics of the other components, it is impossible to make sense of the observed dynamics of the potential, except in contrived simple situations. A similar problem afflicts chemical wave experiments, in which the video camera records at each point a color which, in principle, could convey information about as many chemical species as there are distinct color detectors (three, for humans and human technology), but, in fact, only one of them absorbs in the visible. In both situations, knowledge obtained through indirect experiments about the basic dynamical laws can be deployed to infer the values of a second variable from the one observable variable and its observed time derivative and Laplacian. This is illustrated below [lecture 5]. But it solves The Fundamental Problem only if *just* two variables are dominantly important. Strictly speaking, this is not the case either in chemistry or in electrophysiology, but it often turns out that a consortium of subsidiary variables function as one, en bloc.

OTHER WAVE EQUATIONS

The familiar nondispersive scalar wave equation of physics, written in dimensioned terms, is

$$\frac{\partial^2 \phi}{\partial t^2} = c^2 \nabla^2 \phi, \tag{4}$$

where ϕ is a real-valued potential. This represents local pressure in sound waves or electric potential in light waves. It does not look much like Eq. (3) for the dynamics of a typical excitable medium, which has two *first* time derivatives rather than one second time derivative, has at least two local state variable in place of the single potential here, and has a state-dependent source term lacking here. At second glance, you recognize that were β, γ, δ, all $= 0$ (which is perfectly compatible with excitability), then Eq. (3b) would be simply $u \propto \partial v/\partial t$ and then Eq. (3a) *could* be written as a single equation with a second time derivative, $\partial^2 u/\partial t^2$. But there would still be a source term, now entailing time derivatives, and the Laplacian would now be operating on the *time derivative of* u. These two kinds of wave equation are fundamentally different.

Another famous wave equation is the time-dependent complex Ginzburg-Landau (CGL) equation for generic behavior of a dynamical continuum near a Hopf bifurcation to limit-cycle oscillation of low amplitude (normalized to unity here). After lots of normalizing and making things dimensionless for mathematical (unphysical) transparency:

$$\frac{\partial \mathbf{z}}{\partial t} = (1 + ic_0)\mathbf{z} + (1 + ic_1)\nabla^2 \mathbf{z} - (1 + ic_2)|\mathbf{z}|^2 \mathbf{z} \tag{5}$$

where $\mathbf{z}$ is a complex number representing two interacting variables or (in polar coordinates) the amplitude and phase of oscillation in a two-dimensional center manifold. Coefficient c_0 can be zeroed without loss of generality (moving to a rotating reference frame). This is sometimes called the "Kuramoto-Tsuzuki equation" because "CGL" has by now proliferated into too many meanings. It make waves in space, as shown by Kopell and Howard[118] in the special case with $c_1 = 0$, so it resolves into a pair of reaction-diffusion equations representing a perfectly sinusoidal limit-cycle oscillator spatially distributed with diffusion coupling: the "Λ - Ω equation" or "radial isochron clock" or "Poincaré oscillator" in various literature streams. This was also shown by Kuramoto and Koga[128] and by Yamada and Kuramoto[241] in the "phase model" or "ring device" or "simple clock" limit of strong adherence to the limit cycle (the "Kuramoto-Sivashinski equation"). This model does make spiral waves around phase singularities: if $c_2 \neq 0$, there is an amplitude dependence of rotation frequency so contour lines tend to wind up around the phase singularity, but only up to a certain pitch, limited by the diffusion term. It is not at all clear that the reasons for the CGL equation's behavior are comparable to the reasons for superficially similar behavior in excitable media.

(Anyway, so I thought 20 years ago, and so for publication of my first rotor computation, I abandoned the CGL model in favor of an electrophysiologically motivated, nonoscillatory excitable medium. In the following years I steadfastly advised others also to beware that seductive path...but maybe they had better sense. Much more has since come of the CGL medium in connection with spiral waves, their rotor sources, and the interactions among rotors, not only in two dimensions but also in three dimensions. In a certain parameter range ($1 + c_1c_2 < 0$), it also develops turbulence,[126,127] a feature we will value below in connection with fibrillation. However, I repeat that this field of smooth oscillators is *not* an *excitable* medium (unless periodically driven at a period close to the edge of its ability to entrain). Almost all of this lecture series is about *excitable* media and therefore *not* about these sinusoidal oscillators.)

(Here is yet another aside on "the way things work in the discovery business." Also in 1973 I laid aside cellular automaton models, e.g., Reshodko,[172] as being too unfaithful to the behavior of real excitable media—e.g., the rotor, the spiral's core, is structureless in CA simulations—preferring the partial differential equations of electrophysiology and of physical chemistry. Every 2–3 years thereafter, a conspicuous publication appeared in which they were rediscovered as models of two-dimensional spiral waves simpler than PDE models. I used them again in 1984, before it was feasible to solve PDE's on large enough *three*-dimensional grids, to make the first three-dimensional organizing centers of various topological genres...but again soon set them aside as inadequate for realistic dynamics. Meanwhile, others persevered and created marvelously enhanced CA models which now serve more of the purposes of simulation [e.g., see Gerhardt et al.[72] and Weimar et al.[197]], for those lacking access to the hardware needed to solve three-dimensional reaction-diffusion equations.)

The nonlinear Schrödinger wave equation is another special case of the complex Ginzburg-Landau equation (5), with $c_0 = 0$ and c_1 and c_2 very large, i.e., with the "1"s removed and $c_1 = c_2 = 1$ or functions of position, or of time also in the nonautonomous case. This is another kind of wave equation, with a complex-valued local state $\mathbf{z} = \Psi = (u+iv)$ and the familiar time- and position-dependent potential function $V(\mathbf{r}, t)$ replaced by a state-dependence $|\Psi|^2$. Neglecting units involving h and m,

$$i\frac{\partial\Psi}{\partial t} = |\Psi|^2\Psi - \nabla^2\Psi \tag{6}$$

or

$$\frac{\partial(u+iv)}{\partial t} = i(-(u^2+v^2)(u+iv) + \nabla^2(u+iv))$$

which, separated into real and imaginary parts, reads:

$$\frac{\partial u}{\partial t} = (u^2 + v^2)v - \nabla^2 v,$$
$$\frac{\partial v}{\partial t} = -(u^2 + v^2)u + \nabla^2 u.$$

This resembles the wave equation (2) in oscillatory reaction-diffusion media, but look at the significant differences: one of the diffusion coefficients is negative, u and v are interchanged in the diffusion terms, the amplitude is implicitly very small, and the local kinetics depicts a harmonic oscillator rather than excitability. Vortex solutions exist in two dimensions and they make vortex filaments in three dimensions, which move in ways analogous to hydrodynamic vortex filaments.[134,173] But again, these are not the ways of vortex filaments in excitable media.

The typical wave equation (2) for an excitable medium is thus not much like physically familiar wave equations, so no one need be surprised that it makes waves with distinctive properties.

SECOND LECTURE

Figure 4 shows a vortex in the excitable medium's wave equation (3), calculated on a fine grid of sample points connected by diffusion. The top and bottom boundaries happen to be impermeable, but the side walls are connected: this is a cylinder, slit open and laid flat. The spiral wave front is radiating anticlockwise outward from the tiny central vortex core called the *rotor*. What are the main properties of this distinctive "rotor" solution? When first discovered, they were thought to be few and simple: the rotor has a unique period (the shortest possible) in a given medium, turns at the corresponding uniform angular velocity about a fixed pivot, and radiates a periodic wave train in the form of an involute spiral. For years I accepted most of these beliefs. All four are mistaken. Rotors of discretely different periods can coexist in some media, the period in any case exceeds the minimum response time of the medium by a goodly fraction or even a large multiple, rotors tend to move around spontaneously so the period waxes and wanes as observed from a chosen direction, and the spiral is not exactly an involute even when the rotor pivots rigidly. In fact, as more and more people came into this area of study, there came to be more and more well-studied equations like Eq. (3), and many had unexpected idiosyncrasies. Was there to be no end to the proliferation of diverse rotors and wave behaviors? This possibility is particularly important in clinical cardiology. It was guessed decades ago, and proved experimentally several years ago, that spiral waves underlie an important class of cardiac arrhythmias, including ones that promptly lead to fibrillation and sudden cardiac death. So pharmaceutical modification of

FIGURE 4 An anticlockwise rotor and surrounding spiral wave, computed from Eq. (3) on a cylinder, here slit open and laid flat. From Courtemanche et al.[37] with permission.

the electrical properties of heart muscle by so-called antiarrhythmic drugs might be profitably understood as parametric adjustment affecting the properties of spiral waves. Well, to do that rationally you would need to know what are the present properties, and what properties you would like to have, and what parameter adjustment would effect the change, and what drugs in what amounts would adjust those parameters in the required way. No such rational scheme exists today, but it might become possible if one could understand the parameter dependence of spiral wave behaviors.

To understand that, the most direct approach is to pick one generic excitable medium, nominate a couple of seemingly generic parameters, and vary them systematically and exhaustively, while watching the results. Then do it again for a different medium, and see how much of the first pattern is apparent in modestly distorted form in the second pattern. Well, in the past three years, this has been done in four media, and there does seem to be a common pattern.

Before you can perceive that pattern, you have to see through some of the sources of confusion that obstructed its perception in the past:

1. Equations (2) of a given excitable medium are conventionally written in at least four distinct formats, often using the same canonical variable and parameter

names (e.g., t, x, ε), but with differently scaled meanings that I distinguish by superscript prefixes:

$$\text{format } 0: \quad \frac{\partial u}{\partial\, {}^0t} = f(u,v) + D\frac{\partial^2 u}{\partial\, {}^0x^2}$$

$$\frac{\partial v}{\partial\, {}^0t} = {}^0\varepsilon g(u,v) + \delta D\frac{\partial^2 u}{\partial\, {}^0x^2}$$

$$\text{format } 1: \quad \frac{\partial u}{\partial\, {}^1t} = \frac{f(u,v)}{{}^1\varepsilon} + D\frac{\partial^2 u}{\partial\, {}^1x^2}$$

$$\frac{\partial v}{\partial\, {}^1t} = {}^1\varepsilon g(u,v) + \delta D\frac{\partial^2 u}{\partial\, {}^1x^2}$$

$$\text{format } 2: \quad \frac{\partial u}{\partial\, {}^2t} = \frac{f(u,v)}{{}^2\varepsilon} + D\frac{\partial^2 u}{\partial\, {}^2x^2}$$

$$\frac{\partial v}{\partial\, {}^2t} = g(u,v) + \delta D\frac{\partial^2 u}{\partial^2 x^2}$$

$$\text{format } 3: \quad \frac{\partial u}{\partial\, {}^3t} = \frac{f(u,v)}{{}^3\varepsilon} + {}^3\varepsilon\frac{\partial^2 u}{\partial\, {}^3x^2}$$

$$\frac{\partial v}{\partial\, {}^3t} = g(u,v) + \delta\, {}^3\varepsilon\frac{\partial^2 u}{\partial\, {}^3x^2}$$

The format makes some important differences. For example, in "format 0," the spiral's period is a U-shaped function of ε, while in the other formats it is monotone increasing or decreasing. It is possible to translate among formats by appropriate change of variables. I will present all results in "format 1." There are also several forms of $f()$ and $g()$ in common use, thus some additional translation is needed among usages of "u," "v," and the parameters, before the ostensibly diverse findings of diverse published papers can be unified as an overlapping mosaic of findings. Let it suffice here to just say that they do unify when properly compared.

2. The word "excitability" has distinct meanings to different people. For some, "excitability" is the solitary wave speed or the threshold (the minimum disturbance required to excite from equilibrium). Or the threshold normalized by the range of the ensuing disturbance. For others it means the initial or maximum rate of excitation, or sometimes that relative to the rate of recovery after excitation, in either case, dominated by ε. I use both dimensionless ratios as independent parameters. Also for some people, perhaps thinking of SNIPER bifurcations (see below), "excitability" connotes "inability to oscillate spontaneously," while for others, perhaps thinking of Hopf bifurcations, oscillation is not incompatible with excitability, which then only means "able to take a flying leap across state space when nudged in the right way." I use it in the latter sense, so a medium's ability to excite on cue is independent of its ability or lack of ability to cycle spontaneously if left alone.

3. In some "reaction diffusion" models, only the excitor variable can spread through the medium, e.g., electric potential in electrophysiological models, while the inhibitor variable is strictly local. In others, both diffuse equally, e.g., most chemical reaction models. (And in others the inhibitor diffuses more freely than the excitor: this leads to wave front instabilities and Turing patterns, not discussed here.)
4. There may be qualitatively distinct classes of excitable media, with only one major class thus far examined in any detail. It is not yet clear that this class is fairly representative of all. What I describe below is found in equations (and corresponding laboratory systems) resembling item (3) in that there are two variables, a Z-shaped nullcline for the fast and freely diffusing variable, and a single, globally attracting uniform equilibrium.

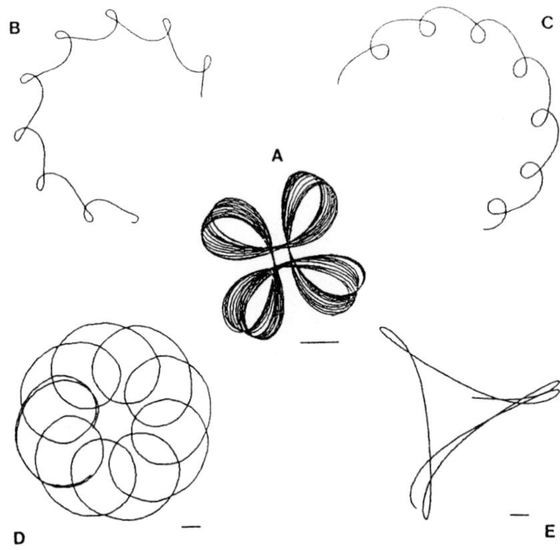

FIGURE 5 With various settings of equation (3)'s parameters β and ε, the tip of the spiral wave in Figure 4 traces out paths like these. The scale bars (accidentally erased during the original publication) represent one "space unit." From Winfree[225] with permission.

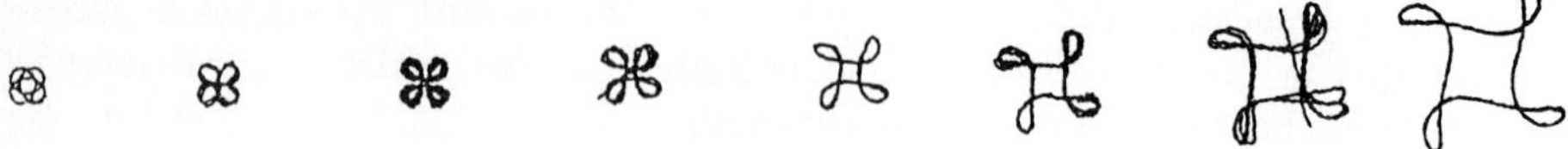

FIGURE 6 Figure 5(a) is one member of this series, obtained by selecting β at each ε to obtain the same (four-fold) symmetry. The scale bar (erased in the original publication) represents 30 "space units." From Winfree[225] with permission.

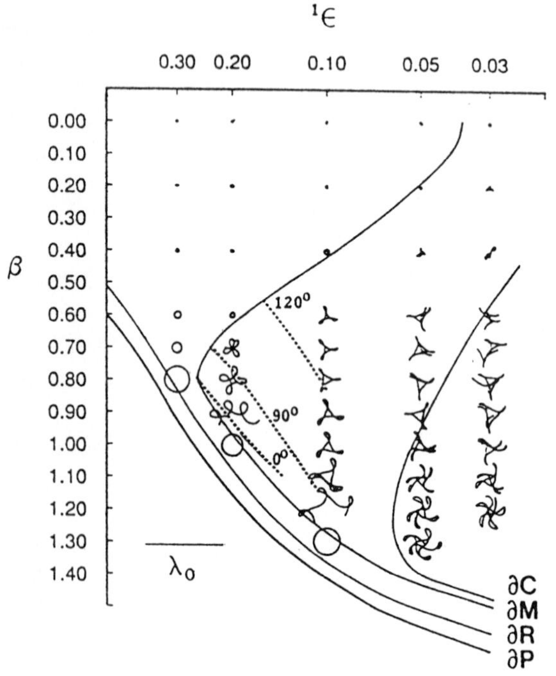

FIGURE 7 The (β, ε) parameter plane, showing the larger context of "flower" shapes from which Figures 5 and 6 were selected. To the lower left from bifurcation boundary ∂P there are no waves; from ∂R there are no rotors or spiral waves; from ∂M there is no meander, no flowers; from ∂C no complex "hyper-meander." Note that the scale bar is now one wavelength, not one space unit: each flower is normalized to its own wavelength. From Winfree[225] with permission.

MEANDER

It turns out that the isolated vortex in a uniform unbounded medium is typically not motionless like those seen in computations during the first several years. Rather, it typically executes a flowery dance. I dubbed this "meander,"[202,248] suggesting randomness, thought to derive from non-uniformity of the experimental arrangements. This was before the discovery that ideal, numerical vortices also can "meander" and that this is not a numerical artifact (my belief until 1985) and that it does not require strange and ungeneric adjustment of the model and before it was appreciated that the most common kind of "meander" is quite regularly biperiodic.[247]

Figure 5 shows several of the biperiodic meander flowers traced by the tip of the spiral wave in the FitzHugh-Nagumo medium (Eq. (3)). Though they vary greatly in absolute size, all are confined to the "core" of the spiral: a central disk of perimeter equal to one wavelength...with an interesting exception that we will come to shortly.

This diagram happens to hold $\gamma = 1/2$ in Eq. (3) while varying ε and β, but slopes 0 and 1 were also explored, with similar results. At slope 1 it is possible to obtain three intersections with the Z-shaped nullcline. This also seems to make little difference for rotor behavior.

Figure 6 shows a series of flowers with four petals, all with greatest curvature to the outside so the petals appear to stick outward. At one end of the series, they are small, nearly perfect circles, but with a little wobble at the right period to exactly retrace an excursion with four-fold symmetry. At the opposite end they consist of long straight segments punctuated by tight turns through -270°. This series corresponds to a one-dimensional locus in Figure 7, along which the angle between petals is 90°. Along adjacent loci, the flower does not quite close. Somewhat further away lies the 60° "isogon" (6 petals), and the 30° isogon (12 petals), etc. Ultimately the 0° isogon is reached, beyond which petals show their maximum curvature on the *in*side (Figure 5(d)). In the neighborhood of the 0° isogon, the wave tip is looping through the medium like a pencil point on a rolling disk. The rotor has a linear momentum (gradually turning unless parameters are exactly on the 0° contour) in an arbitrary direction (Figure 5(b) and (c)). The flower in such cases is not confined to a disk of small perimeter. At about the same time as my study,[225] Dwight Barklay (unpublished) found all the foregoing patterns also in his excitable medium[16,17,18] while varying its equivalents of γ and β. Thus we learn that rotors have an unforeseen property beyond position and phase: they also have intrinsic orientation or direction, which changes on a schedule independent of the familiar rotation.

I am not sure how wide a range of angles the isogons span. Are there 180° and -180° isogons? The former would lie far to the right in Figure 7 and would be populated by tip paths which almost run back and forth along a nearly fixed, but very slowly rotating, slit. Myocardial spiral waves have sometimes been described so, but it is not clear whether this is an artifact of the anisotropic fibrous structure of their substrate, possibly including an actual slit along which fibers fail to communicate laterally. I have never seen them in the FHN medium, except when mistakenly running computations on overly coarse grids. However, Krinsky et al.[123] have reported them in a related model of excitability. The putative -180° isogon would lie under and to the left of the 0° isogon, against ∂M. It would be populated by flowers of a shape not yet seen.

These isogons do not fill the whole parameter space. At the end where the wave tip's path scarcely deviates from a circle, we find ∂M, the parameter-space Hopf bifurcation locus delimiting the domain of two-period meander. The other end has not yet been found, but it might encounter ∂C, a recently discovered locus delimiting more complex meander, sometimes called "hyper-meander." Figure 8 shows two examples: 8(a) from Eq. (3) and 8(b) from the Oregonator model discussed below.

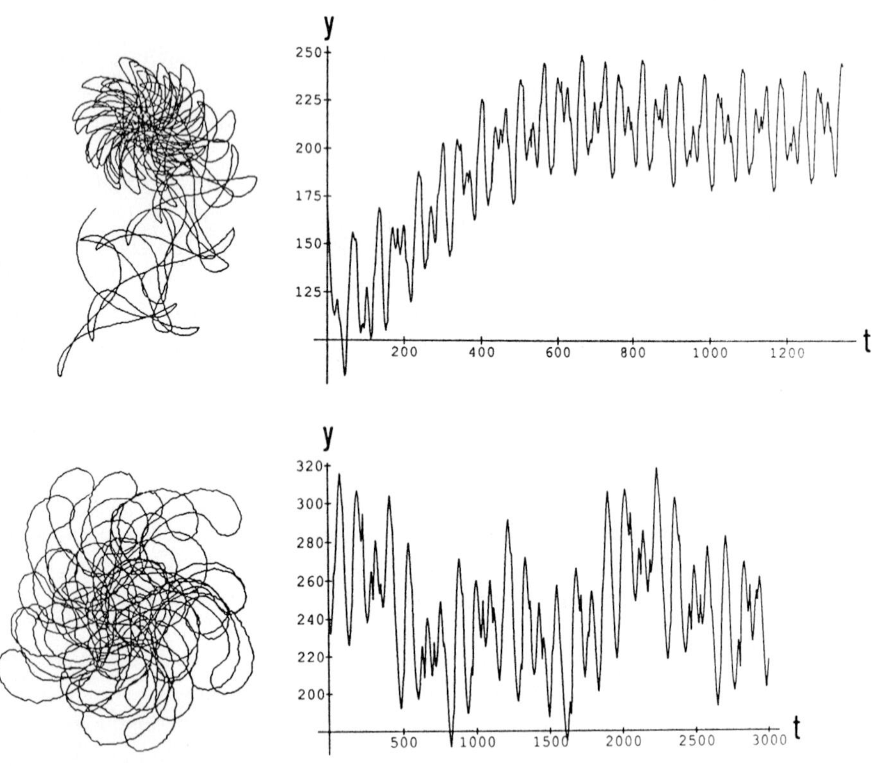

FIGURE 8 To the right from Figure 7's ∂C flowers cannot be described by concatenation of two periodic processes. The upper pannel shows such a flower alongside its y(t), with $\beta = 1.2$ and $\varepsilon = 0.03$. The lower panel shows complex meander in the Oregonator model of the Belousov-Zhabotinsky reaction. Both from Winfree[225] with permission.

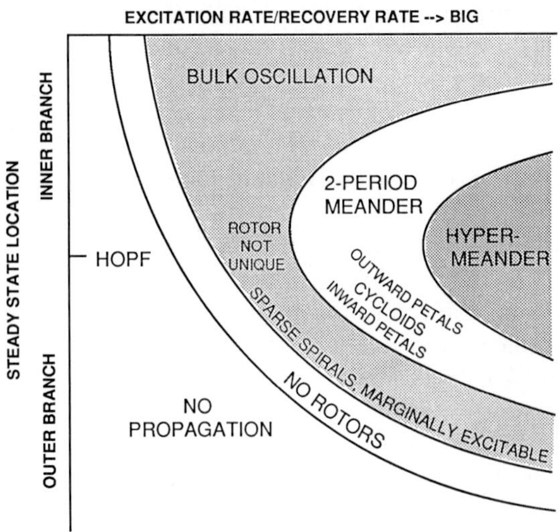

FIGURE 9 This is a guess at the generic landscape of rotor types in excitable media, as two dimensionless parameters are varied: horizontally, the ratio of excitation to recovery rates (called ε above), and vertically the ratio of threshold to full amplitude of excitation (correlated with β above). From left to right the bifurcation loci are ∂P (for which there seems little theory), ∂R (where most Russian work concentrates), ∂M (where Barkley proved a Hopf bifurcation), and ∂C (not understood at all). Work using singular perturbation techniques covers the far right. Symmetric spirals lie along the top edge, where media with smooth, round limit cycles occupy the left part. The tick mark on the vertical axis marks the Hopf bifurcation of the local kinetics, which leads to smooth oscillations only near equilibrium. Note that the Ginzburg-Landau reaction-diffusion model has no home on this plane, nor are there domains of wave-breaking turbulence.

For three-dimensional filaments these boundaries shift slightly unless curvature and twist are zero. In particular, the 0° isogon is delicately sensitive to curvature.[224]

There are many tantalizing regularities about this numerical landscape, which I imagine mathematicians will describe in due course; efforts began in earnest while the publication was still being refereed. One of the most striking is the meander of rotors with small ratio of excitation/recovery rates (small ε). This shows that the assumption of rigid rotation cannot be used in the singular perturbation limit, calling into question the conclusions of much thinking in this area. In general, it is my impression that analytical theory in this area has been excellent for systematizing things already discovered, but of surprisingly little use for discovery. It seems too often that feasible approximations point in many contradictory directions or represent limiting cases remote from the center of scientific attention, or only re-describe in symbols with adjustable parameters what was already established in the laboratory or by numerical experiments. These lectures accordingly emphasize experiments, mostly numerical. The student should be skeptically aware that conclusions from such spotty evidence may also go astray.

Figure 9 gives my "artist's impression" of a generic bifurcation diagram for rotor behavior in excitable medium, as a function of threshold and rate ratios. It indicates the domains of various behaviors and corresponding approximate mathematical methods. People have studied spiral waves analytically and computationally in various approximations, not always clearly distinguished, with results whose overlapping or disjoint domains of validity are not always clearly fenced in, so results often seem contradictory. Experimentalists are forever discovering spiral waves in new biological and physical situations, and making up a model of the putative mechanism, and confirming it by demonstration that the model makes spiral waves with behavior similar to the observations. This tactic relies on the implicit assumption that the idiosyncratic behaviors of spiral waves reflect idiosyncrasies of the underlying mechanisms. But the possibility exists that *any* excitability model is capable of *all* these behaviors, just depending on the tuning of some generic parameters. So then, if you pick the wrong mechanism, you can nonetheless match observed spiral wave behavior by choice of parameters, if nothing else constrains that choice. Thus these exercises are often remarkably successful and seldom exclude an initial mistaken model. If diagrams like Figure 9 can be improved in accuracy and generality, such difficulties will be reduced.

Another curious feature of excitable media is that there is only one kind of spiral wave: it has a characteristic wavelength and rotation period for a given local kinetics. If you create disturbances from which spiral waves emerge, they are all identical except for position and phase: all have the same period and wavelength. Anyway, that what was thought for 20 years. Since everyone knew that, but no one had ever *proved* it, I went in search of a counterexample by recording the rotational period of the spiral wave while varying one parameter of Eq. (3), and while holding the other two fixed in a range guess to favor peculiar behavior. As ε is increased, the period hardly changes until a critical ε_2 at which it jumps to a much longer period, then continues to lengthen as ε is further increased. Backing up this evolution, the long-period solution persists past the former jump point, and drops to the short period solution only at $\varepsilon_1 < \varepsilon_2$. There is an intermediate *range* of parameter values in which *both* types of rotor are viable. Choosing a medium in this range, you thus find both kinds of rotor coexisting. One has tighter spirals than the other, and turns quicker, so it eventually dominates. But until then, or in the absence of the faster solution, the slower solution is perfectly stable.

Incidentally, one of the deepest mysteries in this business is the persistent absence of completely different kinds of reaction-diffusion structures. Why do diverse models keep turning up only Turing patterns (periodic ripples) and rotors? For 20 years I have expected the advent of a marvelous diversity of modes in which diffusion and arbitrarily chosen local kinetics might interact to generate dynamical structures such as one sees, for example, in living cells. But they have not turned up. Is this a problem, or does it only reflect the absence of a conveniently fast visual utility with real-time manual control of parameters and initial conditions, for intuitive exploration of the possibilities? Such a facility constructed for the Silicon Graphics Iris revealed that the most common alternative to rotors and Turing patterns is utter turbulence.

Returning to rotors, the dependence of spiral wavelength, λ_0, and period, τ_0, on small parameter ε is an encouraging exception to my impression that in this area the route to discovery has more often been numerical experiments than analytical mathematics. In the limit as $\varepsilon \to 0$ (the singular perturbation limit), both λ_0 and τ_0 blow up (or vanish, depending on "format" choice). En route, they follow the cube-root laws anticipated by Fife[58,59,60] and by Kessler and Levine[111] more accurately than the square-root laws estimated in the Russian literature: in format 1, e.g., λ_0 is proportional to the inverse sixth root of ε and τ_0 is proportional to the inverse cube root. Fife's scaling was mostly ignored amid contradictory alternatives until it was also discovered numerically.[225] It has swiftly become the basis of newer analytical work.[17,24,101,102,112,113] The basic reason for the cube-root laws is that as ε becomes smaller, the reaction-diffusion system jumps in the u-direction from the lower branch of the Z-nullcline to the upper, and from the upper to the lower, at smaller and smaller $|v|$. It does not wait to jump at the turnaround points of the Z (at $u = \pm 1$, $v = \pm 2/3$). Thus the range of v traversed becomes narrower and narrower as ε decreases, so τ_0 becomes briefer, and the wavelength shorter. (No such ε-dependent abbreviation and attenuation of the cycle is expected in an excitability model like Figure 21 below. This is why I imagine that comparable numerical experiments with that model might reveal limitations to the generality of behaviors explored up to now in excitability models like Eq. (3).)

In contrast to the rotor period τ_0, the shortest period at which the medium can respond in 1:1 fashion, $\tau_{\min}$, apparently does follow a square-root law on ε, so the ratio $\tau_0/\tau_{\min} \to \infty$ as $\varepsilon \to 0$, leaving a wide gap of excitability between the spiral's refractory wake and the next turn of its activation front. This conflicts sharply with a conviction common among cardiologists, that rotors spin nearly at period $\tau_{\min}$ and so they are only precariously viable and have no zone in which an extra stimulus could incite the medium to affect (and perhaps undo) them. The facts seem to be otherwise, both for rotors in typical excitable media and for those in heart muscle: they are perniciously stable, and they can be entrained or even extinguished by properly timed small stimuli.

One consequence of these power law dependencies is that the ratio λ_0^2/τ_0 is independent of (small) ε, and so it is a fixed and dimensionless multiple of the propagator's diffusion coefficient. This multiple, Q, does depend on other details of the local kinetics, e.g., the threshold, but not on ε. Q less than about 6π has never been seen in numerical experiments, nor in the laboratory unless the value of D is very much in doubt. Q seldom differs more than about four-fold from $8\pi^2$, except in media of marginal excitability. This not a theorem but merely an empirical observation about all known continuous reaction-diffusion-based excitable media. But it has its uses, for example in guessing the nature of the communication that binds local reactivity into a spatial pattern. There is a skin disease[83,149,181] resembling an excitable process, even to the extent of making rotating spirals, with wave length about 1 cm and period about 12 hours. If Q is around 80 as in other excitable media, then the propagator's diffusion coefficient is about 3×10^{-7} cm^2/sec, which suggests a protein or a virus struggling from cell to cell. A better-studied case is the catalytic oxidation of carbon monoxide to carbon dioxide on the surface of hot platinum crystals in your car's catalytic converter. This reaction is excitable and propagates typical waves, including spirals, the first of which to be published had wave length about 0.005 cm and period about 10 sec, suggesting propagator diffusion at 420° K about 3×10^{-8} cm^2/sec,[95,224] thus suggesting migration of an adsorbed gas across the surface rather than gas-phase diffusion. As it turned out, that first photograph was atypical: the spiral was apparently stuck on some kind of hole, giving it abnormally long period and wavelength. Nettesheim et al.[151] give a corrected wave length for the free spiral (at higher temperature, 448° K) about 0.0013 cm and period about 8 sec so QD is 20×10^{-8} cm^2/sec, so if D is as above, then $Q \sim 5$: about three times smaller than the record minimum for all known excitable media and mathematical models. This might suggest a closer look at D. Supposing Q within a factor of 4 or so around 80 as usual, the predicted propagator diffusion coefficient would be within a factor of 4 or so around 0.25×10^{-8} cm^2/sec. When the CO diffusion coefficient was measured, it proved to be about 0.27×10^{-8} cm^2/sec;[176] but this was at 406° K and we really want it at 448° K. Nettesheim et al.[151] suggest 4 to 9×10^{-8} cm^2/sec at that temperature, but note that estimates vary widely. Regardless of the residual ambiguity and the need for more refined measurement, the value predicted from familiarity with rotors and the values observed experimentally are as much as three orders of magnitude smaller than values used in recently published mathematical theories about this reaction. So much can yet be expected to change in this area.

Another excitable medium that has been modeled successfully by simple kinetic equations similar to Eq. (3) is the BZ reaction. In silica gel preparations of this reaction, the "controller" quantity v (the colorful catalyst, ferroin) is bound and does not diffuse, but, in other gels or the ungelled liquid, it diffuses freely. Does this make a qualitative difference for rotor behavior in the Oregonator model of the BZ medium? As in the case of Eq. (3), it turns out not to, though there are substantial quantitative differences. With γ diffusing freely we found computationally that the pivot point of the vortex does not sit still, but slowly migrates along a flowerlike path. As chemical parameters ε and f are varied, the wavelength, period, flower symmetry, and flower petal shape all vary. The pattern of these dependencies is about the same as in the FHN model with similar parameters. Something of the sort was shown 20 years ago in the BZ reagents (Winfree,[215] p. 181) but flower varieties more like the computations were discovered by Jahnke et al.,[92] and at least two other laboratories then confirmed these observations.[167,185] Though many flowers like Figure 5 have since been observed in chemically excitable media, two of the most striking predictions have not yet been confirmed in the laboratory. I believe they will be, because diagrams like Figure 7 have turned up in every excitable medium examined to date. One is "hyper-meander," which cannot be described in terms of two distinct periods; Figure 8(b) shows an example. This occurs to the right of ∂C. Another is "linear looping" meander, which occurs along a certain locus (the "0° isogon" between ∂M and ∂C): the center of the spiral in a perfectly uniform isotropic medium glides at fixed speed in a direction determined only by the manner in which the rotor was initiated (or subsequently perturbed). In such media the wave tip traverses a cycloid (at large ε) or (at small ε) straight runs of length less than λ_0 punctuated by a 360° turn about the end of each.

This puts me in mind of the confident assertion of a nineteenth century mathematician and philosopher, Auguste Comte[35] (in translation): "Every attempt to refer chemical questions to mathematical doctrines must be considered, now and always, profoundly irrational and contrary to the spirit of chemistry.... If the employment of mathematical analysis should ever become so preponderant in chemistry—an aberration which is happily almost impossible —it would occasion a vast and rapid and widespread degeneration of that science."

ROTORS IN THE ELECTROPHYSIOLOGIST'S MYOCARDIAL MEMBRANE MODEL

The pattern of bifurcations shown in Figure 7 seems to illuminate diverse excitable media. Thus it seems that real phenomena can be anticipated from principles as basic as merely reaction kinetics in an idealized motionless continuum. In the chemical instance it led directly to several discoveries in the laboratory. Might similar benefits be derived from the equations of electrophysiology? The FHN model that was explored in some detail just above is the simplest half-way plausible caricature of electrical activity in nerve and muscle membrane. It summarizes the accepted principles of electrophysiology that are thought to apply on a large scale in heart muscle. So just as we checked the Oregonator computations in the laboratory to discover strikingly regular and symmetric flowerlike meander in a chemically excitable gel, we should check the FHN computations in the laboratory to see how much can be anticipated that might have a bearing on cardiology.

A caution: whoever designed the human heart ("Mother Nature") would get a D for engineering physics if she left it essentially like any generic excitable medium. Why? Because such excitable media are susceptible to infection by rotors. Figure 9 shows that (for media like Eq. (3)) the only parameter domain in which propagation is reliable yet rotors fail is just a fringe near the propagation boundary, where the medium is only marginally excitable. This would not seem a safe operating regime for hearts. So if hearts manage to evade such infection most of the time, it must be because they are *not* generic excitable media. There must be some special gimmick. What it is I don't know, but these mathematical studies indicate that it should be looked for. This caveat should be borne in mind while examining heart muscle for comparison and contrast with generic excitability.

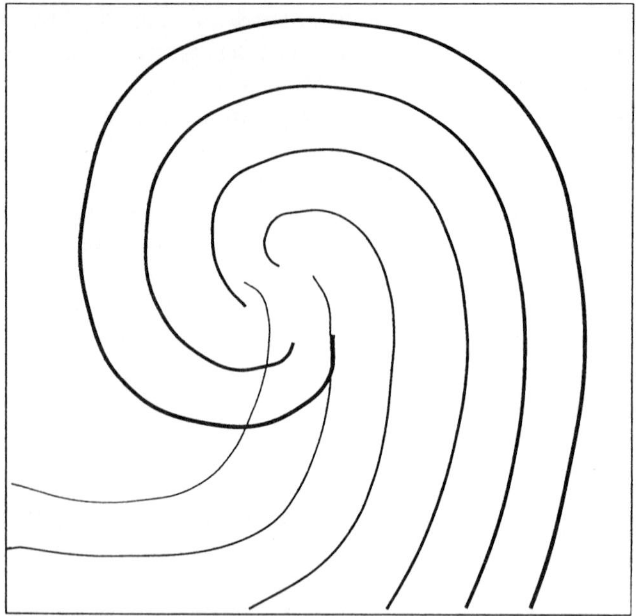

FIGURE 10 The inner turn of a spiral wave in a 16-cm square (200 × 200 grid points) of Beeler-Reuter membrane, slightly modified for external potassium concentration elevated from normal 4.5 mM to 7 mM to simulate heart muscle's excitability in rapidly paced tissue. The contours are hand-sketched activation fronts at intervals of 40 ms in order of line thickness. The inner tip of the activation front traces the border of a 1-cm disk. From Figure 2 of Winfree[221] with permission.

To bring laboratory data to bear on this question, two approaches come to mind. One uses a computational summary of decades of diverse experiments and the other is directly experimental. We will consider both in turn.

The computational approach consists of checking whether *better* electrophysiological equations quantitatively more faithful to heart muscle do indeed support rotors, and whether they confirm the observed size and period. So my graduate students William Skaggs in 1987 and then Marc Courtemanche in 1990 set up arrays of about 40,000 idealized ventricular cells in one or another supercomputer, using a version of the so-called Beeler-Reuter membrane model once considered to give a fair representation of the electrical behavior of the myocardial cell membrane. These equations were formulated using the best description available 15 years ago, while today's productive electrophysiologists were just beginning their careers. Most contemporary users modify it one way or another. It represents calcium kinetics rather sluggishly, for example, so we cut the associated time constants in half. We asked whether it supports two-dimensional vortices similar to those known in all other excitable media, and quantitatively whether it gets the size and period *right*. It almost does. The diameter of the vortex core is about 1 cm and it rotates about 5 times/sec (Figure 10). The main point is that an appropriate stimulus starts a persistent high-frequency rotating excitation. Others have since repeated this kind of computation, also in other models of myocardium, and all obtain vortices of roughly the same size and period. If diverse electrophysiologically motivated membrane models agree on at least these qualitative features, then it seems worthwhile to go looking for them in the real medium that inspired all the models.

ROTORS IN LIVING MYOCARDIUM

These vortices, discovered computationally as an implication of the accepted dynamical equations of heart muscle, had not been seen in the clinic nor in the laboratory. So it seemed worthwhile to find out whether they are real or not. Does real myocardium support them? Are they vastly larger than the hearts in which we would (erroneously) invoke them as interpretation of familiar arrhythmias? Or possibly so much smaller that they would span only a few cells, proving that the continuum approximation used to predict their properties is utterly inappropriate? Or possibly just the size below which small hearts are incapable of sustained arrhythmias and fibrillation?

The question here is about ventricular muscle and "rotors" as conceived under the rubric of cable-equation electrophysiology (Eq. (2), Gulko and Petrov[77] and Winfree[205,206]). In the 1970s it had already been settled that atrial muscle does support them[6,7] though electrophysiologists still prefer to describe them in terms of an *ad hoc* model[8] rather than in terms of electrophysiological equations. There remained doubt as to their role in ventricular (therefore potentially lethal) arrhythmias. This doubt stemmed partly from the difficulties of observation and partly from the failure of any model to predict or account for observations quantitatively, at least to order-of-magnitude, in terms of basic physics. Thus the effort of the 1980s focused on making theory quantitatively predictive and extending it to the ventricular myocardium.

Dog experiments were designed in the format of the computer experiments, but it proved impossible for several years to interest epicardial mapping laboratories in trying the different protocol required for inducing fibrillation in this controlled way. When it was finally tried in about 1986, exactly such vortices were created and observed and found to be about the right size: about 1 cm, the same in all mammalian species tested up to now. They also turn out to have roughly the characteristic frequency of pernicious arrhythmias and fibrillation: around 10 Hz.

The shortest stably sustainable period of rapidly beating mammalian and avian hearts is shorter the smaller is the heart; this period varies as the fourth root of body mass, and heart mass stays about 0.6% of body mass. But all such hearts are made of similar membranes and cells, with similar electrical properties and susceptibility to similar rotors. More quantitatively, it turns out that hearts capable of beating faster than rotors are all too small to accommodate a 1-cm rotor, and are incapable of sustained fibrillation.

These figures can be understood roughly from the observation that dimensionless ratio Q is usually about 100 (order of magnitude) in rotors, regardless of their particular physical, chemical, or mathematical mechanisms. Q can be written as $\lambda_0^2/D\tau_0$ or as $\lambda_0 c_0/D$. Ventricular D is about $1\,\mathrm{cm}^2/\mathrm{sec}$, and speed λ_0/τ_0 is about $30\,\mathrm{cm/sec}$. Speed can be "derived" as $\sqrt{D = \text{excitation time}} = \sqrt{1cm^2/sec/0.001sec}$. This is, of course, the speed of a solitary excitation into virgin territory not of the spiral wave train, but for myocardium the difference is not large. It follows that λ_0 should be expected to be about 1 cm and τ_0 should be expected to be about 0.1 sec (encouragingly, like the dominant period of the EKG during fibrillation).

Rotors seem to be one of the main avenues to fibrillation and death. They have now been observed also in the human heart during potentially lethal arrhythmias. Now in more detail:

The ultimate way to ask the same questions posed above numerically is the experimental approach, exposing real hearts to the same initial conditions that conjured rotors from the equations of electrophysiology, to see if they do the same in normal, healthy heart muscle. This is necessary because the only thing we really know for sure about the purported equations of almost anything biological is that they aren't right. They are *sort of* right, but they *are* necessarily idealized. No matter how detailed, they are still only simplified abstractions, and even though they may capture the simply describable essence of the known phenomena, they might not predict *new* phenomena in even a qualitatively convincing way. Models can only be relied on for hints. The hinted experiment has to be tried.

It was already well known and used in other areas of physiology. Called the pinwheel experiment, it was first used to clarify topological aspects of the dynamics of the (still-unknown) circadian clock mechanism. Then it was used for similar purposes involving the biochemical oscillator that regulates sugar metabolism in yeast cells, then to clarify the dynamics of the human circadian clock, then in the BZ chemically excitable medium to create spiral waves. Starting about 1980 I tried to persuade cardiac physiologists to execute the same procedure on normal heart muscle, but there were no takers. So I just published the ideas in *Scientific American* in 1983, and wrote a more detailed book about it in 1985. In 1986, while the book was still in press, rotors were created in normal heart muscle by this experimental protocol. So it is worthwhile to explain the principle of the protocol:

In a pinwheel experiment a geographical gradient of stimulus intensity is applied across a geographical gradient of timing. A geographical gradient of stimulus intensity is automatically created around any DC stimulus electrode because current density is high nearby and falls off with increasing distance from the electrode. A geographical gradient of timing is automatically established by passage of an action potential, turning on cells in orderly succession. According to the theory and practice of pinwheel experiments, any appropriate crossing of two such gradients should initiate a reentrant vortex in a medium capable of sustained periodic activity, therefore in myocardium.

It should be understood that there is not very much innovation in this. Wiener and Rosenblueth[200] had presented a cogent theory of spiral waves on myocardium, including a mechanism resembling Figure 3 for their creation in mirror-image pairs by a discrete stimulus in the wake of a prior wave. There were doubts about the realism of this model, and the spirals it predicted were enormously too large, and they had no stabilizing "excitable gap" and were liable to spontaneous extinction, but much of the theoretical difficulty was resolved by the improved axiomatic models of V. I. Krinsky,[121,122] co-awardee in the 1980 Lenin Prize. This work included the first real theory, not just of spirals, but also of their deterioration to fibrillation. Yet there remained substantive doubts about this whole approach, since the manner of intercellular coupling used for modeling convenience has implications that differ markedly in some respects from the "cable equation" of electrophysiology or the equivalent diffusion equation of physical chemistry. Some of these problems were resolved independently in Russia and the USA by the first computations of spiral waves using the latter mechanisms,[77,205,206] and by their laboratory demonstration in a chemical reaction-diffusion medium.[202,244] Allessie et al.[6,7,8] then demonstrated rotors in *atrial* myocardium. The trick for initiating them works only at special sites discovered by trail, and its mechanism remains inscrutable. In the experiments described in these lectures, the stress is not on the mere existence of rotors, but more on their demonstration in *ventricular* myocardium as the prelude to fibrillation, and on their creation by a mechanism that is quantitatively understandable in electrophysiological terms.

My present purpose is only to motivate an interest in the theory (which is better presented elsewhere, e.g., Winfree[215,216]), so it will suffice in this lecture to just remind you of the two-stones-into-the-pond image in lecture 1 (Figure 11). The first creates a radial gradient of timing in the wake of the first wave, and the second, landed in that wake, produces a transverse gradient of stimulus intensity which has different effects in different places. On the side opposite from the first wave, the second stimulus creates a new wave front, while between stimulus center and the first wave, that is, in the refractory region, it cannot create a wave front. In-between, on each side, there must be an endpoint of the new wave front. These are the places where the timing gradient and the stimulus gradient cross transversely through critical values, and where rotors arise. The experimental question is: what becomes of the endpoint? Does it move along a hyperbolic arc? Does it retract and vanish, or somehow become diffuse? Or does it evolve into a stable, standard, compact rotor? And, if so, what are its size and period?

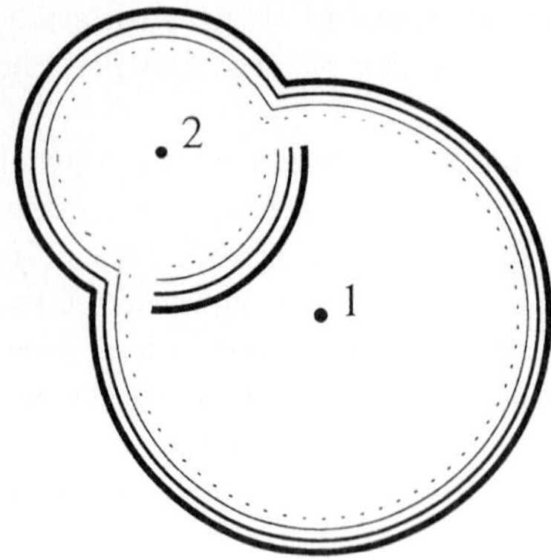

FIGURE 11 Detail of the third panel of Figure 3 to show the interruption of ring 2 by the dotted refractory wake of ring 1. Rotors materialize out of the transverse gradients created near the termini of the segment of ring 2 propagating toward stimulus site 1.

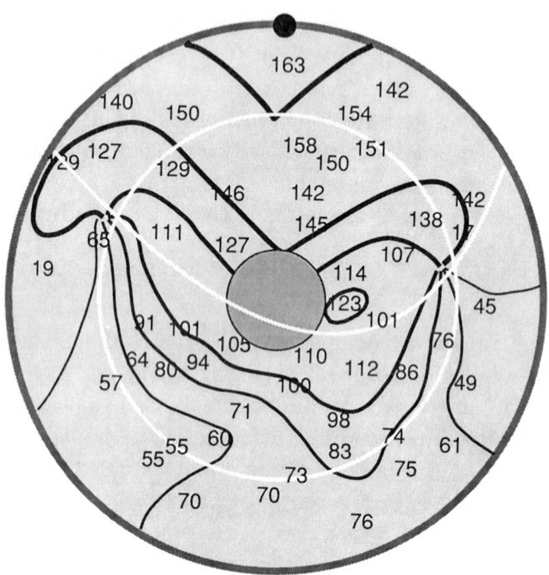

FIGURE 12 The southern hemisphere of a dog's heart, showing successive positions (thin to thick contours) of the excitation front during one rotor period of about 120 msecs. Numbers are excitation times in milliseconds at each electrical pickup site. This mirror-image pair of rotors was initiated by "the pinwheel experiment": when locus T^* had reached roughly the white U-shaped locus while propagating downward from the stimulator at the top (the black dot), locus S^* was created around the gray electrode in the center. This was supposed to create rotors at T^* S^* intersections, and it did. From Winfree[223] with permission.

The easiest way to think about the geometry of these crossed gradients is to focus attention on two special contour lines on the myocardium. One is the *moving* contour along which cells are reaching a certain time of recovery, called T^*. In heart muscle this critical time T^* falls soon after the membrane repolarizes behind the moving activation front. The other is the fixed contour around the stimulus source, along which a certain intensity of local excitation is produced. This critical local stimulus intensity is called S^*. According to theory in heart muscle, this critical stimulus corresponds to an extracellular potential gradient several times exceeding the pacing threshold, which we know to be about 1 volt/cm. When S^* was measured,[68,88] the factor "several" turned out to be about 5. (See below.)

A good thing about these two contours is that both are readily observable and measurable. Where they cross, a reentrant vortex is supposed to be initiated. It may not stay there, particularly if the creating gradients are much deformed relative to the actual shape of a stable rotor, but it might not bounce far from its birthplace. If this is right, then a lot of the horrendously complex subtlety of cardiac electrophysiology has been boiled down to something much more tractable, viz., the locations of two special contours that are observable in heart muscle.

Another way to think about this criterion (the original way[210]) comes from noticing that the phase portrait of myocardial membrane (like Figure 1) resembles that of a limit-cycle oscillator, and that myocardial cells need only very slight encouragement to take up spontaneous oscillation. So let's think of them in approximation as though they already were oscillators, but with period too long to be of practical interest. It is well known that any limit-cycle oscillator has a critical phase, T^*, at which a stimulus of critical size, S^*, will pop it to or beyond the boundary of its attractor basin: it then cannot spontaneously resume its prior normal mode of behavior. Only two experimental parameters are needed to accomplish this, regardless of the complexity of the mechanism[201]; this "pinwheel experiment" has since been used to characterize the "phaseless" states of the circadian clock in several kinds of organism, including humans,[39,99,214] in oscillatory sugar metabolism,[203] and in the BZ oscillator.[76,85,86] In geographical context, in a continuous field of oscillators, such a stimulus generically creates a neighborhood in which the center cell got exactly T^*S^*, and those around it end up with their timing reset in a cycle around the singular center point. This is the perfect initial condition for creating a rotor, as illustrated for example in the BZ medium.[235]

The creation of a rotor supposedly requires that some patch of tissue is near the singular time T^* while it is being excited by nearly the critical electrical stimulus, S^*. This can be arranged in the laboratory if the pertinent quantities are first estimated within narrow enough limits for experimental design. This pinwheel experiment was carried out in dog heart at Duke University Medical School, in the Basic Arrhythmias Laboratory of R. E. Ideker.[68,88,182] The disk in Figure 12 is a flat projection of the southern hemisphere of the dog's heart. From a stimulus electrode at the top of this map (black dot), an action potential was sent forth to sweep across the ventricles, establishing a vertical timing gradient. (Under normal conditions this would be provided by the heart's atrial pacemaker.) When the horizontal midline, roughly, (the white arc) was at singular phase, a second stimulus was provided, this time from the large gray stimulus electrode at the center. Current density was, of course, greatest close to the center, grading to practically negligible at the edges of the disk, establishing a stimulus gradient transverse to the phase gradient. Along the white circle of about half the disk's diameter, the local stimulus had the singular intensity $S^* = 5\,\text{V/cm}$. Wherever the critical time contour crosses the critical stimulus circle, theory predicts creation of a rotor. So we expect two, oppositely rotating, because of the mirror symmetry of the experimental arrangements. The numbers written at sites of pickup electrodes report the timing of local activations, and the hand-drawn contours (thin to thick) segregate those times in 20-msec bunches. After 120 msec, excitation is back to its starting place and it goes around again, continuing in this way cycle after cycle. In short: this healthy dog's heart does exactly what the equations of electrophysiology said it would. The heart is now embarked upon an irreversible course toward fibrillation, which erupts one or two seconds after this 120-ms vortex is started, i.e., after 10–20 cycles. This result is consistently obtained in all dogs, in diverse rearrangements of the experiment. It shows that the conceptual principles used to design the pinwheel experiment are sound in this application, and that the half-century-old electrophysiologist's cable equation is sufficiently reliable for this class of phenomena and has surprises packed away inside it still. It shows that normal healthy heart muscle does support rotors of the kind seen in other excitable media, instigated by the same kinds of stimuli. So the ones seen in humans may be understood as a reflection of understandable normal mechanism, not necessarily requiring poorly characterized aberrations of the individually diseased heart.

Here are four further testable consequences of these principles:

1. If the stimulus gradient is *parallel* to the timing gradient, so that the critical contours are *parallel* and do not cross, then there should be no reentry induced. This was tried by Frazier et al.,[68] and turned out just so. As a caveat: the electrode arrangement of Davidenko et al.[41,42,43] creates rotors, but it looks as though it might also create essentially parallel T and S gradients. (They were not measured, so this scrupulous preservation of a possible counterexample might turn out to be mere worry.).
2. If the stimulus is given too early or too late, then the T^* contour would be far from the S^* ring when the stimulus is given, so again there is no possibility of crossing. One can calculate beforehand what should be "too early" and what should be "too late," from the known mechanism of excitability. Or even more simply, the difference between those times is the diameter of the T^* ring divided by the speed of the activation front. That provides a quantitative prediction of vulnerable phase *duration*. This has been measured, and the estimate seems good to within a factor of 2 or so.

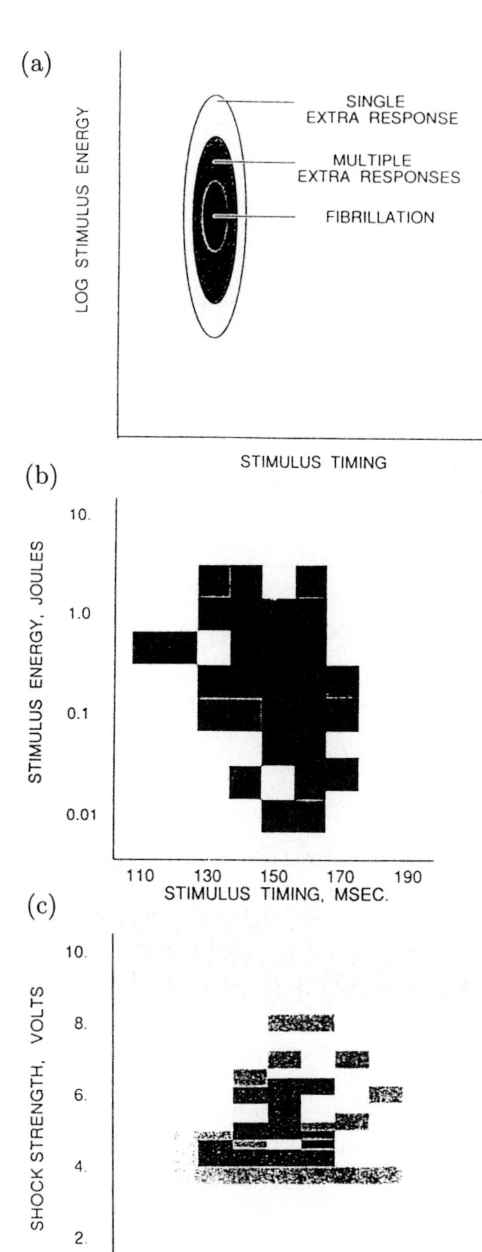

FIGURE 13 (a) The theory behind the pinwheel experiment predicts closed concentric contours of (successively lighter) persistent rotors [which therefore have time to incite fibrillation], ephemeral rotors, a single extra contraction, or no effect, depending on stimulus magnitude (vertically) and timing (horizontally). [From Winfree[223] with permission, adapted from Winfree.[210]] (b) The result of the corresponding experiment in the dog, in vivo [From Winfree[223] with permission, derived from the data of Shibata et al. 1988, here converted into format (a)]. (c) The result in a two-dimensional layer of sheep epicardium *in vitro*. "Fibrillation" is replaced by "sustained vortices" (the black patch near 4,230). As in (a) and (b), dark shading represents "multiple extra responses" and light shading represents "single extra response." [Adapted from Davidenko et al.[42]]

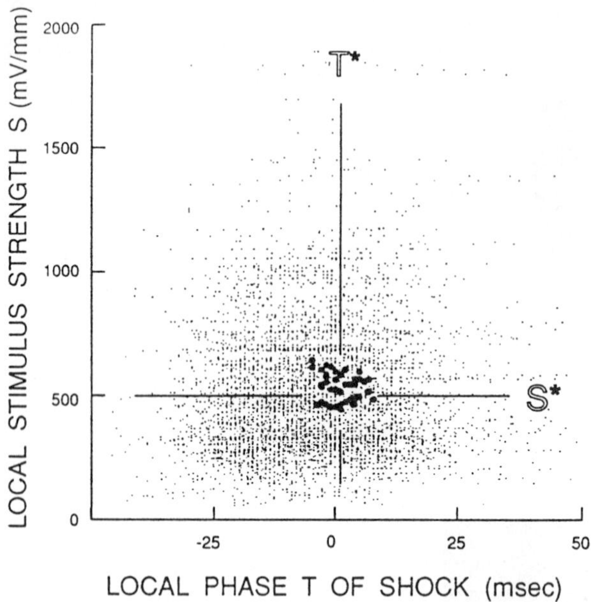

FIGURE 14 Combinations of local stimulus timing and magnitude where (large dots) a rotor is created, or (small dots) not. [From Winfree[223] with permission, adapted from Frazier et al.[68]]

3. If the stimulus is too weak, then the S^* circle will not be large enough to accommodate two mirror-image rotors side by side. If it is too strong, then the S^* contour will not even be in the medium. Thus there is no possibility of crossing and there is no reentry. Again, one can calculate beforehand from simple principles of mechanism what should be "too weak" and what should be "too strong." In marginal cases, contours crossing almost tangentially *twice*, very close together, we should get two mirror-image vortices so close together that they spin only briefly before recombining: just a few rapid beats are expected in these marginal cases. The requirement that the stimulus not be too weak is pretty obvious, because without it we would all be spontaneously breaking into vortices all the time. But the idea provided by the theory of rotors, that the stimulus must also *not be too strong*, was widely regarded as paradoxical or nonsense before the experiment was tried a few years later. And there was no quantitative estimate of either limit.

   Figure 13(a) shows the anticipated results, classified by timing and size of stimulus at the electrode. I was unable to convince experimentalists that this was worth testing, but R. E. Ideker saw this figure in my 1983 *Scientific American* article and with his postdocs ran the experiment in 1986 at Duke University

Medical School using living dog heart. The result (Figure 13(b)) is coarsely resolved, but the main features seem to be there. A core of fibrillation onsets only within a narrow range of stimulus timing, and only when the stimulus is neither too weak nor too strong, and around it a fringe of brief tachycardia episodes, and outside that, no problem. This experiment was repeated in a quite different preparation, the postage-stamp-sized epicardial slices of sheep heart in a petri dish. Davidenko et al.[42] found the same pattern (Figure 13(c)), with only this intriguing difference: where Figures 13(a) and (b) have "fibrillation," Figure 13(c), using a two-dimensional heart, has persistent rotors. Rotors seem to be the gateway to fibrillation in three dimensions, but not in two dimensions.

This bull's-eye structure, and the exact values of the top and bottom of the target, have direct engineering implications for the design of electronic defibrillators. Incidentally, we later learned that the basic result (existence of an "upper limit of vulnerability") had already been put into the literature, but it was apparently so surprising that in the absence of a persuasive theoretical interpretation, the fact was promptly forgotten and left uncited.[131] How many other wonders might linger unfindable in the uncited literature?

These results can also be used to provide a clear measurement of the numerical values of T^* and of S^*. Data from the Duke experiments are gathered in Figure 14 to show the combinations of *local* stimulus timing and strength realized at all points in the hearts of all dogs. Those points that became vortex centers are darkened. They cluster as you can see around special values, providing the clearest measurement obtained to date for T^* and of T^*—and confirmation of their existence and significance.

4. The same principles, plus one extra, provide the first quantitative suggestion for *why* the fibrillation threshold is what it is, rather than 1000 times less or 1000 times more. The extra principle is that in normal ventricular myocardium, rotors promptly incite fibrillation, so the VF threshold cannot be higher than the threshold for creating rotors. Rotors necessarily arise in mirror-image pairs, which will recombine before they can incite VF unless they are initially far enough apart. What is needed to create vortex cores far enough apart? The vortex cores observed in the past few years in many preparations all seem to be about 2/3 cm in diameter, whether in dogs, pigs, sheep, or humans. So we need a T^* surface to cross an S^* surface at sites at least 2/3 cm apart. The T^* surface behind most kinds of activation front is locally similar to a plane, and the S^* surface around a solitary unipolar electrode on the epicardium or endocardium is something like a hemisphere. If T^* cuts the hemisphere along a 2/3-cm diameter, then the area of the S^* hemisphere around the electrode tip is $2\pi(1/3\,\text{cm})^2$. All along this area, the current density S^* is 6 V/cm divided by the electrical resistivity of muscle, about 330 Ω-cm: that is, 18 mA/cm^2. The total over the hemisphere's area is thus 13 mA. Is this close to the observed VFT? The VF threshold had been repeatedly measured on thousands of dogs since it was defined by Wiggers in 1940. The results from direct current, single pulse experiments are predominantly in the range about 12–22 mA. The lower values

are from widely spaced bipolar electrodes, closest to the unipole approximation used here. The higher values come from narrow bipoles, for which a slightly more elaborate theory predicts correspondingly higher VFT's. This simple theory thus suggests the first rough quantitative prediction/interpretation for the known facts in this area.

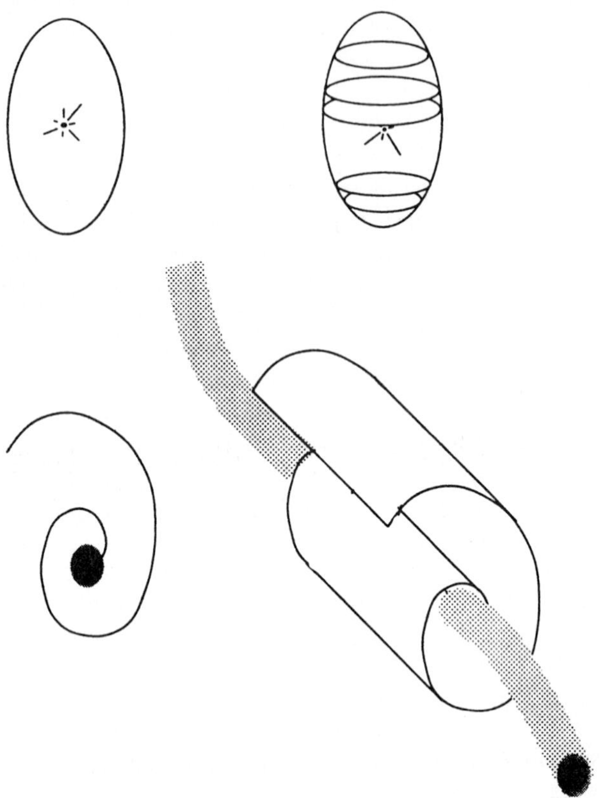

FIGURE 15 Two-dimensional and three-dimensional excitable media (continuous, uniformly anisotropic, unbounded, simply connected) each sustain two kinds of wave: the single elliptical closed ring of activation in response to a central stimulus, and the spiral or scroll. From Winfree[223] with permission.

(a)

(b)

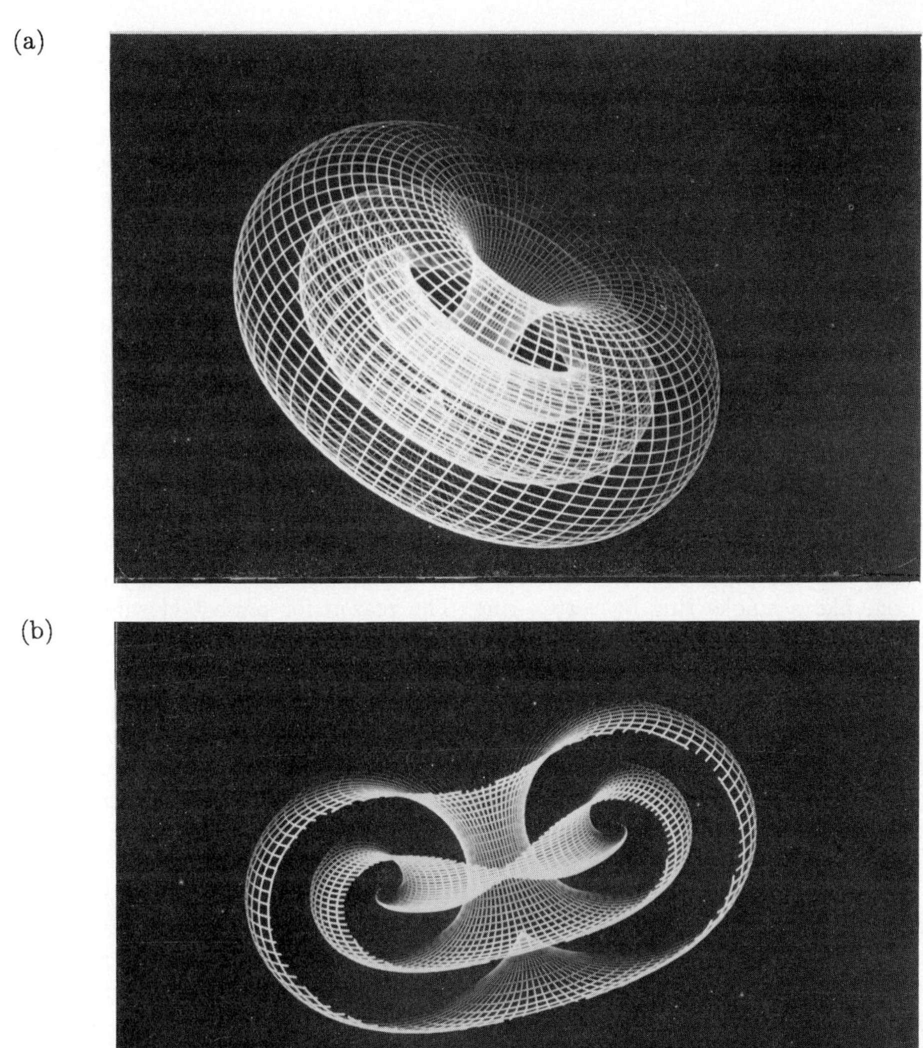

FIGURE 16 (a) The scroll's axis or filament closes in a ring along the axis of this bagel. (b) Each cross section perpendicular to it reveals an outgoing spiral wave in this cut-away of the same wave front. From Winfree[212] with permission.

THIRD LECTURE

We can now generalize from two- to three-dimensional studies. While estimating the VFT threshold, we already ventured into three dimensions by integrating a current flow over a hemisphere, but there was nothing essentially different from a cylindrical or two-dimensional circular version of the problem. But there might be something fundamentally different about three-dimensional excitable media. For example, rotors are stable in two dimensions but spontaneously progress to fibrillation in three dimensions. The beginnings of a theory of three-dimensional vortex filaments has been created, but not tested, during the past decade. That has been my main preoccupation during the last several years and that is what I will present in the second two lectures.

In *three* dimensions the vortex center of the spiral wave becomes a vortex *filament*. In idealized excitable media in three dimensions just as in two dimensions, we still have two distinct modes of activation (Figure 15). The reentrant mode is drawn here as a straight scroll, but more generically its axis curves. Eventually it may encounter the boundaries of the medium. Otherwise, it perfectly closes in a ring, like a smoke ring or vortex ring. The reason for perfect closure is that the vortex center, if motionless, has a certain concentration of substance u, so it lies on a certain iso-u surface...and a certain concentration of v, so it lies on the corresponding iso-v surface...and two smooth surfaces can only intersect in rings (possibly interrupted by the boundaries). At least, this is a "reason" for models involving only two local state variables. For real excitable media which need not reduce in practical approximation to two-variable dynamics, vortex geometry might be more subtle than the visions proffered here.

Such a vortex ring would look like Figures 16(a) and 16(b). The vortex filament is a circle and the pulses radiating away from it form surfaces of revolution. The spirals fit together in three dimensions to make closed egg-shaped wave fronts like those that surround a "pacemaker point" or "leading focal center" of repeated excitations. But the mechanism of *this* pacemaker is three-dimensional reentry, not space-independent pointlike oscillation at an essentially arbitrary period. Only its unique *period* betrays that the mechanism involves rotors rather than abnormal cells or a patch of parametrically different medium.

Vortex *rings* of reentry in normal heart muscle could be as small as 2–3 cm in diameter, according to computations using typical electrophysiological parameters. (In case of prolonged ischemia or infarction, the cells come uncoupled and all activity patterns become smaller, or even change their character entirely if the uncoupling is so extreme as to abolish continuum properties.) Such a vortex ring would fit comfortably in whale myocardium and it might also in healthy dog or pig or even adult human heart muscle if the propagation speed along the 1-cm thickness is much less than in the directions visible on the surface. One has to be a little careful here about the meaning of "cm," since heart muscle is not isotropic. Waves in heart muscle propagate on the epicardium about three times slower transversely to the cellular "grain" of heart muscle, than they do in along the longitudinal direction, the way the cells point. They might be even slower in the other transverse direction, so the 1-cm thickness of the human ventricular wall (if it can be regarded as a continuum at all) might be effectively several centimeters for this purpose. This is not yet clear. In any case it is clear that *bits and pieces* of a scroll ring certainly would fit. Shibata and Frazier at Duke set up an experiment to elicit the simplest geometry, the three-dimensional scroll predicted to touch both epicardium to endocardium perpendicularly. They documented it quite clearly.

The notion that scroll rings might exist in thick myocardium has been kicking around since at least my 1980 book. Medvinsky et al.[141] in Puschino near Moscow were the first to look for vortex rings in myocardium, using the protocol that demonstrated them for the first time in a chemically excitable three-dimensional medium.[207] Their results in the atrium were strongly suggestive, but remain to be repeated or tried in the ventricles.

That is about as far as the three-dimensional vortex story has evolved in normal, healthy myocardium. It might have further to go if the 1-cm thickness of human LV wall is functionally equivalent to several centimeters, due to slowness of conduction in that direction. Or it might not if mid-myocardial tissue is fragmented by cleavage planes so that continuum approximations are less serviceable than they seem to be on the epicardium. It will be hard to know until electrophysiologists figure out how to record from three-dimensional electrode arrays at 1-mm spacing without altering the otherwise normal tissue in the process. Meanwhile understanding of three-dimensional vortex filaments is developing quickly in the experimentally more convenient context of chemically excitable media, and in numerical experiments and mathematics.

CHEMICAL ORGANIZING CENTERS

Vortex *filaments* forming closed rings commonly include topologically distinct knots and links. Compact arrangements of linked rings are called *organizing centers.* Each kind of organizing center radiates closed spherical excitation fronts at the characteristic interval of rotors (or slightly more often due to an effect of "twist"—see below). In the presence of organizing centers, there is always some part of the medium in an excited and contagious state. Without them, the medium promptly reverts to uniform equilibrium, or if it is stimulated, it exhibits one response to each stimulus with intervals of global quiescence between stimuli. This is the normal situation in human heart muscle, for example. But heart muscle, like any other excitable medium, is also susceptible to periodic modes of *self*-organization. The study of periodic self-organized activity in generic excitable media is largely the study of rotors in two dimensions, and in three dimensions, of organizing centers made of vortex rings. It is only a conjecture that this is also the case in the particular excitable medium called heart muscle, but it is quite clear in several other cases.

Since 1974 such vortex rings have been observed and their motions have been studied quantitatively in the laboratory, using the chemical excitable medium. If, as in heart muscle, you could not see into and through this chemical medium—if it were visually opaque—then the nature of the source would only be given away by the characteristic short period of the rotor where the wave front erupts onto the observable boundary as elliptical activation fronts ostensibly radiating from a focus. That unique period would suffice to betray the reentrant mechanism at its source. But in the chemical gels, the entire structure is also visible. In this chemical kinetics the millimeter-size rotor has a period about 1000 times slower than in heart muscle, near 100 seconds rather than 100 msec, and it is a hundred times smaller, about 1/10 mm across rather than 1 cm.

Numerically solving equations like Eq. (3) is not difficult in two dimensions. Stable rotors were first obtained from such equations numerically 20 years ago using a Hewlett-Packard 9830 desk calculator with 1 kilobyte of RAM driving an audiocassette tape "hard disk" for several days. Computation in *three* dimensions only requires more memory and more operations. Twenty years later we use 10^8 bytes of RAM and 10^{11} updates of u and v, but the principles are the same.

Appropriate initial conditions are required, to get anything to happen in a three-dimensional medium that is more interesting than propagation of a single excitation front to the walls of the box, followed by relaxation to uniform quiescence. We need to induce self-organization, and (curiously) the only known mode is the rotor. Thus a filament must be created along which every cross section is a rotor or something sufficiently similar to become a rotor. There is an easy way to do this. The arrangement of u and v concentrations that constitutes a rotor necessarily includes a core of transversally crossing gradients. Let's center a coordinate grid in this disk of (u, v) combinations so that $U = u - \langle u \rangle$ and $V = v - \langle v \rangle$. Then we can refer to each part of the rotor by a complex number, $\mathbf{C} = U + iV$. We want to map each (x, y, z) point of the three-dimensional medium to this plane in such a way that $\mathbf{C} = 0$ along the wanted filament. There are recipes for prescribing z as one or another rational polynomial $\mathbf{C}(x, y, z)$ whose roots lie along variously linked and knotted paths through (x, y, z). We pick such a polynomial, use it to prescribe u and v throughout the cube, and we have our initial conditions. Then we turn on local reactions and diffusion by updating all u and v values as though a short dt had elapsed, and repeat many times.

It is less easy to look at the results. You are confronted by an array of about a million concentrations. Most of them represent waves propagating through the bulk of the medium. These waves are little different from those found in one-dimensional pulse propagation so they are not very interesting. (Their speed depends interestingly on local mean curvature, but there seems to be not much else to study about them.) They obscure your view unless windows are laboriously cut through them. From my perspective, the interesting part of an organizing center is the slowly moving vortex filament that constitutes the source of these waves. There are several ways to extract it numerically, throwing away the larger volume occupied merely by waves radiating from the filament. One way is to select all those grid points whose $U^2 + V^2$ lies sufficiently close to 0. This collection of grid points resembles a long thin swarm of bees which can be threaded by a smooth curve representing the vortex filament.

Having done this we then sample the local geometry at each of 100 equispaced stations along the vortex filament. Then the $u(x, y, z)$ and $v(x, y, z)$ arrays are updated through one more rotor period, and a similar snapshot is obtained again. Each arc between stations on the original filament is found to have moved. In these numerical experiments we ask whether any of the topologically distinct organizing centers are asymptotically stable in the sense that they rebound to a preferred shape after mild perturbation, and otherwise merely tumble through space like rigid particles. We have run across several such kinds, perhaps half of the kinds thus far initiated. And we ask under what circumstances can the motions of vortex filaments be understood just in terms of local geometry in the neighborhood of each moving segment. In those circumstances, what *are* the laws of motion (if they are simple enough to recognize from numerical experiments)? For example, it can be proved that if both u and v diffuse at the same rate and kinetic parameters are such that there is no spontaneous meander, and the filament's torsion and twist (see below) are identically 0 and the radius of curvature greatly exceeds the radius of the rotor, then each segment of filament "should" move toward its center of curvature with a speed equal to the diffusion coefficient over the local radius of curvature.

The first observations of curved three-dimensional filaments in this reagent, 20 years ago, showed that curved segments do indeed contract toward their centers of curvature, leading to the demise of small rings (a couple mm in diameter) within an hour or less. In computations the rate adheres nicely to a linear dependence on curvature of the filament (superimposed on periodic meander). Integrating that rate law, we find that the square of the radius should fall off linearly with increasing time, and the slope of the decline should be twice the diffusion coefficient of the two equally diffusing substances. That is exactly confirmed numerically. The corresponding laboratory measurements of r^2 vs. t later confirmed these computations in the sense that the slope is constant and compatible with a plausible diffusion coefficient. Actually one does not know quite what to use for a diffusion coefficient, since the two substances have about two-fold different coefficients even in water, and in silica gel one of them is immobilized.

Numerical experiments show that things get more interesting when the two diffusion coefficients are not conveniently equal. Figure 17 shows the shrinkage rate of a planar circular vortex ring in the FHN medium (Eq. (3)) plotted against curvature). At *small* curvature the slope is equal to the diffusion coefficient as foreseen in that limit. But as the curvature becomes significant, the shrinkage rate grows slower than linearly, then even decreases, eventually to 0 at a certain curvature. In other words, shrinkage need not continue all the way to extinction. Shrinkage stops about when the hole inside the ring has been squeezed shut, but it is not clear

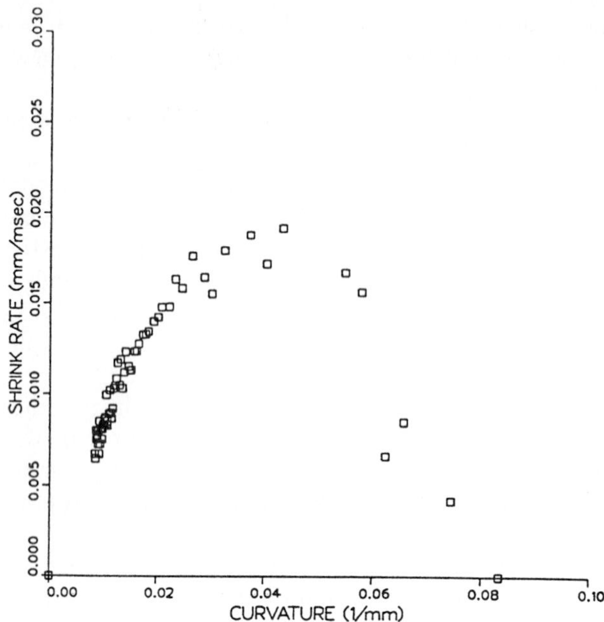

FIGURE 17 The rate at which a uniformly curved filament spontaneously moves toward its center of curvature can be a nonlinear function of curvature. In this computation (Eq. (3) with carefully chosen parameters $\varepsilon = 0.3$, $\beta = 0.7$, $\gamma = 0.5$), there is a zero of shrinkage rate at a certain finite radius. From Courtemanche et al.[37] with permission.

whether this is due to nearness to the other side of the doughnut or simply to local curvature of the filament. In any case, the result is a stable vortex ring, which goes drifting through the medium like a particle with momentum. It can be deflected by ephemeral chemical (numerical) perturbations, after which it regains it shape and continues in the new direction. This is the first discovered stable organizing center.

Not much is yet known about the interactions among such particles. I know that they do not always destroy one another upon mutual encounter, and can combine and transmute into topologically more elaborate stable organizing centers. The possibility is not excluded that such particles, like atoms and molecules, might support a chemistry, in principle permitting mutation, natural selection, and evolution. Perhaps someday it will be ridiculous to contemplate chemistry, biology, and life in excitable media (based on lasing in the solar corona-sphere? based on optical entrainment of nonlinear solid state oscillators?). But at this writing it still seems as realistic a possibility as chemical evolution from the medium described by the partial differential equations of quantum mechanics would have seemed to a contemporary physicist contemplating the universe of 10 billion years ago.

Single, flat, uniformly curved, untwisted organizing centers are not very generic. Their wave fronts are all surfaces of revolution, their vortex filaments are perfect circles lying in a plane, there is no torsion, nothing looks twisted, and so on. In this context the only property possessed by the vortex ring, which suffices to completely characterize it, is its radius. So a dynamical theory need only consist of an expression for the rate of change of radius as a function of radius, and it becomes hard to distinguish mechanistic theory from mere description. Correction: the ring also has a position along the axle of symmetry, so it has a velocity in that direction, and that also depends on the radius (unless all diffusion coefficient are equal: then this perpendicular velocity is identically zero in theory). So now you find yourself contemplating two velocities: one at right angles to the filament's local tangent vector in the direction of the filament's center of curvature, and one at right angles to those two directions. Since we are now talking of local directions and local pieces of filament, not about the whole ring, the idea comes naturally to mind that the local velocities might be determined by local curvature, in case that were not the same all along the ring. Thus arose the notion that V_n in the *n*ormal direction (toward the center of curvature) is a linear or maybe quadratic function of local curvature, at the same time that V_b in the *b*inormal direction (perpendicular to the plane of curvature) is also some such function. This turns out to work fairly well in many cases, at least as a description. But there is a new problem. Notice now, that if curvature is not the same everywhere; i.e., if the ring is not a flat circle, then the binormal velocity is not uniform, and parts of the ring drift out of the plane at different rates, so pretty soon it is not planar any more. It now has nonzero torsion and we have a new ingredient for the theory. You might expect that no more ingredients are needed beyond these two, since any curve in three dimensions can be described by specifying its local curvature and torsion as functions of arc length. So it might suffice to describe V_n and V_b through those two independent variables, possibly mediated by differential operators or functions. This is a mistake, because vortex filaments in excitable media have (at least) one other crucially important local property.

TWIST

In fluid dynamics there is a dynamical law derived from the Navier-Stokes equations for conservation of mass and momentum in a "dry" fluid (no viscosity) that tells just how each element of arc length along the filament moves under the integrated influence of all the vorticity associated with all the other elements of the filament.[115,116,134] This Biot-Savart law is not a *local* rule, but for filaments of slight curvature the excellent "localized induction approximation" is, and something like it should be derivable from Eq. (3). Unfortunately, it is not to be had by straightforward copy-cat procedures. The resemblance of vortex filaments in excitable media to those in fluids is only superficial. First of all, their key feature is that they must close in a ring—unless interrupted by boundaries—and that is a feature of fluid vortex filaments *only* in the nineteenth-century "dry water" idealization, lacking viscosity. Moreover, in the localized induction approximation, a circular filament drifts perpendicular to its plane but does not shrink, and more generally curvature does not change in the absence of torsion (both contrary to the behavior of scroll rings in excitable media) and torsion does not change if curvature is uniform (contrary to the behavior of helical scrolls in excitable media).[173] In the twentieth century, liquid helium provides a sort of "dry water," but its vortex rings are quantized in a way quite different from vortex rings in excitable media. Also unlike any fluid, excitable media have a distinct temporal period associated with every point pervaded by spiral waves from a given organizing center. But most importantly, the substitution of physical motion by chemical activity endows vortex filaments in excitable media with a geometrical property unknown in hydrodynamics. This additional property is called "twist." It measures the rotor's change of phase with arc length along the filament in a still snapshot.

The filament is made of spatial distributions of the local state variables. Those variables have spatial gradients transverse to the filament. Those gradients point in some direction relative to a fixed coordinate frame. That direction changes with arc length along the filament (and, of course, it rotates once in each rotor period). If, in a snapshot at fixed time, it changes a lot in a short distance, then the filament is very "twisted." This twist rate, $w(s)$, is a real-valued function which should not be confused with the integer W denoting a winding number on a closed curve. It represents the "turning to the right" of the rotor with distance (in either direction) along the filament, in a snapshot at fixed time. It is computed from the triple product of three normalized vectors: the unit tangent vector T and the normal component of the u or v gradient (either should serve the same purpose in this idealization) evaluated at adjacent sites along the filament. This is the volume of a parallelepiped of unit height, whose base, projected onto the plane normal to T, is a parallelogram of area equal to the sine of the angle θ between its two edges. With gradients multiplied in the correct order, θ is the clockwise (rightward) rotation of the gradient vector with increasing distance along the filament. Since only the component of the gradient perpendicular to the tangent contributes anything to the triple product, it is sufficient to use the unresolved gradient, normalizing the wanted component by $|T\times\nabla u|$:

$$\sin\theta = \frac{\nabla u(s+\Delta s)\bullet T\times\nabla u(s)}{|T\times\nabla u(s)|\ \Delta s\ |T\times\nabla u(s+\Delta s)|}$$

so

$$w(s) = \frac{\Delta\theta}{\Delta s} = \frac{[d/ds\nabla u(s)]\bullet T\times\nabla u(s)}{|T\times\nabla u(s)|^2}\ .$$

With s oppositely oriented (or dot and cross interchanged), this product remains the same, but interchanging any two of the three vectors reverses its sign.

Nonzero twist arises quite naturally and inevitably. Suppose you had a vortex filament with no twist. But it is curved more in some places than in others. Curvature slows down vortex rotation. Faster-spinning arcs then creep ahead in phase relative to slower-spinning arcs: the accumulating difference is *twist*. Also a curved filament moves in the binormal direction, at rates proportional to local curvature, so if curvature is not uniform then even an initially planar filament acquires torsion, which distorts an initially uniform phase distribution, creating twist. Or even without curvature, real excitable media are not 100% uniform in their local properties, so in some regions the vortex period is a little shorter than in others: if the filament passes through such regions, it becomes more twisted with each successive rotation.

So then it seemed that we want a theory capable of predicting V_n and V_b from local curvature, torsion, and twist.[228] But it was already becoming evident that rotors move spontaneously even in the absence of curvature, torsion, or twist. This spontaneous motion undercuts the present basis for study of vortex filament motion, according to which the filament is a two-dimensional rotor in every perpendicular slice, and its motion is zero when curvature, torsion, and twist are zero, as when isolated in the plane. Because meander is typical in two dimensions, except in an

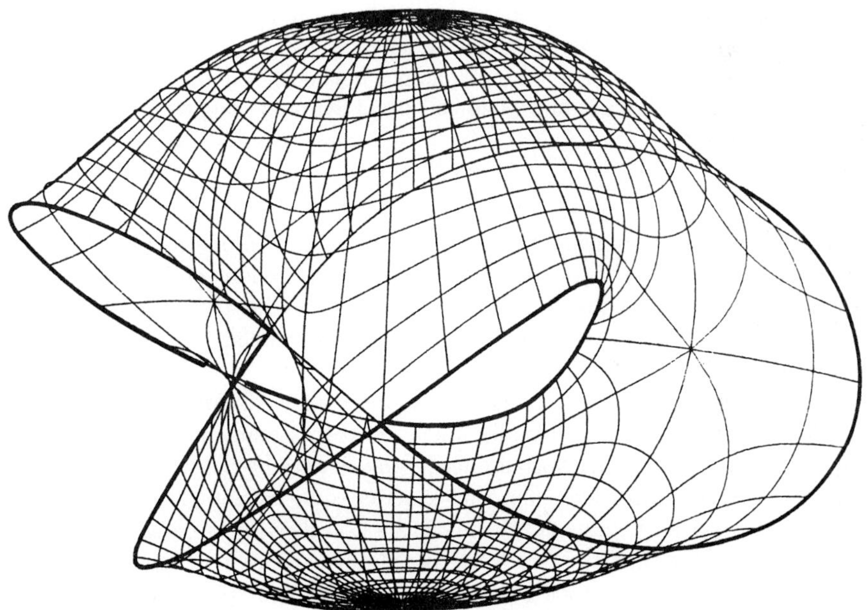

FIGURE 18 From such an initial wave front (implicitly oriented: one side faces quiescent medium, the other, refractory medium) a knotted scroll ring evolves. This figure was drawn by Timothy Poston in 1985; it represents one of the complex-valued polynomials used as initial conditions for organizing centers.[171]

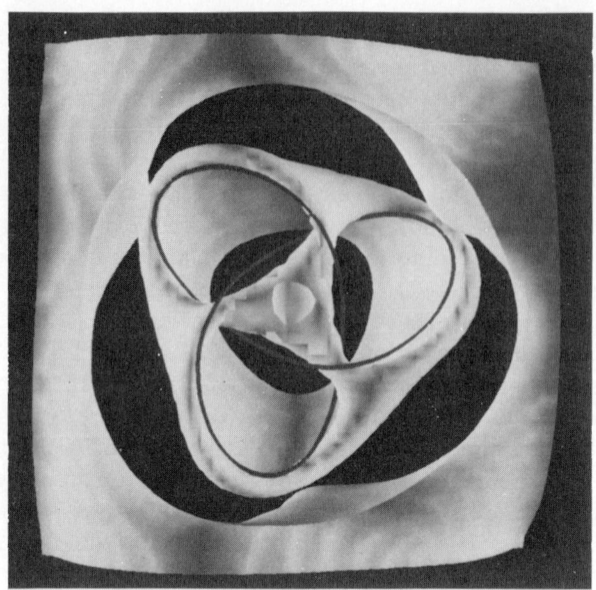

FIGURE 19 At a later stage of evolution, the knotted scroll ring has shrunk to a stable shape, slowly rotating and drifting through the medium (perpendicular to the plane of the page). This snapshot shows only the excitation fronts radiating away from it. The front and back walls of the box are transparently connected but the side walls are all absorbing. (This is a black-and-white version of the organizing center examined in Henze and Winfree.[81])

atypical fringe area of parameter space, chemical and numerical experiments intended to isolate quantitative effects of curvature, torsion, and twist on filament motion must be restricted to that fringe and have little hope of generality. Nonetheless, it is interesting to know qualitatively what new behaviors are introduced by allowing nonzero twist.

A surprising consequence of twist was discovered computationally in 1989.[80] Consider the evolution of an initially uncurved vortex filament given uniform twist. In the initial condition every horizontal x, z plane contains the same two-dimensional vortex computed from the chemical kinetic equations, e.g., of the Belousov-Zhabotinsky reaction. Planes each bearing a spiral wave are stacked up, slightly rotating about a common vertical y-axis. The wave radiating from this axis may be viewed as a skewered stack of rotationally staggered spirals. By repeating this experiment with different twists imposed, the effect of twist can be systematically determined. For example, it shortens the rotation period of the vortex, enabling the twisted segment to compete successfully with waves arriving less frequently from less twisted segments elsewhere. It also turns out (in two of the three different excitable media examined) that if the twist rate exceeds a certain threshold, then the filament sproings into *helical* form. This helix then grows until either stabilizing at finite radius or hitting the no-flux walls of the vessel, where its expansion may or may not be arrested, depending on the kinetics chosen. In these helix experiments, all properties of the filament are uniform along the filament: arc length derivatives of curvature, torsion, twist, and so on are all identically zero.

But suppose we used initial conditions like Figure 18. Here you see one three-dimensional iso-u surface—a wave front because the way the v field is arranged, not visibly here—adjusted to have its edge along a knotted ring. That edge turns into a trefoil-knotted vortex ring, buried in the structure seen in Figure 19. It is confusing to look at the wave fronts, and simpler to look just at the vortex tube from which the waves radiate: Figure 20 shows the core tube of another knotted organizing center, viewed along the vertical axle of the previous figure. It resembles a three-bladed propeller. The propeller tips move at several percent of pulse propagation speed. The entire knot is rigidly precessing, two orders of magnitude slower than vortex rotations, about the vertical axle of Figure 19. This vortex knot acts like a particle gliding through the medium at about 1% of wave speed while it sedately rotates comparably slower than the rotor turns. It is stable in that if you deform it, it gradually reverts to its original shape and speed and precession rate, but now with a phase shift and with its parallel linear and angular momentum vectors turned together to some other direction. Using six adjustable parameters, the motions

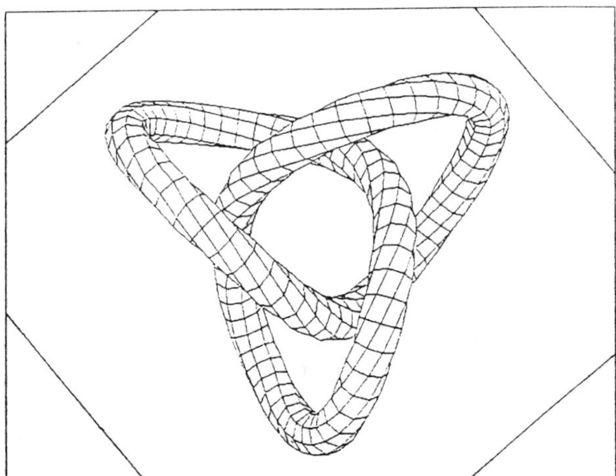

FIGURE 20 The singular filament of a knotted organizing center like Figure 18 (but opposite handed and a different kinetics) is isolated in this view.

of this stable organizing center can be (as foreseen by Keener[104]) described very closely in terms of linear combinations of local filament curvature, twist, and arc-length derivative of twist.[81] I find this surprising because I believe its stability is determined by nonlocal factors: by waves from more rapidly spinning, more-twisted arcs impacting on less-twisted arcs, and by the mutual obstruction of filaments encountering one another transversely at very close quarters.

THE LOCAL GEOMETRY APPROXIMATION

Let's consider the adjective "local." In fluid dynamics local dynamical laws such as the nonlinear Schrödinger equation (6) are rather poor approximations to the Biot-Savart integral for filament motion. Why should a local theory be any better in excitable media? There are two reasons: in excitable media there is no fluid motion induced by remote vortex lines, and in excitable media causation goes outward from wave sources that are protected from incoming influences by shells of outgoing shock waves. But there are situations in which this "local geometry hypothesis," as I dubbed it at the outset of this project in 1986, breaks down. They are important situations. I think they are mainly responsible for the numerically observed stability of some compact organizing centers. I am not going to go into that here, but for those interested, these situations are discussed in Winfree and Guilford,[228] Henze et al.,[80] and Henze and Winfree.[81] There probably are situations in which nothing more than local geometry dominates filament motion, though they might not be convenient for numerical simulation. What would a theory of such motion look like?

What one wants from the local geometry hypothesis is an explicit formulation of local motion in terms of local chemical kinetics and diffusion. Local motion consists of V_n, V_b, and the local period of the rotor, which, as we saw just above, is also affected by local curvature and twist. Those three items would constitute a complete description of the filament's motion if it were determined merely by local geometry—and if, in the absence of curvature, etc., there were no meander—and if the filament could be described in terms of position and phase, like a thread through the pivots of stacked standard rotors. This is an intimidating list of "ifs," but the job was courageously taken on.

For nonmeandering filaments the local geometry hypothesis was implemented by Keener in 1988. The bottom line of Keener's mathematics is that in the case of small curvature and twist, smoothly changing along the filament, the departure of the rotor's angular velocity from the two-dimensional baseline, and V_n, V_b are each compounded by linear superposition of curvature, twist, and—instead of torsion—the arc-length derivative of twist. Thus there are nine undetermined coefficients that in principle depend only on local kinetics and diffusion, but at present they still have to be measured empirically. Without knowing those coefficients we don't yet have an easily testable theory, but several things are clearly right about it, so long as curvature and twist do not become substantial. First of all, it corrects my early erroneous notion that torsion could be an important variable, and replaces torsion by the derivative of twist. Henze and Winfree[81] confirmed Keener's prediction that this is an essential variable. (It hard to imagine an important role for torsion, since a straight line has any torsion you please, inherited from whatever helix you may have made thinner and thinner without changing torsion, to approach straightness.) Secondly, in the utter absence of twist, it agrees with the accepted rule that V_n and V_b depend linearly on (slight) curvature. If the nine linear coefficients could be predicted from the local kinetics of excitability, it might become possible to make a much stronger case.

Meanwhile, I suspect that something is fundamentally wrong with the present incarnation of this theory, since it has no place for motion in the absence of curvature or twist (meander) and since it predicts finite speed of an uncurved uniformly twisted filament, in proportion to the squared twist, even though the direction is indeterminate. It should also be remembered that its domain of validity may be not much wider than what is explicitly claimed in its derivation (the limit of small curvature and twist, and far from other filaments):

1. The observed dependence of the planar ring's shrinkage rate on its size is not always even roughly linear.[37] Maybe these linear terms are just the front end of some Taylor expansions, whose further terms come into play when filaments get close to one another or curvature and twist become severe.
2. The observed motions (especially V_b) of twisted filaments have not yet been successfully fitted closely even after adding several new terms and corresponding fitted parameters.[80,224]
3. Even though V_n and V_b in the one knot examined exhaustively does appear describable in linear terms with only six arbitrary parameters,[81] that knot clearly is not an example of a low-curvature, low-twist, local-geometry situation.

Thus it seems to me possible that we are still at an early stage in this exploration, and that experiments are making indispensable contributions.

WHAT ORGANIZING CENTERS ARE NOT

I think it is worth noting explicitly that organizing centers in excitable media are not examples of a phenomenon long awaited: chemical reaction-diffusion instability. Two-dimensional Turing patterns were anticipated by 40 years, and half-way through that wait, theorists were almost bursting to recognize an example in the real world. My photographs of spiral waves in the Belousov-Zhabotinsky medium were widely reprinted to illustrate such symmetry-breaking instabilities of a uniform steady state. However, it was clear from the start that they are nothing of the sort. The uniform steady state is perfectly stable to small perturbations. And rotating spirals are not among the anticipated eigenfunctions of the Turing instability. In three dimensions a wider variety of patterns occur: the organizing centers classified by Strogatz and Winfree with "quantum numbers" identifying integer linkages and twists. These are neither alternative eigenfunctions of any reaction-diffusion system, nor successive bifurcations on a tree of solutions. They do not arise by Hopf bifurcation from an unstable uniform condition. Many of them are stable, and they are discrete alternatives to the stable uniform state, arrived at only by large-scale disturbances required to place the system in the distinct attractor basins of the distinct organizing centers.

At this stage I leave off the line of development informally followed above with minimal intrusion of citations, and run through about two-thirds of the same material again from another direction, tying it to another one-third of related thoughts and experiments. This time I will try to be more scholarly. I hope the redundancy and the change of perspective will help to illuminate the many intriguing questions that await attention.

FOURTH LECTURE

The main ideas behind these lectures were:

- that normal heart muscle is an excitable medium;
- that some aspects of the dynamics of excitable media are generic, notably their vortex excitations;
- that these aspects are inherently three-dimensional and remain but poorly understood;
- that pure computation from equations of mechanism should be complemented by experiments;
- that in ventricular myocardium these vortex excitations are responsible for the most pernicious arrhythmias, but are almost impossible to study three-dimensionally in living heart muscle;
- that they *can* be studied computationally and in chemically excitable media, both of which support action-potential-like propagation in three dimensions;
- that the utility of these metaphors is attested by verification in the laboratory (both chemical and physiological) of conclusions (which seemed surprising until confirmed) from prior two-dimensional numerical studies; and
- that quantitative experiments might be implemented by adapting existing computational procedures in optical tomography to existing computational procedures for analysis and display of the data we currently obtain by supercomputer solution of the mechanistic equations of cardiac electrophysiology and of the analogous chemical kinetics.

TWO-DIMENSIONAL PHYSIOLOGY

In 1970 I discovered "rotors," the self-organizing sources of spiral waves, in aqueous films catalytically oxidizing malonic acid. This reaction was discovered by Belousov during his study of the Krebs Cycle of cellular metabolism; it is now called the Belousov-Zhabotinsky excitable medium.[211] In Russia Zhabotinsky independently found rotors in the same year. Several years later, the chemical kinetics and reaction-diffusion mechanism of rotors had been deciphered in detail and I noticed that the equations have the same format and properties as those describing action-potential propagation in excitable membranes. It was already widely conjectured (by Wiener and Rosenblueth,[200] followed up mostly in the Russian literature, notably by Krinsky[121]) that spiral waves play a role in cardiac arrhythmias. If so, then the rotor mechanism[77,205,206] might be involved. (In electrophysiological contexts "rotors" are now often called "vortices"; I use the two interchangeably in these lectures.) I ventured some specific predictions on this basis[210] which then seemed counterintuitive, e.g., that there should exist a (then unobserved) "upper limit of vulnerability" (so named by Chen et al.[32]) for a DC electrical stimulus capable of inducing fibrillation during the vulnerable phase. This was checked in dog myocardium and found true (see Chen et al.[32]; see Figure 13(b) here, from Shibata et al.[182]; Frazier et al.[68]; see Figure 13(c) here, from Davidenko et al.[42]). Then physiologists felt motivated to check half a dozen other predictions, all of which were confirmed (summarized in Winfree[216,217]). These were only qualitative implications of the theory summarized in Winfree.[215] With this much encouragement, the theory was made quantitative in Winfree[219,220] by applying in cardiological context the notion of wave front curvature from physical chemistry and other recent understanding of the mechanism of rotors. It thus proved possible for the first time to quantitatively predict optimum conditions and the corresponding electrical thresholds for myocardial stimulation and minimum-energy pacemaking, for induction of tachycardia and fibrillation, and for defibrillation.[220] These thresholds proved quantitatively correct, according to the most recent measurements. This is significant for two reasons:

1. The values predicted were in some cases orders of magnitude lower than expected for pacemaking and defibrillation. Their confirmation has direct and beneficial consequences for the engineering of implanted pacemakers and defibrillators.
2. Cardiac physiology now has a quantitative theory of such things. This should catalyze more experimental tests, providing a platform for debunking or refinement of parts of the theory.

TWO-DIMENSIONAL PHYSIOLOGY FALLS SHORT OF UNDERSTANDING FIBRILLATION, DUE TO IGNORING THE THIRD DIMENSION OF MYOCARDIUM

Though the foregoing two-dimensional theory for DC, single-shock stimulation of normal dog myocardium during its vulnerable period did accurately foresee the mechanism of fibrillation onset in terms of mirror-image rotors, and did for the first time quantitatively derive the measured threshold from basic principles of electrophysiology, nevertheless, in retrospect I see this as partly accidental. What was really predicted was the existence, stability, and main properties of rotors in normal ventricular myocardium, together with necessary and sufficient stimulus conditions for their creation as counter-rotating pairs of predictable size and location (Figure 12). It was a gratuitous accident that they proved unstable and precipitated fibrillation. The mechanism of this transition is still completely unknown. But there is a tantalizing clue.[227] The key reason for expanding the cardiological investigation to three dimensions is a surprising discovery (not yet widely recognized as such) suggested by results from several cardiological laboratories during the past several years: the transition from spiral waves to fibrillation occurs spontaneously only in functionally three-dimensional preparations! The transition from spiral waves to fibrillation does not occur spontaneously in experimental preparations from which the endocardium has been removed to such an extent that the surviving epicardium is essentially two-dimensional (see Dillon et al.[50] and Kavanagh et al.[103] using an infarct, Schalij[179] and Allessie et al.[9,10] using liquid nitrogen, and Davidenko et al.[41,42,43] using dermatomed thin layers of myocardium). The cause does *not* seem to be mere removal the sub-endocardial lining of Purkinje fibers without seriously compromising the three-dimensionality of propagation (see Damiano et al.[40] using iodine). In ischemic tissue similar treatment completely eliminated the otherwise spontaneous transition from tachycardia to fibrillation (see Janse et al.[97,98] using phenol). My synthesis of these disparate observations is that ventricular fibrillation is an intrinsically three-dimensional process. It is therefore essential to understand in what ways three-dimensional excitable media differ fundamentally from now-familiar two-dimensional media. One essential difference is that vortex filaments in three dimensions, unlike rotors in two dimensions, have curvature and twist. The consequences of twist seem to include interesting possibilities for the advent of fibrillation.[219,220,222]

PAST TWO-DIMENSIONAL VORTEX COMPUTATIONS AND PHYSICAL CHEMISTRY

The theory of rotors diversified elaborately during the 1980s; for summaries see Winfree,[215,216,225] Zykov,[246] Keener and Tyson,[110] Meron,[144] and Mikhailov and Zykov.[146] However, it was not shown until about 1990 that the electrophysiologist's best quantitative summary of the ionic mechanisms of myocardial excitability (the Beeler-Reuter model) does in fact support rotors, and that their size and period and wavelength are compatible with their imputed role as a prominent mechanism of ventricular tachycardia. This has recently been established (see Courtemanche and Winfree[38] and citations therein, and Fishler and Thakor[61]; see Figure 10).

In the 20 years since rotors were first computed from a reaction-diffusion mechanism,[77,205,206] the published literature about their behavior has exploded (doubling every five years since about 1970, to about one paper per week currently) into such unanticipated diversity of examples that by the late 1980s there seemed an unlimited variety of vortexlike behaviors of diverse excitable media. Unbelieving, I tested the alternative proposition that for any excitable medium characterized by certain generic parameters (e.g., the ratio of excitation rate to recovery rate, the ratio of excitation threshold to excitation amplitude, the ratio of diffusion coefficients), all the newly familiar behaviors occur somewhere in a landscape coordinated by those parameters. An exhaustive survey was conducted of rotor properties in the Oregonator model of chemical excitability with all reactants diffusing equally[94]—and with only the electrical potential diffusing in the FitzHugh-Nagumo model[225] and Barkley's variant of it[16,17,18] and in Zykov's model of electrophysiological excitability.[133,225,247] Such a landscape was mapped and proved to be qualitatively similar for all four. It reveals the existence of four bifurcation boundaries, in order of increasing "excitability" of the medium (Figures 7 and 9): ∂P at which propagation fails, ∂R at which rotors blow up to infinite size and diffuseness, ∂M at which the compact rotor begins to "meander" in doubly periodic patterns, and ∂C at which meandering becomes more complex. The possibility of chaotic meander was foreseen computationally[175] but has not yet been sought in the laboratory. Computed myocardial rotors seem to belong inside this boundary and there is reason to think real myocardial rotors exhibit similar behavior. Discovery of ∂M provoked our experiments in which doubly periodic meander was first observed experimentally in the BZ chemical medium.[92] This has since become a popular subject for experiment and theory.[17,18,19,100,101,102,112,143,144,167,185]

TABLE 1 Notations for Two Periods

	'Earth period'	'Moon period'
Jahnke et al.[92]	$(N-1)\tau_0$	τ_0
Lugosi[133]	$2\pi/\omega_{\mathrm{tail}}$	$2\pi/\omega_{\mathrm{tip}}$
Skinner & Swinney[185]	$1/f_2$	$1/f_1$
Karma[100]	$1/F_2 = 1/(f_{1q} - f_{2q})$	$1/f_{1q}$
Barkley[16]	$2\pi/(\omega_2 - \omega_1)$	$2\pi/\omega_1$
Meron[144]	$2\pi/\Delta\omega$	$2\pi/\omega_1$

Various notations are used for the two periods or spectral lines ω_2 and ω_1 in a Fourier spectrum of wave trains from a meandering rotor, or the two frequencies f_1 and f_2 involved in tracing the trajectory of the meandering spiral wave's tip: where τ_0 is the rotor period assayed as the average interval between excitations far from the rotor. N (integer-valued only in exhibition flowers) is the reciprocal angle, measured as fraction of a circle, between successive petals of the flower. This is taken positive for outward-petal flowers like Figure 5(a), (b), and (e), and negative for inward-petal flowers like Figure 5(c) and (d). $\omega_2/2\pi$ is the frequency of curvature modulation in the tip's path. (Don't be misled by subscript "2": $\omega_2/2\pi = f_{2q} = f_1+f_2$, not $= f_2$). These relationships are described in Figure 5(a) of Jahnke et al.[92] The "earth-moon" notation is an artifact of our initial impression that two-period meander flowers were "compound *circular* motion," like spirograph tracings composed of two circular orbits, one (the moon, the rotor) carried by the other (the earth's orbit around the sun). The meander bifurcation was identified with "earth orbit amplitude" bifurcating from 0. Skinner and Swinney corrected this to "compound rotation" and Barkley emphasized that there is no reason for one of the closed paths to be circular even right at the Hopf bifurcation, and extracted its actual shape. Both "earth orbit amplitude" and the amplitude of path curvature fluctuations bifurcate from 0 at ∂M, but depend differently on parameters; both can become very large, but this happens to curvature amplitude where flowers are very pointy, and to "earth orbit amplitude" where they are not, along the 0° isogon.

Along the 0° isogon, N and $f_2 = 0$ and $\omega_2 = \omega_1 = 2\pi/\tau_o$: the curvature of the tip's path fluctuates at the period of the rotor. The amplitude of that fluctuation determines the linear speed or momentum of the "linear looping" rotor. Elsewhere in the meander domain, the flower has a center, which is excited at intervals $(1 - 1/N)\tau_o$: positive unless $N < 1$ (which makes no sense), and less than τ_o in outward-petal flowers. Thus, in a triangular flower like Figure 5(e), excitation goes through center three times while the rotor turns (the front moves eastward) $3-1=2$ times: the center is excited at intervals $2\tau_o/3$. This can happen only if this interval exceeds $\tau_{\min}$, which requires small ε, thus outward-petal flowers of lower symmetry are found further to the right in Figures 7 and 9 (until the onset of hyper-meander at sufficiently small ε or large interpetal angle.)

Though the FHN study was published with only the fast "propagator" or "excitor" variable diffusing, this study has since been redone with the slow "recovery" or "controller" variable equally diffusing, with similar results. The published study also fixed the third parameter of the FitzHugh-Nagumo model ($\gamma = 0.5$). It has since been redone with γ changed to 1 and to 0 with similar results (for comparison with analytical approximations subsequently completed.[100,101,102,106]

The pertinence of these results in present context is that without them parameters for numerical experiments and chemical experiments can be chosen only blindly, perchance in unrepresentative regions of parameter space or precariously close to bifurcation loci. The diverse published recipes for BZ media (e.g., many gathered in Jahnke and Winfree[93]) give the impression of unsystematic diversity of rotor behaviors. I believe it should be possible (but I have not yet tried it) to locate each of these recipes on the generic landscape, together with myocardium and its various pharmaceutical modifications, and other biological excitable media such as *Dictyostelium discoideum*[64,148,193] and oocyte cortex,[129] thus putting the whole literature into perspective. For example, the American theoretical literature is mostly in the singular perturbation limit, far to the right on Figures 7 and 9, while much of the Russian literature (theoretical, computations, and experimental) together with some recent American literature (e.g., Karma[101]) is concentrated along the lower part of ∂R. Their contradictory conclusions thus need not be seen as incompatible. The middle region, largely unexplored by theory, seems to be where myocardial rotors live.

Before going on, let me pause to remark that the foregoing characterization of "excitability" might not be completely generic. There is at least one other kind of differential-equation model of excitability, which might turn out to have significantly different properties.[73,78,114,124,139,186] For example, its range of v seems unlikely to contract toward 0 as $\varepsilon \to 0$; but that property underlies much of the behavior of FHN-like excitable media. In the format of Eq. (2), one such example uses

$$
\begin{aligned}
f(u,v) &= \frac{(1-u^2-v^2)}{\varepsilon} + (1+bv)v, \\
g(u,v) &= \frac{(1-u^2-v^2)}{\varepsilon} - (1+bv)u,
\end{aligned}
\tag{7}
$$

in a medium with equal diffusion of both u and v.

Such media have three fixed points instead of just one. In case (7) one is at the center of the unit circle in the (u,v) plane and two occur where it intersects line $1+bv=0$. Such media make the transition from excitability to spontaneous oscillation at $b=1$ by a saddle-node bifurcation through infinite period ("SNIPER"), e.g., as revealed by the period and amplitude of the BZ reagent's bulk oscillations.[69,70,138,153,180,207] In such media, spontaneous oscillation and excitability are indeed mutually exclusive alternatives as commonly supposed (Figure 21), rather than independent, typically coexisting properties of the medium (as in the most profusely modeled chemical and electrophysiological excitable media, exhibiting Hopf bifurcation).

Suppose the single, attracting fixed point of FHN-like models is supplemented by a saddle and a repelling focus which, if the first two would fuse and vanish, would create a limit cycle (Figure 21(a)). The Japanese literature of rotors in excitable media is mostly about a limiting case of this model (Figure 21(b)), "phase models" (called "ring devices" in Winfree,[209] or "active rotators" in Shinomoto and Kuramoto[183]: $d\phi/dt = f(\phi)$, e.g., $1 + b\cos\phi$. These have but a single state variable, ϕ. The class used in the present survey, in contrast, have two or more state variables. The one-variable models can be seen as a limiting case of two-variable

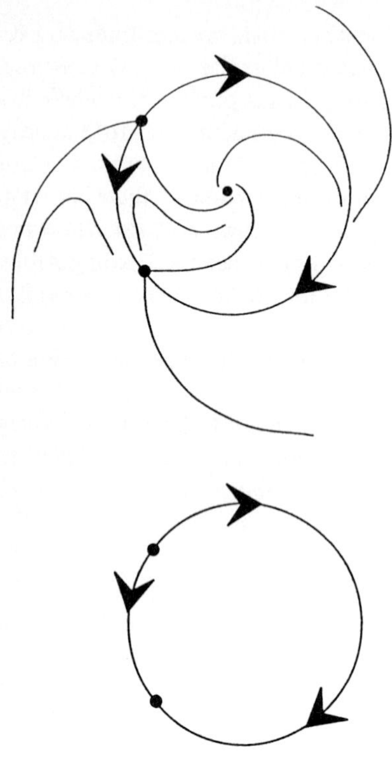

FIGURE 21 (a) The phase portrait of an excitability model like Figure 1, but with an extra pair of fixed points [adapted from Winfree,[209] Figure 9-1]. (b) In a limiting case, the system is confined to a closed trajectory bearing two of the fixed points (the attractor and the saddle).

models, in which the state flows so rapidly onto a one-dimensional ring that the "phase" description suffices. This limiting case has peculiarities that might disqualify it from serious attention (e.g., its spiral wave necessarily has a discontinuity of ϕ near its pivot) except that short of the limit, with rapid but finite radial flow, behavior may in many respects resemble that found in the unrealistic limit.

This model is of particular interest as a representation of any nonlinear dynamical system entrained by a periodic signal. In the entrained steady state, it holds a stable phase relation to the driver, but if perturbed sufficiently, it can briefly lose entrainment, changing relative phase until it approaches that same relative phase again in the next or prior cycle. If it is only marginally entrained (the driver frequency being near one or the other limit of the 1:1 entrainment band), then a relatively small phase displacement suffices to initiate excursion through the complement of one cycle back to the stable relative phase. This is excitability in a dynamical system which, by itself (apart from the dynamic of its entrainment), exhibits nothing of the sort. For example, a photosensitive chemical system may be driven by a spatially uniform cycle of light and dark, or by regular cycle of temperature change, at a period which it is barely able to follow. Appropriate perturbation at any point causes a local excursion en route to re-entrainment one cycle ahead or behind. In a spatial continuum with diffusion this event would trigger the like in neighbors and propagate like the waves familiar in this kind of abstract excitable medium. It is not yet known whether its variety of possible rotor behaviors is equivalent to that already found (e.g., Figures 5–7) in single-fixed-point models.

Supposing the results prove qualitatively the same as before, it may then be possible to characterize all known excitable media (covering a considerable range of biological experiments) by locating them on this two-parameter landscape. Their modifications by pharmaceutical treatment, electrical biasing, aging, mutation, and so on should then become comprehensible as departures in one or another direction on this generic map. If such a classification is really possible, to my mind its clarification would be the most satisfying and widely useful outcome of this whole project.

The next important step is a computational check on the pertinence of this ostensibly "generic" parameter plane to the eight-variable Beeler-Reuter model of ventricular myocardium. This was done independently in at least three laboratories.[38,54,61,216] In the earliest three of these citations, parameters were found at which a more complicated "turbulent" behavior occurs. The fourth, varying parameters systematically outside that range, produced flowers similar to those shown in these lectures. In all, the Beeler-Reuter model was adjusted in diverse ways: these ways evidently make a big difference. It might be that any generic version of Figure 7, like Figure 9, should also contain another bifurcation locus, to "turbulence."

The importance of these studies is that anti-arrhythmic pharmaceuticals are thought to act partly by altering the excitability properties of the myocardium to make rotors less viable (or more: medication does not always help). Knowing the implications for rotor behavior would seem an essential component of rational prescription. An improvement on this plan might use the DiFrancesco-Noble[47] model, which adds to the Beeler-Reuter[23] model the longer term consequences of short-period activation during arrhythmia (the refractory period shortens and propagation speed decreases due to accumulation of interstitial $[K^+]$, etc.). Can the consequences of "fatigue" in a model replete with Na/K pumps and cylindrical gradients be represented by a change in the generic excitability parameters? In what direction relative to the bifurcation boundaries described above? The DFN model is optimized for Purkinje fibers, not for myocardium, so results will only give a hint, but revision to represent myocardium is expected and will be easy to plug into the same code.

Is fibrillation related to such "turbulence" in models? Is it a two-dimensional process, like the models, or does it fundamentally require three dimensions? Fibrillation in the acetylcholine-treated atrium clearly *is* two-dimensional (atrial myocardium being so thin). But maybe there is a special reason for atrial fibrillation which does not apply to normal ventricular myocardium, viz., the greater morphological and electrical inhomogeneity of such atrial tissue. Almost all thinking about *ventricular* fibrillation descended with little if any modification from the non-quantitative model of Moe et al.[147] based on "nonuniform dispersion of refractoriness" in the *atrium*. The relevance of these assumptions to ventricular myocardium has never been demonstrated quantitatively,[218] while that of quite different features of ventricular myocardium has been (viz., the theory of rotors, based on the continuous cable equation of electrophysiology.) But the theory of rotors in nonoscillating excitable media goes only as far as tachycardia. Rotors seem stable in normal two-dimensional ventricular myocardium (e.g., dog right ventricular wall[245]), but always turn into fibrillation in normal three-dimensional ventricular myocardium (e.g., dog left ventricular wall[245]). Existing theory does not illuminate that transition.

The possibility also remains unexcluded that in two-dimensional myocardium or models thereof, wave fronts fragment and rotors multiply in part because, beyond a certain abundance of rotors per unit area of epicardial surface, they interact in complicated ways not yet discovered or described.

In smoothly oscillatory and not-necessarily-excitable perfectly continuous and uniform media even a *single* phase singularity (rotor, vortex) serves as a seed from which ever more spontaneously arise in pairs.[126,127,128] This seems just what a modeler would want as bridge between creation of a rotor and eruption of turbulence. Similar behavior has not yet been observed in continuous models with a single, attracting, fixed point. As in myocardium, except perhaps in three-dimensional BZ media.[80,166]

Coarsely discrete excitable media like cellular automata have long been known to go turbulent once seeded with a rotor. Media can be discrete in space, in state, or in time; in "coupled maps" space or time is so coarsely discretized that turbulence arises much as in discrete-state cellular automaton models.[71,90,159,164] But these can only be related to severely diseased myocardium. One can also use a kinetics with multiple fixed points, one of which has complex conjugate eigenvalues with positive real part. Then we can get Kuramoto's turbulence: If another fixed point is an attractor (as in simpler excitable media), this need not create a limit cycle, yet a rotor in such a medium can provide the seed of turbulence. An example was pointed out to me by M. Baer and M. Eiswirth two weeks after the Santa Fe lectures.[14,15] But such models still differ qualitatively from present visions of heart muscle. A bit more like myocardium are models with discrete nodes connected by continuous one-dimensional paths along which excitation propagates continuously. For example, Suzuki et al.[190] (in Japanese, discussed in Winfree[209] and Suzuki[189]) made movies of spirals breaking up on a mesh of wires each capable of propagating a chemical excitation. A related numerical model exhibits similar behavior.[90]

Another version of reaction-diffusion turbulence occurs in the most realistic myocardial model currently available.[38] This is the "stationary fronts" phenomenon in the original Beeler-Reuter model of ventricular membrane and in some simple modifications of it. The idea in caricature is that myocardium can activate in either of two alternative modes: in both, the sodium channels open, but the calcium channels might or might not, depending on their stage of recovery from prior activation

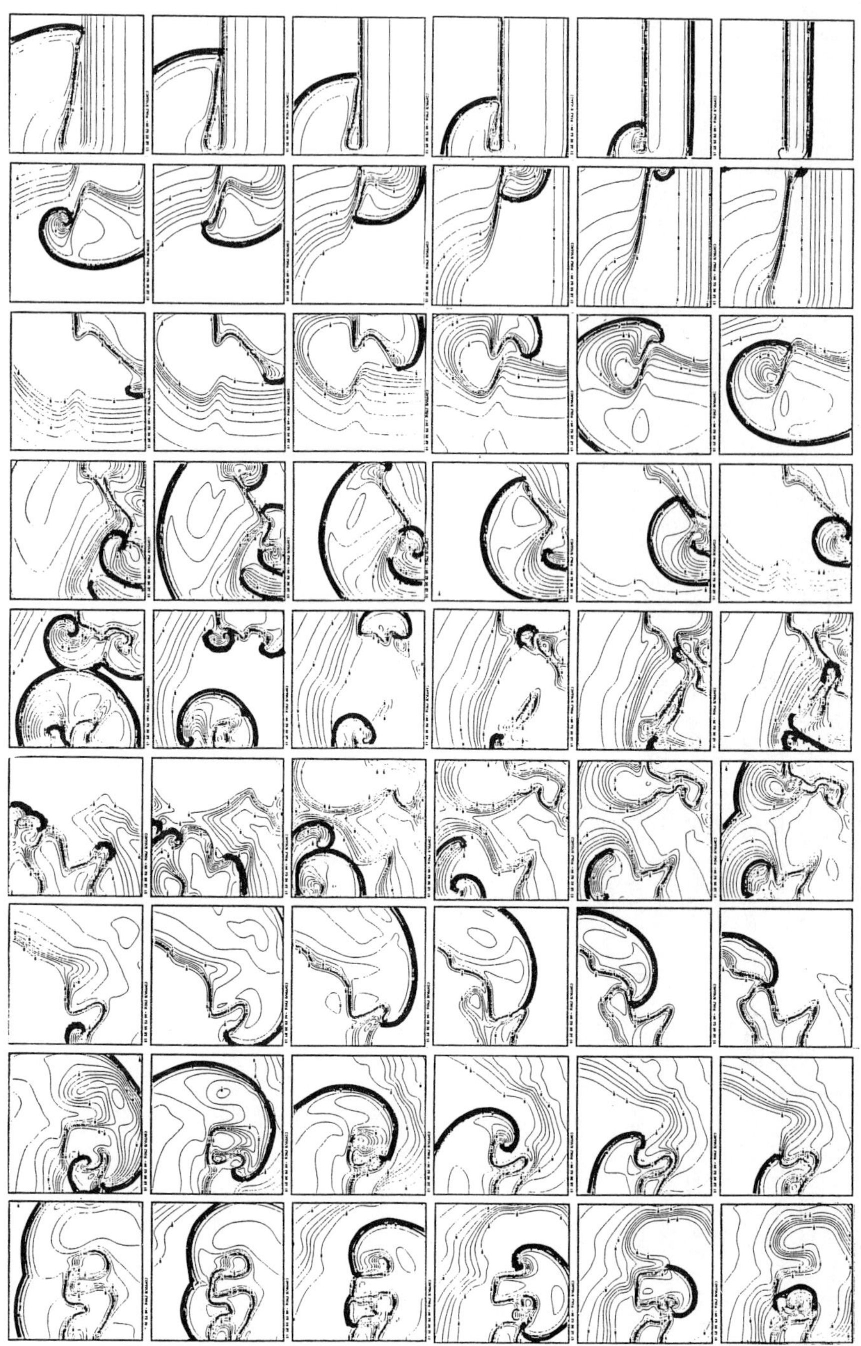

FIGURE 22 The unmodified Beeler-Reuter membrane model, executed again as in Figure 10 on a uniform near-continuum (160,000 grid points) 10 cm on edge, seen at intervals of 40 ms (25 frames per second). Read the 54 frames from top to bottom, then left column to right column. Each panel contours the calculated membrane potential. Contours pack densely where activation or repolarization occur; the activation front is painted red in the original, here darker and thicker. Initial conditions 20 ms before the upper left panel provided an east-west bar of depolarization north of a bar of refractoriness, both stopping slightly inside the east wall: at 20 ms we see the activation front propagating north and around the east end of the bar. At 260 ms (top of second column), it is puncturing the west end of the refractory bar, now recovered. Columns 2 and 3 suggest the rounding-up of a clockwise rotor like Figure 10. But then arcs of wave front fail where encountering regions still depolarized from prior activation, creating new wave tips until the activation map resembles turbulence. In the last column two oppositely-moving wavetips fuse and a single activation front exits the south wall. Afterwards all subsided to quiescence. These are the contour maps from which the poorly printed color relief maps were made for Figures 1 and 2 of Courtemanche and Winfree.[38] (From Winfree[227] with permission.)

at the moment when a new activation front intrudes. If they do, then the local action potential will be much more prolonged than in an adjacent area where (because they were just slightly more recently activated) they do not. If an activation front encounters that relatively motionless boundary before both sides have fully recovered, then once again the two sides will respond differently and recover at different times, renewing the stationary front. New fronts are created wherever activation crosses zones that have repolarized (thus permitting re-activation), in which the calcium-mediated events are ready to recur in one part but not in another: such parts develop action potentials of different durations, separated by a stationary repolarization front. Complicated and abruptly changing mosaics of stationary fronts (and their moving descendants, after both sides have recovered) might exist in any such medium, e.g., heart muscle.

That verbal description is a cartoon-like simplification: really the duration of the calcium-mediated refractory period is a smooth function of the timing of activation (see Courtemanche and Winfree,[38] Figures 3, 6, and 8). Along the wavefront, anticipated duration varies from 60 to 240 ms quite smoothly. Also the "stationary fronts" are not perfectly stationary. But the cartoon remains the best description contrived so far. The numerical data are seen in Figure 22. As in Winfree[216,217,221] and Fishler and Thakor,[61] arcs of conduction block occur "unexpectedly," even in this absolutely uniform, nearly continuous medium. This turbulence is all self-generated, once it is seeded by creation of one short-period reentrant source (wave tip, rotor). Rotors proliferate and recombine at random during this turbulence. Adjacent stationary fronts sometimes channel the excitation front, like walls forming a tunnel. Possibly the curvature-dependent unidirectional blocks and reentrant patterns described by Winfree[213] and by Kogan et al.[117] in models of infarcted media also occur by this mechanism even in perfectly uniform media. This ought to be further explored, both by alertness to such patterns in epicardial maps of fibrillation and by examination of simpler conceptual models than the Beeler-Reuter summary of archaic electrophysiological lore.

Videotapes of epicardial maps during VF in the dog[89,239] dramatize the sharp temporal periodicity of VF at the rotor period and its lack of corresponding spatial periodicity. Instead, pieces of wave front propagate normally, presumably pirouetting about broken ends and occasionally fragmenting to produce more. Such recordings will provide a standard for comparison of the similarly displayed implications of various hypotheses, e.g., of Winfree,[216] Courtemanche and Winfree,[38] and Fishler and Thakor[61] that in sparsely coupled or coarsely discretized versions of the BR model (and maybe in perfectly smooth implementations, too) something like fibrillation *does* spontaneously develop from two-dimensional rotors. The final outcome of these studies is not yet 100% clear. An alternative possibility still needs study: that in normal tissue, continuity is well maintained, and rotors do not spontaneously occur or multiply, yet when the tissue is significantly three-dimensional, rotors (as vortex filaments) *do* serve as the catalysts for the turbulence called fibrillation.

PAST THREE-DIMENSIONAL VORTEX COMPUTATIONS

Three-dimensional vorticity, while in many respects analogous to two-dimensional vorticity, is also fundamentally different in important respects. For example, in two dimensions, the only organizing center for periodic activation fronts (i.e., source of such waves) is the isolated single-armed vortex and it is quite persistent in either of two mirror-image forms (clockwise, anticlockwise). But in three dimensions the vortex filament generically closes in a ring which has no handedness and either expands until hitting boundaries or shrinks and vanishes unless parameters are adjusted with exquisite care[37] or it is knotted or threaded by a linking ring.[81] Several chemistry papers[3,4,91,92,107,135,136,137,166,198,199,204,206,207,208,213,229] describe three-dimensional vortex filaments both in the computer and in the laboratory. The basic idea from 1973 to 1990 was that the vortex filament is a stack of two-dimensional rotors, with some second-order modifications on account of the stacking (e.g., the rotor may now move. . .this was before it was realized that even in perfectly uniform two-dimensional media it moves anyway, in the meander trajectory). If the excitable medium also oscillates spontaneously, then the place of the vanished scroll ring is taken by a radial phase gradient of that oscillation, which looks like a pacemaker (see Winfree[206,207] in three dimensions, independently rediscovered by Muller et al.[150] in two-dimensional cross section.) That is about as far as experiments went up to the late 1980s.

By solving the chemical reactions and the molecular diffusion kinetics of the Oregonator model of BZ medium, we ran numerical simulations of intended experiments in two dimensions and in three dimensions. This resulted in some interesting ostensible "discoveries" that could be checked by implementing the same experiments at the laboratory bench. In the case of vortex rings, their shrinkage at a rate proportional to local filament curvature was already familiar from earlier computations and analytical approximations.[27,28,51,160,162,163] Experiments provided quantitative vindication of twist-free models.[3,4,91,107,229] But twist itself was still ignored.

No one has yet explored the consequences of twist in a generic experimental setting, but computational precedents are accumulating. Awareness of "twist" dates from 1980,[209] but its topological implications were not made explicit until 1983.[230,231,232,233,234,235] Panfilov et al.[161] and Mikhailov et al.[145] drew attention to its implications for spin rate and for catastrophic shortening of wave spacing, but its effects on filament motion were not contemplated until the "local geometry hypothesis"[228] motivated Keener[104] and Biktashev[25] to an explicit mathematical formulation that could be solved analytically in certain limiting cases for nonmeandering vortices. Twist was first given explicit operational definition in numerical experiments only three years ago[222] and measured with demonstrable reliability only in Henze et al.[80] and Henze and Winfree.[81] All analyses of chemical experiments up to present day were limited to geometries in which twist was artificially eliminated with the single exception of the ingenious experiment of Pertsov et al.,[166] which instead constrained filament curvature to zero. This limitation to un-generic limiting cases was essential for getting started but at the present stage it begins to seem like restricting analyses of biological pattern formation to dots on a straight line, or trying to interpret electrical activation of heart muscle by analogy to action potential propagation on a linear nerve fiber. Moreover, inability to quantify most three-dimensional laboratory experiments to precision better than a factor of 2 or 3 leaves the theoretical literature rapidly proliferating and mutating with no natural selection to prune it.

One consequence of realistically admitting nonzero twist to analyses of vortex filaments is that their development need no longer tend to one of two trivial steady states: the straight scroll wave (a mere projection of the trivial two-dimensional spiral perpendicular to its plane) or a shrinking planar circular scroll that will collapse and vanish in finite time. With nonzero twist, computations show that vortex filaments can link and knot in ways that evolve into stable organizing centers. Organizing centers are three-dimensional sources of periodic waves in a parametrically uniform excitable medium. They were predicted mathematically and a "periodic table" of discrete alternative variants was outlined from topological constraints.[230,231,232,233,234,235] But nothing was known of their (theoretically allowed) transmutations, nor about their dynamics, especially stability, and no defense could then be raised against the withering criticism that they all might just be transients en route to uniform quiescence. Since 1989 we know better (though only in the computational world). At least eight qualitatively distinct and remarkably stable organizing centers evolve from generic initial conditions when one solves the known partial differential equations of several different excitable media (see Winfree,[222] Henze et al.,[80] Henze and Winfree,[81] and recent unpublished numerical experiments).

There is no single key to this work in this area; there are three keys, all necessary. The first is facility in writing the initial conditions. Winfree et al.[235] described a general method, but it was not easy to implement except in cellular automaton models. No one figured out a better way until Poston recognized that the problem was equivalent to classifying complex-valued functions on R^3 according to the linkage of their zero loci, and that these zeros could be most succinctly thought of as roots of complex-valued polynomials of two complex variables. By attending only to an S^3 subset of that R^4, and mapping S^3 to the R^3 target space by stereographic projection the one-dimensional root loci (vortex filaments) are made to cluster around the origin (south pole) as the seed of a compact organizing center inside a computationally manageable box, and remote places (north pole) tend to initially uniform composition. This method is briefly described in Winfree[215] and in Henze and Winfree[81]; a more detailed exposition (the promised "Poston and Winfree, 1987" of Winfree[215]) with new IRIS graphics will be Poston, Henze, and Winfree, 1993.[171]

The second key resource is facility in solving the evolution equations of local reaction and diffusion. This is now widely available to many laboratories. The third key resource is the ability to make something instructive from the results of extensive computation. Integrating the partial differential equations is not much use without elaborate facilities for reliably quantitative geometric analysis and display of the avalanche of data resulting. At present this analytical geometry toolkit is best developed in my lab at the University of Arizona in Tucson. The 1990 version comprising about 28,000 lines of Fortran and Pascal is the culmination of efforts by myself, Pramod Nandapurkar, William Guilford, William Skagg, Erzebet Lugosi, Marc Courtemanche, Michael Wolfson, and Chris Henze under 5 years' support from the National Science Foundation. Its adaptation to the Silicon Graphics IRIS initially by Wolfson then especially by Henze during 2 years' further generous support rendered facile the production of such miraculous slides and videos as presented during these lectures (and in Henze and Winfree[81]) and those to appear in Poston, Henze, and Winfree[171] and Henze's dissertation. This was the fifth lecture, unwritten but given by Henze after I returned to Tucson. I fill in a fifth lecture by elaborating on the next intended use of the vintage-1992 Tucson Toolkit.

A newly contemplated possibility is the acquisition of data on vortex filament motion from three-dimensional chemical experiments, using computer-assisted optical tomography. Such data could in principle play the role formerly occupied by PDE solutions as input to the analytical geometry toolkit.

FIFTH LECTURE

THREE-DIMENSIONAL INVESTIGATIONS

Among over 1000 pertinent cardiology reprints copiously marked up in many colors of pen in my files, only two dozen give explicit attention to three-dimensional aspects of cardiac[29,30,33,45,52,53,56,66,67,79,87,96,119,140,141,142,165,169,170,187,188,236,237] propagation. This is not because three-dimensional aspects are unimportant or trivially equivalent to two-dimensional propagation, but because three-dimensional experimental technology is only now coming into being. About half of these concern ischemic or infarcted tissue in which patterns are fragmented and unclear due to regional variegation of substrate properties, and of the remainder only a few even mention rotors.[66,67,140,165] This might be because the reality of rotors could be confirmed and their role in ventricular arrhythmias could be investigated only after computer-controlled multielectrode epicardial mapping at the required resolution became feasible in two dimensions: rather recently, perhaps about 1986.

Electrophysiologists have resorted to analytical mathematics and computational models for previews of what might be seen and what should be watched for in three-dimensional mapping experiments.[11,34,55,84,120,152,154,174,191,192,196] This literature mostly concerns the role of anisotropy in three dimensions and overlooks the role of rotors. We now know that rotors play a crucial role in arrhythmias, though their three-dimensional aspect has been explored only slightly, notably by Chen et al.[33] As noted above, beginnings have been made on exploration of three-dimensional vortex dynamics in excitable media, both in chemical experiments and in numerical experiments, but not often with explicit attention to the special circumstances of their mechanism in myocardium.

Propagation of the activation front in many other three-dimensional numerical models of the heart is represented not by the electrophysiologically more realistic cable (reaction-diffusion) equation, but by the computationally more expedient Huyghens' construction or eikonal equation. According to this algorithm, a cell becomes excited (so the activation front progresses to its location) if it was quiescent and appropriate neighbors are firing; the wave front progresses in each increment of time by a fixed distance perpendicular to itself. This rule was used in the first three decades of two-dimensional modeling, starting with Wiener and Rosenblueth,[200] e.g., in all the "axiomatic" models and in "cellular automaton" models prior to about 1991 (but see Weimar et al.[197] for the newer sense of "cellular automaton"). Though its numerical simplicity makes it a favorite for three-dimensional computations, it is unacceptable for quantitative work, as explained with examples in Plonsey and Barr.[168]

It is important to note that the numerical and chemical experiments discussed in this lecture all focus on *continuous* media. This is deliberate. There is currently much debate (well summarized in Roth[177]) over the roles in ventricular tachycardia (VT), in the transition from VT to ventricular fibrillation (VF), and in sustained VF of the many and diverse discontinuities so evident even in normal myocardium, and *a fortiori* prominent in ischemically damaged myocardium. It is important to know how the electrical behavior of myocardium is *modified* by the presence, quantitative character, and abundance of discontinuities of both structure and function. This question has been hard to answer convincingly, not only because such discontinuities are poorly described and quantified, but also because we still lack a sound appreciation of the electrical behavior of myocardium and analogous excitable media even in their *absence*, i.e., in the continuum case. For example, the two-dimensional epicardial mapping experiments and numerical experiments cited above showed that many features of VT and VF that have for decades been assumed to derive from discontinuities and "non-uniform dispersion of refractoriness" in fact occur in essentially the same form in perfectly continuous and uniform media and in normal myocardium that approximates such media.[218] The three-dimensional experiments considered here are intended to provide the needed background for such theorizing, by quantitatively exposing the dynamics of continuous excitable media in three dimensions. Such media might or might not, for example, be capable of something like "fibrillation" after three-dimensional vortices are initiated; no one yet knows.

Efforts to construct a predictive analytical theory of nonmeandering vortex filament behavior[25,104,110,166,222] have been based on the "local geometry hypothesis."[80,81,104,110,228] This postulates that the motions of the filament can be analyzed on a local basis in terms of the filament's local shape and the local twist of the concentration fields. (Thus, mutual attraction or repulsion of vortex cores, impact of wave trains from afar, spontaneous meander, etc. are overlooked.) Existing analytical models work best for planar filaments with uniform curvature and no twist, or in the limiting case of curvature and twist too slight to represent the compact stable organizing centers that we find evolving from our initial conditions. The behavior of filaments of general shape with substantial curvature, torsion, and twist is still full of surprises. Thus exploration and discovery still depend mostly on computation. The Tucson facility for such numerical experiments was debugged on two published computations:

1. Henze et al.[80] described the consequence of uniform twist on a straight filament: the filament buckles into a helix which expands up to a stable radius or continues unrestrained to explode through the walls of the container, or is confined by repulsion from the walls. The mathematical theory of a different case (the limit of small curvature and twist[104]) was found to predict outcomes substantially different from our experimental results. Attempts to amend it with more terms and corresponding adjustable parameters[108,109] still did not enable an impressive fit to these compact organizing centers, especially for V_b.[224] Their numerically observed behavior remains bewildering, especially in the apparent sensitivity of behavior to slight amounts of twist.
2. Henze and Winfree[81] described in an electrophysiologically motivated excitable medium the stable anatomy of a vortex ring with nonuniform curvature, torsion, and twist (a knot). This experiment tested the local geometry hypothesis and found it badly wanting in that the collision boundary does *not* shield the filament from impact of wave fronts emanating from remote segments of the same filament. The ring is kept from collapsing only by the pressure of other segments of filament obstructing its motion. Moreover, vortex core anatomy in planes normal to the filament looks quite different at various sites. This unexpected fact undercuts the foundation of all contemporary mathematical analyses.

The fact is that the vortex filament is *not* basically a pancake stack, somewhat bent and twisted, of normally functioning two-dimensional rotors. Nonetheless, in these numerical experiments the filament's motions could be described with startling precision by adjusting the six parameters of a linear dependence on local curvature, twist, and arc-length derivative of twist: the conclusion of Keener's[104] mathematics. In a way this might seem unsurprising: the only properties by which the vortex filament is distinguished from a two-dimensional rotor are its local twist and the local curvature and torsion that uniquely define the shape of any space curve. To first order, one might expect their contributions to be additive and proportional to the departures of all three from their values (0) in the comparison case of the uncurved, untwisted filament or two-dimensional rotor (which by choice of the medium's parameters we take to be motionless, nonmeandering). But there is more to it than that: it had to be shown that the arc-length derivative of twist plays an important role, and that torsion *per se* does not. Moreover, this formulation of the dynamics seems a little strange when it is recognized that "motion" means motion perpendicular to the tangent vector, in the local plane of curvature and normal to it. When curvature is nearly 0, these directions become undefined or liable to violent vacillation, yet such a formulation gives a finite (possibly large) speed to the filament in that vacillating direction, if its twist or its derivative are not negligible. This seems physically unrealistic, yet in the one case of general shape tested to date, in which that telltale case is excluded from the data, the linear model statistically fits the remainder quite well.[81] The obtained coefficients seem independent of curvature, twist, and its derivative; i.e., those that can be assayed with any precision seem about the same as in two-dimensional rotors. Is this an accident, a fortuitous consequence of other factors not present in contemporary theory, e.g., the close apposition of adjacent filaments? This surprising ostensible linearity still needs interpretation. The best-fitting parameters cannot yet be compared to values derived from the local kinetics because such values are at present merely assumed to be unique functions of local kinetics. Since these parameters remain purely descriptive, they might be entirely different for a ring of different shape, and since only one shape was observed, the local geometry descriptors (curvature, twist, torsion, etc.) are inevitably confounded with one another and with others not considered.

What is really needed is an examination of diverse filament shapes in one medium, thus breaking the correlation matrix that confounds variables in the regression of motion data on local geometry descriptors. This effort is in progress as the Ph.D. thesis of Chris Henze. Being motivated more by physiology than chemistry, we chose the FitzHugh-Nagumo model of excitable membrane, whose parameter space I had surveyed with attention to two-dimensional rotor behavior.[225] This revealed only a tiny range of parameters in which the medium has reasonable excitability, no spontaneous oscillations, and a stable, nonmeandering vortex core. This is the range in which Skaggs et al.[184] and in Courtemanche et al.[37] discovered the first stable three-dimensional organizing center. Using this medium, we can readily quantify vortex filament motion and relate it quantitatively to local geometry, using a variety of geometries from the "periodic table" of Winfree and Strogatz.[230,231,232,233,234] These numerical experiments will be essential for understanding the laboratory experiments outlined below and their inevitable analogs in myocardium.

No one has yet discovered (nor looked for) any stable organizing centers in any experimental system. Three years ago there was no reason to imagine they exist, but now we have so many computational examples that it seems appropriate to look. How could one look? The simplest way would be to stir up random initial conditions in a volume of BZ liquid, then watch for persistent wave sources with distinctive periods. Thus far only one organizing center has been found and it is not stable in the BZ medium: this is the shrinking scroll ring, with zero twist or torsion and very nearly the same period as the two-dimensional rotor.[204,206,207] But it is easy to imagine that mere stirring of the liquid is inadequate to accidentally provide initial conditions for linked, knotted, twisted organizing centers such as were numerically created in computation (2) above. A more systematic study of filament anatomy is called for.

COMPUTER-ASSISTED TOMOGRAPHY FOR NEW THREE-DIMENSIONAL LABORATORY EXPERIMENTS

Laboratory experiments with rotors in two-dimensional excitable media up to now have done little more than confirm prior numerical discoveries. But there remain two ways that laboratory experiments can now leapfrog to preeminence.

One is already working: multielectrode recording and fluorescent-dye scanning of electrical activity in heart muscle.[41,42,43,48,49,82,178] Numerical experiments played a key role in bringing this about, by showing that there are surprising phenomena to seek which, if they did in fact play the foreseen role in real myocardium, would constitute a usefully new viewpoint on cardiac arrhythmias. In the past five years, it has been demonstrated experimentally that they exist and they do, giving a strong impetus to further improvement of both experimental and numerical techniques in this area.

But there is a second and less widely appreciated way that laboratory experiments can now play a decisive role in discovery, not just in confirmation. This concerns the role played by the three-dimensionality of heart muscle,[169,170] which remains to be completely deciphered by direct electrophysiological experiment. And it might never be, since a sufficient density of metal electrodes in depth might seriously compromise normal electrophysiological function. What goes on in chemically excitable media may have close analogies in myocardium, and if the experience with two-dimensional rotors is any guide, those phenomena will be almost impossible to recognize until they are first perceived in the more tractable context of numerical experiments and numerically analyzed chemical experiments. For example, the "scroll filament" was first detected in three-dimensional dog ventricle[33] by an electrophysiological experiment explicitly contrived to check the analogy to BZ reagent[204,206] and its theoretical generalizations. The next designed experiment, to detect intramural scroll rings,[217] has not yet been carried out in myocardium due to the extreme difficulty of placing enough electrodes three-dimensionally. With sufficient knowledge of scroll ring anatomy and behavior, this technical obstacle might be bypassed in the following way. Scroll rings are known to drift perpendicular to the ring's plane, thus gradually revealing themselves in cross section as they pass through any single observation plane.[207] Thus contemporary high-resolution *epicardial* mapping technique would suffice, were the ring created intramurally in the right orientation.[217] Medvinsky and Pertsov[140] and Medvinsky et al.[141] tried something of the sort, but the experimental technique still needs refinement.

More generally, three-dimensional reentry is an important aspect of myocardial behavior during tachycardias, but almost nothing is known of it, even qualitatively. Partly because of the inevitably low resolution of three-dimensional electrophysiological observations, it is essential to design myocardial experiments to look for specific phenomena. The anatomical aspect of three-dimensional reentrant phenomena seems fairly well understood now (Winfree[222]), but the dynamical aspect has greater importance for myocardium and is still understood only from theory[104,105,108] and computation.[80,81,91,222,229] These two approaches agree well only in limiting cases of little pertinence to feasible experiments, and neither has been adequately compared to experimental reality. Tomographic experiments might provide that comparison and provide a more direct model for electrophysiological observations in myocardium.

Three-dimensional experiments are not new in chemically excitable media,[2,3,4,5,91,107,125,132,166,195,198,199,204,205,207,208,229,242,243] but it has always been hard to see what transpires inside. The first attempts to expose the three-dimensional anatomy of activation fronts in this excitable medium used the following approaches:

1. dissecting microscope observation[204,207,208];
2. absorption of BZ medium into an opaque nitrocellulose block that could be quickly fixed or frozen and sliced into 140-micron fixed sections for one-time reconstruction of three-dimensional anatomy[207];
3. video recording and photography of un-gelled liquid in a test-tube[198,199]; and
4. similar observations in various gels.[2,3,4,91,125,132,166,229,243]

For different reasons in each case, none of these are capable of revealing the three-dimensional anatomy of the wave fronts and vortex filaments, except in the most starkly simple cases (e.g., a perfect twist-free and torsion-free vortex ring.). Most pictures obtained remain unpublished for want of unique interpretation. The only better-than-qualitative experimental studies have necessarily avoided generic combinations of curvature and twist. As noted above, all restrict themselves to situations in which torsion = 0 and twist = 0 (Winfree,[207] Agladze et al.,[3] Jahnke et al.,[91] Keener and Tyson,[107] Winfree and Jahnke[229]) or curvature = 0 (Agladze et al.[3]).

Optical techniques in myocardium[41,42,43,48,49,82,178] are still limited to surface excitations because the tissue is not fully transparent to the lights needed for excitation and fluorescent signal recording. However, BZ reagent is fully transparent, and its oxidation-reduction "action potential" is marked by a dramatic change of absorption in the blue-green. It has already been shown two-dimensionally that much can be learned about myocardium by attention to activation fronts in this "analog computer." I conjecture that three-dimensional optical experiments in this medium will also preview the results of similar inquiries (not yet experimentally practical) in myocardium. The method I have in mind depends almost as heavily upon computation as does numerical solution of the electrophysiologist's cable equation in three dimensions, but instead of *being a numerical* analog of myocardium, it allows us to *see an experimental* analog of myocardium. This technique is computer-assisted tomography. By photographing a three-dimensional volume of BZ reagent simultaneously from many directions, one obtains the two-dimensional projections needed as input to a three-dimensional tomographic reconstruction algorithm.[1,19,20,21] The resulting three-dimensional distribution of optical densities or chemical concentrations has the same format as familiar partial differential equation solutions and can be examined by the same graphics utilities that served well in numerical experiments.

Diverse artifacts must first be eliminated from the BZ reagent itself, e.g., convection currents in the three-dimensional volume, growing CO_2 bubbles, and parameter gradients, notably of temperature and of oxygen. The oxygen gradient afflicts only the surface 1–2 mm layers and can be eliminated by immersing the reaction volume in a bath of the same solutions, lacking catalyst. This expedient also eliminates optical refraction effects that would otherwise interfere with tomographic reconstruction. The temperature gradient comes from released reaction heat (several calories/ml in typical recipes); it is minimized by keeping the reaction volume unconfined in a thermostatted bath and by using so little catalyst that free-energy release is quite slow. This is necessary also to keep optical density low enough for transmission through 1–2 cm thickness. To obtain a uniform convection-free medium free of gas bubbles, I gel the medium with silicic acid. This gel is so hard that it does not permit nucleation/expansion of bubbles. This was the first thing I tried in 1971 while working up the recipes and procedures that became standard in this field. But the necessary alkalinity of silica gel ($pH > $ about 8) was incompatible with the necessary acidity of BZ reagent ($pH <$ about 1.5) and I gave it up. Yamaguchi et al.[242] have since used two-dimensional slabs of silica gel by first preparing it free of BZ reagent, washing it to neutral pH leaving only silicon dioxide, then perfusing with acid BZ reagent. However, for three-dimensional experiments, one cannot wait for reagents to diffuse into an 1–2 cm gel block and equilibrate, and gradients of any kind are completely incompatible with intent to measure vortex filament motion in a uniform field. So the gel must set *in* the reagent. This requires modification of BZ reagent to work well in an environment slightly more alkaline than used heretofore, and discovery of some catalyst (compatible with BZ reagent) by which to set the gel in an environment too acid for uncatalyzed gelation. By tiresome trial and error, the following solution was converged upon at a compromise pH.

The following procedure for a total volume of 8.5 ml, as in the recipes standardized by Jahnke and Winfree,[93] uses those stock solutions. It differs mainly by using five-fold less ferroin (since the optical path will be five times longer) and including the catalyzed silicate gel and sodium phosphate buffer:

MIXTURE A. Dissolve 630 mg colloidal silicon dioxide (e.g., Cab-o-sil) in 2.3 ml of hot 10% NaOH solution plus 3.2 ml of stock (1420 mM) $NaBrO_3$. Filter through millipore. The 6 mmoles of (univalent) NaOH will later be more than neutralized by 15 mmoles of (trivalent) phosphoric acid. pH is 11.

MIXTURE B. Prepare bromomalonic acid by mixing 0.2 ml of $NaBrO_3$ stock plus 0.2 ml stock (3260 mM) H_2SO_4 plus 1.2 ml. stock (1041 mM) malonic acid plus 0.45 ml stock (972 mM) NaBr. When the yellow color clears, this will contain nearly equimolar malonic and bromomalonic acids, plus sodium sulfate. Add 0.2 ml of stock (25 mM) ferroin, 0.6 ml of 85% phosphoric acid to overneutralize the NaOH, and 0.1 ml of 1 gm/25 ml NaF to catalyze gelation even at the unfavorably low pH that will result. pH is now 1.5.

Quickly squirt A (5.5 ml) into B (3.0 ml), filter (through a metal aviation-fuel screen), and allow to gel: about 8 minutes at 25°C with 10 mM NaF and 7.4% SiO_2 at pH 1.7. Rotors in this medium have period 72 sec and wavelength 2.6 mm. The catalyst/indicator ("v," ferroin) is firmly adsorbed onto the silica: it does not diffuse.

Now with a very hard, glass-clear gel free of bubbles, it is necessary to quantify the dynamics of its activation fronts. This could be done by tomography as used in hospitals on a one-meter scale with X-rays, but now on 1-cm scale with blue-green light. In tomography one takes photographs simultaneously from 50–100 different angles. Each snapshot is a different shadowgraph of the object of interest: the "Radon Transform" of the three-dimensional optical density distribution.[19] These can be used to mathematically reconstruct the three-dimensional object thus shadowed (e.g., the inside of one's brain, to study the shape and connections of a tumor without ever opening the head). The mathematical procedures involve "projection" (photography), Fourier transformation of those projections, convolution filtering of those transformations, and "backprojection" to assemble a three-dimensional array of optical density values in computer memory.[1,19,20,21,22] The computational aspects of this procedure have been tested. We have three-dimensional vortex filaments in the CRAY-YMP, obtained by solving the pertinent partial differential equations of chemical reaction and molecular diffusion. By projecting one of these in 50 directions, research assistant Chris Henze obtained "photographs" that were delivered as input to the numerical procedures that will later accept real photographs for reconstruction of the original object. This worked beautifully,[226] with little loss of resolution, even when various experimentally realistic noises were introduced. For example, we let the waves move while the reaction-diffusion equations were solved into forward time between one photograph and the next, as they will do in the real gel while the table turns and the camera scans. While rotating the gel on a turntable, 50 views spanning 180° can be harvested into a video recorder within 2 seconds. Meanwhile the spiral waves move three-dimensionally. With the chosen BZ recipe at room temperature, they move about 2/100 cycle (and much less if chilled). Returning to the numerical backprojection algorithm, we allowed the partial differential equation solution to progress this fast (or several times faster) while consecutively projecting the three-dimensional array in 50 directions. Backprojections obtained from these "smeared" pictures were not seriously degraded until the two rotation speeds became comparable.

Tomographic resolution is theoretically limited mainly by the number of projections taken (small relative to the number of distinguishable pixels in each scan line). One theoretical estimate, presupposing perfect optics, perfect resolution in the two-dimensional projections, optimal filtering, perfect alignment of all projections for the backprojection, etc. is resolution = field width times π divided by the number of projections. "Resolution" here refers to the central disk in Fourier space containing unattenuated harmonics corresponding to the longer wavelengths. Shorter wavelengths outside this disk are unrepresented or underrepresented. Of course, if there are none (if the object is smooth), nothing is lost, so "resolution," though limited in this sense, is nonetheless perfect. Supposing a BZ recipe with wavelength 3 mm, a 10-mm cube would suffice to present typical organizing centers; viewed from 100 directions, resolution would be 1/10 wavelength. Nothing more refined than this may be necessary since wave fronts are visibly quite smooth; the only detail lost would be the sharp cusps where wave fronts collide obliquely. Our numerical dress-rehearsals do indeed indicate that for smooth objects the theoretical worst-case estimate of resolution given above is unduly pessimistic.

This result is a three-dimensional array of optical densities corresponding, in this case, to [ferroin]. This array, like those now obtained from PDE solvers, can then be fed to graphics display programs and geometrical analysis programs. There is one fundamental difference however: data acquired by tomography represent only the single *colored* reactant, whereas PDE solutions provide the spatial distributions of *all* reactants. In the case of the Oregonator model of the BZ reaction, "all" are just two: the concentrations of bromous acid and of ferroin. The wave tip or the rotor consists of those places where both concentrations simultaneously take on average values, inside the cycle of ups and downs experienced anywhere else during the passage of periodic waves. Thus we need pixels (x, y) with ([bromous acid], [ferriin])(x, y) within a small window inside the excitation-recovery loop. . .but distressingly, only [ferriin] is visible to the camera (as a logarithmic transform). Or is that all that is visible? The spatial distribution, thus its Laplacian, is implicitly visible. And the local time derivative is implicitly visible. Both can be made explicit in real time if the experiment is run under the attention of a video camera attached to an image processor.[226] The reaction-diffusion equation says that the difference between the time derivative and the Laplacian is the source term, the local kinetics as described by the ([bromous acid], [ferriin])-dependent net rate of oxidation or reduction. If we know that rate law and we know [ferriin], then we can infer [bromous acid]. So I set up that experiment and took the appropriate derivatives, using the biologist's standard image processor, NIH Image on the Macintosh. According to the two-variable Oregonator with Tyson's parameters, the needed rate equation is quite simple: d/dt[ferriin] = [bromous acid] - [ferriin]. Then in reaction-diffusion context we can solve two successive images for [bromous acid](x, y), taking the optical density ferriin_color(x, y, t) as linear indicator of [ferriin] where the contrast does not change too much:

$$\begin{aligned}[\text{bromous acid}](x,y) =& d/dt \text{ ferriin\_color}(x,y,t) \\ &+ \text{ferriin\_color}(x,y,t) \\ &- D\nabla^2 \text{ ferriin\_color}(x,y,t)\end{aligned}$$

with appropriately scaled discrete approximations to the two linear operators. [Bromous acid](x, y) is a squarish pulse about 1/2 mm wide riding somewhat ahead of the [ferriin] pulse. Near the wave tip (in fact, defining the wave tip) iso-concentration contours of ferriin cross transversely through those of bromous acid. Everywhere else, all the contours are parallel spirals. But near the tip, the steep transverse gradients of two substances create unique combinations of the two which occur nowhere else. This range of combinations defines wave tips or rotors. This method should work as well in three-dimensional context as shown here in two-dimensional.

Now, given that such data *can* be obtained straightforwardly, what good are they? First of all, there is now a substantial volume of (often contradictory) theory about the motion of activation fronts and their sources (twisted vortex filaments) in three-dimensional excitable media.[25,44,74,75,104,105,108,109,110,145,158,164,165,194] Much of this is not testable due to a profusion of undetermined parameters, and much of the rest has been tested only in a vague way, to order of magnitude, so it remains to be tested quantitatively. Four distinct kinds of test are needed:

1. Test of the topological essentials predicted by Winfree and Strogatz.[194,230,231,232,233,234] Are the mutual linkages and knottedness of vortex filaments related as foreseen to the integral twist along each filament? Do filaments fuse and hybridize, or do they pass through one another (or resist such passage), and do the results in any case conform to the "understood" topological transmutation rules?
2. Another piece of quantitative theory that can be tested by experiment is the dependence of activation front speed on local front curvature, expected to be linear when the activation front's radius of curvature greatly exceeds its thickness. (If it does not, there is too much ambiguity involved in defining "front" and "curvature.") This rule has important consequences in cardiology; for example, the coefficient involved determines the threshold stimulus or critical size of a nucleus of depolarized cells sufficient to initiate an ectopic beat.[219,220] The dependence has been confirmed two-dimensionally in BZ reagent and in the slime mold[62,63,64,65] by experiments which seem to exhibit its accuracy even at curvatures far tighter than intended in its derivation and perhaps even too tight to permit definition of "wave front" and "curvature." Its three-dimensional generalization has not been experimentally tested, even numerically.
3. Tests of the various putative laws of motion of vortex filaments which partly determine the stability or instability and the periodicity (in milliseconds and in

centimeters) of reentrant activations. Candidates can be tested by careful numerical experiments, e.g., Henze et al.,[80] and Henze and Winfree.[81] However, existing theory is not really about vortex filaments of the kind that can be reliably computed within a grid of manageable size. Such filaments have substantial curvature and twist. Rather, existing theory is about transients (not stable periodic steady states) in barely curved (huge) and barely twisted filaments which evolve very slowly. But these cannot be economically investigated numerically even in the best contemporary supercomputers, except by using a computational mesh too coarse for believable quantitation. Most such computations have been too coarse and we find that such computations are reliable only in the absence of twist.[81] This is where the chemically excitable "analog computer" provides a usefully complementary perspective. We can image a much larger volume at resolution adequate for measurement of wave front and filament motions on a long time scale, without having to manage the high-resolution mechanics ourselves: that part becomes Mother Nature's job, leaving us the part of the computational job that is within the capacities of the CRAY. Then existing mathematical theories can finally be tested.

Limiting-case theories (of which Keener's[104] still seems the best) for transient processes in large nonmeandering vortex filaments dominate the theoretical effort at present and should be tested before progress can be made toward understanding more realistic situations. Jahnke et al.,[91] Winfree and Jahnke,[229] and Agladze et al.[4] experimentally determined the curvature-dependent inward motion of such curved filaments, and it agrees nicely with theory. The perpendicular "drift" of the filament has been observed[92,229] but not yet measured outside computer simulations. As theory is in excellent shape for this motion and all theorists agree, this should be the second test case for experiment after verifying the known rate law for inward motion. Various versions of theory diverge when it comes to nonzero twist, and no experimental test has been done. The chemical medium is a convenient source of transient and slowly moving vortex filaments of slight curvature and twist. Techniques for initiating them were outlined in Winfree[213] and in Pertsov et al. [166] Some of these techniques have been shown to work well in the laboratory.[91,92,166,229]

4. Tests of the bottom line: are there persistent organizing centers in three-dimensions? They might not be "stable" in any simple sense, e.g., if they glide and tumble through the medium, if their component filaments are meandering, if waves of twistedness circulate along the constituent rings. But do they persist? Does any version of the BZ medium support persistent particle-like solutions even vaguely resembling those anticipated by theorists?[12,37,75,80,81,222]

Such laboratory set-up has another possible application. Since 1952 theoretical biologists and some experimentalists have vigorously pursued the idea of Alan Turing that biological pattern formation might be a consequence of the spatially unstable interplay of local reaction kinetics with local transport processes, notably molecular diffusion or its larger scale equivalent in turbulent convection of cytoplasm. A malonic acid/chlorite/iodide reaction with a starch indicator recently produced the first Turing patterns.[13,31,46,130,156,157] Their three-dimensional aspect has not been explored, but might be, using optical tomography.

ACKNOWLEDGMENTS

This work was supported by the U.S. National Science Foundation and the hard work of recent graduate students William Guilford, Wolfgang Jahnke, Michael Wolfson, Chris Henze, and Marc Courtemanche.

REFERENCES

1. Agard, D. A. "Optical Sectioning Microscopy: Cellular Architecture in Three Dimensions." *Ann. Rev. Biophys. Bioeng.* **13** (1984): 191–219.
2. Agladze, K. I., A. V. Panfilov, and A. N. Rudenko. "Nonstationary Rotation of Spiral Waves: Three-Dimensional Effect." *Physica D* **29** (1988): 409–415.
3. Agladze, K. I., V. I. Krinsky, A. V. Panfilov, H. Linde, and L. Kuhnert. "Three-Dimensional Vortex with a Spiral Filament in a Chemical Active Medium." *Physica D* **39** (1989): 38–42.
4. Agladze, K. I., R. A. Kocharyan, and V. I. Krinsky. "Direct Observation of Vortex Ring Collapse in a Chemically Active Medium." *Physica D* **49** (1991): 1–4.
5. Aliev, R. R., and K. I. Agladze. "Critical Conditions of Chemical Wave Propagation in Gel Layers with an Immobilized Catalyst." *Physica D* **50** (1991): 65–70.
6. Allessie, M. A., F. I. M. Bonke, and F. J. G. Schopman. "Circus Movement in Rabbit Atrial Muscle as a Mechanism of Tachycardia: I." *Circ. Res.* **33** (1973): 54–62.
7. Allessie, M. A., F. I. M. Bonke, and F. J. G. Schopman. "Circus Movement in Rabbit Atrial Muscle as a Mechanism of Tachycardia: II. The Role of Nonuniform Recovery of Excitability." *Circ. Res.* **39** (1976): 168–177.
8. Allessie, M. A., F. E. M. Bonke, and F. J. G. Schopman. "Circus Movement in Rabbit Atrial Muscle as a Mechanism of Tachycardia: III. The Leading Circle." *Circ. Res.* **41** (1977): 9–18.
9. Allessie, M. A., M. J. Schalij, C. J. Kirchoff, L. Boersma, M. Huybers, and J. Hollen. "The Role of Anisotropic Impulse Propagation in Ventricular Tachycardia." In *Cell-to-Cell Signalling: From Experiments to Theoretical Models*, edited by A. Goldbeter, 565–575. New York: Academic Press, 1989.
10. Allessie, M. A., M. J. Schalij, C. J. H. J. Kirchoff, L. Boersma, M. Huybers, and J. Hollen. "Electrophysiology of Spiral Waves in Two Dimensions. The Role of Anisotropy." "Mathematical Approaches to Cardiac Arrhythmias," edited by J. Jalife. *Ann. N.Y.A.S.* **591** (1990): 247–256.
11. Aoki, M., Y. Okamoto, T. Musha, and K. Harumi. "Three-Dimensional Computer Simulation of Depolarization and Repolarization Processes in the Myocardium." *Jap. Heart J.* **27** suppl. (1986): 225–234.
12. Aranson, I. S., K. A. Gorshkov, A. S. Lomov, and M. I. Rabinovich. "Stable Particle-Like Solutions of Multidimensional Nonlinear Fields." *Physica D* **43** (1990): 435–453.
13. Arneodo, A., J. Elezgaray, J. Pearson, and T. Russo. "Instabilities of Front Patterns in Reaction-Diffusion Systems." *Physica D* **49** (1990): 141–160.
14. Baer, M., M. Falcke, and M. Eiswirth. "Dispersion Relation and Spiral Rotation in an Excitable Surface Reaction." *Physica A* **188** (1992) 78–88.

15. Baer, M., and M. Eiswirth. "Turbulence Due to Spiral Breakup in a Continuous Excitable Medium." Preprint, October, 1992 [1992b].
16. Barkley, D. "A Model for Fast Computer Simulation of Waves in Excitable Media." *Physica D* **49** (1991): 61–70.
17. Barkley, D. "Linear Stability Analysis of Rotating Spiral Waves in Excitable Media." *Phys. Lett.* **68** (1992): 2090–2093.
18. Barkley, D., M. Kness, and L. S. Tuckerman. "Spiral Wave Dynamics in a Simple Model of Excitable Media: Transition from Simple to Compound Rotation." *Phys. Rev. A* **42** (1990): 2489–2492.
19. Barrett, H. H. "The Radon Transform and its Applications." *Prog. in Optics* **21** (1984): 217–286.
20. Barrett, H. H. "Image Reconstruction and the Solution of Inverse Problems in Medical Imaging." Preprint, December, 1991 [1993].
21. Barrett, H. H., J. N. Aarsvold, and T. J. Roney. "Null Functions and Eigenfunctions: Tools for the Analysis of Imaging Systems." In *Information Processing and Medical Imaging*, edited by A. Ortendahl and J. Llacer, 212–226. New York: Wesley-Liss, 1991.
22. Barrett, H. H., and W. Swindell. *Radiological Imaging, Radiological Imaging.* New York: Academic Press, 1981.
23. Beeler, G. W., and H. Reuter. "Reconstruction of the Action Potential of Ventricular Myocardial Fibres." *J. Physiol.* **268** (1977): 177–210.
24. Bernoff, A. "Spiral Wave Solutions for Reaction-Diffusion Equations in a Fast-Reaction/Slow-Diffusion Limit." *Physica D* (1992).
25. Biktashev, V. N. "Evolution of Twist of an Autowave Vortex." *Physica D* **36** (1989): 167–117.
26. Brambilla, M., F. Battipede, L. A. Lugiato, V. Penna, F. Prati, C. Tamm, and C. O. Weiss. "Transverse Laser Patterns. I. Phase Singularity Crystals." *Phys. Rev. A* **43** (1991): 5090–5113.
27. Brazhnik, P. K., V. A. Davydov, V. S. Zykov, and A. S. Mikhailov. "Vortex Rings in Distributed Excitable Media." *Zh. Exsp. Teor. Fiz.* **93(11)** (1987): 1725–1736. (English transl. Sov. Phys.-JETP)
28. Brazhnik, P. K., V. A. Davydov, and A. S. Mikhailov. "Scroll Vortex in Excitable Media." *Radiofizika* **32** (1989): 289–293.
29. Brusca, A., and E. Rosettani. "Activation of the Human Fetal Heart." *Am. Heart J.* **86** (1973): 79–87.
30. Burgess, M. J., B. M. Steinhaus, K. W. Spitzer, and P. R. Ershler. "Nonuniform Epicardial Activation and Repolarization Properties of in Vivo Canine Pulmonary Conus." *Circ. Res.* **62** (1988): 233–246.
31. Castets, V., E. Dulos, J. Boissonade, and P. de Kepper. "Experimental Evidence of a Turing Stationary Structure." *Phys. Rev. Lett.* **64** (1990): 2953–2956.
32. Chen, P. S., N. Shibata, E. G. Dixon, R. O. Martin, and R. E. Ideker. "Comparison of the Defibrillation Threshold and the Upper Limit of Ventricular Vulnerability." *Circulation* **73** (1986): 1022–1028.

33. Chen, P. S., P. D. Wolf, E. G. Dixon, N. D. Danieley, D. W. Frazier, W. M. Smith, and R. E. Ideker. "Mechanism of Ventricular Vulnerability to Single Premature Stimuli in Open Chest Dogs." *Circ. Res.* **62** (1988): 1191–1209.
34. Cohn, R. L., S. Rush, and E. Lepeshkin. "Theoretical Analyses and Computer Simulation of ECG Ventricular Gradient and Recovery Waveforms." *IEEE Trans. Biomed. Engr.* **BME-29** (1982): 413–422.
35. Comte, A. *The Positive Philosophy.* London: Trubner, 1875. (Freely translated and condensed by Harriet Martineau.)
36. Coullet, P., L. Gil, and L. Rocca. "Optical Vortices." *Opt. Commun.* **73** (1989): 403–408.
37. Courtemanche, M., W. Skaggs, and A. T. Winfree. "Stable Three-Dimensional Action Potential Circulation in the FitzHugh-Nagumo Model." *Physica D* **41** (1990): 173–182.
38. Courtemanche, M., and A. T. Winfree. "Re-entrant Rotating Waves in a Beeler-Reuter Based Model of Two-Dimensional Cardiac Conduction." *Intl. J. Bif. & Chaos* **1** (1991): 431–444.
39. Czeisler, C. A,. R. E. Kronauer, J. S. Allan, J. F. Duffy, M. E. Jewett, E. N. Brown, and J. M. Ronda. "Bright Light Induction of Strong (Type 0) Resetting of the Human Circadian Pacemaker." *Science* **244** (1989): 1328–1333.
40. Damiano, R. J., P. K. Smith, H. Tripp, T. Asano, K. W. Small, J. E. Lowe, R. E. Ideker, and J. L. Cox. "The Effect of Chemical Ablation of the Endocardium on Ventricular Fibrillation Threshold." *Circulation* **74** (1986): 645–652.
41. Davidenko, J. M., P. F. Kent, D. R. Chialvo, D. C. Michaels, and J. Jalife. "Sustained Vortex-Like Waves in Normal Isolated Ventricular Muscle." *Proc. Natl. Acad. Sci. USA* **87** (1990): 8785–8789.
42. Davidenko, J. M., P. F. Kent, and J. Jalife. "Spiral Waves in Normal Isolated Ventricular Muscle." *Physica D* **49** (1991): 182–197.
43. Davidenko, J. M., A. M. Pertsov, R. Salomonsz, W. Baxter, and J. Jalife. "Stationary and Drifting Spiral Waves of Excitation in Isolated Cardiac Muscle." *Nature* **355** (1992): 349–351.
44. Davydov, V. A., A. S. Mikhailov, and V. S. Zykov. "Kinematical Theory of Autowave Patterns in Excitable Media." In *Nonlinear Waves in Active Media,* edited by A. Crighton and Yu. Engelbricht, 38–51. Berlin: Springer-Verlag, 1989.
45. de Bakker, J. M. T., B. Henning, and W. Merx. "Circus Movement in Canine Right Ventricle." *Circ. Res.* **45** (1979): 374–378.
46. de Kepper, P., V. Castets, E. Dulos, and J. Biossonade. "Turing-Type Chemical Patterns in the Chlorite-Iodide-Malonic Acid Reaction." *Physica D* **49** (1991): 161–169.
47. DiFrancesco, D., and D. Noble. "A Model of Cardiac Electrical Activity Incorporating Ionic Pumps and Concentration Changes." *Phil. Trans. Roy. Soc. London B* **307** (1984): 353–398.

48. Dillon, S. M. "Optical Recordings in Rabbit Heart Show that Defibrillation-Strength Shocks Prolong the Duration of Depolarization and the Refractory Period." *Circ. Res.* **69** (1991): 842–856.
49. Dillon, S. M. "Synchronized Repolarization After Defibrillation Shocks." *Circulation* **85** (1992): 1865–1878.
50. Dillon, S. M., M. A. Allessie, P. C. Ursell, and A. L. Wit. "Influence of Anisotropic Tissue Structure on Reentrant Circuits in the Epicardial Border Zone of Subacute Canine Infarcts." *Circ. Res.* **63** (1988). 182–206.
51. Ding, D. F. "A Plausible Mechanism for the Motion of Untwisted Scroll Rings in Excitable Media." *Physica D* **32** (1988): 471–487.
52. Downar, E., I. D. Parson, L. L. Mickleborough, L. C. Yao, D. A. Cameron, and W. B. Waxman. "On-Line Epicardial Mapping of Intraoperative Ventricular Arrhythmias: Initial Clinical Experience." *J. Am. Coll. Card.* **4** (1984): 703–714.
53. Durrer, D., R. T. van Dam, G. E. Freud, M. J. Janse, F. L. Meijler, and R. C. Arzbaecher. "Total Excitation of the Isolated Human Heart." *Circulation* **41** (1970): 899–912.
54. Efimov, I. R., V. I. Krinsky, and J. Jalife. "Dynamics of Rotating Vortices in the Beeler-Reuter Model of Cardiac Tissue." *J. Nonlin. Sci.*, preprint, 1992.
55. Eifler, W. J., and R. Plonsey. "A Cellular Model for the Stimulation of Activation in the Ventricular Myocardium." *J. Electrocard.* **8** (1975): 117–128.
56. El-Sherif, N., R. Mehra, W. B. Gough, and R. H. Zeiler. "Ventricular Activation Patterns of Spontaneous and Induced Ventricular Rhythms in Canine One-Day-Old Myocardial Infarction." *Circ. Res.* **51** (1982): 152–166.
57. El-Sherif, N., W. B. Gough, R. H. Zeiler, and R. Hariman. "Reentrant Ventricular Arrhythmias in the Late Myocardial Infarction Period. 12. Spontaneous vs. Induced Reentry and Intramural vs. Epicardial Circuits." *J. Am. Coll. Card.* **6** (1985): 124–132.
58. Fife, P. C. "Propagator-Controller Systems and Chemical Patterns." In *Non-Equilibrium Dynamics in Chemical Systems*, edited by C. Vidal and A. Pacault, 76–88. Berlin: Springer-Verlag, 1984.
59. Fife, P. C. "Understanding the Patterns in the BZ Reagent." *J. Stat. Phys.* **39** (1985): 687–703.
60. Fife, P. C. "Dynamics of Internal Layers and Diffusive Interfaces." *CBMS-NSF Regional Conf. Series in Appl. Math.* **53** (1988): 1–93.
61. Fishler, M. G., and N. V. Thakor. "A Massively Parallel Computer Model of Propagation Through a Two-Dimensional Cardiac Syncytium." *PACE* **14** (1991): 1694–1699.
62. Foerster, P., S. C. Muller, and B. Hess. "Curvature and Propagation Velocity of Chemical Waves." *Science* **241** (1988): 685–687.
63. Foerster, P., S. C. Muller, and B. Hess. "Critical Size and Curvature of Wave Formation in an Excitable Chemical Medium." *Proc. Natl. Acad. Sci. USA* **86** (1989): 6831–6834.

64. Foerster, P., S. C. Muller, and B. Hess. "Curvature and Spiral Geometry in Aggregation Patterns of Dictyostelium discoideum." *Development* **109** (1990): 11–16.
65. Foerster, P., S. C. Muller, and B. Hess. "Temperature Dependence of Curvature-Velocity Relationship in an Excitable Belousov-Zhabotinskii Reaction." *J. Phys. Chem.* **94** (1991): 8859–8861.
66. Frazier, D. W., W. Krassowska, P. S. Chen, P. D. Wolf, N. D. Danieley, W. M. Smith, and R. E. Ideker. "Transmural Activations and Stimulus Potentials in Three-Dimensional Anisotropic Canine Myocardium." *Circ. Res.* **63** (1988): 135–146.
67. Frazier, D. W., W. Krassowska, P. S. Chen, P. D. Wolf, E. G. Dixon, W. M. Smith, and R. E. Ideker. "Extracellular Field Required for Excitation in Three-Dimensional Anisotropic Canine Myocardium." *Circ. Res.* **63** (1988): 147–164.
68. Frazier, D. W., P. D. Wolf, J. M. Wharton, A. S. L. Tang, W. M. Smith, and R. E. Ideker. "Stimulus-Induced Critical Point: Mechanism of Electrical Induction of Reentry in Normal Canine Myocardium." *J. Clin. Invest.* **83** (1989): 1039–1052.
69. Gaspar,V., and P. Galambosi. "Bifurcation Diagram of the Oscillatory Belousov-Zhabotinskii System of Oxalic Acid in a Continuous Flow Stirred Tank Reactor." *J. Phys. Chem.* **90** (1986): 2222–2226.
70. Gaspar, V., and K. Showalter. "Period Lengthening and Associated Bifurcations in a Two-Variable Flow Oregonator." *J. Chem. Phys.* **88** (1988): 778–791.
71. Gerhardt, M., H. Schuster, and J. J. Tyson. "A Cellular Automaton Model of Excitable Media Including the Effects of Curvature and Dispersion." *Science* **247** (1990): 1563–1566.
72. Gerhardt, M., H. Schuster, and J. J. Tyson. "A Cellular Automaton Model of Excitable Media IV. Untwisted Scroll Rings." *Physica D* **50** (1991): 189–206.
73. Glass, L., and A. T. Winfree. "Discontinuities in Phase-Resetting Experiments." *Am. J. Physiol.* **246** (1984): R251–258.
74. Gomatam, J., and P. Grindrod. "Three-Dimensional Waves in Excitable Reaction-Diffusion Systems." *J. Math. Biol.* **25** (1987): 611–622.
75. Gorshkov, K. A., A. S. Lomov, and M. I. Rabinovich. "Three-Dimensional Particle-Like Solutions of Coupled Nonlinear Fields." *Phys. Lett. A* **137** (1989): 250–254.
76. Graae Sorensen, P., and F. Hynne. "Amplitudes and Phases of Small-Amplitude Belousov-Zhabotinskii Oscillations Derived from Quenching Experiments." *J. Phys. Chem.* **93** (1987): 5467–5474.
77. Gulko, F. B., and A. A. Petrov. "Mechanism of the Formation of Closed Pathways of Conduction in Excitable Media." *Biofizika* **17** (1972): 261–270.
78. Hanusse, P., V. Perez-Munuzuri, and C. Vidal. "Phase Dynamics and Spatial Patterns in Oscillating and Excitable Media." In *Wave Processes in Excitable Media*, edited by A. V. Holden. New York: Plenum, 1991.

79. Harumi, K., C. R. Smith, J. A. Abildskov, M. J. Burgess, R. L. Lux, and R. F. Wyatt. "Detailed Activation Sequence in the Region of Electrically Induced Ventricular Fibrillation in Dogs." *Jap. Heart J.* **21** (1980): 533–544.
80. Henze, C., E. Lugosi, and A. T. Winfree. "Stable Helical Organizing Centers in Excitable Media." *Canad. J. Phys.* **68** (1990): 683–710.
81. Henze, C., and A. T. Winfree. "A Stable Knotted Singularity in an Excitable Medium." *Intl. J. Bif. & Chaos* **1** (1991): 891–922.
82. Hill, B. C., and K. R. Courtney. "Design of a Multi-Point Laser Scanned Optical Monitor of Cardiac Action Potential Propagation: Application to Microreentry in Guinea Pig Atrium." *Ann. Biomed. Engr.* **15** (1987): 567–577.
83. Holt, P., and M. Davies. "Erythema Gyratum Repens—An Immunologically Mediated Dermatosis?" *Brit. J. Dermatol.* **96** (1977): 343–347.
84. Hunter, P. J., and B. H. Smaill. "The Analysis of Cardiac Function: A Continuum Approach." *Prog. Biophys. Mol. Biol.* **52** (1988): 101–164.
85. Hynne, F., P. Graae Sorensen, and K. Nielsen. "Quenching of Chemical Oscillations." *J. Phys. Chem.* **91** (1987): 6573–6575.
86. Hynne, F., P. Graae Sorensen, and K. Nielsen. "Quenching of Chemical Oscillations: General Theory." *J. Chem. Phy.* **92** (1990): 1747–1757.
87. Ideker, R. E., G. H. Bardy, S. J. Worley, L. D. German, and W. M. Smith. "Patterns of Activation During Ventricular Fibrillation." In *Tachycardias: Mechanisms, Diagnoses, and Treatment,* edited by M. E. Josephson and H. J. J. Wellens, 519–536. Philadelphia, PA: Lea & Feberger, 1984.
88. Ideker, R. E., D. W. Frazier, W. Krassowska, N. Shibata, P.-S. Chen, K. M. Kavanagh, and W. M. Smith. In *Mathematical Approaches to Cardiac Arrhythmias*, edited by J. Jalife. *Ann. NYAS* **591** (1990): 208–218.
89. Ideker, R. E., P. D. Wolf, E. Simpson, E. E. Johnson, S. M. Blanchard, and W. M., Smith. "The Ideal Cardiac Mapping System." In *Proceedings of First International Workshop on Cardiac Mapping*, edited by M. Borgreffe, G. Breithardt, and M. Shenasa. (September 1992, Playa de Ayo). Mount Kisco, NY: Futura Press, 1993.
90. Ito, H, and L. Glass "Spiral Breakup in a New Model of Discrete Excitable Media." *Phys. Rev. Lett.* **66** (991): 671–674.
91. Jahnke, W., C. Henze, and A. T. Winfree. "Chemical Vortex Dynamics in Three-Dimensional Excitable Media." *Nature* **336** (1988): 662–665.
92. Jahnke, W., W. E. Skaggs, and A. T. Winfree. "Chemical Vortex Dynamics in the Belousov-Zhabotinsky Reaction and in the 2-Variable Oregonator Model." *J. Phys. Chem.* **93** (1989): 740–749.
93. Jahnke, W., and A. T. Winfree. "Recipes for Belousov-Zhabotinsky Reagents." *J. Chem. Educ.* **68** (1991): 320–324.
94. Jahnke, W., and A. T. Winfree. "A Survey of Spiral Wave Behavior in the Oregonator Model." *Intl. J. Bif. & Chaos* **1** (1991): 445–466.
95. Jakubith, S., H. H. Rotermund, W. Engel, A. von Oertzen, and G. Ertl. "Spatiotemporal Concentration Patterns in a Surface Reaction: Propagating and

Standing Waves, Rotating Spirals, and Turbulence." *Phys. Rev. Lett.* **65** (1990): 3013–3016.

96. Janse, M. J., F. J. van Capelle, H. Morsink, A. G. Kleber, F. Wilms-Schopman, and R. Cardinal. "Flow of Injury Currents and Patterns of Excitation During Early Ventricular Arrhythmias in Acute Regional Myocardial Ischemia in Isolated Hearts." *Circ. Res.* **47** (1980): 151–165.

97. Janse, M. J., F. Wilms-Schopman, R. J. Wilensky, and J. Tranum-Jensen. "Role of the Subendocardium in Arrhythmogenesis during Acute Ischemia." In *Cardiac Electrophysiology and Arrhythmias*, edited by D. P. Zipes and J. Jalife, 353–362. Orlando, FL: Grune & Stratton, 1985.

98. Janse, M. J., A. G. Kleber, A. Capucci, R. Coronel, and F. Wilms-Schopman. "Electrophysiological Basis for Arrhythmias Caused by Acute Ischemia." *J. Mol. Cell. Cardiol.* **18** (1986): 339–355.

99. Jewett, M. E., R. E. Kronauer, and C. A. Czeisler. "Light-Induced Suppression of Endogenous Circadian Amplitude in Humans." *Nature* **350** (1990): 59–62.

100. Karma, A. "Meandering Transition in Two-Dimensional Excitable Media." *Phys. Rev. Lett.* **65** (1990): 2824–2827.

101. Karma, A. "Universal Limit of Spiral Wave Propagation in Excitable Media." *Phys. Rev. Lett.* **66** (1991): 2274–2277.

102. Karma, A. "The Scaling Regime of Spiral Wave Propagation in Single-Diffusive Media." *Phys. Rev. Lett.* **68** (1992): 397–400.

103. Kavanagh, K. M., J. S. Kabas, D. L. Rollins, S. B. Melnick, W. M. Smith, and R. E. Ideker. "High Current Stimuli to the Spared Epicardium of a Large Infarct Induce Ventricular Tachycardia." *Circulation* **85** (1992): 680–698.

104. Keener, J. P. "The Dynamics of Three-Dimensional Scroll Waves in Excitable Media." *Physica D* **31** (1988): 269–276.

105. Keener, J. P. "Knotted Scroll Wave Filaments in Excitable Media." *Physica D* **34** (1989): 378–390.

106. Keener, J. P. "The Core of the Spiral." *SIAM. J. Appl. Math.* **52** (1992): 1370–1390.

107. Keener, J. P., and J. J. Tyson. "The Motion of Untwisted Untorted Scroll Waves in Belousov-Zhabotinsky Reagent." *Science* **239** (1988): 1284–1286.

108. Keener, J. P., and J. J. Tyson. "Helical and Circular Scroll Wave Filaments." *Physica D* **44** (1990): 191–202.

109. Keener, J. P., and J. J. Tyson. "The Dynamics of Helical Scroll Waves in Excitable Media." *Physica D* **53** (1991): 151–161.

110. Keener, J. P., and J. J. Tyson. "The Dynamics of Scroll Waves in Excitable Media." *SIAM Rev.* **34** (1992): 1–39.

111. Kessler, D., and H. Levine. "Effect of Diffusion on Patterns in Excitable Belousov-Zhabotinskii Systems." *Physica D* **39** (1989): 1–14.

112. Kessler, D., H. Levine, and W. N. Reynolds. "Spiral Core Meandering in Excitable Media." *Phys. Rev. A* **46** (1992): 5264–5268..

113. Kessler, D., H. Levine, and W. N. Reynolds. "The Spiral Core in Singly Diffusive Media." *Phys. Rev. Lett.* **68** (1992): 401–404.

114. Ketnerova, L., H. Sevcikova, and M. Marek. "Periodic Forcing of a Spatially One-Dimensional Excitable Reaction-Diffusion System." In *Nonlinear Wave Processes in Excitable Media*, edited by A. V. Holden, M. Markus, and H. G. Othmer. New York: Plenum, 1991.
115. Kida, S. "A Vortex Filament Moving Without Change of Form." *J. Fluid Mech.* **112** (1981): 397–409.
116. Kida, S. "Stability of a Steady Vortex Filament." *J. Phys. Soc. Japan* **51** (1982): 1655–1662.
117. Kogan, B. Y., W. J. Karplus, B. S. Billett, and W. G. Stevenson. "Excitation Wave Propagation Within Narrow Pathways: Geometric Configurations Facilitating Unidirectional Block and Reentry." *Physica D* **59** (1992): 275–296.
118. Kopell, N., and L. Howard. "Plane Wave Solutions to Reaction-Diffusion Equations." *Stud. Appl. Math.* **52** (1973): 291–328.
119. Kramer, J. B., J. E. Saffitz, F. X. Witkowski, and P. B. Corr. "Intramural Reentry as a Mechanism of Ventricular Tachycardia During Evolving Canine Myocardial Infarction." *Circ. Res.* **56** (1985): 736–754.
120. Krassowska, W., D. W. Frazier, T. C. Pilkington, and R. E. Ideker. "Finite Element Approximation of Potential Gradient in Cardiac Muscle Undergoing Stimulation." *Proc. Sixth Intl. Conf. Math. Modelling.* St. Louis, Missouri, 1987.
121. Krinsky, V. I. "Spread of Excitation in an Inhomogeneous Medium (State Similar to Cardiac Fibrillation) (in Russian)." *Biofizika* **11** (1966): 776–784.
122. Krinsky, V. I. "Fibrillation in Excitable Media." *Prob. in Cyber.* **20** (1968): 59–80 (in Russian) and in *Systems Theory Research* **20** (1969): 46–65 (in English).
123. Krinsky, V. I., I. R. Efimov, and J. Jalife. "Vortices with Linear Cores in Excitable Media." *Proc. Roy. Soc. Lond. A* **437** (1992): 645–655.
124. Krischer, K., M. Eiswirth, and G. Ertl. "Periodic Perturbations of the Oscillatory CO Oxidation on Pt(110): Model Calculations." *J. Chem. Phys* . **97** (1992): 307–391.
125. Kuhnert, L. "Chemische Strukturbildung in Festen Gelen auf der Basis der Belousov-Zabotinskij-Reaktion." *Naturwissenschaften* **70** (1983): 464–466.
126. Kuramoto, Y. "Diffusion-Induced Chemical Turbulence." In *Dynamics of Synergetic Systems*, edited by H. Haken, 134–146. Heidelberg: Springer, 1980.
127. Kuramoto, Y. *Chemical Oscillations, Waves, and Turbulence.* Berlin: Springer-Verlag, 1984.
128. Kuramoto, Y., and S. Koga. "Turbulized Rotating Chemical Waves." *Prog. Theor. Phys.* **66** (1981): 1081–1085.
129. Lechleiter, J., S. E. Girard, E. G. Peralta, and D. E. Clapham. "Spiral Calcium Wave Propagation and Annihilation." *Science* **252** (1991): 123–126.
130. Lee, K. J., W. D. McCormick, Z. Noscticzius, and H. L. Swinney. "Turing Patterns Visualized by Index of Refraction Variations." *J. Chem. Phys.* **96** (1992): 4048–4049.
131. Lesigne, C., B. Levy, R. Saumont, P. Birkui, A. Bardou, and B. Rubin. "An Energy-Time Analysis of Ventricular Fibrillation and Defibrillation Thresholds with Internal Electrodes." *Med. Biol. Engr.* **14** (1976): 617–622.

132. Linde, H., and H. Engel. "Autowave Propagation in Heterogeneous Active Media." *Physica D* **49** (1991): 13–20.
133. Lugosi, E. "Analysis of Meandering in Zykov-Kinetics." *Physica D* **40** (1989): 331–337.
134. Lund, F. "Defect Dynamics for the Nonlinear Schrödinger Equation Derived from a Variational Principle." *Phys. Lett. A* **159** (1991): 245–251.
135. Markus, M., and B. Hess. "Isotropic Cellular Automaton for Modelling Excitable Media." *Nature* **347** (1990): 56–58.
136. Markus, M., and B. Hess. "Isotropic Automata for Simulations of Excitable Media: Periodicity, Chaos, and Reorganization." In *Dissipative Structures in Transport Processes and Combustion*, edited by D. Meinkuhn, 197–214. Berlin: Springer-Verlag, 1990.
137. Markus, M., B. Krafczyk, Hess. "Randomized Automata for Isotropic Modelling of Two- and Three-Dimensional Waves and Spatiotemporal Chaos in Excitable Media." In *Nonlinear Wave Processes in Excitable Media*, edited by A. V. Holden, M. Markus, and H. G. Othmer. New York: Plenum, 1991.
138. Maselko, J. "Determination of Bifurcation in Chemical Systems: An Experimental Method." *Chem. Phys.* **67** (1982): 17–26.
139. McCormick, W. D., Z. Noszticzius, and H. L. Swinney. "Interrupted Separatrix Excitability in a Chemical System." *J. Chem. Phys.* **94** (1991): 2159–2167.
140. Medvinsky, A. B., and A. M. Pertsov. "Initiation Mechanism of the First Extrasystole in a Short-Lived Atrial Arrhythmia." *Biofizika* **27** (1982): 895–899.
141. Medvinsky, A. B., A. V. Panfilov, and A. M. Pertsov. "Properties of Rotating Waves in Three Dimensions. Scroll Rings in Myocardium." In *Self Organization: Autowaves and Structures far from Equilibrium,* edited by V. I. Krinsky. Berlin: Springer-Verlag, 1984.
142. Mehra, R. "Mechanisms of Initiation and Termination of Tachyarrhythmias." *IEEE Engr. Med. Biol.* **3** (1984): 34–35.
143. Meron, E. "The Roles of Curvature and Wavefront Interactions in Spiral Wave Dynamics." *Physica D* **49** (1991): 98–106.
144. Meron, E. "Pattern Formation in Excitable Media." *Phys. Rep.* **218** (1992): 1–66.
145. Mikhailov, A. S., A. V. Panfilov, and A. N. Rudenko. "Twisted Scroll Waves in Active Three-Dimensional Media." *Phys. Lett. A* **109** (1985): 246–250.
146. Mikhailov, A. S., and V. S. Zykov. "Kinematical Theory of Spiral Waves in Excitable Media: Comparison with Numerical Simulations." *Physica D* **52** (1991): 379–397.
147. Moe, G. K., W. C. Rheinboldt, and J. A. Abildskov. "A Computer Model of Atrial Fibrillation." *Am. Heart J.* **67** (1964): 200–220.
148. Monk, P. B., and H. G. Othmer. "Wave Propagation in Aggregation Fields of the Cellular Slime Mold *Dictyostelium discoideum*." *Proc. Roy. Soc. Lond. B* **240** (1990): 555–589.
149. Moore, H. J. "Does the Pattern of *Erythema gryatum repens* Depend on a Reaction-Diffusion System?" *Brit. J. Dermatol.* **107** (1982): 723.

150. Muller, S. C., O. Steinbock, and J. Schutze. "Autonomous Pacemaker of Chemical Waves Created by Spiral Annihilation." *Physica A* **188** (1992): 47–54.

151. Nettesheim, S., A. von Oertzen, H. H. Rotermund, and G. Ertl. "Reaction Diffusion Pattern in the Catalytic CO-Oxidation on Pt(110): Front Propagation and Spiral Waves." Preprint, October, 1992 [1993].

152. Nicholson, P. W. "Experimental Models for Current Conduction in an Anisotropic Medium." *IEEE Trans. Biomed. Engr.* **BME-14** (1967): 55–56.

153. Noszticzius, Z., P. Stirling, and M. Wittmann. "Measurement of Bromine Removal Rate in the Oscillatory BZ Reaction of Oxalic Acid." *J. Phys. Chem.* **89** (1985): 4914–4921.

154. Okajima, M., T. Fujino, T., Kobayashi, and K. Yamada. "Computer Simulation of the Propagation Process in Excitation of the Ventricles." *Circ. Res.* **23** (1968): 203–211.

155. Ouyang, Q., W. T. Tam, P. De Kepper, W. D. McCormick, Z. Noszticzius, and H. L. Swinney. "Bubble-Free Belousov-Zhabotinskii-Type Reactions." *J. Phys. Chem.* **91** (1987): 2181–2184.

156. Ouyang, Q., H. L. Swinney, V. Dufiet, and J. Boissonade. "Transition to Hexagonal and Striped Turing Patterns." *Nature* **352** (1991): 610–612.

157. Ouyang, Q., and H. L. Swinney. "Transition to Chemical Turbulence." *Chaos* **1** (1992): 411–420.

158. Panfilov, A. V. "Three-Dimensional Vortices in Active Media." In *Nonlinear Wave Processes in Excitable Media*, edited by A. V. Holden, M. Markus, and H. G. Othmer. New York: Plenum, 1991.

159. Panfilov, A. V., and A. V. Holden. "Self-Generation of Turbulent Vortices in a Two-Dimensional Model of Cardiac Tissue." *Phys. Lett. A* **151** (1990): 23–26.

160. Panfilov, A. V., and A. M. Pertsov. "Vortex Ring in Three-Dimensional Active Medium Described by Reaction Diffusion Equation." *Dokl. Akad. Nauk. SSSR* **274** (1984): 1500–1503.

161. Panfilov, A. V., A. N. Rudenko, and A. M. Pertsov. "Twisted Scroll Waves in Active Three-Dimensional Media." *Dokl. Acad. Nauk. SSSR* **279** (1984): 1000–1002.

162. Panfilov, A. V., A. V. Rudenko, and V. I. Krinsky. "Vortex Rings in Three-Dimensional Active Media with Diffusion by 2 Components." *Biofizika* **31** (1986): 850–854.

163. Panfilov, A. V., and A. N. Rudenko. "Two Regimes of the Scroll Ring Drift in Three-Dimensional Active Media." *Physica D* **28** (1987): 215–218.

164. Panfilov, A. V., and A. V. Holden. "Spatio-Temporal Chaos in a Model of Cardiac Electrical Activity." *Intl. J. Bif. & Chaos* **1** (1991): 219–225.

165. Pertsov, A. M., and A. K. Grenadier. "The Autowave Nature of Cardiac Arrhythmias." In *Self Organization: Autowaves and Structures far from Equilibrium*, edited by V. I. Krinsky. Berlin: Springer-Verlag, 1984.

166. Pertsov, A. M., R. R. Aliev, and V. I. Krinsky. "Three-Dimensional Twisted Vortices in an Excitable Chemical Medium." *Nature* **345** (1990): 419–421.

167. Plesser, T., S. C. Muller, and B. Hess. "Spiral Wave Dynamics as a Function of Proton Concentration in the Ferroin-Catalyzed Belousov-Zhabotinsky Reaction." *J. Phys. Chem.* **94** (1990): 7501–7507.
168. Plonsey, R., and R. Barr. "Mathematical Modeling of Electrical Activity of the Heart." *J. Electrocard.* **20** (1987): 219–226.
169. Pogwizd, S. M., and P. B. Corr. "Reentrant and Nonreentrant Mechanisms Contribute to Arrhythmogenesis During Early Myocardial Ischemia: Results Using Three-Dimensional Mapping." *Circ. Res.* **61** (1987): 352–371.
170. Pogwizd, S. M., and P. B. Corr. "Electrophysiologic Mechanisms Underlying Arrhythmias due to Reperfusion of Ischemic Myocardium." *Circulation* **76** (1987): 404–426.
171. Poston, T., C. Henze, and A. T. Winfree. *Intl. J. Bif. & Chaos* (1993): In preparation.
172. Reshodko, L. V. "Automata Models and Machine Experiment in the Investigation of Biological Systems." *J. Gen. Biol.* **1** 80–87.
173. Ricca, R. L. "Rediscovery of the Da Rios Equations." *Science* **352** (1991): 561–562.
174. Roberts, D. E., and A. M. Scher. "Effect of Tissue Anisotropy on Extracellular Potential Fields in Canine Myocardium in situ." *Circ. Res.* **50** (1982): 342–351.
175. Rössler, O. E. and C. Kahlert. "Winfree Meandering in a Two-Dimensional 2-Variable Excitable Medium." *Z. Naturforsch.* **34** (1979): 565–570.
176. Rotermund, H. H., S. Nettesheim, A. von Oertzen, and E. Ertl. "Surface Science Letters." **275** (1992): L645–L649.
177. Roth, B. J. "The Bidomain Model of Cardiac Tissue: Predictions and Experimental Verification." In *Neural Engineering*, edited by Y. Kim and N. Thakor, 1993.
178. Salama, G., R. Lombardi, and J. Elson. "Maps of Optical Action Potentials and NADH Fluorescence in Intact Working Hearts." *Am. J. Physiol.* **252** (1987): H384–394.
179. Schalij, M. "Anisotropic Conduction and Ventricular Tachycardia." Ph.D. Thesis, University of Limburg, 1988.
180. Sciegosz, H., and S. Pokrzywnicki. "The Belousov-Zhabotinsky Reaction Under External Periodic Influence near the SNIPER Bifurcation Point." *Acta Chem. Scand.* **43** (1989): 926–931.
181. Segel, L. A., A. S. Perelson, J. M. Hyman, and S. N. Klaus. "Rash Theory." In *Theoretical and Experimental Insights into Immunology*, edited by A. S. Perelson. Berlin: Springer-Verlag, 1992.
182. Shibata, N., P. S. Chen, E. G. Dixon, P. D. Wolf, N. D. Danieley, W. M. Smith, and R. E. Ideker. "Influence of Shock Strength and Timing on the Induction of Ventricular Arrhythmias in Dogs." *Am. J. Physiol.* **255** (1988): H891–H901.
183. Shinomoto, S., and Y. Kuramoto. "Phase Transitions in Active Rotors." *Prog. Theor. Phys.* **75** (1986): 1105–1110.
184. Skaggs, W., E. Lugosi, and A. T. Winfree. "Stable Vortex Rings of Excitation in Neuroelectric Media." *IEEE Trans. Circ. Sys.* **CAS-35** (1988): 784–787.

185.Skinner, G. S., and H. L. Swinney. "Periodic to Quasiperiodic Transition of Chemical Spiral Rotation." *Physica D* **48** (1990): 1–16.

186.Smoes, M. L. "Chemical Waves in the Oscillatory Zhabotinskii System: A Transition from Temporal to Spatio-Temporal Organization." *Dynamics of Synergetic Systems*, edited by H. Haken, vol. 6, 80–95. Springer Series in Synergetics. Berlin: Springer-Verlag, 1980.

187.Spach, M. S., and J. M. Kootsey. "The Nature of Electrical Propagation in Cardiac Muscle." *Am. J. Physiol.* **244** (1983): H3–H22.

188.Spielman, S. R., E. L. Michelson, L. N. Horowitz, J. F. Spear, and E. N. Moore. "The Limitations of Epicardial Mapping as a Guide to the Surgical Therapy of Ventricular Tachycardia." *Circulation* **57** (1978): 666–670.

189.Suzuki, R. "Electrochemical Neuron Model." *Adv. Biophys.* **9** (1976): 115–156.

190.Suzuki, R., S. Sata, and J. Nagumo. "Electrochemical Active Network." *Notes of ICEC Japan Professional Group on Nonlinear Theory*, 1963. (In Japanese.)

191.Swenne, C. A., H. A. Bosker, and N. M. van Hemel. "Computer Simulation of Compound Reentry." *Computers in Cardiol.* (1986): 445–448.

192.Swenne, C. A., N. M. van Hemel, and E. O. Robles de Medina. "ECG Criteria for Assessement of Mechanisms of Arrhythmias: A Review." *Eur. Heart J.* **8** (1987): 800–812.

193.Tyson, J. J., and J. D. Murray. "Cyclic-AMP Waves During Aggregation of Dictyostelium Amoebae." *Development* **106** (1989): 421–426.

194.Tyson, J. J., and S. H. Strogatz. "The Differential Geometry of Scroll Waves." *Intl. J. Bif. & Chaos* **1(4)** (1991): 723–744.

195.Vidal, C., and A. Pagola. "Observed Properties of Trigger Waves Close to the Center of the Target Pattern in Oscillating BZ Reagent." *J. Phys. Chem.* **93** (1989): 2711–2716.

196.Wach, P., R. Killamn, F. Dienstl, and Ch. Eichtinger. "A Computer Model of Human Ventricular Myocardium for Simulation of ECG, MCG, and Activation Sequence Including Reentry Rhythms." *Basic Res. Cardiol.* **84** (1989): 404–413.

197.Weimar, J. R., J. J. Tyson, and L. T. Watson. "Third Generation Cellular Automaton for Modeling Excitable Media." *Physica D* **55** (1992): 328–339.

198.Welsh, B. J., J. Gomatam, and A. Burgess. "Three-Dimensional Chemical Waves in the Belousov-Zhabotinsky Reaction." *Nature* **304** (1983): 611–614.

199.Welsh, B. J., and J. Gomatam. "Diversity of Three-Dimensional Chemical Waves." *Physica D* **43** (1990): 304–317.

200.Wiener, N., and A. Rosenblueth. "The Mathematical Formulation of the Problem of Conduction of Impulses in a Network of Connected Excitable Elements, Specifically in Cardiac Muscle." *Arch. Inst. Cardiol. Mex.* **16** (1946): 205–265.

201.Winfree, A. T. "The Temporal Morphology of a Biological Clock." *Lect. Math. Life Sci.*, edited by M. Gerstenhaber. Providence, RI: Am. Math. Soc., **2** (1970): 109–150.

202.Winfree, A. T. "Spiral Waves of Chemical Activity." *Science* **175** (1972): 634–636.

203. Winfree, A. T. "Oscillatory Glycolysis in Yeast: The Pattern of Phase Resetting by Oxygen." *Arch. Biochem. Biophys.* **149** (1972): 338–401.
204. Winfree, A. T. "Scroll-Shaped Waves of Chemical Activity in Three Dimensions." *Science* **181** (1973): 937–939.
205. Winfree, A. T. "Rotating Solutions to Reaction/Diffusion Equations." In *S.I.A.M./A.M.S.* Proc. Vol. 8, edited by D. Cohen, 13–31. Providence, RI: Am. Math. Soc., 1974.
206. Winfree, A. T. "Rotating Chemical Reactions." *Sci. Am.* **230(6)** (1974): 82–95.
207. Winfree, A. T. "Two Kinds of Wave in an Oscillating Chemical Solution." *Far. Symp. Chem. Soc.* **9** (1974): 38–46.
208. Winfree, A. T. "Spatial and Temporal Organization in the Zhabotinsky Reaction." *Adv. Biol. Med. Phys.* **16** (1977): 115–136. (1973 Katchalsky Symposium, edited by J. H. Lawrence, J. W. Gofman, and T. L. Hayes)
209. Winfree, A. T. *The Geometry of Biological Time.* New York: Springer-Verlag, 1980.
210. Winfree, A. T. "Sudden Cardiac Death—A Problem in Topology." *Sci. Am.* **248(5)** (1983): 144–161.
211. Winfree, A. T. "The Prehistory of the Belousov-Zhabotinsky Reaction." *J. Chem. Educ.* **61** (1984) 661–663.
212. Winfree, A. T. "Wavefront Geometry in Excitable Media: Organizing Centers." *Physica D* **12** (1984): 321–332.
213. Winfree, A. T. "Organizing Centers for Chemical Waves and 2 and 3 Dimensions." In *Oscillations and Travelling Waves in Chemical Systems*, edited by R. Field and M. Burger, 441–472. New York: Wiley, 1985.
214. Winfree, A. T. *The Timing of Biological Clocks.* New York: Scientific American Books, 1986.
215. Winfree, A. T. *When Time Breaks Down.* Princeton: Princeton University Press, 1987.
216. Winfree, A. T. 1989 "Electrical Instability in Cardiac Muscle: Phase Singularities and Rotors." *J. Theor. Biol.* **138** (1989): 353–405.
217. Winfree, A. T. "Ventricular Reentry in Three Dimensions." In *Cardiac Electrophysiology From Cell to Bedside*, edited by D. P. Zipes and J. Jalife, 224–234. Philadelphia, PA: W. B. Saunders, 1989.
218. Winfree, A. T. "Vortex Reentry in Healthy Myocardium." In *Cell-to-Cell Signalling: From Experiments to Theoretical Models*, edited by A. Goldbeter, 609–624. New York: Academic Press, 1989.
219. Winfree, A. T. "Estimating the Ventricular Fibrillation Threshold." In *Theory of Heart*, edited by L. Glass, P. Hunter, and A. McCulloch, 477–531. Berlin: Springer-Verlag, 1990.
220. Winfree, A. T. "The Electrical Thresholds of Normal Ventricular Myocardium." *J. Cardiovasc. Electrophys.* **1** (1990): 393–410.
221. Winfree, A. T. "Vortex Action Potentials in Normal Ventricular Muscle." "Mathematical Approaches to Cardiac Arrhythmias," edited by J. Jalife. *Ann. N.Y.A.S.* **591** (1990): 190–207.

222. Winfree, A. T. "Stable Particle-Like Solutions to the Nonlinear Wave Equations of Three-Dimensional Excitable Media." *SIAM Rev.* **32** (1990): 1–53.
223. Winfree, A. T. "Rotors in Normal Ventricular Myocardium." *Proc. Kon. Ned. Akad. v. Wetensch.* **93(4)** (1990): 513–536, and **94(2)** (1991): 257–280.
224. Winfree, A. T. "Vortices in Motionless Media." *Appl. Mech. Rev.* **43** (1990): 297–309.
225. Winfree, A. T. "Varieties of Spiral Wave Behavior in Excitable Media." *Chaos* **1(3)** (1991): 303–334.
226. Winfree, A. T. "Numerical and Chemical Experiments on Filament Motion." In *Spatio-Temporal Organization in Nonequilibrium Systems*, edited by S. C. Müller and T. Plesser, 270–273. Dortmund: Project Verlag, 1992.
227. Winfree, A. T. "Mapping in Three Dimensions and Future Directions: How Does VT Decay into VF?" In *Proceedings of First International Workshop on Cardiac Mapping*, Ch. 41 (September 1992, Playa de Ayo), edited by M. Borgreffe, G. Breithardt, and M. Shenasa. Mount Kisco, NY: Futura Press, 1993.
228. Winfree, A. T., and W. Guilford. "The Dynamics of Organizing Centers: Numerical Experiments in Differential Geometry." In *Biomathematics and Related Computational Problems*, edited by L. M. Riccardi, 697–716. Dordrecht: Kluwer, 1988.
229. Winfree, A. T., and W. Jahnke. "Three-Dimensional Scroll Ring Dynamics in the Belousov-Zhabotinsky Reagent and in the 2-variable Oregonator Model." *J. Phys. Chem.* **93** (1989): 2823–2832.
230. Winfree, A. T., and S. H . Strogatz. "Singular Filaments Organize Chemical Waves in Three Dimensions: 1. Geometrically Simple Waves." *Physica D* **8** (1983): 35–49.
231. Winfree, A. T., and S. H. Strogatz. "Singular Filaments Organize Chemical Waves in Three Dimensions: 2. Twisted Waves." *Physica D* **9** (1983): 65–80.
232. Winfree, A. T., and S. H. Strogatz. "Singular Filaments Organize Chemical Waves in Three Dimensions: 3. Knotted Waves." *Physica D* **9** (1983): 333–345.
233. Winfree, A. T., and S. H. Strogatz. "Singular Filaments Organize Chemical Waves in Three Dimensions. 4: Wave Taxonomy." *Physica D* **13** (1984): 221–233.
234. Winfree, A. T., and S. H. Strogatz. "Organizing Centers for Three-Dimensional Chemical Waves." *Nature* **311** (1984): 611–615.
235. Winfree, A. T., E. M. Winfree, and H. Seifert. "Organizing Centers in a Cellular Excitable Medium." *Physica D* **17** (1985): 109–115.
236. Wit, A. L., M. A. Allessie, F. I. M. Bonke, W. Lammers, J. Smeets, and J. J. Fenoglio. "Electrophysiologic Mapping to Determine the Mechanism of Experimental Ventricular Tachycardia Initiated by Premature Impulses." *Am. J. Cardiol.* **49** (1982): 166–185.
237. Wit, A. L., S. M. Dillon, J. Coromilas, A. E. Saltman, and B. Waldecker. "Anisotropic Reentry in the Epicardial Border Zone of Myocardial Infarcts."

"Mathematical Approaches to Cardiac Arrhythmias," edited by J. Jalife. *Ann. N.Y.A.S.* **591** (1990): 86–108.

238. Witkowski, F. X., and P. A. Penkoske. "Activation Patterns During Ventricular Fibrillation." "Mathematical Approaches to Cardiac Arrhythmias," edited by J. Jalife. *Ann. N.Y.A.S.* **591** (1990): 218–231.

239. Witkowski, F. X., and P. A. Penkoske. "Epicardial Activation Times After Defibrillation in Open-Chest Dogs Using Unipolar DC-Coupled Activation Recordings." *J. Electrocard.* **23** (Suppl.) (1990): 39–45.

240. Wright, F. J., and M. V. Berry. "Wave-Front Dislocations in the Sound Field of a Pulsed Circular Piston Radiator." *J. Acoust. Soc. Am.* **75** (1984): 733–747.

241. Yamada, T., and Y. Kuramoto. "Spiral Waves in a Nonlinear Dissipative System." *Prog. Theor. Phys.* **55** (1976): 2035–2036.

242. Yamaguchi, T., L. Kuhnert, Zs. Nagy-Ungvarai, S. C. Muller, and B. Hess. "Gel Systems for the Belousov-Zhabotinsky System." *J. Phys. Chem.* **95** (1991): 5831–5837.

243. Yamaguchi, T., and S. C. Muller. "Front Geometries of Chemical Waves Under Anisotropic Conditions." *Physica D* **49** (1991): 40–46.

244. Zhabotinsky, A. M. "Investigations of Homogeneous Chemical Auto-Oscillating Systems" (in Russian). Ph.D. Thesis, Puschino, USSR, 1970.

245. Zipes, D. P., J. Fischer, R. M. King, A. Nicoll, and W. W. Jolly. "Termination of Ventricular Fibrillation in Dogs by Depolarizing a Critical Amount of Myocardium." *Am. J. Cardiol.* **36** 37–44.

246. Zykov, V.S. *Simulation of Wave Processes in Excitable Media* (in Russian), Nauka, Moscow, 1984. Also in translation under same title, edited by A. T. Winfree. Manchester: Manchester University Press 1988.

247. Zykov, V. S. "Cycloidal Circulation of Spiral Waves in an Excitable Medium." *Biofizika* **31** (1986): 862–865.

Jonathan S. Yedidia
Lyman Physics Laboratory, Harvard University, Cambridge, MA 02138

Quenched Disorder: Understanding Glasses Using a Variational Principle and the Replica Method

OUTLINE

1. Introduction and Elements of Statistical Mechanics

   A. Introduction: the goal of these lectures is to learn about the new techniques that have been developed to compute thermodynamic properties (such as the free energy, specific heat, and correlation functions) of physical systems with quenched disorder.

   B. Physical Systems: a physical system is characterized by the states it can be in, and the observable quantities which are a function of the state. The Hamiltonian (energy function) is a very important observable quantity. The partition function, if known exactly, gives complete information about the thermodynamics of a physical system.

   C. Two Exactly Soluble Models: we compute the free energy exactly for two kinds of physical systems: noninteracting Ising spin systems, and particle systems with quadratic interactions.

   D. The Variational Approach: we can define a "trial free energy" which is a quantity that will be greater than the true free energy for any "trial

Hamiltonian," and which will approach the true free energy as the "trial Hamiltonian" approaches the true Hamiltonian.

E. Trial Hamiltonians: our exactly soluble systems provide a source of trial Hamiltonians which permit the analytical computation of trial free energies.

F. Toy Example of the Variational Approach: we use the variational approach for a simple one-dimensional model, and interpret the results.

2. Variational Approach Applied to Proteins and Magnetic Spin Systems

A. The Protein-Folding Problem: we want to predict the shape of a protein, knowing only the sequence of amino acids. We assume that the energy function is known and is a sum of two-body terms.

B. Quadratic Trial Hamiltonian: we assume a trial Hamiltonian in which every monomer is linked to every other monomer by a spring, and try to optimize the spring constants to minimize the trial free energy.

C. Heteropolymer Self-Consistent Equations: we derive a set of self-consistent equations which, when solved, predict the position and correlated fluctuations of all the monomers, given the temperature and two-body potentials.

D. The Ising Spin Glass Hamiltonian: this is a model of Ising spins on the lattice in which the interactions between neighboring spins can be either ferromagnetic or antiferromagnetic. The interactions are chosen from a probability distribution, and are quenched.

E. Mean-Field Theory: when applied to ferromagnetic spin systems, the variational approach generates a self-consistent equation (mean-field theory) which becomes exact when the number of dimensions of space approaches infinity.

F. Corrections to Mean-Field Theory: an alternative derivation of mean-field theory (an expansion in powers of the inverse temperature at fixed magnetization) provides a way to systematically approach the true free energy. The first term correcting ordinary mean-field theory is particularly important for spin glasses, and gives the Thouless-Anderson-Palmer free energy for the Sherrington-Kirkpatrick model.

3. Averaging over Disorder and the Replica Method

A. Averaging over Disorder: we are interested in computing the average free energy for an ensemble of systems (or samples) where each sample has a Hamiltonian which is chosen from a probability distribution.

B. A Very Simple Toy Model: we compute directly disorder averages for a very simple ensemble of systems, one in which each sample consists of a single quadratic well, with its center at a random position. We distinguish between correlation functions measuring thermal and disorder fluctuations.

C. The Replica Method: we explain the mathematical identities underlying the replica method, and work out some interesting results for $n \times n$ replica-symmetric matrices.

D. Check of the Replica Method: we rederive disorder averages for our very simple toy model using the replica method. The replica method involves averaging over disorder first, and leaving any additional computations for later. A quadratic effective replica Hamiltonian with a replica-symmetric Green's function has a straightforward interpretation in terms of the original ensemble of samples.

E. Random Potentials: we learn how to mathematically describe rough random potentials in terms of Gaussian probability distributions. The first two moments of the probability distribution provide complete information about it.

F. Averaging over Random Potentials with the Replica Method: we derive an effective replica Hamiltonian resulting from an average over a random potential.

4. The Variational Replica Approach and Replica Symmetry Breaking (RSB)

   A. Variational Approach to a Toy Model: we derive the replica-symmetric, trial free energy for a toy model of a particle in a rough random potential. The result for the average fluctuations is pathological; the explanation is that a replica-symmetric trial Hamiltonian describes a particle in a single well, and the toy model describes a particle in a potential with many metastable minima.

   B. One-Step RSB: the idea of one-step RSB is that the off-diagonal elements representing correlations between different replicas need not be identical. The replicas can be grouped into families, and intrafamily matrix elements will have different values than interfamily matrix elements.

   C. Mathematics of One-Step RSB: we show how to manipulate one-step RSB matrices, starting with their multiplication.

   D. Physical Interpretation of One-Step RSB: one-step RSB has a straightforward interpretation in terms of an ensemble of samples which have intrasample disorder including metastable minima.

   E. Full RSB: the infinite-step generalization of replica symmetry breaking actually has a very convenient mathematical form in terms of a function of a variable which ranges from 0 to 1. It can be interpreted in terms of each sample being constructed as a infinite hierarchy of wells within wells.

F. The Full RSB Solution of the Toy Model: we show how the full RSB solution of the toy model cures its pathologies, and describes a "freezing" phenomenon as the temperature is lowered.

5. The Variational Replica Approach to Impure Superconductors in a Magnetic Field

A. Type-II Superconductors and the Abrikosov Crystal: a type-II superconductor in a magnetic field will exhibit an intermediate phase in which the magnetic flux penetrates the sample as a triangular lattice of flux lines.

B. Perfect Elastic Crystals: the Abrikosov Crystal of flux lines can be described in terms of a perfectly quadratic classical Hamiltonian. We discuss the microscopic and continuum versions of this model, and introduce the most general Hamiltonian consistent with the triangular symmetry.

C. Random Pinning Potentials: we discuss oxygen vacancies in cuprate superconductors as an example of a quenched defect which could give rise to a rough random potential.

D. Trial Free Energy and Self-Consistent Equations: we average over disorder and introduce a quadratic trial Hamiltonian, with full RSB. The solution for the Green's function is a function of both momentum and the replica variable.

E. Physical Correlation Functions: we compute various physical correlation functions using the variational replica method. The agreement with Bitter pattern decoration experiments is good.

1. INTRODUCTION AND ELEMENTS OF STATISTICAL MECHANICS

In this chapter, I will be discussing the statistical mechanics of various disordered physical systems which have been called "glasses" because of some similarities of their properties to those of more familiar glasses. I will only cover a small portion of the subject, governed partly by my own personal idiosyncratic tastes and mainly by a desire to give a pedagogical introduction to some remarkable theoretical ideas which may at first sight appear overly intimidating. This chapter will concentrate on some general methods of calculation that have proven to be especially useful for systems with quenched disorder. The advantage of focusing on general methods is that when one understands them, one can use them on many different problems in the future. This chapter will be unashamedly technical—we will be striving for a mathematical understanding of the physical problems we consider. On the other hand, because this chapter was delivered at a summer school where many of the students were not physicists, it will also be unashamedly pedagogical and will assume no mathematical background beyond calculus and matrix algebra. I will discuss technical subjects, but I will try my best to introduce all the technical matter in as gentle and comprehensible a way as possible, assuming no previous exposure to the subject of this chapter at all. As you shall see, it will still be possible to address problems at the frontiers of current research.

The general goal in this chapter will be to learn how to compute thermodynamic properties of disordered physical systems. Physicists have long understood how to compute properties like the specific heat of a classical crystal or the magnetic susceptibility of a ferromagnet, or long-distance correlation functions in either system, as long as the crystal or ferromagnet is perfectly regular. They have taken advantage of the symmetries in these systems to invent such important theoretical concepts as phonons and spin waves. In a glass, however, the randomness is intrinsic; each atom in the system is in a different complicated environment, and it seems at first like an impossible goal to compute, say, the specific heat or correlation functions, as precisely as we are used to for ordered systems. Fortunately, since 1975, there has been substantial progress in learning how to make analytical computations for glasses; this chapter is devoted to teaching you about some of the exciting new ideas and concepts that have been invented.

We will make combined use of two major tools for these computations: a variational principle and the replica method. In the first two sections, we will introduce the variational principle without using any replicas, and then in the remaining sections explain, in order of increasing complexity, the ideas behind the replica method. Variational principles are very well established in physics, but the power of the replica method is still not as generally appreciated. For a nontechnical introduction to the history of the replica method and its manifold applications, the reader can consult P. W. Anderson's series of articles on the spin glass in *Physics Today.*[1] For a much more technical treatment, together with a collection of reprints of the more important replica articles up to 1987, see Mézard et al.[17] In this chapter, I will not cover the many interesting applications of the replica method to a variety of problems like neural network theory or optimization problems. Instead, I will concentrate on trying to explain the method itself as simply as possible.

In this first section we will review some fundamentals of statistical mechanics and solve exactly two completely trivial models. The reason that we care about these models is that as we go on, we will be studying much more complicated models which we have no hope of solving exactly. The trivial models will prove useful as building blocks for powerful methods which give us approximate results about the more complicated models. We will also introduce one of these approximation methods in this first section. Even experienced physicists should find it worthwhile to review these simple models, because I will be presenting them in a way that will ultimately make esoteric theories like the TAP equations for spin glasses and Gaussian replica field theory much more transparent. In future sections, we will be applying the approximation techniques to such complex physical systems as proteins, spin glasses, and impure superconductors in a magnetic field.

Enough generalities; let's begin studying the statistical mechanics of some physical systems. A physical system will be characterized by the different states it can be in, and various "observable" quantities which can be measured and which are a function of the state of the system. One very important observable quantity is the energy, or "Hamiltonian." Our first trivial example of a physical system is a single *Ising spin* in a magnetic field. An Ising spin, denoted by the variable S_1, can be in two states: "up," when $S_1 = 1$, or "down," when $S_1 = -1$. (The subscript 1 in "S_1" is just there to indicate that it is our first spin. If we had two spins, we would label them S_1 and S_2.) For a system consisting of a single Ising spin in a magnetic field, the Hamiltonian is

$$H = -h_1 S_1 \tag{1}$$

where h_1 is a magnetic field which tends to align the spin to point "up" if the field is positive, and "down" if the field is negative. The fundamental principle of statistical mechanics is that the probability that a system in each possible state is proportional to the Boltzman weight $e^{-H(\text{state})/T}$ of that state, where T is the temperature. The central object of study in statistical mechanics, from which we can compute all thermodynamic quantities of interest, is the partition function Z, which is the sum of the Boltzman weights of all the states of the system:

$$Z \equiv \sum_{\text{states}} e^{-H(\text{state})/T} \,. \tag{2}$$

For the model of a single Ising spin with Hamiltonian given by Eq. (1), we easily find that

$$Z = e^{-h_1/T} + e^{h_1/T} \,. \tag{3}$$

As I mentioned, when one knows the partition function exactly, as we do for this trivial model, one can then calculate all the thermodynamics exactly. Thus, the *free energy* F is defined by

$$F = -T \ln Z \,, \tag{4}$$

the *entropy* S is given by

$$S = -\frac{\partial F}{\partial T} \,, \tag{5}$$

the *internal energy* U is given by

$$U = \frac{T^2}{Z}\frac{\partial Z}{\partial T} = F + TS \,, \tag{6}$$

and the *specific heat* C is given by

$$C = \frac{\partial U}{\partial T} = -T\frac{\partial^2 F}{\partial T^2} \,. \tag{7}$$

Notice that if we add an overall constant to the Hamiltonian, it is also added to the free energy and the internal energy, but does not affect the entropy or the specific heat.

The probability that the system is in any of its states is just equal to the Boltzman weight of that state divided by the sum of the Boltzman weights of all the states:

$$p_{\text{state}} = \frac{1}{Z} e^{-H(\text{state})/T} . \tag{8}$$

We can define the *thermal expectation value* of any state-dependent quantity as the average (weighted by the state's probability) of that quantity. For example, you can check that the internal energy is actually just the thermal expectation value of the Hamiltonian itself, while the entropy is the thermal expectation value of the negative of the logarithm of the probability of the state, and is thus a measure of how "spread out" the system is between its possible states:

$$U = \sum_{\text{states}} p_{\text{state}} H(\text{state}) = \langle H \rangle \tag{9}$$

$$S = \sum_{\text{states}} -p_{\text{state}} \ln(p_{\text{state}}) = \langle -\ln(p_{\text{state}}) \rangle \tag{10}$$

where the angular brackets are a convenient short-hand notation for the thermal expectation value.

There are other thermal expectation values that we might be interested in. For example, the *magnetization* m_1 of the spin is just the average value of the spin, which for this model we can easily relate to the magnetic field and temperature:

$$m_1 \equiv \langle S_1 \rangle = \tanh(h_1/T) . \tag{11}$$

Finally, the *susceptibility* χ_1 is defined as the response of the magnetization to a change in the magnetic field:

$$\chi_1 = \frac{\partial m_1}{\partial h_1} . \tag{12}$$

So far, it has been natural to think of the free energy and all the other thermodynamic quantities as functions of the magnetic field h_1. But Eq. (11) gives a simple relation between the field and the magnetization, so if we prefer, it is a simple matter to replace the magnetic field by the magnetization and define a magnetization-dependent free energy. We will see later that this is often a convenient thing to do.

Now let us introduce another trivial model. In this model, we consider a particle which moves in one dimension and which can be located at any position r_1 from negative to positive infinity. (The subscript "1" is again a label indicating that this is the first particle.) The Hamiltonian will be a function of the position r_1. The partition function is

$$Z = \int_{-\infty}^{\infty} dr_1 \; e^{-H(r_1)/T} . \tag{13}$$

To make this model exactly soluble, we restrict ourselves to a Hamiltonian function $H(r_1)$ which is quadratic in r_1:

$$H(r_1) = \frac{1}{2G}(r_1 - a_1)^2 . \tag{14}$$

Using the well-known formula for Gaussian integrals (in the appendix, I give a table of Gaussian integrals which will be useful in this chapter), we find that

$$Z = \sqrt{2\pi TG} . \tag{15}$$

Using formulae (4)–(7), we can again compute the free energy, entropy, internal energy, and specific heat. We can also compute (again using the appendix) a couple of thermal expectation values which are particularly relevant for this model. Namely, the *average position* is given by

$$\langle r_1 \rangle = \frac{1}{Z} \int_{-\infty}^{\infty} dr_1 r_1 e^{-(r_1-a_1)^2/2GT} = a_1 \tag{16}$$

while the *average fluctuation* in the position is given by

$$\langle (r_1 - a_1)^2 \rangle = \frac{1}{Z} \int_{-\infty}^{\infty} dr_1 (r_1 - a_1)^2 e^{-(r_1-a_1)^2/2GT} = TG . \tag{17}$$

Physically, one can imagine that the particle described by this model is attached to a spring which is nailed to the position a_1. The amount that it bounces around its average position a_1 is determined by the combination of the temperature T and the softness of the spring G.

In statistical mechanics, we are usually most interested in systems which have a very large number of degrees of freedom, rather than just one as in the examples given. So how can we generalize these models so that they are still exactly soluble but concern a large number of spins or particles? For the Ising spin system, there is not much that we can do beyond considering the system of many *noninteracting* spins, each under the influence of its own private magnetic field, for which the Hamiltonian would be:

$$H = -\sum_{i=1}^{N} h_i S_i \tag{18}$$

where N is the total number of spins. The free energy for this model is just the sum of the free energies for all the individual spins.

We can generalize the other model in a somewhat more interesting way. A Gaussian integral over many *interacting* degrees of freedom will still be soluble as long as all the interactions are quadratic. Thus, we can consider the generalized Hamiltonian for N particles:

$$H = \frac{1}{2}\sum_{i=1}^{N}\sum_{j=1}^{N}(r_i - a_i)\left(G^{-1}\right)_{ij}(r_j - a_j) \tag{19}$$

where $(G^{-1})_{ij}$ is a symmetric matrix of the spring constants connecting all the particles together. We have written this matrix as an inverse matrix to agree with the common convention. This generalized model represents N particles, each attached with a spring to a nail at its own position, but now also connected by springs to every other particle. We have left the values of all the spring constants as general parameters. The wonderful thing about Gaussian integrals is that we can still compute our partition function; we find, using a formula from the appendix, that

$$Z = \sqrt{(2\pi T)^N \det G} \tag{20}$$

where $\det G$ is the determinant of the G matrix. Another extremely useful result is that the expectation value of the correlated fluctuations of two particles around their average positions is simply related to the G matrix:

$$\langle (r_i - a_i)(r_j - a_j)\rangle = TG_{ij}\,. \tag{21}$$

The G matrix is often referred to as "Green's function."

Now you may be thinking that these exactly soluble models are great, but most interacting systems we know of do not have these Hamiltonians, so what use are they? We shall see that these models can actually be used as inputs for a couple of different approximation schemes.

The first of these approximation schemes is based on a general mathematical inequality. Assume that we have some arbitrary physical system which can be in, say, K different states. The probability of each state is some number p_α ($\alpha = 1, 2, ..., K$) where

$$\sum_{\alpha=1}^{K} p_\alpha = 1\,. \tag{22}$$

Let us also imagine that there is some observable quantity X (like the energy) which depends on which state the system is in. We label the different possible values of the quantity by X_α. By our previous notation, the thermal average of X is

$$\langle X\rangle \equiv \sum_{\alpha=1}^{K} p_\alpha X_\alpha\,. \tag{23}$$

The mathematical inequality that I will assert (without proof) is that

$$\left\langle e^{-X}\right\rangle \geq e^{-\langle X\rangle}\,, \tag{24}$$

or written out more explicitly

$$\sum_{\alpha=1}^{K} p_\alpha e^{-X_\alpha} \geq e^{-\sum_{\alpha=1}^{K} p_\alpha X_\alpha} . \tag{25}$$

You can check for yourself that this inequality follows from the convexity of the exponential function.

Let us now return to our model of a single particle whose energy depends upon its position. We want to compute the partition function

$$Z = \int_{-\infty}^{\infty} dr_1 e^{-H(r_1)/T} . \tag{26}$$

Before, we had an especially convenient quadratic function for $H(r_1)$, but imagine that we now have some more complicated function which makes the integral difficult or impossible to compute analytically. We can use our inequality to obtain an approximate solution as follows. The partition function is obviously equal to

$$Z = \int_{-\infty}^{\infty} dr_1 e^{-H(r_1)/T} \frac{\int_{-\infty}^{\infty} dr_1 e^{-H_0(r_1)/T}}{\int_{-\infty}^{\infty} dr_1 e^{-H_0(r_1)/T}} \tag{27}$$

where $H_0(r_1)$ is *any* function at all. We can rewrite this as

$$Z = \frac{\int_{-\infty}^{\infty} dr_1 e^{-[H(r_1)-H_0(r_1)]/T} e^{-H_0(r_1)/T}}{\int_{-\infty}^{\infty} dr_1 e^{-H_0(r_1)/T}} \int_{-\infty}^{\infty} dr_1 e^{-H_0(r_1)/T} \tag{28}$$

or

$$Z = \left\langle e^{-(H-H_0)/T} \right\rangle_0 \int_{-\infty}^{\infty} dr_1 e^{-H_0(r_1)/T} \tag{29}$$

where the notation $\langle X \rangle_0$ means the thermal average of X using the function $H_0(r_1)$ as a so-called *trial Hamiltonian.* We can now use our inequality to assert that

$$Z \geq e^{-\langle (H-H_0)/T \rangle_0} \int_{-\infty}^{\infty} dr_1 e^{-H_0(r_1)/T} \tag{30}$$

for *any* function $H_0(r_1)$. In terms of the free energy $F \equiv -T \ln Z$, we can equivalently assert that

$$F \leq -T \ln \int_{-\infty}^{\infty} dr_1 e^{-H_0(r_1)} + \langle H - H_0 \rangle_0 \equiv \tilde{F} \tag{31}$$

where we define the quantity on the right-hand side of the inequality as the *trial free energy* $\tilde{F}$ corresponding to the trial Hamiltonian H_0. This is our fundamental variational principle. (R. P. Feynman was one of the first physicists to make use of this principle in his famous treatment of the polaron problem.[10] Another interesting application of the variational principle is to the excluded volume problem in polymer physics.[3,7]) It says that the true free energy will always be less than the trial free energy no matter what trial Hamiltonian we choose. Thus, if from some class of trial functions $H_0(r_1)$, we find one that gives a minimal trial free energy, we know that that is our best estimate of the free energy. Note that if H_0 is equal to H, the trial free energy is automatically equal to the true free energy.

To be able to use this variational principle in practice, of course, we must restrict ourselves to a class of trial functions H_0 for which we can analytically compute $\tilde{F}$. The best trial Hamiltonian will be the one in this class which is closest to the real Hamiltonian. Finding a class of analytically tractable trial Hamiltonians is precisely where our previously analysed exactly soluble models become useful. Let us look at a relatively simple example. Consider the Hamiltonian function (we drop the subscript "1" on r_1 for brevity)

$$H(r) = Cr^2 + r^4 \,. \tag{32}$$

If $C \geq 0$, then the function has the form of a "single well" potential, while if $C < 0$, $H(r)$ has the form of a "double well" potential with a barrier of height $C^2/4$ separating the two valleys. In either case, the free energy $F \equiv -T \ln \int_{-\infty}^{\infty} dr e^{-H(r)}$ is some perfectly well-defined function of T and C, but this function is rather difficult to compute analytically. We will compute the function approximately using our variational principle with a class of trial Hamiltonian functions of the now familiar form

$$H_0(r) = \frac{1}{2G}(r-a)^2 \tag{33}$$

where G and a are now arbitrary variational parameters that we will vary in order to minimize the trial free energy.

The trial free energy is

$$\tilde{F} = -T \ln \int_{-\infty}^{\infty} dr e^{-H_0(r)/T} + \langle H - H_0 \rangle_0 \tag{34}$$

$$= -T \ln \int_{-\infty}^{\infty} dr e^{-(r-a)^2/2GT} + C \left\langle r^2 \right\rangle_0 + \left\langle r^4 \right\rangle_0 - \frac{1}{2G} \left\langle (r-a)^2 \right\rangle_0 \tag{35}$$

$$\begin{aligned} = & -T \ln \int_{-\infty}^{\infty} dr e^{-(r-a)^2/2GT} + C \int_{-\infty}^{\infty} dr r^2 e^{-(r-a)^2/2GT} \\ & + \int_{-\infty}^{\infty} dr r^4 e^{-(r-a)^2/2GT} - \frac{1}{2G} \int_{-\infty}^{\infty} dr (r-a)^2 e^{-(r-a)^2/2GT} \,. \end{aligned} \tag{36}$$

Now comes the key step, where we take advantage of the fact that our trial Hamiltonian has such a convenient form. Because it is quadratic, all these integrals are simple to compute. Using the integrals in the appendix, we find that

$$\tilde{F} = -\frac{T}{2} \ln(2\pi TG) + C(a^2 + TG) + (a^4 + 6a^2TG + 3T^2G^2) - \frac{T}{2} \,. \tag{37}$$

It is interesting to note that adding a constant to the trial Hamiltonian would not change the trial free energy. The constant would have been added in the

$$-T \ln \int_{-\infty}^{\infty} dr e^{-H_0(r)/T}$$

piece of the trial free energy but subtracted in the $\langle -H_0 \rangle_0$ piece.

Minimizing the trial free energy with respect to a, we find that $a = 0$ or

$$a^2 = -\left(\frac{C}{2} + 3TG\right) \tag{38}$$

while minimizing with respect to G gives

$$-\frac{1}{2G} + C + 6a^2 + 6TG = 0 \,. \tag{39}$$

Let us examine the solution in more detail. When $C \geq 0$ (the single-well case), we find that the only solution is $a = 0$ and

$$G = \frac{1}{12T}\left[\sqrt{C^2 + 12T} - C\right] . \tag{40}$$

Returning to the trial free energy, which is our estimate for the true free energy, we finally find that

$$\begin{aligned}\tilde{F} = &- T\left(\frac{1}{4} + \frac{1}{2}\ln\left(\frac{\pi}{6}\right)\right) - \frac{T}{2}\ln\left(\sqrt{C^2 + 12T} - C\right) \\ &+ \frac{C}{24}\left(\sqrt{C^2 + 12T} - C\right) .\end{aligned} \tag{41}$$

On the other hand, in the double-well case when $C < 0$, there can be two possible solutions, depending on the temperature. For temperatures higher than some critical temperature T_c, we have a functionally identical solution to that for $C \geq 0$, with $a = 0$, G given by Eq. (40), and $\tilde{F}$ given by Eq. (41). For temperatures below T_c, there is another solution with a lower trial free energy. The variational parameters corresponding to this other solution are

$$a^2 = -\frac{1}{8}\left(5C + \sqrt{C^2 + 12T}\right) \tag{42}$$

and

$$G = \frac{1}{24T}\left[\sqrt{C^2 + 12T} + C\right] , \tag{43}$$

and the trial free energy is

$$\tilde{F} = -T\left(\frac{5}{8} + \frac{1}{2}\ln\left(\frac{\pi}{12}\right)\right) - \frac{T}{2}\ln\left(\sqrt{C^2 + 12T} + C\right) - \frac{5C}{48}\left(\sqrt{C^2 + 12T} + C\right) - \frac{C^2}{4}. \tag{44}$$

The critical temperature itself is simply determined by setting the high-temperature trial free energy of Eq. (41) equal to the low-temperature trial free energy of Eq. (44).

What is the physical meaning of this transition between two solutions? First, it should be made clear that the real free energy does *not* have the "cusp" that our trial free energy exhibits, so that is an artifact of our approximation. Nevertheless, it is not a stupid artifact, because there really are two temperature regimes for the true free energy—it is just that the crossover between them is a gradual one. The Hamiltonian represents a ball in a double-well potential, and that ball is jiggled around by random hits from some background "stuff." When the temperature is high, the jiggling is strong, while when the temperature is low, the ball will just sit at the bottom of the potential, as the jiggling will be weak. Clearly, there will be a high-temperature regime, when the particle bounces back and forth between the two valleys easily, because there is enough thermal energy to get over the barrier. There will also be low-temperature regime, when the particle tends to spend a very long time in one hill before it bounces over the barrier. These two regimes are represented in our solutions. In our low-temperature solution, the ball has some average position at a, with a fluctuation G which represents how much the ball jiggles around that. In the high-temperature solution, $a = 0$ which means the ball is bouncing back and forth between the two wells, and G is much larger than in the low-temperature solution, which again corresponds to the larger fluctuations.

In general, one should be aware that in any variational method, the results one gets for the quantity one is minimizing over (in our case, the free energy) can be very accurate, but the results for other quantities which one deduces from the minimization (like in our case, the size of the fluctuations) will not be as accurate. To understand this, let us imagine that we are minimizing the quantity F over some multidimensional space which we represent by the vector $\vec{x}$. The minimum of $F(\vec{x})$ will always be quadratic in $\vec{x}$. That means that if we miss the optimal $\vec{x}$ by some small amount $\delta\vec{x}$ because we have restricted ourselves to a certain portion of $\vec{x}$-space where we can compute $F(\vec{x})$ analytically, our estimate for the value of F will only be wrong by $(\delta\vec{x})^2$. This does not mean that one should ignore the variational solution except for the upper bound it gives for the free energy (unless you are a rigorous mathematical physicist, in which case it means precisely that), as it is still true that the best $\vec{x}$ found will be closest in the subspace chosen to the optimal $\vec{x}$. It just means that one should beware that a poorly chosen subspace (or class of trial functions) can produce misleading results.

2. VARIATIONAL APPROACH APPLIED TO PROTEINS AND MAGNETIC SPIN SYSTEMS

In this section, we will begin by applying the variational method we learned about in the last section to the important problem of protein folding. The work that I will describe was done in collaboration with Jean-Philippe Bouchaud and Marc Mézard[4] at the Ecole Normale Supérieure in Paris. I will be discussing a highly simplified model of "proteins" in this section, but the ideas presented here could be generalized to a more realistic model.

A protein is a polymer which can be specified by the sequence of amino acids which make up its monomeric units. This sequence is stored biologically in the DNA segment which is ultimately translated into the protein. The sequence of amino acids determines the three-dimensional shape of the protein, and that shape in turn determines how well the protein performs its biological function. The "protein folding problem" is the problem of predicting the three-dimensional shape of a protein given only the sequence of amino acids. It is attracting considerable interest because present technology makes it much easier to sequence proteins than to determine their shape.

The shape of a protein can be specified at varying levels of precision. One could, for example, specify bond angles between neighboring amino acids, or one could specify the position of every atom in the protein. Let us, for the purposes of simplicity, consider a generic model of an N-monomer linear heteropolymer in which the position of the ith monomer in the chain is given by the D-dimensional vector $\vec{r}_i$. (In ordinary space, $D = 3$ of course, but there is no particular difficulty caused by keeping the dimension of space arbitrary.) We will make the huge assumption that the Hamiltonian for our heteropolymer is known and can be reduced to a sum of two-body monomer-monomer potentials; i.e.,

$$H = \sum_{1 \le i < j \le N} V_{ij}(|\vec{r}_i - \vec{r}_j|) \,. \tag{45}$$

The effect of the solvent is taken into account in this Hamiltonian only insofar as the two-monomer potentials are affected by it.

Now suppose that at the temperature we are interested in (physiological temperatures for a protein) the heteropolymer has a shape that is well defined. That is, up to global rotations and translations, each monomer has an average position and some typical fluctuation around that position. This is certainly the case for globular proteins at physiological temperatures, although it is interesting that the physiological temperature is usually not much less than the temperature at which proteins undergo a transition to a fluctuating "coil state" with no definite shape, which suggests that thermal fluctuations are rather significant. If the heteropolymer does have a definite shape, it should not be too bad an approximation to consider a quadratic trial Hamiltonian which assumes that each monomer has some average position with Gaussian fluctuations around that position. Thus we reintroduce the trial Hamiltonian that we first mentioned in the last section:

$$H_0 = \frac{1}{2}\sum_{i=1}^{N}\sum_{j=1}^{N}(G^{-1})_{ij}(\vec{r}_i - \vec{a}_i).(\vec{r}_j - \vec{a}_j) \tag{46}$$

where $\vec{a}_i$ and G_{ij} are two sets of variational parameters which have straightforward interpretations: from our discussion of this Hamiltonian in the first section, we know that $\vec{a}_i$ is the average position of the ith monomer, while G_{ij} is proportional to the correlated fluctuation of monomer i and monomer j. As usual, we will eventually choose these variational parameters to minimize the trial free energy. Because we have so many variational parameters to vary, and because our trial Hamiltonian describes a system which is physically close to the true state of our system, the results we derive from our variational approach should be rather reliable. In fact, one can introduce an even more general and realistic trial Hamiltonian with a tensorial structure for G:

$$H_0 = \frac{1}{2}\sum_{i=1}^{N}\sum_{j=1}^{N}\sum_{\alpha=1}^{D}\sum_{\beta=1}^{D}(G^{-1})_{ij}^{\alpha\beta}(r_i^\alpha - a_i^\alpha)(r_j^\beta - a_j^\beta) \tag{47}$$

where α and β are spatial indices. Such a trial Hamiltonian allows the protein to have anisotropic fluctuations, which is obviously desirable. For the sake of simplicity, we shall keep here to the form of Eq. (46) which is isotropic in space.

We now compute the trial free energy. The following derivation may seem complicated, but keep in mind that it is actually a straightforward generalization of the example from the last section. We have

$$\tilde{F} \equiv F_0 + \langle H - H_0\rangle_0 \tag{48}$$

where

$$F_0 = -T\ln Z_0 \equiv -T\ln\int_{-\infty}^{\infty} d\vec{r}_1 d\vec{r}_2,\ldots,d\vec{r}_N e^{-H_0/T} \tag{49}$$

and $\langle X\rangle_0$ denotes the expectation value of the observable X with respect to the Boltzman measure $\frac{1}{Z_0}\exp(-H_0/T)$:

$$\langle X\rangle_0 \equiv \frac{1}{Z_0}\int_{-\infty}^{\infty} d\vec{r}_1 d\vec{r}_2,\ldots,d\vec{r}_N e^{-H_0/T} X\,. \tag{50}$$

Some of these integrals can actually be done very easily. Using the appendix, we find that

$$F_0 = -T\ln\left([(2\pi T)^N \det G]^{D/2}\right)\,. \tag{51}$$

Using the matrix identity $\ln\det G = \mathrm{Tr}\ln G$, we find

$$F_0 = -\frac{NDT}{2}\ln(2\pi T) - \frac{DT}{2}\mathrm{Tr}\ln G\,. \tag{52}$$

For $\langle -H_0\rangle_0$, we have

$$\langle -H_0\rangle_0 = -\frac{1}{2}\sum_{ij}(G^{-1})_{ij}\,\langle(\vec{r}_i - \vec{a}_i).(\vec{r}_j - \vec{a}_j)\rangle_0 \tag{53}$$

$$= -\frac{DT}{2}\sum_{ij}(G^{-1})_{ij}G_{ij} = -\frac{NDT}{2}\,. \tag{54}$$

Finally, the most interesting and difficult integral to compute is

$$\langle H\rangle_0 = \sum_{1\le i<j\le N}\langle V_{ij}(|r_i - r_j|)\rangle_0\,. \tag{55}$$

To compute it, it helps to first learn a general fact about Gaussian integrals. We clearly need to be able to compute quantities like $\langle f(\vec{r}_i - \vec{r}_j)\rangle_0$, where f is some arbitrary function. $\vec{z} \equiv \vec{r}_i - \vec{r}_j$ is a *Gaussian variable*, which technically means that you can determine all of its higher moments from the first two. Its first two moments are

$$\langle \vec{z}\rangle_0 = \vec{a}_i - \vec{a}_j \tag{56}$$

and

$$\left\langle(\vec{z} - \langle\vec{z}\rangle_0)^2\right\rangle_0 = \left\langle[(\vec{r}_i - \vec{a}_i) - (\vec{r}_j - \vec{a}_j)]^2\right\rangle_0$$
$$= DT(G_{ii} + G_{jj} - 2G_{ij})\,. \tag{57}$$

The important general fact that we need to know is that when one computes the average of some function of a Gaussian variable over all the original N variables in the problem, one can replace the measure over all the original variables with a new measure over only the Gaussian variable, as long as the new measure gives the same first two moments. Perhaps it is clearer if I just write it in equations; if $\vec{z}$ is a D-dimensional Gaussian variable with first two moments $\vec{M}_1$ and DM_2, then

$$\langle f(\vec{z})\rangle_0 = \int_{-\infty}^{\infty}\frac{d\vec{z}}{[2\pi M_2]^{D/2}}\exp\left[-\frac{(\vec{z} - \vec{M}_1)^2}{2M_2}\right]f(\vec{z})\,. \tag{58}$$

Similarly, although we will not need it here, an average of a function of two Gaussian variables can be reduced to an integral over just those two variables if one uses a measure which gives the same first two moments. By the way, we know $\vec{z}$ is a Gaussian variable because linear combinations of Gaussian variables are themselves

Gaussian and the original variables of a quadratic Hamiltonian are always Gaussian variables.

Given the general fact described above, the computation of $\langle H\rangle_0$ follows immediately. We find

$$\langle H\rangle_0 = \sum_{1\le i<j\le N} \int_{-\infty}^{\infty} \frac{d\vec{z}}{[2\pi T(G_{ii}+G_{jj}-2G_{ij})]^{D/2}} \exp\left[-\frac{(\vec{z}-(\vec{a}_i-\vec{a}_j))^2}{2T(G_{ii}+G_{jj}-2G_{ij})}\right] V_{ij}(|\vec{z}|)\,. \tag{59}$$

To summarize our computations, we have found that

$$\tilde{F} = -\frac{NDT}{2}(\ln(2\pi T)+1) - \frac{DT}{2}\mathrm{Tr}\ln G \tag{60}$$
$$+ \sum_{1\le i<j\le N} \int_{-\infty}^{\infty} \frac{d\vec{z}}{[2\pi T(G_{ii}+G_{jj}-2G_{ij})]^{D/2}} \exp\left[-\frac{(\vec{z}-(\vec{a}_i-\vec{a}_j))^2}{2T(G_{ii}+G_{jj}-2G_{ij})}\right] V_{ij}(|\vec{z}|)\,.$$

Minimizing the trial free energy with respect to all the $\vec{a}_i$, we find that for all $i=1,2,\ldots,N$

$$0 = \sum_{j=1,N(j\neq i)} \int_{-\infty}^{\infty} \frac{d\vec{z}}{(2\pi)^{D/2}[T(G_{ii}+G_{jj}-2G_{ij})]^{(D/2+1)}} [\vec{z}-(\vec{a}_i-\vec{a}_j)] \exp\left[-\frac{(\vec{z}-(\vec{a}_i-\vec{a}_j))^2}{2T(G_{ii}+G_{jj}-2G_{ij})}\right] V_{ij}(|\vec{z}|) \tag{61}$$

while minimizing the trial free energy over all the G_{ij} tell us that for all $i\neq j$,

$$(G^{-1})_{ij} = 2\int_{-\infty}^{\infty} \frac{d\vec{z}}{(2\pi)^{D/2}[T(G_{ii}+G_{jj}-2G_{ij})]^{(D/2+1)}} V_{ij}(|\vec{z}|) \exp\left[-\frac{(\vec{z}-(\vec{a}_i-\vec{a}_j))^2}{2T(G_{ii}+G_{jj}-2G_{ij})}\right]\left[1-\frac{(\vec{z}-(\vec{a}_i-\vec{a}_j))^2}{DT(G_{ii}+G_{jj}-2G_{ij})}\right] \tag{62}$$

and for $i=j$,

$$(G^{-1})_{ii} = -\sum_{j\neq i}(G^{-1})_{ij}\,. \tag{63}$$

These equations are obviously complicated, and the only way to solve them for some arbitrary set of two-body potentials V_{ij} would be numerically on the computer. (They simplify considerably and can be dealt with analytically in the case when the potentials V_{ij} are identical, corresponding to an ordinary homopolymer.[3,7]) At a given temperature, there may well be more than one solution to these equations, which would correspond to the different possible metastable states of the heteropolymer. The advantage of this approach is that it naturally acounts for thermal fluctuations while giving a great deal of useful information: given the input of the temperature and the two-body potentials, one gets as output the positions and correlated fluctuations of all the monomers. It would certainly be interesting if someone went ahead and used these equations for a relatively small system, and then compared the results to a more conventional, but time-consuming, Monte Carlo simulation.

The protein is our first example of a system with "quenched disorder." The disorder in this case simply comes from the two-body potentials, which depend in some complicated way on the precise amino acids that the chain is made out of. The disorder is "quenched" in the sense that the two-body potentials are fixed once and for all for any given protein. Of course, different proteins will be made out of different amino acids, and therefore will have different two-body potentials and ultimately different shapes, but for a given protein, the potentials are quenched. The concept of "quenched disorder" is easy to understand, but it is important to continue to learn how to deal with it on a technical level, as it is an intrinsic aspect of many physical systems. For our next example, we will first apply our variational approach, and then introduce a new and potentially even more powerful technique.

The next system that we will consider for which quenched disorder is important is the *Ising spin glass.* Imagine that we have some lattice (for concreteness we will restrict ourselves to D-dimensional hypercubic lattices like the linear, square, or cubic lattice) of N points, and on each point of the lattice, we put an Ising spin which can point up or down. In the first section, we only considered Ising spin systems in which each spin was independent of every other spin, but let us now consider what happens if each spin influences its nearest neighbors. In a *ferromagnet* each spin will tend to make its nearest neighbors point in the same direction that it is pointing. A commonly used Hamiltonian for the ferromagnet is

$$H = -J \sum_{(ij)} S_i S_j \tag{64}$$

where $J > 0$ and the (ij) notation means that the sum is over nearest neighbors in the lattice. This Hamiltonian clearly favors configurations in which all the spins point in the same direction. In an *antiferromagnet,* each spin will tend to make its nearest neighbor point in the direction opposite to its own. The same Hamiltonian as the one used for the ferromagnet will also describe an antiferromagnet if $J < 0$. A *spin glass* is a system in which the interaction between any pair of nearest neighbors is fixed and randomly chosen to be either ferromagnetic or antiferromagnetic. The canonical Hamiltonian for a spin glass is

$$H = -\sum_{(ij)} J_{ij} S_i S_j \, . \tag{65}$$

In practice, if one wants to make a computer simulation of a spin glass, for example, one chooses each J_{ij} to be equal to $+1$ or -1 with equal probability, or chooses the J_{ij}'s from some other probability distribution. One very popular probability distribution is the Gaussian one, for which

$$p(J_{ij}) = \frac{e^{-(J_{ij}-J_0)^2/(2\tilde{J}^2)}}{\sqrt{(2\pi\tilde{J}^2)}} \tag{66}$$

where J_0 is the average value of a bond and $\tilde{J}$ is the standard deviation in the bond strengths. Once the J_{ij}'s are chosen form whichever probability distribution being used, they are quenched and cannot be changed. They play a role analogous to that of the V_{ij}'s in the protein Hamiltonian, while the S_i's play a role analogous to the r_i's as the degrees of freedom in the problem. The ferromagnetic and antiferromagnetic Hamiltonians are obviously just special cases of the more general spin glass Hamiltonian, although in practice, a system is only called a "spin glass" if the J_{ij}'s are chosen at random and from a probability distribution which contains both positive and negative J's. (In 1975, Sherrington and Kirkpatrick introduced their famous model of a spin glass with the above Hamiltonian on the very special lattice in which each spin is a nearest neighbor of every other spin.[24] A spin glass on such a lattice turns out to have thermodynamic properties that are identical to those of a spin glass on a $D = \infty$-dimensional hypercubic lattice.)

Faced with this Hamiltonian, I would hope that you would first consider trying the variational approach on it. Let's see how that would work. We are interested in computing

$$F = -T \ln \mathrm{Tr} \exp(-H/T) \tag{67}$$

where Tr is a shorthand notation for a sum over all possible states of the system:

$$\mathrm{Tr} \equiv \sum_{S_1=\pm1} \sum_{S_2=\pm1} \cdots \sum_{S_N=\pm1} . \tag{68}$$

We know that we can make exact computations with a trial Hamiltonian consisting of noninteracting spins, so we take

$$H_0 = -\sum_{i=1}^{N} h_i S_i \tag{69}$$

where the h_i variables are now variational parameters that we will try to optimize.

The trial free energy is as usual

$$\tilde{F} \equiv F_0 + \langle H - H_0 \rangle_0 \tag{70}$$

where

$$F_0 = -T \ln Z_0 \equiv -T \ln \mathrm{Tr} \exp(-H_0/T) \tag{71}$$

and the expectation value $\langle X \rangle_0$ is taken with respect to the trial Hamiltonian H_0:

$$\langle X \rangle_0 \equiv \frac{1}{Z_0} \mathrm{Tr} X \exp(-H_0/T) \,. \tag{72}$$

We did the trivial computation of F_0 last section:

$$F_0 = -T \sum_i \ln \cosh(h_i/T) \,. \tag{73}$$

The other pieces of $\tilde{F}$ are almost as simple to compute:

$$\langle -H_0 \rangle_0 = \sum_i h_i \langle S_i \rangle_0 = \sum_i h_i \tanh(h_i/T); \tag{74}$$

$$\begin{aligned} \langle H \rangle_0 &= -\sum_{(ij)} J_{ij} \langle S_i S_j \rangle_0 \\ &= -\sum_{(ij)} J_{ij} \langle S_i \rangle_0 \langle S_j \rangle_0 = -\sum_{(ij)} J_{ij} \tanh(h_i/T) \tanh(h_j/T) \,. \end{aligned} \tag{75}$$

We have used the fact that the spins are independent in the trial Hamiltonian to factorize the correlation function $\langle S_i S_j \rangle_0$.

Putting it all together, we find

$$\tilde{F} = -T \sum_i \ln \cosh(h_i/T) - \sum_{(ij)} J_{ij} \tanh(h_i/T) \tanh(h_j/T) + \sum_i h_i \tanh(h_i/T) \,. \tag{76}$$

Minimizing the trial free energy with respect to h_i, we find

$$\begin{aligned} &-\tanh(h_i/T) - \frac{1}{T} \sum_{j(i)} J_{ij} \tanh(h_j/T)(1 - \tanh^2(h_i/T)) \\ &\qquad + \tanh(h_i/T) + \frac{1}{T} h_i (1 - \tanh^2(h_i/T)) = 0 \end{aligned} \tag{77}$$

or

$$h_i = \sum_{j(i)} J_{ij} \tanh(h_j/T) \,. \tag{78}$$

where the notation $j(i)$ means all spins j neighboring spin i.

Notice that these self-consistent equations are considerably simpler than the corresponding equations we derived for the protein problem. We can rewrite them in a different and slightly more conventional way by changing variables to the local magnetizations $m_i \equiv \langle S_i \rangle_0 = \tanh(h_i/T)$. In terms of the magnetizations, we have the trial free energy

$$\tilde{F} = -T \sum_i \left[\frac{1+m_i}{2} \ln \left(\frac{1+m_i}{2} \right) + \frac{1-m_i}{2} \ln \left(\frac{1-m_i}{2} \right) \right] + \sum_{(ij)} J_{ij} m_i m_j \,. \tag{79}$$

(The first term on the right is the entropy, while the second term is the internal energy of the system in this approximation.) In these variables, the self-consistent equations are

$$m_i = \tanh\left(\sum_{j(i)} \frac{J_{ij}m_j}{T}\right). \tag{80}$$

Let us now consider the special case of the ferromagnet where all J_{ij} are equal. We will choose the particular scaling $J_{ij} = 1/(2D)$, where D is the dimension of our hypercubic lattice, in order that the ground state energy density (when all spins point in the same direction) will be $E/N = -1$ irrespective of the dimension. Since all J_{ij}'s are equal, we expect the magnetization at each site to be equal, since there is nothing to distinguish one site from another. Setting $m_i = m$, we find the famous *mean field* equation for the magnetization of an Ising ferromagnet:

$$m = \tanh(m/T). \tag{81}$$

According to this "variational," or "mean field," approximation, the ferromagnet will have a transition at $T = 1$. For $T > 1$, $m = 0$, but for $T < 1$, the magnetization is nonzero, with a magnitude approaching 1 at $T = 0$. We shall eventually see that the mean-field approximation becomes exact for the ferromagnet when D approaches infinity.

To demonstrate this fact, we will need to develop a new and different technique for calculating the free energy. This technique is conceptually very simple—it is based on the idea of expanding the magnetization-dependent free energy in powers of the inverse temperature. We shall see that the form for the trial free energy given in Eq. (79) actually corresponds to just the first two terms in such an expansion. Thus, by computing the higher order terms in this expansion, we can systematically approach the true free energy. My collaborator on the work that I am about to describe was Antoine Georges at the Ecole Normale Supérieure in Paris.

The starting point of this technique is the *magnetization-dependent free energy.* As I have defined it so far, the free energy is just the logarithm of the partition function, and the magnetization will have some equilibrium value at any temperature. The magnetization has not been a free variable—we have been given the temperature and we have computed the magnetization. One can turn the magnetization into a free variable and define a magnetization-dependent free energy by adding to the physical system a set of external auxiliary fields which are used to insure that the magnetizations are at their desired values. Of course, when the magnetizations are at their *equilibrium* values, no auxiliary fields will be necessary. Let's see how this works in equations, using our spin glass Hamiltonian. The magnetization-dependent free energy is

$$-\beta F(\beta, m_i) = \ln \operatorname{Tr} \exp\left(\beta \sum_{(ij)} J_{ij}S_iS_j + \sum_i \lambda_i(\beta)(S_i - m_i)\right) \tag{82}$$

where $\beta \equiv 1/T$ is the inverse temperature. The $\lambda(\beta)$ are our auxiliary fields (or Lagrange multipliers). Note that they depend explicitly on the inverse temperature, which is just a reflection of the important (and obvious) fact that the fields necessary to fix a certain set of magnetizations will change as the temperature changes. As usual, the magnetizations m_i are defined as $\langle S_i \rangle$, where the expectation value is taken with respect to an effective Hamiltonian which is the sum of the original Hamiltonian and the auxilary fields: if X is some observable, then

$$\langle X \rangle \equiv \frac{\mathrm{Tr} X \exp(\beta \sum_{(ij)} J_{ij} S_i S_j + \sum_i \lambda_i(\beta)(S_i - m_i))}{\mathrm{Tr} \exp(\beta \sum_{(ij)} J_{ij} S_i S_j + \sum_i \lambda_i(\beta)(S_i - m_i))} . \tag{83}$$

Eventually, we are going to minimize the free energy with respect to the magnetizations (set $\partial F / \partial m_i = 0$). You can work out that this condition, when combined with the constraint that $m_i = \langle S_i \rangle$, ensures that the auxiliary fields $\lambda_i(\beta) = 0$, precisely as they should at equilibrium.

We are going to expand $-\beta F(\beta, m_i)$ around $\beta = 0$ using a Taylor expansion. You can already see that this trick will be useful because at $\beta = 0$, the spins will be entirely controlled by their corresponding auxiliary fields, and we will thus have again reduced our problem to one of independent spins. Since m_i is fixed equal to $\langle S_i \rangle$ for any inverse temperature β, it is in particular equal to $\langle S_i \rangle$ when $\beta = 0$, which gives us the important relation

$$m_i = \langle S_i \rangle_{\beta=0} = \frac{\mathrm{Tr} S_i \exp(\lambda_i(0) S_i)}{\mathrm{Tr} \exp(\lambda_i(0) S_i)} = \tanh(\lambda_i(0)) . \tag{84}$$

We now expand the $-\beta F(\beta, m_i)$ around $\beta = 0$ using a Taylor expansion:

$$-\beta F(\beta) = -(\beta F)_{\beta=0} - \left.\frac{\partial(\beta F)}{\partial \beta}\right|_{\beta=0} \beta - \left.\frac{\partial^2(\beta F)}{\partial \beta^2}\right|_{\beta=0} \frac{\beta^2}{2} - \cdots \tag{85}$$

where we have temporarily suppressed the dependence of F on m_i. From the definition of $-\beta F(\beta, m_i)$ given in Eq. (82), we find that

$$-\beta F(\beta, m_i)_{\beta=0} = \sum_i \ln[\cosh(\lambda_i(0))] - \lambda_i(0) m_i . \tag{86}$$

At this point, we can choose to work with either the variables m_i or the variables $\lambda_i(0)$, which are directly related to the m_i through Eq. (84). We will choose to eliminate the $\lambda_i(0)$ (note that the formal manipulations are very similar to some of those we did previously when using the variational approach, but the meanings of our variables are somewhat different), and thereby recover

$$-\beta F(\beta, m_i)_{\beta=0} = -\sum_i \left[\frac{1+m_i}{2} \ln\left(\frac{1+m_i}{2}\right) + \frac{1-m_i}{2} \ln\left(\frac{1-m_i}{2}\right) \right] \tag{87}$$

which is the entropy of noninteracting Ising spins constrained to have magnetizations m_i. (Compare with the formula

$$S = \sum_{\text{states}} -p_{\text{state}} \ln(p_{\text{state}}) \tag{88}$$

from last section.) Considering next the first derivative in Eq. (85), we find that

$$-\beta \frac{\partial(\beta F)}{\partial \beta}\bigg|_{\beta=0} = \beta \Big\langle \sum_{(ij)} J_{ij} S_i S_j \Big\rangle_{\beta=0} + \beta \left\langle S_i - m_i \right\rangle_{\beta=0} \frac{\partial \lambda_i}{\partial \beta}\bigg|_{\beta=0}. \tag{89}$$

At $\beta = 0$, the spin-spin correlation functions factorize so we find that

$$-\beta \frac{\partial(\beta F)}{\partial \beta}\bigg|_{\beta=0} = \beta \sum_{(ij)} J_{ij} m_i m_j \,. \tag{90}$$

This is, of course, the "variational" internal energy, so we see that as claimed, the first two terms in our expansion give the variational trial free energy.

Naturally, we can continue our expansion, and to arbitrarily high order. If you are interested in some formal details and tricks which make the computation easier, you can refer to Georges and Yedidia.[11] To order β^4, one finds that

$$\begin{aligned}
-\beta F(\beta, m_i) = &- \sum_i \left[\frac{1+m_i}{2} \ln\left(\frac{1+m_i}{2}\right) + \frac{1-m_i}{2} \ln\left(\frac{1-m_i}{2}\right) \right] \\
&+ \beta \sum_{(ij)} J_{ij} m_i m_j \\
&+ \frac{\beta^2}{2} \sum_{(ij)} J_{ij}^2 (1-m_i^2)(1-m_j^2) \\
&+ \frac{2\beta^3}{3} \sum_{(ij)} J_{ij}^3 m_i (1-m_i^2) m_j (1-m_j^2) \\
&+ \beta^3 \sum_{(ijk)} J_{ij} J_{jk} J_{ki} (1-m_i^2)(1-m_j^2)(1-m_k^2) \\
&- \frac{\beta^4}{12} \sum_{(ij)} J_{ij}^4 (1-m_i^2)(1-m_j^2)(1+3m_i^2+3m_j^2-15m_i^2 m_j^2) \\
&+ 2\beta^4 \sum_{(ijk)} J_{ij}^2 J_{jk} J_{ki} m_i (1-m_i^2) m_j (1-m_j^2)(1-m_k^2) \\
&+ \beta^4 \sum_{(ijkl)} J_{ij} J_{jk} J_{kl} J_{li} (1-m_i^2)(1-m_j^2)(1-m_k^2)(1-m_l^2) + \ldots
\end{aligned} \tag{91}$$

where the notation (ij), (ijk), or $(ijkl)$ means that one should sum over all distinct pairs, triplets, or quadruplets of spins.

For the ferromagnet on a hypercubic lattice, all these terms can be reorganized according to their power in $1/D$. It is easy to show that only the zeroth- and first-order term contribute in the limit $D \to \infty$, and to generate $1/D$ expansions for all the thermodynamic quantities, including the magnetization.[11] In 1977, Thouless, Anderson, and Palmer (TAP) pointed out that the "mean field" theory for the Sherrington-Kirkpatrick spin glass model should include also include the second-order term (in β).[25] Unfortunately, solving the N equations obtained from minimizing the TAP free energy is still no easy task for the spin glass problem, although they were nevertheless able to make a number of interesting deductions based on their equations. The interested reader is referred directly to their paper for the details. In the next section, we will finally begin studying an even more powerful technique for dealing with quenched disorder, the famous replica method.

3. AVERAGING OVER DISORDER AND THE REPLICA METHOD

In this section, we will introduce a new subject, *averaging over the disorder* of a physical system, and a technique to do it, the *replica method.* The idea of averaging over disorder may have already occured to you when you saw the results of the variational method for the spin glass Hamiltonian. For N spins, there were N self-consistent equations to solve! This compared very unfavorably to the single self-consistent equation that we needed to solve in the ferromagnetic mean-field theory. Of course, if one is interested in some very specific system, for example the shape of a specific protein, then it makes sense that one will have to solve a lot of equations to get the answers—if you want a lot of detailed information, you need to do a lot of work. But, if instead, one is satisfied to know the *average* value of thermodynamic quantities for some typical system with quenched disorder chosen from some probability distribution, then it makes sense that one can simplify the problem. Sam Edwards and Philip Anderson were the first to attempt to compute such average quantities using the replica method for the spin glass in 1975,[8] and Giorgio Parisi first gave the correct solution of the replica mean-field theory for spin glasses in 1981.[19,20,21,22] We will introduce the replica method on models which are somewhat simpler than the spin glass, so that the ideas will be clearer. In fact, we will begin with a model which is sufficiently simple that the average over disorder can easily be done without replicas, so that we can check that the replica method does indeed give the correct answer.

Suppose that we have a single particle governed by the Hamiltonian

$$H = \frac{r^2}{2} - f\, r \tag{92}$$

where f is a quenched random force field. Or, rather, we actually have an ensemble of such physical systems, each with a different value of f. A single examplar system from the ensemble is called a *sample*. Suppose that the samples are assigned values of f chosen from the probability distribution

$$p(f) = \frac{e^{-f^2/2f_0^2}}{\sqrt{2\pi f_0^2}}\,. \tag{93}$$

For this ensemble of systems, we can compute thermodynamic quantities like the free energy F_f for any particular sample with a certain value of f, but what is more, we can compute the free energy $\overline{F}$ averaged over the entire ensemble, with each value of f weighted by its probability:

$$\overline{F} \equiv \int_{-\infty}^{\infty} df\; p(f)\; F_f\,. \tag{94}$$

Let's see how this works. We have

$$H = \frac{r^2}{2} - f\,r = \frac{1}{2}(r-f)^2 - \frac{f^2}{2} \tag{95}$$

so we see that if the force field is f, the particle will actually be in a quadratic well centered at $r = f$ with minimum $H = -f^2/2$. We have

$$Z_f = \int_{-\infty}^{\infty} dr\; e^{-\frac{1}{2T}[(r-f)^2 - f^2]} \tag{96}$$

$$= e^{f^2/2T}\int_{-\infty}^{\infty} dr\; e^{-\frac{1}{2T}(r-f)^2} = e^{f^2/2T}\sqrt{2\pi T}\,. \tag{97}$$

The free energy F_f for a sample with force field f is

$$F_f = -T\ln Z_f = -\frac{f^2}{2} - \frac{T}{2}\ln(2\pi T)\,. \tag{98}$$

This is as we would expect: the free energy is shifted from the $f = 0$ free energy by the same amount as the Hamiltonian was shifted. Now we can compute the desired average free energy:

$$\overline{F} = \int_{-\infty}^{\infty} \frac{df}{\sqrt{2\pi f_0^2}}\, e^{-f^2/2f_0^2}\left(-\frac{f^2}{2} - \frac{T}{2}\ln(2\pi T)\right), \tag{99}$$

$$= -\frac{f_0^2}{2} - \frac{T}{2}\ln(2\pi T)\,. \tag{100}$$

We can compute other disorder-averaged quantities. For example, the thermal average of the position for a given force field f is

$$\langle r \rangle_f = \frac{\int_{-\infty}^{\infty} dr\; r\; e^{-\frac{1}{2T}(r^2-2fr)}}{\int_{-\infty}^{\infty} dr\; e^{-\frac{1}{2T}(r^2-2fr)}} = f\,. \tag{101}$$

Again this makes sense given that our quadratic well is centered at $r = f$. If we now do the disorder average, we find that the disorder average of the thermal average of the position is zero:

$$\overline{\langle r \rangle} \equiv \int_{-\infty}^{\infty} df\; p(f)\, \langle r \rangle_f = 0\,. \tag{102}$$

This result is the consequence of the fact that our probability distribution for f is even, so that the samples with negative and positive force fields cancel each other out. Nevertheless, the disorder average of the *square* of the average position will be nonzero:

$$\overline{\langle r \rangle^2} = \overline{f^2} = \int_{-\infty}^{\infty} \frac{df}{\sqrt{2\pi f_0^2}} e^{-f^2/2f_0^2} f^2 = f_0^2\,. \tag{103}$$

You can easily work out other averages; for example, the disorder average of the thermal average of the squared position is

$$\overline{\langle r^2 \rangle} = T + f_0^2\,. \tag{104}$$

Note that this correlation function, which measures how much the particles fluctuate around the origin, actually has two contributions. The disorder contribution, $\overline{\langle r \rangle^2} = f_0^2$, does not depend on the temperature and is proportional to the strength of the disorder. On the other hand, the thermal part of the fluctuation, given by

$$\overline{\langle (r - \langle r \rangle)^2 \rangle} = \overline{\langle r^2 \rangle - \langle r \rangle^2} = T \tag{105}$$

does not depend on the disorder and is proportional to the temperature. This thermal part has the form of a so-called *connected* correlation function; note how it automatically subtracts away the shift in the average position caused by the random force field.

We will now rederive these results using the replica method. For this problem, using replicas is certainly overkill, but, of course, the point of the method is that it will work for many other problems where a direct computation is impossible. Our direct computation relied on the fact that we could compute the free energy for any particular sample. Of course, we have seen previously that for other problems, computing the free energy of a particular sample meant solving a large number of self-consistent equations. The idea of the replica method is to reverse the order of the computations—we want to average over disorder *first*, leaving any additional computations for later. The additional computations may still be possible, or even actually simpler, if we do them in this reverse order. In our example, we have

$$\overline{F} = -T \int_{-\infty}^{\infty} \frac{df}{\sqrt{2\pi f_0^2}} \, e^{-f^2/2f_0^2} \, \ln \int_{-\infty}^{\infty} dr \, e^{-\frac{1}{T}\left(\frac{r^2}{2} - fr\right)} . \tag{106}$$

Instead of doing the r integration first, we want to start with the disorder average represented by the f integration. This would be possible if the logarithm were not there—the f integral would then be a soluble Gaussian. The replica method is based on the mathematical identity

$$\ln Z = \lim_{n\to 0} \frac{Z^n - 1}{n} \tag{107}$$

which enables us to pull the logarithm out of the way. Let us rewrite our average free energy as

$$\overline{F} = -T \int_{-\infty}^{\infty} df \; p(f) \; \ln Z_f \, . \tag{108}$$

We use the mathematical identity

$$x = \lim_{n\to 0} \frac{1}{n} \ln(1 + nx) \tag{109}$$

to rewrite this as

$$\overline{F} = -T \lim_{n\to 0} \frac{1}{n} \ln\left(1 + n \int_{-\infty}^{\infty} df \; p(f) \ln Z_f\right) . \tag{110}$$

Using the fact that $\int_{-\infty}^{\infty} df \; p(f) = 1$, we have

$$\overline{F} = -T \lim_{n\to 0} \frac{1}{n} \ln \int_{-\infty}^{\infty} df \; p(f) \; (1 + n \ln Z_f) \, . \tag{111}$$

Finally, using the identity (107), we have

$$\overline{F} = -T \lim_{n\to 0} \frac{1}{n} \ln \int_{-\infty}^{\infty} df \; p(f) \; Z_f^n \tag{112}$$

which is a form that we can work with; we now will be able to perform the f integration. One way to think of the term Z_f^n is to imagine a new physical system consisting of n identical replicas of the old system; then Z_f^n is just the partition function of the new system. If we label the replicas by the index a, where a can run from 1 to n, then

$$Z_f^n = \int_{-\infty}^{\infty} \prod_{a=1}^{n} dr_a \; \exp\left(-\frac{1}{T} \sum_{a=1}^{n} \left(\frac{r_a^2}{2} - fr_a\right)\right) . \tag{113}$$

Finally, we have transformed our original Eq. (106) for $\overline{F}$ into the mathematically equivalent expression

$$\overline{F} = -T \lim_{n\to 0} \frac{1}{n} \ln \int_{-\infty}^{\infty} \frac{df}{\sqrt{2\pi f_0^2}}\, e^{-f^2/2f_0^2} \int_{-\infty}^{\infty} \prod_{a=1}^{n} dr_a \exp\left(-\frac{1}{T}\sum_{a=1}^{n}\left(\frac{r_a^2}{2} - f r_a\right)\right). \tag{114}$$

The advantage of this horrible-looking expression is that we can now complete the square, perform the Gaussian f integration, and arrive at the formula

$$\overline{F} = -T \lim_{n\to 0} \frac{1}{n} \ln \int_{-\infty}^{\infty} \prod_{a=1}^{n} dr_a \exp\left(-\frac{1}{T}\left(\sum_{a=1}^{n}\frac{r_a^2}{2} - \frac{f_0^2}{2T}\sum_{a=1}^{n}\sum_{b=1}^{n} r_a r_b\right)\right). \tag{115}$$

By averaging over disorder, we have converted our original problem into the mathematically equivalent problem of a system of n particles with no disorder, and interacting according to the effective Hamiltonian

$$H_{eff} = \sum_{a=1}^{n}\frac{r_a^2}{2} - \frac{f_0^2}{2T}\sum_{a=1}^{n}\sum_{b=1}^{n} r_a r_b\,. \tag{116}$$

Now we have to do the "additional computations" that we have postponed—that is, we have to compute the free energy for this new system. Fortunately, in this case the integration over the r_a variables is Gaussian so it can be performed exactly. (For other problems this next step can only be done approximately, for example, by a variational approximation; we shall see how this works in the next section.) We write the effective Hamiltonian in the form

$$H_{eff} = \frac{1}{2}\sum_{a=1}^{n}\sum_{b=1}^{n}\left(G^{-1}\right)_{ab} r_a r_b \tag{117}$$

where $(G^{-1})_{aa} = 1 - f_0^2/T$ and $(G^{-1})_{a\neq b} = -f_0^2/T$. Doing the Gaussian integral, we find

$$\overline{F} = -T \lim_{n\to 0} \frac{1}{n} \ln(\sqrt{(2\pi T)^n \det G}) = -\frac{T}{2}\ln(2\pi T) - \frac{T}{2}\lim_{n\to 0}\frac{1}{n}\mathrm{Tr}\ln G\,. \tag{118}$$

Now we have to take to logarithm of the determinant of our G matrix, or equivalently the trace of the logarithm of that matrix. At this point, it is worthwhile to make a digression to study some needed matrix algebra. We are interested in $n \times n$ matrices of the form

$$\begin{pmatrix} \tilde{a} & a & a & a \\ a & \tilde{a} & a & a \\ a & a & \tilde{a} & a \\ a & a & a & \tilde{a} \end{pmatrix}$$

We call these matrices *replica-symmetric matrices*, and we will adhere to the convention that the diagonal elements of the replica-symmetric matrix A_{ab} will be denoted by $\tilde{a}$, while the off-diagonal elements will be denoted by a. We want to derive rules for multiplying such matrices which are correct for arbitrary n, so that we can take the $n \to 0$ limit. We will then be able to formally deal with 0×0 matrices!

If we multiply two replica-symmetric matrices A_{ab} and B_{ab}, the result is a new replica-symmetric matrix C_{ab} with

$$\tilde{c} = \tilde{a}\tilde{b} + (n-1)ab \tag{119}$$

and

$$c = \tilde{a}b + a\tilde{b} + (n-2)ab\,. \tag{120}$$

Notice that we derive these formulas by thinking of n as an integer, but that we can then extend their validity to all real n. By requiring that $\tilde{c} = 1$ and $c = 0$, we get the conditions which must be satisfied if A_{ab} is the inverse matrix of B_{ab}. In the $n \to 0$ limit, these conditions are

$$\tilde{a} - a = \frac{1}{\tilde{b} - b}\ , \quad a = \frac{-b}{(\tilde{b} - b)^2}\ , \quad \tilde{a} = \frac{\tilde{b} - 2b}{(\tilde{b} - b)^2}\,. \tag{121}$$

Applying these formulae to our G_{ab} matrix, of which we only knew the inverse until now, we find that $\tilde{g} = 1 + f_0^2/T$ and $g = f_0^2/T$.

The trace of the logarithm of a replica-symmetric matrix can be worked out using the Taylor expansion

$$\ln(1+X) = X - \frac{X^2}{2} + \frac{X^3}{3} - \frac{X^4}{4} + \ldots \tag{122}$$

in its matrix form. We leave it as an exercise for the reader to work out that for the replica symmetric matrix A

$$\lim_{n\to 0} \frac{1}{n}\ \mathrm{Tr}\ln A = \ln(\tilde{a} - a) + \frac{a}{\tilde{a} - a}\,. \tag{123}$$

Using this result in Eq. (118), we finally obtain

$$\overline{F} = -\frac{T}{2}\ln(2\pi T) - \frac{f_0^2}{2} \tag{124}$$

in agreement with our previous direct computation.

Other disorder averages can also be worked out using the replica method. Again, one does the average over disorder *first*, leaving any additional computations until the end. In this way, one can show that

$$\overline{\langle r^2 \rangle} = \left\langle \sum_{a=1}^{n} r_a^2 \right\rangle_{\text{eff}} \tag{125}$$

where the "effective" expectation value is taken with respect to the replica Hamiltonian: we define an effective partition function

$$Z_{\text{eff}} \equiv \int_{-\infty}^{\infty} \prod_{a=1}^{n} dr_a \exp\left(-\frac{H_{\text{eff}}}{T}\right) \tag{126}$$

and the effective expectation value of some observable X is then

$$\langle X \rangle_{\text{eff}} \equiv \lim_{n\to 0} \frac{1}{n} \frac{1}{Z_{\text{eff}}} \int_{-\infty}^{\infty} \prod_{a=1}^{n} dr_a \; X \; \exp\left(-\frac{H_{\text{eff}}}{T}\right). \tag{127}$$

Completing the computation, we find

$$\overline{\langle r^2 \rangle} = \lim_{n\to 0} \frac{1}{n} \sum_{a=1}^{n} TG_{aa} = T\tilde{g} = T + f_0^2 \,. \tag{128}$$

This result is very important—it tells us that the diagonal element of our Green's function matrix in replica space is proportional to the correlation function measuring the combined fluctuations caused by thermal and disorder effects. Similarly, we can show that

$$\begin{aligned} \overline{\langle r^2 \rangle - \langle r \rangle^2} = \left\langle \sum_{a=1}^{n} \sum_{b=1}^{n} r_a r_b \right\rangle_{\text{eff}} &= \lim_{n\to 0} \frac{1}{n} \left(n \; TG_{aa} + n(n-1) \; TG_{a\neq b}\right) \\ = T(\tilde{g} - g) = T \,. \end{aligned} \tag{129}$$

This tells us that the purely thermal fluctuations given by the "connected" correlation function are proportional the difference of the diagonal and off-diagonal elements of the Green's function matrix.

We can summarize what we have learned by saying that an effective replica Hamiltonian of the form

$$H_{\text{eff}} = \frac{1}{2} \sum_{a=1}^{n} \sum_{b=1}^{n} \left(G^{-1}\right)_{ab} r_a r_b \tag{130}$$

with a replica-symmetric G matrix represents an exact description of an ensemble of physical systems, each one of which is a particle in a quadratic well with identical thermal fluctuations equal to $T(G_{aa} - G_{a\neq b})$, but with the center of the well assigned a different position in each sample, with fluctuations in the position of the well equal to $TG_{a\neq b}$. The total combined fluctuations in the position averaged over temperature and from sample to sample will be equal to TG_{aa}.

We will now move on to a much more challenging example of averaging over quenched disorder, for which a direct computation is impossible. Our first example was not really so "disordered," as each sample was still just a single particle in a perfectly quadratic well. The disorder was just sample-to-sample disorder, not disorder within a given sample. Much more interesting would be an ensemble of physical systems such that even within a given sample, there exists intrinsic disorder. This is really what most physicists have in mind when they think of glasses. To be specific, let us imagine an ensemble of systems, each one again consisting of a single particle in a potential. The Hamiltonian for the particle will again consist of a quadratic piece plus a random term:

$$H = \frac{r^2}{2} + V(r), \tag{131}$$

but now the random term $V(r)$ represents a whole random potential landscape, looking something like the profile of a one-dimensional mountain landscape, or perhaps the graphical history of a (purely random) stock market. A particle in such a potential will certainly feel intrinsic disorder within the sample, and if each sample has a *different* random potential, we will still have sample-to-sample disorder. Before we can try to solve for the average thermodynamics of such an ensemble of systems (which we will eventually do), we first have to understand how one can even describe such random potentials in a mathematically precise way, which is the problem we turn to now.

One convenient (and realistic) way to make an ensemble of random potentials is to imagine a potential landscape which is actually very slowly fluctuating according to some Hamiltonian at the effective temperature T_{eff}, so that the potential chosen for a particular sample is just a snapshot of the fluctuating potential at some time. We assume that the fluctuations of the particle on the potential are very much faster than the fluctuations of the potential, so that on time scales relevant to the particle, the potential still appears quenched. Technically, this means that we do the thermal average over the Hamiltonian of the particle first, and only afterwards do the disorder average, which is interpreted as a thermal average over the slow Hamiltonian of the potential. One reasonable "slow Hamiltonian" is

$$H_{\text{slow}} = \frac{T_{\text{eff}}}{2} \int_{-\infty}^{\infty} dr \, \left(\frac{\partial V}{\partial r} \right)^2 . \tag{132}$$

Such a Hamiltonian favors potential landscapes for which nearby points are correlated. A more general quadratic slow Hamiltonian is

$$H_{\text{slow}} = \frac{T_{\text{eff}}}{2} \int_{-\infty}^{\infty} dr \int_{-\infty}^{\infty} dr' \left(K^{-1}\right)_{rr'} V(r) V(r') . \tag{133}$$

This slow Hamiltonian will give us a Gaussian probability distribution

$$p(V(r)) = \frac{1}{Z_0} \exp\left(-\frac{1}{2} \int_{-\infty}^{\infty} dr \, dr' \, \left(K^{-1}\right)_{rr'} V(r) V(r') \right) \tag{134}$$

where Z_0 is a normalization constant. (To make sense of these expressions, it may help you to think of discretizing space so that the number of points where the particle can sit is finite. Then, for example, $K_{rr'}$ will become a finite matrix. At the end, one can go back to the continuum limit.) One can take this probability distribution as a starting definition of a random potential—the description in terms of a slow Hamiltonian was merely to help give some qualitative understanding. This Gaussian probability distribution has for its first two moments

$$\overline{V(r)} = 0 \quad , \quad \overline{V(r)V(r')} = K_{rr'} \, . \tag{135}$$

Clearly, the specifications of a Gaussian probability distribution by its slow Hamiltonian or by its first two moments are equivalent, as one can switch from one to the other by simply inverting the K matrix.

Normally, $K_{rr'}$ is just a function of $|r - r'|$:

$$K_{rr'} = K(|r - r'|) \, . \tag{136}$$

In that case,

$$\overline{(V(r) - V(r'))^2} = 2\left(K(0) - K(|r - r'|)\right) \, . \tag{137}$$

It is reasonable that $K(|r|)$ should monotonically decrease from its value at $|r| = 0$, as that implies from Eq. (137) that the closer two points are, the more closely their potentials are correlated. One very reasonable and technically convenient form that for $K(|r|)$ is a Gaussian decay:

$$K(|r - r'|) = W \exp\left(-\frac{(r - r')^2}{2\Delta^2}\right) \, . \tag{138}$$

Such a form describes a random potential with typical magnitude W and correlation length Δ. In the "mountain landscape" analogy, W corresponds to the typical height of the mountain peaks, while Δ is the typical distance between mountain peaks. Another very popular form is a linear decay:

$$K(|r - r'|) = W - f|r - r'| \, . \tag{139}$$

In contrast to the Gaussian decay, the linear form has no characteristic spatial or energy scale. Instead, the typical squared difference in potential will grow linearly with distance at all scales; such a form corresponds to a Brownian random walk for the potential.

We will now begin the computation of the average free energy for an ensemble of systems with a Gaussian random potential by averaging over the disorder with the replica method. In this case, we have no hope of succeeding with the "direct approach" of computing the free energy for an arbitrary sample, and then averaging over the disorder. We must try to average over the disorder first, and leave any additional computations for later. I will first tell you the result of the average over disorder, and then give a derivation of this result.

The result is: if we have an ensemble of systems with Hamiltonian

$$H = \frac{r^2}{2} + V(r) \tag{140}$$

where $V(r)$ is a Gaussian random potential with first two moments $\overline{V(r)} = 0$ and $\overline{V(r)V(r')} = K_{rr'}$, then

$$\overline{F} = -T \lim_{n \to 0} \frac{1}{n} \ln \int_{-\infty}^{\infty} \prod_{a=1}^{n} dr_a \exp\left(-\frac{H_{\text{eff}}}{T}\right) \tag{141}$$

where

$$H_{\text{eff}} = \frac{1}{2}\sum_{a=1}^{n} r_a^2 - \frac{1}{2T}\sum_{a=1}^{n}\sum_{b=1}^{n} K_{r_a r_b} . \tag{142}$$

The derivation: we have (all the initial steps are the same as our previous example)

$$\overline{F} = -T \lim_{n \to 0} \frac{1}{n} \ln \overline{Z_V^n} \tag{143}$$

where Z_V represents the partition function of the system with a given potential $V(r)$. We take the nth power of the partition function by replicating the system n times:

$$Z_V^n = \int_{-\infty}^{\infty} \prod_{a=1}^{n} dr_a \; \exp\left(-\frac{1}{T}\sum_{a=1}^{n}\left(\frac{r_a^2}{2} + V(r_a)\right)\right) . \tag{144}$$

Since we have n identical replicas of the system with potentials $V(r_a)$, the correlation function of the disorder will not depend on which two replicas we choose, so that

$$\overline{V(r_a)} = 0 \;\; , \;\; \overline{V(r_a)V(r_b')} = K_{r_a r_b'} \tag{145}$$

and we find

$$\overline{F} = \\ -T \lim_{n \to 0} \frac{1}{n} \ln \frac{1}{Z_0^n} \int_{-\infty}^{\infty} DV(r) \exp\left(-\frac{1}{2}\sum_{a=1}^{n}\sum_{b=1}^{n}\int_{-\infty}^{\infty} dr_a \, dr_b' \left(K^{-1}\right)_{r_a r_b'} V(r_a)V(r_b')\right) \\ \int_{-\infty}^{\infty} \prod_{a=1}^{n} dr_a \; \exp\left(-\frac{1}{T}\sum_{a=1}^{n}\left(\frac{r_a^2}{2} + V(r_a)\right)\right) \tag{146}$$

where the notation $DV(r)$ represents a *functional integral* over the function $V(r)$. (Again, if you have trouble with this, just imagine that space is discretized into a finite number of points. The functional integral then becomes a multiple integral over the values of the potential at the discretized points.) The integral over $V(r)$ is Gaussian and can be done (using the functional integral extension of the formulae in the appendix) yielding the desired result of Eqs. (141) and (142). In the next section we will start from this point and use the variational method to try to complete, at least approximately, the calculation of the average free energy of an ensemble of systems with Gaussian random potentials. We shall discover that very interesting problems will arise in such a computation.

4. VARIATIONAL REPLICA APPROACH AND REPLICA SYMMETRY BREAKING

In this section we shall attempt an approximate computation of the average free energy of the ensemble of physical systems consisting of a single particle in a Gaussian random potential. To be specific, we choose the Hamiltonian to be

$$H = \frac{r^2}{2} + V(r) \tag{147}$$

where $\overline{V(r)} = 0$ and

$$\overline{V(r)V(r')} = W - f|r - r'|\,. \tag{148}$$

As we learned in the last section, the average over disorder can be done by the replica method, yielding the expression

$$\overline{F} = -T \lim_{n\to 0} \frac{1}{n} \ln \int_{-\infty}^{\infty} \prod_{a=1}^{n} dr_a \exp\left(-\frac{H_{\text{eff}}}{T}\right) \tag{149}$$

where in this case, the effective replica Hamiltonian is

$$H_{\text{eff}} = \frac{1}{2}\sum_{a=1}^{n} r_a^2 - \frac{1}{2T}\sum_{a=1}^{n}\sum_{b=1}^{n} W - f|r_a - r_b|\,. \tag{150}$$

Notice that the average over disorder induces an effective *attractive* interaction between particles from different replicas.

Unfortunately, the form of the interaction makes an exact integration over the r_a variables impossible, so we must resort to approximate methods. Fortunately, our variational method can still be applied (although because of the $n \to 0$ limit, the trial average free energy becomes a lower bound on the true average free energy rather than an upper bound). Recalling the formalism that we learned in the first two sections, we define a trial average free energy $\tilde{F}$ by

$$\tilde{F} \equiv \overline{F_0} + \langle H_{\text{eff}} - H_0 \rangle_0 \tag{151}$$

where H_0 is a trial replica Hamiltonian and the thermal average $\langle\ \rangle_0$ is taken with respect to it:

$$\overline{F}_0 \equiv -T \lim_{n\to 0} \frac{1}{n} \ln Z_0; \tag{152}$$

$$Z_0 \equiv \int_{-\infty}^{\infty} \prod_{a=1}^{n} dr_a \exp\left(\frac{-H_0}{T}\right); \tag{153}$$

$$\langle X \rangle_0 \equiv \lim_{n\to 0} \frac{1}{n} \frac{1}{Z_0} \int_{-\infty}^{\infty} \prod_{a=1}^{n} dr_a\ X\ \exp\left(\frac{-H_0}{T}\right). \tag{154}$$

As the class of trial replica Hamiltonians, we use the quadratic form that we solved exactly in the last section:

$$H_0 = \frac{1}{2} \sum_{a=1}^{n} \sum_{b=1}^{n} \left(G^{-1}\right)_{ab} r_a r_b \tag{155}$$

where G is a replica-symmetric matrix with diagonal elements $\tilde{g}$ and off-diagonal elements g. These matrix elements are now the variational parameters which we will vary to optimize the trial free energy.

(The idea of using quadratic trial replica Hamiltonians [with a more general replica-symmetry-broken form which we will learn about later] in a variational approach was first suggested in a paper by Shaknovich and Gutin[23] on the replica approach to the heteropolymer problem. Mézard and Parisi[15] significantly developed the ideas of this approach and, in a recent preprint,[16] applied it directly to the problem we are studying today. I have chosen to discuss their work because it is a particularly illuminating example of the replica method from the pedagogical point of view.)

Fortunately, we have already done most of of the work necessary to compute the various pieces of the trial average free energy. For example, from our computation of the free energy in the last section, we have (see Eq. (118))

$$\overline{F_0} = -\frac{T}{2} \ln(2\pi T) - \frac{T}{2} \lim_{n\to 0} \frac{1}{n} \operatorname{Tr} \ln G \tag{156}$$

and using the result we derived last section for the trace of the logarithm of a replica-symmetric matrix (Eq. (123)), we find

$$\overline{F_0} = -\frac{T}{2} \ln(2\pi T) - \frac{T}{2} \left(\ln(\tilde{g} - g) + \frac{g}{\tilde{g} - g} \right). \tag{157}$$

For $\langle -H_0 \rangle_0$, we have

$$\langle -H_0 \rangle_0 = -\frac{1}{2} \left\langle \sum_{a=1}^{n} \sum_{b=1}^{n} \left(G^{-1}\right)_{ab} r_a r_b \right\rangle_0 \tag{158}$$

$$= -\frac{T}{2} \lim_{n \to 0} \frac{1}{n} \sum_{a=1}^{n} \sum_{b=1}^{n} \left(G^{-1}\right)_{ab} G_{ab} = -\frac{T}{2} . \tag{159}$$

$\langle H_{\text{eff}} \rangle_0$ is itself composed of a few pieces:

$$\left\langle \frac{1}{2} \sum_{a=1}^{n} r_a^2 \right\rangle_0 = \frac{T}{2} \lim_{n \to 0} \frac{1}{n} \sum_{a=1}^{n} G_{aa} = \frac{T\tilde{g}}{2} . \tag{160}$$

The $a = b$ piece of the interaction term is

$$\left\langle -\frac{W}{2T} \sum_{a=1}^{n} 1 \right\rangle_0 = -\frac{W}{2T} \lim_{n \to 0} \frac{1}{n} n = -\frac{W}{2T} . \tag{161}$$

For the $a \neq b$ piece of the interaction term, we note that $z \equiv r_a - r_b$ is a Gaussian variable, with

$$\langle z \rangle_0 = 0 \,, \quad \langle z^2 \rangle_0 = B_{ab} \equiv T(G_{aa} + G_{bb} - 2G_{ab}) = 2T(\tilde{g} - g) \,. \tag{162}$$

Using what we learned about Gaussian variables in the second section (see the discussion around Eq. (58)), that means

$$\frac{f}{2T} \left\langle \sum_{a \neq b} |r_a - r_b| \right\rangle_0 = \frac{f}{2T} \lim_{n \to 0} \frac{1}{n} \sum_{a \neq b} \int_{-\infty}^{\infty} \frac{dz}{\sqrt{2\pi B_{ab}}} \exp\left(-\frac{z^2}{2B_{ab}}\right) |z| \tag{163}$$

$$= -\frac{f}{2T} \int_{-\infty}^{\infty} \frac{dz}{\sqrt{2\pi[2T(\tilde{g} - g)]}} \exp\left(-\frac{z^2}{2[2T(\tilde{g} - g)]}\right) |z| \tag{164}$$

$$= -f \sqrt{\frac{(\tilde{g} - g)}{\pi T}} . \tag{165}$$

Note that the sign of this term switched when taking the $n \to 0$ limit because there are $n(n-1)$ off-diagonal matrix elements. Collecting all the terms, we find the trial free energy

$$\tilde{F} = -\frac{T}{2}(\ln(2\pi T) + 1) - \frac{W}{2T} - \frac{T}{2}\left(\ln(\tilde{g} - g) + \frac{g}{\tilde{g} - g}\right) + \frac{T\tilde{g}}{2} - f\sqrt{\frac{(\tilde{g} - g)}{\pi T}} . \tag{166}$$

Setting the derivatives with respect to $\tilde{g}$ and g equal to zero, we find the two equations

$$\frac{T}{2}\left(-\frac{1}{\tilde{g}-g}+\frac{g}{(\tilde{g}-g)^2}+1\right)-\frac{f}{2\sqrt{\pi T(\tilde{g}-g)}}=0; \tag{167}$$

$$\frac{T}{2}\left(\frac{1}{\tilde{g}-g}-\frac{g}{(\tilde{g}-g)^2}-\frac{1}{\tilde{g}-g}\right)+\frac{f}{2\sqrt{\pi T(\tilde{g}-g)}}=0\,. \tag{168}$$

Adding these two equations, we find $\tilde{g}-g=1$, and plugging that back into the second equation, we find $g=f/\sqrt{\pi T^3}$. We recall from the last section (Eq. (130)) that the physical correlation function corresponding to the thermal fluctuations is

$$\overline{\langle r^2\rangle-\langle r\rangle^2}=T(\tilde{g}-g)=T \tag{169}$$

which tells us that the thermal fluctuations are proportional to temperature, as we expect. The correlation function measuring the average of both thermal and disorder-induced fluctuations is

$$\overline{\langle r^2\rangle}=T\tilde{g}=T+\frac{f}{\sqrt{\pi T}} \tag{170}$$

which is...a disaster! Our result suggests that the typical displacement caused by disorder diverges as the temperature approaches zero, which makes no sense. We know, in fact, that at zero temperature, the particle will sit at the bottom of the lowest well in each sample, and the lowest well should always be some finite distance from the origin, given the quadratic term we have included in the Hamiltonian. Therefore, we should have gotten a finite answer for the fluctuations at zero temperature.

What went wrong? In fact, it should have been obvious that we were headed for trouble given what we learned in the last section about the physical meaning of the replica-symmetric trial Hamiltonian. As we learned, a replica-symmetric Hamiltonian gives an exact description of an ensemble of systems, each consisting of a particle in a *single* quadratic well, with the position of the well distributed from sample to sample according to a Gaussian distribution. This is very far from our ensemble of systems, for which every sample has many metastable minima. We should not be surprised that we get nonsense from an approach which approximates rough random potentials by a single quadratic well. Garbage in, garbage out.

What is not so obvious is how to make a better approximation. We need an approximation which includes the possibility of disorder within a sample, as well as sample-to-sample disorder. In fact, such an approximation is possible, even retaining the quadratic form of our trial replica Hamiltonian, using an amazing idea due originally to Giorgio Parisi, which he proposed in the context of a replica approach to the spin glass problem.[19,20,21,22] The idea, called *replica symmetry breaking*, technically amounts to widening the class of $n\times n$ replica matrices G_{ab} considered in the trial replica Hamiltonian to include matrices for which the off-diagonal elements are not necessarily equal. We will first explain how to do this technically, and then describe the physical meaning of an ensemble of systems represented by a trial quadratic replica Hamiltonian with replica-symmetry-broken matrices. As we will see, replica symmetry breaking is the key ingredient to describing ensembles of systems in which each sample has many metastable states.

We will begin our description of replica-symmetry-broken (RSB) matrices with the simplest such possibility, called *one-step replica symmetry breaking.* In a one-step RSB matrix, the n replicas are grouped into n/m families of m members each. There are three kinds of matrix elements: the diagonal elements, off-diagonal elements for which the row and column replicas belong to the same family, and off-diagonal elements for which the row and column replicas belong to different families. For example, in the following one-step RSB matrix A_{ab}, $n = 4$, while $m = 2$:

$$\begin{pmatrix} \tilde{a} & a_1 & a_0 & a_0 \\ a_1 & \tilde{a} & a_0 & a_0 \\ a_0 & a_0 & \tilde{a} & a_1 \\ a_0 & a_0 & a_1 & \tilde{a} \end{pmatrix} .$$

We have adopted the convention that for a one-step RSB matrix A_{ab}, the n diagonal elements are denoted $\tilde{a}$, the $n(m-1)$ "intrafamily" matrix elements are denoted a_1, and the $n(n-m)$ "interfamily" matrix elements are denoted by a_0.

If we multiply 2 one-step RSB matrices which share the same values of n and m, the result is also a one-step RSB matrix with the same n and m values. In fact, it is easy to work out that if

$$C_{ab} = \sum_{c=1}^{n} A_{ac} B_{cb} \,, \tag{171}$$

then

$$\tilde{c} = \tilde{a}\tilde{b} + (n-m)a_0 b_0 + (m-1)a_1 b_1 \,, \tag{172}$$

$$c_0 = \tilde{a} b_0 + (m-1)a_1 b_0 + \tilde{b} a_0 + (m-1)b_1 a_0 + (n-2m)a_0 b_0 \,, \tag{173}$$

$$c_1 = \tilde{a} b_1 + a_1 \tilde{b} + (m-2)a_1 b_1 + (n-m)a_0 b_0 \,. \tag{174}$$

We derived these equations thinking of n and m as positive integers with $n \geq m$, but we can consider the equations to be valid for arbitrary real n and m. We can take the $n \to 0$ limit of these equations, and then use them as we did last section to find the set of conditions that must be satisfied if A_{ab} is the inverse of B_{ab}. We can also use them to compute the trace of the logarithm of a one-step RSB matrix A_{ab}; the result is

$$\lim_{n \to 0} \frac{1}{n} \mathrm{Tr} \ln A = \ln(\tilde{a} - \langle a \rangle) + \frac{a_0}{\tilde{a} - \langle a \rangle} - \frac{m-1}{m} \ln \left(\frac{\tilde{a} - a_1}{\tilde{a} - \langle a \rangle} \right) \tag{175}$$

where

$$\langle a \rangle \equiv m a_0 + (1-m) a_1 \,. \tag{176}$$

We recover our previous replica-symmetric results in either of the limits $m \to n$ or $m \to 1$, corresponding to all the off-diagonal elements being "intrafamily" or "interfamily" respectively.

We have developed enough "replica technology" to be able to use a trial Hamiltonian of the same quadratic form as in Eq. (155) but with a one-step RSB G_{ab} matrix. The one-step RSB trial Hamiltonian has four variational parameters: $\tilde{g}$, g_0, g_1, and m, and must give a trial free energy at least as good as the replica-symmetric Hamiltonian, because it is more general. When we compute thermal replica expectation values and encounter sums over off-diagonal elements, we must remember to break the sum into intrafamily and interfamily parts. This has implications for the computation of disorder averages. For example, the connected correlation function measuring thermal fluctuations is within a one-step RSB trial Hamiltonian (compare Eq. (155))

$$\begin{aligned}\overline{\langle r^2 \rangle - \langle r \rangle^2} &= \left\langle \sum_{a=1}^{n} \sum_{b=1}^{n} r_a r_b \right\rangle_0 \\ &= \lim_{n\to 0} \frac{1}{n} (nT\tilde{g} + n(n-m)Tg_0 + n(m-1)Tg_1) \\ &= T(\tilde{g} - \langle g \rangle) \,.\end{aligned} \tag{177}$$

In the computation of the trial free energy, the only terms that change from their replica-symmetric forms are the term giving the trace of the logarithm of the G_{ab} matrix and the term from the $a \neq b$ piece of the interaction expectation value. The average square of the Gaussian variable $z \equiv r_a - r_b$ depends on whether replica a and b are in the same family or not; $\langle z^2 \rangle_0 = 2T(\tilde{g} - g_1)$ if they are in the same family, and $\langle z^2 \rangle_0 = 2T(\tilde{g} - g_0)$ if they are in different families. We thus find (compare with the replica-symmetric result of Eq. (165)) that

$$\frac{f}{2T} \left\langle \sum_{a \neq b} |r_a - r_b| \right\rangle_0 = -\frac{fm}{\pi} \sqrt{\frac{(\tilde{g} - g_0)}{T}} - f(1-m) \sqrt{\frac{(\tilde{g} - g_1)}{\pi T}} \,. \tag{178}$$

If we collect all the pieces together, the one-step RSB trial free energy is

$$\begin{aligned}\tilde{F} = &-\frac{T}{2} \left(\ln(2\pi T) + 1 - \tilde{g} + \ln(\tilde{g} - \langle g \rangle) + \frac{g}{\tilde{g} - \langle g \rangle} + \frac{m-1}{m} \ln \left(\frac{\tilde{g} - g_1}{\tilde{g} - \langle g \rangle} \right) \right) \\ &- \frac{W}{2T} - fm \sqrt{\frac{(\tilde{g} - g_0)}{\pi T}} - f(1-m) \sqrt{\frac{(\tilde{g} - g_1)}{\pi T}} \,.\end{aligned} \tag{179}$$

Now, of course, we can optimize the free energy with respect to the variational parameters $\tilde{g}$, g_0, g_1, and m. It is important to realize that in the $n \to 0$ limit, the optimal values of the parameters will give a *maximum* of the free energy, rather than a minimum.[17]

The physical meaning of these variational parameters has been worked out by Mézard and Parisi in their paper.[15] It can be understood in much the same way that we understood the physical meaning of the variational parameters $\tilde{g}$ and g in the replica-symmetric case: by comparing results of a "direct" computation of some disorder average with the results of the replica computation. Specifically, Meźard and Parisi showed that if we have an ensemble of physical systems for which the average free energy is given by the expression

$$\overline{F} = -T \lim_{n \to 0} \frac{1}{n} \ln \int_{-\infty}^{\infty} \prod_{a=1}^{n} dr_a \exp\left(-\frac{H_{\text{replica}}}{T}\right) \tag{180}$$

with

$$H_{\text{replica}} = \frac{1}{2} \sum_{a=1}^{n} \sum_{b=1}^{n} \left(G^{-1}\right)_{ab} r_a r_b \tag{181}$$

with a one-step RSB G_{ab} matrix, then the ensemble is equivalent to one constructed in the following way:

1. For each sample we determine a "central point" r_0 by choosing it from the probability distribution

$$p(r_0) = \frac{1}{\sqrt{2\pi g_0}} \exp\left(-\frac{r_0^2}{2g_0}\right). \tag{182}$$

2. Around the central point of each sample, we generate an infinite number of quadratic wells. The positions r_α chosen for the center of each quadratic well are uncorrelated with each other; they are chosen from the probability distribution

$$p(r_\alpha) = \frac{1}{\sqrt{2\pi(g_1 - g_0)}} \exp\left(-\frac{(r_\alpha - r_0)^2}{2(g_1 - g_0)}\right). \tag{183}$$

The potential V_α at the minimum of each well is also a random variable uncorrelated with the position r_α or with any of the other wells. The V_α are chosen from a probability distribution such that the average fraction of wells with minima below the level V is e^{mV}. (That is, all the wells have minima below $V = 0$; while a fraction e^{-m} have minima below $V = -1$, and so on. m is normally between 0 and 1.) In the total Boltzman sum for the sample, each well will have a weight

$$W_\alpha = \frac{e^{-V_\alpha/T}}{\sum_\alpha e^{-V_\alpha/T}}. \tag{184}$$

3. The thermal fluctuations within each well are equal to $T(\tilde{g} - g_1)$; that is, the effective well Hamiltonian is

$$H_{\text{well}} = \frac{1}{2}\frac{(r - r_\alpha)^2}{(\tilde{g} - g_1)} . \tag{185}$$

As I mentioned, this "interpretation" was arrived at by Mézard and Parisi[15] by demonstrating that the disorder averages that one computes using the interpretation agree with those obtained from the replica method.

An ensemble such as that described above clearly exhibits the kind of intrasample disorder that we want. The many quadratic wells mimic the many metastable minima in our original problem, and our various variational parameters give a clear quantitative measure of the disorder.

Nevertheless, one can do better than the one-step RSB scheme. In the one-step scheme, the replicas are organized into a very simple hierarchy of families. Parisi[19,20,21,22] proposed an even more general scheme called *k-step replica symmetry breaking* in which the n replicas are organized into n/m_1 "1-families" of m_1 elements each. The m_1 replicas in a "1-family" are then further organized into m_1/m_2 "2-families" of m_2 elements each, and so on until we reach the level of "k-families," which are ordinary families consisting of m_k elements, and "$k+1$-families" which are simple elements. We use the convention that all replicas belong to the same "0-family" so that $m_0 = n$ and $m_{k+1} = 1$. Off-diagonal elements of the k-step RSB matrix A_{ab} for which replicas a and b belong to the same "l-family" (the maximal l being chosen) are labeled a_l.

As an example, we give a two-step RSB A_{ab} matrix, with $m_0 \equiv n = 8$, $m_1 = 4$, and $m_2 = 2$:

$$\begin{pmatrix} \tilde{a} & a_2 & a_1 & a_1 & a_0 & a_0 & a_0 & a_0 \\ a_2 & \tilde{a} & a_1 & a_1 & a_0 & a_0 & a_0 & a_0 \\ a_1 & a_1 & \tilde{a} & a_2 & a_0 & a_0 & a_0 & a_0 \\ a_1 & a_1 & a_2 & \tilde{a} & a_0 & a_0 & a_0 & a_0 \\ a_0 & a_0 & a_0 & a_0 & \tilde{a} & a_2 & a_1 & a_1 \\ a_0 & a_0 & a_0 & a_0 & a_2 & \tilde{a} & a_1 & a_1 \\ a_0 & a_0 & a_0 & a_0 & a_1 & a_1 & \tilde{a} & a_2 \\ a_0 & a_0 & a_0 & a_0 & a_1 & a_1 & a_2 & \tilde{a} \end{pmatrix} .$$

Two k-step RSB matrices with the same values of m_i will, when multiplied together, give another k-step RSB matrix with the same values of m_i. As we did above for one-step RSB matrices, we can write down the multiplication rules for k-step RSB matrices, and then derive formulas for the inverse and the trace of the logarithm of a k-step RSB matrix. We can thus compute the trial free energy, and optimize it with respect to all the variational parameters in the problem ($\tilde{g}$, the $k+1$ off-diagonal elements g_i, and the k parameters m_i). Clearly, a trial Hamiltonian based on a k-step RSB matrix will always be more general than one based on a $k-1$-step RSB matrix, and should thus provide a trial free energy that is at least as good.

The physical interpretation of a such a Hamiltonian is a rather straightforward generalization of the interpretation of a one-step RSB Hamiltonian. For example, in the case of a two-step RSB Hamiltonian, in each sample, one first determines a "central point" from a Gaussian probability distribution. Around that central point one distributes an infinite number of "wells" as as we did in the one-step case. Now, however, the position of the center of each well is just a "central point" around which one distributes an infinite number of quadratic "sub-wells," again with spatial and energy displacements determined by probability distributions using the g_i and m_i parameters. Finally, the particle fluctuates thermally in the sub-wells. In three-step RSB, the sub-wells are further broken down into sub-sub-wells, and so on. For more details, see Mézard and Parisi.[15]

The best we could do within the k-step RSB scheme would be to take the $k \to \infty$ limit. Incredibly enough, this limit can be taken and is referred to as *full replica symmetry breaking.* Actually, in order to make sense of this limit, we need the following fact, which has been empirically observed in all known examples of k-step replica symmetry breaking with finite k. The parameters m_i should, in the $n \to 0$ limit, obey the inequalities

$$0 \leq m_1 \leq m_2 \leq .... \leq m_{k-1} \leq m_k \leq 1\,. \tag{186}$$

We can "justify" these inequalities in the following way. In each row of a RSB matrix, there are $n-1$ off-diagonal elements, which in the $n \to 0$ limit, equals -1. Say, for example, that we had the two-step RSB matrix A_{ab}. Then $n - m_1$ of the off-diagonal elements on each row would equal a_0, $m_1 - m_2$ of the off-diagonal elements would equal a_1, and $m_2 - 1$ of those elements would equal a_2. If we want a certain fraction of the elements to have each of the possible values, and if all the fractions should be between 0 and 1, then we require, for example, that $n - m_1$ should be between 0 and -1 (so that the number of elements equal to a_0 be a fraction of -1), which means (for $n \to 0$) that $0 \leq m_1 \leq 1$. The other inequalities similarly follow. If the above inequalities are obeyed, then we can represent all the parameters m_i and a_i in terms of a single function $a(x)$, where x ranges from 0 to 1. We construct the function piece by piece, with $a(x) = a_i$ for $m_i \leq x < m_{i+1}$. For any finite k, the function will $a(x)$ will thus consist of a series of steps.

The variable x in the function $a(x)$ is a measure of the relatedness of the two replicas denoted by the row and column indices of a matrix element of A_{ab}. When $x \to 1$, the relatedness is high, while when $x \to 0$, the relatedness is low. x is the generalization of the one-step variable m, which you recall was physically interpreted in terms of the difference in potentials between the different wells. Physically, therefore, the small x regime refers to correlations between wells that have a big potential difference.

The rules for multiplying two full RSB 0×0 matrices A_{ab} and B_{ab} to get a new matrix C_{ab} can be written in terms of the parameter representing the diagonal element and the function representing all the off-diagonal elements.[19,20,21,22] The following formulae are taken from Mézard and Parisi,[15] Appendix II:

$$\tilde{c} = \tilde{a}\tilde{b} - \langle ab \rangle \tag{187}$$

$$c(u) = (\tilde{b} - \langle b \rangle)a(u) + (\tilde{a} - \langle a \rangle)b(u) - \int_0^u dv\ (a(u) - a(v))(b(u) - b(v)) \tag{188}$$

where

$$\langle a \rangle \equiv \int_0^1 du\ a(u)\,. \tag{189}$$

In order that B_{ab} be the inverse of A_{ab}, we require that $\tilde{c} = 1$ and $c(u) = 0$. The resulting conditions are written out in detail in reference Mézard and Parisi.[15] They have also worked out the formula for the trace of the logarithm of a full RSB matrix:

$$\lim_{n \to 0} \frac{1}{n} \mathrm{Tr} \ln A = \ln(\tilde{a} - \langle a \rangle) + \frac{a(0)}{\tilde{a} - \langle a \rangle} - \int_0^1 \frac{du}{u^2} \ln \left(\frac{\tilde{a} - \langle a \rangle - [a](u)}{\tilde{a} - \langle a \rangle} \right) \tag{190}$$

where

$$[a](u) \equiv -\int_0^u dv\ a(v) + ua(u)\,. \tag{191}$$

You can check that our previous formulae in the replica symmetric or one-step RSB cases are just special cases of these more general formulae.

We can now employ a quadratic trial Hamiltonian with a full RSB G_{ab} matrix. The variational parameters will be $\tilde{g}$ and the full function $g(x)$. Note that whenever we encounter a sum over off-diagonal elements, the sum can be replaced by an integral over x (with a minus sign to account for the fact that there are -1 off-diagonal elements per row.) For example, the connected correlation function measuring thermal fluctuations is

$$\begin{aligned} \overline{\langle r^2 \rangle - \langle r \rangle^2} = \left\langle \sum_{a=1}^{n} \sum_{b=1}^{n} r_a r_b \right\rangle_0 &= \left\langle \sum_{a=1}^{n} r_a^2 + \sum_{a \neq b} r_a r_b \right\rangle_0 \\ = T\left(\tilde{g} - \int_0^1 dx\ g(x) \right) &= T(\tilde{g} - \langle g \rangle)\,. \end{aligned} \tag{192}$$

With this information, it is easy to recompute the trial free energy assuming a full RSB G matrix; the final result is

$$\begin{aligned} \tilde{F} = \\ &-\frac{T}{2}\left(\ln(2\pi T) + 1 - \tilde{g} + \ln(\tilde{g} - \langle g \rangle) + \frac{g(0)}{\tilde{g} - \langle g \rangle} - \int_0^1 \frac{du}{u^2} \ln\left(\frac{\tilde{g} - \langle g \rangle - [g](u)}{\tilde{g} - \langle g \rangle} \right) \right) \\ &- \frac{W}{2T} - f \int_0^1 du \sqrt{\frac{\tilde{g} - g(u)}{\pi T}}\,. \end{aligned} \tag{193}$$

We can now optimize this free energy with respect to $\tilde{g}$ and $g(u)$. Actually, Mézard and Parisi used the equivalent and technically simpler procedure of differentiating with respect to the G_{ab} matrix first, before taking the $n \to 0$ limit, and then breaking replica symmetry on the saddle point equations. Because the inverse of G naturally arises in this procedure, they chose to write the trial Hamiltonian in terms of the "self-energy" matrix σ_{ab}:

$$H_0 = \frac{1}{2}\sum_{a=1}^{n} r_a^2 - \frac{1}{2}\sum_{a=1}^{n}\sum_{b=1}^{n} \sigma_{ab} r_a r_b \tag{194}$$

so that

$$G_{ab} = \left((1-\sigma)^{-1}\right)_{ab} . \tag{195}$$

Mézard and Parisi[16] found the following solution: For $t > 1$, where t is the "reduced temperature"

$$t \equiv T\left(\frac{2\sqrt{\pi}}{f}\right)^{2/3} \tag{196}$$

the replica-symmetric solution given previously is valid, with $g(u) = \sigma(u) = 2/t^{3/2}$. For $t \leq 1$, the result is

$$\sigma(u) = \begin{cases} 2/t, & 0 < u < 3t/4, \\ 32u^2/9t^3, & 3t/4 < u < 3/4, \\ 2/t^3, & 3/4 < u < 1, \end{cases} \tag{197}$$

and

$$\tilde{\sigma} - \langle\sigma\rangle = 1 . \tag{198}$$

Their final result for the average of combined thermal and disorder-induced fluctuations is

$$\overline{\langle r^2 \rangle} = \begin{cases} T\left(1 + 2t^{-3/2}\right), & t > 1, \\ \frac{3T}{t} = 3\left(\frac{f}{2\sqrt{\pi}}\right)^{2/3}, & t < 1, \end{cases} \tag{199}$$

which makes much more sense than our previous replica-symmetric result. According to this solution, the system "freezes" at the critical temperature $T_c = (f/(2\sqrt{\pi}))^{2/3}$ into its low-temperature configuration, and at zero temperature, the fluctuations are finite. Of course, we learned in the first section that we should not trust a variational approach to the extent of believing that there is actually a sharp transition as described here. Indeed, Mézard and Parisi[16] performed numerical simulations of this system and showed that in the true system, the crossover from the low- to high-temperature regimes is actually smoothed out. Nevertheless, their comparison does show that the predictions of the full RSB variational approach are not too far from reality.

5. REPLICAS AND THE IMPURE SUPERCONDUCTOR IN A MAGNETIC FIELD

In this last section, we will apply the ideas we have learned about to a real physical system which has attracted considerable experimental interest, especially in the last five years—the impure superconductor in a magnetic field. The work that I will describe was done in collaboration with Jean-Philippe Bouchaud and Marc Mézard at the Ecole Normale Supérieure in Paris.[5,6]

As you may know, a superconductor has the property that sufficiently small magnetic fields cannot penetrate inside of it. Nevertheless, for all superconductors, there is a critical magnetic field (which will depend on the temperature) above which the superconductor cannot maintain itself. In type I superconductors, there is a single transition from low magnetic fields when the sample will superconduct, to high magnetic fields when it will not. In type II superconductors, the type which we will discuss today (the new high-T_c superconductors are all type II), there exists an intermediate regime in which the magnetic field partially penetrates into the sample. In this intermediate regime, the magnetic field inside the superconductor is organized into a triangular lattice of flux lines (called an "Abrikosov lattice") parallel to the direction of the external field. As the magnetic field is increased from zero in a type II superconductors, two transitions occur. At the first transition, called H_{c1}, the magnetic field first begins to penetrate and the Abrikosov lattice is formed. As the magnetic field is further increased, the density of flux lines will continually increase, until at a second transition called H_{c2}, the superconductivity is finally destroyed.

The Abrikosov lattice can be experimentally identified in a *Bitter decoration experiment.* The experimenter lays nickel filings on the superconductor, and they are attracted to the flux lines. A photograph is taken of the filings, and they indicate the structure of the lattice. A striking feature of the new high-temperature superconductors has been the extent of the disorder in the Bitter patterns of some samples, disorder which perhaps is caused by intrinsic quenched impurities in the sample. In this section, we will be using our replica methods to compute the properties of a model that assumes such impurities, and we shall see that one can make detailed predictions that can be compared to the Bitter decoration experiments. We will begin, however, by constructing a model of an Abrikosov lattice in a perfectly pure superconductor.

An Abrikosov lattice is much like any other classical crystal, except that it is constructed from flux lines rather than atoms. Nevertheless, it will have thermal fluctuations which can be understood using classical statistical mechanics. The simplest model that describes a classical crystal is the perfectly quadratic elastic solid. One can think of an elastic solid in terms of either a microscopic atomic picture or a more macroscopic continuum picture. In the two-dimensional version of the microscopic atomic picture, the atoms are arranged in a triangular array, with a lattice spacing a, and each atom is linked to its nearest neighbors by a spring.

The springs are taken to be perfectly quadratic harmonic oscillators, with possibly different spring constants for longitudinal and transverse fluctuations. Each atom has some unperturbed equilibrium position $\vec{x}$. You can call $\vec{x}$ the "label" of the atom, because no matter how much it jiggles around, its unperturbed equilibrium position $\vec{x}$ will never change. At any given time, each atom will actually be at some position $\vec{r}(\vec{x})$. The displacement of each atom from its equilibrium position is $\vec{u}(\vec{x}) \equiv \vec{r}(\vec{x}) - x$. The energy of the system is just the sum of all the two-body terms corresponding to stretching each spring.

This microscopic picture is useful to keep in mind, especially when discussing experiments that can measure the microscopic structure of the lattice. But it is often more useful and general to think in terms of a continuum description—such a description will be appropriate when we work at length scales very long compared to the atomic spacing. In this description, we imagine following some point $\vec{x}$ in the solid as we stretch it—again it gets displaced by an amount $\vec{u}(\vec{x})$ to the new position $\vec{r}(\vec{x})$. Quite generally, we know that when we make a perturbation of a system around its minimum, the energy will grow quadratically with the perturbation—in this case, $\vec{u}(\vec{x})$. In fact, we expect that the continuum generalization of the atomic Hamiltonian will look like

$$H_{\text{elastic}} = \int d^2\vec{x} \left[(C_{11} - C_{66}) \left(\sum_{\alpha} \frac{\partial u_\alpha}{\partial x_\alpha} \right)^2 + C_{66} \sum_{\alpha\beta} \left(\frac{\partial u_\alpha}{\partial x_\beta} \right)^2 \right]. \tag{200}$$

The Greek indices α and β refer to the x and y directions, and the derivatives are the generalizations of nearest-neighbor energy cost terms. This is actually the most general quadratic form consistent with the symmetries of a triangular lattice. C_{11} is called the "bulk modulus" and tell you how hard it is to squeeze the solid; C_{66} is called the "shear modulus" and tell you how difficult it is to shear the solid. For a solid like rubber which is easy to shear, $C_{66} \ll C_{11}$. This will also be true for superconductors in the regime that we are interested in.

For a three-dimensional solid of triangular lines, the continuum elastic Hamiltonian is

$$H_{\text{elastic}} = \int d^2\vec{x}dz \left[(C_{11} - C_{66}) \left(\sum_{\alpha} \frac{\partial u_\alpha}{\partial x_\alpha} \right)^2 + C_{66} \sum_{\alpha\beta} \left(\frac{\partial u_\alpha}{\partial x_\beta} \right)^2 + C_{44} \sum_{\alpha} \left(\frac{\partial u_\alpha}{\partial z} \right)^2 \right] \tag{201}$$

where C_{44} is the "tilt modulus" which measures how hard it is to tilt the lines as they travel in the z direction. This is the standard Hamiltonian used to describe the Abrikosov lattice of vortex lines in type II superconductors. It should generally give a good description when the displacements $\vec{u}$ are not too large—that is, when the temperature is low enough.

Because these Hamiltonians are quadratic, we can solve for all the thermodynamics exactly. In particular, we can calculate any correlation function that we want. One interesting example is the "translational correlation function" or "Debye-Waller factor"

$$g_{\vec{K}}(\vec{x}) \equiv \left\langle e^{i\vec{K}(\vec{u}(\vec{x})-\vec{u}(\vec{0}))}\right\rangle \tag{202}$$

where $\vec{K}$ is some reciprocal lattice vector. We can calculate this correlation function exactly; in two dimensions we find the interesting result[18]

$$g_{\vec{K}}(\vec{x}) \sim x^{-\eta_{\vec{K}}(T)} \tag{203}$$

where $\eta_K(T)$ is a temperature-dependent exponent:

$$\eta_{\vec{K}}(T) = \frac{T(2C_{66}+C_{11})|\vec{K}^2|}{4\pi C_{66}(C_{66}+C_{11})}\,. \tag{204}$$

In three dimensions, this correlation function does not decay to zero for large $\vec{x}$, but only goes to a constant. This correlation function measure density-density correlations; it can be related to the probability that if we have an atom at some position, there is an atom exactly x lattice spacings away. The fact that it goes to a constant at long distances in three dimensions means that the crystal has "long-range order." We will be able to compare the above results with our computations in the disordered case. Even more interestingly, we will be able to compare with experiments where these correlation functions are measured in real systems.

So far, our Hamiltonians have described the Abrikosov lattice in a perfectly pure superconductor. We will now add a term which models the effect of defects through a random pinning potential. I will motivate the form of this potential using the example of a thin film cuprate superconductor in a magnetic field that points in a direction perpendicular to the film. In this example, the triangular lattice of flux points (flux points in two dimensions; flux lines in three dimensions) will sit on square lattice of copper and oxygen attoms. These atoms never move, but the flux points do. The spacing of the lines will be on the order of 1000 lattice spacings of the square copper oxide lattice. We assume that there are some "point" defects in the copper oxide lattice, like oxygen vacancies. We assume that whatever the defects are, their typical correlation length Δ_{xy} will be much less than the lattice spacing of the flux lines a. Note that while the flux points move around, the defects in the copper oxide lattice are "quenched": they are frozen in and never move.

The defects in the copper-oxide lattice will attract flux points. To properly describe the lattice of flux points including their attraction to the defects, we should add to our elastic Hamiltonian a pinning potential

$$H = H_{\text{elastic}} + \sum_{\vec{x}} V(\vec{r}(\vec{x}))\,. \tag{205}$$

We idealize the pinning potential a little by assuming that it is given by a Gaussian probability distribution similar to those we have been using in the last two sections; in the two-dimensional case, for example, the first two moments are chosen to be

$$\overline{V(\vec{r})} = 0 \ , \ \overline{V(\vec{r})V(\vec{r}')} = U_{\rm pin}^2 \exp\left(-\frac{(\vec{r}-\vec{r}')^2}{2\Delta_{xy}^2}\right) . \tag{206}$$

Such a form implies that the typical magnitude of the potential is the pinning energy $U_{\rm pin}$ and the correlation length is Δ_{xy}.

The reasonableness of our assumptions is, in fact, bolstered by a quick examination of the results of some recent Bitter decoration experiments on three-dimensional cuprate superconductors.[12] In the 69 Gauss experiment of the Bell Labs group, one can see that a perfect triangular topological structure is maintained within the camera's field of view. The physical idea behind our calculation is that the visible local distortions of the triangular lattice are caused by the attraction flux lines feel for microscopic defects like oxygen vacancies. This hypothesis can be checked by comparing the quantitative predictions that we derive for correlation functions with those obtained by the experimentalists for their samples. We shall see that the agreement is quite good.

We are interested in computing the average free energy of an ensemble of systems described by the above Hamiltonian, and we begin by averaging over the disorder using the replica method. As usual, the average over disorder converts our problem into a mathematically equivalent one of n identical crystals, for which the atoms no longer feel the random potential, but instead feel an inter-replica attraction. In particular, we have

$$\overline{F} = -T \lim_{n\to 0} \frac{1}{n} \exp\left(-\frac{H_{\rm eff}}{T}\right) \tag{207}$$

with (in two dimensions)

$$\begin{aligned} H_{\rm eff} = & \frac{1}{2}\sum_{a=1}^{n} \int d^2\vec{x} \left[(C_1 - C_6)\left(\sum_{\alpha} \frac{\partial u_\alpha^a}{\partial x_\alpha}\right)^2 + C_6 \sum_{\alpha\beta}\left(\frac{\partial u_\beta^a}{\partial x_\alpha}\right)^2 \right] \\ & - \frac{W}{2T}\sum_{a=1}^{n}\sum_{b=1}^{n}\sum_{\vec{x}\vec{x}'} \delta(\vec{r}_a(\vec{x}) - \vec{r}_b(\vec{x}')) \end{aligned} \tag{208}$$

where we have taken the limit $\Delta_{xy} \ll a$ to convert our Gaussian into a delta-function and where now all distances are written in units of the lattice spacing a. We define

$$W \equiv \frac{2\pi U_{\rm pin}^2 \Delta_{xy}^2}{a^2} \tag{209}$$

and

$$C_1 \equiv C_{11} a^2, \quad C_6 \equiv C_{66} a^2 . \tag{210}$$

The first part of the effective Hamiltonian is just the elastic crystal replicated n times. In the other part, one has an effective attraction between all the atoms in the system. Of course, they will not all sit on top of one another because the elastic term keeps them apart.

We cannot compute the average free energy exactly because the effective Hamiltonian is not quadratic. As usual, we will use the variational approach with a trial quadratic Hamiltonian which is as general as possible. Thus, we choose

$$H_0 = \frac{1}{2} \sum_{ab} \sum_{\alpha\beta} \sum_{\vec{x}\vec{x}'} \left(G^{-1}\right)^{ab}_{\alpha\beta} (\vec{x} - \vec{x}')u^{\alpha}_{a}(\vec{x})u^{\beta}_{b}(\vec{x}'). \tag{211}$$

Physically we are coupling the fluctuation of every atom (or flux point) in the system with every other atom, taking into account with the α and β indices the difference between longitudinal and transverse fluctuations. Because the system is translationally invariant, we can diagonalize the spatial part of the G matrix by going into Fourier space, so that

$$H_0 = \frac{1}{2} \sum_{ab} \sum_{\alpha\beta} \int \frac{d^2\vec{q}}{(2\pi)^2}\, u^{\alpha}_{a}(\vec{q}) \left(G^{-1}\right)^{ab}_{\alpha\beta} (\vec{q})\ \ u^{\beta}_{b}(-\vec{q}). \tag{212}$$

We can diagonalize the spatial indices of the Green's function by breaking it down into longitudinal and transverse fluctuations

$$G_{\alpha\beta}(\vec{q}) \equiv \left(\delta_{\alpha\beta} - \frac{q_\alpha q_\beta}{q^2}\right) G_T(q) + \left(\frac{q_\alpha q_\beta}{q^2}\right) G_L(q). \tag{213}$$

Finally, we can compute the trial free energy $\tilde{F}$ and try to optimize it with respect to the Green's functions $G^{ab}_{L,T}(q)$. I will not write out the trial free energy or the saddle-point equations here; they are long and not too enlightening and you can find them in Bouchaud et al.[5] Of course, when we optimize with respect to the replica indices, we use a full RSB matrix. That means that we ultimately optimize with respect to the Green's functions $G_{L,T}(q,v)$, where v is a real number ranging from 0 to 1. In fact, if one is lazy and tries to use a replica-symmetric ansatz, one runs into precisely the same type of trouble as we saw in the last section with a one-particle problem—all the fluctuations seem to diverge at zero temperature. Again a full RSB approach is necessary to account for the fact that the many possible metastable configurations of the atoms.

The final form of the Green's functions are (for $C_{66} \ll C_{11}$, which is reasonable for the superconductors)

$$G_{L,T} = \frac{1}{q^{2+2\nu}} g_{L,T}\left(\frac{v}{q^{\omega}}\right) \tag{214}$$

where $\omega = 2\nu$ in two dimensions ($\omega = 2\nu + 1$ in three dimensions) and $g_{L,T}(x)$ is a complicated function given explicitly in Bouchaud et al.[5] ν is an exponent whose value and physical significance we will discuss shortly. Of course, the Green's function in and of itself is not too interesting. We are more interested in using the Green's function to compute various disorder averages which have a more obvious physical and experimental significance.

Imagine, for example, that we take two atoms some distance x apart along the x axis and look at their typical squared longitudinal and transverse fluctuations. These are given by the disorder average correlation functions $\tilde{B}_L(x)$ and $\tilde{B}_T(x)$:

$$\tilde{B}_L(x) = \overline{\left\langle (u_x(\vec{x}) - u_x(\vec{0}))^2 \right\rangle}\,, \tag{215}$$

$$\tilde{B}_T(x) = \overline{\left\langle (u_y(\vec{x}) - u_y(\vec{0}))^2 \right\rangle}\,. \tag{216}$$

We find that these correlation functions increase with distance with a power-law form (with possible logarithmic corrections)

$$\tilde{B} \sim x^{2\nu}\,. \tag{217}$$

In particular, a full variational calculation gives the result (in two dimensions)

$$\tilde{B}_L(x) = \frac{3}{5}\tilde{B}_T(x) = \frac{3\Gamma(2/3)^2}{8\pi^{2/3}2^{1/3}}\left(\frac{x}{\xi}\right)^{2/3} \simeq 0.25\left(\frac{x}{\xi}\right)^{2/3} \tag{218}$$

for $1 \ll x \ll \xi = C_6/\sqrt{W}$ and

$$\tilde{B}_T(x) = \frac{1}{2}\tilde{B}_T(x) = \frac{\sqrt{\pi}\,2^{1/2}}{6\;\;3^{1/4}}\left(\frac{x}{\xi\sqrt{\ln x}}\right) \simeq 0.32\left(\frac{x}{\xi\sqrt{\ln x}}\right) \tag{219}$$

for $x \gg \xi$. In three dimensions, we find

$$\tilde{B}_L(x) = \frac{3}{4}\tilde{B}_T(x) = \frac{3\Gamma(2/3)}{7\pi^{2/3}}\left(\frac{x}{\xi}\right)^{1/3} \simeq 0.27\left(\frac{x}{\xi}\right)^{1/3} \tag{220}$$

for $1 \ll x \ll \xi = C_4^{1/2}C_6^{3/2}/W$ and

$$\tilde{B}_L(x) = \frac{2}{3}\tilde{B}_T(x) = \frac{2^{1/2}4}{3^{1/4}5}\left(\frac{x}{\xi\ln x}\right)^{1/2} \simeq 0.43\left(\frac{x}{\xi\ln x}\right)^{1/2} \tag{221}$$

for $x \gg \xi$. Notice that the ratio of transverse to longitudinal fluctuations is always equal to $2\nu + 1$. Of course, we do not expect these variational results to be exact, but there are various arguments, including other more qualitative approaches,[13,14,9] which indicate that these results are reasonable.

The correlation length ξ is the distance at which fluctuations become equal to a full lattice spacing. Until that length scale, all the atoms see essentially independent potentials, but beyond that length scale, two different atoms can take advantage of the same attractive pinning potential. ξ can actually be quite large, so that for many experiments, the relevant result is intermediate distance regime $1 \ll x \ll \xi$.

Finally, we get to the promised density correlation function $g_{\vec{K}}(\vec{x})$. We find

$$g_{\vec{K}}(\vec{x}) = \exp\left(-\frac{K^2}{2}(\tilde{B}_L(x)\cos^2\theta + \tilde{B}_T(x)\sin^2\theta)\right) \tag{222}$$

where θ is the angle between $\vec{K}$ and $\vec{x}$. Compared to the behavior in a pure system—power law in two dimensions or decay to a constant in three dimensions—the decay of $g_{\vec{K}}(\vec{x})$ is quicker; it is a stretched exponential with

$$g_{\vec{K}}(\vec{x}) \sim e^{-x^{2\nu}} \tag{223}$$

radial behavior. This prediction appears to agree quite well with experimental results,[12] with values of ν that are also consistent with our predictions.

For a much more detailed account of these results, the reader is again referred to Bouchaud et al.[5] Some extensions of these ideas to more complicated physical situations have also been worked out.[2] The main message of this section is that the replica technology that we have learned about in previous sections really can be a working tool of physicists, who can thereby make theoretical predictions which can be compared with experiment.

APPENDIX

Some useful Gaussian integrals:

$$\int \frac{\prod_{i=1}^{n} dx_i}{\sqrt{(2\pi)^n \det A}} \exp\left(-\frac{1}{2}\sum_{i=1}^{n}\sum_{j=1}^{n} x_i \left(A^{-1}\right)_{ij} x_j + \sum_{i=1}^{n} x_i J_i\right)$$

$$= \exp\left(\frac{1}{2}\sum_{i=1}^{n}\sum_{j=1}^{n} J_i A_{ij} J_j\right) \tag{224}$$

$$\int_{-\infty}^{\infty} \frac{\prod_{i=1}^{n} dx_i}{\sqrt{(2\pi)^n \det A}} x_i x_j \exp\left(-\frac{1}{2}\sum_{i=1}^{n}\sum_{j=1}^{n} x_i \left(A^{-1}\right)_{ij} x_j\right) = A_{ij} \tag{225}$$

$$\int_{-\infty}^{\infty} \frac{dx}{\sqrt{2\pi A}} e^{-x^2/2A} = 1 \tag{226}$$

$$\int_{-\infty}^{\infty} \frac{dx}{\sqrt{2\pi A}} x e^{-x^2/2A} = 0 \tag{227}$$

$$\int_{-\infty}^{\infty} \frac{dx}{\sqrt{2\pi A}} |x| e^{-x^2/2A} = \sqrt{\frac{2A}{\pi}} \tag{228}$$

$$\int_{-\infty}^{\infty} \frac{dx}{\sqrt{2\pi A}} x^2 e^{-x^2/2A} = A \tag{229}$$

$$\int_{-\infty}^{\infty} \frac{dx}{\sqrt{2\pi A}} x^4 e^{-x^2/2A} = 3A^2 \tag{230}$$

ACKNOWLEDGMENTS

I thank my longtime friends and collaborators, Jean-Philippe Bouchaud, Antoine Georges, and Marc Mézard, with whom I have enjoyed exploring the issues described in this chapter. I thank Dan Stein for inviting me to give these lectures and the members of the Santa Fe Institute for their hospitality during my visit. My work has been supported by the Harvard Society of Fellows, where I am a Junior Fellow, and by the National Science Foundation, through Grant No. DMR91-15491.

REFERENCES

1. Anderson, P. W. "Reference Frame" articles appearing in *Physics Today* from January, 1988 to March, 1990.
2. Bouchaud, J.-P., and A. Georges. "Competition Between Lattice Pinning and Impurity Pinning: Variational Theory and Physical Realizations." *Phys. Rev. Lett.* **68** (1992): 3908–3911.
3. Bouchaud, J.-P., M. Mézard, G. Parisi, and J. S. Yedidia. "Polymers with Long-Ranged Self-Repulsion: A Variational Approach." *J. Phys. A* **24** (1991): L1025–L1030.
4. Bouchaud, J.-P., M. Mézard, and J. S. Yedidia. "Some Mean-Field-Like Equations Describing the Folding of Heteropolymers at Finite Temperature." LPTENS preprint 91/27.
5. Bouchaud, J.-P., M. Mézard, and J. S. Yedidia. "Variational Theory for Disordered Vortex Lattices." *Phys. Rev. Lett.* **67** (1991): 3840–3843.
6. Bouchaud, J.-P., M. Mézard, and J. S. Yedidia. "A Variational Theory for the Pinning of Vortex Lattices by Impurities." LPTENS preprint 92/3. *Phys. Rev. B* (1992): to be published.
7. des Cloizeaux, J. "The Statistics of Long Chains with Non-Markovian Repulsive Interactions and the Minimal Gaussian Approximation." *J. Physique* **31** (1970): 715–736.
8. Edwards, S., and P. W. Anderson. "Theory of Spin Glasses." *J. Phys. F* **5** (1975): 965–974.
9. Feigel'man, M., V. B. Geshkenbein, A. Larkin, and V. Vinokur. "A Theory of Collective Flux Creep." *Phys. Rev. Lett.* **63** (1989): 2303–2306.
10. Feynman, R. P. "Slow Electrons in a Polar Crystal." *Phys. Rev.* **97** (1955): 660–665.
11. Georges, A., and J. S. Yedidia. "How to Expand Around Mean-Field Theory Using High-Temperature Expansions." *J. Phys. A* **24** (1991): 2173–2192.
12. Grier, D. G., C. A. Murray, C. A. Bolle, P. L. Gammel, D. J. Bishop, D. B. Mitzi, and A. Kapitulnik. "Translational and Bond-Orientational Order in the Vortex Lattice of the High-T_c Superconductor $Bi_{2.1}$ $Sr_{1.9}Ca_{0.9}Cu_2O_{8+\delta}$." *Phys. Rev. Lett.* **66** (1991): 2270–2273.
13. Halpin-Healy, T. "Diverse Manifolds in Random Media." *Phys. Rev. Lett.* **62** (1989): 442–445
14. Halpin-Healy, T. "Disorder-Induced Roughening of Diverse Manifolds." *Phys. Rev. A* **42** (1990): 711–722.
15. Mézard, M., and G. Parisi. "Replica Field Theory for Random Manifolds." *J. Physique I* **1** (1991): 809–836.
16. Mézard, M., and G. Parisi. "Manifolds in Random Media: Two Extreme Cases." LPTENS preprint 92/12, 1992.
17. Mézard, M., G. Parisi, and M. A. Virasoro. *Spin Glass Theory and Beyond.* Singapore: World Scientific, 1987.

18. Nelson, D. R. "Defect-Mediated Phase Transitions." In *Phase Transitions and Critical Phenomena*, edited by C. Domb and M. S. Green, Vol. 7, 5. London: Academic Press, 1983.
19. Parisi, G. "Toward a Mean-Field Theory for Spin Glasses." *Phys. Lett.* **73A** (1979): 203–205.
20. Parisi, G. "A Sequence of Approximated Solutions to the S-K Model for Spin Glasses." *J. Phys. A* **13** (1980): L115–L121.
21. Parisi, G. "Order Parameter for Spin Glasses." *J. Phys. A* **13** (1980): 1101–1112.
22. Parisi, G. "Magnetic Properties of Spin Glasses in a New Mean-Field Theory." *J. Phys A* **13** (1980): 1887–1895.
23. Shakhnovich, E. I., and A. M. Gutin. "Frozen States of a Disordered Globular Heteropolymer." *J. Phys. A* **22** (1989): 1647–1659.
24. Sherrington, D., and S. Kirkpatrick. "Solvable Model of a Spin Glass." *Phys. Rev. Lett.* **35** (1975): 1792–1796.
25. Thouless, D. J., P. W. Anderson, and R. G. Palmer. "Solution of 'Solvable Model of a Spin Glass.'" *Phil. Mag.* **35** (1977): 593–601.

Robert H. Austin
Department of Physics, Princeton University, Princeton, NJ 08544

Complexity in Biological Molecules

1. INTRODUCTION

I was terrified when Dan Stein asked me to give a series of lectures at the Santa Fe summer school on complexity because I regard myself as a blue-collar experimentalist—a yeoman's work—trying to figure out the dynamics of these complex biomolecules. By some bizarre twist of fate, I am at Princeton University, the Land of Eternal Theory, and I am fully aware of how difficult it can be to communicate with students who only do abstract theory and are proud of it. I stumbled through my lectures and I still haven't worked up the courage to look at the student evaluations which I am sure were not exactly rave reviews. But I tried to communicate some of the excitement I feel as I do experiments that probe the complex dynamics of biomolecules, and I listened to many student comments and insightful observations. I don't think I communicated much to the students, but I learned some things from them. The following summary of some of the topics I discussed may prove useful.

1992 Lectures in Complex Systems, Eds. L. Nadel & D. Stein, SFI Studies in the Sciences of Complexity, Lect. Vol. V, Addison-Wesley, 1993

Typically, when people talk about complexity in biological systems, they talk about the complexity that occurs at relatively high levels of organization, such as the immune system or neural networks. The individual molecules that make up the system are regarded as bits that form the system, but one need not concern oneself with the details of how they work; it is only necessary to know their basic transfer function, as it were, between input and output. In fact, it is sort of low class to worry about how the molecules actually work—details, details, details! But, I am well aware of my virtues and my weaknesses. I am pretty good in the lab and have an ability to pull off experiments because I can speak to the little people in the equipment who like to screw up experiments (a talent many theorists *definitely* don't have because of a bad attitude!), while my mathematical abilities are definitely second rate. I can't imagine myself living up on the theorist's floor in a physics department in a spartan office nervously wrestling with a set of coupled differential equations, trying to ignore the pipe smoke from the bearded, brilliant but caustic mathematical physicist next door. Give me the lab and good rock-and-roll.

However, it is possible that good physics can be found in trying to understand how a complex biological polymer goes about its business. I suspect that it is important to have people who worry about high-level complexity and people who worry about low-level complexity, and hopefully insights will come from both kinds of endeavors.

2. A WHIRLWIND INTRODUCTION TO PROTEINS AND NUCLEIC ACIDS

I have noticed, as has my biophysicist wife Shirley Chan, that *many* physicists do not know the difference between a protein and a nucleic acid, literally. This has always surprised me—you would think that there would be a bit of interest in how the incredible miracle "life" works. Here is a thumbnail sketch of some basic ingredients for life.

What does a generic protein look like? It starts out as a polymer consisting of strings of amino acids; the amino acids come in 20 different chemical flavors. I am no chemist and I will leave it to you to read a good textbook like Stryer's *Biochemistry* to get a good overview of the amino acids. However, it is important to note that, chemically, the 20 different amino acids cover a full spectrum of chemical properties, from highly hydrophobic aliphatic groups such as alanine to strongly negative hydrophilic groups such as aspartic acid. Since most water-soluble proteins use the full range of amino acids, one could expect that the problem of predicting the structure formed by a given sequence of amino acids could be a formidable problem.

We will discuss—I am by no means an expert in the field—the protein folding problem a bit later. Instead, we will *assume* that in fact the protein has folded into a three-dimensional structure consisting of some complicated trists and turns of the polymer. In fact, proteins seem to form reproducible structures and diffract to subangstrom resolution in the best of cases. The crystal structures show that the protein molecules are close-packed and have nearly the same density as simple crystals of single amino acids. However, the question remains if the protein folds to a *unique* structure or if there is a *distribution* of conformational configurations that are thermally accessible. The answer to the last question is probably "yes," since one can easily reduce the definition of a thermally reachable conformation to any very small displacement of an amino acid group. A more interesting but not easily answered question is can a protein be "put" into a metastable conformation that is thermally very unlikely to occur yet has a very large activation energy separating it from a deeper lying conformation?

DNA is also a polymer although structurally very different from a protein. Instead of 20 different amino acids, a DNA polymer is composed of only 4 different nucleotides (Adenine, Guanine, Cytosine, and Thymine) that are similar to each other chemically. The nucleotides are strung together in a chain separated by a phosphate-ester bond that produces a negative charge between every base. Ordinarily you wouldn't expect such a negatively charged polymer to bind to another similar negatively charged line, but the single-stranded DNA molecule actually lines up with another DNA single-stranded molecule because of the very unexpected way in which the adenine base can form hydrogen bonds with a thymine base and the guanidine base can form hydrogen bonds with a cytosine base. This only works if the single-stranded molecules (which are chiral and therefore have a sense of direction) run antiparallel to each other.

The DNA double-helix molecule is very stiff because the negative charges impart a large internal tension. The persistent length of the molecule, roughly defined as the mean radius curvature induced by thermal fluctuations, is known to be about 60 nm for DNA so, unlike a protein, DNA forms a very open structure. In the absence of proteins DNA forms a classic semi-stiff polymer which can be treated analytically. Of course there are complications: the basepairs are not identical and the hydrogen bonding that occurs across the bases is not the only interaction. Basepair stacking interactions strongly influence the thermodynamics of double-helix formation and possibly hydrogen bonds form diagonally across the basepairs to adjacent ones. These nearest-neighbor interactions can also have a major influence on DNA structure. Further, in closed circular DNA and when DNA is being processed by proteins, major topological considerations seem to play an important role in the way that DNA expression is controlled. This is the end of the whirlwind tour. It is time to look at experiments.

The choice of topics will basically be historical in nature, hopefully paralleling my growth in biophysics. I am slowly moving away from trying to understand detailed microscopic aspects of biological molecules and toward more global aspects for a practical reason: life is short, and these molecules are so large and intricate that I suspect that only a more global view will lead to some insight into how the system integrates itself.

3. A SPIN GLASS ANALOGY TO PROTEIN DYNAMICS

Via some rather indirect physical probes, it is possible to measure the recombination of small molecules such as carbon monoxide in a protein such as myoglobin after a photon is used to break the bond between CO and an iron atom. These experiments allow one to ascertain how rapidly the molecule moves through the protein and the rate at which the small molecule recombines with the iron atom.

I want to explore the complexities of living systems, not some physical chemistry remotely connected to the diffusion of small molecules. My mentor Hans Frauenfelder would make a distinction between biophysics and biological physics. If you do biophysics, then you explore the fundamental physical questions concerning the dynamics of complex molecules as proteins. If you do biological physics, then you use the tools of physics to explore biological systems, but you don't expect or look for new insights into physics in the work—you look for new insights into biology.

The idea behind the biophysics of the ligand recombination experiments is that after photolysis the ligand can diffuse from the original binding site and either escape from the original protein where it was bound or rebind. If these proteins have a complex conformational landscape in the sense that within a given sample many molecules with different conformations exist, then the kinetics of recombination might be quite different in different protein conformations.

My early recombination work was done at the University of Illinois, under the guidance of Hans Frauenfelder and with my fellow crew of terrific graduate students.[7,8,9] This was a simple experiment involving the observation of a chemical reaction in myoglobin at cryogenic temperatures, from 300 K to 4.2 K. We observed that below 200 K in a glass of glycerol water the rate of the reaction—i.e., the recombination of carbon monoxide with iron—could not be characterized by a single rate constant but, instead, seemed to be due to a *spectrum* of rates.

That is, if a reactant has to surmount a *single, mono-energetic* barrier of height E_a, then the rate k for the reaction should proceed as:

$$k = A \exp\left(\frac{-E_a}{kT}\right) \tag{1}$$

and the actual number of molecules $N(t)$ surviving after a time t is given by:

$$N(t) = N(0)\exp(-kt)\,. \tag{2}$$

Please note that the deviations from an exponential are basically believed to be due to: (1) a *heterogeneous* distribution of occupied states and (2) the distribution in sites is *continuous* and *not* discrete. This is a supposition at this point—do the data stand up to these two critical requirements?

The next question is: can we understand the functional form of the distribution? Three major attempts have been made to understand the form of the distribution; one by Stein[93] directly used the spin glass analogy. The other two come from rather different camps. The model of Bowne and Young[105] explicitly uses an adiabatic formulation to model the low-temperature recombination process[43] and assumes anharmonic potential wells based upon the temperature-dependent x-ray scattering data of Petsko and coworkers.[44] From Agmon and Hopfield[1] the third model, used an ansatz based upon a distribution arising from Franck-Condon factors and harmonic potential wells. In any event, all models do agree on one thing: there is a distribution of activation energy barriers!

Casual inspection of the data indicates about 6 exponentials would be needed to fit the data; expanded data by Frauenfelder's group that covers 12 decades of time would need on the order of 12 exponentials. As the number of exponentials goes to infinity, we arrive at a continuous distribution of states. *We thus assume that the activation energies for recombination are given by a probability distribution* $g(E)$ *for finding a molecule with activation energy* E *or, equivalently, a distribution* $g(k)$ *for finding a molecule with rebinding rate constant* k. The kinetics then become:

$$N(t) = N(0) \int g(k) e^{-kt} dk \,. \tag{3}$$

Basically, this experiment revealed that at low temperatures the protein seemed to have a time-invariant and temperature-invariant *spectrum* of activation energies. Other workers have proposed that the distribution of rates is due to internal dynamics of a single molecule[21] or that a small number (four or less) of exponentials can be used to fit the data,[80] but several experiments have ruled out those possibilities. Acceptance of the continuous distribution explanation of our data is critical for the following material.

In the previous section we have seen that at least myoglobin reveals a complex kinetic recombination pattern at low temperatures. Other physical systems show such kinetic complexity; in particular, spin glasses show glass transitions and distributed kinetics at low temperatures. Dan Stein has edited a book on the subject of spin glasses and their analogy to biological systems.[94] In the spirit of the Santa Fe "Lecture Notes," I have included an abbreviated account of my contribution to Dan's book,[94] *Spin Glasses in Biology.*

Now, let's see if we can contrast a spin glass with a globular protein and find similarities. Twenty different amino acids commonly found in a globular protein are of highly variable chemical composition in contrast with another common biological polymer, DNA, which is chemically much more homogeneous (although, structurally, it reveals considerably more variety than has been assumed). Thus, while under physiological conditions DNA is in a B-helix form, under physiological conditions proteins assume a staggering variety of conformational shapes, depending upon the amino acid composition.

Even within one configuration of N amino acid sequences, there are probably exponentially ($e^{\alpha N}$) many ways to fold the protein into a globular configuration. Even in the absence of the elastic interaction mentioned above, perhaps one can say that, because of the complex array of amino acids present, the interaction J_{ij} between the ith amino acid and the jth amino acid is likely to be random in sign, in analogy to the RKKY potential in spin glasses, if that is a legitimate thing to say.

Of course, the x-ray crystallographers present very nice pictures of "the structure" for a given protein. While proteins have a unique structure and don't look like a glass at all, we are more concerned with the small structural variations within the broad set of a given framework. These variations can be hidden within the highly massaged structures that computers spit out. Possibly crystallization of a protein selects out a subset of the total protein structures in solution, and crystallization forces can drive the system into a common structure. Computer folding simulations indicate that the structures fold into many separate energy minima that are quite distinct from one another. Thus, even at present, we do not know if "the structure" seen by x-ray crystallography is a global minimum structure or a selected one. The level to which the reader wishes to draw the spin glass analogy resides on several possible levels.

At this point it is appropriate to raise the question of the size of the protein molecule. As Fisher et al. made very clear in a review article,[42] theory in the true spin glass systems assumes an infinite size sample. In proteins, the molecules are definitely of finite size. For example, a small protein such as myoglobin has a molecular weight of approximately 18,000 daltons and a radius of approximately 25 Å. Such a small radius immediately says that many of the approximations used by the theorists will not work.

Does this then mean that there are no phase transitions in proteins? Even small proteins have well-defined denaturation temperatures where, over the range of only a degree or so, the globular protein changes to a random coil.[26] These changes of state, in my opinion, are sharp enough to merit the label of a "phase transition." As is always true in the biological physics of macromolecules, we must work by analogy and try to accommodate the physics as best we can. To demand perfect rigor is to abandon the field altogether.

Below this denaturation temperature we believe that the globular protein can be in a very large number of structurally different spatial configurations. At room temperature in solution (that is, fully hydrated), the protein is believed to jump rapidly between the thermally accessible conformation states. We would expect (hope) that below some temperature T_c the protein molecules no longer can jump between these conformational states, and each protein molecule becomes "frozen" in some particular conformation. Note that there is no good reason from what we have said so far that (a) a conformational distribution exists, (b) the protein can rapidly jump between these conformations, or (c) a glass transition can occur. The evidence for this will come later.

However, if such a glass transition can occur, then the entire conformational space formed by the ensemble of protein molecules with a given amino acid sequence and general structure and once rapidly sampled by each protein molecule is now time invariant in the sense that the protein molecule can no longer sample the states. The space can be considered to be a glass in the spin glass sense if the two criteria of *randomness* and *frustration* have been satisfied. We hasten to point out that to our knowledge no one has ever cleanly demonstrated that frustration occurs in a protein structure.

A great deal of theory has been written concerning the consequences of frustration and disorder on the magnetization M of a spin glass vs. temperature T. When kT is considerably greater then the mean interaction energy $\langle J^2\rangle^{1/2}$ between the spins, we expect that the magnetization of the spin glass should obey a simple Curie-Weiss law[63]:

$$M(t) \propto \frac{1}{T}\,. \tag{4}$$

This simple temperature dependence is called free-spin paramagnetism and, of course, is due to a single spin S interacting with the magnetic field B. Imagine, however, that there is spin-spin interaction. If the interaction is of a ferro- or antiferro-magnetic type, then below the Curie temperature the magnetization becomes very large due to the net alignment of the spins along some K vector. Since all the spins point in the same direction, the system can be said to be in one macroscopic state.

However, for a spin glass there is no net alignment of the spins although the spins are no longer free to point any direction given the statistics of the Boltzmann relation. Since there is no structural change in the system if any thermodynamic phase transition exists, we would expect that the transition will not be like a simple first- or second-order phase transition. Indeed, while above the critical temperature in spin glasses the magnetization does follow a $1/T$ susceptibility below the glass transition there is a *cusp* in the magnetization followed by a roughly linear T-dependence of the magnetization vs. temperature.[74]

The presence of a cusp in the susceptibility followed by a *decline* is evidence that the spins are no longer free but, instead, have been constrained in some direction. However, since there can be no net alignment of the spins in a frustrated system, the magnetization falls to zero at low temperatures, unlike a ferromagnetic or paramagnetic system.

As Binder and Young express it,[22] we can characterize a spin glass at low temperatures by two equations. A spin has some orientation in space that when averaged over time t, is not zero,

$$\langle \vec{S}_i\rangle_t \neq 0\,, \tag{5}$$

but there is no net alignment of the entire spin ensemble,

$$\frac{1}{N}\sum_i \langle \vec{S}_i \rangle_t \exp(i\vec{K}\cdot\vec{R}_i) = 0\,. \tag{6}$$

Presumably, at the critical temperature T_c the spins undergo a glass transition to a time-invariant disordered phase. The actual definition of the glass transition T_c is rather murky, since there is no discontinuous change in any thermodynamic properties such as susceptibility or specific heat, and the value at which the cusp in the susceptibility is reached depends on the frequency f at which the measurements are made. In general the higher the frequency, the lower the temperature at which the kinetic arrest seems to occur. However, a plot of the cusp vs. frequency f usually yields a well-defined extrapolated T_c at $f = 0$.[40]

Since the temperature T_c can only be obtained by extrapolation to $f = 0$, it is a matter of controversy as to whether a spin glass undergoes a true thermodynamic phase transition or, instead, simply experiences a kinetic runaway where the relaxation times to the true ground state become unreasonably long.

The above reasoning would seem to indicate that the "glass transition" is just a trivial freeze-out of relaxation. However, Kauzmann pointed out quite a while ago[61] that, as the glass transition is approached, the entropy of the system falls so steeply with temperature that at some temperature T_k the entropy of the liquid is *less* than the entropy of the crystal, which is disturbing. However, the glass transition seems to arrive in the nick of time like the cavalry to keep the entropy of the glass greater than the entropy of the liquid. As Stein has pointed out in a popular article,[96] it is unclear at present if this result means that there really is a thermodynamic glass transition or just a kinetic arrest. Of more general concern is the metastability of glasses: if there is no true underlying thermodynamic phase transition, then all glasses can be viewed as metastable systems out of equilibrium.[59] When it comes to "living" molecules, this is matter of supreme importance.

Up to this point we have discussed how frustration in spin-spin interactions can give rise to a glass transition below which the spins are no longer able to respond to the applied magnetic field, and we have very briefly discussed how the relaxation times of the system seem to diverge below the glass transition temperature and are given by a distribution of relaxation times. The most interesting analogy to be drawn to biological systems still awaits us, however. A protein in a glass state is effectively a dead protein. We would like to know the *dynamics* of the relaxation of the system above the glass transition. It is not enough then to state what the distribution of ground state energies are, since the very existence of a glass transition implies that there must be *very large barriers* between the local minima in free energy. In any kinetics problem one wants to know the pathway by which the system can relax to other local minima. We will now show, one would suspect, that this issue is also quite deep in spin glass physics and new concepts as important as frustration emerge.

The first "new physics" we want to very briefly discuss is the concept of a hierarchical space, which is a form of evolutionary tree. You can "prove" that a set of states forms a hierarchical space if a reasonable definition of distance (not Euclidian!) can be found to characterize the path one must take between different spin states. If the distance is very large, then "yuh can't get thar from here." If any state can be reached by crossing one barrier, then the system is not hierarchical since there is no distance.

Intuitively, one way to explain hierarchical states is to look at a Rubik's cube. Suppose you have some random color distribution (state ψ_r) and would like to go back to the ordered color state ψ_c. If you could arbitrarily change any color, going back to the desired state would be trivial and quick. However, because of the construction of the cube, large free-energy barriers exist between states that are not "close" to the one you are in: you must flow back over the *allowed* states in some very slow process in order to arrive where you want to be. This distribution of allowed states close in "distance" and forbidden states, separated by a large "distance," can give rise to a hierarchical distribution of states.

Now, consider a particular spin glass configuration α. The individual spins in the α configuration can be labeled by S_i^α, which in a one-dimensional case can be viewed as a series of +1 (up) or −1 (down). Consider how you would change the spin state to another state β. One could compare the two spin configurations inside randomly flipping spins, and leave alone the spins that pointed in the same direction, but flip spins that point in opposite directions. For example, suppose that we found that we had to flip the fourth spin of state α. Flipping that spin up or down oscillates between two "nearby" lying states. The bifurcation between these two "nearby" states can be seen as a branch point above the two states. If we restrict ourselves to single spin flips, it is possible to construct an "evolutionary tree" via single spin flips to a common ancestor from which we could descend to the state β. Flowing from a common ancestor via intermediate states this structure is called a hierarchy.

As we stated, in any hierarchy you must have some quantitative way to characterize the distance between different states in the hierarchy. In a spin glass, a convenient way to parameterize the similarity between the two different states might be given by the overlap $q^{\alpha\beta}$ between two states:

$$q^{\alpha\beta} = \frac{1}{N}\Sigma_i S_i^\alpha S_i^\beta \,. \tag{7}$$

Essentially this definition of similarity is a measure of the probability that two spin configurations match. States that are close to one other will have a value for $q^{\alpha\beta}$ that is close to unity, while states that are far away will have a value near zero. Thus, the "distance" between two states would be given by:

$$d_{\alpha\beta} = |q_{\alpha\beta}| \,. \tag{8}$$

(We do not distinguish between states related by a global spin flip.)

Next, in a big leap of faith, we also would expect that the energy barrier between the states would be related in some way to the distance between the states. The actual quantitative relationship must depend on the functional form of the spin-spin interaction. It is reasonable, though unjustified, to assume that they should scale together.

It is believed that the Sherrington-Kirkpatrick (SK) model of a spin glass[90] forms a hierarchical space.[78] This seems fairly certain since, for infinite-ranged spin interactions, it is easy to calculate both distances and barriers. Unfortunately, this case may be pathological because of the infinite range of the spin-spin interaction. In fact, Huse and Fisher[57] questioned whether a real spin glass could have any of the qualities that the SK model has. The more physical Edwards-Anderson type of spin glass involves finite-ranged coupling constants between the spins and also has been proposed to form a hierarchical system,[39] but the issue is unfortunately unresolved.

The pay-off from all of these musings comes when we start to consider the origin of observable, nonexponential time dependence of things like remanent magnetization in the spin glass below the phase transition temperature. We explained this in the above by assuming that a multitude of energy barriers existed, but we did not really justify the existence of the distribution. In the hierarchical scheme, it is possible (but necessary) for a distribution to arise from the many branches of the tree and, by considering diffusion from branch to branch, one also has hopes of actually doing a dynamical calculation on this lattice.

For example, Palmer, Stein, Abrahams, and Anderson[77] (hereafter PSAA) have performed an interesting examination of hierarchically constrained dynamics, in an attempt to understand why some sort of power law was often seen in such systems. In hierarchically constrained dynamics all of the states are assumed to have the same ground state energy, but we wish to observe the diffusion of the probability density $P(t)$ give that we start in one particular state. PSAA were able to show that in various situations kinetics could be observed that fit a variety of nonexponential curves, including power-law decays, as we discussed in the relaxation rate section, or the Kohlrausch decay law:

$$\sigma_R \propto \exp(-t^{\beta}) \tag{9}$$

where β is less than 1. As we have pointed out, relaxation dynamics in disordered systems often seem to fit such a decay law.

Given that the spin states of at least some spin glasses form a hierarchy, we can finally ask what are the mathematical and physical consequences of this hierarchy. Suppose we ask how one can pass from one spin state to another? Since the spins act in a cross-coupled way, with attendant frustration "clashes" occurring between certain configurations, randomly flipping the spins is likely to move along high-activation energy paths that are unlikely to occur.

A consistent and logical approach would be to work through the hierarchical tree of states from one state to another. In this way one always goes through states that are closely related to another, and hence presumably travels over minimum energy routes.

If we define the "distance" D between any two states as the number of generations that one must go back to find a common ancestor, then the concept of distance takes on a decidedly non-Euclidian turn. In fact, in such a space the distance between any three points x, y, and z satisfies the inequality:

$$d(x,z) < \max[d(x,y), d(x,z)] \,. \tag{10}$$

A space which satisfies this relationship is called an ultrametric space.[82] In this simple expression lies a great deal of subtle mathematics. It is not our purpose in this brief review to go into the complexities of these mathematical ramifications, since, in fact, the purpose of this paper is to draw the analogy of spin-glass physics to proteins. We will try to address aspects of ultrametricity that may help us understand the dynamics of protein conformational relaxation and the ordering of protein conformational states.

One striking effect of ultrametricity is the granular nature it gives to different spin configurations. For example, one of the major differences between an ultrametric space and a Euclidian metric is its lack of intermediate states. You can't have three points on a straight line since, if points A, B, and C are on a straight line with one meter separation between A and B and one meter between B and C, we will violate our inequality stated above: A and C can be no more than 1 meter apart in an ultrametric space rather than the two meters needed here.

A consequence of the above section is an important theorem, important to us in the protein section, where a ball is defined as all those sets of configurations that are closer than some distance D from each other. The theorem is: any two balls must be either disjoint or contained within the other; that is, no ball can have parts of itself contained in other balls. In the tree analogy, no branch can belong to two separate trunks. This also implies that any two balls of equal radius must be either disjoint or identical. Consider a ball of unity radius. Let point A be at the center of the ball and point B be on the sphere surface. There can be no point outside the ball less than 1 unit from B, because then it would also be at most 1 unit from A, and thus in or on the ball by our definition of distance. Thus, the sets of spheres are necessarily disjoint. We will return to this important point in the section on proteins.

As we mentioned above, it is not yet clear that the metastable low-energy states of a finite-range, three-dimensional spin glass forms an ultrametric space. The criterion for ultrametricity is quite formal and mathematical; hence, even *thinking* of applying such a concept to a protein must seem like sheer folly and may well be. However, as we will see, there are several aspects of ultrametricity that can be "tested" in protein dynamics simulations.

Finally, we will discuss briefly the question of diffusion in an ultrametric space. In our discussion of diffusion in a hierarchical space, we found that we expect to get nonexponential kinetics. It should come as no surprise that the nonexponential kinetics are also expected in the more confined ultrametric space. The paper that most directly addresses this question is by Ogielski and Stein.[75] They assumed that the bifurcation in the space is formed by a hierarchy of activation energy barriers Δ_i which can be ranked in order of increasing magnitude:

$$\Delta_1 \leq \Delta_2 \ldots \leq \Delta_k . \tag{11}$$

In this scheme the ultrametric distance between two sites is related to the total activation energy that must be surmounted in going from one site to another. Several different activation energy rankings in the ultrametric space were studied. The simplest case of equal barriers δ gave the simple result in the limit of an infinite number of sites

$$P_o(t) \sim t^{-T \ln 2/\Delta} \tag{12}$$

where $P_o(t)$ is the probability of finding site 0 occupied at time t, assuming that $P_o(0) = 1$. Note that we expect a temperature-dependent power law in this case, with the slope of the power law linearly dependent on temperature.

We are walking the reader from the cool and abstract beauties of the spin glass to the wet and wild world of the protein. To ease the reader into this cultural shift, we want to discuss how the dielectric glasses resemble spin glasses in terms of experimental observables.

We believe Dan Stein was the first to draw the analogy between the low-temperature distribution of states and the predictions of the spin glass model.[93,95]. In some sense what Stein did was to sense intuitively the physical similarities between the two systems and use an ansatz to map the distribution in energy states seen in one system over to another system. Thus, while what he did was not rigorous, it provided inspiration for others in the field. Stein simply noted the correspondence between spin glasses and orientational glasses, assumed a Gaussian distribution of activation energies arising from a Gaussian distribution of energies within most spin glass models, and used this reasonable approximation to fit the low-temperature recombination data.

Stein used a Gaussian coupling constant between spins:

$$P(J_{ij}) = \frac{1}{\sqrt{2\pi J^2}} \exp(-(J_{ij}^2/2J^2)) \tag{13}$$

and then supposed that the result that the probability distribution of the metastable spin states of energy E is

$$P(E) = \frac{1}{\sqrt{N\pi J^2}} \exp\left(-\frac{(E-E_o)^2}{NJ^2}\right) . \tag{14}$$

This Gaussian distribution in energy levels was then "frozen" at some temperature T_f, presumably the glass transition temperature of the protein, to yield the final energy distribution:

$$D(E) = \exp\left(-\frac{(E-E_o)^2}{NJ^2}\right) \times \exp\left(-\frac{E}{k_b T_f}\right) . \tag{15}$$

This distribution, "predicted" from spin glass physics, was used to fit the recombination data of T-state Hb. Stein's paper showed a comparison of a Gaussian distribution with the actual data. The fit was reasonably good, although the model by Bowne and Young[105] claimed to achieve substantially better χ^2. It is not clear that the two models clash with one another; probably there are a variety of ways to express phenomena in complex systems, although a Gaussian is an expected functional form for a random process. Indeed, although it does not explicitly use any concepts from spin glasses, the model by Agmon and Hopfield[1] also arrives at a Gaussian distribution.

Actually, although Young and Bowne ruled that the simple Gaussian distribution of activation energy barriers did not fit the recombination data, recent results using the so-called "A" state infrared CO stretch bands, split in myoglobin, seem to reveal that the recombination kinetics on a *single* band are actually quite well fit by a Gaussian distribution![19] The story doesn't seem to be over yet. Probably the main point is simply that a straightforward application of the Edwards-Anderson spin glass model to the barrier distribution in heme proteins gives a "reasonable" fit to the data.

In *Protein Structure: Molecular and Electronic Reactivity*,[10] there are a number of experimental articles showing how a protein shows glasslike properties at temperatures below the glass transition. The most direct evidence is from Gol'danskii et al.[49] and Finegold,[38] who studied the specific heat of hydrated proteins at low temperatures. As Gol'danskii points out, the data is unfortunately spotty in the temperature range of interest.[38,41,102] Although the early measurements by Finegold and colleagues attempted to fit the low-temperature specific heats to either a varying-dimensionality model or to computer simulations, the "modern" view of this data is to fit C_p to a semi-empirical formula:

$$C_p(T) = C_1 T + C_2 T^3 + C_E \left(\frac{\Theta_E}{T} \right) , \tag{16}$$

where C_1 is the specific contribution due to amorphous states as we discussed in the spin glass section, C_2 is a T^3 contribution empirically tied to a Debye relaxation process in three dimensions, and C_E is the Einstein coefficient contribution. With so many variables, it is not too surprising that the above equation fits the limited data rather well. The complexity of the equation should not detract from the fact that the interesting term C_1 is dominant at low temperatures.

The conclusion that one can draw from this is that at low temperatures proteins do indeed seem to resemble disordered glasses in analogy to the spin glass or the orientational glasses in at least one respect: the dominant linear-specific heat dependence. Anderson et al.[5] and Phillips[79] would say that the structure is characterized by a large number of two-level tunneling states, which is closely related to the conformational states we have been talking about. Probably more detailed work should be done in this area. The complexity of the equation used to fit the data causes some discomfort, and we wonder if some of the more recent ideas concerning phonons on disordered lattices[103] might be applicable here. More experimental and theoretical work to be done here, but we have to confess it is not exactly the kind of thing to make the blood hot. Best left to graduate students who like to keep tidy and neat desks. The low temperature-specific heat of proteins doesn't give too much information about what happens up at higher temperatures!

A dynamic picture of the glassy state of a protein at low temperatures can be found by using the technique of flash photolysis, as we discussed in the earlier part of this section. Glass transitions can be identified by dramatic changes in both dynamic and thermodynamic quantities.

The time course of rebinding gives important information about the dynamics of the protein myoglobin. In particular, even if the spin glass analogy allowed us to confidently predict that the protein had many conformational states, if the myoglobin freely sampled all of its states on a time scale much faster than the mean recombination time, then the recombination will be a simple exponential, as is indeed observed at room temperature. Now, as the protein is cooled we might hope that a glass transition will occur, as happens in spin glasses and in orientational glasses. Below the glass transition temperature the system will no longer be ergodic: the dynamical divergence of the relaxation times will keep various protein molecules in various states for effectively infinite times. We should then see deviations from simple exponential recombination. Depending on the solvent, the kinetics change from the temperature-invariant distribution of rates at low temperatures to a quasi-narrow single rate at high temperatures. This change occurs over a narrow temperature range that seems to be linked to the glass transition of the solvent,[58] a fact that has given rise to the concept of a "slaved glass" transition in the protein.

Recently, some exciting work by Hans Frauenfelder and his colleagues has resulted in direct measurements of a nuclear coordinate in the vicinity of the "slaved" glass transition.[58] Frauenfelder has exploited the fact that the CO stretch band of iron-ligated CO in most heme proteins is split into a number of sub-bands,[6] called "A" states by Frauenfelder in his historical identification of the bound CO. These bands are sensitive to many external parameters, including temperature and pressure.

Observation of the ratio of the A_0 to A_1 states as a function of temperature at a static atmospheric pressure revealed that the temperature dependence of the ratio of the states stopped at the glass transition temperature of the solvent. Thus, the ability of the conformational distribution to adjust to temperature seems to halt at the external glass transition temperature, as we would expect from the experiments discussed above. Of greater interest is the question: what is the rate at which the protein is able to approach equilibrium as the glass transition is approached from above?

Actually, this is a rather deep question, especially when recast into some of the language we used to discuss the transition in a spin glass. In other words, is there a true thermodynamic phase transition underlying the kinetic slowdown of the glass transition, or do we merely observe a thermally driven fall out of equilibrium? Frauenfelder was able to look at conformational relaxation by cooling the protein under hydrostatic pressure to a given temperature T and then suddenly releasing the pressure. The ratio of the A states relaxes to a new value appropriate to the equilibrium value at the lower pressure. He observed that, as the glass transition was approached, the relaxation kinetics were nonexponential and highly temperature dependent. A loose analogy to this experiment in spin glass lore would be to suddenly decrease the magnetic field on the spin glass sample at some temperature T and observe the relaxation of the magnetization, although the closer analogy of actually changing the hydrostatic pressure on the sample in a fixed magnetic field, to my knowledge, has not been done.

The interesting parameter that Frauenfelder et al. measure here is the temperature dependence of the relaxation process near the glass transition. Frauenfelder et al. choose to fit the relaxation to a Bassler-Zwanzig function

$$k(T) = k_o \exp\left[-\left(\frac{T_o}{T}\right)\right]^2 \tag{17}$$

rather than the Vogel-Tamman-Fulcher (VTF) relation

$$k(T) = k_o \exp\left[-\frac{E}{(k_b(T - T_o))}\right] \tag{18}$$

to fit the data. The question of which function to use maps back to the question of whether the glass transition is some sort of hydrodynamic arrest or has underneath it some sort of a phase transition. A clear summary of the differences between these two pictures can be found in a paper by Bassler.[16] The argument we have is that the VTF relation has a singularity at the critical temperature T_0 rather than the continuity everywhere of the BZ function. Both curves give adequate fits to the relaxation data, but the extrapolation into lower temperatures is, of course, completely different. Unfortunately, *both* equations give impressively good fits to data over nine orders of magnitude![60]

In the case of glass relaxation work of Frauenfelder, the relaxation of the ratio of the peaks was given by a power law

$$\Phi_r(t) = [1 + k_r(T)t]^{-n} \tag{19}$$

and fits were done with n as a variable and k_r was fit to the hydrodynamic law. It is clear that proteins at present offer no clear test of the troubling question of whether phase transitions do or do not exist in glasses (in general) or proteins (in particular). These are excellent experiments, and intriguing results, but the fundamental questions are still not tested.

However, it seems clear that the protein shows kinetic aspects that glasses and spin glasses show: significant slowing down near some sort of a fixed temperature.

Does a protein show hierarchical relaxation? Frauenfelder's group at Illinous wrote the one paper that made a real stab at using these terms to explain some experimental aspects of protein behavior.[6] In this paper the authors attempted a synthesis of experiments from a large range of physical techniques to show that the structural relaxation of the protein myoglobin could be characterized as a hierarchal diffusion through connected states. The main thrust was to show that the sets of states through which the protein relaxed after photolysis could be characterized within a hierarchical scheme of progressively large motions.

Frauenfelder coined the phrase conformational substates, abbreviated as CS, to describe these hierarchically connected substates. Thus, CS0 would be the set of substates closest to the iron atom, while CS4 would be the substates associated with conformationally distinct protein substates that might have different surface configurations. It is a little strained to connect these CS levels with particular spin glass states since the three-dimensional topology of a protein has no easy analog with a spin glass. Instead, the analogy should be to the "distance" between a particular protein conformation and another one. We defined a distance for one particular spin glass state from another one that made sense, yet no one (except maybe Karplus and Elber, as seen later) has really come up with a way to systematize the concept of distance clearly in proteins—and we must keep in mind that distance may not have the intuitive meaning that we are used to. That is, two protein configurations may be rather close to one another as viewed by x-ray diffraction yet the folding path between the two conformations could be very large. This lack of a clear definition of distance in the protein systems will cause us much grief later in the dreaded ultrametric section.

The next step removes us from any idea in spin glasses and separates the physicists from the biophysicists. No one has ever spoken about a functionally important motion in a spin glass and, probably, would be driven from the high holy temple of condensed matter physics if they did. However, in a proteinlike myoglobin there exists two different sets of conformational substates: those associated with no bound ligand and those that are associated with the bound ligand. The connection between these two sets is via what Frauenfelder called a *functionally important motion*, or an FIM. We can't think of an analogy in the spin glass system that would be physically realizable. Frauenfelder viewed the recombination process as consisting of relaxation between the CS's and lateral movements over via the FIM's.

The paper was possibly flawed by the interpretation of the shift in a near-infrared charge transfer band at 760 K. The evolution of the maximum of this band vs. recombination at low temperatures (less than 180 K) was interpreted as evidence for conformational flow of the structure within the CS_2. Several workers[2,25] have subsequently pointed out that in fact what was occurring was a form of reactive hole burning. That is, each different conformational state of the protein has a particular band near 760 nm and, as recombination proceeds, the band appears to shift as the long-wavelength sub-bands combine. The effect of this hole burning is to make the maximum of the band appear to "move," although no conformational relaxation is occurring. Thus, rather than relaxing, the substates in CS_2 are actually temperature invariant, along the lines of a glass transition.

In our view this flaw in no way invalidates the basic ideas of Frauenfelder; in fact, the existence of hole burning would seem to give strong support to the idea of a distribution of conformationally distinct substates! It would be interesting in our opinion to continue pushing on the hierarchical concept especially in the light of kinetic theories to determine the rate of flow of the substates in the protein and verify that hierarchical complexity serves as a kinetic bottleneck in the process.

In general, to us the analogy of hierarchal structures in the protein seems apt, since the compact folding of the protein in what seems to be a directed sequential manner would imply that the structure must flow through different layers of organization to undergo an arbitrary relaxation.

Questions arise as to whether a real three-dimensional spin glass is ultrametric, so things in the protein arena will be much worse. However, we believe that the question is important, since ultrametricity in proteins reflects upon both how the protein is folded into its structure and on the overlap of adjacent structures. As yet, we cannot predict the folding of a protein given the sequence, so even as abstruse a concept as ultrametricity could help us to codify the problem. Perhaps a clean way to put it is: if ultrametricity is operative in protein conformation space, then the conformational substates of a protein do not arise from a "kicking, screaming stochastic walk" as Gregorio Weber has characterized it but, instead, evolve from paths determined by the previous history of the folding of the polymer.

We'll put our cards on the table right here: it is clear since that proteins form history-dependent conformations, the ultrametric idea is of great importance[68] and full credit should go to Frauenfelder for pushing this concept.

In their paper[6] Frauenfelder and his coworkers also were bold enough to claim that the protein space could be ultrametric as well; they also discussed this concept in several review papers.[45,46] The concept of ultrametricity is of very little use unless a very crisp definition of distance exists by which to address the ultrametricity question. On the basis of computer simulations, Karplus and Elber[36] attempted to come up with a workable definition of distance between protein conformations. They decided that ultrametricity was of little use, but the paper was flawed so the issue not so clearly dead.

In their paper, they did a 300-picosecond simulation of myoglobin structural relaxation. A set of randomly chosen structures consistent with x-ray crystallography were allowed to relax over this time range, and the root-mean-square differences between the structures were compared before and after relaxation to determine if two nearby initial structures converged to a common structure or diverged to separate structures separated by an energy barrier. In essence, this is equivalent to the test within spin glass physics for the presence of nearly iso-energetic ground substates separated by energy barriers.

The basic concept of a glasslike structural space was verified by the observation that the configuration space seemed to be made up of many minima with small energy differences. A test was then made to see if the space was ultrametric. The "distance" between different stable configurations K and K' was defined by:

$$D_{K,K'} = \sum_{i,j} \Delta_{ij}(K, K') = \sum_{i,j} R_{ij}(K) - R_{ij}(K') \tag{20}$$

where $R_{ij}(K)$ is the distance between the amino acid units i and j in configuration K. Similar structures have small $D_{K,K'}$, while dissimilar structures have large $D_{K,K'}$.

How we can use this matrix to test for ultrametricity? Suppose we look at, say, $N = 50$ different stable conformations (denoted by K from 1 to 50). If two structures are closer than some given distance apart, then we can define them to be in the same cluster. Ultrametricity occurs if the grouping of similar structures via the distance matrix results in *disjoint* clusters. Of course, the size of the clusters is dependent on the amount of overlap defined for similarity. What does ultrametricity mean physically for the protein substates? It means that protein structures *within a cluster* evolve like a species, retaining their identity and not blurring into a structure that could arise from another disjoint cluster.

Now, in Elber and Karplus' paper one of us (CC) noticed a mistake in the logic. As the paper was written, the authors confused "distance" with "overlap": overlap is effectively 1-distance. Thus, their statement that "There is a rather sharp transition between the range ($D_{K,K'} \geq 1.5$ Å) when all structures are disjoint, and the range ($0 \leq D_{K,K'} \leq 1$ Å) when all the structures belong to the same cluster."[36] This statement makes no sense as we hope we made clear in the above discussion: if all the structures have less than 1 Å difference, there surely will be none with a difference greater than 1 Å! In a personal communication with Dr. Elber, we received the clarification that the offending sentence should have read: "All the clusters form disjoint clusters for $0 \leq D_{K,K'} \leq 1$ Å and a single cluster at $0 \leq D_{K,K'} \leq 1.5$Å"; that is, there are no structures greater than 1.5 Å apart. The corrected statement is the *logical inverse* of the original statement. This clarification makes the ultrametric nature of the conformational substates not as useless a concept as it appeared!

To see why this is true, let's refer to some data from the actual distance matrix that Dr. Elber sent us. Elber and Karplus picked out, presumably at random, 28 converged conformational substates after 300 ps of relaxation. The matrix elements $D_{K,K'}$ were evaluated for all the possible combinations and scored a 1 if $D_{K,K'}$ was *less* than some value, and a 0 if it was *greater* than some value. For a value less than some small number, such as 0.5 Å, we expect only the unit matrix and this indeed is seen. As the distance cut-off increases, we begin to obtain disjoint groups, for example, as seen at $D_{K,K'} \leq 1.5$Å. There was a problem: the system was not rigorously ultrametric. For distances greater than 1.5 Å, we get one large group ball. Should Elber and Karplus have thrown out ultrametricity? We feel not. The fundamental ultrametric nature of the grouping is actually quite impressive, minus a few problems.

As stated in the introduction, our goal in biophysics is not to prove mathematical theorems! Our goal is to take ideas from some of the powerful and beautiful work that has been done on "clean systems" and try to apply these concepts to help categorize the complex molecules we study. To close, one of the most sophisticated ideas to arise from the study of spin glasses—ultrametricity—was shown to be potentially useful when attacking one of the most nagging problems in biophysics: what is the nature of the conformational heterogeneity of protein structure?

4. ENERGY FLOW IN BIOMOLECULES

We have studied to some depth the conformational complexity of proteins and their thermally driven dynamics. But, proteins aren't just little balls that vibrate in solution. They are molecular machines: they perform chemical feats of magic, transforming chemical potential energy in one molecule into highly directed and specific reactions somewhere else. The question is: are there general physical principles to be learned? Let's examine the issues at hand. Proteins are large, highly condensed polymers that are roughly spherical in shape. If we ignore their important structural roles, their mission is to catalyze chemical reactions in living organisms. The catalyzed reactions often run uphill in free energy and hence require an external source of energy such as adenosinetric-phosphate (ATP). Usually these reactions are *extraordinarily* slow to proceed in the absence of the protein. Thus the proteins act as marvelous mesoscopic reactors of highly specific reactions. If we were chemists, which we most assuredly are not, then our approach might be to treat each protein as a wonderful puzzle, complex and quite unique, to be carefully unraveled and explained based on detailed chemical mechanisms. But, we have this training as physicists where we are taught to look for global mechanisms of unifying importance.

In principle, computers might be able to model the dynamics of proteins. The basic problem in any computer dynamics simulation is how to get an overriding view of the dynamics of the process. If you look at a single atom in the biomolecule, it would appear to be oscillating in some random manner with very little correlation with atoms some distance away. Basically, the whole object seems to a quivering chaotic mass of atoms. The question is, as we discussed above, whether there is any collective aspect to the motions in the molecule. Since Bill Bialek is an enthusiastic promoter of neural processing in organisms, we could make the analogy to studying one neuron firing versus looking for a collective response in the entire neural network.

Angel Garcia pointed out in a provocative paper[47] that collective motions in condensed polymers such as proteins cannot be viewed as an analysis problem: the interactions between the different amino acids are highly nonlinear with the distance between the groups and the polymer held together very weakly so that the nonlinearities are strongly present. The result is that the complex motions are highly anharmonic and metastable in nature: the protein can be expected to reside in some particular conformation for some time τ and then to switch to another conformation. Note that, because of the metastability expected of these conformations, even if the system is highly overdamped and dissipative, it is still possible for the system to display anomalously long "excited state" lifetimes.

The protein was simulated in an aqueous environment of 1,315 water molecules. The temperature of this simulation was 300 K. The most striking thing in this simulation was that the motions of the amino acids were not distributed in a simple Gaussian manner around an average position, nor was the motion a random walk in phase space. Rather, the motions were metastable in nature, characterized by confinement to a particular angular range for an extended time followed by a very rapid jump to another metastable minimum. The motions were also correlated, as one would expect in a highly condensed molecule: as one amino acid moves one way, others must move the other way to make room. Interestingly, the average lifetime of these metastable states is on the order of 100 picoseconds, although the transition to a particular state takes about a picosecond, indicating that the actual structural transition time is quite rapid.

Such a simulation is done in a fluid solvent, and no attempt has been made to calculate the oscillator strength connecting the states, so it may be that these collective modes are in no way correlated with the broad continuum of absorbance observed in protein thin films. However, it is intriguing to note the correlation between the metastable lifetimes and the saturation recovery times observed in the experiment discussed here.

We have explored using far-infrared (FIR) photoexcitation of perhaps functionally important collective modes in proteins. FIR is defined in different ways: we will take the semi-arbitrary cut-off of FIR as those excitations lying below 400 cm^{-1}. Looking back at these experiments, we would say that they definitely put the horse before the cart: we tried to see *if* a change in a reaction rate could be observed when we pumped in the FIR. One rationale for doing this reverse-order experimentation is simply to see if there is anything going on: if not, then forget it!

In our case we have seen that FIR pumping at 50 cm^{-1} to 80 cm^{-1} does seem to influence reaction rates in an *athermal* way. What does *athermal* mean anyway? It means that if you pump energy into some particular state, then the energy resides in that state for a "substantially" long period of time. While the energy is in that state and not in other states, the system is not in thermal equilibrium. Hence, it is athermal. A thermal change in the rate would occur if the energy of the absorbed FIR photon thermalizes *very rapidly*, on the order of picoseconds, into all the degrees of freedom of the system. Thus, while a specific mode may be heated to an effective "temperature" 100 K in the absorption of a 100-cm^{-1}photon instantaneously, after relaxation the amount of energy/mode increase is on the order of 10^{-2} K in the end. This is what one of our skeptical colleagues at Bell Labs meant when he characterize the FIR excitation as a blowtorch, quickly heating all modes. However, one can do experiments to find out how long these FIR modes live.

Bacteriorhodopsin (bR) is a very interesting protein that has many important properties. The molecule has a molecular weight of 26,534 daltons and consists of a single polypeptide chain consisting mostly of seven α-helical stretches which repeatedly span the membrane in which the protein is found.[51] bR has a chromophore (retinal), a linear conjugated polyene, which is covalently linked to lysine-216 in the backbone of the protein. The retinal undergoes conformational cis-trans transitions when photo-excited by a visible photon[98] and ultimately the pumping of a proton (a positive charge, that is) across the membrane "uphill" against the chemical potential gradient.

The trans-cis conformational transition of the chromophore in bR leads to conformational transitions in the protein itself.[14,34,70] Usually, these conformational transitions are monitored *indirectly* by observing the perturbations that structure makes upon the chromophore absorbance. The visible work has been very important: it has shown (a) that the trans-cis isomerization occurs very rapidly, in less than a picosecond[71] and (b) the protein cycles through a series of metastable states that can be characterized by the absorbance of the chromophore.[24] The lifetimes of these states are highly temperature dependent and show freeze-outs over a range of temperatures, some as low as 100 K.[76] Since this temperature corresponds to energies on the order of 100 cm^{-1}, it has been speculated that, in fact, collective protein motions are involved in this kinetic freeze-out.[72] It is interesting that at low temperatures the K state (bR_{630}) and the light-adapted bR *trans* ground state (bR_{568}) can be created reversibly by absorption of the appropriate color photon.[106]

The presence of these intermediate states tells us something interesting: in these large structures, metastable conformational states are separated by energetic barriers. It is impossible to tell from either the optical spectra or mid-IR spectra what exactly is the conformation of the macromolecule in these metastable states, but the relatively low energy barrier of the state suggests that the state is a "soft" state consisting of the perturbation of many atoms over small distances rather than a highly localized deformation.

As interesting as these optical measurements may be, ultimately one really wants to probe the protein conformation as a function of time. One way to do this is to probe directly in the infrared where the local vibrational transitions of the molecular elements of the protein are evident. In fact, there is a substantial amount of work in the *mid-IR* (3000 cm^{-1}to 500 cm^{-1}) of structural changes in bR,[48] using time-resolved FTIR. Although many of the features are dominated by chromophore changes and, in fact, can be compared to the resonance Raman results,[62] protein features also can be ascertained. In particular, the backbone amide stretch from 1671–1650 cm^{-1}, which is a sensitive indicator of the α-helix conformation, shows the same time-dependent changes as the chromophore bands do. This seems to indicate that, as Gerwert et al. stress, "all reactions in various parts of the protein are synchronized to each other and no independent cycles exist for different parts."

Although the amide stretch band is a good place to look for semi-local conformational changes, it would be better if indicators could be found for truly large-scale motions in the protein. It would be nice if we had information about the FIR portion of the spectrum where one would hope that the collective modes might have some oscillator strength variations with conformation, but none exists.

The most interesting and experimentally accessible states are the light-adapted bR_{568} and the first excited-state K_{630} states, easily isolated at low temperatures (below 70 K). So-called light-adapted bR_{568}, that is, bR which has been continuously exposed to light levels on the order of milliwatts/cm^2, will have its chromophore in the *trans* configuration at low temperatures if cooled quickly in the dark. As mentioned above, if green light is shone on the sample, the chromophore makes a trans-cis isomerization[98] and the protein is now in the K state, with chromophore absorbance maximum at 630 nm. This trans-cis isomerization is actually photoreversible at low temperatures: illumination with red light drives the system back to the trans state. Both states seem to be very stable at temperatures below 40 K: the thermal relaxation rate between the two states seems to be very slow. However, at 70 K the relaxation rate is quite fast, on the order of milliseconds.

It is known that the trans-cis photo-driven energy storage is highly effective: of the 2 eV carried by the visible photon, about 1 eV is stored as chemical energy.[52] There is something important and puzzling in this last fact: somehow the protein absorbs 2 eV and stores 1 eV by effectively transferring a charge over at least 50 Å. One would guess that maybe energy could be stored locally by the trans-cis isomerization, but how can energy be transferred over a large distance without losing it?

One possibility is that the energy is stored "mechanically" in a strained conformation of the protein. In fact, this is the essence, we think, of John Hopfield's obvious but subtle idea of how the R–T free difference is stored in hemoglobin.

Such an idea still has problems, however. A strained conformation implies that many atoms over a large volume of the protein are slightly moved relative to the conformation before the "event" to a metastable configuration. At least initially, this conformational shape change must occur fast enough to compete with thermal and viscous relaxation, since the pre-event state is a local minimum in free energy by definition. It is easy to see how 2 eV photons could use the Franck-Condon[17] effect to snap a protein into a metastable state, but how can soft energies on the order of hundreds of cm^{-1} do such a thing? Remember that the protein doesn't stay in the initial metastable state after the "event"; it moves to other metastable states with high retention of the initial energy deposition. It stays cocked.

There is one highly disputed answer to this puzzle, namely that the energy is transferred in the same way water waves transfer energy over large scales in a highly viscous medium: by excitation of large-scale collective motions whose group velocity carries energy faster than the relaxation process can remove it. The key is to make the wavelength sufficiently long so that the diffusive relaxation time becomes very large compared to the wave period. Of course, the longest possible wavelength in a protein is its diameter, about 50 Å typically. If the speed of sound c in a protein is about 10^5 cm/sec (typical for liquids), the frequency of the wave is approximately 10^{11} Hz. Can such a wave travel across a protein without attenuation?

What sets the time scale for that relaxation? We roughly can guess this criterion in a protein. Let's imagine that again we have a protein 50 Å in diameter. We assume again that the sound group velocity c in a protein is roughly the same as in water, 1.4×10^5 cm/s.[4] Attenuation (relaxation) of acoustic waves occurs through two basic mechanisms: viscous damping and thermal diffusion. Of course, it isn't clear that the macroscopic equations that came from the Navier-Stokes equation are applicable at the length and frequency scale we wish to discuss: we are at the murky line between mesoscopic and atomic scale phenomena. However, it is interesting to see what the predictions of continuum mechanics are.

The attenuation coefficient α_{thermal} for an acoustic wave due to thermal diffusion is

$$\alpha_{\text{thermal}} = \frac{\gamma - 1}{\gamma} \times \frac{\omega^2 \kappa}{2\rho c^3 c_v} \tag{21}$$

where γ is ratio of specific heat at constant pressure to constant temperature, ω is the frequency of the wave, κ is the thermal conductivity, ρ is the density of the fluid, and c_v is the specific heat/gram at constant volume. If the attenuation length is set to the diameter of the protein and we assume that a protein macromolecule physical properties is approximately like water is, we find that the maximum frequency ω_{max} at which we can expect acoustic transmission of energy is roughly 2×10^{12} Hz. Incorrectly assuming a linear restoring potential, we find that the corresponding phonon energy, is equal to $\hbar\omega_{\text{max}} = 1.1 \times 10^{-3}$eV or 10 cm^{-1}. This corresponds to FIR frequencies if these oscillations are excited by absorbed photons, and we would expect that 10 cm^{-1} photons could excite acoustic modes which could propagate across the molecule. The attenuation coefficient α_{viscous} for viscous damping is given by

$$\alpha_{\text{viscous}} = \frac{2\omega^2 \eta}{3\rho c^3} \tag{22}$$

where η is the viscosity of the medium. For illustration, we will do the calculation for water at 20°C. Our experiments are mostly done at cryogenic temperatures where the protein is undoubtedly a solid, in which case this expression is meaningless. In any event, we find again that if we want transmission across 50Å, the maximum frequency ω_{max} is 3×10^{11} Hz. This would indicate that viscous damping would strongly attenuate FIR modes; in fact, this may be why we observed no signal with FIR excitation at 50 cm^{-1} above approximately 180 K in myoglobin.[11] However, the decoupling frequency of the solvent from the molecule (mid-IR transitions in solution of course are quite sharp) and the internal viscosity of the protein leave many questions unanswered.

In sum, what we have in mind is in this frequency range, collective modes, can have a long enough attenuation length and can live long enough to move the protein to a collectively strained state. These stress-carrying waves should propagate in picoseconds across the 50 Å protein and move it to a strained configuration. They are *not* solitons, Davydov or otherwise[28]—they carry very small amounts of energy. Nor do our ideas agree with a model proposed by Bialek and Goldstein,[21] which basically proposed the idea of a distribution of metastable conformational states. The strained metastable states are the source of reaction-rate heterogeneity at low temperatures. Further, we wish to point out that what we are proposing here is in some sense a version of Hans Frauenfelder's brilliant idea of conformational substates in proteins that have a corresponding distribution in reaction rates.[12] We simply claim here that we can pump between Hans' states with FIR radiation. Can it be done?

The first experiment we did was simply to find if pulsed FIR radiation had any effect on the bR photocycle bR_{568}–K_{630} at low temperatures. (See our papers on FIR effects in electron transport[13] and ligand recombination[11] for details about the technique.) In the case of the bR experiments, thin films of hydrated bR films were deposited on a plastic disk (TPX, a common plastic, has excellent transmission in the optical and FIR regions) via careful deposition of a solution containing bR vesicles. The final optical density of the films was quite high, on the order of 10 optical density units (OD) at the 568-nm absorbance maximum of light-adapted bR (an OD is the base-ten absorbance: $I = I_o 10^{-OD}$). The bR thin films were held in a copper sample holder mounted in a Janis flow cryostat with z-cut quartz optical windows. A quartz-halogen lamp with a 3-inch water filter to remove IR was used to maintain the bR in a cycling state: the green portion of the spectrum drove the bR to the K state, and the red portion of the spectrum drove the K-to-bR cycle. The basic idea is that the system will settle down into an equilibrium concentration of K and bR. If it does allow some of the protein to leak out of the K state pulsed FIR, will leave a perturbed amount of bR and K states immediately after FIR pulse. If the ratio of the bR-to-K is monitored at an appropriate wavelength sensitive to the amount of (in our case) the K state present, then we would expect a *prompt* change in the absorbance of the signal. *Prompt* means that, since we expect that any sort of reasonable lifetime for FIR excitations will be less than a nanosecond, the FIR sets up a new quasi-equilibrium set of rate constants during the laser pulse. When the FIR laser pulse is over, the rates relax back to the normal rates in nanoseconds. In our case, we used a 650-nm filter of 20-nm bandwidth to monitor the amount of K-state absorbance.

Since we have length constraints, we won't go into a detailed description of the experiment. The results show that the absorption changes occur during the 5-μsec duration of the FEL pulse and, in fact, the change rises linearly with time during the pulse. Other than FIR-induced reaction rate changes the only reasonable explanation that we have for such a signal would be a change in the absorption spectrum due to sample heating. However, unlike the case of the reaction centers where the 860 special pair band turned out to be surprisingly temperature dependent,[13] the temperature dependence of the absorption spectra of bR is far less than the 860 band[18] and the same analysis we applied to the reaction center system will show that we cannot explain the observed signals in bR. The FIR-induced absorbance change seen at 70 K has a changed sign and now indicates an absorbance increase with FIR irradiation. In fact, we find that the FIR-induced bR signal for a particular hydrated sample has two temperatures where no signal is seen, at approximately 50 K and 150 K. It is unlikely that any simple temperature induced broadening could give rise to such a complex scenario, but it is possible that the thermal occupation of further states of the bR photocyle and subsequent FIR perturbation of the rates could give rise to a complex signal compared to temperature. Because of this complexity, we will restrict our comments to the signal observed at temperatures below 50 K, where we presumably know the state of the protein. We also have to point out that the sign of the signal seen at room temperature (data not shown) is a function of the hydration of the sample: highly dried samples show an absorbance decrease with FIR, while highly hydrated samples show an absorbance increase. Again, a simple thermal broadening is probably not responsible for this behavior.

A last and highly informative piece of data can be found in a "double pulse" experiment. In this experiment only steady-state red light (650 nm) and no green light illuminated the sample; hence, the sample should be predominantly in the bR_{568} state. A 532-nm doubled YAG pulsed laser (10 nsec) was used to transiently drive the system to the K_{630} state and to increase the absorbance. That was indeed observed. Then, a fixed delay after the 532-nm pulse, the FIR FEL was fired and resultant absorbance change observed. If our analysis of the signal at low temperatures is correct, an *increase* in transmission should be seen—and it is. If the Nd:YAG laser is not fired, then no signal is seen, consistent with the claim that at least at low temperatures the FIR signal comes exclusively from the K state. Both the K-state signal induced by the 532-nm laser pulse and the FIR-induced change are eventually lost due to optical pumping of the red light of the monitoring beam.

It is informative from the Nd:YAG experiment to estimate the quantitative influence of the FIR pulse. Since no signal is seen when the Nd:YAG laser does not fire, there is effectively no base of K state molecules present in the absence of the Nd:YAG laser pulse. At the time of the FEL pulse, the "amount" of K state is approximately 50 mV in terms of ΔI. The FIR shot removes about 4 mV of K state in 12 μsec. Since we have confirmed (data not shown) that the signal is linear with FEL energy, we get simply that the rate of K state loss is $(4/50) \times 1/12 \times 10^{-6}$ sec^{-1}, or 7×10^3 sec^{-1}. If we know the lifetime of the collective modes and the pumping rate of the FIR photons, we can convert this number to a true transition rate in the excited state.

Suppose that collective modes can be excited and that in the excited states the reaction rates are different. The most important question to be answered is: what are the lifetimes of collective modes in proteins? If the lifetime is on the order of period one (about 10 picoseconds from our calculation above), then there is a possibility of significant kinetic steerage of a reaction. Lifetimes on the order of 100 ps to 1 nanosecond would represent considerably less damping and greater efficiency. A one-nanosecond lifetime would be fantastic. What are the relaxation rates?

The brute force way to measure the lifetime of an excited state is to do a pump-probe experiment: put in enough energy to equalize the populations in the ground and excited states, then probe with a weak beam the resulting loss of absorbance with time after the pump. The relaxation rate $\mathcal{K}_r$ of the excited state, also known as the longitudinal relaxation rate or $1/T_1$ in analogy to NMR jargon, is obtained by measuring the recovery of the initial ground-state population. Unfortunately, the UCSB FEL has a 10-μsec FIR pulsewidth T, much too long to do direct time-resolved work. The next best thing is to make a quasi-equilibrium pump-probe experiment: assume that the relaxation rate is fast enough (that is, $1/\mathcal{K}_r \ll T$) and a steady state saturation is obtained during the FIR pump pulse. Measurement of this bleach then can be indirectly back-calculated to obtain the K_{-1}.

There is a clever way to compose an experiment that doesn't require the separation of the FIR beam into a separate pump and probe beam and doesn't require great linearity in the FIR detectors. Let the FIR beam be focussed onto a spot of diameter D in the material of interest. The material is assumed to have an absorbance A_s (the absorbance is the same as OD, as we mentioned above). Let there be an absorbing filter of absorbance A at the frequency of interest. Let the incoming FIR pulse have total energy E and pulse duration T. A FIR detector is placed after the sample. A beam splitter in front of the entire apparatus is used to measure the (possibly variable) beam energy in a separate detector.

If the filter is put in *front* of the sample, then the energy incident on the sample is $E \times 10^{-A}$ and the energy incident on the detector is $E \times 10^{-(A+A_s)}$. Let the filter now be put in *back* of the sample. The sample now sees the full pulse energy E, but the detector still has incidently the same attenuated energy $E \times 10^{-(A+A_s)}$. Clearly, if the incident power $P = E/T$ is insufficient to cause appreciable saturation, the ratio of incident energy to detected energy will be the same, independent of the filter position. However, if the FIR power is sufficiently high to saturate the system, then the detector will record greater energy transmission when the filter is placed *after* the sample than when it is placed *before* the sample. Call this ratio $\mathcal{R}$:

$$\mathcal{R} = E_{\text{after}}/E_{\text{before}} \,. \tag{23}$$

Since we know the OD of the sample at the illuminated FIR wavelength, the number of molecules of the absorbing molecule from the visible OD, the energy/photon, the energy of the FIR laser pulse, and the pulse width, we know the rate at which FIR photons are hitting the protein molecules. Call that rate $\mathcal{K}_p$. Further, we assume that the relaxation rate of the excited state is an unknown number $\mathcal{K}_r$. If the laser pulse is much longer in duration than the mean excited-state lifetime, then the sample will come into a steady-state value of ground N_g and excited-state population N_e. In equilibrium we have

$$\frac{dN_g}{dt} = -\mathcal{K}_p N_g + \mathcal{K}_p N_e + \mathcal{K}_r N_e = \frac{dN_e}{dt} = +\mathcal{K}_p N_g - \mathcal{K}_p N_e - \mathcal{K}_r N_e = 0 \,. \tag{24}$$

This yields

$$\mathcal{K}_r = \mathcal{K}_p \left(\frac{N_g}{N_e} - 1 \right) . \tag{25}$$

Finally, we note that

$$\mathcal{R} = \frac{N_g}{N_g - N_e} \tag{26}$$

in the limit where $N_e \ll N_g$. We finally have

$$\mathcal{K}_r = \mathcal{K}_p \frac{1}{1 - R} . \tag{27}$$

The experiment to determine $\mathcal{K}_r$ was carried out recently at the UCSB FEL. The protein sample on which we made measurements was essentially the same bR thin film as that used in the above experiments. The bR sample was light adapted before cooling to 7 K. A FIR absorption spectrum of the BR sample was measured to have an absorbance at 100 cm^{-1} of 1.0 OD. The sample was mounted in a cryostat so that the temperature could be easily varied. Since we expect that the T_1's for proteins will be rather short, we need the highest possible intensities for the FIR pump beam to see appreciable saturation. The highest flux FIR source at the UCSB CFELS is a CO_2 pulsed laser which drives an alcohol vapor column to produce stimulated Raman emission off the rotational levels of alcohol molecules. The FIR output from this laser is quite intense: typically 1.0 millijoules in a 50-nanosecond pulse width, photon energy 100 cm^{-1}. The FIR beam was focussed to a spot size of 4×10^{-2} cm^2 on the sample. This corresponds to about 10^{19} photons/cm^2. The sample OD at 563 nm was 10 and, using a ϵ of 60 $mM^{-1}cm^{-1}$, gives about 4×10^{15} molecules in the illuminated spot. We then find that the pump rate $\mathcal{K}_p$ is 2×10^9 sec^{-1}. That is, every 0.5 nsec, a bR molecule absorbs a 100-cm^{-1} photon.

After some false starts due the nasty ability of FIR radiation to bounce all over a lab, we found that the ratio of the signal detected on the pyroelectric detector with a 1.0-OD filter in back of the sample (I_b) and 1.2 ± 0.05 in front of the sample (I_f)—with the sample at 7 K. Note that the ratio of I_b/I_f is greater than 1, as expected for a saturation of the absorbing levels in the FIR. Simple tests, such as the use of just the TPX plastic sample holders, yielded nulls with R = 1.00 ± 0.01. The saturation was only evident when the sample BR film was present.

A value of $\mathcal{R} = 1.2$ then yields our desired result, namely that *the relaxation rate* $\mathcal{K}_r \sim 10^{10} sec^{-1}$ *at 100* cm^{-1}! Now you know. You may recall that this is roughly, within an order of magnitude, of what we guessed would be the relaxation rate from simple Navier-Stokes equation noodling.

The reader should be aware that a *100-ps lifetime for a 100-*cm^{-1} *mode* is pretty heretical. That is a reasonably long time for energy to bounce around the protein before it becomes thermalized.

Now that we have $\mathcal{K}_r$, we can continue our calculation to get the excited-state reaction rate. The pump rate $\mathcal{K}_p$ in the Nd:YAG laser experiment has to be modified since the [CO_2 laser was not used, but instead the CFELS FEL was used instead of the CO_2 laser]. The pump rate is modified by the increased pulse energy (16 mJ vs. 1 mJ) and the increased pulse width (12 μsec vs. 0.05 μsec). The net effect is that the pump rate in the Nd:YAG experiment should have been approximately 10^8 sec^{-1}. Thus, FIR photons excited the bR molecules only every 10 nsec in that experiment, and we believe the excitation rattled around the protein for about 0.1 nsec after excitation. The effective rate then has to be multiplied by a "duty factor" of 100. *Finally, when excited by 100* cm^{-1} *photons, the rate of K to "something" is approximately* 7×10^5 sec^{-1}. This number isn't totally crazy: the rate of decay of the K state at room temperature is approximately 10^6 sec^{-1}.

Recently we have become aware of some work concerning the effects of "resonant" activation over a fluctuating barrier.[32,97] Here, if an activation barrier fluctuates at the same rate as the characteristic crossing rate, then the true rate of crossing can be enhanced greatly. Perhaps the FIR oscillations that we create in the protein act as resonant activators, if we assume that the characteristic fluctuations of the barrier are also in this terahertz range. I don't know, but I believe that the signals are real.

5. DNA ELASTICITY

DNA is equally as fascinating as proteins in biological systems—and then there are the RNA's, the lipid systems, the list goes on. For lack of superhuman or even human strength, we will confine ourselves to some studies on DNA that we have carried out.

Of interest is the flexibility of DNA. Since B-DNA consists of stacks of almost-flat basepairs to first order, we can imagine that the flat planes of the basepairs are basically rigid and that the bending of DNA is caused by increasing the distance between adjacent basepairs, and furthermore that the twisting of DNA is due to the shearing, or sliding, of adjacent basepairs over each other. In this model all of the rigidity of DNA is due to basepair *stacking* interactions and not the Watson-Crick hydrogen bonding between the basepairs. If, on the other hand, deformation of the helix also strains the Watson-Crick hydrogen bonding pattern of the basepairs, then the Watson-Crick hydrogen bonding will also play a role. Finally, the negative charge of the phosphate groups on the backbone imparts a substantial contribution to the rigidity, which is most evident at low (less than 1 mM) salt concentrations.

It is not convenient to give the DNA stiffness as a spring constant, since that is an extensive parameter that varies with the length of the spring and the strained area. One characterizes the intrinsic rigidity of an elastic material by the elastic moduli of the material, which are intrinsic parameters independent of the size or shape of the material. There are two elastic moduli that are of interest here—the Young modulus E and the shear modulus G. The Young modulus is responsible for the restoring forces that are felt when an object is simply stretched or bent, while the shear modulus is responsible for the restoring torque that is felt when an object is twisted. Of course, what is actually measured in a material is the net effective stretching spring constant k_s and twisting spring constant k_t. The stretching spring constant k_s of a material of length L and cross-sectional area A is:

$$k_s = \frac{EA}{L}. \tag{28}$$

Bending an object is a considerably more complex deformation than stretching or twisting an object. Internal restoring torques originate from the differential stretching and compression of sections of the material above and below the unstrained neutral plane of the object. The internal torque τ which exists in an object bent into an arc of radius R is

$$\tau = \frac{EI_A}{R} \tag{29}$$

where I_A is the surface moment of inertia measured in the x-y cross section of the rod:

$$I_A = \int x^2 dxdy. \tag{30}$$

If the rod is of length L, the energy U stored in the rod is:

$$U(R) = \frac{-EI_A L}{2R^2} = -\frac{B}{2R^2} \tag{31}$$

where we have defined $B = EI_A L$ as the *bending spring constant.* It is interesting to recast this expression in terms of the angle θ that is formed by tangents to the ends of a rod of length L bent into an arc of radius R. Since $R \times \theta = L$ we have simply

$$U(\theta) = \frac{EI_A}{2L}\theta^2 \tag{32}$$

which is, of course, the expected harmonic response. In thermal equilibrium the angle θ will be distributed therefore in a Gaussian distribution with width depending on temperature.

Now consider twisting the rod of length L through an angle α. The restoring torque τ applied at the free end of the rod is

$$\tau = \frac{GI_p}{L}\alpha \tag{33}$$

where G is the shear modulus and I_p is called the polar moment of inertia

$$I_p = \int r^2 2\pi r dr \tag{34}$$

where r is the distance from the center of mass of the cross section of the object. It is convenient to define the quantity C as the torsional rigidity of the rod:

$$C = GI_p. \tag{35}$$

The torsional spring k_t is then C/L. It is important to distinguish between extrinsic spring constants and intrinsic rigidities when reading the literature! One last point: the potential energy term for strain as a function of angle is quadratic in α, so the torsional angles will have a Gaussian distribution by the Boltzmann relation at temperature T.

The two quantities E and G are related to each other by Poisson's ratio σ:

$$G = \frac{E}{2(1+\sigma)} . \tag{36}$$

Typical values for σ seem to be about 0.5 (negative values of σ occur if objects expand when stretched). The coupling between G and E is a deep subject that needs to be explored since it seems to be clearly related to the complex structures formed in super-coiled DNA.

Finally, we need to define the very important *intrinsic* quantity called the persistence length $\mathcal{P}$ which is related to the bending rigidity of a material. As the name implies, $\mathcal{P}$ is a statistical length over which vectors tangent to the symmetry axis of a polymer are correlated. If one wants to make a simplified calculation of the RMS radius of gyration of a very long polymer (length L) and persistence length $\mathcal{P}$, then the standard random-walk arguments can be used if you let the length of the random step be the Kuhn random flight length, $2\mathcal{P}$. The relation between the persistence length $\mathcal{P}$ and the Young modulus E is

$$\mathcal{P} = \frac{EI_a}{k_b T} \tag{37}$$

where I_a is the surface moment of inertia, as before.

Schurr has written a wonderful paper[85] that summarizes his work with Allison and concerns the anisotropy decay of a deformable molecule with mean local cylindrical symmetry.[3] The rotational diffusion equation has three degrees of freedom corresponding to the three Euler angles. It is possible to write the solution to the diffusion equation in terms of eigenfunctions to the diffusion equation which of course are nothing more than the spherical harmonics Y_{lm}. When the excited state is a simple dipole transition, only the spherical harmonics $\mathrm{Y}_{2,m}$ have nonzero eigenvalues and the anisotropy decay must have the form

$$r(t) = \sum_{m=-2}^{2} I_m F_m(t) T_m(t) \tag{38}$$

where I_m is a factor containing the Y_{2m} spherical harmonics, $F_m(t)$ is a function related to the bending of the polymer, and $T_m(t)$ is related to the twisting of the polymer. A more specific expression is:

$$\begin{aligned} r(t) =& \frac{2}{5}\left(\frac{4\pi}{5}\right) \sum_{m=-2}^{2} \langle Y_{2m}^{*}(\omega_R(0)) Y_{2m}(\Omega_R(t))\rangle \\ &\times \exp\left[-\frac{1}{2}(6-m^2)\langle\Delta_x^2(t)_R\rangle\right] \exp\left[-\frac{1}{2m^2}\langle\Delta_z^2(t)\rangle_R\right] \end{aligned} \tag{39}$$

where (in Schurr's words) $\omega_R(0) = (\mu(0), \nu(0))$ is the instantaneous orientation in polar coordinates of the absorption dipole *in the frame of the binding site on the molecule* at time $t = 0$; $\Omega_R(t) = (\epsilon(t), \zeta(t))$ is the instantaneous orientation of the emission dipole at time t; and the angular brackets $\langle\ \rangle_R$ imply an average over the dye molecules in the body frame of the molecule. Most importantly, $\Delta_x^2(t)$ and $\Delta_z^2(t)$ represent the *fixed* mean *squared* angular displacements of the body along the transverse and symmetry axis, respectively. An intuitive way to view these terms is to call the $\Delta_x^2(t)$ the bending terms and the $\Delta_z^2(t)$ the twisting terms as we stated above. As Schurr points out, the determination of the $\Delta^2(t)$ functions is a separate statistical-mechanical and hydrodynamic problem that has by no means been solved. In fact, they will occupy us for the remainder of this paper.

It often is the case that the absorption and emission moments are parallel to one another in fluorescence work: in the case of singlet depletion techniques, they necessarily are so. In that case the spherical harmonic functions become the Legendre harmonics. The angle θ is the angle that the transition moment makes with the symmetry axis of the polymer. Our expression becomes:

$$\begin{aligned} I_0 &= \frac{(3\cos^2\theta - 1)^2}{2} \\ I_1 &= 3\cos^2\theta\sin^2\theta\,, \\ I_2 &= \frac{3}{4}\sin^4\theta \end{aligned} \tag{40}$$

$$F_m = \exp\left[-\frac{1}{2}(6 - m^2)\Delta_x^2(t)\right], \tag{41}$$

$$T_m = \exp\left[-\frac{m^2}{2}\Delta_z^2(t)\right], \tag{42}$$

where we have listed the separate angular, bending, and twisting functions respectively. The problem is to compute the bending and twisting functions.

The major complication is that DNA is not a rigid molecule. Thus, while the anisotropy decay of rigid objects is well known, the anisotropy decay of *flexible* objects is very difficult to compute and no analytical solutions exist.

The dynamics of a semi-flexible molecule such as DNA as viewed by an intercalated dye is a formidable problem that has not yet been solved analytically. One way to handle such a problem is semi-brute force: do a Brownian dynamics simulation. We will compare the Brownian dynamics simulation to the analytical theories that have been developed but, to our mind, it is probably the Brownian dynamics that have the greatest chance of being "right." Unfortunately, the sheer magnitude of the computing problem precludes using a computer to model dynamics from the sub-nanosecond to the millisecond time scale, which one would really want to do in order to capture the full dynamics. First I will discuss the analytical theories.

We need to compute the Δ^2 functions. Barkeley and Zimm[15] made a serious attempt to model the dynamics of DNA over appropriate time scales. They treated the DNA molecule as a continuous, isotropic elastic rod. The fundamental idea of their theory is to compute the angular position of an arbitrary dipole moment as a function of time after excitation by photoselection. Since one deals with an ensemble of dipoles, the problem boils down to calculating the probability $\psi(\alpha, t; \alpha_o, 0)$ of finding a dipole with orientation α at time t when the initial orientation is α_o at $t = 0$.

The basic equation for twisting motions is

$$\frac{\partial \alpha}{\partial t} = -\left(\frac{C}{\rho L}\right) \frac{\partial^2 \alpha}{\partial z^2} \tag{43}$$

where C is the torsional rigidity of the rod (given by Eq. 8 in terms of the shear modulus), $\alpha(z, t)$ is the dipole angle in the cross-sectional plane of the polymer as a function of position z, and ρ is the rotational frictional coefficient per unit length, such that

$$\rho = \frac{k_b T}{D_{\text{par}} L} = 8\pi\eta b^2 \tag{44}$$

where D_{par} is the twisting diffusion constant used in the rigid-rod discussion above. Note that since twisting around the symmetry axis is essentially a one-dimensional problem, the diffusion constant per unit length is independent of length!

The equation of motion for bending motions is much more difficult to write down since the bending motions can be strongly coupled together by hydrodynamics. This can be seen in the tumbling diffusion constant D_{perp} which is dependent on the *cube* of the length, meaning that, unlike in the twisting motion, no frictional coefficient per unit length exists which is independent of the length of the rod. It is "straight forward" to write down the restoring force $F(x)$ per unit length acting on a rod bent in the x-y plane:

$$F(x) = -EI_a \frac{\partial^4 y}{\partial x^4} . \tag{45}$$

Note that since bending is effectively a two-dimensional displacement, the differential equation is more complex. If we wanted to be simple-minded, the damping force per unit length ρ_o would be expressed in terms of the diffusion constant for tumbling of a rigid rod:

$$\rho_o = \frac{k_B T}{2D_{\text{perp}} L} \sim \frac{4}{3\pi\eta L^2} . \tag{46}$$

It still really bothers me that there is an explicit L term in the frictional force term per unit length. In reality the Oseen-Burgers tensor must be used.

It is good to pause here and reflect upon the complexities and simplicities of our problem, especially the bending equation. The twisting equation is relatively straightforward and at short times represents diffusion in a harmonic (parabolic) potential. However, while on a harmonic surface, the bending equation of motion, however, while on a harmonic surface is nonlinear due to hydrodynamics. Note also that inertial terms play no role since we assume that the motion is totally overdamped.

We still have Gaussian distributions for the angles because of the harmonic surface that the diffusion occurs in. This means that the basic solution of Schurr, which assumes Gaussian distribution functions, should still be correct. The time-dependent width σ of our Gaussian distribution function of angles will not go as $t^{1/2}$ because the diffusion is now constrained by the harmonic surface and would be expected to go with some lower power of time due to that constraint. Although the problem of one-dimensional diffusion in a harmonic potential has been solved by Chandrasekhar long ago,[27] I see no obvious way how his solution for small t has a simple limiting expression. As we will discuss, the limiting values of the decay functions seem to go as t^{-n}, where n must be less than 1.0 since free diffusion gives a value of 1.0, as we discussed above. Because some sort of "kinking" of the helix for example, true nonlinearities in the system will make even the Gaussian approximation for the distribution of angles incorrect.

Our attempts to understand the bending solutions have been a failure. The bending potential is quadratic in the angle of bending θ, so we can guess that the distribution of bending angles should be a Gaussian distribution. However, the higher orders of the derivatives in the equation of motion makes the time-dependent width of the distribution Δ_x^2 very difficult to understand intuitively. I have yet to come up with an intuitive explanation for the dependence of the squared width of the distribution on the fourth (1/4) power of time. The derivation comes from throwing out the odd terms in an expansion using boundary value arguments, I know, "but" the physical origin behind this eludes me. Unfortunately, the bottom line here seems to be that we have to trust the incredible mathematics that are thrown at this problem to get the time-dependent second moments of the distributions.

We have made enough attempts to understand the solutions intuitively; let us be mindless robots and write down the solutions. Barkley and Zimm calculated the time-dependent probability distributions for both bending and twisting. The probability $\psi(\alpha, t; \alpha_o)$ of finding a segment at the body fixed angle α when time $t = 0$ and the segment was at angle α_o is

$$\psi(\alpha, t; \alpha_o) = \frac{1}{(\pi\Gamma)^{1/2}} \exp\left[-\frac{(\alpha - \alpha_o)^2}{\Gamma(t)}\right] \tag{47}$$

where the width of the distribution, the twisting decay function $\Gamma(t)$, is

$$\Gamma(t) = \frac{k_B T}{4\pi\eta b^2 L} t + \frac{4k_B T}{CL} \sum_{k=1}^{\infty} \left(1 - \exp\left[-\frac{\lambda_k^2 \sigma t}{\lambda_k^2}\right]\right) \tag{48}$$

where $\lambda_k = k\pi/L$ and $\sigma = C/8\pi\eta b^2$. This is pretty awful; fortunately, in the limit of $Ct/8\pi\eta b^2 \ll 1$, this simplifies to a mercifully useful expression:

$$\Gamma(t) = 4k_BT\left(\frac{t}{8\pi^2Cb^2\eta}\right)^{1/2} = \left(\frac{t}{\tau_t}\right)^{1/2} \tag{49}$$

where the time constant τ_t is

$$\tau_t = \frac{2\pi^2GI_pb^2\eta}{(k_BT)^2}. \tag{50}$$

This is good—we have a direct expression for the torsional rigidity in terms of the *width* of a Gaussian distribution of angles which is increasing with the square root of time.

The distribution of angles for bending motions is once again a Gaussian:

$$\Psi(\beta, t; \beta_o) = \frac{1}{(\pi\Delta)^{1/2}}\exp(-(\beta-\beta_o)^2/\Delta(t)) \tag{51}$$

where $\Delta(t)$ is the bending decay function, a measure of the width of the Gaussian distribution. I won't give the full expression for $\Delta(t)$ because it is just too depressing (read: complicated and requires solving ugly transcendental equations). However, if $EI_At/L^4\eta \ll 1$, the expression sort of simplifies to

$$\Delta(t) \sim B(t)t^{1/4} \tag{52}$$

where unfortunately $B(t)$ is

$$B(t) = \left[\left(\frac{4}{EI_A}\right)^{3/4}\frac{k_BT\Gamma(3/4)}{\pi^{3/4}\eta^{3/4}}\right] \times (\text{ something}) \tag{53}$$

where (something) is supposedly a slowly varying function of time (I have not been able to confirm this with my own calculation). If we *ignore* (something), we can write the bending function as

$$\Delta(t) = \left(\frac{t}{\tau_b}\right)^{1/4} \tag{54}$$

where

$$\tau_b = \left[\frac{EI_A\pi\eta}{4}\right]^3 \times \left[\frac{1}{k_BT}\right]^4. \tag{55}$$

Now, in reality, both bending and twisting motions occur. Barkley and Zimm used an Euler transform, also used to find the anisotropy decay for arbitrary motions of the helix, but in order to solve the equations they were forced to use a small angle approximation in the Euler angle transform. The result was that the final expression for anisotropy was incorrect since linearization of the equations forces the result to not extrapolate properly to the rigid rod limit.

Schurr not only derived the correct form for the anisotropy decay in terms of the twisting and bending functions as we have discussed but, with Allison, also derived twisting angle distribution functions.[3] For their model, Allison and Schurr used a discrete model of DNA consisting of rigid cylinders connected by torsional springs, rather like the Rouse model of polymer dynamics.[84] Because of the discrete nature of the diffusing object, the solutions to the diffusion equation break up into certain time domains which have quite different functional dependences on time. The troubles lie in the actual functional *form* of the mean angular displacements—they are extremely complex and always involve some simplifications. The relative simplicity and intuitive feel of the continuum rod solution by Barkley and Zimm is the reason we stress their result. Also see the excellent paper by Shibat, Fujimoto and Schurr[91] which has carefully and thoughtfully addressed the issue of the comparison between the theories of Barkley and Zimm versus Shurr and his colleagues. Further, presumably because of the complexities associated with the Oseen-Burgers hydrodynamic tensor, Schurr et al. have not attempted to derive the actual functional dependence of the bending decay function, although they know how to *use* the form once it has been found as we have discussed. The twisting function of Allison and Schurr, as we mentioned above, is quite complex (amazingly so, in fact) and split up into various time ranges. Of most interest to us, the intermediate zone is

$$T_m(t) = \exp\left[-\frac{m^2}{2}\Delta_x(t)^2\right] = \exp\left[\frac{-m^2 k_B T t^{1/2}}{(\pi\alpha\gamma)^{1/2}}\right]. \tag{56}$$

Translation: Barkley and Zimm used C as the torsional rigidity while, in Schurr's notation, α is the effective spring constant between two adjacent basepairs, separated by a distance h. Thus, in terms of the shear modulus I can write:

$$\alpha = \frac{GI_p}{h} = \frac{C}{h}. \tag{57}$$

The parameter γ is the frictional coefficient of one basepair of length h and radius b:

$$\gamma = \rho h = 4\pi\eta b^2 h. \tag{58}$$

We can rewrite the equation as

$$T_m = \exp\left[-\frac{m^2 k_B T t^{1/2}}{(\pi^2 G I_p \rho)^{1/2}}\right] = \exp\left[-m^2 (t/\tau_t)^{1/2}\right] \tag{59}$$

where the twisting time constant is

$$\tau_t = \frac{4\pi^2 G I_p b^2 \eta}{(k_B T)^2}. \tag{60}$$

Finally we see that the twisting anisotropy decay functions derived by both Barkley and Zimm and Schurr are the same, outside of a factor of two which I assume is due to Barkley and Zimm's incorrect use of linearized Euler transforms for Barkley and Zimm. The expected anisotropy decay is a sum of stretched exponential decays for both the bending and twisting anisotropy decay. A stretched exponential is an equation of the form

$$s(t) = s(0) \exp\left[-\left(\frac{t}{\tau}\right)^{\beta}\right] . \tag{61}$$

The twisting modes have a β of 0.5. Schurr and his colleagues did not attempt to find a functional form of the bending function $\Delta_x^2(t)$; the Barkley-Zimm stretched exponential with $\beta = 0.25$ remains the only *model* used by Schurr and his colleagues. However, since electric birefringence decay is also essentially due to the bending modes, Schurr and his colleagues have utilized some birefringence data derived from the work of Eden and his colleagues[37] for 600-bp-long DNA fragments as empirical sources of twisting decay functions. As Schurr himself has pointed out however, birefringence is a coherent phenomena since the phase is first summed over the entire molecule and then squared while, in emission anisotropy, the signal is due to the incoherent sum over the intensities of excited states. One would expect that the forms of the decay laws to be different so it is somewhat of a miracle that the two decay functions should overlap. Further, as we point out in a recent paper[56] the field-free electric birefringence decay is strongly dependent on the *width* of the aligning pulse so it is somewhat of a dicey situation to mix birefringence with excited optical anisotropy.

At the fastest time scale of 0 to 1 nanosecond, the characteristic length scale of motions is on the order of 1 basepair, so that strictly local events are observed in this range unless it is possible for the DNA molecule to decouple from the viscous solvent; that is, one would guess that the motions are highly overdamped and that our equations presented above should still work.

However, since the times are very short a possibility of *decoupling* hydrodynamically from the solvent exists. This whole story has a very controversial history, with several apparently false starts. One unsolved problem that has not been explored very much is the question as how strongly the dye (or the DNA) is coupled dynamically to the solvent. That is, does the characteristic time of motions on the sub-nanosecond timescale scale with the viscosity, as we would expect from our hydrodynamic expressions, or not? For a while this was not a moot point since there was a flurry of excitement about the possibility that underdamped, high-Q acoustic modes could propagate along the helix backbone at frequencies in the 2–10 GHz range (0.5 to 0.1 nanoseconds)[35]; in fact, an editorial in *Nature*[69] included a warning to molecular biologists that physicists just might "high jack" DNA and soon no molecular biology lab would be complete without microwave plumbing. Unfortunately the early tantalizing results were not supported by later experiments,[53] and it would seem that damping turns on at shorter times than 100 picoseconds.

It is interesting that while the earlier theoretical calculation had predicted over-damping, the theory was quickly modified to accommodate these new results.[101] The motions of intercalated dyes should be sensitive probes to the onset of high-Q acoustic modes if the mode involves tilting or twisting of the basepairs, but no one seems to have explored this.

There really have been only three techniques that probe in any systematic way the dynamics of the double helix in this time range: dynamical light scattering, electric birefringence/dichroism, and triplet anisotropy. The first two techniques do not require a label and thus can claim to look at "native" dynamics, especially dynamical light scattering.

If we look once more at the twisting and bending functions, it is clear that out beyond 1 microsecond or so the anisotropy decay is totally dominated by the F_0 bending decay function, which corresponds to the end-over-end tumbling mode if the molecule is a rigid rod. Further, just as the rotation time of the rigid rod is highly sensitive to length, going as $1/L^3$, so too does the decay rate of the bending motions become more sensitive to variations in the rigidity of the helix.

The Barkley-Zimm theoretical derivation of the bending decay function took the Oseen-Burgers tensor into account to find the bending decay function. The long-range nature of the tensor force matrix makes the bending motions strongly dependent on how close segments of the rod are to each other, a hand-waving explanation for why the microsecond bending motions would be more sensitive to the rigidity of the helix than the twisting motions. However, the approximations used in deriving the bending expressions F_m invalidate the expression for large times. As we move into the microsecond time domain the relaxation times are for a length scale on the order of the persistence length of the polymer, which would imply that the nature of the problem shifts over from a local bending problem to the kind of statistical mechanics problem that the Rouse model of polymer dynamics addresses. The Rouse model views the polymer as a collection of spheres that are of diameter b and connected by "springs" of strength $3k_BT/b^2$. These "springs" really are a convenient place to store entropic free energy and are not the internal enthalpic energy springs in Schurr's model. The Rouse model incorrectly throws out the hydrodynamic interaction between spheres; Zimm corrected this by including the Oseen-Burgers tensor. The final step to be taken is to include the Rouse-Zimm formalism, the elastic energy terms due to bending. We are working on this problem at present and it appears that a closed-form expression for the F_o bending function at long wavelengths is possible.

For the truly long modes, with wavelength greater than about 10 persistence lengths, the purely entropic Rouse-Zimm modes without rigidity are the dominant modes. The result of the analysis of the Rouse-Zimm model is that the rotational relaxation time τ_p of the pth mode is

$$\tau_p = \frac{\eta(N^{1/2}b)^3}{(3\pi p^3)^{1/2}k_BT} \tag{62}$$

where N is the number of beads in the polymer. We can recast this, ignoring the numerical factors to convert from beads to rods, by noting that b is $\mathcal{P}$ and $N = L/2\mathcal{P}$ (L is the length of the polymer). We get

$$\tau_p \sim \left(\frac{2L\mathcal{P}}{p}\right)^{3/2} \tag{63}$$

so that the rotational relaxation times go as $\mathcal{P}^{3/2}$. Or, by the fundamental relation between the persistence length $\mathcal{P}$ and the Young modulus E, it follows that the rotational relaxation times go as $E^{3/2}$. The progression should now be clear: twist times go as the $G^{1/2}$, internal bending relaxation goes as $E^{3/4}$, and the Rouse-Zimm tumbling relaxation goes as $E^{3/2}$. It then follows that the most sensitive region to look for variations in the elastic properties of DNA is the *long, many-microsecond* time domain.

Qualitatively this is exactly what the triplet data look like: the anisotropy decay data for different sequences are very similar at the shortest times (nanoseconds), spread apart with increasing time, and are strongly apart in the time range beyond 10 microseconds. Unfortunately, we haven't yet properly mixed in the Rouse-Zimm bending modes with the Barkley-Zimm internal modes in Schurr's equation so there are no solid fits.

It seems somewhat senseless at this point to analyze further our best single depletion anisotropy data in terms of shear and bending rigidities given that Schurr vigorously disputes *any* sequence dependence to the shear modulus and no adequate bending theory exists at present by which one can fit the data. The best, and most honest thing to do, is to report our "best" triplet data at present. These DNA samples were rigorously fractionated to a *common length* of 450 bp—still too short to get the internal bending modes of dG–dC but long enough to see the internal modes all the way out for A–T sequences. These are best sets of data, and a clear difference exists between AT and GC sequences as well as a strong difference between dA–dT and d(AT)–d(TA).

It is most interesting to look at the overlap of the nonpathological d(AT)–d(TA) data and compare it to the d(GC)–d(CG) data, especially at short times. It is *very difficult* to observe differences between the GC and AT sequences over the range 0 to 100 nanoseconds while, at times greater than a microsecond, a clear difference exists between the two sequences. Does this mean that we were incorrect in our previous statements that both G and E are strongly sequence dependent? In my opinion the jury is still out on this issue.

We mentioned earlier that, when the DNA helix forms, the basepair interactions are not strictly across the helix to the neighbor on the other side. Since the bases sit on top of one another in Vander Wall's contact (the spacing between bases is only 3.4 Å) ample opportunity exists for nearest-neighbor interactions between the bases directly above and below a given base and next-nearest neighbor interactions with bases that are diagonally across from a given base.

These interactions can give rise to sequence-dependent conformations of the DNA that involve more than the simple counting of basepairs that we mentioned earlier. Further, one would expect that due to the nonlocal nature of the interactions that structural phase transitions will occur in the system since the strength of the nonlocal interactions will be a function of what has proceeded it. For example, in the Zimm-Bragg theory of the thermodynamics of the helix-coil phase transition a zippering aspect to the phase transition exists: once a single hydrogen bond is formed between the first and third amino acids, the coil is positioned so that it is much easier for the next hydrogen bond to form. Zimm and Bragg long ago were able to derive a partition function that was able to reproduce the observed temperature dependence of the formation of the α helix from a denatured polymer chain. In the case of internal structural changes that occur in DNA, their analysis cannot be applied, since it assumes that the system changes from a disordered polymer to a internally linked polymer. In the case of DNA we have something like a solid-state phase transition or, actually since the system is essentially one-dimensional, something like an Ising spin transition. We don't know.

6. SYNTHETIC COMPLEX ENVIRONMENTS

There is a fascinating possibility that by using the nanotechnology of the semiconductor industry we will be able to develop the ability to manipulate and modify single biological molecules. Visionaries such as Drexler have proposed a brave new world where nanotechnology will produce micromachines that enter the body and do repairs. We have a much more limited but more achievable contribution to the field of nanotechnology here.

We have used optical microlithography to fabricate capped quasi-two-dimensional obstacle courses in SiO_2. We have made observations using epi-fluorescence microscopy of the electrophoresis and length fractionation of large (100 kbase) DNA molecules confined in arrays. Biased reputation theory, based on the work of deGennes,[29] predicts that at low-electric fields, for a polymer of length L much greater than the persistence length p the electrophoretic mobility scales inversely with L.[83] However, elongation of the coil in the matrix at sufficiently strong electric fields[92] results in a length-independent electrophoretic mobility.[54,66] The application of suitably timed pulsed electric fields[87] restores the fractionating power of gels for long molecules,[33] but the protocols of pulsed field electrophoresis are of necessity semi-empirical because the complex and ill-understood gel matrix plays a critical role in fractionation. With their low dimensionality, small volume, and extremely reproducible topography, microlithographically constructed obstacle arrays will make it possible to understand the motion and fractionation of large polymer molecules in complex but well-characterized topologies.

The persistence length of DNA under normal buffer conditions is approximately 0.06 μ,[50] so the DNA is confined to a vertical space slightly more than two persistence lengths or one Kuhn random flight length. DNA molecules were stained by ethidium bromide and then imaged in the array by epi-fluorescence video microscopy.[73,23]

In our epifluorescence work the DNA molecules are in focus over the entire field of view because they are confined to two dimensions. Note that the longer fragments all show substantial elongation, and that many of the longest fragments are clearly "hooked" by single posts. Typically, a 100-kbase fragment hooked three times on crossing the 130-μ-long field of view. Surprisingly, the amount of elongation in this coarse array is comparable to that observed in gels, which are much finer grained.[88] A 100-kbase molecule comes in as a distorted coil, drapes around a post, hooks, and then slides off. The hooked DNA molecule actually is stretched to nearly its full contour length since the observed stretched length (30 μ) is the expected contour length of a 100-kbase molecule. The molecule has greater fluorescence intensity at its ends farthest from the post. This is presumably due to the randomization of the ends of the molecule where the internal tension is very small.[86]

Our arrays correspond to "unphysical" agarose gels. The effective pore size of 1.0 μ corresponds roughly to a physically unstable 0.05% agarose gel.[55] Such a large pore spacing is comparable to the 1.5 μ radius of gyration of the largest DNA molecules. The amount of elongation and hooking that occurs in the arrays is encouraging but not explained by simple models in such an open lattice. Of course, the applied electric field of 1 V/cm is not responsible for the alignment of the polymer in the absence of the lattice, as is quantitatively seen by evaluation of the dimensionless number $\kappa_a = -QEp/kT$ (k is Boltzmann's constant, T is the absolute temperature, and Q is the net charge of a persistence length).[31] If κ_a is less than 1, than the entropy of the polymeric chain dominates and the coil is random while, if κ_a is greater than 1, the potential energy gained in translation of the persistance length along the field direction dominates and the chain is elongated. Since the applied electric field E is 1 V/cm and the net charge per unit length should be 0.1 e^- per 3.4 Å,[67] we have $\kappa_a = 0.005$.

This small value for κ_a is not consistent with the observations of the elongation of the hooked molecules. The characteristic distance L' over which this thinning from random coil at the ends to fully stretched polymer has been calculated by Schurr and Smith to be $L' \sim 2p/\kappa_a$. In our case $\kappa_a = 0.005$ gives a value for L' of approximately 30 μ or 100 kbases, while the measurements reveal a much smaller L' of approximately 5 kbases (about 1.5 μ). This may be due to electroosmosis[64] in our relatively open array. The hydrodynamic drag from electroosmosis causes a larger force per unit length on the molecules, effectively increasing κ_a. The influence of two-dimensional statistics in the array may also play a significant role in the dynamics of the polymers.

The binned data indicate for up to a length of approximately 100 kbases the array is capable of length fractionation in a DC field. This is at the limits of fractionation for conventional DC electrophoresis in agarose gels,[89] giving hope that our arrays can extend the present hard-won limits on gel electrophoresis. Noolandi first suggested the importance of hooking in gels[65] and Song and Maestre[99] have done a preliminary analysis of unhooking times. We believe that fractionation in our arrays occurs because of dispersion in the time it takes a hooked polymer to thermally free itself from a post.

In fact, it is rather easy to compute the motion of a polymer as it moves off the post. Consider a fully stretched polymer hung over a single post. Let x represent the difference in lengths for a polymer of contour length L hung asymmetrically over a post, and we assume zero coefficient of friction over the post contact region for simplicity. We will show that the present experiments within our error bars are adequately fit without assumption of a coefficient of friction.

The net force acting along the polymer due to the applied electric field is γEx, and the polymer thus moves on an inverted harmonic potential surface of the form $U(x) = -\frac{1}{2}\gamma Ex^2$. In the absence of any frictive forces due to the contact of the hung polymer on the post, the damping force acting on the polymer is simply the viscous drag F_D on a rod of length L and diameter a moving along its length in a solvent of viscosity ν:

$$F_D = v_{cm} \frac{2\pi\eta L}{\ln(L/a)} \tag{64}$$

where v_{cm} is the velocity of the center of mass of the polymer.

We first examine the deterministic motion of the polymer in the absence of any thermal fluctuations. Since the viscous drag forces are much larger than any inertial terms (the Reynolds number[81] $\mathcal{R} = \rho v L/\eta$, where ρ is the mass density of the polymer, is exceedingly small, $\sim 10^{-10}$), the equation of motion for the stretched polymer as it slides off the post determined by the combination of the electric and viscous forces is:

$$\gamma Ex = \frac{\pi\eta L}{\ln(L/a) k_B T} \frac{dx}{dt} \tag{65}$$

where it should be understood that electro-osmotic hydrodynamic forces may supplement or even be greater than the columbic force γEx. Eq. (3) yields the simple exponential solution:

$$x(t) = x_o \exp\left(\frac{\gamma E \ln(L/a)}{2\pi\eta L} t\right) \tag{66}$$

where x_o is the position of the particle at $t = 0$, when it becomes fully extended after uncurling on the post. Since a is on the order of 20 Å and $L \gg a$ in all of our samples, the simple conclusion is that the polymer should slide off the post with x increasing exponentially with a time constant τ proportional to L:

$$\tau = \frac{2\pi\eta L}{\gamma E \ln(L/a)} . \tag{67}$$

In reality, the predicted linearity of τ with length L is good as long as the length L is much greater than κ. We predict from our value of $\gamma E = 7 \times 10^{-9} N/m$, assuming $\eta = 1 \times 10^{-3}$ Pa-s, that a rod of length 30μ should have a time constant τ of approximately 5 seconds, compared to our measured value of 7 ± 1 seconds.

The above analysis ignores thermal fluctuations. Consequently, Eq. (4) incorrectly predicts that, if $x_o = 0$, the system is infinitely metastable; thus, Eq. (4) cannot be used to correctly predict polymer retention times on a post for arbitrary starting positions. In the high-friction limit, the correct description of the Brownian motion of the molecule is provided by the Smoluchowski equation.[27] This equation states that for a system diffusing in a potential $U(x)$ (here, $-\frac{1}{2}\gamma E x^2$) with diffusion coefficient D (here, $k_B T \ln(L/a)/2\pi\eta L$), the probability density $p(x,t)$ satisfies the following equation[100,104]:

$$\frac{\partial p(x,t)}{\partial t} = \frac{\partial}{\partial x} D \left(\frac{\partial}{\partial x} + \frac{1}{k_B T} \frac{\partial U}{\partial x} \right) p(x,t) \, . \tag{68}$$

The second moment $x^2(t)$ has been shown[100] to be

$$\langle x^2(t) \rangle = x^2(0) e^{2t/\tau} + \frac{1}{\lambda^2} \left(e^{2t/\tau} - 1 \right) \tag{69}$$

where $\lambda^2 = \gamma E / k_B T$. This result can be used to find a useful estimate for the *mean first passage time* $\langle t_o \rangle$ for $x(t)$ to reach a value of $\pm L$. The limits $x = \pm L$ represent perfectly absorbing boundaries because the DNA irreversibly falls off a post when $x = \pm L$ if $(\lambda L) \gg 1$; i.e., $U(\pm L) \ggg k_B T$. The mean first passage times $\langle t_o \rangle$ for a particle initially at x_o to reach $\pm L$ must satisfy the following equation:

$$D \frac{d}{dx_o} e^{-U(x_o)/k_B T} \frac{d}{dx_o} \langle t(x_o) \rangle = -e^{-U(x_o)/k_B T} \, . \tag{70}$$

For an inverted potential it is straightforward to show that:

$$\langle t(x_o) \rangle = 2\tau [I(\lambda L / 2^{1/2}) - I(\lambda x_o / 2^{1/2})] \tag{71}$$

where

$$I(w) = \int_0^w e^{-z^2} \int_0^z e^{y^2} \, dz dy \tag{72}$$

$$= \int_0^\infty e^{-a^2} \sin^2(wa) \frac{da}{a} \tag{73}$$

$$\sim \frac{1}{4} \ln(1 + 2w^2) + \frac{w^4}{\pi(2 + w^4)} \, . \tag{74}$$

The simple approximate expression is exact as $w \to 0$ or $x \to \infty$ and errs by less than 7% for all w. Note that aside from a weak logarithmic dependence on L, the mean first passage time is proportional to L for small x_o with no singularity at $x_o = 0$. For $x_o = 0, L = 30\mu, \lambda = 1.3\times10^6\ m^{-1}$, $\tau = 7$ seconds we predict that $\langle t(0)\rangle \sim 30$ seconds, in comparison to our measured value of 40 ± 10 seconds.

Further, work will be necessary to test some of the conclusions of this analysis, since at present our statistics are constrained by the small number of DNA molecules that have been studied. Further, the matrix has enough other posts along the length of the molecule to possibly perturb the motion. However, we believe that this work has demonstrated that a key aspect of electrodiffusion of very long polymers in a synthetic lattice can be understood quantitatively and that the prospects are excellent that this new environment can be understood and optimized for any desired molecular-processing task.

7. OPEN SPECULATIONS

We usually aim wide of our targets and we are sure that once again we have missed one. Instead of giving some achingly formal and elegant view of the way that biological molecules interact to form complex biological structures, I have instead gotten pretty well buried in the chemical physics of individual molecules.

Is it necessary to know any of this stuff? Can young mathematical types, assured by all how very brilliant and clever you are, safely ignore the nasty and unclean facts about how these molecules actually work? We would guess yes and no. Yes, you can probably ignore their internal complexity if you are content to treat them as black boxes with arbitrary properties. No, you probably cannot ignore the facts if you wish to do something really useful. We suppose the best analogy is to Hopfield's original neural network model: it assumed symmetrical $J_{ij} = J_{ji}$ neuronal couplings. His model made an enormous impact, but real neurons I have been told do *not* have symmetrical couplings, throwing many aspects of the biological relevance of the model into doubt.

In the area of proteins, the one area where all this talk about conformational complexity may be useful some day will be understanding the immune system, particularly how antibodies can be directed with high specificity towards a particular antigen. How does that happen? Can an antibody change with time¿ Can it be directed in a given ammino acid sequence towards a different antigen? I don't know.

In the area of DNA elasticity, it is clear that we are only beginning to understand the connection between basepair composition and DNA structure and the control of information flow in the cell. We know very little, and much more quantitative work along the lines we have outlined will be essential in learning the rules. We suspect that many aspects of the DNA sequence, particularly the 90% of the genome that is not expressed and sometimes considered to be obsolete evolutionary baggage, may turn out to be coding for the *structure* of the chromosome and may be very important.

The microlithography work will have a rich future. We will learn how to manipulate single DNA molecules, how to perform specific chemical reactions on DNA molecules, separate the parts, and proceed to do other operations on the molecule like a little microfactory. Further uses of two-dimensional arrays will be to move micro-organisms through molecules, challenge them with different environments, and find out the rules by which they navigate, find food, or solve puzzles. Ultimately one could also imagine using three-dimensional lithographic structures to realize true neural nets.

There is much to do in biophysics. Almost anything you grapple with is interesting and involves dealing with highly complex, nonlinear systems. What you really should do boils down to a matter of talent and taste. We hope that the really powerful people out there, the creative ones, may find something useful in the stew we presented here.

REFERENCES

1. Agmon, N., and J. J. Hopfield *J. Chem. Phys.* **78** (1983): 6947.
2. Agmon, N. *Biochem.* **27** (1988): 3507.
3. Allison, S. A., and J. M. Schurr. *Chem. Phys.* **41** (1979): 35–59.
4. Anderson, H., ed. *Physics Vade Mecum.* New York: American Instit. Physics, 1989.
5. Anderson, P. W., B. L. Halperin, and C. M. Varma. *Phil. Mag.* **25** (1971): 1.
6. Ansari, A., J. Berendzen, S. F. Bowen, H. Frauenfelder, I. E. T. Iben, T. B. Sauke, E. Shyamsunder, and R. D. Young.*Biophys. Chem.* **26** (1985): 337.
7. Austin, R., K. Beeson, L. Eisenstein, H. Frauenfelder, I. C. Gunsalus, and V. Marshall. *Science* **181** (1973): 541.
8. Austin, R., K. Beeson, L. Eisenstein, H. Frauenfelder, I. C. Gunsalus, and V. Marshall. *Phys. Rev. Lett.* **32** (1974): 403.
9. Austin, R., K. Beeson, L. Eisenstein, H. Frauenfelder, I. C. Gunsalus, and V. Marshall. *Biochem.* **14** (1975): 5255.
10. Austin, R., ed. *Protein Structure, Molecular and Electronic Reactivity.* New York: Springer-Verlag, 1987.
11. Austin, R. H., M. W. Roberson, and P. Mansky. *Phys. Rev. Lett.* **62** (1989): 1912–1915.
12. Austin, R. H., K. W. Beeson, L. Eisenstein, and H. Frauenfelder. *Biochem.* **14** (1975): 5355–5373.
13. Austin, R. H., M. K. Hong, C. Moser, and J. Plombon. *Chem. Phys.* **158** (1991): 473–486.

14. Bagley, K., V. Balogh-Nair, A. A. Croteau, G. Dollinger, T. G. Ebrey, L. Eisenstein, M. K. Hong, K. Nakanishi, and J. Vittitow. *Biochem.* **24** (1985): 6055–6071.
15. Barkley, M. D., and B. H. Zimm. *J. Chem. Phys.* **70** (1979): 2991–3007.
16. Bassler, H. *Phys. Rev. Lett.* **58** (1987): 766.
17. Becker, R. S. *Theory and Interpretation of Fluorescence and Phosphorescence.* New York: Wiley, 1969.
18. Becher, B., F. Tokunaga, and T. G. Ebrey. *Biochem.* **17** (1980): 2293–2300.
19. Berendzen, Joel. Personal Communication.
20. Bialek, W., and J. N. Onuchic. *Proc. Natl. Acad. Sci. USA* **85** (1988): 5908.
21. Bialek, B., and R. Goldstein. *Protein Structure, Molecular and Electronic Reactivity*, edited by R. Austin et al. New York: Springer-Verlag, 1987.
22. Binder, K., and A. Young. *Rev. Mod. Phys.* **58** (1986): 846.
23. Bustamante, C. *Ann. Rev. Biophys. Biophys. Chem.* **20** (1991): 415–446.
24. Callender, R., and B. Honig. *Ann. Rev. Biophys. Bioeng.* **6** (1977): 33–55.
25. Cambell, B. C., M. R. Chance, and J. M. Friedman. *Science* **38** (1987): 373.
26. Cantor, C., and P. Schimmel. *Biophysical Chemistry*, Vol. III, 1041. New York: W. H. Freeman, 1980.
27. Chandrasekar, S. *Rev. Mod. Phys.* **15** (1943): 1–89.
28. Christiansen, P. L., and A. C. Scott, eds. *Davydov's Soliton Revisited.* New York: Plenum Press, 1990.
29. DeGennes, P. *Scaling Concepts in Polymer Physics*, 4th Ed. Ithaca, NY: Cornell University Press, 1991
30. Degiorgio, V., T. Bellini, R. Piazza, F. Mantegazza, and R. E. Goldstein. *Phys. Rev. Lett.* **64** (1990): 1043–1046.
31. Deutsch, J. M., and T. L. Madden. *J. Chem. Phys.* **90** (1989): 2476–2485.
32. Doering, C. R., and J. C. Gadoua. *Phys. Rev. Lett.* **69** (1992): 2318.
33. Doi, M., T. Kobayashi, Y. Makino, M. Ogawa, G. W. Slater, and J. Noolandi. *Phys. Rev. Lett.* **61** (1988): 1893–1896.
34. Draheim, J. E., and J. Y. Cassim. *Biophys. J.* **47** (1985): 497–507.
35. Edwards, G. C., C. C. Davis, J. D. Saffler, and M. L. Swicord. *Phys. Rev. Lett.* **53** (1984): 1284–1287.
36. Elber, R., and M. Karplus. *Science* **235** (1987): 318.
37. Elias, J. G., and D. Eden. *Macromolecules* **14** (1981): 410–419.
38. Fanconi, B., and L. Finegold. *Science* **190** (1975): 458.
39. Feigelman, M. V., and L. B. Ioffe. *J. de Physique Lettres* **45** (1984): 475.
40. Ferre', J., M. Ayadi, R. V. Chamberlin, R. Orbach, and N. Bontemps. *J. Magn. Mater.* **54-57** (1986): 211.
41. Finegold, L., and J. L. Cude. *Nature* **238** (1972): 38.
42. Fisher, D. S., G. M. Grinstein, and A. Khurana. *Physics Today* **14** (1988): 56.
43. Frauenfelder, H., and P. Wolynes. *Science* **229** (1985): 337.
44. Frauenfelder, H., G. A. Petsko, and D. Tsernoglou. *Nature* **280** (1979): 558.

45. Frauenfelder, H. *Amorphous and Liquid Materials*, edited by E. Luscher, G. Fritsch, and G. Jacucci. Proceedings NATO Summer School, 3–18. Holland: Martinus Nijhoff, 1987.
46. Frauenfelder, H. *Physics in Living Matter*, Lecture Notes in Physics, vol. 284, 1–14. Berlin: Springer-Verlag, 1987.
47. Garcia, A. *Phys. Rev. Lett.* **68** (1992): 2696–2699.
48. Gerwert, K., G. Souvignier, and B. Hess. *Proc. Natl. Acad. Sci. USA* **87** (1990): 9774–9778.
49. Gol'danskii, V. I., Y. F. Krupyanskii, and V. N. Fleurov. *Protein Structure: Molecular and Electronic Reactivity*, edited by R. Austin et al., 95. New York: Springer-Verlag, 1987.
50. Hagerman, P. J. *Ann. Rev. Biophys. Biophys. Chem.* **17** (1988): 265–286.
51. Henderson, R., J. M. Baldwin, T. A. Ceska, F. Zemlin, E. Beckmann, and K. H. Downing. *J. Mol. Biol.* **213** (1990): 899–929.
52. Honig, B., T. G. Ebrey, R. H. Callender, U. Dinur, and M. Ottolenghi. *Proc. Natl. Acad. Sci. USA* **76** (1979): 2503–2507.
53. Gabriel, C., E. H. Grant, R. Tata, P. R. Brown, B. Gestblom, and E. Noreland. *Nature* **328** (1987): 145–146.
54. Hervet, H., and C. P. Bean. *Biopolymers* **26** (1987): 727–742.
55. Holmes, D., and N. Stellwagen. *Electrophoresis* **11** (1990): 5–15.
56. Hong, M. K., O. Narayan, R. E. Goldstein, E. Shyamsunder, R. H. Austin, D. S. Fisher, and M. Hogan. *Phys. Rev. Lett.* **68** (1992): 1430.
57. Huse, D. A., and D. S. Fisher. *J. Phys. A: Math. Gen.* **20** (1987): L997.
58. Iben, I. E. T., D. Braunstein,W. Doster, H. Frauenfelder, M. K. Hong, J. B. Johnson, S. Luck, P. Ormos, A. Shulte, P. J. Steinbach, A. H. Xie, and R. D. Young. *Phys. Rev. Lett.* **62** (1989): 1916.
59. Jackle, J. *Rep. Prog. Phys.* **49** (1986): 171.
60. Jeong, Y., S. R. Nagel, and S. Bhattacharya. *Phys. Rev. A* **34** (1986): 602.
61. Kauzmann, W. *Chem. Rev.* **43** (1948): 219.
62. Kitagawa, T., and A. Maeda. *Photochem. Photobiol.* **50** (1989): 883–894.
63. Kittel, C. *Introduction to Solid State Physics*. New York: John Wiley.
64. Kozak, M. W., and E. J. Davis. *Langmuir* **6** (1990): 1585–1590.
65. Lalande, M., J. Noolandi, C. Turmel, R. Brousseau, J. Rousseau, and G. W. Slater. *Nucl. Acids Res.* **16** (1988): 5427–5437.
66. Lumkin, O. J., P. Dejardin, and B. H. Zimm. *Biopolymers* **24** (1985): 1573–1593.
67. Manning, G. S. *J. Chem. Phys.* **51** (1969): 924–933.
68. Maddox, J. *Nature* **324** (1986): 205.
69. Maddox, J. *Nature* **324** (1986): 11.
70. Marrero, H., and K. J. Rothschild. *Biophys. J.* **52** (1987): 629–635.
71. Mathies, R. A., S. W. Lin, J. B. Ames, and W. T. Pollard. *Ann. Rev. Biophys. Chem.* **20** (1991): 491–518.
72. Milder, S. J., and D. S. Kliger. *Biophys. J.* **49** (1986): 567–570.
73. Morikawa, K., and M. Yanagida. *J. Biochem.* **89** (1981): 693–696.

74. Mulder, C. A. M, A. M. van Duyneveldt, and J. A. Mydosh. *Phys. Rev. B* **23** (1981): 1384.
75. Ogielski, A, and D. L. Stein. *Phys. Rev. Lett.* **55** (1985): 1634.
76. Ormos, P., D. Braunstein, M. K. Hong, S. Lin, and J. Vittitow. *Biophysical Studies of Retinal Proteins*, edited by T. G. Ebrey et al. Urbana: University of Illinois Press, 1987.
77. Palmer, R. G., D. L. Stein, E. Abrahams, and P. W. Anderson. *Phys. Rev. Lett.* **53** (1984): 958.
78. Parisi, G. *Phys. Rev. Lett.* **53** (1979): 1754.
79. Phillips, W. A. *J. Low Temp. Phys.* **7** 351.
80. Powers, L., and W. Blumberg. *Biophysical J.* **54** (1988): 181.
81. Purcell, E. M. Special Issue: Physics and Our World: A Symposium in Honor of Victor Weisskopf. *AIP Conf. Proceedings* **28** (1976): 49.
82. Rammal, R., G. Toulouse, and M. A. Virasoro. *Rev. Mod. Phys.* **58** (1986): 765.
83. Rodbard, D., and A. Chrambach. *Proc. Natl. Acad. Sci. USA* **65** (1970): 970–977.
84. Rouse, P. E. *J. Chem. Phys.* **21** (1953): 1272–1280.
85. Schurr, J. M. *Chem. Phys.* **84** (1984): 71–96.
86. Schurr, J. M., and S. B. Smith. *Biopolymers* **29** (1990): 1161–1165.
87. Schwartz, D. C., and C. R. Cantor. *Cell* **37** (1984) 67–75.
88. Schwartz, D. C., and M. Koval. *Nature* **338** (1989): 520–522.
89. Serwer, P. *Electrophoresis* **10** (1989): 327–331.
90. Sherrington, D., and S. Kirkpatrick. *Phys. Rev. Lett.* **35** (1975): 1972.
91. Shibata, J. H., B. S. Fujimoto, and J. M. Schurr. *Biopolymers* **24** (1985): 1909–1930.
92. Smith, S. B., P. K. Aldridge, and J. B. Callis. *Science* **243** (1989): 203–206.
93. Stein, D. L. *Protein Structure: Molecular and Electronic Reactivity*, edited by R. Austin et al., 85. New York: Springer-Verlag, 1987.
94. Stein, D. L. ed. *Spin Glasses in Biology.* Singapore: World Scientific, need year.
95. Stein, D. L. *Protein Structure: Molecular and Electronic Reactivity*, 85. New York: Springer-Verlag, 1987.
96. Stein, D. L. *The Sciences* **Sept.-Oct.** (1988): 22
97. Stein, D. L., R. G. Palmer, J. L. van Hemmen, and C. R. Doering. *Phys. Rev. A* **136** (1989): 353.
98. Stoeckenius, W., and R. A. Bogomolni. *Ann. Rev. Biochem.* **51** (1982): 587–616
99. Song, L., and M. F. Maestre. *J. Biomol. Struc. & Stero Dynamics* **9** (1991): 87–99.
100. Szabo, A., K. Schulten, and Z. Schulten. *J. Chem. Physics* **72** (1980): 4350.
101. Van Zandt, L. L. *Phys. Rev. Lett.* **57** (1986): 2085–2087.
102. Verkin, B. I., V. Suharevskii, Y. Telezhenko, A. V. Alapina, and N. Y. Vorob'eva. *Fizika Nizk. Temp* **3** (1977): 252.

103. Webman, I., and G. S. Grest. *Phys. Rev. B* **31** (1985) 1689.
104. Weiss, G. H. *Adv. Chem. Phys.* **13** (1967): 1.
105. Young, R. D., and S. F. Bowne. *J. Chem. Phys.* **81** (1984): 3730.
106. Zimanyi, L., P. Ormos, and J. K. Lanyi. *Biochem.* **28** (1989): 1656–1661.

Raymond E. Goldstein
Department of Physics, Jadwin Hall, Princeton University, Princeton, NJ 08544; e-mail gold@puhep1.princeton.edu

Nonlinear Dynamics of Pattern Formation in Physics and Biology

This chapter summarizes recent work on the dynamics of pattern formation in dissipative and Hamiltonian systems in two dimensions carried out in collaboration with S. A. Langer, D. P. Jackson, and D. M. Petrich. Some unifying geometric aspects are illustrated in the context of three distinct problems: the motion of shapes governed by bending elasticity, labyrinthine pattern formation of dipolar domains, and vorticity dynamics in ideal two-dimensional fluids. Following a discussion of perhaps the simplest dynamical formalism for shape evolution with global geometric constraints, the dynamics of some simple two-dimensional systems are considered, with emphasis on the complex configuration space which exists when short-range and long-range interactions compete in the presence of constraints. We review recent experimental and theoretical work on patterns formed by quasi-two-dimensional drops of magnetic fluid in a magnetic field, again illustrating that the space of configurations has a large number of local minima and that kinetic and energetic effects compete in selecting patterns. Finally, the mathematics of integrable systems related to the Korteweg-de Vries equation are shown to be fully equivalent to a

1992 Lectures in Complex Systems, Eds. L. Nadel & D. Stein, SFI Studies in the Sciences of Complexity, Lect. Vol. V, Addison-Wesley, 1993

hierarchy of chiral shape dynamics of closed curves in the plane, a result shedding light on the geometry of Euler's equation.

INTRODUCTION

In fields as diverse as developmental biology and fluid dynamics, there is a recurring notion of a link between *form* and *motion*. The present section is a summary of recent theoretical[13,14,19] and experimental work[6] which has endeavored to shed light on this general question of the interplay of geometry and dynamics of pattern formation in a class of systems. It is a common feature in nature that patterns arise from a competition between thermodynamic driving forces and global topological and geometrical constraints. While much is known concerning the static aspects of problems of this type (e.g., the elastic basis for the shapes of red blood cells), little is known about the *dynamics* by which arbitrary shapes and patterns relax to the global or local minimum of free energy under such constraints.

In the first section, we elaborate on a central theme for this investigation, the notion that patterns of interest are often well represented as d-dimensional surfaces (or interfaces) which are governed by a configurational energy and remark on the rather general occurrence of a competition between short-range and long-range interactions between points on the surface. The necessary differential geometry for discussing the simplest problems, those involving curve motion in the plane, is then described, followed by a variational principle for the dissipative evolution of boundaries of two-dimensional domains subject to global geometric constraints. The formalism is then applied to the motion of incompressible domains which relax to an accessible energetic minimum driven by line tension, elasticity, or nonlocal interactions as found in polymers, amphiphilic monolayers, and magnetic fluids. We seek to answer such questions as: Is an observed time-independent shape a unique energetic ground state or does the energy functional contain multiple metastable local minima? Are these minima roughly equivalent in energy? How can they be organized and classified? What kinetic considerations force a relaxing system into a metastable local minimum instead of the true ground state? Such questions are, of course, not confined to these particular examples of pattern formation but also arise in systems such as spin glasses[3] and in protein folding.[9]

Finally, we close with a very brief discussion of how aspects of the differential geometry of curve dynamics are shown to provide a new interpretation of the mathematics of integrable Hamiltonian systems. In particular, certain hierarchies of integrable systems are shown to be equivalent to a hierarchy of chiral shape dynamics of closed curves in the plane. These purely local dynamics conserve an infinite number of global geometric properties of the curves, such as perimeter and enclosed area. They in turn are related to the motion of iso-vorticity surfaces in ideal incompressible flow in two dimensions.

ENERGETICS AND CONSTRAINTS OF SURFACES

Here we describe some typical examples of patterns defined by surfaces and their energetics and constraints. Perhaps the simplest is a drop of incompressible fluid surrounded by a second immiscible fluid. The energy of interest is just that associated with surface tension γ, written as

$$\mathcal{E}_\gamma = \gamma \oint dS\,, \tag{1}$$

where dS is the differential of surface area. Such a drop moves with conserved volume, but nonconserved surface area. A lipid membrane, like that of a biological cell or artificial vesicle, is described by an elastic energy of the form[16]

$$\mathcal{E}_k = \oint dS \frac{1}{2} k_c (H - H_0)^2 + \oint dS \bar{k}_c K, \tag{2}$$

where $H = 1/R_1 + 1/R_2$ is the mean curvature and $K = 1/R_1 R_2$ is the Gaussian curvature, where R_1 and R_2 are the local principle radii of curvature of the surface. By hypothesis, this surface moves with fixed total area and may be impermeable to the flow of fluid (and hence have conserved enclosed volume) or leaky (with a volume determined by the osmotic pressure difference across the membrane).

More complicated and nonlocal potential energies arise when surfaces are allowed to interact with themselves. For instance, a membrane may have a long-range van der Waals attraction which leads to adhesion, described by some pairwise interaction ϕ between points on the surface,

$$\mathcal{E} = \mathcal{E}_k + \frac{1}{2} \oint dS \oint dS' \phi\left(|\mathbf{r}(S) - \mathbf{r}(S')|\right). \tag{3}$$

Such two-body interactions may be more complex, involving not just the positions of points on the surface but also their orientation. Indeed, we shall see below both for two-dimensional dipolar domains and ideal fluids that these interactions may involve the *tangent vectors* $\hat{\mathbf{t}}$ to a curve in the plane,

$$\mathcal{E} = \mathcal{E}_\gamma + \oint ds \oint ds' \hat{\mathbf{t}}(s) \cdot \hat{\mathbf{t}}(s') \Phi\left(|\mathbf{r}(s) - \mathbf{r}(s')|\right). \tag{4}$$

If we turn to a molecule like DNA that has an internal helical structure, an additional internal degree of freedom must enter the energy,

$$\mathcal{E} = \oint ds \left[\frac{1}{2} k_c \kappa^2 + \frac{1}{2} g \left(\omega - \omega_0\right)^2 \right], \tag{5}$$

where κ is the local curvature, ω describes the local rate of *twist* of the helix, and ω_0, the natural twist rate. A closed (e.g., circular) DNA molecule may adopt a complex three-dimensional *supercoiled* shape as a consequence of internal elastic strains associated with an excess or deficiency of windings of one edge of the helix about the other, a conserved *topological* quantity known as the "linking number deficit." The length is, of course, fixed and, barring the action of certain enzymes, so is the *knottedness* of the molecule.

Finally, turning to hydrodynamics, and in particular ideal inviscid flows, we find again the appearance of surface motion. Classical examples are the dynamics of vortex lines[2] and patches.[28] The important conservation laws in such systems include, of course, energy and momentum but also quantities related to the vorticity through the Kelvin circulation theorem,

$$\oint_{\mathcal{C}} d\mathbf{l} \cdot \mathbf{v} = \text{const.}, \tag{6}$$

where $\mathcal{C}$ is a contour in the fluid flow and $\mathbf{v}$ is the local velocity.

With these examples in mind, the central question we would like to address is: Given some energy functional $\mathcal{E}$, what is the motion it determines for the shape? More specifically, we may ask how a shape relaxes to some local (or global) minimum of that energy if it is prepared initially in some nonequilibrium state. To begin, we require some basic results from differential geometry and Lagrangian mechanics.

DYNAMICS OF CURVES IN THE PLANE

GEOMETRICAL PRELIMINARIES

For the remainder of this discussion, we will restrict our attention to motion of the simplest kind of surfaces, closed curves in the plane. Such geometrical objects, along with space curves, possess "arclength" as a natural parametrization unlike their generic higher dimensional counterparts. Nevertheless, it is often pedagogically useful to imagine an arbitrary variable $\alpha \in [0,1]$ labelling points $\mathbf{r}(\alpha)$ along the curve. Using a subscript to denote differentiation, the metric factor $\sqrt{g} = |\mathbf{r}_\alpha|$ is the Jacobian of the transformation between α and distance along the curve. Thus, for the *arclength parameterization* $s(\alpha)$ of the curve, there is the differential relation $ds = \sqrt{g}\,d\alpha$. The primary global geometrical quantities of interest, the length L and area $\mathcal{A}$, are then given by

$$L = \int_0^1 d\alpha\,\sqrt{g}, \qquad \mathcal{A} = \frac{1}{2}\int_0^1 d\alpha\,\mathbf{r} \times \mathbf{r}_\alpha. \tag{7}$$

Here, $\mathbf{a} \times \mathbf{b} \equiv \epsilon_{ij} a_i b_j$.

At each point along the curve, there is a local coordinate system defined by the unit tangent vector $\hat{\mathbf{t}}$ and normal $\hat{\mathbf{n}}$. The former is defined as $\hat{\mathbf{t}}(\alpha) \equiv g^{-1/2}\mathbf{r}_\alpha$, the latter is rotated by $-\pi/2$ with respect to it. Traversing the shape in a counter-clockwise direction, the curvature κ is defined through the arclength derivatives of these vectors by the Frenet-Serret equations,

$$\partial_s \begin{pmatrix} \hat{\mathbf{t}} \\ \hat{\mathbf{n}} \end{pmatrix} = \begin{pmatrix} 0 & -\kappa \\ \kappa & 0 \end{pmatrix} \begin{pmatrix} \hat{\mathbf{t}} \\ \hat{\mathbf{n}} \end{pmatrix}. \tag{8}$$

The curvature in turn is related to the angle $\theta(s)$ between the tangent vector and some arbitrary fixed axis by $\kappa = \theta_s$.

KINEMATIC CONSTRAINTS

A dynamics for the shape may be specified in terms of the components of the velocity in the local Frenet-Serret frame as

$$\mathbf{r}_t = U\hat{\mathbf{n}} + W\hat{\mathbf{t}}, \tag{9}$$

where normal and tangential velocities U and W are arbitrarily complicated local or nonlocal functions of $\mathbf{r}(\alpha)$. For closed curves, these functions must be periodic functions of s. Constraints such as length or area conservation now may be recast as constraints on the velocities U and W. Let us remark, however, that whatever the particular forms of U and W, the time evolution of κ and θ follow from Eq. (9) as[4]

$$\theta_t = -U_s + \kappa W, \tag{10}$$

and

$$\kappa_t = -\left(\partial_{ss} + \kappa^2\right) U + \kappa_s W. \tag{11}$$

The intrinsically nonlinear nature of these evolution equations reflects the interdependence of the internal coordinate s and the vector $\mathbf{r}(s)$.

If we consider length first among the global conservation laws, we observe that it may be conserved either globally or locally, the latter implying the former but not vice versa. Local conservation means that the distance along the curve between two points labelled by α and α' remains constant in time. Since this distance is determined by the metric, local length conservation means $\partial_t \left|\mathbf{r}_\alpha\right| = 0$, or, equivalently, $\hat{\mathbf{t}} \cdot \partial_s \mathbf{r}_t = 0$. Using Eq. (9) we obtain the local arclength conservation constraint in differential form, $W_s = -\kappa U$, or integral form,

$$W(s) = -\int^s ds' \kappa U \equiv -\partial^{-1} \kappa U \qquad \text{(local)}. \tag{12}$$

If we demand only the global conservation condition $L_t = 0$, then only the integral over s of $\partial_t \sqrt{g}$ need vanish. This can be shown to yield the global constraint

$$\oint ds\, \kappa U = 0 \qquad \text{(global)}. \tag{13}$$

Note that the local constraint relates both components U and W of the velocity, whereas the global constraint leaves the tangential component free. Given a normal velocity U, the local constraint determines W only up to an additive time-dependent function independent of s. A nonzero value of this function simply reparameterizes the curve, without changing its shape.

We appeal to reparameterization invariance to choose a convenient tangential velocity W. For systems with conserved *total* arclength the natural choice would be one which conserves local arclength, i.e., the metric $\sqrt{g}$. When the total arclength is not constant, a useful choice is still that which maintains uniform spacing of points on the curve, the *relative arclength* gauge. The condition $\partial_t(s/L) = 0$ determines W as

$$W(s) = \frac{s}{L}\oint ds'\, \kappa U - \int_0^s ds' \kappa U. \tag{14}$$

The surface of an incompressible two-dimensional domain moves with fixed enclosed area, a constraint which is again nonlocal in U,

$$\oint ds\, U = 0, \tag{15}$$

the form of which is clear on an intuitive level. In general, a *local* normal velocity $U(s)$, that is, one that depends only on the local geometry of the curve at the point s, will not satisfy these global integral constraints, so the motion will be intrinsically nonlocal (and mathematically complex!). Exceptions to this occur if, for instance, U and κU are total derivatives with respect to s. In the dissipative dynamics formulation, just as in equilibrium statistical mechanics, the mathematical freedom to satisfy these constraints arises from Lagrange multipliers present in an augmented free energy, as we now describe.

DISSIPATIVE DYNAMICS

A VARIATIONAL FORMALISM

In many of the systems of physical and biological interest, the surface motion is dominated by viscous drag—inertial effects are unimportant. Thus, we shall focus on strongly overdamped dynamics and, hence, seek *first-order* equations of motion that relax a shape to a minimum of an energy functional $\mathcal{E}$. It is convenient to derive the motion from an action principle using Lagrange's formalism for dissipative processes.[12] In constructing a Lagrangian $\mathcal{L}$, the generalized coordinates q_α are the positions of the points $\mathbf{r}(\alpha)$ on the curve and the potential energy is just the energy functional $\mathcal{E}$. In general, the equations of motion are

$$\frac{d}{dt}\frac{\partial \mathcal{L}}{\partial \dot{q_\alpha}} - \frac{\partial \mathcal{L}}{\partial q_\alpha} = -\frac{\partial \mathcal{F}_d}{\partial \dot{q_\alpha}}, \tag{16}$$

where the Rayleigh dissipation function $\mathcal{F}_d$ is proportional to the rate of energy dissipation by the viscous forces. For the typical viscous forces linear in the velocity, $\mathcal{F}_d$ is quadratic in $\mathbf{r}_t$, and so its derivative is linear in $\mathbf{r}_t$.

In order to study the interplay of geometry, dynamics, and constraints, we make a model for $\mathcal{F}_d$ that assumes *local dissipation with isotropic drag*, and write

$$\mathcal{F}_d = \frac{1}{2}\eta \int_0^1 d\alpha \sqrt{g} \left| \mathbf{r}_t - \Theta(\alpha, t)\hat{\mathbf{t}} \right|^2 , \tag{17}$$

where η is some friction coefficient. For motions that must be invariant under arbitrary time-dependent reparameterizations, we need the "gauge function" $\Theta(\alpha, t)$ to ensure that the reparameterizations do not contribute to the dissipation. Under the transformation $\alpha \to \alpha'(\alpha, t)$, the velocity transforms as $\mathbf{r}_t \to \mathbf{r}_t + \mathbf{r}_{\alpha'}\alpha'_t$, and the dissipation function is unchanged if we let $\Theta \to \Theta + \sqrt{g}\alpha'_t$.

In the viscous limit, we neglect the kinetic energy terms in the Lagrangian, so $\mathcal{L} = -\mathcal{E}[\mathbf{r}]$ and, by absorbing η into a rescaled time, we may rewrite Eq. (16) in terms of functional derivatives as

$$\mathbf{r}_t = -\frac{1}{\sqrt{g}} \frac{\delta \mathcal{E}}{\delta \mathbf{r}} + \Theta \hat{\mathbf{t}}. \tag{18}$$

The gauge function Θ is a tangential velocity, showing that it is indeed a reparameterization of the curve. Equation (18) has the appearance of the time-dependent Ginzburg-Landau equation[17] of dynamic critical phenomena and is also a version of the Rouse model of polymer dynamics.[7]

CONSTRAINED DYNAMICS

To include the possibility of imposing global conservation laws, we introduce time-dependent Lagrange multipliers Π and Λ conjugate to the area and length in an augmented energy functional

$$\mathcal{E} = \mathcal{E}_0 - \int_0^1 d\alpha \sqrt{g} \Lambda(\alpha) - \Pi A, \tag{19}$$

where $\mathcal{E}_0$ is the energy of the unconstrained system. Π and Λ are determined as follows. Let U_0 and W_0 be the velocities derived from $\mathcal{E}_0$,

$$-\frac{1}{\sqrt{g}} \frac{\delta \mathcal{E}_0}{\delta \mathbf{r}} = U_0(\alpha, t)\hat{\mathbf{n}} + W_0(\alpha, t)\hat{\mathbf{t}}. \tag{20}$$

Differentiation of the augmented free energy in Eq. (19), with L and A given in Eq. (7), yields dynamics with

$$U(s) = U_0 + \Lambda\kappa + \Pi \qquad \text{and} \qquad W(s) = W_0 - \Lambda_s + \Theta. \tag{21}$$

It is now necessary to consider two classes of motion, distinguished from each other in the way in which the unknown functions Θ and Λ are determined. In the *reparameterization-invariant* (RI) class, only the curve itself has physical meaning; the points α are simply labels. Given dynamics that conserves global arclength, we may always find a reparameterization Θ so that local arclength is conserved as well. A consistent value of $\Lambda(s)$ is then the constant determined at each instant of time by the global length constraint. This implies

$$\oint ds \kappa U_0 + \Lambda \oint ds \kappa^2 + 2\pi \Pi = 0, \tag{22}$$

where we have used $\oint ds\kappa = 2\pi$. In the *nonreparameterization-invariant* (NRI) class, we require $\Theta = 0$ and local arclength conservation is accomplished by choosing Λ to satisfy a differential equation

$$\left(\partial_{ss} - \kappa^2\right) \Lambda(s,t) = \kappa U_0 + \partial_s W_0 + \kappa \Pi. \tag{23}$$

If area is also conserved, then Λ and Π also satisfy

$$\oint ds U_0 + \oint ds \kappa \Lambda + \Pi L = 0. \tag{24}$$

We can see from Eqs. (22) and (24) that even though the dissipation function was taken to be local, global constraints ultimately do lead to nonlocality through the Lagrange multipliers.

EXAMPLES

Here we illustrate the dissipative dynamics formalism developed above in the context of three systems: tense interfaces, elastic polymers, and dipolar domains. In the discussion of the third class we will also summarize the salient results from recent experiments.

TENSE INTERFACES. Consider again the incompressible fluid drop mentioned in the introduction. In the two-dimensional problem, the motion is driven by the *line tension* γ. The energy $\mathcal{E}_0 = \gamma L$ is clearly minimized by shapes having the smallest perimeter consistent with the prescribed area, i.e., circles. We find, by functional differentiation, that the normal velocity is $U_0(s) = -\gamma\kappa$, and W_0 vanishes. Since the perimeter is clearly not conserved, we need only determine the area Lagrange multiplier Π. It is $\Pi = (2\pi/L)\gamma$. The entire dynamics reduces to two coupled differential equations for the length and curvature,

$$L_\tau = \frac{(2\pi)^2}{L} - \oint ds \kappa^2, \qquad \kappa_\tau = \kappa_{ss} + \kappa^3 - \frac{2\pi}{L}\kappa^2 + \kappa_s W, \tag{25}$$

with W given by Eq. (14), $U = U_0 + \Pi$, and $\tau = \gamma t$. Apart from the nonlocality associated with the tangential velocity, Eq. (25) is an area-conserving version of the well-known "curve-shortening equation."[11] If we neglect the tangential velocity and the area constraint, the dynamics is just

$$\mathbf{r}_\tau = \mathbf{r}_{ss}, \tag{26}$$

a diffusion equation. This linearity is deceptive since, as mentioned earlier, $\mathbf{r}$ and s are not independent. The full dynamics in Eq. (25) relaxes the shape to one of uniform curvature $\kappa = 1/R_0$, with $L = 2\pi R_0$, where πR_0^2 is the area of the initial shape. This is illustrated in Figure 1, where we see the curvature evolution associated with an ellipse relaxing to a circle. The corresponding shape evolution is shown in Figure 2, along with the perimeter relaxation. In this simple example, it is plausible that the circle is the unique minimum in the energy functional and that all initial conditions will relax to it.

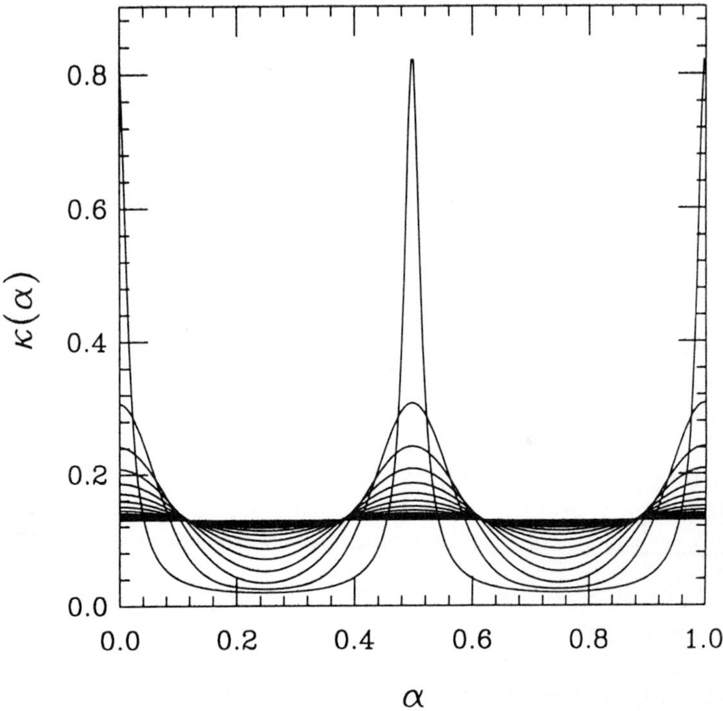

FIGURE 1 Curvature evolution, for the relaxation of an ellipse to a circle, according to the dissipative dynamics formalism.

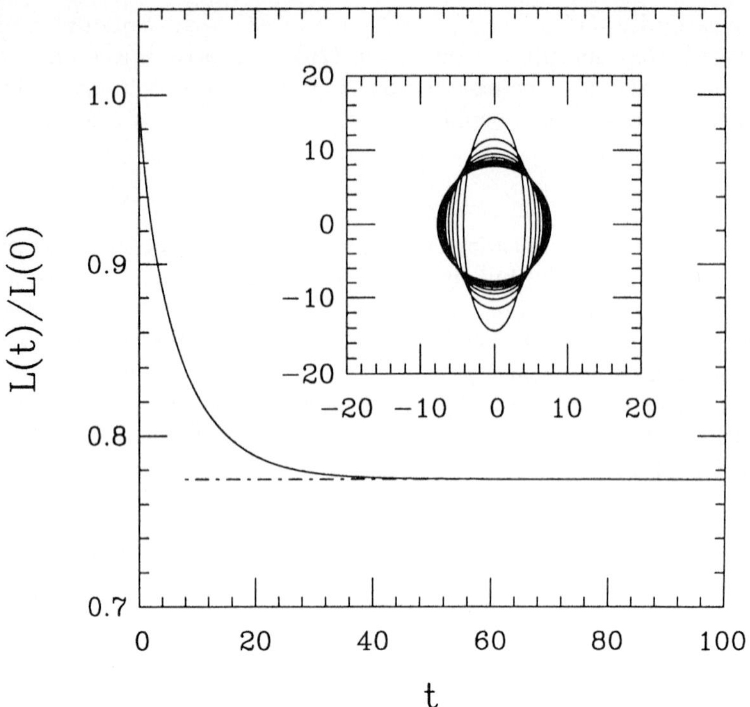

FIGURE 2 Shape evolution, associated with Figure 1, and perimeter relaxation.

ELASTIC MEMBRANES AND CLOSED POLYMERS. If we now endow the planar curves with an elastic energy, we have a simple model of two-dimensional vesicles or ring polymers. Those with constrained enclosed area are "impermeable," whereas permeable vesicles have an area set by specifying an osmotic pressure Π. We may then ask the question: How does a closed polymer or vesicle in two dimensions "fold" itself into an energetic minimum without self-intersections?

The conventional elastic energy is simply quadratic in the curvature[16]:

$$\mathcal{E}_{cur} = \frac{1}{2}k_c \oint ds\, \kappa^2, \tag{27}$$

where k_c is the rigidity of the membrane. The functional derivative of Eq. (27) combines with the Lagrange multipliers to provide the normal velocity

$$U = k_c\left(\kappa_{ss} + \frac{1}{2}\kappa^3\right) + \Lambda\kappa + \Pi, \tag{28}$$

and the bare tangential velocity W_0 vanishes. Interestingly, the dynamics in Eq. (28) appears as an expansion in the curvature and its derivatives, like the "geometrical" models of pattern formation used in the study of crystal growth.[4] Its presence here, however, is a consequence of a variational formulation not envisioned in the nonequilibrium crystal growth process.

In the presence of pairwise interactions as in Eq. (3), the normal velocity has an additional contribution

$$U_0 = \hat{\mathbf{n}}(s) \cdot \oint ds' \frac{(\mathbf{r}(s) - \mathbf{r}(s'))}{|\mathbf{r}(s) - \mathbf{r}(s')|} \phi' \left(|\mathbf{r}(s) - \mathbf{r}(s')|\right), \tag{29}$$

which is simply the (nonlocal) normal force at point s due to the rest of the chain. A repulsive core to the potential ϕ will prevent self-crossings and combines with the attractive tail to produce a minimum in ϕ.

Figure 3 illustrates how a random initial shape relaxes under these dynamics to a local energetic minimum. We note the appearance of "hairpin loops." These are the natural compromise between the attractive membrane interactions, favoring local parallelism of the chain segments, and the curvature energy, which disfavors the bends necessitated by the constraint of closure.

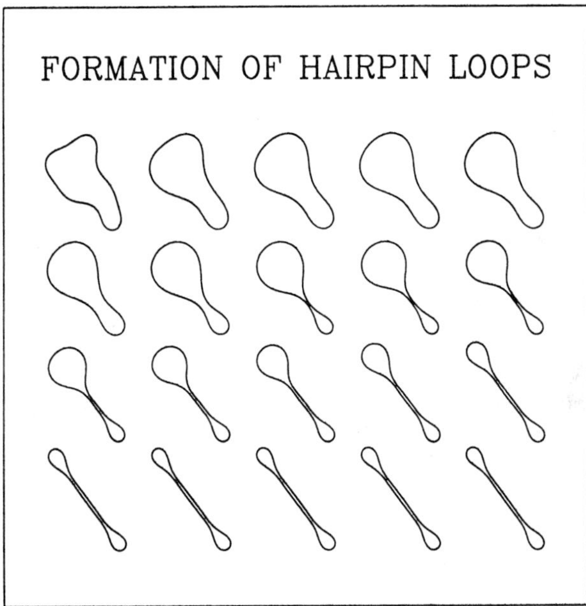

FIGURE 3 Intermediate stages in the "folding" of a closed elastic polymer in two dimensions.

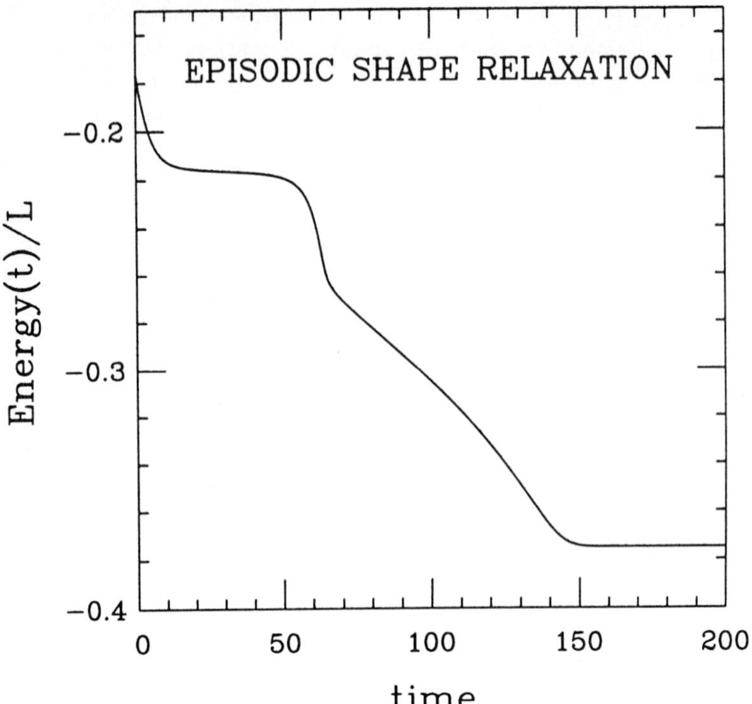

FIGURE 4 Episodic relaxation of the energy corresponding to the shape evolution in Figure 3.

The time evolution of the energy, shown in Figure 4, is found to be rather *episodic*, with periods of rather gradual decay interrupted by rapid relaxation intervals. These may plausibly be associated with configurational "bottlenecks" in the relaxation process. It is of great interest that such complex dynamics can emerge from such a simple energy function. Figure 5 shows how a different initial condition relaxes to the same local minimum in configuration space, albeit by a quite different path. Finally, by reducing the elastic constant k_c, the polymer can tolerate more bending, and fall into a branched local minimum like that shown in Figure 6.

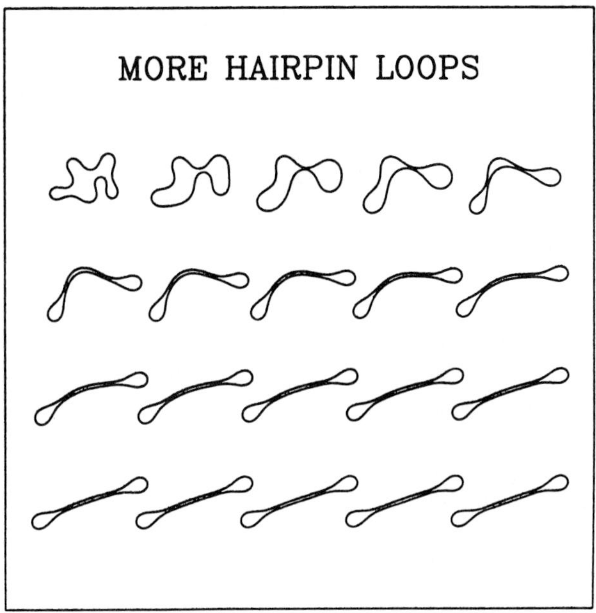

FIGURE 5 Convergence of a different initial condition to the same local minimum as in Figure 3.

MAGNETIC FLUIDS AND DIPOLAR DOMAINS. Motivated by the appearance of "labyrinthine" patterns in several quite distinct physical systems, thin magnetic films,[26] amphiphilic "Langmuir" monolayers,[20,23,25] and Type I superconductors in magnetic fields,[18] we have been led to investigate both theoretically and experimentally the patterns formed by magnetic fluids[24] ("ferrofluids"). These materials are colloidal suspensions of microscopic magnetic particles. When placed between closely spaced parallel glass plates and magnetized by an external magnetic field normal to the plates, they are macroscopic examples of two-dimensional dipolar domains. As such, they constitute a convenient system in which to study the competition between short-range and long-range forces.

We are interested in understanding whether the similarities found in the pattern formation of such distinct physical systems, as mentioned above, do actually reflect a common mechanism. To the extent that the observed patterns reflect the underlying energetics of the shapes, the similarities are indeed understandable. In each case, the labyrinth is formed of the boundary between two thermodynamic phases (up- and down-magnetized domains, expanded and condensed dipolar phases, normal and superconducting regions, and magnetic fluid against water), and has an associated surface tension that favors minimizing the contour length. Each system also possesses long-range bulk interactions of various origins. In the ferrofluid example studied here, the dipole-dipole force between suspended magnetic particles aligned with the applied field is repulsive, tending to extend the fluid along the plates. Similar interactions exist in amphiphilic monolayers, the dipolar molecules of which are aligned perpendicular to the air-water interface. In solid-state magnetic systems the spontaneous magnetization produces the long-range interactions, while in superconducting thin films the in-plane Meissner currents interact via the Biot-Savart force.

Figure 7 illustrates the shape evolution of a ferrofluid domain after the magnetic field is brought rapidly to a fixed value. The pattern evolution lasted approximately 60 s after the application of the field; the figure at the lower right is essentially time-independent and locally stable to small perturbations. In general we find that the branching process displays sensitive dependence on initial conditions in the sense that two initially circular shapes, indistinguishable to the eye, evolve under identical applied fields to trees differing in the shapes, lengths, and connectivity of their branches. There is thus a vast number of topologically different minima that may be reached by the system. The relaxation of the tree to a circular shape during a gradual decrease of the applied field to zero is shown in Figure 8.

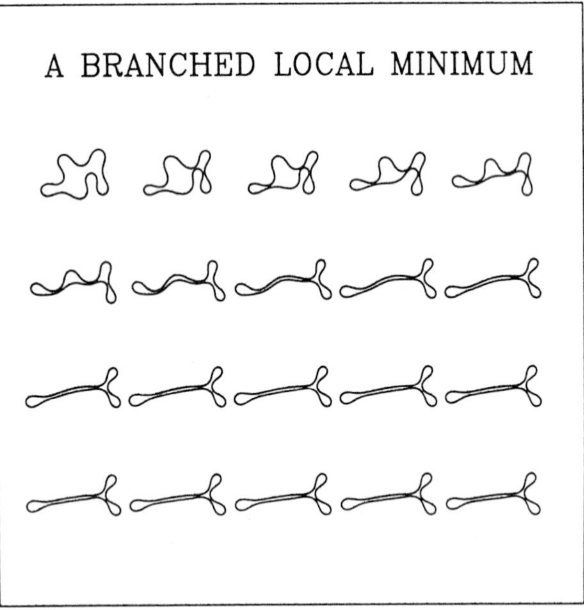

FIGURE 6 Relaxation toward a branched local minimum at lower rigidity.

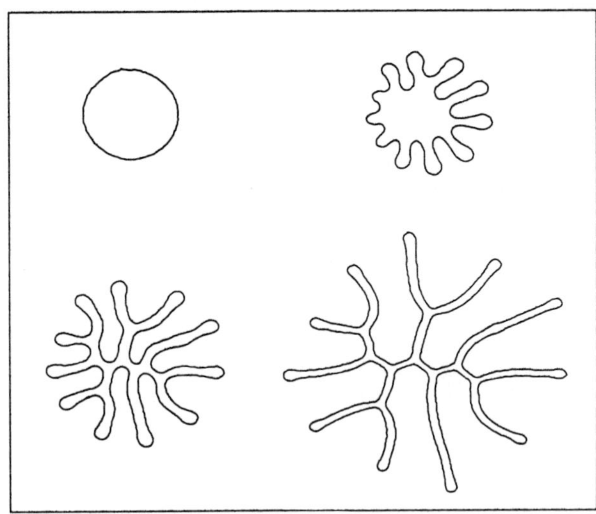

FIGURE 7 Snapshots of the fingering instability of an initially circular ferrofluid domain.

It has been recognized for some time, especially in the context of amphiphilic systems,[1] that the competition between these long-range forces and surface tension can result in a variety of *regular* patterns, such as lamellar stripe domains, hexagonal arrays, etc. The more widely encountered *irregular*, or disordered, patterns such as those in Figure 7 are, however, poorly understood.

These shapes may be classified according to their topology. All observed trees have n free ends and $n-2$ three-fold coordinated nodes, at which branches meet at nearly 120°. The patterns may be arranged in the hierarchy shown in Figure 9 by grouping together trees with a given number of free ends (or, equivalently, a given number of nodes) into a single generation and by linking those members of adjacent generations that are related by the addition or subtraction of a single node. These relationships are shown by the lines in Figure 9. A compact labeling scheme[10] for these is achieved by first labeling the free ends of the pattern in a cyclic fashion $(1, 2, \ldots, n)$. Then, one may construct a vector $\mathbf{v} = (a_1, \ldots, a_n)$ of n integers such that, between the free ends of the pattern labeled i and $i+1$, there are a_i three-fold coordinated nodes. Apart from cyclic permutations depending on choice of the first free end, this vector uniquely identifies the topology. This hierarchical arrangement is reminiscent of proposed phase-space geometries of spin glasses and proteins, which have the additional property of *ultrametricity*.[22] We observe that the arrangement here is *not* ultrametric, since there are closed loops in the hierarchy.

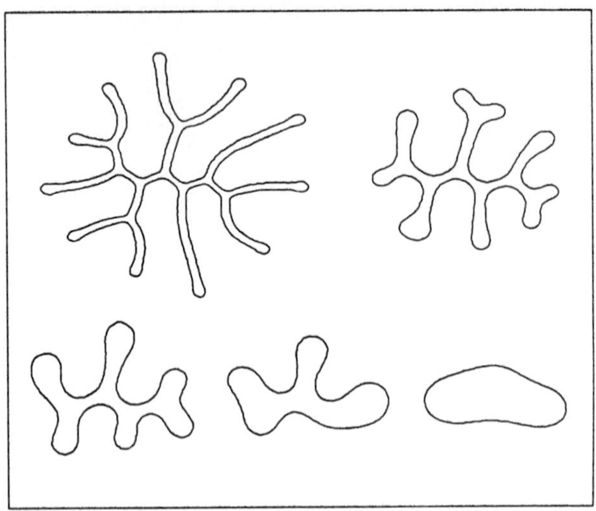

FIGURE 8 Relaxation of the final state in Figure 7 as the magnetic field is ramped slowly back to zero.

Despite the fact that the patterns observed are so different in numbers of arms and in connectivities, they share two very basic geometric quantities: the perimeter L and radius of gyration R_G, defined as $R_G^2 \equiv (1/L) \oint ds\, (\mathbf{r}(s) - \mathbf{r}_c)^2$, where $\mathbf{r}_c \equiv (1/L) \oint ds\, \mathbf{r}(s)$ is the center of mass of the boundary of the pattern and s is the arclength. A convenient method for summarizing the shape evolution is to consider its trajectory in the $L - R_G$ plane, with time varying parametrically. Figure 10 displays trajectories in this space for four shapes found following a rapid ramp to a large field. The scales of the two axes have been normalized by the perimeter L_0 and radius R_0 of the initial circular state. We see that patterns which differ in the details of their branching nevertheless may exhibit a high degree of overlap, both during the evolution and in their final states. This similarity suggests the near energetic equivalence of these patterns.

Turning to the means by which patterns are selected, we find that the degree of branching (and hence the level to which the hierarchy in Figure 9 is descended) is highly correlated with the rate at which the magnetic field is ramped to its final value. Figure 11 shows how the initial mode number of the instability depends on dH/dt, as the rate is varied over three orders of magnitude. Note that the selection has occurred long before the ramping was complete. A detailed understanding of this effect is not yet at hand.

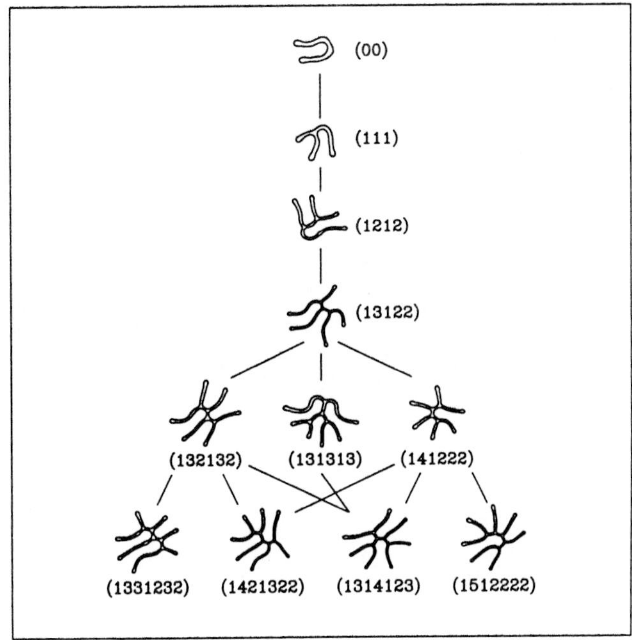

FIGURE 9 Hierarchical arrangement of topologically distinct local minima. Numerical labels distinguish the topologies.

In the simplest theory of this pattern formation, we obtain an energy functional like that in Eq. (4), where the nonlocal term represents the field energy associated with the oriented dipoles. The calculation of this is presented elsewhere.[19] Its form is understandable from the usual association between magnetization and current loops; it is just the self-energy of a ribbon of current flowing around the boundary of the domain. Viewed this way, the scalar product in the energy reflects the attraction (repulsion) between parallel (antiparallel) current-carrying wires. The pair interaction has a complex form which reflects the finite height h of the sample:

$$\Phi(\xi) = -\frac{\mu^2}{h}\left\{\sinh^{-1}(1/\xi) + \xi - \sqrt{1+\xi^2}\right\}. \tag{30}$$

Here, $\xi = R/h$, where $R = |\mathbf{R}| = |\mathbf{r}(s') - \mathbf{r}(s)|$ is the in-plane distance between points at positions s and s' on the boundary and μ is the dipole density per unit area. For $\xi \gg 1$, the function Φ is essentially Coulombic ($\Phi \simeq 1/2\xi$), whereas for $\xi \leq 1$ it is less singular, varying as $\ln(2/\xi)$. This crossover to logarithmic behavior occurs because of the finite thickness h of the slab, and prevents the integrals from diverging without additional cutoffs.

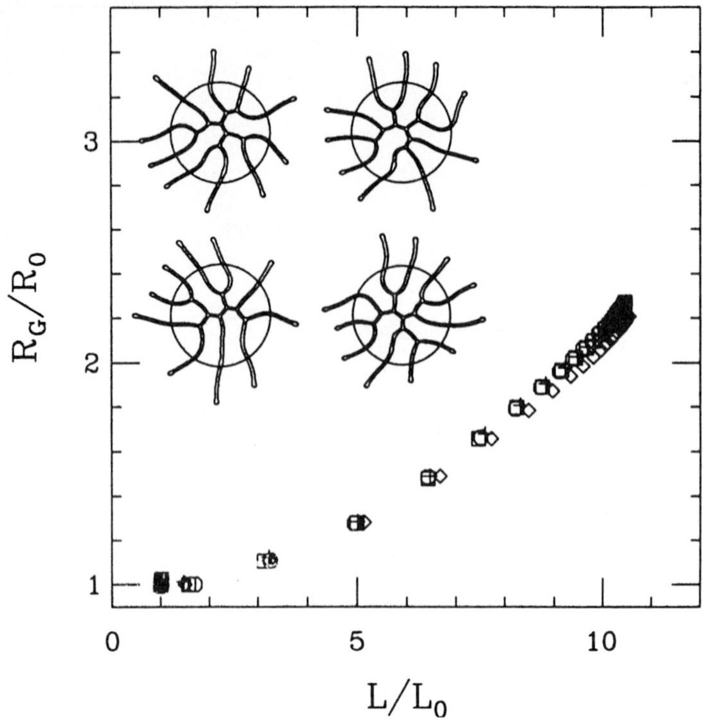

FIGURE 10 Radius of gyration and perimeter evolution for four trees, illustrating the near equivalence of the local minima.

From these results we deduce that the full energy is determined by one dimensionless parameter, the "magnetic Bond number"[24] $N_{\rm Bo} \equiv 2\mu^2/\gamma$, and by the shape of the dipolar region.

The velocity arising from the functional derivative of the total energy is

$$U_0(s) = -\gamma\kappa + \frac{2\mu^2}{h^2}\oint ds'\, \hat{\mathbf{R}} \times \hat{\mathbf{t}}' \left[\sqrt{1 + (h/R)^2} - 1\right], \tag{31}$$

where $\hat{\mathbf{R}} = \mathbf{R}/R$ is the unit vector pointing from the point s towards s'. The nonlocal term is essentially a Biot-Savart force due to a wire (of finite height) carrying an effective current $I = E_0hc/4\pi$ around the boundary. The tangential force $W_0(s)$ vanishes.

Analytic progress can be made in the linear stability analysis about a circular shape,[19,27] but the nonlinear regime requires numerical study; some results are shown in Figure 12. This simple dynamics satisfactorily reproduces the essential

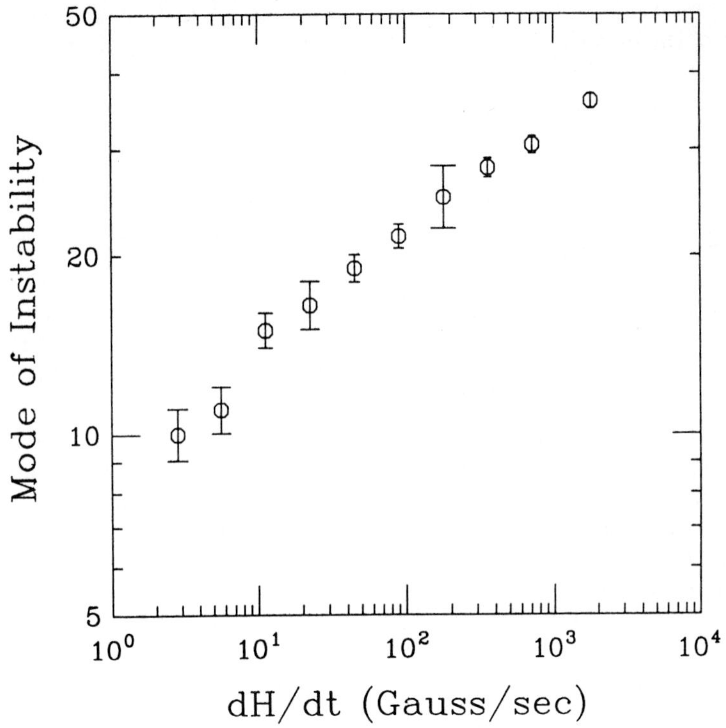

FIGURE 11 Observed mode of instability as a function of magnetic field ramp rate.

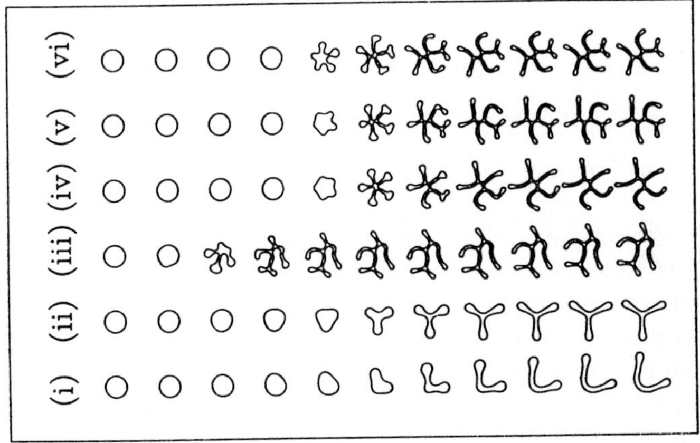

FIGURE 12 Theoretical pattern formation of a dipolar domain,[19] with time increasing downward.

features of the experimental pattern formation, most notably the existence of many local minima, sensitive dependence on initial conditions, and the gross geometric features of the trees.

CURVE DYNAMICS AND SOLITONS

The problems in pattern formation considered thus far have been strongly dissipative. Quite remarkably, however, many of the mathematical issues and techniques brought to bear on those problems also have application in integrable Hamiltonian dynamics, as we now describe.

Returning to the integral constraints required by perimeter and area conservation, Eqs. (13) and (15), one notices that such conservation is automatic if U and κU are total derivatives with respect to arclength of any periodic functions. Moreover, with the tangential velocity determined by local arclength conservation, W is then determined *locally* from U. It follows then that the curvature evolution is determined entirely from the normal velocity as $\kappa_t = -\Omega U$, where $\Omega = \partial_{ss} + \kappa^2 + \kappa_s \partial^{-1}\kappa$. The simplest pair of velocity functions which conserve length and area is $\left(U^{(1)} = 0, W^{(1)} = -c\right)$, with c a constant. The motion is simply a reparameterization, as can be seen by the curvature evolution

$$\kappa_t = c\kappa_s. \tag{32}$$

A second choice,[5] $\left(U^{(2)} = \kappa_s, W^{(2)} = -(1/2)\kappa^2\right)$, due to Constantin, yields

$$\kappa_t = -\kappa_{sss} - \frac{3}{2}\kappa^2\kappa_s. \tag{33}$$

These two curvature dynamics happen to be the first two members of a hierarchy of integrable systems,[8] an infinite set of 1+1-dimensional partial differential equations, each with a common infinite number of conserved quantities. Equation (33) is known as the "modified Korteweg-de Vries (mKdV) equation" related to the KdV equation which describes solitons in narrow channels of fluid.[8] Each of these dynamics is *chiral*, breaking the symmetry between s and $-s$ which we saw in the dissipative dynamics in earlier sections.

In addition to conserving length and area, by construction, the additional conserved quantities of Eq. (33) are of the form $H_k = \oint ds\ h_k$ with h_k obeying a continuity equation $\partial_t h_k + \partial_s j_k = 0$, for some currents j_k. For mKdV, which is already in the form of a continuity equation, the successive conserved quantities are

$$H_1 = \oint ds\kappa, \quad H_2 = -\frac{1}{2}\oint ds\kappa^2, \quad H_3 = \oint ds\left\{-\frac{3}{8}\kappa^4 + \frac{1}{2}\kappa_s^2 - \kappa\kappa_{ss}\right\}, \tag{34}$$

etc. For $k \geq 2$ these are just the tangential velocities of the hierarchy, while H_1 is just the "winding angle"—the angle through which the tangent vector rotates as the curve is traversed. Thus, starting from the conservation of perimeter and enclosed area, we have ended up with dynamics with an infinite number of conservation laws!

The mKdV hierarchy parallels the more familiar KdV hierarchy which is based on the KdV equation itself, $u_t + u_{sss} - 3uu_s = 0$. The two hierarchies are connected by the Miura transformation[21]

$$u = -\frac{1}{2}\kappa^2 - i\kappa_s, \tag{35}$$

such that if $\kappa(s,t)$ satisfies the nth-order mKdV equation, then u satisfies the nth-order KdV equation. Since the variable of the mKdV hierarchy is the curvature, it is natural to inquire about the geometrical significance of u. Consider then the curve in the complex plane given by $z(s,t) = x(s,t) + iy(s,t)$, with $z_s(s,t) = e^{i\theta(s,t)}$ being the tangent vector. Using the associated representation of the curvature, $\kappa = -iz_{ss}/z_s$, we find

$$u = -\left[\left(\frac{z_{ss}}{z_s}\right)_s - \frac{1}{2}\left(\frac{z_{ss}}{z_s}\right)^2\right] \equiv -\{z, s\}. \tag{36}$$

We recognize the quantity $\{f, x\}$ as the Schwarzian derivative of a function f with respect to its argument x. This quantity has the property of being invariant under fractional linear transformations in the complex plane; that is, $\{z, s\} = \{w, s\}$ under transformations of the form $z \to w = (az+b)/(cz+d)$, which takes circles to circles. Thus, not only do the KdV curve dynamics have an infinity of conservation laws, they also have very strong invariance properties under mappings of the complex plane.

In addition to the conservation of enclosed area, which we naturally associate with an incompressible fluid, a second aspect of the KdV dynamics suggests that the curves whose motion is described by them are associated with ideal fluid flow, and hence with solutions of Euler's equation. Among the conserved quantities for each member of the hierarchy is the tangential velocity W. This is actually just the Kelvin circulation theorem in Eq. (6).

We have found[14] that the relationship between the KdV dynamics and the motion of ideal fluids with vorticity mirrors a well-known result in three-dimensional ideal fluid flow, the connection[15] between the nonlinear Schrödinger (NLS) equation, $i\psi_t = -\psi_{xx} - (1/2)|\psi|^2\psi$, and the motion of a vortex filament. The NLS is the geometric evolution equation in a *local approximation* to the full nonlocal dynamics governed by the Biot-Savart law. Unlike the Euler equations themselves, the NLS is an integrable system with an infinite number of conserved quantities.

In two dimensions, the idealized distribution of vorticity analogous to the filament is a *vortex patch*, a bounded region of constant vorticity surrounded by irrotational fluid. The known exact equation of motion[28] for the boundary of such a domain is very nonlocal (again reflecting an underlying Biot-Savart law). Under a local approximation like that used in the NLS, the evolution equation for the curvature of the boundary is the mKdV equation.

Among the most interesting features of these results is that the integrable curve dynamics obeys a variational principle of the form

$$\hat{\mathbf{n}} \cdot \mathbf{r}_t = \partial_s \left(\hat{\mathbf{n}} \cdot \frac{\delta \mathcal{H}}{\delta \mathbf{r}} \right) , \tag{37}$$

remarkably similar to the dissipative dynamics result in Eq. (18). This suggests that it may be possible to develop a common language with which to describe both dissipative and Hamiltonian pattern formation.

CONCLUSIONS

Our emphasis here has been to illustrate issues and techniques that arise in the study of pattern formation in biological and physical systems. Whether for dissipative or Hamiltonian systems, some common aspects of these systems are emerging, including the stucture of variational principles, the competition between short- and long-range interactions, and the unifying point of view stemming from the differential geometry of surface motion. It is hoped that these investigations may provide a framework for deeper study of particular systems of physical or biological interest and for addressing the striking complexity of the patterns seen in nature.

ACKNOWLEDGMENTS

I am indebted to my colleagues Stephen A. Langer, David P. Jackson, Dean M. Petrich, Akiva J. Dickstein, and Shyamsunder Erramilli for ongoing collaborations described in this review. This work has been supported by NSF Grant No. CHE9106240 and the Alfred P. Sloan Foundation. The opportunity to present these lectures at the complex systems summer school and to partake in its exciting atmosphere of interdisciplinary discussion are most appreciated.

REFERENCES

1. Andelman, D., F. Brochard, and J.-F. Joanny. "Phase Transitions in Langmuir Monolayers of Polar Molecules." *J. Chem. Phys.* **86** (1987): 3673–3681.
2. Batchelor, G. K. *An Introduction to Fluid Dynamics*. Cambridge: Cambridge University Press, 1967.
3. Binder, K., and A. Young. "Spin Glasses. Experimental Facts, Theoretical Concepts, and Open Questions." *Rev. Mod. Phys.* **58** (1986): 801–976.
4. Brower, R. C., D. A. Kessler, J. Koplik, and H. Levine. "Geometric Models of Interface Evolution." *Phys. Rev. A* **29** (1984): 1335–1342.
5. Constantin, P. Private communication, 1990.
6. Dickstein, A. J., S. Erramilli, R. E. Goldstein, D. P. Jackson, and S. A. Langer. "Labyrinthine Pattern Formation of Magnetic Fluid." Preprint, Princeton University, 1993.
7. Doi, M. and S. F. Edwards. *The Theory of Polymer Dynamics*. New York: Oxford University Press, 1986.
8. Drazin, P. G., and R. S. Johnson. *Solitons: An Introduction.* New York: Cambridge University Press, 1989.
9. Frauenfelder, H., S. G. Sligar, and P. G. Wolynes. "The Energy Landscape and Motions of Proteins." *Science* **254** (1991): 1598–1603, and references therein.
10. Hwang, F. K., and J. F. Weng. "Hexagonal Coordinate Systems and Steiner Minimal Trees." *Discrete Math.* **62** (1986): 49–57.
11. Gage, M. E. "Curve Shortening Makes Convex Curves Circular." *Invent. Math.* **76** (1984): 357–364.
12. Goldstein, H. *Classical Mechanics*. Reading, MA: Addison-Wesley, 1980.
13. Goldstein, R. E., and D. M. Petrich "The Korteweg-de Vries Hierarchy as Dynamics of Closed Curves in the Plane." *Phys. Rev. Lett.* **67** (1991): 3203–3206.
14. Goldstein, R. E., and D. M. Petrich. "Solitons, Euler's Equation, and Vortex Patch Dynamics." *Phys. Rev. Lett.* **69** (1992): 555–558.
15. Hasimoto, H. "A Soliton on a Vortex Filament." *J. Fluid. Mech.* **51** (1972): 477–485.
16. Helfrich, W. "Elastic Properties of Lipid Bilayers—Theory and Possible Experiments." *Z. Naturforsch.* **28c** (1973): 693–703.
17. Hohenberg, P. C., and B. I. Halperin. "Theory of Dynamic Critical Phenomena." *Rev. Mod. Phys.* **49** (1977): 435–479.
18. Huebner, R. P. *Magnetic Flux Structures in Superconductors.* New York: Springer-Verlag, 1979.
19. Langer, S. A., R. E. Goldstein, and D. P. Jackson. "Dynamics of Labyrinthine Pattern Formation in Magnetic Fluids." *Phys. Rev. A* **46**: 4894–4904.
20. McConnell, H. M., and V. T. Moy. "Shapes of Finite Two-Dimensional Lipid Domains." *J. Phys. Chem.* **92** (1988): 4520–4525.

21. Miura, R. S. "Korteweg-de Vries Equation and Generalizations. I. A Remarkable Explicit Nonlinear Transformation." *J. Math. Phys.* **9** (1968): 1202–1204.
22. Rammal, R., G. Toulouse, and M. A. Virasoro. "Ultrametricity for Physicists." *Rev. Mod. Phys.* **58** (1986): 765–788.
23. Rice, P. A., and H. M. McConnell. "Critical Shape Transitions of Monolayer Lipid Domains." *Proc. Natl. Acad. Sci.* **86** (1989): 6445–6448.
24. Rosensweig, R.E. *Ferrohydrodynamics.* Cambridge: Cambridge University Press, 1985.
25. Seul, M., and M. J. Sammon. "Competing Interactions and Domain-Shape Instabilities in a Monomolecular Film at an Air-Water Interface." *Phys. Rev. Lett.* **64** (1990): 1903–1906;
26. Seul, M., L. R. Monar, L. O'Gorman, and R. Wolfe. "Morphology and Local Structure in Labyrinthine Stripe Domain Phase." *Science* **254** (1991): 1616–1618.
27. Tsebers, A. O., and M. M. Maiorov. "Magnetostatic Instabilities in Plane Layers of Magnetizable Fluids." *Magnetohydrodynamics* **16** (1980): 21–28.
28. Zabusky, N. J., M. H. Hughes, and K. V. Roberts. "Contour Dynamics for the Euler Equations in Two Dimensions." *J. Comp. Phys.* **30** (1979): 96–106.

Joshua M. Epstein
Senior Fellow, The Brookings Institution and Visiting Lecturer, Princeton University

On the Mathematical Biology of Arms Races, Wars, and Revolutions

This chapter and the next one will appear as Lectures 2 and 3 in *Nonlinear Dynamics, Mathematical Biology, and Social Science* by Joshua M. Epstein, to be published as a lecture notes volume in the Santa Fe Institute Studies in the Sciences of Complexity series (Addison-Wesley, 1994).

In the preceding lecture, we developed some powerful mathematics.[1] In subsequent lectures, we will use it to delve more deeply into the dynamics of war, arms racing, and revolution. In this chapter, I attempt a unifying overview of these social phenomena from the perspective of mathematical biology, a field which, in my view, must ultimately subsume the social sciences.[2] Unfortunately, few social

[1]See Lecture 1, "An Introduction to Nonlinear Dynamical Systems," in Joshua M. Epstein, *Nonlinear Dynamics, Mathematical Biology, and Social Science*, forthcoming (Addison-Wesley, 1994).

[2]Edward O. Wilson, in his book *Sociobiology*, has advanced a closely related view. The perspective taken here, however, is quite distinct from that taken by Wilson. Specifically, I do not discuss the role of genes in the control of human social behavior. Rather, the argument is that macro social behaviors such as war, revolution, arms races, and the spread of drugs may conform well to equations of mathematical biology—ecology and epidemiology in particular. That, ultimately,

scientists are exposed to mathematical biology, specifically the dynamical systems perspective pioneered by Alfred Lotka, Vito Volterra, and others. In turn, mathematical biologists—with such notable exceptions as John Maynard Smith and Marcus Feldman—have not considered the application of mathematical biology to problems of human society.

Particularly in areas of interstate and intrastate conflict is there a need to explore formal analogies to biological systems. On the topic of animal behavior and human warfare, the anthropologist Richard Wrangham observes,

> "The social organization of thousands of animals is now known in considerable detail. Most animals live in open groups with fluid membership. Nevertheless there are hundreds of mammals and birds that form semi-closed groups, and in which long-term intergroup relationships are therefore found. These intergroup relationships are known well. In general they vary from benignly tolerant to intensely competitive at territorial borders. The striking and remarkable discovery of the last decade is that only two species other than humans have been found in which breeding males exhibit systematic stalking, raiding, wounding and killing of members of neighbouring groups. They are the chimpanzee (*Pan troglodytes*) and the gorilla (*Pan gorilla beringei*) (Wrangham, 1985). In both species a group may have periods of extended hostility with a particular neighbouring group and, in the only two long-term studies of chimpanzees, attacks by dominant against subordinate communities appeared responsible for the extinction of the latter.
>
> "Chimpanzees and gorillas are the species most closely related to humans, so close that it is still unclear which of the three species diverged earliest (Ciochon & Chiarelli, 1983). The fact that these three species share a pattern of intergroup aggression that is otherwise unknown speaks clearly for the importance of a biological component in human warfare." (Wrangham,[14] p. 78)

Although man has engaged in arms racing, warring, and other forms of organized violence for all of recorded history, we have comparatively little in the way of formal theory. Mathematical biology may provide guidance in developing such a theory. Wrangham writes, "Given that biology is in the process of developing a unified theory of animal behavior, that human behavior in general can be expected to be understood better as a result of biological theories, and that two of our closest evolutionary relatives show human patterns of intergroup aggression, there is a strong case for attempting to bring biology into the analysis of warfare. At present, there are few efforts in this direction."[14] I would like to see more effort, specifically more mathematical effort, in this direction and hope to stimulate some interest among you. To convince you that there might conceivably be some "unified field

there is a genetic component to all of this seems beyond doubt. But I do not attempt to gauge it. Perhaps "socioecology" would be a suitable name for this level of analysis.

theory" worth pursuing, I want to share some observations with you. To set them up, a little background is required.

The fundamental equations in the mathematical theory of arms races are the so-called Richardson equations, named for the British applied mathematician and social scientist Lewis Frye Richardson, who first published them in 1939.[9,10] The fundamental equations in the mathematical theory of combat (warfare itself, as against peacetime arms racing) were published in 1916 by Frederick William Lanchester.[7] The formal theory of interstate conflict, to the extent there is one, rests on these twin pillars, if you will. Meanwhile, the classic equations of mathematical ecology are the Lotka-Volterra equations.

In light of the remarks above, I find the following fact intriguing: The Richardson and Lanchester models of human conflict are, mathematically, specializations of the Lotka-Volterra ecosystem equations.

Before proceeding, I must make one point unmistakably clear. I do not claim that any of these models is really "right" in a physicist's sense. They are illuminating abstractions. I think it was Picasso who said, "art is a lie that enables us to see the truth." So it is with these simple models. They continue to form the conceptual foundations of their respective fields. They are universally taught; mature practioners, knowing full-well the models' approximate nature, nonetheless entrust to them the formation of the student's most basic intuitions. And this because, like idealizations in other sciences—idealizations that are ultimately "wrong"—they efficiently capture qualitative behaviors of overarching interest. That these ecosystem and, say, arms race equations should look at all alike is unexpected. That, on closer inspection, they are virtually identical is, to me, really quite interesting. Let me go a bit further.

Under yet other parameter settings, the Lotka-Volterra equations yield standard models of epidemics. And, in other Lectures, I will argue that social revolutions, riots, and illicit drugs may well spread in a strictly analogous way or—at the very least—that an epidemiological perspective on such social processes is promising. Once more, the point is simply that social science might learn a lot from mathematical biology and, conceivably, might inherit some of its apparent unity.

Let me now introduce the Lotka-Volterra equations and show how the classic arms race and war models fall out as special cases. Then, I will explore the analogy between revolutions and epidemics. In subsequent Lectures, we will move beyond these simple—too simple—models.

THE LOTKA-VOLTERRA WORLD

The Lotka-Volterra equations are as follows:

$$\begin{aligned} \dot{x}_1 &= x_1(r_1 - a_{11}x_1 + a_{12}x_2), \\ \dot{x}_2 &= x_2(r_2 + a_{21}x_1 - a_{22}x_2). \end{aligned} \tag{1}$$

In discussing these equations, I will freely invoke nonlinear dynamical systems terminology presented in the preceding lecture. Turning now to Eqs. (1), $x_i(t)$ is the species i population at time t; the a's and r's are real parameters.

If all a_{ij}'s equal zero and $r_1, r_2 > 0$, we have unbounded exponential—so-called Malthusian—growth. Since, ultimately, there are limits, for instance, environmental carrying capacities, the terms $a_{11}, a_{22} > 0$ are preceded by a negative sign. Then, in the language of the preceding lecture, the species are self-inhibiting. Leaving r_1 and r_2 positive and still assuming $a_{12} = a_{21} = 0$, this assumption yields a logistic approach for each species to the positive phase plane equilibrium

$$(\overline{x}_1, \overline{x}_2) = \left(\frac{r_1}{a_{11}}, \frac{r_2}{a_{22}}\right),$$

a node sink.

Now, life really gets interesting only when species interact, and this involves the cross-terms a_{12} and a_{21}.

MUTUALISM

Leaving everything else as is, let us now assume $a_{12}, a_{21} > 0$. In that case our species are said to be in a relationship of mutualism, or reciprocal activation; the population level of one feeds back positively on the growth rate of the other. Bees and flowers—pollinators and pollinatees, if you will—provide examples. There are many others.

Setting $\dot{x}_1 = \dot{x}_2 = 0$, the interior equilibrium conditions are

$$\begin{aligned} r_1 - a_{11}x_1 + a_{12}x_2 &= 0, \\ r_2 + a_{21}x_1 - a_{22}x_2 &= 0. \end{aligned} \tag{2}$$

Of course, these are also the equilibrium conditions for the linear system:

$$\begin{aligned} \dot{x}_1 &= r_1 - a_{11}x_1 + a_{12}x_2, \\ \dot{x}_2 &= r_2 + a_{21}x_1 - a_{22}x_2. \end{aligned} \tag{3}$$

But this is exactly the famous Richardson model of an arms race! The more bees, the more flowers, and vice versa. It's the same in Eqs. (3), but not quite as idyllic. The more weaponry my adversary has, the more I want, and vice versa, up to some economic—or ecological—limit or carrying capacity.

Richardson's basic idea is that a state's arms race behavior depends on three overriding factors: the perceived external threat, the economic burden of military competition, and the magnitude of grievances against the other party. These are discussed at greater length in Lecture 4. Suffice it to say here that $r_1, r_2 > 0$ represent fundamental grievances; $a_{12}, a_{21} > 0$ are the reciprocal activation coefficients (the rates at which each arsenal grows in response to the other); and a_{11}, a_{22} are the self-inhibiting, or damping, terms which Richardson identified with economic fatigue.

Mathematical biologists have long asked how mutualistic populations avoid exploding in what Robert May called an "orgy of mutual benefaction."[8] Likewise, we can ask what mechanism damps the upward action-reaction military dynamic represented in the Richardson model. In each case, self-inhibitory effects must somehow dominate reciprocal activation effects if a stable species equilibrium—or military "balance of power"—is to emerge. Stability analysis bears this out.

Clearly, we can write (2) in matrix form $r + Ax = 0$, $x \in \mathbf{R}^2$. The positive (or interior) equilibrium of Eqs. (1) and the sole equilibrium of Eqs. (3) is therefore given by $\bar{x} = -A^{-1}r$. For each model, the stability of $\bar{x}$ can be evaluated by the methods of Lecture 1.

By a simple translation, the Richardson equations (3) are globally asymptotically stable at $\bar{x}$ if and only if $\dot{y} = Ay$ is globally asymptotically stable at the origin, where $y = x - \bar{x}$. From Lecture 1, we have the well-known stability criterion

$$\operatorname{Tr} A < 0 \text{ and } \det A > 0. \tag{4}$$

Now, Richardson's economic fatigue *means* $a_{11}, a_{22} > 0$. So, we have

$$\operatorname{Tr} A = -a_{11} - a_{22} < 0.$$

And we will have $\det A > 0$ precisely when $a_{11}a_{22} > a_{12}a_{21}$, which is to say that inhibition ($a_{11}a_{22}$) outweighs activation ($a_{12}a_{21}$), confirming our intuition.

One can demonstrate[4] that the eigenvalues of the Jacobian of Eq. (1) at $\bar{x}$ have negative real parts (indeed, are negative reals) when the same condition is met. An isocline analysis is also revealing. You recall that an isocline is a curve—here a line—where one side's rate of growth is zero; clearly, an equilibrium is a point where isoclines intersect. From Eq. (2), the isoclines are given by:

$$\begin{aligned} \phi_1(x_1) &= \frac{a_{11}}{a_{12}}x_1 - \frac{r_1}{a_{12}} \quad \text{(the } x_1\text{-isocline)}, \\ \phi_2(x_1) &= \frac{a_{21}}{a_{22}}x_1 + \frac{r_2}{a_{22}} \quad \text{(the } x_2\text{-isocline)}. \end{aligned} \tag{5}$$

For local stability of the equilibrium $\bar{x}$, we require the configuration of Figure 1. But, this occurs only if the slope of ϕ_1 exceeds the slope of ϕ_2, which is to say $a_{11}/a_{12} > a_{21}/a_{22}$, or

$$a_{11}a_{22} > a_{21}a_{12}.$$

Our intuition is again confirmed: stability requires self-inhibition to exceed reciprocal activation in this sense.

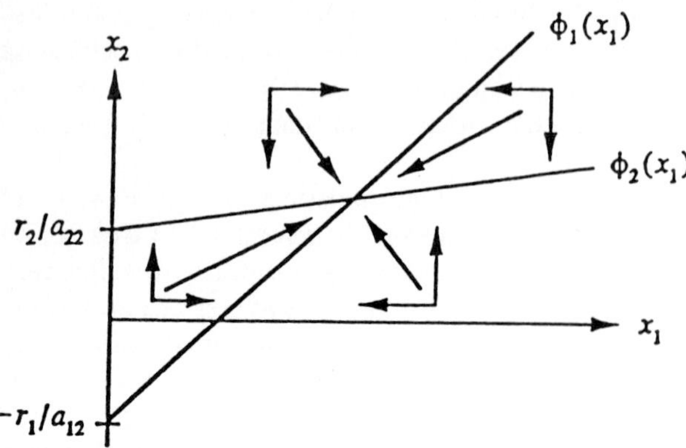

FIGURE 1 Mutualistic stability.

The main point, however, is that the classic Lotka-Volterra model of mutualistic species interaction embeds, in its equilibrium behavior, the classic Richardson arms race model.

AN ASIDE ON COEVOLUTION

In the models above, of course, the "phenotypes" do not change. In fact, ecosystem dynamics select against certain phenotypes. Roughly speaking, phenotypic frequencies and population levels have interdependent trajectories. This is very clear, for example, in immunology, where antigens and antibodies coevolve in a so-called "biological arms race." But, of course, real arms races work this way, too. Ballistic missiles beget antiballistic missile defenses, which beget various evasion and defense suppression technologies. The machine gun makes cavalry obsolete, giving rise to the "iron horse"—the tank—which begets antitank weapons, which beget special armor, and so on. Michael Robinson's analogy between moth-bat coevolution and the coevolution of World War II air war tactics is apposite.

> "Moths and their predators are in an arms race that started millions of years before the Wright brothers made the Dresden raids possible. Butterflies exploit the day, but their 'sisters' the moths dominate the insects' share of the night skies. Few vertebrates conquered night flying. Only a small fraction of bird species, mostly owls and goatsuckers, made the transition. Bats, of course, made it their realm. Many species of bats are skilled 'moth-ers': they pursue them at speed after detecting them with their highly attuned echolocation system. Some moths, however, have developed 'ears' capable of detecting the bat's ultrasonic cries. When they hear a bat coming, the moths take evasive action, including dropping below the bat's track. The

> parallels of the response of Allied bombers to the radar used by the Germans in World War II are interesting. If we visualize the bombers as the moths, and radars on the ground and in the night-fighter aircraft as bats (a reversal of sizes), the situation is similar. Bombers used rearward-listening radar to detect enemy night fighters. When they detected a fighter, they took evasive action. But heavy bombers, heavily laden, were not very maneuverable. They couldn't dodge about quite as well as moths. Some pilots tried to drop their aircraft into a precipitous dive. Moths also do this; it is easy for them to fold their wings and drop. The next stage in the night-battle escalation is predictable. The night fighter's radar was eventually tuned to detect the bomber's fighter-detector, and thus the bomber itself. Bats have not yet tuned in on moths' ears.
>
> "Bombers also used technological disruption. Night fighters came to be guided to bombers by long-distance radars on the ground. The fighters started winning. But nothing remains static. The ground radars could be jammed by various kinds of radio noise. The technological battle swung the other way. Then the fighters acquired radar. Much like a bat, a fighter emitted and listened to radar signals of its own. These, too, proved to be susceptible to countermeasures, however. The RAF could jam the fighters' radar or 'clutter' it with strips of aluminum foil. Each bomber in a formation dropped one thousand-strip bundle per minute, so that huge clouds of foil foiled the radar. Amazingly, there may be a similar counter-weapon among moths. Some moths can produce ultrasonic sounds that fall within the bats' audio frequency. The moths' voice boxes are paired, one on each side of the thorax; double voices must be particularly confusing. Alien sounds in their waveband could confound the bats, exactly in the same way the foil confounded the fighters.
>
> "The next steps in the bat-versus-moth war may simply be awaiting discovery by some bright researcher; after all, we did not know a lot about echolocation in bats until after World War II. My guess would be that the detector will get more complex to meet the defenses. This may already have happened; bats specializing in moths with ears may have moved to a higher frequency sound outside the moths' hearing range!"[11]

Quite clearly, *levels* of armament (in the international system) and *levels* of population (in an ecosystem) interact, as in the Lotka-Volterra and Richardson models, but *phenotypes* themselves are also changing. In biology, there is a mathematical theory of coevolution.[12] In social science, there isn't. There probably could be, so I simply mention it as a promising direction.

Now, let's shift gears from the mutualistic/arms race variant of Eqs. (1). Specifically, instead of assuming that all $a_{12}a_{21}$ are positive, assume that they are negative.

COMPETITION

Rearranging slightly, Eqs. (1) take the form

$$\begin{aligned} \dot{x}_1 &= a_{12}x_1x_2 + r_1x_1\left(1 - \frac{x_1}{k_1}\right), \\ \dot{x}_2 &= a_{21}x_1x_2 + r_2x_2\left(1 - \frac{x_2}{k_2}\right), \end{aligned} \tag{6}$$

where $k_i \equiv (r_i/a_{ii}) > 0$ is the carrying capacity of the environment for each species. These equations were published in 1934 by the great Russian mathematical biologist G. F. Gause in his book *The Struggle for Existence*. Indeed, he termed a_{12} and a_{21} "coefficients of the struggle for existence."[3]

Now, examining Eq. (6), each species would exhibit logistic growth to its respective carrying capacity but for these interaction—struggle—terms. Including them, Eq. (6) gives a picture of uniform mixing of the populations x_1 and x_2, with contacts proportional to the product x_1x_2. Now, however, since the interaction coefficients are negative, each contact *kills* species 1 at rate a_{12} and species 2 at rate a_{21}. Quite clearly, a parallel to combat is suggested. But more is true.

In fact, unbeknownst to Gause, Eq. (6) is an exact form of the famous—and to this day ubiquitous—Lanchester[7] model of warfare!

The transition from arms race to war, then, might be seen as a transition from the case of $a_{12}, a_{21} > 0$ to the case of $a_{12}, a_{21} < 0$. In the latter context, the well-known biological "principal of competitive exclusion" simply maps to the military principle that, usually, one side wins and the other side loses. Both these competitive exclusion behaviors reflect the mathematical fact that the interior $(x_1, x_2 > 0)$ equilibrium of Eq. (6) is a saddle. The stable equilibrium in the mutualistic—peacetime arms race—case was a node. To the extent these models are correct, then, we can say (pacem Poincaré) that war is topologically different from peace; the outbreak of war is a bifurcation from node to saddle.

Thus far we have been exploring a mathematical biology of interstate relations; what about intrastate dynamics? Is there a Lotka-Volterra perspective on revolution, for instance? And, to what biological process might such social dynamics correspond?

REVOLUTIONS AND EPIDEMICS

Consider the following specialization of Eqs. (1):

$$a_{12} = a_{21} > 0; \quad r_1 = r_2 = a_{11} = a_{22} = 0.$$

Then Eqs. (1) become

$$\begin{aligned}\dot{x}_1 &= -\,a_{12}x_1x_2,\\ \dot{x}_2 &= a_{12}x_1x_2,\end{aligned} \tag{8}$$

which is the simplest conceivable epidemic model. Now, rather than armament levels, x_1 represents the level of susceptibles, and x_2 the level of infectives, while the parameter a_{12} is the infection rate, expressing the contagiousness of the infection. Ideal homogeneous mixing, once more, is assumed. If population is constant at P_0, then $x_1 = P_0 - x_2$ and we obtain

$$\dot{x}_2 = a_{12}x_2(P_0 - x_2), \tag{9}$$

our familiar friend the logistic differential equation. Here, $x_2 = 0$ is an unstable equilibrium; the slightest introduction of infectives, and the disease whips through the whole of society.

A traditional tactic for combatting the spread of a disease is removal of infectives. Sometimes, nature does the removing, as with fatal diseases; often, society removes infectives from circulation by quarantine. The simplest possible assumption is that removal is proportional to the size of the infective pool, yielding the following variant of Eqs. (1):

$$\begin{aligned}\dot{x}_1 &= -\,a_{12}x_1x_2,\\ \dot{x}_2 &= a_{12}x_1x_2 - r_2x_2,\end{aligned} \tag{10}$$

with $r_2 > 0$. This is the famous Kermack-McKendrick (1927) *threshold* epidemic model,[6] so-called because it exhibits the following behavior.

By definition, there is an epidemic outbreak only if $\dot{x}_2 > 0$. But this is to say $a_{12}x_1x_2 - r_2x_2 > 0$, or

$$x_1 > \frac{r_2}{a_{12}}. \tag{11}$$

The initial susceptible level $x_1(0)$ must exceed the threshold $\rho \equiv r_2/a_{12}$, sometimes called the relative removal rate, for an epidemic to break out. The fact that epidemics are threshold phenomena has important implications for public health policy and, I will argue below, for social science.

The public health implication, which was very controversial when first discovered, is that *less than* universal vaccination is required to prevent epidemics. By the threshold criterion (1), the fraction immunized need only be big enough that the unimmunized fraction—the actual susceptible pool—be below the threshold ρ. "Herd immunity," in short, need not require immunization of the entire herd. For instance, diphtheria and scarlet fever require 80 percent immunization to produce herd immunity.[1] Hethcote and Yorke argue that "a vaccine could be very effective in controlling gonorrhea...for a vaccine that gives an average immunity of 6 months, the calculations suggest that random immunization of 1/2 of the general population each year would cause gonorrhea to disappear."[5]

Mathematical epidemic models are discussed more fully in Lecture 5. With the above as background, let us now consider the analogy between epidemics (for which a rich mathematical theory exists) and processes of explosive social change, such as revolutions (for which no comparable body of mathematical theory exists). Again, a more careful and deliberate development is given in Lecture 5. Here, we simply offer the main idea. It will facilitate exposition to re-label the variables in Eqs. (10). If $S(t)$ and $I(t)$ represent the susceptible and infective pools at time t and if r and γ are the infection and removal rates, the basic model is:

$$\begin{aligned} \dot{S} &= -rSI, \\ \dot{I} &= rSI - \gamma I, \end{aligned} \tag{12}$$

with epidemic threshold

$$S > \frac{\gamma}{r} = \rho. \tag{13}$$

The basic mapping from epidemic to revolutionary dynamics is direct. The infection or disease is, of course, the revolutionary idea. The infectives $I(t)$ are individuals who are actively engaged in articulating the revolutionary vision and in winning over ("infecting") the susceptible class $S(t)$, comprised of those who are receptive to the revolutionary idea but who are not infective (not actively engaged in transmitting the disease to others). Removal is most naturally interpreted as the political imprisonment of infectives by the elite ("the public health authority").

Many familiar tactics of totalitarian rule can be seen as measures to minimize r (the effective contact rate between infectives and susceptibles) or maximize γ (the rate of political removal). Press censorship and other restrictions on free speech reduce r, while increases in the rate of domestic spying (to identify infectives) and of imprisonment without trial increase γ.

Symmetrically, familiar revolutionary tactics—such as the publication of underground literature, or "samizdat"—seek to increase r. Similarly, Mao's directive that revolutionaries must "swim like fish in the sea," making themselves indistinguishable (to authorities) from the surrounding susceptible population, is intended to reduce γ.

GORBACHEV, DETOQUEVILLE, AND THE THRESHOLD

Interpreting the threshold relation (13), if the number of susceptibles S_0 is, in fact, quite close to ρ, then even a slight reduction (voluntary or not) in central authority can push society over the epidemic threshold, producing an explosive overthrow of the existing order. To take the example of Gorbachev, the policy of Glasnost obviously produced a sharp increase in r, while the relaxation of political repression (e.g., the weakening of the KGB, the release of prominent political prisoners, the dismantling of Stalin's Gulag system) constituted a reduction in γ. Combined, these measures evidently depressed ρ to a level below S_0, and the "revolutions of 1989" unfolded. Perhaps DeToqueville intuited the threshold relation (13), describing this phenomenon, when he remarked that "liberalization is the most difficult of political arts."

As a final element in the analogy, systematic social indoctrination can produce herd immunity to potentially revolutionary ideas. We even see "booster shots" administered at regular intervals—May 1 in Moscow; July 4 in America—on which occasions the order-sustaining myths ("The USSR is a classless workers' paradise"; "Everyone born in America has the same opportunities in life") are ritually celebrated.

Now, as I said before, all these analogies are doubtlessly terribly crude. I certainly do not claim either that any of the models are right or that the dynamical analogies among them are exact. Yet, the very fact that a single ecosystem model—the Lotka-Volterra equations—could specialize to equations that even caricature, however crudely, such basic and important social processes as arms racing, warring, and rebelling is, I believe, very interesting and serves to reinforce the larger point with which I began: social science is ultimately a subfield of biology.

Finally, let me conclude with an admission. I was surprised when I began to notice these connections. But why should we be surprised? In certain non-Western cultures, where our species is seen as a "part of nature," where gods—like the sphinx—can be part man and part lion, all these connections between ecosystems and social systems might appear quite unremarkable. But in Western cultures shaped by the Old Testament, where God creates *only* man—not the fishes, birds, and bushes—in *his* own image, man is seen as "apart *from* nature." And, accordingly, we are surprised when our models of fish—or worse yet, of viruses—turn out to be interesting models of man. Perhaps we are true Darwinians more in our heads than in our hearts. Creatures of habit, we are captive to a transmitted and slowly evolving culture. But, of course, this too is "only natural."

ACKNOWLEDGMENTS

I am grateful to Robert L. Axtell, Samuel David Epstein, and Elaine McNulty for their thoughtful comments. I also thank Dan Stein for organizing the Summer School and the students who made it a learning experience for all involved.

REFERENCES

1. Edelstein-Keshet, Leah. *Mathematical Models in Biology*, 255. New York: Random House, 1988.
2. Epstein, Joshua M. *The Calculus of Conventional War: Dynamic Analysis Without Lanchester Theory*. Washington, DC: Brookings, 1985.
3. Gause, G. F. *The Struggle for Existence*, 47. Baltimore: Williams & Wilkins, 1934.
4. Goh, B. S. "Stability in Models of Mutualism." *Am. Natur.* **113(2)** (1979): 261–275.
5. Hethcote, Herbert W., and James A. Yorke. *Gonorrhea Transmission Dynamics and Control*, 47. New York: Springer-Verlag, 1980.
6. Kermack, W. O., and A. G. McKendrick. "Contributions to the Mathematical Theory of Epidemics." *Proc. Roy. Stat. Soc. A* **115** (1927): 700–721. For a contemporary development, see Waltman.[13]
7. Lanchester, F. W. *Aircraft in Warfare: The Dawn of the Fourth Arm*. London: Constable, 1916. For a contemporary discussion with references, see Epstein.[2]
8. May, Robert M. "Models for Two Interacting Populations." In *Theoretical Ecology*, edited by Robert M. May. London: Blackwell Scientific Publications, 1981.
9. Richardson, L. F. "Generalized Foreign Policy." *Brit. J. Psych. Monograph Suppl.* **23** (1939).
10. Richardson, Lewis F. *Arms and Insecurity: A Mathematical Study of the Causes and Origins of War*. Pittsburgh: The Boxwood Press, 1960.
11. Robinson, Michael H. "Nature's Game of Attack and Defense." *Smithsonian* **23(1)** (1992): 77–79.
12. Roughgarden, Johnathan. *Theory of Population Genetics and Evolutionary Ecology: An Introduction*. New York: Macmillan, 1979.
13. Waltman, Paul. *Deterministic Threshold Models in the Theory of Epidemics*. Lecture Notes in Biomathematics, Vol. 1. New York: Springer-Verlag, 1974.
14. Wrangham, Richard W. "War in Evolutionary Perspective." In *Emerging Syntheses in Science*, edited by David Pines. Santa Fe Institute Studies in the Sciences of Complexity, Proceedings Volume I. Reading, MA: Addison-Wesley, 1988.

Joshua M. Epstein
Senior Fellow, The Brookings Institution and Visiting Lecturer, Princeton University

The Adaptive Dynamic Model of Combat

This chapter will appear as Lecture 3 in *Nonlinear Dynamics, Mathematical Biology, and Social Science* by Joshua M. Epstein, to be published as a lecture notes volume in the Santa Fe Institute Studies in the Sciences of Complexity series (Addison-Wesley, 1994). All lectures referred to in the present chapter are from this forthcoming book.

In this chapter I would like to give an introduction to some simple mathematical models of combat, including my own Adaptive Dynamic Model. Here, we are concerned with the *course* of war, rather than the arms races or crises that may precipitate war. Before discussing specifics, it may be well to consider the basic question: What are appropriate goals for a mathematical theory of combat at this point?

First and foremost, we need to be humble. Warfare is complex. Outcomes may depend, perhaps quite sensitively, on technological, behavioral, environmental, and other factors that are very hard to measure before the fact. Exact prediction is really beyond our grasp.

But, that's not so terrible. Theoretical biologists concerned with morphogenesis —the development of pattern—are, in some cases, situated similarly. For the particular leopard, we certainly cannot predict the exact size and distribution of spots. But, certain classes of partial differential equations—reaction diffusion equations—will generate generic animal coat patterns of the relevant sort. So, we feel that this is the right body of mathematics to be exploring. The same sort of point holds for epidemiologists. Few would claim to be able to predict the exact onset point or severity of an epidemic. Theoreticians seek simple models that will generate a reasonable menu of core qualitative behaviors: threshold eruptions, persistence at endemic levels, recurrence in cycles, perhaps chaotic dynamics. The aim is to produce transparent, parsimonious models that will *generate the core menu of gross qualitative system behaviors.* This, it seems to me, is the sort of claim one would want to make for a mathematical theory of combat.

Now, in classical mechanics, the crucial variables are mass, position, and time. In classical economics, they are price and quantity. War, traditionally, is about territory and, unfortunately, death, or mutual attrition. A respectable model, at the very least, should offer a plausible picture of the relationship between the fundamental processes of attrition and withdrawal (i.e., territorial sacrifice). I will discuss attrition first.

LANCHESTER'S EQUATIONS

The big pioneer in this general area was Frederick William Lanchester (1868–1945). The eclectic English engineer made contributions to diverse fields, including automotive design and the theory of aerodynamics.[6] He is best remembered for his equations of war, appropriately dubbed the Lanchester equations. First set forth in his 1916 work, *Aircraft in Warfare*, these have a variety of forms, the most renowned of which is called—for reasons that will be given shortly—the Lanchester "square" model.[7] With no air power and no reinforcements, the Lanchester square equations are

$$\begin{aligned} \frac{dR}{dt} &= -\,bB, \\ \frac{dB}{dt} &= -\,rR. \end{aligned} \tag{1}$$

Here, $B(t)$ and $R(t)$ are the numbers of "Blue" and "Red" combatants—each of which is an idealized fire source—and $b, r > 0$ are their respective firing effectiveness per shot. Qualitatively, these equations say something intuitively very appealing, indeed, seductive: *The attrition rate of each belligerent is proportional to the size of the adversary.* The system (1) is, of course, soluble exactly. With $B(0) = B_0$ and $R(0) = R_0$,

$$\begin{aligned} R(t) =& \frac{1}{2}\left[\left(R_0 - \sqrt{\frac{b}{r}}B_0\right)e^{\sqrt{rb}t} + \left(R_0 + \sqrt{\frac{b}{r}}B_0\right)e^{-\sqrt{rb}t}\right], \\ B(t) =& \frac{1}{2}\left[\left(B_0 - \sqrt{\frac{r}{b}}R_0\right)e^{\sqrt{rb}t} + \left(B_0 + \sqrt{\frac{r}{b}}R_0\right)e^{-\sqrt{rb}t}\right], \end{aligned} \tag{2}$$

with various trajectories for R and B over time. In the phase plane, the origin is obviously the only equilibrium of (1) and the Jacobian of (1) at $\overline{x}$ is

$$DF(\overline{x}) = \begin{pmatrix} 0 & -r \\ -b & 0 \end{pmatrix}.$$

The eigenvalues are clearly $\pm\sqrt{rb}$. Hence, the origin is a saddle, though the positive quadrant is all we care about. Clearly, depending on the parameters (b, r) and the initial values (B_0, R_0), either side can start ahead and lose, or start behind and win, as is observed historically.[1]

The most celebrated result of the theory is the so-called Lanchester Square Law, which is obtained easily. From (1), we have

$$\frac{dR}{dB} = \frac{bB}{rR}. \tag{3}$$

Separating variables and integrating from the terminal values $(R(t), B(t))$ to the higher initial values,

$$r\int_{R(t)}^{R_0} RdR = b\int_{B(t)}^{B_0} BdB,$$

we obtain the state equation

$$r(R_0^2 - R(t)^2) = b(B_0^2 - B(t)^2) \tag{4}$$

or, after a bit of rearranging,

$$bB(t)^2 - rR(t)^2 = bB_0^2 - rR_0^2.$$

[1] Indeed, the numerically smaller force was the victor in such notable cases as Austerlitz (1805); Antietam (1862); Fredericksburg (1862); Chancellorsville (1863); the Battle of Frontiers (1914); the fall of France (1940); the invasion of Russia (Operation Barbarossa, 1941); the battle of Kursk (1943); the North Korean invasion (1950); the Sinai (1967); the Golan Heights (1967 and 1973); and the Falklands (1982), to name a few.

The left-hand side is a Hamiltonian of the system. Of course, stalemate occurs when $B(t) = R(t) = 0$, which yields the Lanchester Square Law:

$$
\begin{aligned}
bB_0^2 &= rR_0^2 \qquad \text{or} \\
B_0 &= \sqrt{\frac{r}{b}}R_0.
\end{aligned} \tag{5}
$$

This equation is very important. It says that, to stalemate an adversary three times as numerous, it does not suffice to be three times as effective; you must be nine times as effective! This presumed heavy advantage of *numbers* is deeply embedded in virtually all Pentagon models. For decades, it supported the official dire assessments of the conventional balance in Central Europe, giving enormous weight to sheer Soviet numbers and placing a huge premium on western technological supremacy. That, of course, had budgetary implications. But, the presumption of overwhelming Soviet *conventional* superiority also shaped the development of so-called theater-nuclear weapons and produced a widespread assumption that their early employment would be inevitable, which drove the Soviets to seek preemptive offensive capabilities, and so on, in an expensive and dangerous military coevolution (see the preceding chapter).

The whole dynamic while driven by myriad political and military-industrial interests on all sides, was certainly supported by Lanchester's innocent-looking *linear* differential equations, (1). But, the linearity itself implicitly assumes things that are implausible on reflection and it mathematically precludes phenomena that, in fact, are observed empirically. Moreover, anyone exposed to mathematical biology would have found the Lanchester variant (1) to be suspect immediately.

DENSITY

The equations, once again, are

$$\frac{dB}{dt} = -rR, \tag{6}$$

$$\frac{dR}{dt} = -bB. \tag{7}$$

In this framework, increasing density is a *pure benefit.* If the Red force R grows, a greater volume of fire is focused on the Blue force B, and in Eq. (6), the Blue attrition rate dB/dt grows proportionally. At the same time, however, *no penalty* is imposed on Red in Eq. (7) when, in fact, if the battlefield is crowded with Reds, the Blue target acquisition problem is eased and red's attrition rate should grow.

In warfare, each side is at once *both predator and prey.* Increasing density is a benefit for an army as predator, but it is a cost for that same army as prey. The

Lanchester square system captures the predation benefit but completely ignores the prey cost of density. The latter, moreover, is familiar to us all. For instance, if a hunter fires his gun into a sky black with ducks, he is bound to bring down a few. Yet if a single duck is flying overhead, it takes extraordinary accuracy to shoot it down. For ducks, considered as prey, density carries costs.

And, as any ecologist would expect, the effect is indeed observed. Quoting Herbert Weiss, "the phenomenon of losses increasing with force committed was observed by Richard H. Peterson at the Army Ballistic Research laboratories in about 1950, in a study of tank battles. It was again observed by Willard and the present author [Weiss] has noted its appearance in the Battle of Britain data."[8] The work referred to is D. Willard's statistical study of 1500 land battles.[9]

To his credit, Lanchester actually offered a second, nonlinear variant of these equations, which is much more plausible in this ecological light. Here,

$$\frac{dR}{dt} = (-bB)R, \tag{8}$$

$$\frac{dB}{dt} = (-rR)B. \tag{9}$$

In parentheses are the Lanchester square terms reflecting the "predation benefit" of density, but they are now multiplied by a term (the prey force level) reflecting "prey costs," as it were. The Red attrition rate, dR/dt in Eq. (8), slows as the Red population goes to zero, reflecting the fact that, as the prey density falls, the predator's search ("foraging") requirements for the next kill increase. Equivalently, red's attrition rate grows if, like the ducks in the analogy, its density grows. In summary, a density cost is present to balance the density benefit reflected in the parenthesized term.

If we now form the casualty-exchange ratio

$$\frac{dR}{dB} = \frac{b}{r},$$

separate variables, and integrate as before, we obtain the state equation

$$r\big(R_0 - R(t)\big) = b\big(B_0 - B(t)\big)$$

and the stalemate requirement

$$rR_0 = bB_0.$$

Now, as against the Lanchester Square Law, it *does* suffice to be three (rather than nine) times as good to stalemate an adversary three times as numerous.

AMBUSH AND ASYMMETRY

Further, asymmetrical, variants of the basic Lanchester equations have been devised. For example, the so-called ambush variant imputes the "square law" fire concentration capacity to one side (the ambushers) but denies it to the other (the ambushees). Here,

$$\frac{dB}{dt} = -\, rR,$$
$$\frac{dR}{dt} = -\, bBR,$$

so that

$$\frac{dB}{dR} = \frac{r}{bB},$$
$$b\left(B_0^2 - B(t)^2\right) = r\left(R_0 - R(t)\right).$$

Now assuming a fight to the finish ($R(t) = B(t) = 0$) and equal firing effectiveness ($r = b$), a Blue force of B_0 can stalemate a Red force numbering B_0^2—a hundred can hold off ten thousand. It's Thermopolae.

REINFORCEMENT

Thus far the discussion has concentrated on the dynamics of *engaged* forces. Often, however, there is some flow of reinforcements to the combat zone proper. But, there are limits to the number of forces one can pack into a given area—there are "force to space" constraints. One might therefore think of the combat zone as having a carrying capacity and, accordingly, posit logistic reinforcement. Attaching such a term to the Lanchester nonlinear attrition model produces

$$\begin{aligned} \frac{dR}{dt} &= -\, bRB + \alpha R\left(1 - \frac{R}{K}\right), \\ \frac{dB}{dt} &= -\, rBR + \beta B\left(1 - \frac{B}{L}\right), \end{aligned} \tag{10}$$

where α, β, K, and L are positive constants. As observed in the preceding chapter, this is *exactly* Gause's (1935) famous model of competition between two species, itself a form of the general Lotka-Volterra ecosystem equations.

Equations (10) admit four basic cases, corresponding to different "war histories." These are shown in the phase portraits in Figure 1.

Cases (a) and (b) are clear instances of the biological "principle of competitive exclusion," or military principle that one or the other side usually wins. Case

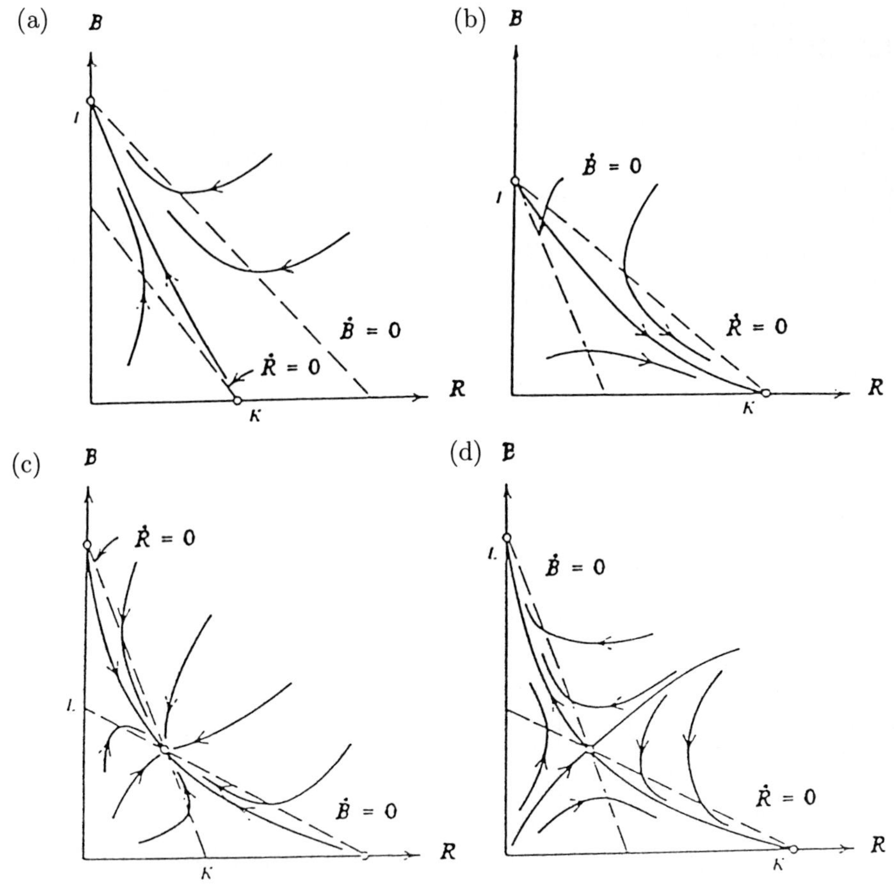

FIGURE 1 Phase portraits for Lanchester/Gause Model. Adapted from Colin W. Clark's *Mathematical Bioeconomics* (New York: John Wiley, 1990), p. 194.

(c) shows the horrific stable node—the "permanent war" that neither side wins. Finally, we have case (d), a saddle equilibrium. Any perturbation (off the stable manifold) sends the trajectory to a Red or Blue triumph. There is, however, the interesting and important region below both isoclines. Each side feels encouraged in this zone; reinforcement rates exceed attrition rates so the forces are growing. But, for instance, as the trajectory crosses the $\dot{B} = 0$ isocline, matters start to sour for Blue; $\dot{B}$ goes negative while Red forces continue to grow. Expectations of Blue defeat may set in, Blue morale may collapse, and, as a result, the Blue force can "break" long before it is physically annihilated. Indeed, the general phenomenon of "breakpoints" is common.

BREAKPOINTS

Literal fights to the finish are actually rare. Normally, there is some level of attrition at which one belligerent "cracks." Suppose Blue breaks if $B(t) = \beta B_0$ and Red breaks if $R(t) = \rho R_0$, with $0 < \rho, \beta \leq 1$ and ρ not necessarily equal to β. Clearly, breakpoints divide phase space into four zones, as shown in Figure 2.

In Zone III, each side exceeds its breakpoint, so there is combat. Red wins if a trajectory crosses from Zone III to Zone II. All's quiet in Zone I, and so forth.

Substituting the stalemate conditions, $B(t) = \beta B_0$ and $R(t) = \rho R_0$ into, for illustration, the Lanchester square state equation (4) yields

$$r\left[R_0^2 - (\rho R_0)^2\right] = b\left[B_0^2 - (\beta B_0)^2\right],$$

which implies the (with breakpoints) stalemate condition

$$R_0\sqrt{r(1-\rho)^2} = B_0\sqrt{b(1-\beta^2)}.$$

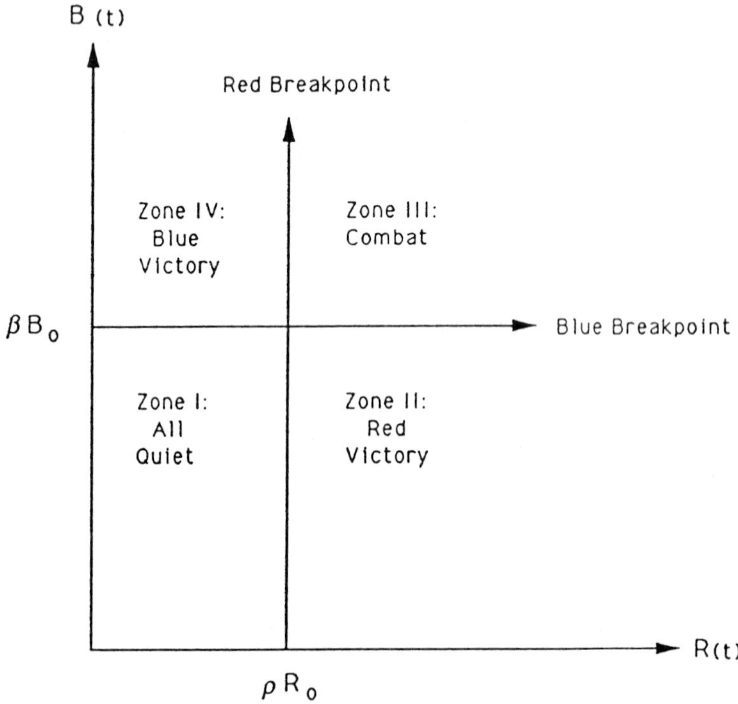

FIGURE 2 Breakpoints.

GENERALIZED EXCHANGE RATIO

As discussed in Epstein[2,3] these variants are all special cases of the general system

$$\frac{dR}{dt} = -\, bB^{c_1}R^{c_2}, \tag{11}$$
$$\frac{dB}{dt} = -\, rR^{c_3}B^{c_4}. \tag{12}$$

The corresponding casualty-exchange ratio is

$$\frac{dR}{dB} = \frac{b}{r}\frac{B^{c_1-c_4}}{R^{c_3-c_2}},$$

where c-values are simply reals in the closed interval $[0, 1]$.

Clearly, from Eq. (9), c_1 is Blue's predation benefit from increasing density while from Eq. (10), c_4 is Blue's prey cost of increasing density. Hence the exponent $c_1 - c_4$ might be thought of as the *net predation benefit of increasing density*, which is net fire concentration capacity in Lanchester's sense. The Red exponent $c_3 - c_2$ is analogously interpreted. Therefore, let us define

$$\lambda_b = \text{Blue's net predation benefit} = c_1 - c_4,$$
$$\lambda_r = \text{Red's net predation benefit} = c_3 - c_2.$$

Then,

$$\frac{dR}{dB} = \frac{b}{r}\left(\frac{B^{\lambda_b}}{R^{\lambda_r}}\right). \tag{13}$$

Again separating variables and integrating from terminal to (higher) initial values, we have

$$b\int_{B(t)}^{B(0)} B^{\lambda_b}dB = r\int_{R(t)}^{R(0)} R^{\lambda_r}dR.$$

With stalemate defined as $B(t) = R(t) = 0$, we obtain the stalemate condition

$$\frac{b}{1+\lambda_b}B_0^{1+\lambda_b} = \frac{r}{1+\lambda_r}R_0^{1+\lambda_r}, \qquad \text{or}$$
$$B_0 = \left[\frac{r}{b}\left(\frac{1+\lambda_b}{1+\lambda_r}\right)R_0^{1+\lambda_r}\right]^{\frac{1}{1+\lambda_b}}, \tag{14}$$

which specializes to all the cases discussed earlier (e.g., $\lambda_d = \lambda_a = 1$ implies square law), and many more.

Equation (13) is the algebraic form of the exchange ratio $\rho(t)$, used in my own Adaptive Dynamic Model. On separation of variables and integration, it also yields the measure of net military advantage used in the nonlinear arms race models of Lecture 4.[2] For further discussion of the λs, see Epstein.[3]

Of course, mere casualty-exchange ratios do not necessarily determine actual outcomes. Even defenders with favorable exchange ratios in engagements may run out of room or run out of time (e.g., popular support may collapse before the attacker's breakpoint is reached). Duration and territory—space and time—can loom every bit as large as physical attrition in determining outcomes. And this brings us to the topic of movement.

MOVEMENT

Historically, war has been about territory. On a map of the modern world, the jagged borders are often simply the places where battle lines finally came to rest. It is interesting to compare these with the straight borders arrived at more contractually, peacefully—say, the borders between Nebraska and Kansas or between the U.S. and Canada. This is the reason that mountain ranges are such common borders: they were natural lines of military defense. The Alps, Himalayas, Pyrenees, and Caucases are examples. The same obviously holds for major bodies of water, like the English channel, and rivers, like the Yalu. In short, political borders reflect military technology. In any event, movement is a central aspect of war. And, as I argued at the outset, a plausible model should capture the basic connection between the fundamental processes: attrition and movement.

Lanchester himself had nothing to say about this and offered no model of movement. Contemporary extensions of Lanchester all handle it in essentially the same way: they posit that the velocity of the front—that is, the rate of defensive withdrawal—is some function of the force ratio. So, if

$$x \equiv \frac{R(t)}{B(t)},$$

these models posit a withdrawal rate, a velocity, $W(x)$ with $W(1) = 0$; $W'(x) > 0$ if $x > 1$ and eventually $W''(x) < 0$, implying some asymptote. (The direction of movement is always "forward" for the larger force.) One published example[5] is

$$W(x) = \frac{W_{\max}}{\sqrt{e^{(4/x)^2}}}.$$

The basic setup, then, is this: forces grind each other up via the attrition equations; the force ratio changes accordingly; and, as a function of that changing force ratio, the front's velocity changes, as depicted in the flow diagram of Figure 3.

[2] See Epstein, *Nonlinear Dynamics, Mathematical Biology, and Social Science.*

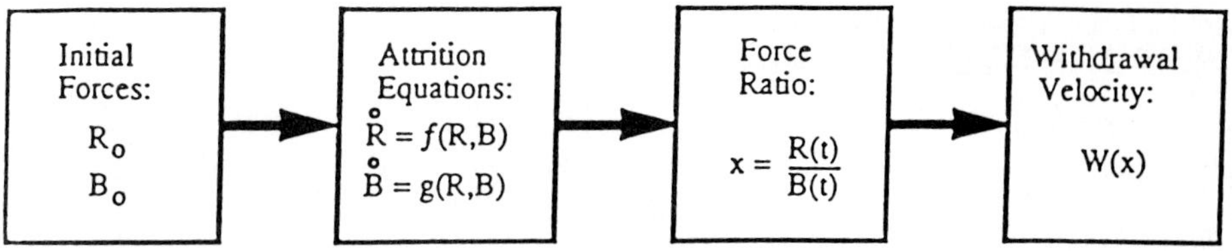

FIGURE 3 Flow diagram for the standard model.

The framework is very neat indeed. The only problem is that any combat model with this basic structure is fundamentally implausible, and for one basic reason: movement of the front—defensive withdrawal—is anomalous! For a given pair of attacking and defending forces, the course of attrition on the defender's side, as calculated in this framework, is *exactly the same* whether he withdraws or not. The course of attrition on the attacker's side is also unchanged whether the defender withdraws or not. In short, defensive withdrawal neither benefits the defender nor penalizes the attacker. So, why in the world would the defender ever withdraw? The framework itself mathematically eliminates any rationale, or incentive, for the very behavior—withdrawal—it purports to represent. Movement is influenced by attrition, but not conversely. The movement of the front (withdrawal) is not *fed back* into the ongoing attrition process, when the entire point of withdrawal was presumably to affect that process—in the prototypical case, the point is to reduce one's attrition. Surely, it is contradictory to assume some benefit in withdrawal (otherwise, why would anyone withdraw?) and then to reflect *no benefit* whatsoever in the ongoing attrition calculations. Yet, all the contemporary Lanchester variants of which I am aware suffer this inconsistency.[3]

In turn, because defensive withdrawal cannot slow the defender's attrition (or, for that matter, the attacker's), the sacrifice of territory cannot prolong the war. And so, the most fundamental tactic in military history—the trading of space for time—is mathematically precluded. But, this tactic saved Russia from Napoleon and, later, from Hitler. A plausible model should certainly permit it.

[3]It is interesting to note that the battle of Iwo Jima—an island, where movement of the front was all but impossible—is the only case (to my knowledge) in which there is any statistical correspondence between events as they unfolded and as hypothesized by the Lanchester equations. Even if the statistical fit were good, there would be no basis for extrapolation to cases where substantial movement is possible. And, in fact, the fit is marred by insufficient data. On this issue, see Epstein.[2]

THE ADAPTIVE DYNAMIC MODEL[3]

So, how do I fix it—how do I build in a *feedback* from movement *to* attrition? As simply as possible. The key parameters are the "equilibrium" attrition rates, α_{dT} and α_{aT}. The first, α_{dT}, is defined as the daily attrition rate the defender is willing to suffer in order to hold territory. The second, α_{aT}, is defined as the daily attrition rate the attacker is willing to suffer in order to take territory. I assume $0 < \alpha_{dT}, \alpha_{aT} < 1$.

War, in addition to being a contest of technologies, is a contest of wills. So it is not outlandish to posit basic levels of pain (attrition rates) that each side comes willing to suffer to achieve its aims on the ground. If the defender's attrition rate is less than or equal to α_{dT}, he remains in place. If his attrition rate exceeds this "pain threshold," he withdraws, in an effort to restore attrition rates to tolerable levels, an effort that may fail dismally depending on the adaptations of the attacker, a similar creature. If the attacker's attrition rate exceeds tolerable levels, he cuts the pace at which he prosecutes the war; if his attrition rate is below the level he is prepared to suffer, he increases his prosecution rate.[5]

It is the interplay of the *two adaptive systems, each searching for its equilibrium, that produces the observed dynamics, the actual movement that occurs and the actual attrition suffered by each side.* Indeed, in its most basic form, withdrawal might be thought of as an attrition-regulating servomechanism. The pain thresholds α_{dT} and α_{aT} play the roles of homeostatic targets, in other words. The introduction of these thresholds struck me—and still strikes me—as the most direct mathematical way to *permit* defensive withdrawal to affect attrition and, thus, to permit the trading of space for time. Their introduction also generates the fertile analogy between armies and a broad array of goal-oriented, feedback-control (cybernetic) systems.

Before delving into the mathematics, one possible misconception about these "pain" thresholds should be addressed. I do not claim, nor does my model imply, that battlefield commanders are necessarily *aware* of the numerical values of α_{dT} and α_{aT}. Humans in the eighteenth century were not "aware" that they were sweating and shivering depending on the error: "body temperature minus 98.6 degrees, Fahrenheit." But the homeostatic behavior was there nonetheless.

OVERVIEW OF THE MODEL

Let me now turn to the Adaptive Dynamic Model itself. The full apparatus includes air power as well as air and ground reinforcements, factors I will not discuss here.[3]

[3]For earlier versions see Epstein.[2,3]

[5]These parameters represent daily *rates* of attrition, not total or cumulative attrition *levels*, as discussed above in connection with breakpoints.

The model is a system of delay equations where the unit of time is usually interpreted as the day. If $A(t)$ and $D(t)$ are the attacker's and defender's ground forces surviving at the start of the tth day and $\alpha_a(t-1)$ is the attacker's attrition rate over the preceding day, we have the accounting identity

$$A(t) = A(t-1) - \alpha_a(t-1)A(t-1). \tag{15}$$

The attacker's force on Tuesday is his force on Monday, minus total losses Monday. Likewise, it must be true that

$$D(t) = D(t-1) - \big(\text{Defender's losses on day } (t-1)\big).$$

What are these losses? Well, if we define the casualty-exchange ratio as

$$\rho(t-1) \equiv \left(\frac{\text{Attackers Lost on day } t-1}{\text{Defenders Lost on day } t-1}\right),$$

the defender's losses must be

$$\frac{\alpha_a(t-1)A(t-1)}{\rho(t-1)},$$

since the numerator is the attackers lost on $(t-1)$. Thus, we have the second accounting identity

$$D(t) = D(t-1) - \frac{\alpha_a(t-1)A(t-1)}{\rho(t-1)}. \tag{16}$$

Obviously, once we attach specific functional forms to $\alpha_a(t)$ and $\rho(t)$, we no longer have accounting identities; we have a model. Above we discussed $\rho(t)$ and argued that a plausible and relatively general functional form is

$$\rho(t) = \rho_0 \frac{D(t)^{\lambda_d}}{A(t)^{\lambda_a}}, \tag{17}$$

where $\lambda_a, \lambda_d \in [0,1]$ are parameters. The real action—all feedback from movement to attrition—is inside $\alpha_a(t)$. Here is where the *interplay of adaptive belligerents* unfolds. As mentioned, this interplay is between the attacker's prosecution rate (reflecting the pace at which he chooses to press the attack) and the defender's withdrawal rate, both of which are attrition-regulating servomechanisms, in effect. The defender is, in some respects, simpler. We discuss him first.

ADAPTIVE WITHDRAWAL AND PROSECUTION

The defender's withdrawal rate for day t is assumed to depend on the difference between his actual and his equilibrium attrition rate for the preceding day, day $(t-1)$. The functional form of that dependence should satisfy some basic requirements:

1. As the actual attrition rate for day $(t-1)$ approaches 1, the withdrawal rate for day t should approach the maximum feasible daily rate, $W_{\max}$.
2. If the actual attrition rate for day $(t-1)$ is greater than the equilibrium rate α_{dT}, the withdrawal rate for day t should be greater than for day $(t-1)$.
3. If the actual attrition rate for day $(t-1)$ is less than or equal to the equilibrium rate α_{dT}, then the withdrawal rate for day t is zero.

It may not be correct, but the simplest functional form I can think of that satisfies these requirements is

$$W(t) = \begin{cases} 0 & \text{if } \alpha_d(t-1) \leq \alpha_{dT}, \\ W(t-1) + \left(\frac{W_{\max} - W(t-1)}{1-\alpha_{dT}}\right)\left(\alpha_d(t-1) - \alpha_{dT}\right) & \text{otherwise,} \end{cases} \tag{18}$$

where

$$\alpha_d(t) = \frac{D(t) - D(t+1)}{D(t)}. \tag{19}$$

While in particular cases, there may be departures, exceptions,[3] and so forth, as a first-order idealization, the notion that, *ceteris paribus*, the aim of withdrawal is to reduce one's attrition rate seems fairly compelling. It also enjoys a certain biological plausibility. If the heat is too great, we yank our hand from the fire; Ashby's cat comes to mind. Surely, flight is a basic mechanism of defense for all species. One of the more famous experiments in this connection was conducted by our friend Gause and is known as his "flour beetle" experiment. He began with two beetle species competing in an environment of flour. He found competitive exclusion to be operative; left alone, one species consistently exterminated the other. But, when Gause inserted small lengths of glass tubing into the flour, the weaker species was able to retreat into the tubing, establish refuges, and survive—they could "trade space for time," as it were. So can the defenders in the Adaptive Dynamic Model. As we will see, they may choose to forego that option. But, a reasonable model should not preclude it.

Turning to the attacker, the model assumes that the pace at which he presses the attack, his prosecution rate for day t, which we denote $P(t)$, depends on the difference between his actual and his equilibrium attrition rates for the preceding day, day $(t-1)$. The functional form of that dependence should satisfy some basic requirements:

1. As the attacker's actual attrition rate for day $(t-1)$ approaches 1, the prosecution rate for day t should approach zero.
2. If the actual attrition rate for day $(t-1)$ is greater than (less than) the equilibrium rate α_{aT}, the prosecution rate for day t should be less than (greater than) for day $(t-1)$.
3. If the actual attrition rate for day $(t-1)$ equals the target, or equilibrium, rate, then there is no change in the prosecution rate.

It may not be correct, but the simplest functional form I can think of that satisfies these requirements is[6]

$$P(t) = P(t-1) - \left(\frac{P(t-1)}{1-\alpha_{aT}}\right)\left(\alpha_a(t-1) - \alpha_{aT}\right). \tag{20}$$

As I said earlier, it is the *interplay* of these adaptive agents that shapes the dynamics; they are linked in the formula for $\alpha_a(t)$, the attacker's attrition rate for day t. This functional form should satisfy some basic requirements:

1. *Ceteris paribus*, the higher is the attacker's prosecution rate, the higher should be his attrition rate;
2. *Ceteris paribus*, the higher is the defender's withdrawal rate, the lower should be the attacker's attrition rate.
3. As the defender's withdrawal approaches full flight ($W(t) \to W_{\max}$), the attacker's attrition rate should approach zero.

It may not be correct, but the simplest functional form I can think of that satisfies these requirements is

$$\alpha_a(t) = P(t)\left(1 - \frac{W(t)}{W_{\max}}\right). \tag{21}$$

Once the initial conditions and parameter values are specified, these equations produce the dynamics. And, as noted above, it is the coadaptation of these agents, each searching for its equilibrium, that determines the actual movement that occurs and the actual attrition that is suffered by each side.

In a nutshell, the attacker makes an opening "bid" on the pace of war, the rate at which his own forces are consumed (of course, he can set his rate at zero by not attacking). He may want to press the attack at an extremely high pace and may be willing to suffer extremely high attrition rates, if—for operational, strategic, or

[6]I am grateful to Mike Sobel for pointing out to me that the functional form for $P(t)$ that I originally published in Epstein[2] actually fails requirement (2). Subsequent to our discussion, I noticed that it also fails (1).

political reasons—a quick decision is paramount.[7] Via the casualty-exchange ratio (defenders killed per attacker killed), this imposes an attrition rate on the defender. The latter may elect to hold his position and accept this attacker-dictated rate, or he may choose to reduce his attrition rate by withdrawing at a certain speed.

The mathematical mechanism whereby the defender's withdrawal reduces his attrition is not obvious. From Eq. (16), the attacker's attrition rate over day t, $\alpha_a(t)$, produces, via the inverse exchange ratio $1/\rho$, a defensive attrition rate over day t, $\alpha_d(t)$. If this exceeds the defender's movement threshold α_{dT}, then on the next day the defender withdraws at a rate $W(t+1)$. This action reduces (that is, feeds back negatively on) the *attacker's* attrition rate $\alpha_a(t+1)$. In turn, this decrease in the attacker's attrition rate produces (again via $1/\rho$) a reduction in the defender's attrition rate $\alpha_d(t+1)$, whose size relative to α_{dT} determines the rate of any subsequent withdrawal. If $\alpha_d(t+1)$ is less than α_{dT}, no subsequent withdrawal occurs. The front then remains in place unless and until the attacker—by attempting to force the combat at his chosen pace—imposes on the defender an attrition rate exceeding his withdrawal threshold, and so on. One might think of the defender as an adaptive system, with withdrawal rates as an attrition-regulating servomechanism.

All the while, the attacker, too, is adapting; the prosecution rate $P(t)$ is his servomechanism. Just as there is some threshold α_{dT} beyond which the defender will withdraw, so the attacker possesses an "equilibrium" attrition rate α_{aT}. If on day $(t-1)$ he records an attrition rate exceeding α_{aT}, the attacker reduces the pace at which he prosecutes the combat. If he records an attrition rate lower than α_{aT}, he accelerates by raising $P(t)$. The magnitude of these changes in $P(t)$ approach zero if the attacker's attrition rate approaches α_{aT}, the equilibrium rate. Each side's adaptation may damp or amplify, penalize or reward, the adaptation of the other.

The adaptations are perhaps more sophisticated than meets the eye. Specifically, a primitive type of learning can occur. Suppose that on Monday, the defender's attrition rate exceeds his threshold α_{dT} by some amount X. In response, the defender withdraws at a rate $W(t)$ on Tuesday. Suppose, however, that—because his own attrition rate on Monday was below his threshold α_{aT}—the attacker increases his prosecution rate on Tuesday and that, as a result, the defender's attrition rate on Tuesday again exceeds his threshold by the same amount X. Only a defender unable to learn would withdraw at $W(t)$ again, since that rate *already failed* to solve his problem. A more deeply adaptive defender would withdraw at a rate greater than $W(t)$; in the Adaptive Dynamic Model, he does. To me, this makes a certain amount of biological sense. If walking slowly away from a swarm of attacking bees

[7]As an operational matter, a quick decision can circumvent logistical problems that could prove telling in a prolonged war. Strategically, the attacker may seek a decision before the defense has a chance to mobilize superior industry, superior reinforcements, or superior allies. An attacker with unreliable allies of his own may seek a quick win lest they begin to defect. An attacker may also choose to press the attack at a ferocious pace to secure a decision before the defender's nuclear options can be executed. A classic strategy of states facing enemies on multiple fronts has been to win quickly through offensive actions on one front and then switch forces to the second.

does not reduce the sting rate, we try jogging. If jogging doesn't reduce the sting rate, we run, and so on, until we are running as fast as we can (W_{max}). Of course, in the bee case we actually are free to pick something close to W_{max} as a first "trial retreat rate" because we are not concerned with territorial sacrifice. Analogous points apply to the attacker and his learning behavior in adjusting his prosecution rate, $P(t)$, as we will illustrate in the simulations below.

By setting the two fundamental thresholds α_{dT} and α_{aT} in various ways, the model will generate a reasonable spectrum of war types—bellotopes—from the war of entrenched defense, à la Verdun, to guerrilla war. I will discuss the four extreme settings and then present some simulations.

CASE 1: $\alpha_{aT} \approx 1$

The British at the Somme (1916) offer perhaps the great example of an attacker with no apparent pain threshold. Considering the extraordinary pain involved, we can ask with Jack Beatty, "What made them do it?"

> "'It' was to march, in an orderly way, rank by rank, column by column, to their death. That is what 20,000 British soldiers did on July 1, most of them falling between 7:30 and 8:30 A.M., the taste of tea and bacon still fresh on their lips. They got out of their trenches and marched to their death, or to some other form of mutilation.... Methodically, these [German] gunners raked the British formations. Methodically new formations set out, were shot down in no-man's-land, were replaced by other formations, and so on, turn and turn about, through the long day." (Beatty,[1] p. 112–114)

Long indeed. Here, perhaps, is a case of $\alpha_{aT} \approx 1$. Along similar lines, one thinks of the fateful Argonne Forest offensive of 1918 and, in particular, of Pershing's order to "push ahead *without regard to losses* and without regard to the exposed condition of the flanks." Surely, for Pershing, α_{aT} was close to 1. And, as Beatty notes, "It is no wonder that the cemetery at Romagne-Sous-Montfaucon, deep in the Argonne, is the largest American military cemetery in Europe, containing the remains of 14,246 soldiers."[1]

CASE 2: $\alpha_{dT} \approx 1$

The defensive analogue of the British at the Somme is undoubtedly the French at Verdun, also in 1916—not a good year, as Beatty recounts:

> "The French rotated seven tenths of their army though the meat grinder of Verdun. A colonel's order to his regiment gives the death-heavy flavor of the battle: 'You have a mission of sacrifice.... On the day they want to, they will massacre you to the last man, and it is your duty to fall.' The losses

> on both sides were appalling—perhaps a million and a quarter casualties in all. (The *ossuaire* at Verdun is full of the bones of the 150,000 unidentified and unburied corpses.) In short, Verdun was a demographic catastrophe for France. Yet, following Pétain's famous order, *'Ils ne passeront pas!'* the French Army held Verdun for the ten months of the battle—an epic of courage and endurance but not of victory. The standoff of Verdun, in the words of Alistair Horne, 'was the indecisive battle in an indecisive war; the unnecessary battle in an unnecessary war; the battle that had no victors in a war that had no victors.'" (Beatty,[1] p. 117)

Perhaps this is the terrible stable node I spoke of above—the sink of all sinks and, I would argue, a case of $\alpha_{dT} \approx 1$.

CASE 3: $\alpha_{dT} \approx 0$

Diametrically opposed to the French at Verdun are guerrilla defenders; their withdrawal threshold α_{dT} is close to zero. In guerrilla wars, like Vietnam, larger "superior" forces seeking direct engagements find themselves frustrated by defenders who withdraw—"vanish into the brush"—at the slightest attrition, the extreme case of trading space for time. Indeed, the entire strategy of the guerrilla—his only real hope—is precisely to *prolong* indecisive hostilities until domestic support for the war disintegrates, as it did for the United States in Vietnam.

CASE 4: $\alpha_{aT} \approx 0$

The fourth and final "pure" variant is the case where the attacker's equilibrium rate α_{aT} is close to zero. The natural example here is the so-called "fixing operation." The classic case is where an attacker is attempting a concentrated breakthrough in some sector of the battle front. He wants to prevent the defender from shifting forces from neighboring sectors to reinforce the breakthrough sector. Standard procedure for the attacker is to "pin," or "fix," these neighboring defensive forces by applying some pressure, but not enough to incur serious losses.

By specializing these two parameters, α_{dT} and α_{aT}, the model will produce the "pure" forms, shown in Table 1, as well as myriad mixed cases.

TWO SIMULATIONS

For illustrative purposes, I offer two simulations representing mixed cases. The numerical settings are given in Table 2. In the first, I posit a ferocious attacker, with an equilibrium attrition rate of $\alpha_{aT} = 0.6$. The defender's withdrawal threshold attrition rate is set at $\alpha_{dT} = 0.3$, respectably stalwart. Though not shown in Figure 4, the forces are initially equal (at half a million). What coadaptive story, then, is this picture telling?

TABLE 1 Adaptive Dynamic Model

Thresholds	
α_{dT}	Defender's Threshold
α_{dT}	Attacker's Threshold
Qualitative Range	
$\alpha_{dT} \longrightarrow 1$	Trench War (Verdun)
$\alpha_{dT} \longrightarrow 0$	Guerrilla War
$\alpha_{aT} \longrightarrow 1$	The Somme
$\alpha_{aT} \longrightarrow 0$	Fixing Operations

TABLE 2 Numerical Settings for Figures 4 and 5.

	Setting	
Variable	Figure 4	Figure 5
α_{aT}	0.6	0.1
α_{dT}	0.3	0.1
$P(1)$	0.1	0.2
$A(1)$	5×10^5	—[1]
$D(1)$	5×10^5	—
$W(1)$	0	—
W_{max}	20.0	—
$\rho(t)$	1.1	—

[1] Dash indicates "same as in Figure 4."

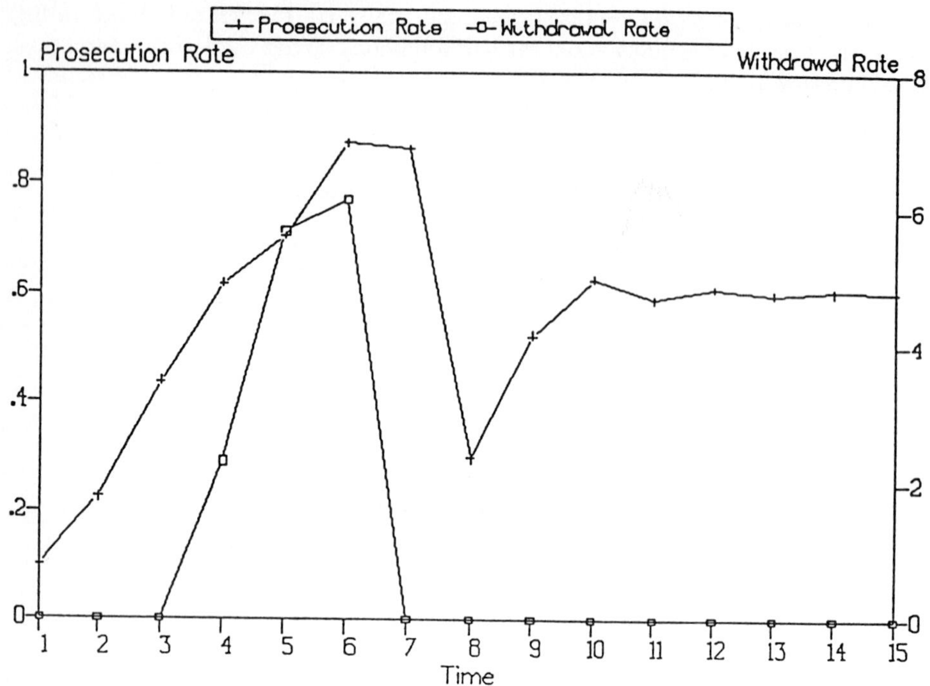

FIGURE 4 High unequal thresholds.

The attacker's opening "bid" on the pace of war, his opening prosecution rate, is $P(1) = 0.1$. At this low level, the resulting attrition rate for the attacker is well below the 0.6 level he is, in fact, prepared to suffer. And so, as shown, he begins raising his prosecution rate. But, this must climb to around 0.4 (on day 3) before it produces a defensive attrition rate above the defender's threshold of $\alpha_{dT} = 0.3$, which induces withdrawal.[8] Both curves then rise to day 6. In this phase, the defender's withdrawals (partial disengagements) are thwarting the attacker's effort to attain his "ideal" attrition rate of $\alpha_{aT} = 0.6$, so the attacker prosecutes with increasing vigor, which efforts induce successive withdrawals at increasing rates.

Now, all the while in this simulation, the casualty exchange ratio (attackers killed per defender killed on day t) has been constant at a rate favoring the defender. And, by day 6, he has whittled down the attacker to such an extent that, even at high prosecution, the attacker cannot exact defensive attrition sufficient to induce withdrawal—so, withdrawal stops, the defender halts, on day 7.

[8]The computer has simply connected the dots in these pictures.

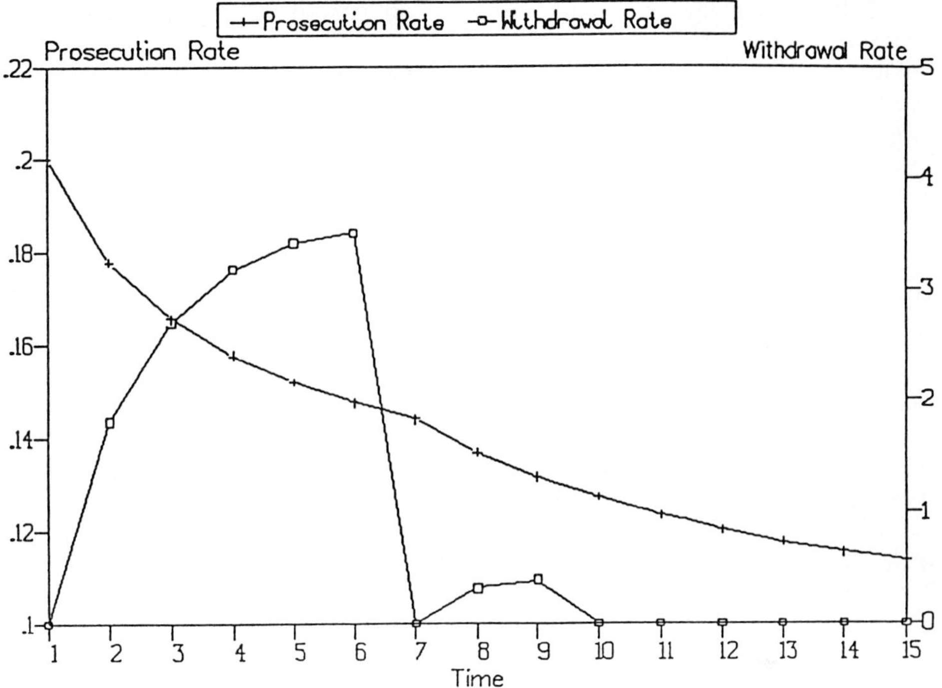

FIGURE 5 Low equal thresholds.

In effect, the attacker "slams into" the now stationary defender on that day, producing attacker attrition well in excess of the attacker's tolerance α_{aT}; "ouch," in other words. The attacker reacts to this extraordinary pain by cutting his prosecution rate sharply on day 8—too sharply, it turns out. He has overshot, as evidenced by his subsequent increases in $P(t)$ which ultimately levels off at around $P(t) = 0.6$.

A rather different history is portrayed in Figure 5. The prosecution rate decreases monotonically, while the withdrawal rate rises and falls twice. In this case, the attacker's equilibrium, and defender's threshold, attrition rates are set equal at $\alpha_{aT} = \alpha_{dT} = 0.1$, considerably lower than in the preceding case. Initial force levels are as before.

Here, the attacker's opening prosecution rate *exceeds* his equilibrium rate: $P(1) = 0.2$. This opening rate imposes on the defender an attrition rate that exceeds his withdrawal threshold. Over the first six days, *both* sides are above tolerance; the defender withdraws at a growing (though diminishing marginal) rate, while the attacker decreases his prosecution rate.

These coadaptations (plus a casualty-exchange ratio favoring the defender) gradually depress the defender's attrition rate to a level below his withdrawal threshold; so, on day 7, he halts. Though the attacker is steadily reducing his prosecution rate, the weight of his impact on the stationary defender is sufficiently painful to drive the latter from his position once more until, on day 10, the front stabilizes. The attack nonetheless persists, though at a declining level of ferocity, $P(t)$.

SUMMARY

In Lanchester Theory—by which I mean the original equations and their contemporary extensions—these *behavioral* dimensions of combat are ignored. Mere opposing numbers and technical firing effectiveness completely determine the dynamics: there is no adaptation. In the Adaptive Dynamic Model, the parameters α_{dT} and α_{aT} allow one to reflect the different ways in which given forces can behave. As we have seen, with a given force, an attacker may prosecute the offensive at a ferocious pace, virtually unresponsive to losses. The British at the Somme in 1916 come to mind. Or, an attacker may operate *the same forces* at a more restrained pace, as in fixing operations. A high value of α_{aT} will produce the former type of attacker; a low value of α_{aT} will generate the latter.

Similarly, the tactical defender may be more or less stalwart in holding his positions. Guerrilla defenders may withdraw—"disappear"—when even slight attrition is suffered. For such tactical defenders, the withdrawal-threshold attrition rate α_{dT} is close to zero. At Verdun, by contrast, no attrition rate was high enough to dislodge the defenders from their entrenched positions. Pètain's famous order—"Ils ne passeront pas!"—effectively set α_{dT} equal to one.

These strategic and human realities are captured, however crudely, in the Adaptive Dynamic Model. And they are captured by a mechanism that permits movement to affect attrition, a *feedback* that is not possible in any version of Lanchester's equations. So, I feel some confidence in claiming that my equations present a *less crude* caricature of combat dynamics. But, given the complexity of the process, that is all I claim.

ACKNOWLEDGMENTS

I am grateful to Robert L. Axtell for his thoughtful comments.

REFERENCES

1. Beatty, Jack. "Along the Western Front." *The Atlantic Monthly* **258** (1986): 112–115.
2. Epstein, Joshua M. *The Calculus of Conventional War: Dynamic Analysis Without Lanchester Theory.* Washington, DC: Brookings, 1985.
3. Epstein, Joshua M. *Conventional Force Reductions: A Dynamic Assessment,* 92–93, 98–99. Washington, DC: Brookings, 1990.
4. Gauss, G. F. *The Struggle for Existence,* 47. Baltimore: Williams & Wilkins, 1934.
5. Kaufmann, William W. "The Arithmetic of Force Planning." In *Alliance Security: NATO and the No-First-Use Question,* edited by J. D. Steinbruner and L. V. Sigal, 214. Washington, DC: Brookings, 1983.
6. Lanchester, F. W. "Mathematics in Warfare." In *The World of Mathematics,* edited by James R. Newman, vol. 4, 2136–2137. New York: Simon & Schuster, 1956.
7. Lanchester, F. W. *Aircraft in Warfare: The Dawn of the Fourth Arm.* London: Constable, 1916. The same model was apparently developed independently by the Russian M. Osipov in 1915. See "The Influence of the Numerical Strength of Engaged Forces on Their Casualties" by M. Osipov. Originally published in the Tzarist Russian journal *Military Collection,* June–October, 1915. Translated by Robert L. Helmbold and Allan S. Rehm, U.S. Army Concepts Analysis Agency, CAA-RP-91-2, 1991.
8. Weiss, Herbert K. "Combat Models and Historical Data: The U.S. Civil War." *Oper. Res.* **14** (1966): 788.
9. Willard, D. *Lanchester as Force in History: An Analysis of Land Battles of the Years 1618–1905.* Technical Paper RAC-TP-74. Bethesda, MD: Research Analysis Corp., 1962.

Seminars

E. Atlee Jackson
Santa Fe Institute and Department of Physics, Center for Complex Systems Research, Beckman Institute, University of Illinois at Urbana-Champaign, Urbana, IL 61801

Chaos Concepts

The term "chaos" has become a much-used word in recent years, appearing in numerous articles, and on the cover of books in a variety of fields. It is a term that catches the imagination of people in general, and students in particular. It may well be that your interest in complex dynamic systems was sparked by this term.

Like any term that attempts to deal with the complex dynamics in the real world, "chaos" actually represents a variety of distinct features. Presumably we all know that it has something to do with "sensitivity to initial conditions," but there remains a general lack of appreciation concerning its fundamental significance to science, and how this relates to the future studies of "complexity." This search for the fundamental aspects of complexity is illustrated by the recent article that appeared on the editorial pages of *Physics Today*,[2] where Philip Anderson, a Nobel Laureate and External Faculty Member to SFI, wrote the article "Is Complexity Physics? Is it Science? What is it?" I recommend that you read this article, to see the point of view of an enlightened physicist[1] to this changing scene in science. He referred to it as "this infinitely quiet revolution," and indeed this would have been so if people like Joe Ford had not been beating a lonesome "chaos drum" for over ten years.[18,19] Even with these efforts, it certainly has been a very quiet revolution, whose significance remains largely unrecognized. I will attempt, in this very short time, to discuss some of the basic issues.

1992 Lectures in Complex Systems, Eds. L. Nadel & D. Stein, SFI Studies in the Sciences of Complexity, Lect. Vol. V, Addison-Wesley, 1993

This very quiet revolution began a century ago, led by Poincaré, and was totally overshadowed by the quantum revolution in the first half of this century. Only the mathematician Birkhoff kept the enquiry of chaotic dynamics alive until around 1950, when Cartwright and Littlewood[7] and Levinson[35] proved that chaotic solutions exist in simple equations related to physical systems (Poincaré correctly conjectured, around 1890, that this occurs in astronomical systems but could not prove it). Around that time the digital computer was invented, opening an entirely new method for uncovering the wonders of our dynamic world. However, as Anderson's article illustrates, there is great confusion about where science is going with these new concepts. This issue goes much deeper than just the topic of chaos, but I do not have time to explore these more general aspects of this metamorphosis of science. In any case, chaos occupies a special position in the general area of complex behavior because it is readily appreciated on one level, yet it contains some fundamental messages for science that are appreciated by very few scientists. Moreover, chaos has many possibly practical applications in chemistry, medicine, neurology, and engineering.

Before discussing aspects of chaos, I need to make some general remarks about the study of any complex phenomenon in dynamic systems. We always need to keep in mind that we know what we know only when we know how we know (or, at least, think we know!). This is an obvious statement, but great confusion exists in the areas of chaos and complexity simply because the bases of statements are not clearly defined. I suggest that you keep asking "How do I know this is true?"

Presently we use three distinct operational methods for obtaining quantitative information about phenomena in Nature. They are:

1. Physical Experiment (PE): yields finite data, over a finite duration of time, and with finite accuracy. In such experiments one seeks to establish correlations between a few observables.
2. Mathematical Model (MM): operates in the formal world of real numbers, with the infinite precision of variables, infinite time durations, and various infinite limiting processes. In this formal world, logical rules of inference are applied to arrive at deductions.
3. (Digital) Computer Experiment (CE): yields finite data, as in the case of physical experiments, but operate by the same logical rules that apply to mathematical models.

These operational methods do not form a scientific method until they are linked together by various inductive processes (e.g., the invention of differential equations, or algorithms, that are intended to predict new physical situations), and the encoding and decoding of the finite/infinite precision numbers, required to connect the physical observations with the MM or CE results. This is obviously a very large topic, and I only have time to sensitize you to these issues. I need to do this because the more profound messages that chaos has to offer to science can only be understood in the above context. I will return to this point after we see some of the technical features of chaos.

While chaos is one of the least structured components of complex dynamics, it has considerably more structure than the name may convey. "Chaos," in the nonscientific context, is often considered to be any condition of "total, utter, and extreme disorder and confusion." Let's refer to this concept as "total chaos," and denote it by TC. Commonly such terms as "noise," "stochastic process," or "random process" are also used to describe this TC. Whatever terminology is used, it is meant to convey the total lack of any known deterministic feature. This contrasts with the modern studies of deterministic chaos (DC), to be discussed here. But it should be kept in mind that yesterday's TC, or noise, may become tomorrow's deterministic chaos!

Before considering deterministic chaos, consider the following idea:

TC + constraints may yield "structures" of scientific interest.

There is possibly no more impressive application of this idea than in equilibrium statistical mechanics. When all of the sophistry is distilled away, the basis of this remarkably successful model of equilibrium molecular behavior (the Boltzmann-Gibbs probability structure) rests upon our nearly complete ignorance of the detailed dynamics of this system (TC), save for our knowledge of a few additive constants of the motion of the assumed basic equations of motion for these molecules (the constraints). An elementary presentation of how these ideas can be joined to yield the Boltzmann-Gibbs distribution can be found in my 1968 book on equilibrium statistical mechanics.[29] It is a grand example of the above idea, and surprisingly it works for many systems.

The first dynamic application of this principle may well have been Boltzmann's use of TC in his assumption of statistical independence prior to the collision of molecules, which yielded his famous Boltzmann equation, and irreversibility predictions. While Boltzmann had to continually defend his insights against numerous attacks (based on faulty understanding of the limitation of the operational methods of PEs and MMs), the proper defense of his insight can only be understood on the basis on modern chaotic theory, as I will point out later.

Other examples of this principle can be found in many physical systems that exhibit "order out of chaos" ("chaos" = TC), as has been expressed in the title of a book by I. Prigogine and I. Strenger.[47] Famous examples involve the formation of coherent vortices in both the Taylor and Rayleigh-Bénard fluid systems. Here, despite the TC of the molecular motion, when appropriate constraints are imposed at the boundaries (momentum and energy, respectively), these orderly structures can appear.

Finally, it might be remarked that a variant of the above idea has been applied to the area of the basic laws of physics by John Archibald Wheeler. In an article "Law Without Law,"[12,58] he expressed this principle as

"higgledy-piggledy" + regulating principles $\rightarrow$ EVERYTHING (no "law" required).

If you would like to discover an interesting structure for yourself, generate the numbers (1,2,3) randomly on a computer (your TC). Construct an equilateral triangle with vertices labeled (1,2,3) on the computer screen. Start a point anyplace in this triangle, and when you generate the number k, move the point half the distance toward the vertex k (the constraint). Continue this process for a large number of steps. You will see a structure emerge (an attractor) that has some interesting "fractal" properties (EAJ; 2.6; Schroeder[49]), and which you should be able to predict (in retrospect at least!).

Fortunately, we do not need to search for the concept of Chaos (i.e., deterministic chaos) in the complicated systems noted above. The first surprise was that Chaos can be found in "simple" deterministic systems (MMs). Please note carefully that "determinism" in this context is a mathematical concept; it decidedly does not apply directly to PEs, as will be made clear in what follows. Historically (around 1890) the first example, where Poincaré correctly suspected this phenomena occurs, was the system of three bodies that gravitationally attracted each other. Since this three-body system had defied all the efforts of mathematicians to obtain analytic solutions, Poincaré invented a variety of new and general methods to extract some information about the dynamic properties of such complex systems.

To appreciate the concept of Chaos at a MM level, it is necessary to outline some of Poincaré's tools. For more details on these points, let me suggest my friendly set of books, *Perspectives of Nonlinear Dynamics.*[31] I will reference topics in them by "(EAJ; section)." For the real devotee, perhaps the most extensive, and generally accessible presentation of chaos and fractals can be found in Peitgen et al.[45]

We begin with the MM of dynamical systems, given by the differential equations for some dynamic variables, say $(x_1(t), x_2(t))$:

$$\frac{dx_1}{dt} = F_1(x_1, x_2, t), \qquad \frac{dx_2}{dt} = F_2(x_1, x_2, t).$$

When there are more variables, it is useful to use the vector notation

$$\frac{dx}{dt} = F(x,t) \qquad (x, F \in R^n).$$

If $F(x,t) = F(x)$ (i.e., does not depend explicitly on t), the system is called autonomous (otherwise, nonautonomous). For the present we will consider only autonomous systems,

$$\frac{dx}{dt} = F(x) \qquad (x, F \in R^n). \tag{1}$$

The solution of such an equation, $x(t) = g(t; x_0)$, depends on the initial conditions $x(0) = x_0$, and at any time it can be pictured as a point in an n-dimensional space, whose coordinates are $(x_1, x_2, \ldots, x_n)$. This space was introduced by Poincaré, and is called the phase space of the system. As time changes, the point $x(t)$ traces

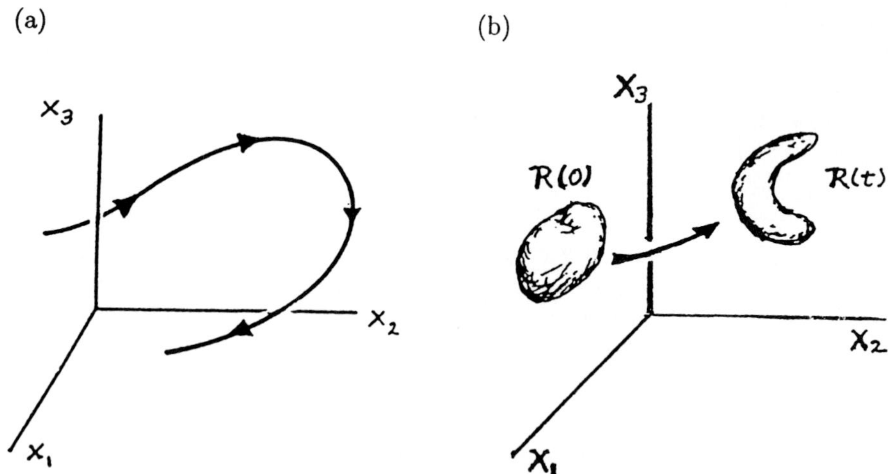

FIGURE 1 (a) A solution, $x(t)$, traces out a curve in phase space as time increases. (b) The dynamics of a region $R(t)$, representing a continuum of solutions.

out a curve in this space, passing through x_0. If we attach an arrow indicating its direction as time increases, we get its orbit in this space, as illustrated in Figure 1(a). If we consider a collection of solutions passing through different initial points, then we obtain a family of such orbits (EAJ; 2.1), and none of them intersect if the solutions are unique (EAJ; 2.2). The idea of considering the properties of families of solutions, rather than only individual solutions, was another important contribution of Poincaré. Its importance will become clearer as we explore chaos.

But first consider a region, $R(t)$, in the phase space which moves in such a way that all the initial points in the region, $R(0)$, remain in the region $R(t)$ at time t, when their dynamics are given by Eq. (1). This is illustrated in Figure 1(b). Poincaré considered the volume of this region as a function of time

$$V(t) = \int \cdots \int_{R(t)} dx_1 dx_2 \ldots dx_n. \tag{2}$$

Actually he considered more general integrals of this type (EAJ; 2.4), but we will limit our discussion to Eq. (2). One can show without much difficulty (EAJ, Appendix C) that the time derivative of $V(t)$ is given by

$$\frac{dV(t)}{dt} = \int \cdots \int_{R(t)} \nabla \cdot F(x) dx_1 dx_2 \ldots dx_n. \tag{3}$$

To illustrate what this tells us, consider a damped nonlinear oscillator.

$$\dot{x} = v, \qquad \dot{v} = -\mu v - x.$$

We find that

$$\nabla \cdot F = \frac{\partial v}{\partial x} + \frac{\partial}{\partial v(-\mu v - x)} = -\mu,$$

and substituting this into Eq. (3), we obtain

$$\frac{dV(t)}{dt} = -\mu V(t).$$

In other words, the volume is decreasing at an exponential rate. What this tells us is that all of these solutions are crowding together as time increases—they are being attracted toward each other. Since the volume is not conserved in time, this is called a nonconservative system (in the present case it is caused by the damping, perhaps due to friction). By contrast, a conservative system satisfies

$$\frac{dV(t)}{dt} = 0 \qquad \text{(conservative)}. \tag{4}$$

Now what has all this to do with chaos? Well, as I pointed out, chaos comes in several forms. Chaos in a conservative system is quite different from that in a nonconservative system, so we need to distinguish these types of systems. More on this later.

First, it is useful to know a little about constants of the motion. $K(x,t)$ is a constant of the motion of $dx/dt = F(x)$, provided that

$$\frac{dK(x,t)}{dt} = 0 \qquad \text{(for all solutions)}. \tag{5}$$

If $K(x,t)$ does not depend explicitly on time, it is called a time-independent constant of the motion.

A simple example illustrates these points. Consider a particle acted on by a constant force, F. Its equations of motion are

$$\frac{dx}{dt} = v \text{ and } \frac{dv}{dt} = F \qquad \text{(unit mass)},$$

with the solution

$$v = v_0 + F \times t; \qquad x = x_0 + v_0 \times t + 0.5 \times F \times t^2.$$

Two constants of the motion are the two initial conditions

$$K_1 = v_0 = v - F \times t; \qquad K_2 = x_0 = x - v \times t + 0.5 \times F \times t^2.$$

We can obtain a time-independent constant of the motion by eliminating t between these, yielding

$$K_3 = K_2 - \frac{K_1^2}{(2 \times F)} = x - 0.5 \times \frac{v^2}{F}.$$

K_3 is, of course, a multiple of the total energy of this system. The importance of time-independent constants of the motion is that equations of the form

$$K(x) = K_0 \qquad \text{(a constant)}$$

define a fixed "surface" (manifold) in the phase space that has a dimension $(n-1)$. An initial state, which begins on this manifold, remains on the manifold for all time. Thus, in the above example, $K_3(x, v) = K_0$ defines a line in the two-dimensional phase space (x, v).

The knowledge of such "integral manifolds" is very useful in assessing some the characteristics of the dynamics of a system. In particular, if the dynamics of a system move on a smooth two-dimensional integral manifold, it cannot be chaotic. In other words, in order to exhibit chaotic behavior, a system needs to have more dynamical freedom than a smooth two-dimensional "surface" provides. It needs at least three dimensions.

Now returning to the three-body problem, around 1890, no general solutions had been found. Indeed, no convergent perturbation methods could be found (that is, methods where one could systematically get better and better approximate solutions). The only time-independent constants of the motion that were known were ten classic results, despite the fact that there are $3 \times 6 = 18$ constants, corresponding to the 18 initial conditions. Indeed, it was proven in the 1890s that the remaining unknown constants of the motion could not be any rational functions of the variables. This strongly suggests that they must be complicated functions that do not represent "smooth' and "predictable" forms of dynamics. Thus, it seems likely that most constants of the motion of most MMs (not just the three-body problem) are inherently "uncontrollable" from the point of view of the finite accuracies of PEs. As we will see, this is also one of the hallmarks of chaotic systems, and the need to distinguish between MMs and PEs.

Fortunately Poincaré introduced a variety of new methods to help science extricate itself from this quagmire. Building again on the above picture of orbits in a phase space, he introduced the idea we should not look at the details of an orbit; instead, we consider it only when it passes through a surface in the phase space (EAJ; 2.5). Let S_0 be such a surface, through which an orbit passes an infinite number of times. This surface is called a surface of section, and it is illustrated in Figure 2(a) for two different orbits. One orbit is a periodic orbit that passes through the point p_0, over and over again. The other orbit passes through S_0 at the point x_0 and the next time at point x_1. Because x_1 is the first return after the point x_0, the association of $x_0 \to x_1$ is called Poincaré's first return map. In other words, all of points near p_0 "map" to other points in their first return to the surface S_0. This concept of a map has become one of the staples for the study of chaos.

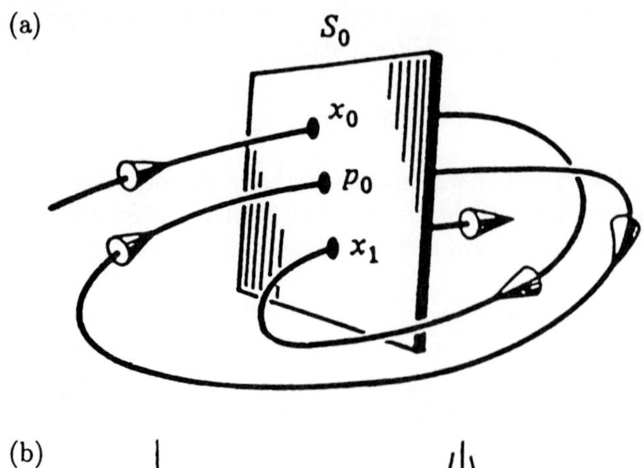

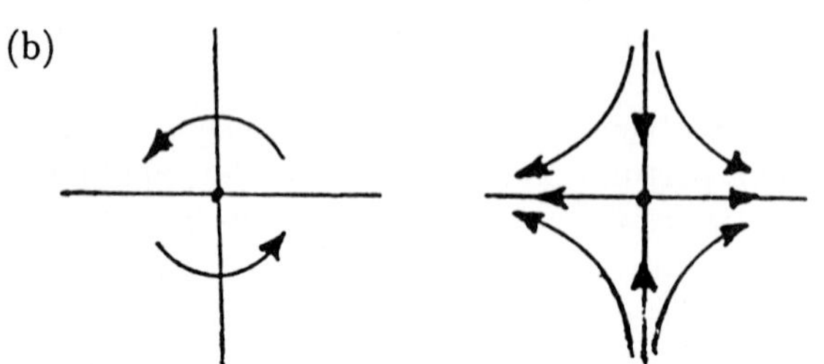

FIGURE 2 (a) The intersection of a periodic orbit, and a nearby orbit, with a surface of section, S_0. (b) The dynamics near an elliptic (left) and a hyperbolic fixed point.

Now the three-body system is a conservative system, and as a consequence it turns out that any little region near p_0 will have a first-return region that has the same area as the original region (EAJ; 6.1, 6.12). Thus these maps are called area-preserving maps, and that property turns out to be crucial in proving that such systems are generally chaotic. The reason that this is so is due to the very limited repertoire such maps possess near fixed points. They can either rotate around the fixed point of map, or else they can map inwards in one direction and outward in another direction (to preserve the area of any region). These are called elliptic and hyperbolic fixed points respectively, and are illustrated in Figure 2(b).

All of these ideas of Poincaré are readily applied to the so-called restricted three-body problem. In this system two heavy masses rotate around each other in some plane, under the influence of their gravitational attraction. A third very light mass is set into motion along this same plane, attracted by the other heavy masses but not influencing them. The phase space for this light mass is four-dimensional, with axes $(x, y, dx/dt, dy/dt)$. Since the total energy is conserved, and since the small mass doesn't change the energy of the large masses, the energy of the small mass is also conserved. This constant of the motion defines a three-dimensional manifold in this phase space along which the mass moves (just enough room for Chaos!). If now one can find a surface of section in this manifold, it will have only two dimensions, and it will be "relatively easy" to study the first-return map in

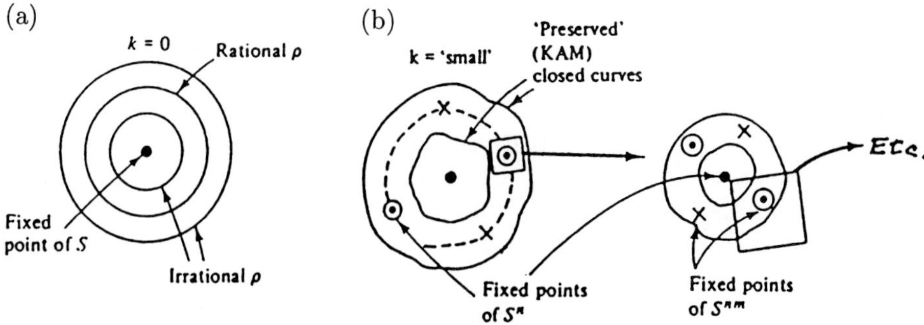

FIGURE 3 (a) When $k = 0$, each circle around an elliptic fixed point is mapped onto itself by the standard (or Chirikov[8]) map, S. (b) When $k \neq 0$ those circles that had a rational rotation rate (at $k = 0$) break up into a finite number of hyperbolic-elliptic pairs of fixed points of S^n. See the text for details.

this plane. That's what Birkhoff did around 1913, proving a theorem that Poincaré had conjectured but was unable to prove.

To appreciate some of the aspects of this chaos in a conservative system, as uncovered by the Poincaré-Birkhoff theorem, consider again the Poincaré first-return map near an elliptic periodic point, P_0. This is illustrated in the Figure 3(a), where different rotation rates typically occur at different distances from P_0. Moreover, an essential feature is that the map is area-preserving. Such a map is frequently referred to as a twist map, because of the different rotation rates (EAJ; 6.6).

An explicit example of such a map, which is widely referred to these days, is the so-called Standard (or Chirikov) Map, S:

$$r(n+1) = r(n) + K \times \sin(\theta(n)); \quad \theta(n+1) = \theta(n) + r(n+1). \tag{6}$$

This is an area-preserving map for any value of K. When $K = 0$, the collection of points on a circle simply map along this circle at a rotation rate denoted by ρ (and equal to r, the radius of that circle; see Figure 3(a)). This type of dynamics is, of course, very boring, and one might think we are a long way from discovering Chaos. But, when K is not zero, all of the circles that correspond to a rational rotation rate break up into a most remarkable collection of dynamics (depending on the initial state). A few points remain periodic after n iterations (denoted S^n). They are illustrated by a cross and a circle on the dashed curve in Figure 3(b). However, most points are no longer periodic. Around the elliptic fixed points of S^n, the dynamics are again of the form (6), but on a much smaller scale (having their own periodic point of S^{nm}!). The dynamics around the "cross" points are that of the hyperbolic fixed points, noted above in Figure 2(b). Those orbits that tend inward and outward at the hyperbolic points can be traced further away and generally intersect (remember this is a map, so that's okay). This implies that there are an infinite number of intersections, because these intersection points both came about from being mapped outward and mapped inward to the fixed point (think about it, or see EAJ; 6.6). The final hooker is that the loops that exist because of two intersections enclose some area, and this area must be preserved when the loop area is mapped, and there's an infinite number of these, all getting "squeezed" by intersection points that have the hyperbolic point as their limit point! Wow!! It is no wonder that Poincaré refused to attempt to illustrate this complicated situation. Only much later, less cautious individuals attempted to illustrate this "Poincaré tangle," as shown in Figure 4(a). This is a caricature of the mathematical results. Figure 4(b) illustrates what one obtains from a CE, with its finite data set (here $K = 1.2$), and this difference from that inferred from an MM is noteworthy. Note that in Figure 4(b) the axes are the cylindrical coordinates of Eq. (6), whereas Figure 4(a) is in the $(x, \dot{x})$ space. It appears in this figure as the chaos is localized in the phase space, since some circles (distorted as they may be) remain intact. These come from the set of circles that had irrational rotation rates when $K = 0$ in Eq. (6). They are called KAM surfaces, after Kolmogorov-Arnold-Moser, who proved that, as $K \to 0$, the measure of these "preserved tori" (in the three-dimensional manifold) goes to one—in other words, "most" of the dynamics remain nonchaotic. On the other hand, since the rational circles are dense, the chaotic dynamics is also dense for any nonzero K. All very mathematical!!

To obtain results which are more significant for physical phenomena, with their finite-K nonlinearities, it requires careful CEs to determine finite-K effects, such as when there is no longer any preserved regular dynamics. Indeed, CEs have been essential for exploring Chaos with finite nonlinearities (e.g., studies by Greene[22] and Chirikov[8]).

The Poincaré tangle is the Chaos of classic conservative systems. The Poincaré-Birkoff theorem expresses this Chaos in approximately the following form:

THEOREM In any neighborhood of a periodic point of a nonintegrable conservative system, there are an infinite number of periodic points with different periods, and an uncountable number of aperiodic solutions.

Now what does this tell us as scientists? It tells us that there are mathematical solutions of our MM that are not related to any PE in a predictable manner for long periods of time. There are solutions of our MM that are physically "forever unusable," in the sense that we cannot set up a physical system with initial conditions that will yield any specific periodic solution. This expression was used by Duheim[14] in 1914 in connection with a rather abstract dynamical result obtained by Hadamard. Indeed, there are many physical systems whose dynamics are related to the standard map, some of which are illustrated in Figure 5 (EAJ: 6.5). When one finds these results in these much more physically realistic MMs, it makes it very clear that it is necessary to pay attention to the reasons we believe we know something. The world of solutions of MMs is a metadynamic world that transcends what we can observe in our PEs, and we must pay attention that we give it scientific credence only after careful examination of the quantitative encoding/decoding possibilities.

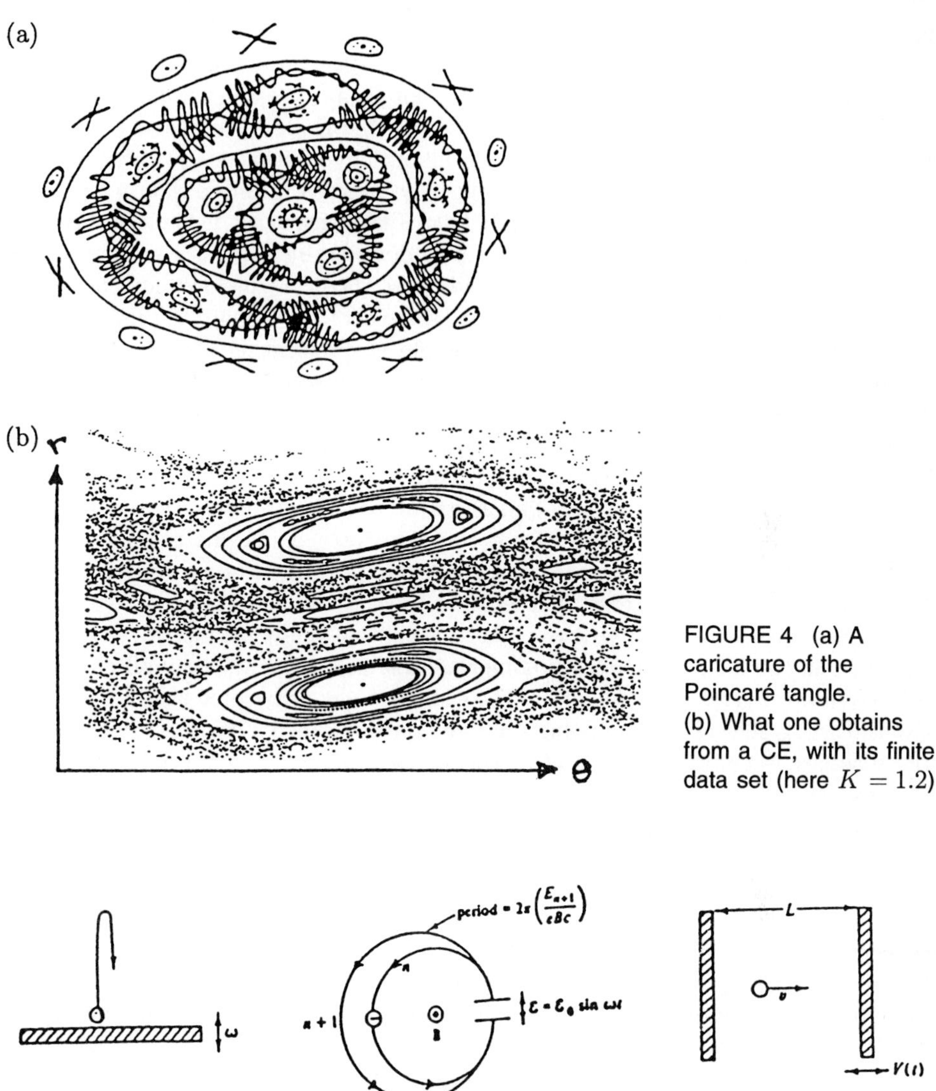

FIGURE 4 (a) A caricature of the Poincaré tangle. (b) What one obtains from a CE, with its finite data set (here $K = 1.2$)

FIGURE 5 Examples of three physical systems whose dynamics are related to the standard map.

One of the important physical examples of such conservative Chaos is found in the chaotic motions in our solar system. A nice discussion of these effects can be found in Wisdom's article.[59] What is also of interest, considering the long history of these results, is how slow their recognition has been (e.g., see Sir J. Lighthill, FRS[36]). Indeed, there apparently has not been a careful re-examination of Boltzmann's stosszahlansatz in light of our new appreciation of the sensitivity of molecular dynamics to arbitrarily small perturbations. In 1914 Borel pointed out this sensitivity, as discussed in Brillouin's book.[6] He noted that if a 1-gram mass is moved 1 cm on the star Sirius, the gravitational field on the Earth's surface changes by 1 part in 10^{100}. If there is a relative error in the initial conditions of 10^{-100}, then the trajectory of a gas particle cannot be accurately followed for more than a nanosecond (a very conservative estimate). In light of such sensitivity, it is clear that the statistical assumption made by Boltzmann in deriving his famous equation is the only defensible assumption that can be made. The classic Zermelo (recurrence) and Loschmidt (reversibility) "paradoxes" (e.g., see Kac[34] or Tolman[54]), which Boltzmann had to fight against all his life, have no physical relevance. They are examples of confusing the (assumed!) results of MMs and the observations in PEs (i.e., both the lack of appreciation of chaotic effects, and the all-important encoding/decoding connections). But I don't have time to digress down this important road.

The above deterministic chaos that occurs in conservative systems has a very "delicate" (unpredictable, uncontrollable) nature to it. Fortunately that is not true of many systems, because they are not conservative. This is due to some strong interaction with their environment; simple examples are friction and metabolic processes in biological systems. The essential feature of nonconservative systems is that they have dynamic attractors. This means that many initial states can tend to behave in the same dynamical manner as time goes on; they may all go to the same equilibrium state, or oscillate in the same fashion, or become chaotic. The set of initial states that all end up doing the same thing is called a basin of attraction of that ultimate dynamics. A basin of attraction in dynamics is somewhat analogous to a river basin, namely, all of the initial locations where the raindrops end up in a particular river.

Let's consider an example that is easy to understand. Figure 6(a) illustrates a double-well surface on which a point mass can slide, being pulled down by gravity and acted on by the friction. The motion is rather simple; depending on where the mass is started, it will ultimately end up at the bottom of one well or the other. We say that this system has two "fixed-point" (equilibrium-point) attractors, and the basins of attraction of these fixed points are illustrated as shaded and unshaded regions in the Figure 6(b). It should be noted, once again, that this is a MM. Do you think that either a PE or a CE can determine these basins of attraction far from the origin? Just a little thing to think about (you don't know until you know how you know!). This system certainly is not chaotic, but what would happen if a large periodic force were applied to this system? One might visualize the resulting "indecisive" motion by sliding Figure 6(b) periodically up and down, and trying to see where the particle might go (which "basin" is it in?).

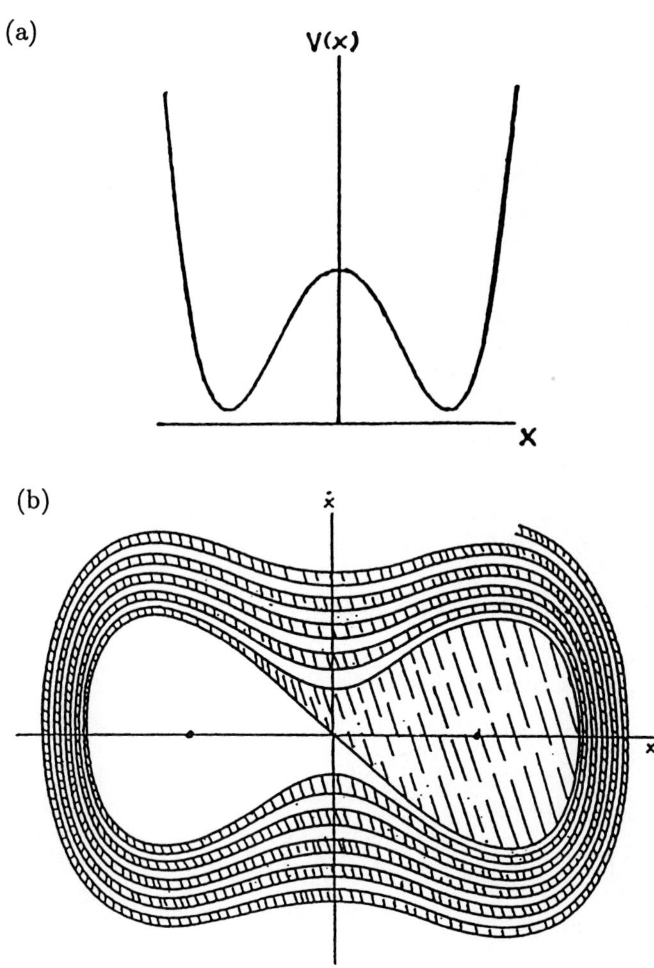

FIGURE 6 (a) A double-well surface on which a point mass can slide, being pulled down by gravity and acted on by the friction. (b) The system in (a) has two "fixed-point" (equilibrium-point) attractors, and the basins of attraction of these fixed points are illustrated as shaded and unshaded regions.

(a)

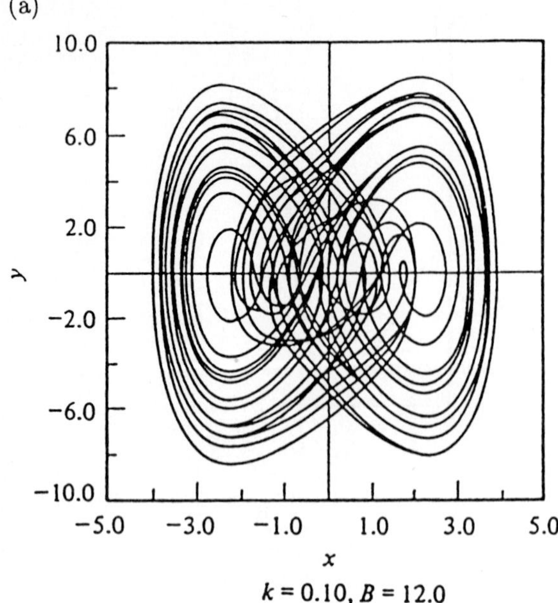

(b)

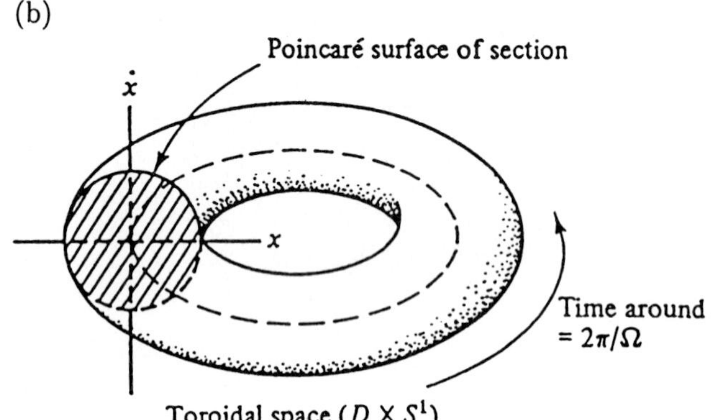

FIGURE 7 (a) Chaotic dynamics in the phase space. (b) The extended phase space. The intersections with a surface of section records oly a small portion of the complications in (a).

As I already mentioned, chaos cannot occur in two-dimensional motion, but if we take systems and shake them like this, with some applied periodic force, chaos can often result. An illustration of this is shown in Figure 6(a) (CE), which comes from a CE of the forced nonlinear oscillator

$$\ddot{x} + k\dot{x} + x^3 = B\cos t$$

studied by Ueda (EAJ; 5.14). While this clearly appears chaotic, it also does not convey much insight into its structure. Taking a leaf from Poincaré's notebook, we

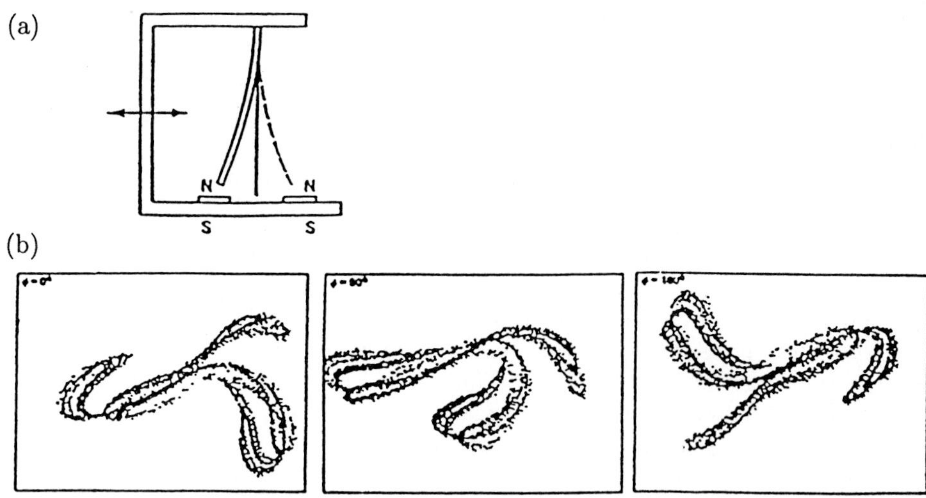

FIGURE 8 (a) Moon's experiment. (b) Moon's Poincaré maps. Copyright © 1987 by John Wiley; reprinted by permission.

can generate a new type of Poincaré map, using a so-called extended phase space. This is illustrated in Figure 6(b). The idea is to make use of the periodicity of the force, to extend the phase plane $(x, dx/dt)$ along the time axis, and reconnect it after one period (because then the equations of motion repeat). We can either use one Poincaré surface of section, or several surfaces around this "toroidal" space, which allows us to capture more details of this highly complex motion.

In some very pretty physical experiments made by Moon,[40] he studied the motion of a metal strip, attracted by two periodically displaced magnets (Figure 8(a)). Three of his Poincaré maps at 0°, 90°, and 180° in the phase of the periodic force are illustrated in Figure 8(b) (so each picture is the map of one dynamic state, passing many times through each of these surfaces in succession). The highly complex character of this fractal attractor is much more clearly seen than in the usual phase plane (e.g., Figure 7(a)). Such fractal attractors are known as strange attractors, being very different in character from simple fixed points, or periodic attractors (limit cycles. EAJ; 5.6). For many other nice experimental results involving chaos and fractals, see Moon.[41]

Perhaps the most famous strange attractor, which we will not have time to explore, is the Lorenz attractor (EAJ; 7.3–7.9). Edward Lorenz is a meteorologist at MIT, and his interests in the difficulties in weather prediction led him to his famous discovery of the first strange attractor in 1963. The discovery was only made possible with CEs. It may be that the PEs carried out by van der Pol and van der Mark around 1927, involving a periodically forced relaxation oscillator, was the first physical strange attractor, but they described it as noise, and simply "a subsidiary phenomena" (EAJ; 5.14). In any case, it was Lorenz who first appreciated the novelty of this strange-attractor dynamics. Since then many other examples have been found. One physical example that is presently widely studied is Chua's electrical oscillator.[9]

Now let me describe some ways that we can characterize chaos:

- The information that we can "predict" about the future of the system in a PE is essentially the same amount of information that we know about its initial state. Is that a "prediction"?
- The fact that one can find a solution of an MM that can be put into correspondence with any sequence of coin tosses (a Bernoulli sequence).
- The sensitive dependence of the solutions to their initial conditions. On the average this can be measured by the system's "Lyapunov exponents"; a concept that is not identical in a PE, MM, or CE.
- The fractal dimension of the points in some Poincaré map; again dependent on the operational basis.
- The "stretch-fold-squeeze" view of the dynamics of strange attractors.

In the limited time remaining, let me outline these ideas with the help of several maps, beginning with the famous logistic map. One of the simplest contexts in which to think of this map is in the area of ecology. Let $x(n)$ be the fractional population of some bugs at generation n, normalized by their maximum value, so $0 < x(n) < 1$. This population at the next generation, $x(n+1)$, will change due to reproduction, but it will be limited by the finite resources in its environment. A simple model incorporating these ideas is

$$x(n+1) = c\,x(n)\,(1 - x(n)), \tag{7}$$

where c is a reproduction rate, and the factor $(1 - x(n))$ limits the population. For very obscure reasons, Eq. (7) is called the logistic map (EAJ; 4.2–4.7).

The dynamics of the logistic map can be very complicated. A classic introductory discussion of some of these wonders can be found in May's article.[37] We do not have time to discuss most of this, but will focus on some of the chaotic features of Eq. (7). One of the most revealing cases is when $c = 4$, where Eq. (7) has its most chaotic behavior. It also can be solved exactly, with the help of a famous transformation due to Ulam and von Neumann. Let $z(n)$ be given by

$$x(n) = \sin^2(\pi z(n)). \tag{8}$$

Substituting this into Eq. (7) $(c = 4)$, and using a little trigonometry, we obtain

$$\sin^2(\pi z(n+1)) = \sin^2(2\pi z(n)).$$

From this we conclude that $z(n+1)$ equals $2z(n) + $ (any integer). But any integer part of $z(n)$ doesn't change $x(n)$ in Eq. (8), and hence can be ignored. Thus we can write the dynamics of $z(n)$ in the form

$$z(n+1) = 2z(n) \bmod (1), \tag{9}$$

where mod (1) means that we discard any integer part of $z(n+1)$. If $z(0)$ is the initial value of z, the solution of Eq. (9) is simply

$$z(n) = z(0)\, 2^n \bmod (1), \tag{10}$$

and this can be substituted into Eq. (8) to give the general solution of Eq. (7) $(c = 4)$

$$x(n) = \sin^2(\pi z(0) 2^n). \tag{11}$$

Note again that this is an exact solution of an MM of chaos—an uncommon event.

Now the solution, particularly in the form (10), sheds a lot of light on the nature of this extreme chaos. To see this most clearly, assume that we represent the number $z(n)$ in binary notation, so

$$z = \sum_{k=1}^{\infty} a(k)\, 2^{-k} \tag{12}$$

where the $a(k) = (0 \text{ or } 1)$. Note that the sum starts at $k = 1$, since we only need to consider numbers less than 1, because of the mod (1) in Eq. (10). We can then represent $z(n)$ in binary notation as

$$.a(1)a(2)a(3)a(4)\ldots \qquad (a(k) = 0 \text{ or } 1). \tag{13}$$

If we multiply $z(n)$ by 2, as in Eq. (9), it can be seen from Eq. (12) that the new binary representation of $z(n+1)$ is

$$.a(2)a(3)a(4)\ldots.$$

In other words, the $a(k)$'s in Eq. (13) are simply shifted to the left one space, and the left-most $a(k)$ is discarded (due to the mod (1)).

So much for the beautiful, infinite world of the MM. Now consider a PE such that, at $n = 0$, we know $z(0)$ to some finite accuracy involving N terms $a(k)$, so

$$z(0) \sim .a(1)a(2)\ldots a(N)???????. \tag{14}$$

Here the ? indicates our ignorance about these higher terms. Then, after five steps (for example), we would have

$$z(5) \sim .a(6)a(7)\ldots a(N)?????,$$

and after N steps,

$$z(N) \sim .????????.$$

Hence, after N steps we have no way to predict the value of $z(N)$. Chaos has limited the duration of prediction to times proportional to the initial information we had about the system. One might say that we have "information output is proportional to information input" in chaotic systems. This is a basic lesson to be learned from chaos—mathematical determinism (such as the solution (11)) is totally distinct from the question of physical predictability (as Boltzmann sensed, and Poincaré pointed out)! It is not a solution of the MMs (like Eq. (11)) that is important; it is the behavior of a FAMILY of solutions (like Eq. (14)) that has physical significance. This is why the regions $R(t)$ in Figure 1(b) are so important.

We can also see that the above dynamics are as random as a coin-toss dynamics, in the sense that we can find solutions of Eq. (9) that can be related to ANY infinite sequence of (Heads,Tails). Note that now we are back in the arena of mathematics—we cannot accomplish this in a PE for the reasons just discussed. Nonetheless, it is interesting to see how this correspondence can be made. To do this, note that Eq. (9) can be represented by a "Bernoulli map"

$$z(n+1) = \begin{cases} 2z(n) & (\text{if } 0 \leq z(n) < \frac{1}{2}), \\ 2(z(n) - \frac{1}{2}) & (\text{if } \frac{1}{2} \leq z(n) < 1), \end{cases} \tag{15}$$

where, given $z(n)$, we can represent $z(n+1)$ by the graphical method shown in Figure 9(a). In this method, we plot $z(n+1)$ vs. $z(n)$ for the function (15), and then draw a 45° line along which $z(n+1) = z(n)$. A few minutes of thought will show that the map dynamics, given by Eq. (15), can be constructed in a manner illustrated in Figure 9(a).

Now divide the z-axis into two regions as shown in Figure 9(b),

$$\text{H (Heads) if } z < 1/2, \qquad \text{T (Tails) if } z > 1/2.$$

If $z(0)$ is in the region H, $z(1)$ will also be in H if $z(0) < 1/4$; otherwise, $z(1)$ will be in T. Therefore, we have found two regions, such that if $z(0)$ is in these regions, we can associate the sequence HH or HT to them. The same can be done for sequences TT and TH. Specifically we have the association

$$\begin{aligned} &\text{HH if } \quad 0 < z(0) < 1/4; \\ &\text{HT if } 1/4 < z(0) < 1/2; \\ &\text{TH if } 1/2 < z(0) < 3/4; \\ &\text{TT if } 3/4 < z(0) < 1. \end{aligned}$$

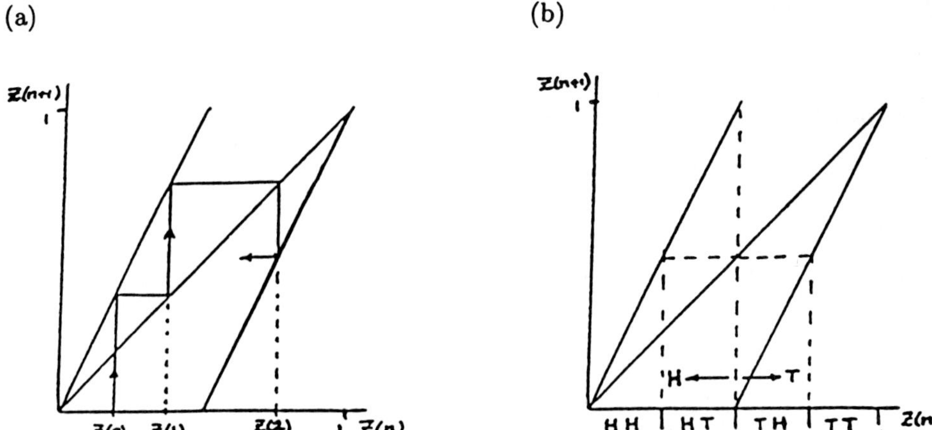

FIGURE 9 (a) A graphical method for representing $z(n+1)$ for a Bernoulli map. (b) Same as (a), but the z-axis has been divided into (H, T) regions.

These regions are shown in Figure 9(b). We can proceed in this fashion to find regions in which $z(0)$ will yield any finite sequence of (H,T). For example, HTH is produced by $1/4 < z(0) < 3/8$, and so on. Thus the chaotic dynamic system (9) has solutions that are just as diverse in character as those obtained from a fair coin toss! That's pretty random! This is a symbolic dynamic way of representing chaos (by showing that the dynamics can be put into correspondence with any so-called "Bernoulli sequence"). The first person to introduce this representation of chaos for physical systems was Levinson,[35] and it subsequently led to the introduction of "horseshoe" maps by Smale[51] (EAJ: Appendix K).

A quantitative method of measuring how nearby solutions diverge from one another (on the average) is given by Lyapunov exponents. Say we have two solutions of some map, $x(n+1) = F(x(n))$, that are separated by a small distance DX about the point $x(0)$. As we see from Figure 10, the separation after one step is given by

$$DX(1) = \left(\frac{dF}{dx}\right)_{x(0)} DX(0). \tag{16}$$

After the next step, we similarly obtain

$$DX(2) = \left(\frac{dF}{dx}\right)_{x(1)} DX(1) = \left(\frac{dF}{dx}\right)_{x(1)} \left(\frac{dF}{dx}\right)_{x(0)} DX(0).$$

After n steps, the separation distance is

$$DX(n) = \left(\frac{dF}{dx}\right)_{x(n-1)} \cdots \left(\frac{dF}{dx}\right)_{x(0)} dDX(0). \tag{17}$$

Now if we assume that this separation distance is increasing exponentially as n increases, then

$$|DX(n)| = |DX(0)| \exp(n\lambda).$$

If we take the logarithm of this, we obtain

$$\lambda = \left(\frac{1}{n}\right) \log \left[\frac{|DX(n)|}{|DX(0)|}\right],$$

and, if we substitute Eq. (17), we obtain the final result

$$\lambda = \left(\frac{1}{n}\right) \sum_{k=0}^{n-1} \log \left| \left(\frac{dF}{dx}\right)_{x(k)} \right|. \tag{18}$$

λ is therefore the average separation rate (per iteration), over n iterations. If one takes the limit of infinite n, this yields the so-called Lyapunov exponent. In all CEs we can only obtain an approximation of this exponent (limited to a finite number of steps), and usually obtained by another approximate method (EAJ; 7.10). In the case of PEs yet other methods must be used to obtain an estimate of the Lyapunov exponents (see Eckmann et al.[15]).

If $\lambda > 0$, nearby orbits rapidly separate from one another (on average). However, if their motion is confined to a bounded region of phase space, then there must be "converging" periods of their motion (for limited times). When these two ingredients are put together, we have chaotic behavior. Thus, in the case of Eq. (15), the slope at every iterated point has magnitude 2. Hence, from Eq. (18), we obtain the Lyapunov exponent $\lambda = \log 2 > 0$, and chaos. Similarly one can see that Eq. (15) is the same dynamics as Eq. (9), with the same chaotic Lyapunov measure. On the other hand, the Lyapunov exponent in the dynamics of Eq. (7), whose solution is Eq. (11), is different (does it depend on $x(0)$?). I do not have time to discuss the fractal dimensions of strange attractors, but an introduction can be found in (EAJ: 2.6). The relationship between these fractal dimensions and the Lyapunov exponents is discussed in (EAJ: 7.12).

This chaos can be usefully visualized by various forms of "stretch-fold-squeeze" dynamics in phase space, as schematically illustrated in Figure 11. This shows a region in phase space going through such a process. If solutions are attracted toward such a region, they continue to "mix," yielding a strange attractor. For other examples see (EAJ; 4.5) (logistic map) and (EAK; 7.11) (Rössler attractors). If more than one Lyapunov exponent is positive, the dynamics are called "hyperchaos." The knowledge of negative Lyapunov-exponent dynamics, and converging regions in the phase space, can be put to good use in influencing/controlling such complex motion (as I'll discuss shortly).

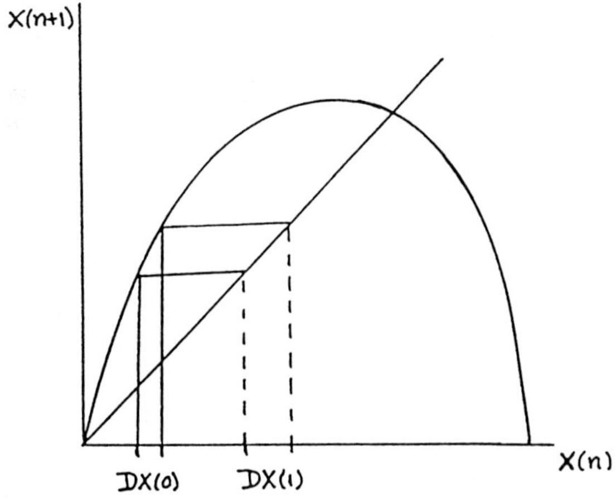

FIGURE 10 The separation of nearby solutions, given by Eq. (16). The average value of this separation rate is measured by the Lyapunov exponent, Eq. (18).

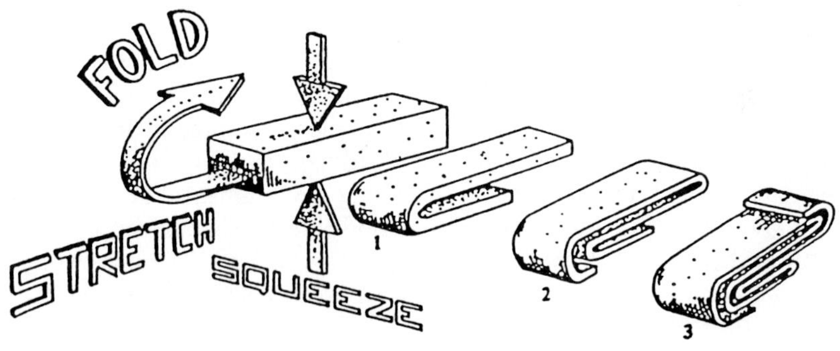

FIGURE 11 Chaotic dynamics visualized as a result of "stretch-fold-squeeze" dynamics in phase space.

Let me make one final comment on chaos as predicted by MMs, particularly when the systems are not conservative. Not all MM chaos is observable, either in a PE or a CE. A famous example is the mathematical result for some one-dimensional maps (e.g., the logistic map), which proves that if the system has a period 3 solution, then it has solutions with any period. This is known as the "period 3 implies chaos" theorem (EAJ; 4.5). However, when one observes period 3 solutions of the logistic map, there is no observable chaos; all observable solutions tend to the attracting period 3 solution. This is another example where we have to be aware of the source of our information. MMs do not necessarily give the same information as PEs or CEs. Moreover, there are many situations (e.g., in the Lorenz system and others), where all solutions may tend to a nonchaotic behavior, but they have very long periods of transient chaos from the point of view of a PE or a CE. Such "transient" chaos may be just as important, because it may persist over the entire observational period.

Let me close with some brief comments concerning a number of other topics which I do not have time to discuss in any detail, but can give you some references.

- One very interesting area concerns spatio-temporal chaos. This occurs in many physical systems, but perhaps most PEs have been done in systems with Rayleigh-Bénard (thermally driven) "turbulence" or Taylor (momentum-driven) "turbulence." Many beautiful experiments have been done by Swinney and Gollub,[52] and other talented people. See the references for some surveys of spatio-temporal chaos.[17,21,60,61,56,57] Some MM exploration of these phenomena have been attempted with coupled cellular dynamics, and more abstractly, using coupled map dynamics. The former research attempts to maintain some direct connection with physical principles, while the latter is more abstract. What is badly needed in this area are some good quantitative measures that relate to physical concepts.
- There are many suggestions that chaos may be an important positive aspect of the dynamics of living systems. One idea is that chaos may be an effective way for a system to explore sensory inputs from the environment. In other words, chaos may be part of a search/response mechanism in systems, a way for a system to most readily acclimate to changing environmental situations. Thus, for example, the heartbeat, while quite periodic, has a chaotic component to its rhythm. Is this so that it can better respond to stresses? Neural networks have highly complex EEG patterns. Is this similarly caused?[23,50] The eye constantly makes erratic "searching" motions, obviously for some beneficial purpose. Not all chaos is beneficial, of course. Fluttering airplane wings are not reassuring, nor are fibrillations of the heart. Just as chaos (TC) can give rise to order, order can give rise to DC. If a system is in a DC state, it may be easier for it to "internally recognize" changes in its dynamics caused by environmental changes, and to make adjustments which "improve" its well-being, performance, or whatever—exactly how, generally speaking, is anybody's guess. Said differently, if a system is in a DC state, its dynamic change may be more easily internally decipherable for purposes of adjustment. One idea along these lines has been suggested by W. J. Freeman[21] (also see Conrad[10]).
- An important aspect of DC is that, since we know something about its origins, we can use this knowledge to influence it in various fashions. Several methods have been proposed. The first direct attack on chaos was done by Hübler in 1989,[27] and this idea has been generalized and refined in several respects since then.[46] Many of these ideas (and others) are discussed by Hübler in his extended article.[28] Another approach, which draws a chaotic system to one of its unstable

periodic orbits, has been used extensively by Ott, Grebogi, and Yorke,[43] and implemented experimentally by Ditto et al.[13] and others. Other applications have focused on the permanent transferral of a chaotic system to another stable attractor (this can, of course, only be done if the system has multiple attractors, as is frequently the case). Thus, knowledge of the features of DC can be used in a variety of important applications in the future. Many of these applications do not depend on explicit MMs, but can be accomplished by the combined use of PEs and CEs (e.g., see Breeden[5] or Hübler[28]). One of the main points to note is that DC is not uniformly chaotic in the phase space, and this can be used to many advantages.[31,32,33]

- The subject of quantum chaos, even its very existence, is not one where there is general agreement. The eigenfunction/eigenvalue formalism of bounded systems, with their almost-periodic structure, does not seem to offer the dynamic freedom required of chaotic motion.[20] On the other hand, it does not appear that anybody has done a general, finite-time analysis of localized particle separations, which are the hallmark of classical chaos. I suspect that the mathematical formalism is again being confronted with physical observations in which the encoding and decoding is more sophisticated than presently recognized. We'll have to wait and see!
- A very active area of research, to which Santa Fe is no stranger, concerns extracting predictable components from chaotic dynamics.[5,39] The "components" may involve very time limited and very time specific predictions, which one tries to differentiate from the total chaos (TC) discussed at the beginning. Such predictions necessarily involve bounded and selective aspects, some of which can be gleaned from the references. Typically it takes a considerable amount of data to distill out some deterministic component, and if the system is not stationary during this collection period, then all bets are off. It's all very exciting, and possibly quite profitable when applied to financial markets (we hope so for our friends!). There have been careful studies searching for DC in economics, but the search may require considerable "filtering." Modifying my opening remarks, I should probably say "yesterday's TC may have at least some small component of DC."

And so we have come full circle! Hopefully I have conveyed the message that distinct forms of "chaos" arise in each of the operational methods of science (PE, CE, and MM). The clarion call of chaos is that it is necessary to evaluate carefully how these methods can be joined in a scientific method of the future, in order to understand complex systems. This is your great challenge and excitement! Thank you for your attention.

REFERENCES

1. Anderson, P. W. "More is Different." *Science* **177** (1972): 393.
2. Anderson, P. W. "Is Complexity Physics? Is It Science? What Is It?" *Physics Today* **July** (1991): 9–10.
3. Berry, M. V. "Quantum Chaology." *Proc. Roy. Soc. Lond. A* **413** (1987): 183–198.
4. Berry, M. V., I. C. Percival, and N. O. Weiss, eds. *Dynamical Chaos.* Princeton: Princeton University Press, 1987.
5. Breeden, J. "Optimal Representation of Experimental Data." Ph.D. Thesis, Department of Physics, University of Illinois at Urbana-Champaign, 1992.
6. Brillouin, L. *Scientific Uncertainty and Information*, chapters 9 and 10. New York: Academic Press, 1964.
7. Cartwright, M. L., and J. E. Littlewood. "On Nonlinear Differential Equations of the Second Order. I. The Equation $\ddot{y} + k(1 - y^2)\dot{y} + y = b\lambda k \cos(\lambda t + a), k$ large." *J. Lond. Math. Soc.* **20** (1945): 180–189.
8. Chirikov, B. V. "Patterns in Chaos." *Chaos, Solitons & Fractals* **1** (1991): 79–103.
9. Chua, L. O., M. Komuro, and T. Matsumoto. "The Double Scroll Family," parts I and II. *IEEE Trans. Circuits Sys.* **CAS-33** (1986): 1073–1117.
10. Conrad, M. "What is the Use of Chaos?" In *Chaos*, edited by A. V. Holden, 1–14. Princeton: Princeton University Press, 1986.
11. Cvitanovic, P. ed. *Universality in Chaos*, 2nd edition. New York: Adam Hilger, 1989.
12. Deutsch, D. "On Wheeler's Notion of 'Law without Law' in Physics." *Found. Phys.* **16(6)** (1986): 583–590.
13. Ditto, W. L., S. N. Rauseo, and M. L. Spano. "Experimental Control of Chaos." *Phys. Rev. Lett.* **65** (1990): 3211–3214.
14. Duheim, P. M. M. *La Theorie Physique, Son Objet—Sa Structure*, 206. Paris: Marcel Riviere, 1914.
15. Eckmann, J-P., S. O. Kamphorst, D. Ruelle, and S. Ciliberto. "Liapunov Exponents from Time Series." *Phys. Rev.* **Q34** (1986): 4971–4979.
16. Farmer, J. D., and J. J. Sidorowich. "Predicting Chaotic Time Series." *Phys. Rev. Lett.* **59** (1987): 845–848.
17. Fenstermacher, P. R., H. L. Swinney, and J. P. Gollub. "Dynamical Instabilities and the Transition to Chaotic Taylor Vortex Flow." *J. Fluid Mech.* **94** (1979): 103–128.
18. Ford, J. "How Random is a Coin Toss?" *Phys. Today* **36(4)** (1983): 40–47.
19. Ford, J. "What is Chaos, that We Should Be Mindful of It?" In *The New Physics*, edited by P. Davis, 348–372. Cambridge: Cambridge University Press, 1989.

20. Ford, J., G. Mantica, and G. H. Ristow. "The Arnold Cat: Failure of the Correspondence Principle." In *Chaos/Xaoc: Soviet-American Perspective of Nonlinear Science*, edited by D. K.Campbell. New York: American Institute of Physics, 1990.
21. Freeman, W. J. "Simulation of Chaotic EEG Patterns with a Dynamic Model of the Olfactory System." *Biol. Cyber.* **56** (1987): 139–150.
22. Green, J. M. "A Method of Determining a Stochastic Transition." *J. Math. Phys.* **20** (1979): 1183–1201.
23. Guevara, M. R., L. Glass, M. C. Mackey, and A. Shrier. "Chaos in Neurobiology." *IEEE Trans. Sys., Man, & Cyber.* **SMC-13:5** (1983): 790–798.
24. Gutzwiller, M. C. *Chaos in Classical and Quantum Mechanics.* Berlin: Springer-Verlag, 1990.
25. Hassell, M. P., H. N. Comins, and R. M. May. "Spatial Structure and Chaos in Insect Population Dynamics." *Nature* **353** (1991): 255–258. Also, commentary: A.R. Ives, ibid., 214–215.
26. Holden, A. V. *Chaos.* Princeton: Princeton University Press, 1986.
27. Hübler, A., and E. Lüscher. "Resonant Stimulation and Control of Nonlinear Oscillators." *Naturwissenschaften* **76** (1989): 67.
28. Hübler, A. "Modeling and Control of Complex Systems: Paradigms and Applications." In *Modeling Complex Phenomena*, edited by L. Lam and V. Naroditsky, 5–65. Berlin: Springer-Verlag, 1992.
29. Jackson, E. A. *Equilibrium Statistical Mechanics*, 86–87. Englewood Cliffs, NJ: Prentice-Hall, 1968.
30. Jackson, E. A., and A. Hübler. "Periodic Entrainment of Chaotic Logistic Map Dynamics." *Physica D* **44** (1990): 404–420.
31. Jackson, E. A. *Perspectives of Nonlinear Dynamics*, vols. 1,2. Cambridge: Cambridge University Press, 1991.
32. Jackson, E. A. "Controls of Dynamic Flows with Attractors." *Phys. Rev. A* **44** (1991): 4839–4853.
33. Jackson, E. A., and A. Kodogeorgiou. "Entrainment and Migration Controls of Two-Dimensional Maps." *Physica D* **54** (1992): 253–265.
34. Kac, M. *Probability and Related Topics in Physical Sciences.* New York: Wiley Interscience, 1959.
35. Levinson, N. "A Second-Order Differential Equation with Singular Solutions." *Ann. Math.* **50** (1949): 123–153.
36. Lighthill, J. "The Recently Recognized Failure of Predictability in Newtonian Dynamics." *Proc. Roy. Soc. Lond. A* **407** (1986): 35–50.
37. May, R. M. "Simple Mathematical Models with Very Complicated Dynamics." *Nature* **261** (1976): 459–467
38. Meyer, T. P., F. C. Richards, and N. H. Packard. "A Learning Algorithm for the Analysis of Complex Spatial Data." *Phys. Rev. Lett.* **63** (1989): 1735–1738.

39. Meyer, T. P. "Long-Range Predictability of High-Dimensional Chaotic Dynamics." Ph.D. Thesis, Department of Physics, University of Illinois at Urbana-Champaign, 1991.
40. Moon, F. C. *Chaotic Vibrations.* New York: John Wiley, 1987.
41. Moon, F. C. *Chaotic and Fractal Dynamics.* New York: John Wiley, 1992.
42. Orszag, S., and L. Sirovich, eds. *New Perspectives in Turbulence.* New York: Springer-Verlag, 1991.
43. Ott, E., C. Grebogi, and J. A. Yorke. "Controlling Chaos." *Phys. Rev. Lett.* **64** (1990): 1196–1199.
44. Pecora, L. M., and T. L. Carroll. "Driving Systems with Chaotic Signals." *Phys. Rev. A* **44** (1991): 2374–2383.
45. Peitgen, H.-O., H. Jürgens, and D. Saupe. *Chaos and Fractals; New Frontiers of Science.* Berlin: Springer-Verlag, 1992.
46. Plapp, B. B., and A. W. Hübler. "Nonlinear Resonances and Suppression of Chaos in the rf-Biased Josephson Junction." *Phys. Rev. Lett.* **65** (1990): 2302–2305.
47. Prigogine, I., and I. Strengers. *Order Out of Chaos.* New York: Bantam Books, 1984.
48. Ruelle, D. *Chance and Chaos.* Princeton: Princeton University Press, 1991.
49. Schroeder, M. *Fractals, Chaos, Power Laws.* San Francisco: W. H. Freeman, 1991.
50. Skarda, C. A., and W. J. Freeman. "How Brains Make Chaos in Order to Make Sense of the World." *Behav. & Brain Sci.* **10(2)** (1987): 161–195.
51. Smale, S. "Diffeomorphisms with Many Periodic Points." In *Differential and Combinatorial Topology*, edited by S. S. Cairns, 63–80. Princeton: Princeton University Press, 1963.
52. Swinney, H. L., and J. P. Gollub, eds. *Hydrodynamic Instabilities and the Transition to Turbulence.* Berlin: Springer-Verlag, 1985.
53. Thompson, J. M. T., and H. B. Stewart. *Nonlinear Dynamics and Chaos.* Chichester, UK: John Wiley, 1988.
54. Tolman, R. L. *The Principles of Statistical Mechanics*, sections 45–49. Oxford: Oxford University Press, 1962.
55. Tsuda, I. "Dynamic Link of Memory—Chaotic Memory Map in Nonequilibrium Neural Networks." *Neur. Nets.* **5** (1992): 313–326.
56. Tsuda, I., T. Tahara, and H. Iwanaga. "Chaotic Pulsation in Human Capillary Vessels and Its Dependence on Mental and Physical Conditions." *Intl. J. Bif. & Chaos* **2** (1992): 313–324.
57. Vorhra, S., M. Spano, M. Shlesinger, L. Pecora, and W. Ditto, eds. *Proceedings of the First Experimental Chaos Conference.* Singapore: World Scientific, 1992.
58. Wheeler, J. A. "On Recognizing 'Law Without Law,'" *Amer. J. Phys.* **51(5)** (1983): 398.
59. Wisdom, J. "Chaotic Behavior in the Solar System." *Proc. Roy. Soc. Lond. A* **413** (1987): 109–129.

David H. Wolpert
Santa Fe Institute, 1660 Old Pecos Trail, Suite A, Santa Fe, NM, 87501, USA;
e-mail: dhw@santafe.edu

Combining Generalizers by Using Partitions of the Learning Set

For any real-world generalization problem, there are always many generalizers that could be applied to the problem. This chapter discusses some algorithmic techniques for dealing with this multiplicity of possible generalizers. All of these techniques rely on partitioning the provided learning set in two, many different times. The first technique discussed is cross validation, which is a winner-takes-all strategy (based on the behavior of the generalizers on the partitions of the learning set, it picks one single generalizer from the set of candidate generalizers and tells you to use that generalizer). The second technique discussed, the one this chapter concentrates on, is an extension of cross validation called stacked generalization. As opposed to cross validation's winner-takes-all strategy, "stacked generalization" uses the partitions of the learning set to combine the generalizers, in a nonlinear manner, via another generalizer (hence the term "stacked generalization"). This chapter ends by discussing some possible extensions of stacked generalization.

1. INTRODUCTION

This chapter concerns the problem of inferring a function f from a subset of $\mathbf{R}^n$ to a subset of $\mathbf{R}^p$ (the *target* function), given a set of m samples of that function (the *learning set*). The subset of $\mathbf{R}^n$ is the *input space*, labeled X, and the subset of $\mathbf{R}^p$ is the *output space*, labeled Y. A *question* is an input space (vector) value. A generalizer is an algorithm that guesses what the target function is, and bases that guess only on a learning set of m $\mathbf{R}^{n+p}$ vectors read off of that target function. It guesses an appropriate output for a question via the target function that it infers from the learning set. Colloquially, we say that the generalizer is "trained," or "taught," with the learning set and then "asked" a question.

For any real-world generalization problem, there are always many possible generalizers. Accordingly, one is *always* implicitly presented with the problem of how to address the multiplicity of possible generalizers. One possible strategy is to simply choose a single generalizer according to subjective criteria. As an alternative, this chapter discusses the objective (i.e., algorithmic) technique of "stacking." However, to provide some context and nomenclature, we first present a cursory discussion of the cross-validation procedure, the traditional method for objectively addressing the multiplicity of possible generalizers.

2. CROSS VALIDATION

Perhaps the most commonly used "objective technique for addressing the multiplicity of generalizers" is cross validation.[4,8,9,10,13] It works as follows:

Let $L = \{(x_i, y_i)\}$ be the learning set of m input-output pairs. The (leave-one-out) *cross-validation partition set* (CVPS) is a set of m partitions of L. It is written as $\{L_{ij}\}, 1 \le i \le m, 1 \le j \le 2$. For fixed i, L_{i2} consists of a single input-output pair from L, and L_{i1} consists of the rest of the pairs from L. The input component of L_{i2} is written as $in(L_{i2})$, and the output component is written as $out(L_{i2})$. Varying i varies which input-output pair constitutes L_{i2}; since there are m pairs in L, there are m values of i.

We have a set of generalizers $\{G_j\}$. Indicate by $G_j(L'; q)$ the guess of generalizer G_j when trained on the learning set L' and when asked the question q. For each generalizer G_j, use the CVPS to compute the (leave-one-out) *cross-validation error*, $\sum_{i=1}^{m} [G_j(L_{i1}; in(L_{i2})) - out(L_{i2})]^2/m$. This is the average (squared) error of G_j for guessing one pair of $L\,(L_{i2})$ when trained on the rest of $L\,(L_{i1})$. If we interpret the cross-validation error for G_j as an estimate of the generalization error of G_j when trained on all of L, then the technique of cross validation provides a winner-takes-all rule: choose the generalizer that has the lowest cross-validation error on the learning set at hand, and use that generalizer to generalize from the entire learning set.

There are other partition sets besides the leave-one-out cross-validation partition set. Two of the most important are the J-fold cross validation partition set and the bootstrap partition set. In J-fold cross validation, i ranges only from 1 to J. For all i, L_{i2} consists of m/J input-output pairs from L, and L_{i1} consists of the rest of L. The input-output pair indices comprising L_{i2} are disjoint from those comprising L_{j2} for $i \neq j$ (so, if no input-output pair is duplicated in L, $L_{i2} \cap L_{j2} = \phi$ for $i \neq j$). Accordingly, the set of all L_{i2} covers L, assuming J is a factor of m. Using this partition set, one computes the cross-validation error exactly as in leave-one-out cross validation, and then chooses the generalizer with the lowest error. Since it only requires the training of a generalizer J times (rather than m times, as in leave-one-out cross validation), J-fold cross validation is less computationally expensive than leave-one-out cross validation. Sometimes it is also a more accurate estimator of generalization accuracy.[2]

As another alternative to leave-one-out cross validation, one can use a *bootstrap* partition set.[4] Such a partition set is found by stochastically creating the L_{i1}: each L_{i1} is formed by sampling L, m times, in an i.i.d. manner (according to a uniform distribution over the elements of L). Other variations of these basic ideas exist, e.g., generalized cross validation,[6] stratification, etc. For a discussion of (some) such variations, see Weiss-Kulikowski.[11]

Although it is not based on subjective judgments for addressing the multiplicity of generalizers, cross validation (and its variations) would be pointless if it didn't work well in the real world. Fortunately, it does work, usually quite well. For example, see Wolpert[13] for an investigation in which cross-validation error almost perfectly correlates with generalization error for the NETtalk data set.

3. STACKED GENERALIZATION

Cross validation is a winner-takes-all strategy. As such, it is rather simpleminded; one would prefer to *combine* the generalizers rather than choose just one of them. Interestingly, this goal of combining generalizers can be achieved by exploiting the partition sets employed by cross validation and its variations. Consider Figure 1. We have a learning set L and a set of two candidate generalizers, G_1 and G_2. We want to infer (!) an answer to the following question: if G_1 guesses g_1 and G_2 guesses g_2, what is the correct guess?

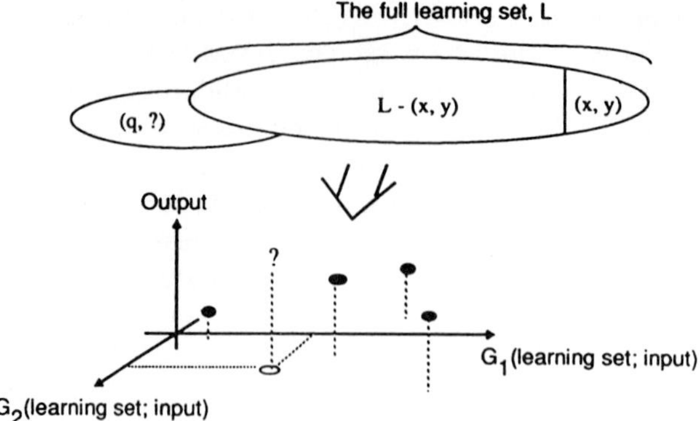

FIGURE 1 A stylized depiction of how to combine the two generalizers G_1 and G_2 via stacked generalization. A learning set L is symbolically depicted by the full ellipse. We want to guess what output corresponds to the question q. To do this we create a CVPS of L; one of these partitions is shown, splitting L into $\{(x, y)\}$ and $\{L - (x, y)\}$. By training both G_1 and G_2 on $\{L - (x, y)\}$, asking them the question x, and then comparing their guesses to the correct guess y, we construct a single input-output pair (indicated by one of the small solid ellipses) of a new learning set L'. This input-output pair gives us information about how to go from guesses made by the two generalizers to a correct output. The remaining partitions of L give us more of such information; they give us the remaining elements of L'. We now train a generalizer on L' and ask it the two-dimensional question $\{G_1(L; q), G_2(L; q)\}$. The answer is our final guess for what output corresponds to q.

To answer this question we make the same basic assumption underlying cross validation: generalizing behavior when trained with proper subsets of the full learning set correlates with generalizing behavior when trained with the full learning set. To exploit this assumption, the first thing one must do is choose a partition set L_{ij} of the full learning set L. For convenience, choose the CVPS.[1] Now pick any partition i from L_{ij}. Train both G_1 and G_2 on L_{i1} and ask them both the question $in(L_{i2})$. They will make the pair of guesses g_1 and g_2. In general, since the generalizers were not trained with the input-output pair L_{i2}, neither g_1 nor g_2 will equal the correct output, $out(L_{i2})$. Therefore we have just learned something: when G_1 guesses g_1 and G_2 guesses g_2, the correct guess is $out(L_{i2})$.

From such information we want to infer what the correct guess is when both G_1 and G_2 are trained on the full learning set and are asked a question q. The most natural way to carry out such inference is via a generalizer. To do this, first cast the information gleaned from the partition i as an input-output pair in a new space

[1]In general we will want to pick a partition set in which L_{i1} is a proper subset of L, lest we "generalize how to learn" rather than "generalize how to generalize" (see Wolpert[16]). Indeed, one could make a case for the opposite extreme, in which one picks the partition set so that as few as possible of the actual input-output pairs in L_{i2}'s are also found in L_{i1}'s.

(the new space's input being the guesses of G_1 and G_2, and the new space's output being the correct guess). Repeat this procedure for all partitions in the partition set. Different partitions gives us different input-output pairs in the new input-output space; collect all these input-output pairs and view them as a learning set in the new space.

This new learning set tells us all we can infer (using the partition set at hand) about the relationship between the guesses of G_1 and G_2 and the correct output. Now we can use this new learning set to "generalize how to generalize"; we train a generalizer on this new learning set and ask it the two-dimensional question $\{G_1(L;q), G_2(L;q)\}$. The resulting guess serves as our final guess for what output corresponds to the question q, given the learning set L. Using this procedure, we have inferred the biases of the generalizers with respect to the provided learning set (loosely speaking), and then collectively corrected for those biases to get a final guess.

Procedures of this sort where one feeds generalizers with information from other generalizers are known as "stacked generalization."[1,15] The original learning set, question, and generalizers are known as the "level 0" learning set, question, and generalizers. The new learning set, new question, and generalizer used for this new learning set and the new question are known as the "level 1" learning set, question, and generalizer.

Some important aspects of how best to use stacked generalization become apparent when the learning set is large and the output space is discrete, consisting of only a few (k) values. Assume that we are using the architecture of Figure 1 to combine three generalizers, using a bootstrap partition set. Assume further that all $k^3 \times k$ possible combinations of a level 1 input and a level 1 output can occur quite often (i.e., assume the learning set consists of many more than k^4 elements).

View the level 1 learning set as a histogram of (number of occurrences of the various possible) level 1 inputs and associated outputs. Assuming the statistics within the level 0 learning set mirrors the statistics over the whole space, this histogram should approximate well the true joint probability distribution Pr(output, guesses of the three generalizers). (In particular, since the learning set is large, finite sample effects should be small.) Therefore, if we guess by using that histogram, we will be guessing according to (a good approximation of) the true distribution Pr(output | guesses of the three generalizers). In general, such guessing cannot give worse behavior than guessing the value of any single one of the three generalizers. (Leo Breiman[3] has proven a more formal version of this statement.) Therefore, one should expect that for large learning sets, the error of stacking with a "histogram" level 1 generalizer is bounded below by the error of any winner-takes-all technique like cross validation.

Even when the learning set is large enough that we can ignore finite sample effects, so that the statistics inside L mirrors the statistics across the whole space etc., it might still be that stacking with a histogram level 1 generalizer will not do better than using one of the level 0 generalizers by itself. For example, this is the case if the following condition always holds: no matter what the guesses by the three level 0 generalizers, the best guess (i.e., the made output value of the histogram for the given level 1 input) is always equal to the guess of the same one of those three level 0 generalizers. For such a scenario, the stacking is simply telling you to always use that guess.

A somewhat more illuminating example arises when the three generalizers not only have the same cross-validation error, but actually make the same guesses when presented with the same L_{i1} and L_{i2}. When the generalizers are synchronized this way (synchronized as far as L is concerned), combining them gains nothing; one might as well run the stacking using only a single one of the generalizers, as described by Wolpert.[15] (Intuitively, if the generalizers behave identically as far as (partitions of) the data are concerned, then combining them gains nothing.) This example suggests that when combining generalizers one should find generalizers that behave very differently from one another, which are in some sense "orthogonal," so that their guesses are not synchronized. Indeed, if for a particular learning set the guesses are synchronized even for generalizers that are usually considered to be quite different from one another, the suspicion arises that this is a data-limited situation; in a sense, there is nothing more to be milked from the learning set.

For these kinds of reasons, it might be that best results arise if the $\{G_j\}$ are not very sensible as stand-alone generalizers, so long as they exhibit different generalization behavior from one another. (After all, in a very loose sense, the $\{G_j\}$ are being used as extractors of high-level, nonlinear "features." The optimal extractors might not make much sense as stand-alone generalizers.) As a simple example, one might want one of the $\{G_j\}$ to be a shallow decision-tree generalizer and another one a generalizer that makes very deep decision trees. Neither generalizer is particularly reasonable considered by itself (intermediate depth trees are usually best), but they might operate quite well when used cooperatively in a stacked architecture.

A good deal of evidence supports stacking in several scenarios. It appears to systematically improve upon both ridge and subset regressions.[1] Moreover, after learning about stacked generalization partitions, Zhang et al. have used it to create the current champion at protein folding.[17] See also Gustafson[5] and Wolpert.[15]

In addition, it should be possible to use stacked generalization in combination with other schemes designed to augment generalizers. For example, one might use stacking to improve the "boosting" procedure developed recently in the COLT community.[7] The idea would be to train several versions of the same generalizer, exactly as in boosting. However, rather than training them with input-output examples chosen from all L as in conventional boosting, one trains each of them with examples chosen from one part of a partition set pair (i.e., from an L_{i1}). One then uses the other part of the partition set pair (L_{i2}) to see how to combine the generalizers (rather than just using a fixed majority rule, as in standard boosting). This might improve performance more than boosting used by itself. It also naturally extends boosting to situations with nonbinary (and even continuous-valued) outputs, in which a simple majority rule makes little sense.

4. VARIATIONS OF STACKED GENERALIZATION

There are many variations of the basic version of stacked generalization outlined above. (See Wolpert[15] for a detailed discussion of some of them.) One variation is to have the level 1 input space contain information other than the outputs of the level 0 generalizers. For example, if one suspects a strong correlation between {the guesses of the level 0 generalizers, together with the level 0 question} and {the correct output}, then one might add a dimension to the level 1 input space (several dimensions for a multidimensional level 0 input space), for the value of the level 0 question.

Another useful variation is for the level 1 output space to be an estimate for the error of the guess of one generalizer rather than a direct estimate for the correct guess. In this version of stacked generalization the level 1 learning set has its outputs set to the error of one generalizer rather than to the correct output. When the level 1 generalizer is trained on this learning set and makes its guess, that guess is interpreted as an error estimate; that guess is subtracted from the guess of the appropriate level 0 generalizer to get the final guess for the output that corresponds to the question.

There are some particularly nice features of this error-estimating version of stacked generalization: (1) Rather than use it to (try to) improve a guess, one can use it simply to get a confidence estimate for that guess. (2) This estimate can be multiplied by a real-valued constant before being subtracted from the appropriate level 0 generalizer's guess (when one is trying to improve that guess). When this constant is 0, the guessing of the entire system reduces to simply the use of the level 0 generalizer by itself. As the constant grows, the guessing of the entire system becomes less and less like the guess of that level 0 generalizer by itself, and more and more like the guess of a full stacked generalization architecture; that constant provides us with a knob determining how conservative we wish to be in our use of stacked generalization. (3) With such an architecture the guess of that level 0 generalizer often no longer needs to be in the level 1 input space, since its information is already incorporated automatically into the final guess (when one does the subtraction). In this way one can reduce the dimensionality of the level 1 input space by one.

Another variation of stacked generalization is suggested by the distinction between "strong" cross validation and "weak" cross validation.[12,16] The conventional form of cross validation discussed so far is "weak" cross validation. Using it to judge amongst generalizers is equivalent to saying, "Given L, I will pick the G_j that best guesses one part of L when trained on another part of it." In strong cross validation, one instead says, "Given a target function f, I will pick the G_j that best guesses one part of f when trained on (samples from) another part of it." Intuitively, with strong cross validation we are saying that we don't want to rely too much on the learning set at hand, but rather want to concern ourselves with behavior related to the target function from which the learning was sampled (presumably randomly).

We want to take into account what would have happened if we had had a different learning set chosen from the same target function.

There are a number of ways to use strong cross validation in practice. One entails using decision-directed learning to create guesses for f, and then measuring strong cross validation over those guessed f. This procedure starts by training all the generalizers on the entire learning set L. Let the resultant guesses for the input-output function be written as $\{h_j\}$ (the index of a generalizer G_j and of its guess h_j have the same value). For each h_j, (1) randomly sample h_j according to the distribution $\pi(x \epsilon X)$ to create a "learning set" and then a separate "testing set"; (2) train the corresponding generalizer G_j on that learning set; and then (3) see how well the trained G_j predicts the elements of the testing set. For all j one does this many times, and tallies the average error. The j that gives the smallest average error for this procedure is the one picked (i.e., one generalizes from L with h_j, where j is the index giving the smallest average error). Other variations involve observing the behavior of generalizers G_j when they are trained on learning sets constructed from function $h_{i \neq j}$. Note that the whole procedure then can be iterated: one uses the original L to create h's which are used to create new L's, then use those new L's to create new h's, and so on.

Strong cross validation makes the most sense if one is in a noise-free scenario. It also does not view cross validation (directly) in terms of generalization error estimation and makes most sense how to use L to perform such estimation. Instead the idea is to view it as an *a priori* reasonable criterion for choosing amongst a set of $\{G_j\}$: choose the G_j that is, loosely speaking, most self-consistent with respect to the learning set. In other words, take the generalizers at their word. If a generalizer guesses h_j, one calls its bluff, and then sees what the ramifications are, what kind of cross-validation errors would have arisen if the target function were indeed h_j and one sampled it to get a different learning set from L.

Such a viewpoint notwithstanding, when using strong cross validation in practice, it often makes sense to bias the distribution $\pi(x)$ used for sampling the h_j—perhaps strongly—in favor of the points in the original learning set. In the extreme, when the (level 0 input space) sampling only runs over the input values in the original L, strong cross validation essentially reduces to the bootstrap procedure (assuming the sampling is done without any noise, and that each G_j acts appropriately and perfectly reproduces the elements of the learning set on which it is trained).

One might use the idea behind strong cross validation to construct a kind of "strong" stacked generalization. For example, one might start by forming the $\{h_j\}$, and then forming learning and testing sets by sampling the $\{h_j\}$, just as in strong cross validation. For each such learning and testing set, one forms new level 1 input-output pairs which are added to the level 1 learning set (i.e., one treats the newly created learning set as an L_{i1} and the newly created testing set as an L_{i2}). As with strong cross validation, in practice one should probably have the sampling used to create the new learning and testing sets heavily weighted towards the points in the original learning set. In the extreme where the sampling only runs over the input values in the original L, one essentially recovers stacked generalization with a bootstrap partition set (assuming the sampling is done without any noise, and that each G_j acts appropriately and perfectly reproduces the elements of the learning set on which it is trained).

Other variations of stacked generalization are based on using the level 1 learning set differently from the way it is used in Figure 1. As an example, one might never train the level 0 generalizers on all of L, but rather use only the guesses of the generalizers when trained on the L_{i1}, those level 0 learning sets directly addressed by the level 1 learning set. (The idea is that the level 1 learning set only directly tells us how to guess when the training is on subsets of L, so perhaps we should try to use it only in concert with such subsets.) To guess what output goes with a question q, one uses a "cloud" consisting of a set of points in the level 1 input space. Each element of the cloud is fixed by a partition set index i and is given by the level 1 input vector $\{G_j(L_{i1}; q)\}$ (j indexes the components of the vector). In other words, each such point is the vector of guesses made by the generalizers when trained on L_{i1}. The "cloud" of such points is formed by running over all i, i.e., over all elements of the partition set. Given this cloud, there are a number of ways to use the level 1 learning set to make the final guess. For example, one might make a guess for each level 1 input value in the cloud, by using the level 1 generalizer and the level 1 learning set. One then could average these guesses over the elements of the cloud, where each guess is weighted according to how close (according to a suitable metric) the associated level 1 input is to an element of the level 1 learning set. (In other words, we would weight a guess of the level 1 generalizer more if we had more confidence in it, based on how close the associated level 1 question is to elements of the level 1 learning set.)

There are a number of possible schemes for automatically optimizing the choice of generalizers and/or stacking architecture. It is easiest to consider them in the context of the architecture of Figure 1. One of the most straightforward of these schemes is to use minimal cross-validation error of the entire stacked generalization structure as an optimality criterion, perhaps together with genetic algorithms as a search strategy.[2] It should be noted that besides minimal cross-validation error of the entire system, there are many other possible optimality criteria, some of which are much more computationally efficient. One of the simplest is the mean-squared error of a least-mean-squared (LMS) fit of a hyperplane to the level 1 learning set. In a similar vein, one might use as criterion the degree to which the guesses being fed to the level 1 input space are not synchronized (see above), or more generally the degree to which the level 1 learning set is both single-valued and spread out in the level 1 input space. (This last criterion is quite similar to the idea behind

[2] As an aside, it is interesting to view such use of a genetic algorithm from an artificial life perspective. Both the constituent level 0 generalizers and the full stacked structure perform the same task (generalization). However, whereas the full structure will be "fit" (have low cross-validation error), the level 0 generalizers in general need not be. The situation is somewhat analogous to forming a fit eukaryote out of not-very-fit prokaryotes.

error-correcting output codes, except that rather than changing the way an output vector is coded, here we are changing the level 1 input space.)

The basic idea of searching over stacking structures is not restricted to changes in discrete quantities like network topologies or choices of generalizers. For example, consider the case where each level 0 generalizer is parametrized by a constant specifying the degree of regularization (or depth of a decision tree, or some such). One could keep the topology and choice of generalizers constant and vary the level 0 generalizers' parameterizing constants (so as to maximize the value of an optimality measure). This might result in level 0 generalizers with very different behavior from one another, a property which, as mentioned above, is often desirable.

As a practical note, if one uses schemes like those just mentioned with cross validation as one's optimality criterion, it is important to bear in mind that one can "over-cross-validate" just as one can "over-train."[14] Accordingly, one might try either to stop the search process early or to "regularize" the cross-validation error somehow (e.g., penalize use of those level 0 generalizers that have many degrees of freedom which the search-over-cross-validation-errors can vary). Similar considerations often apply to other optimality criteria besides cross validation.

As a final point, note that it might be possible to use stacking profitably for purposes other than generalization. For example, consider combining a set of generalizers, as in Figure 1, where each of those generalizers is a Bayesian generalizer. The differences between the generalizers lies in their choice of prior. In other words, there is an implicit hyperparameter α, and, rather than a known prior $P(f)$, there is a known conditional prior, $P(f|\alpha)$ (α indexes the different generalizers). One might want to know $P(\alpha)$. (This is needed, for example, to perform a Bayesian generalization, since $P(f) = U \int d\alpha P(f|\alpha)P(\alpha)$.)

How does one find $P(\alpha)$? One could be a pure Bayesian and (try to) derive $P(\alpha)$ using first-principles reasoning. Or one could be an empirical Bayesian: loosely speaking, one "cheats" (i.e., uses less than fully rigorous reasoning) and sets priors using frequency count (i.e., maximum likelihood) estimates based on past experience. Or one could cheat a different way, by using stacked generalization. One version of this idea is to use a level 1 generalizer that is an LMS fit of a hyperplane to the level 1 learning set, with all the coefficients of the hyperplane restricted to be non-negative. Given such a fit, one might, as a crude heuristic, take the resultant guessing of the full stacked structure to give the maximum *a posteriori* $P(f|L)$. Since $P(f|L) = \sum_\alpha [P(f|L,\alpha) \times P(\alpha|L)]$, and since the guesses of the level 0 generalizers give $P(f|L,\alpha)$ as the coefficients of the hyperplane fit one might estimate $P(\alpha|L)$. (Loosely speaking, for large enough L and a leave-one-out CUPS, one might presume that $P(\alpha|L_{i1}) \simeq P(\alpha|L)$.) One might then estimate $P(\alpha)$ from $P(\alpha|L)$, using the (assumed known) likelihood $P(L|f)$ and conditional prior $P(f|\alpha)$. The idea would be that all the Bayesian generalizers use the same likelihood, so the difference in their utility (as measured by the hyperplane coefficients) must reflect differences in α. Just as with empirical Bayesianism, one is setting priors by means of the data.[3]

ACKNOWLEDGMENTS

This article was supported by the Santa Fe Institute and by NLM grant F37 LM00011.

[3]This idea had its genesis during a discussion I had with Peter Cheeseman and Leo Breiman in August 1992. Of course, any flaws in the idea as presented here are wholly a reflection of my elaboration of it.

REFERENCES

1. Breiman, L. "Stacked Regressions." Technical Report 367, Department of Statistics, University of California at Berkeley, 1992.
2. Breiman, L., and P. Spector. "Submodel Selection and Evaluation—X Random Case." *Intl. Stat. Rev.* (1992): in press.
3. Breiman, L. Personal communication, 1992.
4. Efron, B. "Computers and the Theory of Statistics: Thinking the Unthinkable." *SIAM Rev.* **21** (1979): 460–480.
5. Gustafson, S., G. Little, and D. Simon. "Neural Network for Interpolation and Extrapolation." Report number 1294–40, University of Dayton Research Institute, Dayton, Ohio, 1990.
6. Ker-Chau Li. "From Stein's Unbiased Risk Estimates to the Method of Generalized Cross-Validation." *Ann. Stat.* **13** (1985): 1352–1377.
7. Schapire, R. E. "The Strength of Weak Learnability." *Symposium on Foundations of Computer Science* (1989): 28–33.
8. Stone, M. "An Asymptotic Equivalence of Choice of Model by Cross-Validation and Akaike's Criterion." *J. Roy. Stat. Soc. B* **39** (1977): 44–47.
9. Stone, M. "Asymptotics For and Against Cross-Validation." *Biometrika* **64** (1977): 29–35.
10. Stone, M. "Cross-Validatory Choice and Assessment of Statistical Predictions." *J. Roy. Stat. Soc. B* **36** (1974): 111–120.
11. Weiss, S. M., and C. A. Kulikowski. *Computer Systems that Learn.* San Mateo, CA: Morgan Kaufmann, 1991.
12. Wolpert, D. "A Mathematical Theory of Generalization: Part II." *Complex Systems* **4** (1990): 201–249. Cross validation is a special case of the technique of "self-guessing" discussed here.
13. Wolpert, D. "Constructing a Generalizer Superior to NETtalk via a Mathematical Theory of Generalization." *Neur. Nets.* **3** (1990): 445–452.
14. Wolpert, D. "On the Connection Between In-Sample Testing and Generalization Error." *Complex Systems* **6** (1992): 47–94.
15. Wolpert, D. "Stacked Generalization." *Neur. Nets.* **5** (1992): 241–259.
16. Wolpert, D. "How to Deal with Multiple Possible Generalizers." In *Fast Learning and Invariant Object Recognition*, edited by B. Soucek, 61–80. New York: J. Wiley & Sons, 1992.
17. Zhang, X., J. P. Mesirov, and D. L. Waltz. "A Hybrid System for Protein Secondary Structure Prediction." *J. Mol. Biol.* **225** (1992): 1049–1063.

Robert S. Maier
Department of Mathematics, University of Arizona, Tucson, AZ 87521

Large Fluctuations in Stochastically Perturbed Nonlinear Systems: Applications in Computing

1. INTRODUCTION

Nonlinear dynamical systems often display complex behavior. In this lecture I shall review the behavior of *stochastically perturbed* dynamical systems, which is a field of its own. I shall use this as an opportunity to discuss applications to computer science, though applications to statistical physics, chemical physics, and elsewhere in the sciences are also numerous.

If a deterministic dynamical system has an attractor, by definition the system state approaches the attractor in the long time limit. But if the system is regularly subjected to small stochastic fluctuations (random kicks, or noise), this approach will only be approximate. In the long time limit the system state will typically be specified by a probability distribution (a "noisy attractor") centered on the attractor proper. In the limit as the noise strength tends to zero, this distribution will converge to the attractor.

Even if the system has a single globally stable point as its only attractor, one can pose an interesting question: if the noise strength is very small, what is the probability of finding the system in a specified state macroscopically distant from the attractor? How long must one wait before this occurs? If the system has more than a single stable state, each with its own basin of attraction, one can similarly ask for the time scale on which transitions between the two basins occur. Such questions are really questions about the character of the extreme tail of the noisy attractor and can be answered only by quantifying the probability of *large fluctuations* of the system. The mathematical field dealing with such matters is known as large deviation theory.[4,26]

In scientific applications one would usually like to know not only how frequently atypical fluctuations occur, but also along which trajectory the system state moves during transitions from one stable state to another. It turns out that, in most stochastically perturbed dynamical systems, a single trajectory in the system state space, or at most a discrete set, is singled out in the limit of weak noise as by far the most likely.

This phenomenon has long been known to chemical and statistical physicists, but its importance in other fields that make use of stochastic modeling, such as ecology and evolutionary biology, has only recently become clear.[8,20] In chemical physics the most likely transition trajectory is interpreted as a reaction pathway, since chemical reactions are modeled as transitions from a metastable state to a more stable state.[25] But the mathematical approach I shall sketch is much more general: the dynamical system can be continuous or discrete, and the system dynamics need not obey detailed balance. Some of the strongest results on systems without detailed balance have only recently been obtained.[14,15] The system can even be *distributed*, with nontrivial spatial extent; this includes stochastic cellular automata and those systems specified by stochastic partial differential equations rather than stochastic ordinary differential equations.

The quasi-deterministic phenomena (optimal trajectories, well-defined reaction pathways, *etc.*) which arise in stochastically perturbed dynamical systems can be viewed as *emergent*. They are determined by the stochastic dynamics, but in a rather complicated way, and they manifest themselves only in the weak-noise limit. Their appearance in computer science applications is not well known; I hope the two examples treated in this lecture will correct that. Attempts have recently been made to interpret the behavior of computers, or interacting networks of computers, in dynamical system terms or even ecological terms.[7] But stochasticity is, I think, a crucial part of any such interpretation.

2. A SIMPLE STOCHASTIC MODEL: ALOHAnet

As a first example drawn from computer science, consider a stochastic model that attempts to capture the essential features of a large number of computers communicating with each other across a data network, such as an Ethernet. The model will be idealized, but it will be typical of ("in the same universality class as") models in which a large number of agents share occasional access to a single resource. Here the resource will be the network bus: the ether, which only one computer can use at a time.

You are no doubt familiar with such application programs as **telnet** and **ftp**, which allow a user of one machine to communicate with another. Behind the scenes ("at a lower protocol layer," in telecommunications jargon) these programs work as follows.[24] A connection between two computers consists of a stream of data packets, each typically containing between 10 and 10^3 bytes. (A data packet is simply a train of square waves.) An interactive log-in program like **telnet** normally transmits a packet whenever the user presses a key; the packet contains the typed character. Less interactive programs like **ftp**, which transfers whole files, employ larger packets. There is a scheme known as TCP/IP (Transmission Control Protocol/Internet Protocol) for specifying the destination of packets and for keeping the two communicating computers synchronized. This last task may involve the transmission of additional packets.

Let us suppose that a computer is making substantial use of the network: several users are running **ftp** simultaneously, for example. In this situation a statistical treatment is possible. In the context of a particular stochastic model, it is possible to estimate mean network usage and the probability that data packets are transmitted successfully. That is what I shall now do.

A slight digression is necessary on the issue of *successful* transmission. Ethernet, besides being a trade name, is a multiaccess protocol: a scheme for sharing access to the cable connecting two or more computers. Normally when a computer wishes to transmit a packet, it does so immediately. Therefore it is possible for two machines to transmit colliding packets, in which case both packets are corrupted: the information in both is lost. The Ethernet protocol (a CSMA/CD [Carrier Sense Multiple Access/Collision Detect] protocol) embodies a heuristic for minimizing the probability of collisions, i.e., of unsuccessful transmissions.

A description of the protocol may be found in the book by Bertsekas and Gallager.[1] On grounds of simplicity, I shall model a conceptually similar but simpler protocol known as ALOHAnet. ALOHAnet was one of several Ethernet precursors, developed at the University of Hawaii during the 1970s. Although it has long since been superseded, it lives on in the form of a tractable mathematical model. The stochastic ALOHAnet model is a discrete-time model or Markov chain, unlike the continuous-time models that must be employed in the performance analysis of real-world Ethernets. The following description is standard.[5,9,13,19]

Suppose that N computers are attached to the network; N will eventually be taken to infinity, yielding a continuum limit which (if proper scaling is imposed) can be viewed as a weak-noise limit. At each integer time $j = 1, 2, 3, \ldots$, a packet of data originates with probability p_0 on each computer not currently blocked. When is a computer blocked? When a previously generated packet has failed to be transmitted successfully, and the packet is awaiting retransmission.

Newly generated packets are always transmitted immediately but, of course, they may collide with packets transmitted by other computers at the same integer time. Such collisions are immediately detected, and each of the transmitting computers enters a blocked state (if it was not blocked already). While in the blocked state, at each subsequent integer time a computer will attempt a retransmission with probability p_1. In other words, each of the blocked computers backs off a random amount of time and tries again to transmit its packet. The back-off time is geometrically distributed, with parameter p_1. This random back-off policy facilitates the breaking of the deadlock: if the blocked computers each backed off a *fixed* amount of time, they would simply run into each other again.

This ALOHAnet model has only three parameters: p_0, p_1, and N. If y_j is the number of computers blocked at time j, then $y_1, y_2, y_3 \ldots$ is a Markov chain on the discrete state space $\{0, 1, 2, \ldots, N\}$. Let us analyze this Markov chain.

At any time j, the number of retransmitted packets is binomially distributed, with parameters p_1 and y_j. Similarly, the number of newly generated (and transmitted) packets is binomially distributed with parameters p_0 and $N - y_j$. If X_1 and X_0 denote these two random variables, the total number of packets transmitted at integer time j is $X_1 + X_0$, and

$$\xi \equiv y_{j+1} - y_j = \begin{cases} -1, & \text{if } X_0 = 0,\ X_1 = 1; \\ X_0, & \text{if } X_0 + X_1 > 1; \\ 0, & \text{otherwise.} \end{cases} \tag{1}$$

y_j will decrease by 1 if a previously unsuccessfully transmitted packet (and only that packet) is retransmitted. It will increase by X_0 in the event of a collision, and so forth. From Eq. (1), it is easy to work out the density of the random variable $\xi \equiv \Delta y$.

Since we wish to construct a continuum large-N limit, we define the normalized network state x at any time to be y/N, the fraction of computers that are currently blocked. Necessarily $0 \leq x \leq 1$. Besides scaling the state space in this way, we scale time by defining normalized time t to equal j/N, so that x, if viewed as a function of t, jumps at $t = 1/N, 2/N, \ldots$ by a random quantity $N^{-1}\xi$. The density of the random variable ξ is specified by the current normalized state x; we write ξ as $\xi(x)$ to make this clear.

To get a nontrivial large-N limit, we need to scale the probabilities p_0 and p_1 as well; we take $p_0 = q_0/N$ and $p_1 = q_1/N$, for some N-independent q_0 and q_1. So $q_0 x$ is the expected number of newly generated packets, and $q_1(1-x)$ the expected number of retransmitted packets, at any specified normalized time j/N. It is an easy exercise to verify that in the large-N limit

$$\langle \xi(x) \rangle = q_0(1-x) - [q_0(1-x) + q_1 x] \exp\left[-q_0(1-x) - q_1 x\right] \tag{2}$$

is the expected change in the number of blocked computers, at any specified time j/N. Equation (2) gives us an explicit expression for $\langle \Delta x \rangle$, the mean amount by which the normalized state x changes at any specified time j/N; it is simply $N^{-1}\langle \xi(x) \rangle$. So in the large-$N$ limit the dynamics of our network model are *on the average* completely specified by Eq. (2).

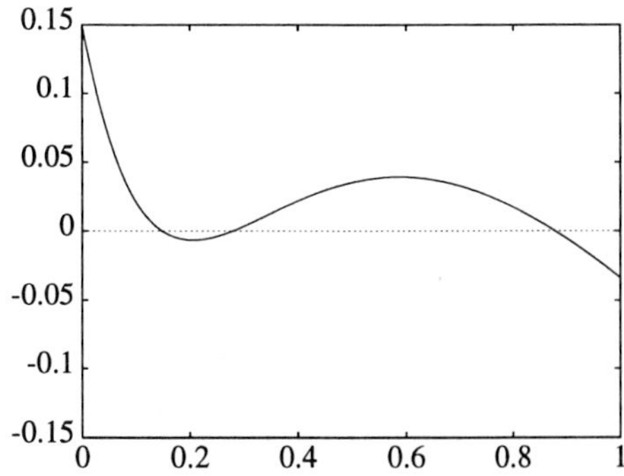

FIGURE 1 The expected drift velocity $\langle\xi(x)\rangle$ of the stochastic ALOHAnet model, as a function of normalized network state x. Model parameters are $q_0 = 0.43$ and $q_1 = 5.0$.

We now can see how the ALOHAnet model can be viewed as a stochastically perturbed dynamical system. In expectation, the large-N ALOHAnet model looks very much like a one-dimensional dynamical system

$$\dot{x}(t) = \langle\xi(x)\rangle, \tag{3}$$

defined on the closed interval $[0, 1]$. Such an associated deterministic dynamical system is called a *fluid approximation* by network performance analysts. Although (as we shall see) it cannot answer the questions about large fluctuations in which we are interested, the fluid approximation says quite a bit about the stability of the network. In Figure 1, the drift field $\langle\xi(x)\rangle$ is plotted as a function of x, for $q_0 = 0.43$ and $q_1 = 5.0$ (parameter values originally chosen by Günther and Shaw[5]). It is clear that for this choice of parameters the system has two point attractors: $x_0 \approx 0.150$ and $x_1 \approx 0.879$. Each has its own basin of attraction and, in the fluid approximation, the network state flows deterministically to one or the other. The two attractors are interpreted as follows. Networks, in particular heavily loaded networks, are prone to *congestion*, and the two attractors are respectively a low-congestion and a high-congestion state.

The presence of more than a single attractor, for certain parameter values, is an unfortunate feature of the ALOHAnet protocol. If at time zero all computers begin unblocked, with these parameter values the fraction of blocked computers will rise swiftly to ≈ 0.150. If, on the other hand, at time zero the computers all begin in the blocked state, the fraction will decrease to ≈ 0.879 and no further. In the latter case very few packets are successfully transmitted or retransmitted, since the probability of more than a single computer transmitting a packet is always very high. (Since $q_1 = 5.0$, when $x \approx 1$ about 5 computers, on average, attempt to retransmit a packet at each time j/N.) The ALOHAnet protocol makes no provision for breaking the deadlock by sharing the network in a sequential or round-robin fashion: in the event of extreme congestion, the computers get in each others' way.

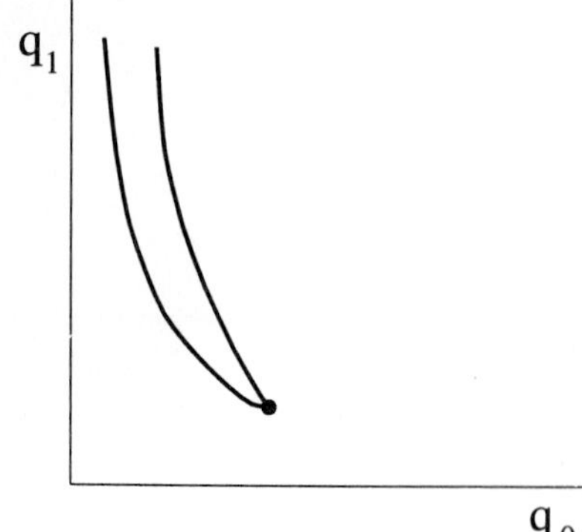

FIGURE 2 An impressionistic sketch of the parameter space of the stochastic ALOHAnet model. Within the horn-shaped region, the network is bistable; outside it, monostable. The tip of the horn is analogous to a statistical-mechanical critical point.

The appearance of more than a single point attractor is actually a bit atypical; it will occur only for certain values of the scaled parameters. (See Figure 2.) The (q_0, q_1)-plane is divided into two regions: a monostable (one-attractor) region and a bistable (two-attractor) region. The equilibrium blocking fraction is a single-valued function of (q_0, q_1) in the former region, and a double-valued function in the latter. Nelson[18] has shown that this phenomenon, which is so suggestive of statistical-mechanical critical behavior, generalizes naturally to multidimensional parameter spaces. The Ethernet protocol modifies the packet retransmission probability each time an unsuccessful retransmission occurs, so a more realistic ALOHAnet model would be specified by a vector $(p_0, p_1, p_2, \ldots)$ of probabilities, with p_k, $k \geq 1$, the probability of transmitting a packet that has failed to be successfully transmitted exactly k times. The corresponding normalized system state would be a vector $(x^{(1)}, x^{(2)}, \ldots)$ of blocking fractions: $x^{(k)}$, $k \geq 1$, would be the fraction of computers that are blocked and that have failed to transmit a stored packet exactly k times. The analogue of Figure 2 would be a multidimensional phase diagram, some regions in which would be characterized by the presence of multiple point attractors in the multidimensional normalized state space.

The preceding treatment has been entirely in the context of the deterministic fluid approximation. The network state does not actually evolve deterministically, except in expectation. The expected increment $\langle \Delta x \rangle$ equals $N^{-1}\langle \xi(x) \rangle$, but the standard deviation of Δx is also proportional to N^{-1}. Δx equals $\langle \Delta x \rangle$ plus $\Delta x - \langle \Delta x \rangle$, and the latter term can be viewed as a stochastic perturbation superimposed on the dynamical system. These stochastic perturbations will broaden the point attractors into noisy attractors and occasionally induce transitions between them.

These transitions are of considerable practical interest, since they are sudden changes in network congestion. A heavily loaded network can suddenly shift from a low-congestion state to a high-congestion state, in which almost no packets are transmitted successfully. (This has rather drastic effects on the computers attached to the network!) But to model such transitions, a fully stochastic treatment is necessary.

3. THE WENTZELL-FREIDLIN THEORY

The techniques employed to estimate the transition time between metastable states, and in general to estimate the probability of unlikely events in the weak-noise limit, go under the name of Wentzell-Freidlin theory.[26] The Wentzell-Freidlin theory is simply the large deviation theory of stochastically perturbed dynamical systems. Many results in this area are due to physicists and chemists,[6,23,25] but Wentzell and Freidlin were the first to put the subject on a sound mathematical footing.[4,27] I shall summarize their main results, and extensions.

Consider a multidimensional random process $\mathbf{x}(t)$ similar to the normalized ALOHAnet process. $\mathbf{x}(t)$ is assumed to jump at times $t = N^{-1}, 2N^{-1}, 3N^{-1}, \ldots$, and the jump magnitude is N^{-1} times a random vector whose distribution depends on the current state $\mathbf{x}$. We write this random vector as $\xi(\mathbf{x})$, so $\Delta\mathbf{x} = N^{-1}\xi(\mathbf{x})$. The $N \to \infty$ limit will be a weak-noise limit.

This random process strongly resembles a diffusion process with drift. In fact the expected drift velocity at any point $\mathbf{x}$ is $\mathbf{u}(\mathbf{x}) \equiv \langle\xi(\mathbf{x})\rangle$, and the diffusion tensor is N^{-1} times $D_{ij}(\mathbf{x}) \equiv \mathrm{Cov}(\xi_i(\mathbf{x}), \xi_j(\mathbf{x}))$, the covariance matrix of the components of $\xi(\mathbf{x})$. A continuous-time diffusion process $\mathbf{x}(t)$ with these parameters would satisfy the stochastic differential equation

$$dx_i(t) = u_i(\mathbf{x}(t)) + \sum_j \frac{\sigma_{ij}(\mathbf{x}(t))}{\sqrt{N}} dw_j(t) \tag{4}$$

where $d\mathbf{w}(t)$ is white noise, and the tensor $\sigma = (\sigma_{ij})$ is related to the tensor $\mathbf{D} = (D_{ij})$ by $\mathbf{D} = \sigma\sigma^t$. But this continuous-time "diffusive approximation" to the underlying jump process is not especially useful for our purposes: the large fluctuations of the jump process turn out to depend crucially on the higher moments of $\xi(\mathbf{x})$.

Suppose that $\mathbf{x}_0$ is an attractor for the expected drift field $\mathbf{u}(\mathbf{x})$. Then in expectation $\mathbf{x}(t)$ will tend to flow toward $\mathbf{x}_0$ if it begins in the basin of attraction of $\mathbf{x}_0$. Thereafter, $\mathbf{x}(t)$ will tend to wander near $\mathbf{x}_0$ for a long time. But statistical fluctuations of all magnitudes will occur; the stochastic perturbations $N^{-1}[\xi(\mathbf{x}) - \mathbf{u}(\mathbf{x})]$ will eventually push $\mathbf{x}$ outside any specified region U surrounding $\mathbf{x}_0$. In other words, the noise will eventually overcome the drift.

Since the effective diffusion coefficient decays as N^{-1}, one expects that the time to exit any specified region U grows (in expectation) exponentially in N. That is correct, and the Wentzell-Freidlin theory provides a technique for computing the asymptotic exponential growth rate. Of course, this will depend on the choice of U.

In most applications U is the entire basin of attraction of the attractor $\mathbf{x}_0$, though a smaller region could be chosen.

The technique is as follows. According to theory the expected exit time $\langle t_{\text{exit}} \rangle$ has weak-noise asymptotics

$$\langle t_{\text{exit}} \rangle \sim \exp(N\mathcal{S}_0), \qquad N \to \infty, \tag{5}$$

where

$$\mathcal{S}_0 = \inf \int L(\mathbf{x}(t), \dot{\mathbf{x}}(t))\, dt \tag{6}$$

is a *minimum action* for exiting trajectories. The infimum is taken over all trajectories $\mathbf{x}(t)$ which begin at $\mathbf{x}_0$ and terminate on the boundary of U. The transit time is left unspecified. Here $L(\mathbf{x}, \dot{\mathbf{x}})$ is a Lagrangian function, dual to a Hamiltonian or energy function constructed from the distribution of $\xi(\mathbf{x})$ by the formula

$$H(\mathbf{x}, \mathbf{p}) = \log\langle \exp(\mathbf{p} \cdot \xi(\mathbf{x}))\rangle. \tag{7}$$

It is clear that the higher moments of $\xi(\mathbf{x})$ enter into the computation of the function H. In fact, $H(\mathbf{x}, \cdot)$ is the cumulant-generating function of the random variable $\xi(\mathbf{x})$.

The sudden appearance of a classical Hamiltonian and its dual Lagrangian is quite remarkable. They are not mere mathematical auxiliaries. The trajectory $\mathbf{x}^*(t)$ minimizing the action (it usually exists, and is unique) is interpreted as the *most probable exit path* (MPEP) in the limit of weak noise. It is not difficult to check, using standard methods of classical mechanics, that the optimization of the action over transit times yields an MPEP which is a *classical trajectory of zero energy*. So the "momentum" $\mathbf{p}$, which has no direct physical interpretation, as a function of position $\mathbf{x}$ along the MPEP must satisfy

$$\langle \exp(\mathbf{p} \cdot \xi(\mathbf{x}))\rangle = 1. \tag{8}$$

If the state space is one-dimensional, this zero-energy constraint alone will determine the MPEP.

The MPEP $\mathbf{x}^*$ is not only a most probable exit path: it is also an exit path of least resistance. Although $\mathbf{x}(t)$ will remain in U for an exponentially long time, it will fluctuate out along the MPEP (and in other directions) an exponentially large number of times before the MPEP is traversed in full and U is exited. The final fluctuation will follow $\mathbf{x}^*$ quite closely in the large-N limit. One can view the equilibrium distribution of the system state $\mathbf{x}$ (the noisy attractor) as being concentrated near $\mathbf{x}_0$ but having a tubelike protuberance stretching out toward the boundary of U along the trajectory $\mathbf{x}^*$. In the large-N limit the tube is exponentially suppressed, and the noisy attractor converges to the point attractor $\mathbf{x}_0$.

$\langle t_{\text{exit}} \rangle$ grows exponentially in N, but the limiting *distribution* of t_{exit} has not yet been specified. It turns out to be an exponential distribution. This is very typical of weak-noise escape problems, where the probability of any single escape attempt is small. (The same exponential distribution is seen in radioactive decay.)

$\mathcal{S}_0$, the weak-noise growth rate of the expected exit time, can be viewed as a *barrier height*: a measure of how hard it is to overcome the drift driving $\mathbf{x}$ toward $\mathbf{x}_0$ and away from the boundary of U. In fact, if extended to conservative continuous-time processes described by Eq. (4), the Wentzell-Freidlin framework yields the familiar Arrhenius law for the growth of the exit time in the limit of weak noise. For such systems $\mathcal{S}_0$ is simply the height of the potential barrier surrounding the attractor.

What is not clear from the Wentzell-Freidlin treatment (and is still not *rigorously* clear, though numerous nonrigorous results have been obtained[14,16,17]) is the subdominant large-N asymptotics of $\langle t_{\text{exit}} \rangle$. In general, one expects

$$\langle t_{\text{exit}} \rangle \sim C N^{\alpha} \exp(N \mathcal{S}_0), \qquad N \to \infty, \tag{9}$$

for some constants C and α, but the Wentzell-Freidlin theory yields only the exponential growth rate $\mathcal{S}_0$. The preexponential factor in Eq. (9) remains to be determined.

The current status of the prefactor problem can be summed up as follows. If U is taken to be the entire basin of attraction of $\mathbf{x}_0$, α is typically zero and C can be obtained by a method of matched asymptotic expansions, i.e., a method of systematically approximating the equilibrium distribution of $\mathbf{x}$. However in multidimensional models there is an entire zoo of possible pathologies, including the appearance of caustics and other singular curves in the state space,[2,14,15] which can induce a nonzero α or hinder a straightforward computation of C. This is the case, at least, for continuous-time diffusion processes defined by stochastic differential equations. The situation for jump processes is expected to be similar.

4. APPLYING THE THEORY

The Wentzell-Freidlin theory, with extensions, can be applied to the stochastic ALOHAnet model, and to other stochastically perturbed dynamical systems arising in computer science. The quantity most readily computed is $\mathcal{S}_0$, the exponential growth rate in the weak-noise limit of the expected time before the system leaves a specified region surrounding a point attractor in the system state space. Recall that in the ALOHAnet model this region is the basin of attraction; a departure from it signals a drastic change in network congestion.

If the system state space is one-dimensional, as in the ALOHAnet model, the classical-mechanical interpretation of $\mathcal{S}_0$ facilitates its computation. $\mathcal{S}_0$ is always the action of a zero-energy trajectory, with energy as a function of position and momentum given by Eq. (7). This Hamiltonian is a convex function of $\mathbf{p}$ at fixed $\mathbf{x}$, so if the state space is one-dimensional (and $\langle \xi(\mathbf{x}) \rangle \neq \mathbf{0}$, which will always be the case within the basin of attraction), the equation $H(\mathbf{x},\mathbf{p}) = 0$ will have only two solutions for $\mathbf{p} = \mathbf{p}(\mathbf{x})$. One of these is $\mathbf{p} \equiv \mathbf{0}$, which is not physical. This solution is not physical because, if $\mathbf{p} = \mathbf{0}$,

$$\dot{\mathbf{x}} = \frac{\partial H}{\partial \mathbf{p}} = \langle \xi(\mathbf{x}) \exp(\mathbf{p} \cdot \xi(\mathbf{x}))\rangle / \langle \exp(\mathbf{p} \cdot \xi(\mathbf{x}))\rangle = \langle \xi(\mathbf{x})\rangle \tag{10}$$

and the $\mathbf{p} \equiv \mathbf{0}$ trajectory simply follows the mean drift, which points *toward* the attractor rather than away from it. The MPEP must be a classical trajectory emanating from the attractor, so in a one-dimensional system it is uniquely characterized by the condition that $\mathbf{p} = \mathbf{p}(\mathbf{x})$ must be the nonzero solution of $H(\mathbf{x},\mathbf{p}) = 0$. Actually there are two such trajectories, one emanating to either side of the attractor; the true MPEP will be the one with lesser action.

In general, to compute $\mathcal{S}_0$, even in higher-dimensional models, one needs only the MPEP and the momentum as a function of position along it. This is because the action of any zero-energy classical trajectory may be written as a line integral of the momentum, so that

$$\mathcal{S}_0 = \int \mathbf{p}(\mathbf{x}) \cdot d\mathbf{x}, \tag{11}$$

the integral being taken along the MPEP from the attractor to the boundary of the region. But only in one-dimensional models is Eq. (11) easily applied. In d-dimensional models, merely finding the MPEP requires an optimization over the $(d-1)$-dimensional family of zero-energy trajectories extending to the boundary. Except in models with symmetry, this optimization must usually be performed numerically.

4.1 THE ALOHANET APPLICATION

In the ALOHAnet model, the expected drift $\langle \xi(x)\rangle$ as a function of normalized network state x is given by Eq. (2). But to study large fluctuations and compute the MPEP, one needs the Wentzell-Freidlin Hamiltonian $\log\langle \exp(p\xi(x))\rangle$. In the large-$N$ limit the random variables X_1 and X_0, in terms of which ξ is expressed by Eq. (1), become a Poisson random variable with parameter $q_1 x$ and a Poisson random variable with parameter $q_0(1-x)$, respectively. A bit of computation yields

$$H(x,p) = \log\left[e^{q_0(1-x)(e^p-1)} + q_0(1-x)e^{-q_0(1-x)-q_1 x}(1-e^p) + q_1 x e^{-q_0(1-x)-q_1 x}(e^{-p}-1)\right] \tag{12}$$

as the Hamiltonian.

If the parameters q_0 and q_1 are known, it is easy to compute the momentum $p = p(x)$ along the MPEP, by numerically solving for the nonzero solution of the implicit equation $H(x,p(x)) = 0$. But the MPEP, and hence $\mathcal{S}_0$, will depend on the choice of basin of attractor. With the parameter values $q_0 = 0.43$ and $q_1 = 5.0$ of Figure 1, the two attractors $x_0 \approx 0.150$ and $x_1 \approx 0.879$ have respective basins of attraction $[0, x_c)$ and $(x_c, 1]$, with $x_c \approx 0.278$ the intermediate repellor. MPEPs extend from x_0 to x_c, and from x_1 to x_c. Numerical integration of $p(x)$ gives

$$\mathcal{S}_0[x_0 \to x_c] \approx 0.00177 \tag{13}$$

$$\mathcal{S}_0[x_1 \to x_c] \approx 0.014 \tag{14}$$

as the growth rates of the expected transition times.

We see that for the stochastically modeled ALOHAnet, in the large-N limit a reduced description is appropriate. Asymptotically, it becomes a *two-state process*. The network is either in a low-congestion state (the basin of attraction of x_0) or a high-congestion state (the basin of attraction of x_1), and the transition rates between them (the reciprocals of the expected transition times) display exponential falloffs

$$\exp\left(-N\mathcal{S}_0[x_0 \to x_c]\right), \qquad \exp\left(-N\mathcal{S}_0[x_1 \to x_c]\right), \tag{15}$$

respectively. With the above choice of parameters, for reasonable-sized N the latter transition rate is much smaller than the former. Once congestion has interfered with the proper performance of the back-off algorithm, the network gets "stuck" for a potentially long time. This is clearly not a good choice of network parameters!

In a real-world N-computer ALOHAnet implementation, q_0 would be the total network load and would be determined by the level of interprocessor computing taking place on the network. The back-off parameter $q_1 = Np_1$, however, would probably be fixed, with p_1 in hard code in a data communications chip installed in each computer. So the Wentzell-Freidlin approach could be employed to determine the likelihood, as a function of network load, of irreversible (or all but irreversible) congestion occurring.

Of course the bistability of the system is itself a function of q_0 and q_1. As noted, for many values of the parameters the network is monostable: there is only a single attractor, which may be characterized by a comparatively low level of congestion. For such a network, one could compute an action $\mathcal{S}_0$ for any specified maximum tolerable congestion level. The associated optimal (i.e., most probable) approach path would be computed much as the MPEP is computed in the bistable case.

4.2 A COLLIDING STACKS APPLICATION

There have been several applications of large deviation theory to the stochastic modeling of *dynamic data structures*.[10,11,12] The memory usage of a program or programs being executed by a computer can be modeled as a discrete-time jump process. In many cases this process may be viewed as a finite-dimensional dynamical system, subject to small stochastic perturbations. Of interest is the amount of time expected to elapse before a particularly large fluctuation away from a deterministic point attractor occurs. This would correspond, in real-world terms, to an atypical string of memory allocations leading to an exhaustion of memory.

The following two-dimensional "colliding stacks" model was first studied by Flajolet,[3] having been first suggested by Knuth. Suppose that N cells of memory, arranged in a linear array, are available for use by two programs. Suppose that at any given time, the programs will require $y^{(1)}$ and $y^{(2)}$ cells of memory respectively. It will be most efficient for them to employ respectively the first $y^{(1)}$ and the last $y^{(2)}$ cells of the array, so as to avoid contention for memory. It is necessary that $y^{(1)} + y^{(2)} \le N$; if this inequality becomes an equality, the two-program system runs out of memory.

A natural model for the evolution of $y^{(1)}$ and $y^{(2)}$ is as follows. At any integer time $j = 1, 2, 3, \ldots$, there are four possibilities: $y^{(1)}$ may increase by 1, $y^{(1)}$ may decrease by 1, $y^{(2)}$ may increase by 1, and $y^{(2)}$ may decrease by 1. These are assigned probabilities $p/2$, $(1-p)/2$, $p/2$, $(1-p)/2$, for p the probability of a net increase in memory usage. Let us take $0 < p < 1/2$, so that deallocations of memory are more likely than new allocations. (Note that, if $y^{(1)} = 0$ or $y^{(2)} = 0$, the assigned probabilities must differ, since neither $y^{(1)}$ nor $y^{(2)}$ can become negative.)

Just as in the ALOHAnet model, it is natural to scale both time and and the state space as the amount of memory N tends to infinity. However, we shall not need to scale the model parameter p. Let $\mathbf{x} = (x_1, x_2) = (y^{(1)}, y^{(2)})/N$ be the normalized state of the two-program system, and let $t = j/N$ be normalized time. $\mathbf{x}$ jumps at $t = 1/N, 2/N, 3/N, \ldots$ by an amount $N^{-1}\xi$, where ξ is a random variable with discrete density

$$\mathbf{P}\{\xi = \mathbf{z}\} = \begin{cases} p/2, & \text{if } \mathbf{z} = (1,0); \\ p/2, & \text{if } \mathbf{z} = (0,1); \\ (1-p)/2, & \text{if } \mathbf{z} = (-1,0); \\ (1-p)/2, & \text{if } \mathbf{z} = (0,-1). \end{cases} \tag{16}$$

As defined, the density of ξ is essentially independent of $\mathbf{x}$. It is useful to relax this assumption, so as to permit more realistic stochastic modeling of dynamic data structures. Let

$$\mathbf{P}\{\xi(\mathbf{x}) = \mathbf{z}\} = \begin{cases} p(x_1)/2, & \text{if } \mathbf{z} = (1,0); \\ p(x_2)/2, & \text{if } \mathbf{z} = (0,1); \\ (1-p(x_1))/2, & \text{if } \mathbf{z} = (-1,0); \\ (1-p(x_2))/2, & \text{if } \mathbf{z} = (0,-1). \end{cases} \tag{17}$$

This is a natural generalization. Here $p(x)$ (assumed to take values between 0 and 1/2 exclusive) specifies the probability of an increase in memory usage by either program, as a function of the fraction of available memory which that program is currently using. We now write ξ as $\xi(\mathbf{x})$, to indicate the dependence of its density on $\mathbf{x}$.

The normalized state $\mathbf{x}$ is confined to the right triangle with vertices $(0,0)$, $(1,0)$ and $(0,1)$. The expected drift

$$\langle \xi(\mathbf{x}) \rangle = \left(p(x_1) - \tfrac{1}{2}, p(x_2) - \tfrac{1}{2}\right) \tag{19}$$

may be viewed as a deterministic dynamical system on this two-dimensional normalized state space. Clearly, the vertex $(0,0)$ is the global attractor. In this model the two programs tend on the average not to use much memory.

Since there is only a single attractor, the quantity of interest is the expected time which must elapse before a fluctuation of specified magnitude occurs. Fluctuations that take the system state to the hypotenuse of the triangle (where $x_1+x_2=1$, or $y^{(1)}+y^{(2)}=N$) are *fatal*: they correspond to memory exhaustion. The rate at which they occur can be estimated in the large-N limit.

This is a two-dimensional system, so the optimal (least-action) trajectories are not determined uniquely by the zero-energy constraint. However, we still have

$$\langle t_{\text{exit}}\rangle \sim \exp(N\mathcal{S}_0), \qquad N\to\infty, \tag{20}$$

with $\mathcal{S}_0$ the action of the least-action trajectory that exits the triangle through the hypotenuse. The action is computed from the Lagrangian dual to the Wentzell-Freidlin Hamiltonian

$$\begin{aligned} H(\mathbf{x},\mathbf{p}) &= \log\langle\exp(\mathbf{p}\cdot\xi(\mathbf{x}))\rangle \\ &= -\log 2 + \log\bigl\{\cosh p_x - [1-2p(x)]\sinh p_x \\ &\qquad\qquad + \cosh p_y - [1-2p(y)]\sinh p_y\bigr\}, \end{aligned} \tag{20}$$

which follows from Eq. (17).

The zero-energy trajectories determined by Eq. (20) are studied at length in Maier[11] where he shows that the MPEP depends strongly on the behavior of the function $p(x)$. (See Figure 3.) If $p(x)$ is a strictly decreasing function, so that the model is "increasingly contractive," with large excursions away from the attractor strongly suppressed, then the MPEP turns out to be directed along the line segment from $(0,0)$ to $(1/2,1/2)$. Its action is

$$\mathcal{S}_0 = 4\int_{x=0}^{1/2} \tanh^{-1}[1-2p(x)]\,dx. \tag{21}$$

If, on the other hand, $p(x)$ is a strictly increasing function, so that the model is decreasingly contractive, with large excursions less strongly suppressed, then there is a twofold degeneracy. MPEPs are directed outward from $(0,0)$ to the two other vertices of the triangle, and

$$\mathcal{S}_0 = 2\int_{x=0}^{1} \tanh^{-1}[1-2p(x)]\,dx \tag{22}$$

is their common action.

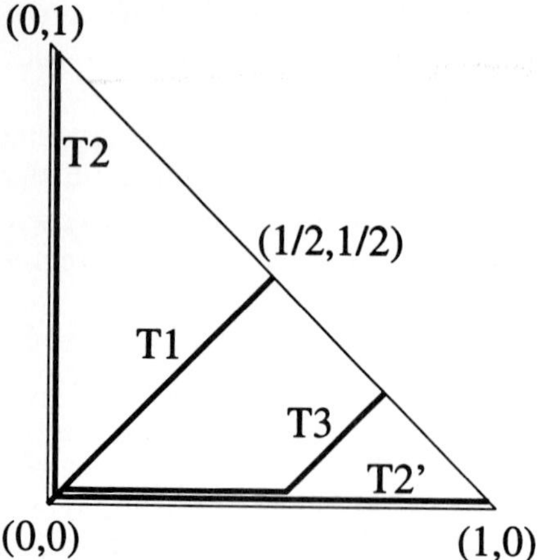

FIGURE 3 The triangular normalized state space of the colliding stacks model. Trajectory T1 is the most probable exit path when the function $p(x)$ is strictly decreasing, but, if $p(x)$ is strictly increasing, then T2 and T2′ are both MPEPs. Trajectory T3 is one of the uncountably many MPEPs that arise when the function $p(x)$ is constant.

So when $p(x)$ is strictly increasing, there is a "hot spot" on the hypotenuse of the triangle at $(1/2, 1/2)$. When the two-program system runs out of memory, as $N \to \infty$ it is increasingly likely that each program will be using approximately $N/2$ memory cells. If, on the other hand, $p(x)$ is strictly decreasing, there are hot spots at the vertices $(0, 1)$ and $(1, 0)$. Exhaustion increasingly tends to occur when one or the other program is using all, or nearly all, of the N memory cells.

If $p(x)$ is neither strictly increasing nor strictly decreasing, the large-N asymptotics may become more complicated. The most easily treated case is that of $p(x) \equiv p$, a constant, i.e., the model of Eq. (16). In this model an *infinite degeneracy* occurs: any trajectory that moves some distance (possibly zero) from $(0, 0)$ toward $(0, 1)$ or $(1, 0)$ and then moves into the interior of the triangle at a 45° angle until it reaches the hypotenuse is a least-action trajectory. Large fluctuations away from the attractor may proceed along any of this uncountable set of MPEPs. As a consequence, there is no hot spot: in the large-N limit, the exit location is uniformly distributed over the hypotenuse. Flajolet[3] first discovered this phenomenon combinatorially, but it has a natural classical-mechanical interpretation. However, it is a bit counterintuitive: it says that when memory is exhausted, the fractions allocated to each program are as likely to be small as large. This is a very sensitive phenomenon.

5. CONCLUSIONS

We have seen that the Wentzell-Freidlin results on scaled jump processes throw considerable light on the fluctuations of stochastically perturbed dynamical systems, in the weak-noise limit. Even if the unperturbed dynamical system is in no sense Hamiltonian, the appearance of a classical Hamiltonian and Lagrangian is quite striking. So is the central importance of zero-energy trajectories.

In this chapter I have focused on jump processes since they are the most relevant to computer science applications. (Computing is inherently discrete.) But they also occur in chemical physics: there is always an integer number of molecules in any given region of space. Attempts are now being made to interpret the stochastic aspects of chemical reactions in terms of optimal trajectories.[21] This is very reminiscent of our focus on most probable exit paths (MPEPs).

There is also a large deviation theory of continuous-time processes,[4,26] such as the diffusion processes specified by the stochastic differential equation (4). Associated to each such process is a Fokker-Planck equation (a parabolic partial differential equation) describing the diffusion of probability. The zero-energy classical trajectories of continuous-time large deviation theory can be viewed as the *characteristics* of this differential equation. Normally one expects only hyperbolic equations to have characteristics, but these characteristics are emergent: they manifest themselves only in the weak-noise limit.

A large deviation theory of spatially extended systems would be an interesting extension but is still under development. Such systems include stochastic partial differential equations and stochastic cellular automata. In such systems an MPEP would be a trajectory in the system state space, describing the most probable *spatially extended* fluctuation leading from one metastable state to another. Much work has been done on this by statistical mechanicians and field theorists (who call such fluctuations "instantons"[22]), but the theory is less complete than the theory I have sketched in this chapter. The theory of extended fluctuations in particular has not been applied to distributed computer systems. There is clearly much work left to be done!

ACKNOWLEDGMENTS

Supported in part by the National Science Foundation under grant NCR-90-16211.

REFERENCES

1. Bertsekas, D., and R. Gallager. *Data Networks.* Englewood Cliffs, NJ: Prentice-Hall, 1987.
2. Chinarov, V. A., M. I. Dykman, and V. N. Smelyanskiy. "Dissipative Corrections to Escape Probabilities of Thermally Nonequilibrium Systems." *Phys. Rev. E* **47** (1993): 2448–2461.
3. Flajolet, P. "The Evolution of Two Stacks in Bounded Space and Random Walks in a Triangle." In *Mathematical Foundations of Computer Science: Proceedings of the 12th Symposium.* Lecture Notes in Computer Science, vol. 233, 325–340. Berlin: Springer-Verlag, 1986.
4. Freidlin, M. I., and A. D. Wentzell. *Random Perturbations of Dynamical Systems.* New York: Springer-Verlag, 1984.
5. Günther, N. J., and J. G. Shaw. "Path Integral Evaluation of ALOHA Network Transients." *Info. Process. Lett.* **33** (1990): 289–295.
6. Hänggi, P., P. Talkner, and M. Borkovec. "Reaction-Rate Theory: Fifty Years After Kramers." *Rev. Mod. Phys.* **62** (1990): 251–341.
7. Huberman, B. A., ed. *The Ecology of Computation.* New York: Elsevier, 1988.
8. Lande, R. "Expected Time for Random Genetic Drift of a Population Between Stable Phenotypic States." *Proc. Nat. Acad. Sci. USA* **82** (1985): 7641–7645.
9. Lim, J.-T., and S. M. Meerkov. "Theory of Markovian Access to Collision Channels." *IEEE Trans. Comm.* **COM-35** (1987): 1278–1288.
10. Louchard, G., and R. Schott. "Probabilistic Analysis of Some Distributed Algorithms." *Random Struc. & Algor.* **2** (1991): 151–186. A preliminary version appeared in *Proceedings of CAAP '90.* Lecture Notes in Computer Science, vol. 431, 177–190. Berlin, Springer-Verlag, 1991.
11. Maier, R. S. "Colliding Stacks: A Large Deviations Analysis." *Random Struc. & Algor.* **2** (1991): 379–420.
12. Maier, R. S. "A Path Integral Approach to Data Structure Evolution." *J. Complexity* **7** (1991): 232–260.
13. Maier, R. S. "Communications Networks as Stochastically Perturbed Nonlinear Systems: A Cautionary Note." In *Proceedings of the 30th Allerton Conference on Communication, Control and Computing*, 674–681. Conference held in Monticello, Illinois. Urbana, IL: University of Illinois 1992.
14. Maier, R. S., and D. L. Stein. "Transition-Rate Theory for Non-Gradient Drift Fields." *Phys. Rev. Lett.* **69** (1992): 3691–3695.
15. Maier, R. S., and D. L. Stein. "The Escape Problem for Irreversible Systems." *Phys. Rev. E* **48** (1993): 941–948.
16. Matkowsky, B. J., Z. Schuss, C. Knessl, C. Tier, and M. Mangel. "Asymptotic Solution of the Kramers-Moyal Equation and First-Passage Times for Markov Jump Processes." *Phys. Rev. A* **29** (1984): 3359–3369.

17. Naeh, T., M. M. Kłosek, B. J. Matkowsky, and Z. Schuss. "A Direct Approach to the Exit Problem." *SIAM J. Appl. Math.* **50** (1990): 595–627.
18. Nelson, R. "The Stochastic Cusp, Swallowtail, and Hyperbolic Umbilic Catastrophes as Manifest in a Simple Communications Model." In *Performance '84*, edited by E. Gelenbe, 207–224. Amsterdam: North Holland, 1984.
19. Nelson, R. "Stochastic Catastrophe Theory in Computer Performance Modeling." *J. Assoc. Comput. Mach.* **34** (1987): 661–685.
20. Newman, C. M., J. E. Cohen, and C. Kipnis. "Neo-Darwinian Evolution Implies Punctuated Equilibria." *Nature* **315** (1985): 400–401.
21. Ross, J., K. L. C. Hunt, and P. M. Hunt. "Thermodynamic and Stochastic Theory for Nonequilibrium Systems with Multiple Reactive Intermediaries." *J. Chem. Phys.* **96** (1992): 618–629.
22. Schulman, L. S. *Techniques and Applications of Path Integration.* New York: Wiley, 1981.
23. Schuss, Z. *Theory and Application of Stochastic Differential Equations.* New York: Wiley, 1980.
24. Stallings, W. *Data and Computer Communications*, 2nd ed. New York: Macmillan, 1988.
25. van Kampen, N. G. *Stochastic Processes in Physics and Chemistry.* Amsterdam: North-Holland, 1981.
26. Varadhan, S. R. S. *Large Deviations and Applications.* Philadelphia: Society for Industrial and Applied Mathematics, 1984.
27. Wentzell, A. D. "Rough Limit Theorems for Large Deviations for Markov Processes." *Theory Probab. Appl.* **21** (1976): 227–242, 499–512.

G.Yagil
The Weizmann Institute of Science, Rehovot, Israel 76100

On the Structural Complexity of Designed Systems

1. INTRODUCTION

Biologists occasionally state that a certain system or phenomenon "is very complex indeed." What exactly does that mean? Does it mean merely that no real understanding of the system concerned is available, or can the term "complex" be assigned a positive and precise meaning? We can consult a dictionary; in addition to unhelpful definitions like "Complex: An irrational attitude (Psych.)" or a "non-real number (Math.)," we also find "Composed of many interconnected parts" or, "Of intricate design."[1] The latter two definitions sound intuitively appropriate because most biosystems are composed of many parts, or components, and certainly have an intricate design. These components are atoms or molecules on the ultimate level, and can be molecular assemblies on the intermediate level, and cells organs on the higher levels of bio-organization. The intricate details of these bioassemblies are coded for, at least partly, by the genome of the organism concerned. This coded information has been accumulated during evolution, is faithfully transmitted from generation to generation, and is decoded precisely (expressed) within each generation.

All this is well known. The questions to be addressed here are whether these complex aspects of biosystems can be given a precise definition suitable for quantitative evaluation, and what such an evaluation can contribute to the understanding of these systems. Several formal approaches to the quantitative definition of complexity are available (for a critical evaluation, see Bennet[2]).

The first approach is based on the tacit assumption that behind many complex phenomena hides a simple mathematical relation, like a set of differential equations or a rule of a cellular automaton. The solution of these within a range of real conditions manifests the complex behavior or pattern observed. The task of the biologist, or biophysicist, is to detect the generating function giving rise to that complex pattern. This approach gives little weight to the fact that bioentities have many of their properties specified by the vast repertoire of instructions encoded in their genome. An example is the generation of biological form—morphogenesis. While a certain amount of symmetry-related features can be beautifully explained by simple growth mechanisms,[7] even quite simple creatures like viruses cannot be generated unless dozens of genes produce their precisely coded product. For instance, in bacteriophage *T*4, at least 49 genes must be precisely expressed in order to generate the three structural components of its virion form—head, tail, and tail fibers—and to produce an infective virion.[12]

A second approach does not insist on the presence of a simple generating function but tries to evaluate complexity by estimating the information content of the states that a biosystem realizes in relation to the information content of all possible states of the system. The complexity of a system is expressed in terms of informational entropy, employing Shannon-Weaver related expressions.[5,11] This approach is particularly suitable for the description of stochastically determined features of processes but, so far, it has not been able to incorporate those aspects of life where a high input of genome-coded instructions is involved.

The third approach, formulated by Solomonoff,[10] Kolmogorov,[8] and Chaitin,[3] characterizes the complexity of a string of symbols by the minimal size of the program that will compute that string; in a real system this means the minimal size of the set of instructions required to obtain a complete description of that system. This approach neither presumes a mechanistic understanding nor requires complete knowledge of all possible states that a system can assume. Complexity thus evaluated has been called algorithmic complexity and has been shown to be able to account for the entropic properties of physical systems.[15]

This last approach is particularly suitable for the characterization of biosystems, because it has the ability to evaluate long sets of coded information inherent in biosystems. For a beautiful descriptive exposition, see Dawkins.[4] Examining (so far) sequenced genomic DNA, it does not seem that we shall discover simple functions that can generate most of the the many million bits of information present in the genetic code. This implies that the complexity of a substantial part of biological information is more likely to be described by approaches that just enumerate features than by approaches that calculate probabilities or presume simple underlying processes. Consequently we have adopted the Kolmogorov-Chaitin approach to the evaluation of the structural features of biocomplexity, applying it to molecular assemblies of interest in biosystems.[13]

2. STRUCTURAL COMPLEXITY

In this section the formalism designed to evaluate the structural complexity of bio- and other systems is described and applied in detail to some fairly simple examples. The basic idea is to express the structural complexity of a system in terms of the size of the shortest instruction set leading to that structure. The formalism was developed for *typed point systems*, i.e., for sets of points that may have a different composition each. Simple molecules and biomolecules and their assemblies can be regarded as typed point systems. In previous papers[13] we analyzed two very simple molecules (methane and ethane) as well as several macromolecules and bioassemblies,[14] and showed a connection between the resulting complexities and the coding requirements of the biomolecules treated.

In this paper, the complexity of three simple organic molecules is described, clarifying both the procedure employed and the assumptions inherent in the proposed treatment. To analyze a molecule, the composing atoms are numbered, and the molecule is put in a suitable coordinate system. Next, each coordinate is examined as to whether it has the same value in every molecule (is ordered), or whether it assumes different values at different times (is random). Complexity of the ordered coordinates then is evaluated by the following set of rules (essentially the same as in Yagil[13]):

1. *Structural complexity* of a system C is the size of the set of specifications describing that system.
2. A *specification* can be the assignment of a numerical value to one or more spatial coordinates of a point in the system or the declaration of the type of that point. A type may be a chemical element, a nucleotide base, a cell type, or any other compositional *element* ε.
3. Coordinate values that can be correlated by a mathematical expression are counted either as a single specification when a single numerical value is involved or by as many new numerical constants as are present in the expression.
4. An ordinal number is not counted as a separate specification.
5. The declaration of the *range* of atom numbers over which an expression is valid is not counted as a separate specification.
6. A simple numerical coefficient like π or $(-1)^i$ is not counted as a separate specification (this rule makes tetrahedral and planar coordination spheres, for example, equally complex).

7. A transformation of the coordinate system adds to the complexity specifications equal in number to the previously unspecified constants present in the transformation matrix; only a single (dummy) transformation can accompany a specification statement.

The criteria by which these rules have been formulated are the extent to which they lead to consistent descriptions using different coordinate and numbering systems. A formal justification is not attempted at present. The crucial rule in determining structural complexity is rule 3, which reduces the number of specifications needed for each k-fold regularity from k statements to a single one:

$$C = \sum_k \left[\frac{c(k)}{k}\right] - c' \tag{1}$$

where $c(k)$ is the number of coordinates sharing a k-fold regularity and c' is the number of the coordinates necessary for placing the system in the external space (usually 5 or 6). Equation (1) represents the intuitive idea that the more *regular*, repetitive features a system has, the lower its complexity will be.

The $C(1)$ term of Eq. (1) gives the contribution of uniquely specified coordinates, while all other terms represent coordinates of some repetition or regularity. These uniquely specified coordinates are not random but *ordered*, because random coordinates have been excluded on the grounds that they are indeterminate and it therefore cannot determine whether they obey any regular relationships. Most natural DNA templates are uniquely specified rather than random, because DNA in every cell of the same organism will have the same base sequence. In summary, the total coordinates of a system c are composed of the random coordinates c_{ran} and the ordered coordinates c_{ord}; c_{ord} is in turn composed of the regular (c_{reg}) and the uniquely specified coordinates (c_{us}), as follows:

$$c = c_{\text{ran}} + c_{\text{ord}} = c_{\text{ran}} + c_{\text{reg}} + c_{\text{us}} \,. \tag{2}$$

Soon we shall see how this distinction between random and ordered elements can be implemented.

3. THE NEOPENTANE MOLECULE

The simple hydrocarbon molecule of pentane will serve as an example. A pentane molecule is composed of $n = 12$ atoms, 5 carbons and 12 hydrogens. Three noncyclic isomers exist: n-pentane, isopentane, and neopentane; the structural formulas are shown in Figure 1. Each pentane molecule is fully specified when its 68 coordinates (one coordinate for type and three in space for each atom) are specified. The values

TABLE 1 Specification Table of Neopentane, $C(CH_3)_4$; $n = 17$.[1]

i	ε	r	ϕ	θ	T
1	C	0	0	0	T_0
2	C	R_{CC}	0	$\Theta_{CC}/2$	T_0
3	C	R_{CC}	$\pi/4$	$-\Theta_{CC}/2$	T_0
4	C	R_{CC}	$2\pi/4$	$\pi - \Theta_{CC}/2$	T_0
5	C	R_{CC}	$3\pi/4$	$\pi + \Theta_{CC}/2$	T_0
6	H	R_{CH}	Any_1	Θ_{CH}	T_1
7	H	R_{CH}	$\text{Any}_1 + 2\pi/3$	Θ_{CH}	T_1
8	H	R_{CH}	$\text{Any}_1 + 4\pi/3$	Θ_{CH}	T_1
9	H	R_{CH}	Any_2	Θ_{CH}	T_2
10	H	R_{CH}	$\text{Any}_2 + 2\pi/3$	Θ_{CH}	T_2
11	H	R_{CH}	$\text{Any}_2 + 4\pi/3$	Θ_{CH}	T_2
12	H	R_{CH}	Any_3	Θ_{CH}	T_3
13	H	R_{CH}	$\text{Any}_3 + 2\pi/3$	Θ_{CH}	T_3
14	H	R_{CH}	$\text{Any}_3 + 4\pi/3$	Θ_{CH}	T_3
15	H	R_{CH}	Any_4	Θ_{CH}	T_4
16	H	R_{CH}	$\text{Any}_4 + 2\pi/3$	Θ_{CH}	T_4
17	H	R_{CH}	$\text{Any}_4 + 4\pi/3$	Θ_{CH}	T_4

[1] R_{CC}, R_{CH} are C-C and C-H bond lengths; Θ_{CC}, Θ_{CH} are the C-C-C and C-C-H bond angles, respectively. The listed coordinate values in each row are valid for the coordinate system T shown in the last column of the row. System T_0 has its origin on the central carbon, with its z-axis bisecting the C_2-C_1-C_4 angle, and the x-axis in the plane of that angle. Systems T_1 to T_4 have their origins on carbons C_2–C_5, with their z-axis along the C-C bonds and their x-axes in planes rotated successively by $\pi/2$. To relate the listed coordinates values (x) to a single system, for example to T_0. The following orthogonal transformation has to be applied[6]:

$$(x') = \mathbf{R}(x) + \mathbf{D} = \begin{pmatrix} \sin\theta\cos\phi & -\cos\theta\cos\phi & \sin\phi \\ \sin\theta\sin\phi & -\cos\theta\sin\phi & -\cos\phi \\ \cos\theta & \sin\theta & 0 \end{pmatrix} + \begin{pmatrix} R_{CC} \\ 0 \\ 0 \end{pmatrix} \tag{2}$$

TABLE 1 (cont'd.)

[1] here $\phi = 0$ when successive z-axes are trans. **R** is the rotational angle matrix and **D** is the displacement vector (polar coordinates). The column vector (x') represents the coordinates based on C_1 and (x) is the vector based on the system listed. For T_1 to T_4 of neopentane, ϕ and θ values are: $\phi_i = (i-1)\pi/2$; $\theta_i = \pi(i-2) - (-1)^i\Theta_{CC}\cdot \pi/2$, Θ_{CC} are pre-specified in rows 1 and 5 and therefore, by rule 7, are not counted as separate specifications for rows 6–17. Rule 7 requires, in addition, that only a single set of ϕ_i, θ_i, R_{CC} values is shared by transformations associated with a statement; otherwise, an increase in complexity is involved. In neopentane, unlike isopentane, only a single set is involved

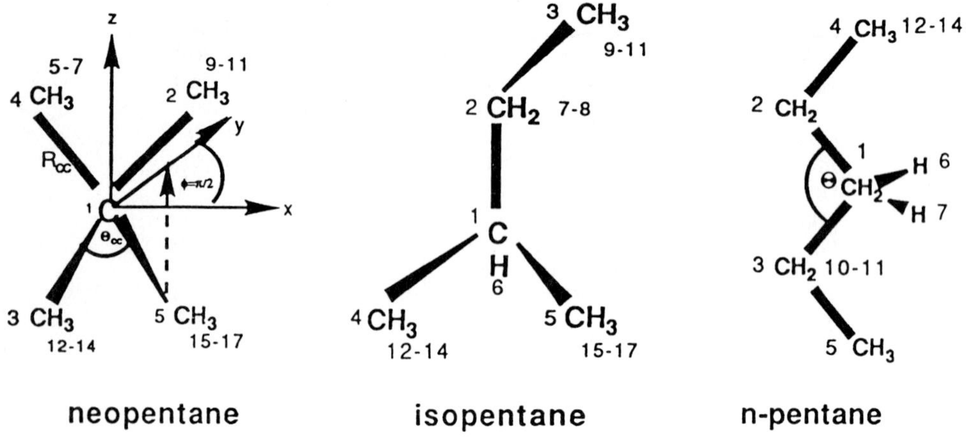

FIGURE 1

that the coordinates of neopentane assume are listed in the *specification table* for neopentane (Table 1). This table contains 68 numerical entries; 68 is consequently an upper limit for the complexity of a pentane molecule.

The actual structural complexity of neopentane is, however, considerably lower than 68, for three reasons:

1. The placement coordinates, $c' = 6$ (four zeros and $\pm\Theta_{CC}/2$), should not be counted, because they fix the position of the molecule in the external space, independent of the internal complexity.
2. More significant, four coordinates in each pentane molecule have no fixed value (at high enough temperatures) because of the free rotations around four of the bonds. This results in four ϕ angles having indeterminate values, designated "Any_1" to "Any_4" in Table 1. In other words, the values of these ϕ angles are different for each molecule in a molecular ensemble as well as at any particular time point. These coordinates cannot be considered as ordered features and, as said, do not contribute to the complexity of the system.
3. Many entries in the table are redundant because of the many specifications that are either equal or inter-related. These specifications can be correlated by short statements like: $r_i = \text{R}_{\text{CC}}, i = 2-5$ (or: $r_{2-5} = \text{R}_{\text{CC}}$) for the four methyl carbons ($i = 2-5$ is a range statement; R_{CC}, is the carbon-carbon bond length). These four r_i values thus form, by rule 3, a single $c(4)$ contribution to Eq. (1).

Points 1 and 2 imply that the *maximal complexity* C_{max} that a pentane analogue with no regular feature can attain is: $C_{\text{max}} = 4n - c' - c_{\text{ran}} = 68 - 6 - 4 = 58$. An examination of Table 1 (and of analogous tables constructed with different coordinate or numbering systems) leads to the conclusion that polar coordinates give the most concise instruction set for neopentane, comprised of the following minimal set of statements:

1. $\varepsilon_1 = \text{C}$
2. $\varepsilon_{2-5} = \text{C}_{\text{pri}}$
3. $\varepsilon_{6-17} = \text{H}$
4. $r_1 = 0$
5. $r_{2-5} = \text{R}_{\text{CC}}$
6. $r_{16-17} = \text{R}_{\text{CH}} \quad (T = T_k); \quad k = \text{INT}(i/3) - 1 \quad (\text{i.e., } 1, 2, 3, \text{or } 4)$
7. $\phi_1 = 0$
8. $\phi_{2-5} = \pi(i-2)/2$
9. $\phi_{6-17} = \text{Any}_k + 2\pi(i-6)/3 \quad (T = T_k)$
10. $\phi_1 = 0$
11. $\phi_{2-5} = \pi i + (-1)^i\Theta_{\text{CC}}/2$
12. $\theta_{6-17} = \Theta_{\text{CH}} \qquad (T = T_k)$

C_{pri} is a primary methyl carbon. R_{CC}; R_{CH} are carbon-carbon and carbon-hydrogen bond lengths. Θ_{CC}; Θ_{CH} are CCC and CCH bond angles.

These 12 statements provide all the information needed to construct a neopentane molecule. Statements 4, 7, and 10 are, however, placement statements, which do not contribute to the complexity of the molecule. The remaining nine statements are required and lead to a value of $C = 9$ for the structural complexity of neopentane. If we want to relate this value to the maximal complexity available for a 17-atom system with four random coordinates, we obtain a *relative complexity* $C_r = C/C_{\max}$ of 9/58, i.e., $C_r = 0.155$ for neopentane. The four indeterminate "Any" are included in statement 9, which in addition contains the constant $2/3\pi$. Relative complexity values can help to relate complexities of differently sized systems.

---

4. n-PENTANE AND isoPENTANE.

Is neopentane more or less complex than n-pentane $(\underset{4}{\mathrm{CH_3}} \,.\, \underset{2}{\mathrm{CH_2}} \,.\, \underset{1}{\mathrm{CH_2}} \,.\, \underset{3}{\mathrm{CH_2}} \,.\, \underset{5}{\mathrm{CH_3}})$ or isopentane $(\underset{3}{\mathrm{CH_3}} \,.\, \underset{2}{\mathrm{CH_2}} \,.\, \underset{1}{\mathrm{CH}} \;.\, \underset{4,5}{(\mathrm{CH_3})_2})$? An effort to answer this question was the incentive to analyze the pentanes. To this end, the specification tables for n- and isopentanes were set up and examined. The following minimal sets of statements resulted (the student is encouraged to do that).

For n-pentane (for numbering see Figure 1):

1. $\varepsilon_{1-3} = \mathrm{C_{sec}}$
2. $\varepsilon_{4-5} - \mathrm{C_{pri}}$
3. $\varepsilon_{6-17} = \mathrm{H}$
4. $r_1 = 0$
5. $r_{2-3} = \mathrm{R_{CC'}}$ $\quad (T_0)$
6. $r_{4-5} = \mathrm{R_{CC''}}$ $\quad (T_{1,2})$
7. $r_{6-11} = \mathrm{R_{CH'}}$ $\quad (T_{0,1,2})$
8. $r_{12-17} = \mathrm{R_{CH''}}$ $\quad (T_{3,4})$
9. $\phi_{1,2} = 0$
10. $\phi_3 = \pi$
11. $\phi_4 = \mathrm{Any}$
12. $\phi_5 = \mathrm{Any}$
13. $\phi_{6-11} = \mathrm{Any}_k + (-1)^i\pi/2 \quad k = \mathrm{INT}(1/2) - 3 \quad (\text{i.e.}, 0, 1, \text{or } 2)(T = T_k)$
14. $\phi_{12-17} = \mathrm{Any}_k + 2\pi i/3 \quad k = \mathrm{INT}(i/3) - 1 \quad (\text{i.e.}, 3 \text{ or } 4)(T = T_k)$
15. $\theta_1 = 0$
16. $\theta_{2-5} = (-1)^i\Theta_{cc}/2$
17. $\theta_{6-11} = \Theta_{\mathrm{CH}} \quad (T = T_k)$
18. $\theta_{12-17} = \pi - \Theta_{\mathrm{CH}} \quad (T = T_k)$

$\mathrm{C_{sec}}$ is a secondary methylenic carbon. $\mathrm{R_{CC'}}$; $\mathrm{R_{CC''}}$ are the distances between primary-secondary and secondary-secondary carbons, respectively. Statements 4, 9, and 15 are placement statements; statements 11 and 12 refer to random coordinates only. This leaves 13 statements to describe the ordered part of the molecule. Consequently, the complexity of n-pentane is $C = 13$ and $C_r = 13/58 = 0.225$, more complex than neopentane on both absolute and relative scales.

For isopentane:

1. $\varepsilon_1 = \mathrm{C_{ter}}$
2. $\varepsilon_2 = \mathrm{C_{sec}}$
3. $\varepsilon_{3-5} - \mathrm{C_{pri}}$
4. $\varepsilon_{6-17} = \mathrm{H}$
5. $r_1 = 0$
6. $r_2 = \mathrm{R_{CC''}}$
7. $r_3 = \mathrm{R_{CC'}}$ (T_1)
8. $r_{4-5} - \mathrm{R_{CC'}}$
9. $r_6 = \mathrm{R_{CH'''}}$
10. $r_{7-8} = \mathrm{R_{CH''}}$ (T_1)
11. $r_{9-17} = \mathrm{R_{CH'}}$ $(T_2, T_{3,4})$
12. $\phi_{1,2} = 0$
13. $\phi_3 = \mathrm{Any}_1$ (T_1)
14. $\phi_{4,5} = -\Phi_{\mathrm{CC}}$
15. $\phi_6 = 0$
16. $\phi_{7-8} = \mathrm{Any}_1 + (-1)^i\ \Phi_{\mathrm{CH''}}$ (T_1)
17. $\phi_{9-17} = \mathrm{Any}_k + 2\pi i/3; k = \mathrm{INT}(i/3) - 1,$ i.e., 2, 3, or 4 $(T_2, T_{3,4})$
18. $\phi_{1,2} = 0$
19. $\theta_3 = \pi - \Theta_{\mathrm{CC''}}$ (T_1)
20. $\theta_{4-5} = \Theta_{\mathrm{CC'}}$
21. $\theta_6 = \Theta_{\mathrm{CH'''}}$
22. $\theta_{7-8} = \Theta_{\mathrm{CH''}}$ (T_1)
23. $\theta_{9-17} = \Theta_{\mathrm{CH'}}$ $(T_2, T_{3,4})$

Statements 5, 12, 15, and 18 are placements and 13 is random. On the other hand, statements 11, 17, and 23 have to be counted twice, because they each involve two different transformations (T_2, which transforms from C_3 to C_1, is different from T_3, T_4). This results in 21 necessary statements; i.e., the structural complexity of isopentane is C = 21 (C_r = 0.36). Isopentane is thus the most complex of the three noncyclic pentanes, as intuitively expected. Note that both isopentane and n-pentane have a plane of symmetry at certain values of "Any$_k$." Symmetry relations so far have not been too helpful; further analysis, nevertheless, might be rewarding.

CONCLUSIONS

The examples analyzed demonstrate that:

a. A value for the structural complexity of a typed point system can be assigned. This assignment is based on a somewhat arbitrary set of rules. Practice shows, however, that changing these rules leads to inconsistencies when the same system is analyzed in different ways. A more rigorous mathematical analysis is needed to determine whether the assignments are indeed unique and whether algorithms can be devised leading uniquely to these assignments.

b. An important step in the complexity analysis of any system is the determination of which coordinates are random and which are ordered (whether regular or uniquely specified). The test is, in principle, simple: Let us examine a certain number of systems in an ensemble, for example, molecules in a specimen. If a certain coordinate assumes the same value in each molecule of the ensemble, then it belongs to the ordered repertoire. On the cellular level, one can compare, for instance, tubuli cells in the kidney to red blood cells. Tubuli cells are arranged in a radial fashion around the kidney tubuli, so that they represent an ordered, fairly regular set, and their contribution to the complexity of the organ can be assessed. In contrast, an erythrocyte (red cell) can be found anywhere in the blood stream; its positional coordinates are random, and no complexity value can be assigned to the arrangement of the erythrocytes in the organism. This randomness test has to be applied for each coordinate before the complexity of any element can be assessed.

c. The formalism permits the assignment of a value not only to the complexity but also to the degree of ordering for a system. This can be done by simply subtracting those coordinates that are indeterminate in the system; for instance, order $= 58/62$ for each of the three noncyclic pentanes. In cyclopentane ($n = 15$), order $= 52/54$, because the ring constraints leave only two indeterminate angles, the "pseudo rotation" and one torsion angle.[9] The distinction between random and ordered coordinates is important because not only biosystems, but most real-world systems, have indeterminate coordinates; consider a point on the rim of a car wheel or the number of twigs on a tree. Most real systems are only partially ordered, just like the pentanes. Structural complexity is relevant and assignable only to the ordered part of a system. The distinction between ordered and random coordinates of a system is a fundamental feature of the treatment presented, separating it from all previous treatments of the subject.

d. Structural complexities are extra thermodynamic quantities, because the structural complexities are determined by the stable molecular bonds and not by the occupancy of internal energy levels (except for possible rotational levels associated with random coordinates). Complexity differences persist at 0°K, where all internal energies are in the ground state and where, according to the third law of thermodynamics, all crystalline (ordered) compounds have a physical entropy of zero. Further, while conventionally measured entropy is an extensive property of systems, structural complexity is an intensive property, the complexity of a single pentane molecule being equal to that of a mol.

e. Structural complexity is low for most natural systems but will assume high values in designed systems—systems that are created with the help of instructions specifying their pattern and composition. In the primitive molecular systems tackled here, instructions are provided by specific catalysts that can direct a chemical reaction towards one isomer (that is, select one pentane isomer in preference to others). The degree of complexity thus achieved is, as we have seen, not too high. Higher degrees of complexity can be achieved when, in addition to a catalyst (enzyme), template molecules participate, like in DNA or protein biosynthesis. The high degree of complexity attained in the bioworld would be unthinkable without participation of replicable templates. In contrast to simple catalysts, templates can store and transmit large amounts of information, and their active presence accounts for the high complexity found in bio-organisms. Even higher degrees of complexity are achieved in artificial systems created by intelligent beings: The creation of a template, blueprint, or a design (all synonyms for the present discussion), whether in the mind of the designer or on paper, is an essential step in making complicated instruments or works of art. Therefore we can expect that the concept of structural complexity will reach its full utility in the physical and chemical analysis of templated and otherwise designed systems.

REFERENCES

1. *Random House Dictionary*, 1987.
2. Bennet, C. H. "Entropy and Information: How to Define Complexity in Physics and Why." In *Complexity, Entropy, and Physics of Information*, edited by W. H. Zurek. Santa Fe Institute Studies in the Sciences of Complexity, Proc. Vol. VII, 137–148. Redwood City, CA: Addison-Wesley, 1990.
3. Chaitin, G. J. *Algorithmic Information Theory.* Cambridge, MA: Cambridge University Press, 1987.
4. Dawkins, R. *The Blind Watchmaker.* New York: W. W. Norton, 1987.
5. Feistel, R., and W. Ebeling. *Evolution of Complex Systems.* Dordrecht: Kluwer, 1989.
6. Flory, P. J. *The Statistical Mechanics of Chain Molecules.* New York: Interscience, 1969.
7. Green, P. B. "Inheritance of Pattern: Analysis from Phenotype to Gene." *Amer. Zoologist* **27** (1987): 657–673.
8. Kolmogorov, A. N. "Three Approaches to the Quantitative Definition of Information." *Prob. in Infor. Trans.* **1** (1965): 4–7.
9. Sanger, W. *Principles of Nucleic Acid Structure.* New York: Springer-Verlag, 1983.
10. Solomonoff, R. J. "A Formal Theory of Inductive Reference." *Infor. & Cont.* **7** (1964): 1–22; 224–254.
11. Wicken, J. S. *Evolution, Thermodynamics, and Information.* Oxford: Oxford University Press, 1987.
12. Wood, W. B. "T4 Morphogenesis." *Quart. Rev. Biol.* **55** (1980): 353–387.
13. Yagil, G. "On the Structural Complexity of Simple Biosystems." *J. Theor. Biol.* **112** (1985): 1–23.
14. Yagil, G. "Complexity Analysis of a Protein Molecule." In *Proceedings, First European Conference on Mathematical and Theoretical Biology*, edited by J. Demongeot. Heidelberg: Springer-Verlag, 1992.
15. Zurek, W. H. "Algorithmic Information Content, Church-Turing Hypothesis, Physical Entropy and Maxwell's Demon." In *Complexity, Entropy, and Physics of Information*, edited by W. H. Zurek. Santa Fe Institute Studies in the Sciences of Complexity, Proc. Vol. VII, 73–89. Redwood City, CA: Addison-Wesley, 1990.

Gottfried Mayer-Kress
Center for Complex Systems Research, Department of Physics, University of Illinois at Urbana-Champaign, 3025 Beckman Institute, 405 N. Mathews, Urbana, IL 61801; gmk@pegasos.ccsr.uiuc.edu

Global Information Systems and Nonlinear Methods in Crisis Management

Crisis management can be seen as one of the major problems of sustainable development in the post-Cold-War world order. Traditional modeling approaches, based on closed descriptions of more or less abstract global systems, do not appear to be adequate for the new challenges. We suggest that new evolutionary, integrated models will make extensive use of a rapidly growing global computer network that will permit direct communication and efficient exchange of information as well as quantitative and conceptual sub-models and simulations.

We present a very incomplete overview of some of the information and modeling tools available today on the internet. We discuss some recent network discussions on the current regional crises in the Balkans and how distributed integrated models on the internet might help to prevent the violent escalation of future crises.

1992 Lectures in Complex Systems, Eds. L. Nadel & D. Stein, SFI Studies in the Sciences of Complexity, Lect. Vol. V, Addison-Wesley, 1993

INTRODUCTION

In this chapter we want to discuss some limited aspects of the development of better tools to evaluate policy decisions in a globally connected, complex world (see Brecke[3] and Isard[11] for an overview of existing tools and problems[21]). Recognizing that a solution to all problems is not feasible in the foreseeable future, we take a pragmatic perspective. Under the assumption that we have access to a well-developed, global computer network,[1] we identify the following steps:

i. Define targets for the solution of important problem areas (population, CO_2 level, violation of human rights, etc.) and assign a relevance weight to each of the problem areas;

ii. Acquire qualitative information on the current status of the problem;

iii. Define sub-areas where a quantitative approach appears to be promising;

iv. For those areas in iii, obtain current quantitative data and identify models that deal with the solution of any of the sub-problems;

v. Create a conceptual model of the integrated system;

vi. Link data and simulation models to an interdependent, distributed network;

vii. Perform simulation, sensitivity analysis; and

viii. Compare the results with the updated information from ii and iii and evaluate them with respect to the targets specified in i (see Hasselman[8] and Forrest et al.[5] for a similar discussion).

From the study of nonlinear and chaotic systems, we know that only short-term predictions are possible if the system exhibits chaos (see, for example, Grossman,[6] Abraham et al.,[1] Campbell et al.,[4] and Mayer-Kress[13,14,15]). Therefore, a typical five-year time scale between formulation and verification of a global model appears to be too long in a world where time scales of, say, eastern Euopean regional conflicts are significantly shorter than one year. Future models will have to be object-oriented with links to other models and information systems and they will have to be adaptive to changing basic conditions. Here we mainly focus on items ii, iv, and v in the above list.

First, we have to realize that we are faced with vast amounts of quantitative data such as that from satellite-based Earth Observing Systems (EOS). EOS will transmit daily an amount of data equivalent to about 100,000 complete works of Shakespeare. But we also have to use less quantitative knowledge based on wisdom and insights that cannot be easily described in tera-bytes. Future modeling

[1]Today this assumption is valid to a degree that depends strongly on factors like geographical location or affiliation with an academic institution.

approaches will have to tap into the global wisdom of informal, anecdotal, and descriptive knowledge as well as into the results of extensive quantitative analysis and supercomputer computations.[2]

Many quantitative models suffer when researchers jump too abruptly from an unspecific conceptual analysis to a highly complicated formal model, a move sometimes impossible or very difficult to justify in detail. This is true especially in disciplines with a tradition of quantitative models, for example, in econometrics where routines of mapping concepts onto categories of models have been developed which make it tempting for the user to focus on mathematical details and the model's implications without questioning thoroughly the validity of its assumptions and approximations. Therefore, we suggest using conceptual models as completely as possible and then using quantitative simulations as necessary.[3] The development of intuitive human-computer interfaces, visualization, and audification of complex structures and processes can be helpful in developing more realistic models and rapidly detecting their problems which would be hidden in a purely quantitative description.

For a successful global modeling approach it is not sufficient to have a good interface with a local computer and database. One of the main challenges will be the interconnection between all distributed computational and informational units. This also includes efficient communication among the researchers who work jointly on a distributed project. Conference calls and fax machines are fairly limited methods of scientific collaborations. We believe that multimedia electronic mail is a more appropriate way to exchange and distribute information within a geographically separated team.[4] The Sequoia 2000 project of the University of California[5] uses distributed data management tools to make about 100 Tera-bytes of global change data available to researchers on the internet.[25]

In a few years we expect to approach a world population of 10 billion ($= 10^{10}$), about the same number of neurons in a brain. It has been speculated that if global communication and the connections between human "units" improves, at some point the qualitative nature of the global human network will change and something like a "global brain" will emerge.[22] While we should not take this analogy too

[2] For example, "Project Gutenberg" transfers classic literature into digital form and makes it available on the internet—from fairy tales to government documents (more information is available from hart@vmd.cso.uiuc.edu).

[3] The use of conceptual models has a long tradition, for example, at IIASA; see, for example, Shaw et al.[24]

[4] Arbitrary documents of formatted text, directory folders, executable programs, graphics, and sound can be sent by modern mail programs (for example, NeXT mail) by simply dragging and dropping the corresponding icon into the mail document. The recipients access the documents in the same way by either dragging the icon into a storage folder or by simply double-clicking the icons.

[5] Reports and graphic materials describing the project are available from a central ftp server (postgres.berkeley.edu).

literally, nevertheless it can serve as a useful paradigm for conceptualizing the nature of future simulation and global modeling approaches.

Today we experience the emergence of a technology that allows an unprecedented degree of communication and information exchange (but also very intimate, person-to-person, conversational discourse)[6] between people located anywhere in the world (that has a connection to the network).[7] In this short paper, we want to give a *necessarily incomplete* description of some of the information, communication, and simulation tools that are already available.

ELECTRONIC INFORMATION SERVERS

While it is foreseeable that all information will be stored in digital form on some (opto-)electronic media, the breathtaking developments in storage capacity and the dramatic reduction of storage cost per giga-byte need not be discussed here. We will mention a few examples of the electronically stored information systems available. In Figure 1 we present schematically a cross section of different services that are available today on the internet. In Appendix A we list the electronic mail addresses of those services.

ELECTRONIC LIBRARY CATALOGS

Electronic library catalogs allow us to get information about literature relevant to our problem directly from our desk. The time and effort it takes to answer a question is directly related to the probability that this question will be answered. Electronic catalogs have decreased the delay between asking a question, finding a relevant publication, and having access to that publication. This is especially true when the library can provide the publication via electronic media: either by fax or by scanning in the text and mailing it electronically.

In Table 1 we have a list of library catalogs currently available through internet; hence, the physical location of the catalog becomes irrelevant. Networking among libraries to optimize service by reducing work duplication poses a considerable challenge. Today the familiarity and convenience of the user interface of the library seems to be one of the determining factors in the choice of a library catalog.

[6] It is interesting to observe the emergence of notation among net users that allows the efficient expression of emotions. For a while, expressions like (smile), (frown), etc. were inserted or appended to messages. More recently a more iconographic notation, such as ":-) 8-) :-) !-)," has become quite popular. The availability of multimedia mail (such as NeXT mail) will increase these possibilities (for example, through sound or personal picture icons).

[7] Psychological resistance to new information technologies is apparent. The enormous spread of fax machines is a phenomenon that probably will be the subject of future psychological/sociological studies.

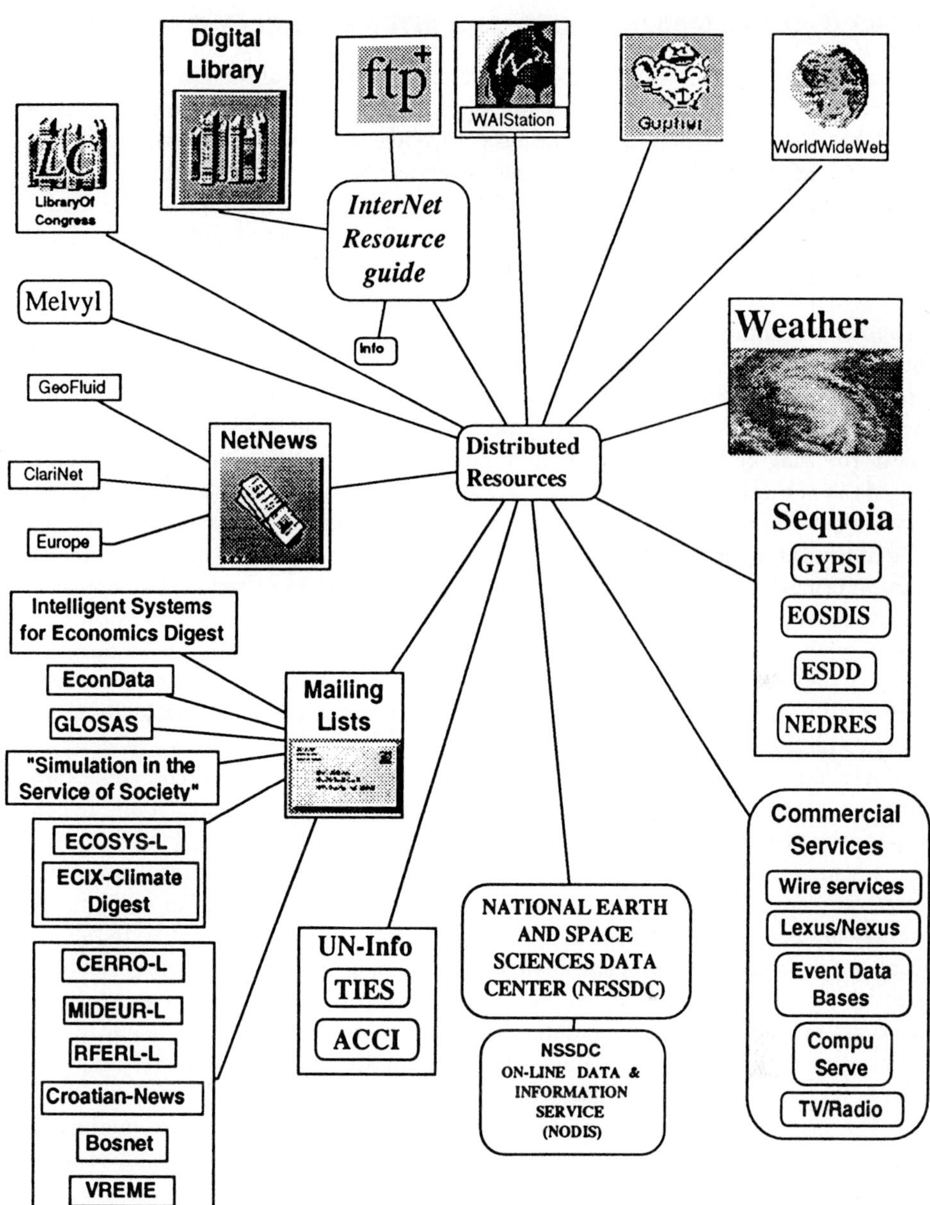

FIGURE 1 Network of distributed information and simulation systems (see Appendix A).

NEWS GROUPS, MAILING LISTS, AND INTERNET RELAY CHAT (IRC)

It is a well-known fact among scientists that direct personal interaction is a tremendous source of information and inspiration. The ability to walk down the hallway and chat with an expert colleague used to make a difference between a first-class research institution and the academic province, where the information from that short conversation has to be painfully retrieved from the literature instead. An electronic analog of this research environment is represented today in the form of network news groups and mailing lists.

In *news groups*, discussions take place through submission to either a news group or a mailing list of contributions covering a specific topic. In the first case, the message will be posted to an electronic bulletin board and then retrieved through special news software that connects a personal computer or workstation with a news server, i.e., a computer that stores a subset of the news items. Probably the most relevant criterion for the usefulness of net news (besides the transmission speed) is the quality of the user interface: the amount of available information is often overwhelming, and the fraction of really interesting items can be very small. Therefore the speed at which one can identify a message with respect to its value can be crucial in deciding if this service will be accepted or not. Filter options are essential since often news groups are dominated by contributors who flood a news group with irrelevant (or worse) contributions. Other time sinks can be discussions on very popular topics that continue over a long stretch of time. Tools that allow the user to filter out messages according to author or topic can dramatically increase the productivity of news sessions.

Mailing lists are a very efficient and flexible way to create the spontaneous, informal exchange of information and ideas by a group of people. While net news groups require a certain amount of administration (there are regulated procedures for setting up new groups, their content is archived, etc.), mailing lists only use electronic mail for the organization of the information exchange. For example, temporary mailing lists can be set up by individual groups as an alternative to telephone conference calls. They have the advantage that an arbitrary large number of readers can be reached within minutes; i.e., quasi-interactive discussions are possible. As opposed to conference calls, mailing lists automatically provide the means for documentation of the discussion. While small mailing lists can be set up very quickly by just collecting the names of users into one mail alias, this is certainly infeasible for large international or global mailing lists. In those cases, automated mailing lists handle most of the administrative work: subscription and cancellation of mailing list memberships is done by sending a "subscribe" or "unsubscribe" note to a mailing list server. Some mailing lists provide not only discussions among participants but also act as a news service and distribute news from UPI, RFE, ClariNet,[8] as

[8]The UPI sources that we quote were available via a mailing list. Problems arise when news from commercial services are redistributed through internet mailing lists.

well as often detailed news from local sources, for example, VREME, a Belgrade-based weekly.[9]

Internet Relay Chat (IRC) is a tool that allows continuous conversation, on selected topics, in a virtual lounge (see, for example, Reid[19] for an extensive report). It is very efficient for rapid answers to simple questions since there is a good chance that someone who knows the answer is listening to the conversation. Also, in this informal environment, people get to know each other and can create a community which is loosely connected all over the world. Physical location becomes less relevant; the fellow from Australia or Finland might turn out to be the better expert in some areas than the colleague from MIT or Berkeley. One problem is the continuous demand for attention from the chat channel, which has led to the development of *cyber-friends*—programs that chat on the net while the real person is busy doing something else.

This raises the issue of misinformation. It is relatively easy to post messages to the network with wrong references to sources, apparently originating from different geographical locations. This will become a serious problem and procedures need to be developed to protect users against planted fake news and data. Users need to develop a critical approach to information from the net. For example, potential sources of abuse might be quotes, from reputable newspapers or wire services, easily falsified for the mailing list. But this seems to be a general problem of public information. On network-based systems, we have the advantage of immediate public questioning of the author and requests for confirmation or alternative opinions.

ANONYMOUS FTP SITES AND WIDE AREA INFORMATION SERVERS

Quantitative information as well as software, sound, images, etc. can be accessed through anonymous ftp sites[10]—storage media on some computer systems that can be publicly accessed through the internet. The number of those publicly accessible sites is so large that manual searches are basically hopeless. Therefore, network tools have been developed that allow searching across the internet for relevant items. To our knowledge the first such tool was "Archie," a network tool that searches through listings of all ftp archives. More sophisticated hyper-text and hyper-media search tools were developed recently. The most common ones are Wide Area Information Servers (WAIS), Gopher, and World Wide Web (WWW). Information servers can be indexed for each of these network tools; i.e., information about the structure and contents of a specific information source can be reported to the information servers.

For these three network tools, we give a brief description and references for more information.

[9] Distributed as "Vreme" News Digest Agency by Croatian-News@bumrl1.bu.edu.

[10] The term "ftp" stands for File Transfer Protocol, an internet standard that is used primarily for data transfer.

GOPHER.

> The Internet Gopher client/server provides a distributed information delivery system around which a world/campus-wide information system (CWIS) can readily be constructed. While providing a delivery vehicle for local information, Gopher facilitates access to other Gopher and information servers throughout the world.
>
> (from: pit-manager.mit.edu:/pub/usenet/news.answers/gopher-faq, Q0: What is Gopher?[11])

Gopher provides an efficient browse capability for a large number of different network services. Search capabilities, if applicable, are embedded in each of the sub-areas.

WAIS.

> Users on different platforms can access personal, company, and published information from one interface. The information can be anything: text, pictures, voice, or formatted documents. Since a single computer-to-computer protocol is used, information can be stored anywhere on different types of machines. Anyone can use this system since it uses natural language questions to find relevant documents. Relevant documents can be fed back to a server to refine the search. This avoids complicated query languages and vendor-specific systems. Successful searches can be automatically run to alert the user when new information becomes available.
>
> (B. Kahle, "Overview of Wide Area Information Servers," April 1991, quake.think.com:wais/doc/overview.txt.[12])

A search request can be addressed to a sub-list[13] of WAIS servers. The search result is a list of servers and documents with a quantitative indicator about their relevance to the request. Unfortunately, the exact nature of the search algorithm is not evident, making it difficult to precisely control the search or anticipate the results.

[11] More Gopher information is available via e-mail: gopher@boombox.micro.umn.edu.

[12] More WAIS information is available from the mailing list "wais-talk@think.com."

[13] A list of all WAIS servers and information about WAIS itself can also be obtained as a search result.

WWW. The World Wide Web has a distributed hyper-text structure; i.e., documents have links attached to specific regions in the text that allow access to new documents anywhere on the network. It seems that many of its general functions have been taken over by WAIS or Gopher, and WWW has concentrated on providing hyper-text information to the worldwide high energy physics community.[14]

These information servers are linked together; i.e., information that is accessed through one server is also available on the others. Since all the systems are still under development, it is not clear how well this works in daily use. Very recently, meta-search tools have been developed, where the search is not done on the ftp-site level but on the level of indexed sites.

VERONICA.

> Very Easy Rodent-Oriented Net-wide Index to Computerized Archives, Veronica offers a keyword search of most gopher-server menus in the entire gopher web. As Archie is to ftp archives, Veronica is to gopherspace. ...Veronica was designed as a response to the problem of resource discovery in the rapidly-expanding gopher web. Frustrated comments in the net news-groups have recently reflected the need for such a service. Additional motivation came from the comments of naive gopher users, several of whom assumed that a simple-to-use service would provide a means to find resources "without having to know where they are."
>
> (foster@cs.unr.edu (Steve Foster), November 17, 1992)

We should also mention the network tools *Prospero*, *Knowbots*, and *Netfind*. Details are available in Schwartz.[23] Besides the internet, there are, of course, many other specialized network and information systems. We only mention the electronic mail system that was installed recently for the member countries of the Conference on Security and Cooperation in Europe, and the Telecom Information Exchange Service of the United Nations (TIES). It seems that these systems have evolved in parallel and only were integrated recently into a globally interconnected network of networks. This parallel and decentralized evolution of sub-nets also resembles the evolution of nerve connections in biological brains.

[14] More information about WWW can be obtained via e-mail from the mailing list "www-talk@nxoc01.cern.ch."

INTEGRATION OF VIDEO AND TELEVISION WITH COMPUTER NETWORKS

A second, parallel development in communication networks can be observed in radio and television networks. In spite (or because) of their popularity, they have several drawbacks with respect to the requirements of information systems: some main disadvantages of television systems are the lack of interactivity and the difficulty in searching for specific information. Efforts exist to make television more interactive through a variety of call-in options. Interactivity would be especially important for "Global Townhall" types of projects, where citizens are encouraged to more actively participate in political discussions and even decisions. Integration of computers in the interactive aspects of television seems to be a natural perspective. The second problem of the current television system is its inefficiency in conveying relevant information according to the individual viewer's needs.

A multitude of television channels allows for multiple choices by the viewer, which is relatively unspecific, like the choice between sports and politics.

If we focus on television news, for example, as provided by CNN, then we observe a vast inefficient information server. This inefficiency could be measured as the ratio between the amount of information that is uploaded via satellite to the television headquarters to the amount of air time that is given to specific news items. From a viewer's perspective, a strong imbalance exists between the news provided by the news service and the fraction of the news items that are actually of interest. Many, if not all, news items are broadcast repeatedly, and viewers who are specifically interested in details of a specific topic will spend too much time scanning the news for new information.

Specifically, scanning and individual storage of relevant news could be handled effectively by an efficient integration of television (TV) and computer networks. We can envision a geographic information system (GIS) that can be programmed to scan television news according to a specific topic and a specific geographical location. For the viewer this would mean that, at any time, (s)he can get an instant overview from the GIS about where new developments in regional crises have arisen during the last 24 hours. By selecting a specific region and topic, the computer then could play back the recorded accumulation of news clips. In Figure 2(a) we show a simple configuration which allows a very easy and efficient example of how such an integration could take place. The main interface is the infrared (IR) receiver/transmitter[15] shown in the figure. It can record IR code from any audio-visual (AV) or TV unit that can be controlled by IR remote control. The computer

[15]Produced by Edmund Ronald, Paris, after some weeks of e-mail exchange based on a message that we had posted to comp.sys.next. Except for one night in Paris (for prototype testing), the entire collaboration took place electronically.

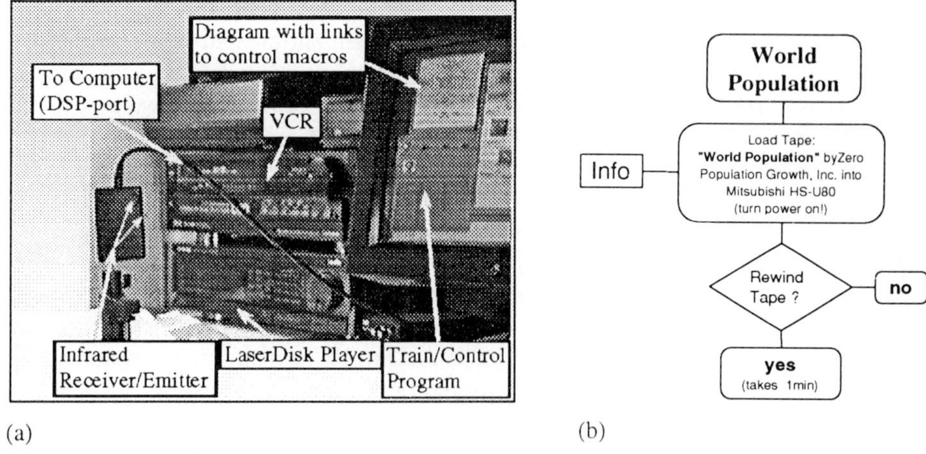

FIGURE 2 (a) Infrared interface and hardware configuration with NeXT monitor, laser disk player, and VCR. (b) Diagram interface that is linked to a node in a network diagram of global problems (see Campbell and Mayer-Kress[4] for a description).

can transmit control sequences from the same application (for example, the diagram in Figure 2(b)) to different AV or TV units. In the configuration shown in Figure 2(a), we control a laser disk player and a VCR simultaneously from a shell script that is linked to a diagram interface (Figure 2(b)). In this example the script will show a computerized animated map of the world from the laser disk player and then will present a video on population growth data.

Note that the integration of computers with the telephone/fax system is already much more advanced and will show some breakthrough developments in the very near future.

SIMULATION SERVERS

Modern computer workstations are used increasingly as mail, information access, and local storage tools. Computation tends to be done remotely on large computers that are accessed through some network. Software becomes more modularized and object-oriented such that many codes either in compiled or in source form are shared through anonymous ftp sites on the internet. Relatively generic programs like algorithms of a mathematical library are mainly shared as source codes. Other software is more specific and depends on the type of computer (mainly for personal computers). More recent software has been categorized according to the window system that is used for the front end: X-windows and NeXTStep are examples.[16] As long as these programs run locally and are not dependent on a special environment (libraries, data, etc.), they can be transferred easily and installed. Mailing lists that discuss different aspects of simulations and global computer networks provide a good source for the most recent developments in that area.[17]

In more complicated cases, the installation of software on a new computer can be considerable work. For those cases, where a port for the software to a different computer is difficult, an alternative seems to be software that can run a computer "kernel" on a remote machine with a "front end" local workstation. Today we have mathematical software available that organizes the "program" in a "notebook" that contains the software and the corresponding documentation as well as the results of the computation in multimedia format.[18] Multiple notebooks can be launched, for example, from a NeXT workstation, each of them linked to a kernel on a computer server anywhere on the network. From the remote kernel, one can then access remote data and libraries without the need for a port to the local machine. Future implementations of distributed models will have to provide similar integrated capabilities.

We should mention that there are already simulation servers on the network with traditional command-line interfaces: the user connects to the simulation server and then follows an interactive command menu. The desired parameters can be inserted, a batch job submitted, and the results of the simulation will be sent out via electronic mail. One elaborate system for geophysical problems[19] can be accessed together with the corresponding data bases. These more traditional simulation servers can be linked to integrated front ends, which could make the connections automatically and could send the appropriate commands to the server. The concept of having computer servers completely transparent to the user on the network is discussed in a meta-computer context.[23] In that scenario, specific programs can be run from any personal computer or workstation connected to the network and can be executed on any computer or supercomputer without the interference of the user who runs the program. For the user the appearance would be that of a single, powerful computing environment.

[16] Standard ftp site is export.lcs.mit.edu for X-based software and cs.orst.edu for NeXTStep-based software.

[17] We just mentioned Simulation in the Service of Society (available from <mcleod@sdsc.bitnet> and also published in the comp.simulation usenet news group). A project that integrates computer mailing lists with satellite links for global lectures is discussed in the Electronic Bulletin of the GLObal Systems Analysis and Simulation Association in the U.S.A. (GLOSAS@vm1.mcgill.ca).

[18] The address of one mathematics mailing list is mathgroup@yoda.physics.unc.edu; a collection of notebooks is archived at mathsource@wri.com. Sending any message to the latter address will result in a response with instructions.

[19] The geophysical models of the National Earth and Space Sciences Data Center (NESSDC) are accessible via nssdca.gsfc.nasa.gov.

COMPLEX ADAPTIVE MODELS FOR CRISIS MANAGEMENT

In the previous sections we gave an overview about methods to access global information, data, and simulation tools. Now we would like to speculate how these tools might be used to develop computational tools for crisis management. In the tradition of Richardson[20] and Lancaster,[12] we used quantitative models to attempt to understand the arms competition among nations or the dynamics of battle. The models are characterized by a small number of global variables (arms expenditures of nations, attrition rates of armies in battle, etc.). Thus, as long as we have well-defined countries and armies, those models will have a chance to describe some relevant aspects of the system. For example, operational planners of NATO forces used Lanchester-type models to estimate the requirements (troops, firepower, logistics, etc) needed to achieve a well-specified military goal, such as how to get Hussein out of Kuwait. One of the main problems for military planners in the Balkan is that there are no well-defined military objectives between the extremes "Drop nuclear bombs onto the area until the fighting stops!" and "Send in enough troops to protect any civilian against any aggressor." In the case of weakly coherent military units, partisans, militias, independent terrorist gangs, and robbers, those concepts do not work. Since it is unlikely (and the UN and EC attempts confirm this assumption) that a complete solution will be found, one can apply the concept outlined in the beginning of this chapter—identify global goals that, in a vague formulation, could include:

- discourage violation of international, humanitarian law,
- minimize suffering of civilians,
- discourage snipers and use of heavy arms,
- encourage and support the supply of humanitarian aid to civilian populations,
- etc.

This would mean a quite different approach to current UN strategies which could be described by these objectives:

- coordinate "peacekeeping" operations,
- provide humanitarian aid to civilians in occupied territories,
- protect refugees,
- avoid confrontations that might endanger the personal safety of UNPROFOR troops,
- etc.

The main difference between the two approaches concerns the basic division of responsibilities between local authorities and the representatives of global institutions like the UN, NATO, and EC. In the latter case the UN, say, takes very low-level responsibility, namely transporting food and medicine, blankets and clothing directly to individual children, even in violation of high-level, international sanctions.[20] The authority regarding police and security issues lies with the local groups that are in power, even when they are without any international or democratic legitimization[21] or even any form of government. The paradox arises that the representatives of the world community have to bribe local gangs for their permission to fulfill humanitarian tasks (like feeding and sheltering their children), tasks traditionally handled by the family or the lowest order of organization like a clan or tribe. This strategy, on a UN level, leads to a very rapid adaptation on the lowest level of anarchistic self-organization: aggressive behavior is rewarded; compliance with international law and civilized behavior dies out since it is not protected by a higher authority or it finds support outside the UN.[22] This local adaptation can even go so far that the result of a specific action is worse than the situation without any intervention. For example, a trade embargo against a whole country that is announced together with the assurance that it will not be enforced sends a very clear and strong message: it will be very profitable (since no risks are attached) to violate the sanctions while it will be very costly for established businesses to honor the sanctions and stop trading. Any business that will not violate the sanctions will go bankrupt and lose the business to black market organizations. Other, probably unintended, effects of the sanctions are described in the following UPI message, quoted as it was posted to cro-news@mph.sm.ucl.ac.uk:

> ...But the U.N. sanctions against Serbia have not forced Milosevic to agree to compromise in the ongoing internationally brokered search for peace.
>
> Instead, import taxes on the embargo-evading flood of fuel has been a major source of badly needed hard currencies by a regime drained by last year's Serb-Croat war in Croatia and the conflict in Bosnia-Herzegovina. The multi-million-dollar trade has also created a new class...experiencing the rapid growth of a syndrome of "the ever-greater bonding of the state power with the economic underground and mafia."
>
> The spread of the underground economy has been accompanied by a massive surge in violent crime. Belgrade's murder rate has hit an all-time high of at least two per day as gangsters armed with machine guns, hand grenades and shoulder-fired rockets vie for control of the lucrative black market.
>
> UPI - Nov, 15 (de2j@uva.pcmail.Virginia.EDU)

[20] "The Bosnian government Monday prepared to return clothing and children's shoes delivered by UNICEF over the weekend because the items came from Serbia, in violation of a U.N. embargo" (UPI, November 1, 1992, per Davor <de2j@uva.pcmail.Virginia.EDU>).

[21] In a similar, or perhaps even more extreme, situation, Somalia has local clans that determine the fate of international aid shipments..

[22] In the case of Bosnia-Herzegovina, this would naturally be Islamic countries, which then could lead to an escalation of the conflict.

Thus we have the interesting situation that, for the population under sanction, the supply is much better than the supply in cities, for which the UN has organized relief efforts. A similar situation occurred in Somalia: by stealing unprotected UN aid supplies and selling them, overpriced, in the markets, local gangs could afford to buy Qat, which is relatively expensive since it has to be flown in from Kenya. Although many factors contribute to this situation, some researchers speculate that it is more efficient to create an *economic* incentive for a humanitarian effort rather than organizing that effort on a UN level. The recommendation then would be to use global or international organizations like the UN or NATO to provide global parameters for the actors in the conflict region: Increased risks for snipers[23] or for crews of heavy weaponry near protected areas, with incentives for, say, private organizations who risk transporting aid into designated areas in response to economic incentives. These general principles of incentives and risks should be applied equally to each of the fighting parties and they should be enforceable without the need for assistance from the local group in power.

Independent of the specific situation, efficient information systems can be used to obtain data for dynamical models that are not integrated representations of nations in an arms race or of armies in a battle but that contain individual actors as elementary units. Especially with the help of supercomputers, the simulation of a few million simplified agents should be feasible and should capture essential elements that would lead to different types of collective behavior. The norms model by Axelrod[2] as well as the educational software toys "SimCity"[26] and "SimEarth"[27] use similar concepts for models of individual interacting agents that exhibit global emergent behavior of coherent dynamical structures. (The application of nonlinear mathematics and chaos theory to control complex and chaotic systems has been discussed in Hübler[9,10] and Ott et al.[17]) In our case these dynamical structures would correspond to collective behavior of aggression, flight, breaking of sanctions, smuggling, etc.

The empirical basis of such a model could be polls on the issues related to individual conditions for behavior (see Figure 3 for a simplistic schematic). In the case of the Serbo-Croat conflict, a psycho-cultural analysis can be found in Grossarth-Maticek.[7] In Figure 3 we have tried to summarize some of the findings of that study. Information systems of the type described in the first part of this chapter could make it possible to monitor those polls over time and use the inputs (in combination with results from more traditional fact-finding missions) for updating the models. The results from the model then can be checked against conditions that

[23]Modern electronic equipment makes it possible to return fire accurately and automatically within a fraction of a second. Snipers whose locations—for example, in Sarajewo—are generally known would experience a strong risk increase if such devices were installed in populated civilian areas.

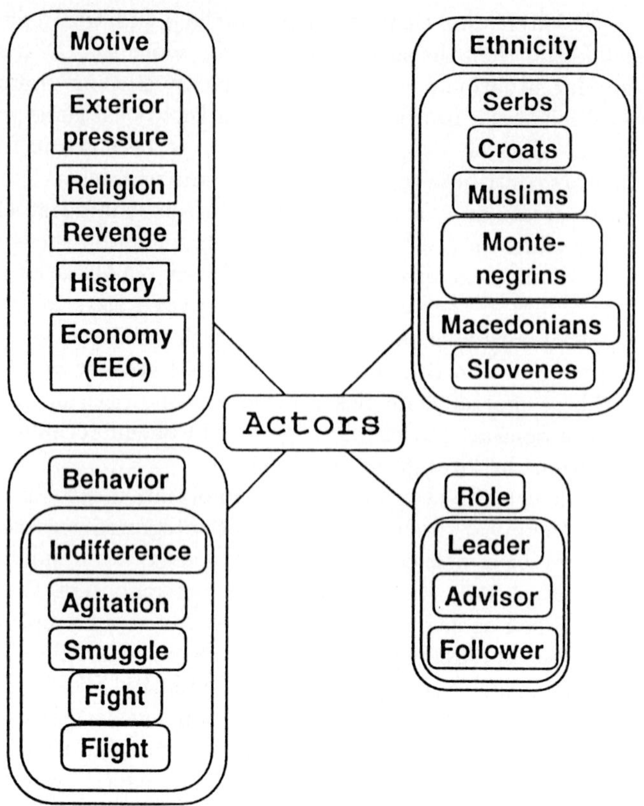

FIGURE 3 Simple diagram of categories for actors in the Balkan conflict.

would indicate, for example, the outbreak of a conflict. There seems to be empirical evidence for indicators that are highly predictive regarding the outbreak of a conflict. Some that were discussed by the German planning staff are[18]:

- image of an enemy
- national identification
- charismatic leader
- regional tensions
- gradient in wealth
- organizational capabilities
- dissatisfaction with the standard of living
- domestic repression

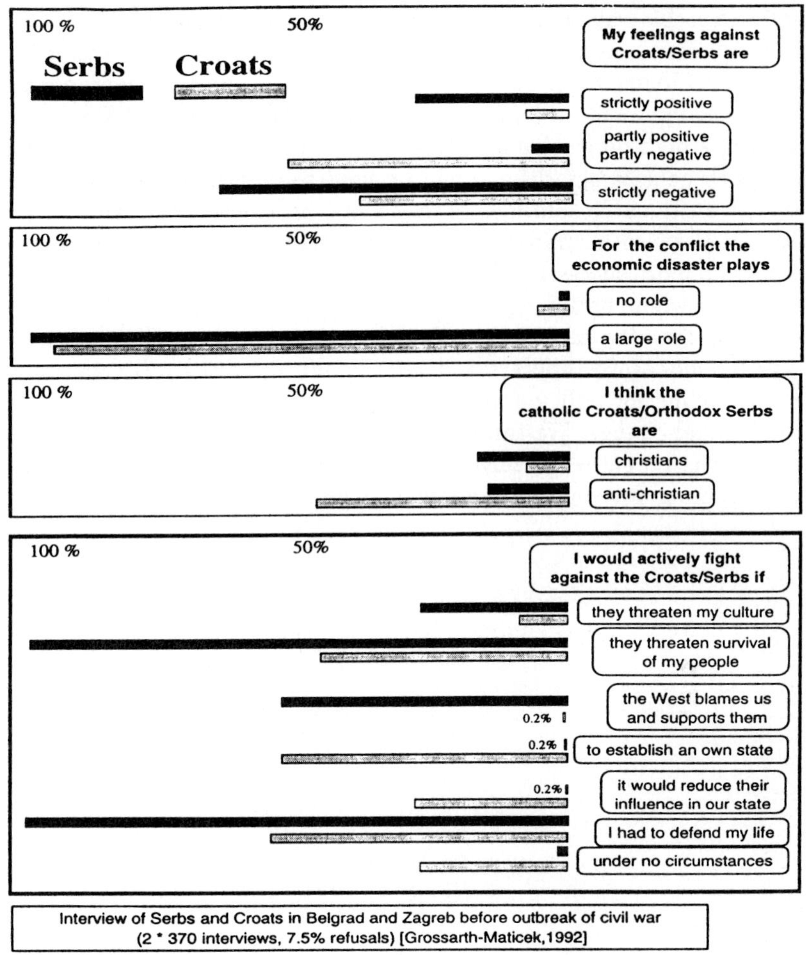

FIGURE 4 Serbo-Croat statistics prior to outbreak of civil war.

- restricted information
- military imbalance
- military-economic potential

With the help of modern tools, we can go much farther: the data presented in Figure 4 integrate across individuals for each of the questions. Thereby we lose most of the information that was gathered in the interviews, namely all the correlations between the answers to different questions by the same person. It is well known, for example, that the person who has strong negative feelings against the Serbs/Croats might often feel this way because (s)he thinks that they are anti-Christian, etc. With modern simulation and information tools, we can start exploiting this information on a large scale. Instead of integrated simple statistics, one could have a mapping of a population sample onto a highly structured (possibly low-dimensional) manifold embedded in a high-dimensional feature space. Exploration of these manifolds (for example, with genetic algorithms) can be used to design very specific, integrated pathways to sustainable solutions of, say, crisis problems. Even with very modest quantities of data with limited accuracy, it is easy to see that such an approach promises to be more successful than global, unspecific methods like economic sanctions against a whole nation.

We think that the integration of the nonlinear science of complex, adaptive systems with modern computers and information networks will create a new era of global modeling for policy evaluations in many different areas, especially in the domain of global change. We think that these methods will provide powerful tools and thereby the potential for misuse. Thus we need to encourage early discussion of these upcoming developments in a wide community.

APPENDIX A: ELECTRONIC MAIL ADDRESSES OF SERVICES REFERRED TO IN FIGURE 1

We are aware that this list is very incomplete and we apologize for all omitted services.

Global Change Information

ECIX CLIMATE DIGEST	larris@igc.apc.org
Weather information	sdm@madlab.sprl.umich.edu
Ecosystem theory and modeling	ECOSYS-L@vm.gmd.de
National Earth and Space Sciences Data Center	jcooper@nssdca.gsfc.nasa.gov
Sequoia 200	claire@postgres.berkeley.edu
EOSDIS Earth Observing System Data Info System	dozier@crseo.ucsb.edu
National Environmental Data Referal Service (NEDRES), Earth Science Data Directory (ESDD)	tgauslin@ridgisd.er.usgs.gov

Global Information

Telecom Information Exchange Service (TIES)	helpdesk@itu.arcom.ch
Simulation in the Service of Society (S3)	mcleod@Sdsc.Edu
GLObal Systems Analysis and Simulation	GLOSAS@vm2.mcgill.ca

Mailing Lists of Events in Eastern Europe

Discussion of Middle Europe topics	MIDEUR-L@ubvm.cc.buffalo.edu
Central European Regional Research Organization	<CERRO-L2AEARN.BITNET>
Radio Free Europe/Radio Liberty Daily Report, Inc.	RFERL-L@UBVM.cc.buffalo.edu
Bosnet	saiti@mth.msu.edu
Croatian-News/Hrvatski-Vjesnik	Croatian-News@andrew.cmu.edu
Cro-News/SCYU-Digest	Cro-News-Request@mph.sm.ucl.ac.uk
VREME	Dimitrije@buenga.bu.edu

Economic Data

Intelligent Systems for Economics Digest	IE-list@cs.ucl.ac.uk
EconData (info.umd.edu)	news@umd5.umd.edu

Network Information Tools

Internet Resource Guide	info-server@nnsc.nsf.net
World Wide Web (WWW)	www-talk@nxoc01.cern.ch
Wide Area Information Server (WAIS)	wais-talk@think.com
Gopher	gopher@boombox.micro. umn.edu
Internet Resource Guide	info-erver@nnsc.nsf.net

Libraries and News Services

Library of Congress	catalog@dra.com
MELVYL—University of California Catalog	melvyl@dla.ucop.edu
Digital Library	info@next.com
ClariNet News, a live electronic newspaper	info@psi.com

REFERENCES

1. Abraham, R., A. Keith, M. Koebbe, and G. Mayer-Kress. "Double Cusp Models, Public Opinion, and International Security." *Intl. J. Bifur. & Chaos* **1(2)** (1991): 417–430.
2. Axelrod, R. M. "An Evolutionary Approach to Norms." *Amer. Pol. Sci. Rev.* **80(4)** (1986).
3. Brecke, Peter. "Integrated Global Models that Run on Personal Computers." Preprint, School of International Affairs, Georgia Institute of Technology, June 1992.
4. Campbell, D., and G. Mayer-Kress. "Chaos and Politics: Simulations of Nonlinear Dynamical Models of International Arms Races." In the Proceedings of the United Nations University Symposium "The Impact of Chaos on Science and Society," Tokyo, Japan, April 15–17, 1991.
5. Forrest, S., and G. Mayer-Kress. "Using Genetic Algorithms in Nonlinear Dynamical Systems and International Security Models." In *The Genetic Algorithms Handbook*, edited by L. Davis. New York: Van Nostrand Reinhold, 1991.
6. Grossmann, S., and G. Mayer-Kress. "The Role of Chaotic Dynamics in Arms Race Models." *Nature* **337** (1989): 701–704.
7. Grossarth-Maticek, R. "Postkommunistische Krisen und Psychokulturerlle Lösungsmodelle am Beispiel des Serbo-Kroatischen Konflikts." Preprint, Institute für Präventive Medizin, April 1992.
8. Hasselman, K. "Wieviel ist der Wald wert?" *Der Spiegel* **41** (1992): 268.
9. Hübler, A., J. Miller, D. Pines, and N. Weber. "Optimal Adaptation in a Randomly Evolving Environment." Preprint, 1991.
10. Hübler, A. "Modeling and Control of Complex Systems: Paradigms and Applications." In *Modeling Complex Phenomena*, edited by L. Lam. New York: Springer, 1992.
11. Isard, W., S. Saltzman, and C. Smith, eds. "Contributions to the Cornell-NCSA Workshop on Large-Scale Social Science Models." Preprint, National Center for Supercomputing Applications, Urbana, IL, 1989.
12. Lanchester, F. W. "[need chapter title]." In *The World of Mathematics*, edited by J. R. Newman, vol. 4, 2138–2157. New York: Simon & Schuster, 1956.
13. Mayer-Kress, G. "Nonlinear Dynamics and Chaos in Arms Race Models." In *Proceedings of the Third Woodward Conference: Modelling Complex Systems*, edited by Lui Lam. Held in San Jose, April 12–13, 1991.
14. Mayer-Kress, G. "EarthStation." In *Out of Control*, edited by K. Gerbel. Ars Electronica 1991. Linz: Landesverlag Linz, 1991.
15. Mayer-Kress, G. "Chaos and Crises in International Systems." Technical Report CCSR-92-15, Center for Complex System Research, Urbana, Illinois, 1992. To appear in the proceedings of SHAPE Technology Symposium on crisis management, Mons, Belgium, March 19–20, 1992.

16. Meadows, D. H., and D. L. Meadows. *Beyond the Limits: Confronting Global Collapse, Envisioning a Sustainable Future.* Post Mills, VT: Chelsea Green, 1992.
17. Ott, E., C. Grebogi, and J. Yorke. "Controlling Chaos." *Phys. Rev. Lett.* **64 N11** (1990): 1196–1199.
18. Papenkort, B. Private communication, 1992.
19. Reid, E. M. "Electropolis: Communication and Community on Internet Relay Chat." Preprint, Department of History, University of Melbourne, 1991. Available from igc.apc.org:pub/ELECTROPOLIS/ELECTOPOLIS).
20. Richardson, L. F. *Arms and Insecurity.* Pittsburgh: Boxwood, 1960.
21. Rosen, L., and R. Glasser, eds. *Climate Change and Energy Policy: Proceedings of the International Conference on Global Climate Change, Its Mitigation Through Improved Production and Use of Energy, Los Alamos National Laboratory, October 21–24, 1991.* New York : American Institute of Physics, 1992.
22. Russell, P. *The Global Brain: Speculations on the Evolutionary Leap to Planetary Consciousness.* Boston: Houghton Mifflin, 1983.
23. Schwartz, M. F., A. Emtage, B. Kahle, and B. C. Neuman. "A Comparison of Internet Resource Discovery Approaches." *Comp. Sys.* **5.4** (1992). (Requests: brewster@Think.COM.)
24. Shaw, R., G. Gallopin, P. Weaver, and S. Öberg. "Sustainable Development, a Systems Approach." Status Report SR-92-6, IIASA, 1992.
25. Stonebraker, M. "An Overview of the Sequoia 2000 Project." Technical Report, Sequoia 2000, 1991. (Requests: claire@postgres.berkeley.edu.)
26. Wilson, J. L. *The SimCity Planning Commission Handbook.* Berkeley: Osborne McGraw-Hill, 1990.
27. Wilson, J. L. *The SimEarth Bible.* Berkeley: Osborne McGraw-Hill, 1991.

Student Contributions

Subbiah Baskaran† and David Noever‡
†Institute for Theoretical Chemistry, University of Vienna, Währingerstrasse 17, 1-1090, Vienna, Austria and ‡Biophysics Branch, ES-76, National Aeronautics and Space Administration, George C. Marshall Space Flight Center, Huntsville, AL 35812, USA

Excursion Sets and a Modified Genetic Algorithm: Intelligent Slicing of the Hypercube

The genetic algorithm (GA) represents a powerful class of search and optimization techniques developed in analogy to genetic laws and natural selection. A consistent picture of GA dynamics and convergence is reached here using ideas central to random field theory. Excursion sets are introduced to parameterize the GA's implicit parallelism and exponential elevation of subthreshold solutions toward optimum. Simulations on trial functions demonstrate this connection between a strong variant of the schema theorem and set theoretic concepts.

INTRODUCTION

In the last fifteen years, many interesting varieties of genetic algorithms (GA) have been designed and implemented.[2,3] The understanding of GAs can be represented on two dimensions: (1) the way user-defined objective functions map to the fitness measure and (2) the way the fitness measure is used to assign offspring to parents.

Along these two dimensions, almost all genetic algorithms exhibit some form of multiple sampling and implicit parallelism. The present work reports the results of a slightly different approach to understanding GA dynamics. By mediating selection through the introduction of excursion sets and random field theory, we achieve stronger conditions for implicit parallelism and simulate better GA performance. We call this new GA the excursion-set-mediated genetic algorithm (ESMGA).

In general, if the excursion sets arise in the *fitness function* space, then it should appear internal to the problem and remain hidden to the user; in contrast, if the excursion sets arise in the *objective function* space, then it becomes transparent to the user. Because one can generate excursion sets equally well, either in the objective function space or in the fitness function space, excursion levels introduce a natural hierarchical structure on the hypercube of available genomes. At increasingly higher excursion levels, population is forced to rise up in fitness above the excursion level, while at the same time distributing the evolved population among the possible solutions in the excursion set. Thus, by introducing the concept of excursion sets, we are able to judiciously balance both internal and external representations and thus to preserve a stronger condition for implicit parallelism.[4] This condition arises directly from the EMSGA's conservative attitude towards disrupting the higher order building blocks. In the remainder of this paper, we describe briefly the notion of excursion sets and explain the ESMGA in detail. Using simple mathematical arguments, we provide a theoretical justification for a strong version of schema theorem from excursion sets. We discuss the application and performance of the ESMGA using a trial function and conclude by indicating the direction of future research.

EXCURSION SETS AND OBJECTIVE FUNCTIONS

Excursion sets provide a natural basis to control the adaptive GA performance in terms of objective functions. We define an excursion set A_u at a given excursion level parameter u for any arbitrary objective function $f(x)$ as follows[1]:

$$A_u = \{x_i : f(x_i) \geq u\}\,. \tag{1}$$

Excursion sets induce a nontrivial hierarchy in search space that is represented in the evolving GA population. Excursion sets and local optima above the level u are closely related entities. For example, if only local optimal solutions of importance turn out to be those above the given level u, then these optimal solutions certainly lie within the excursion set A_u. However, an excursion set need not be in one single connected piece but, in general, it tends to be composed of finitely many components. The sum total of all excursion sets generated from a finite number of increasing excursion levels, $u_0 < u_1 < \ldots < u_k$, form a hierarchical space. Excursion sets at higher levels are contained in the excursion sets of the immediately lower level and so on. Thus we obtain the following strict inclusion:

$$A_{u_0} < A_{u_l} < A_{u_2} < \ldots < A_{u_k} .$$

However since the excursion set at any level contains finitely many disconnected components, we can write

$$A_{u_0} = U_{i=1}^{m} B_i \quad (2)$$

where B_i represents the m components of the excursion set at level u_0.

From the preceding, we observe that in terms of their objective function values, excursion sets at higher levels *strictly dominate* their counterparts at lower levels. While schemas can be thought to subdivide solutions in the genome space, the excursion sets alternatively subdivide the solutions in a hierarchical way within the objective function (/ fitness) space.

EXCURSION SETS AND MODIFIED GA DYNAMICS

FORMULATION OF ESMGA WITH EXCURSION SETS

Figure 1 schematically represents the various stages of the excursion-set-mediated GA dynamics. The procedure involves the following steps: (1) We generate a random population of genomes for an arbitrarily chosen excursion level u, and we partition the possible genome space into upper and lower portions. The upper set contains individuals whose fitness is greater than or equal to the excursion threshold (equivalently, the excursion set at that level). (2) During the selection stage, we retain the upper, excursion set component and fill the lower portion by performing tournament selection upon the entire population (excursion set plus its complement). This step introduces a certain bias against mating similar individuals (incest prevention). (3) The modification stage which follows this selection proceeds by applying the crossover and mutation operators *only to the lower components.* These operations produce the new population at generation 2. Finally, (4) the new population is evaluated; individuals scoring above the excursion level get pushed up into the excursion set and, subsequently, are preserved for future generations.

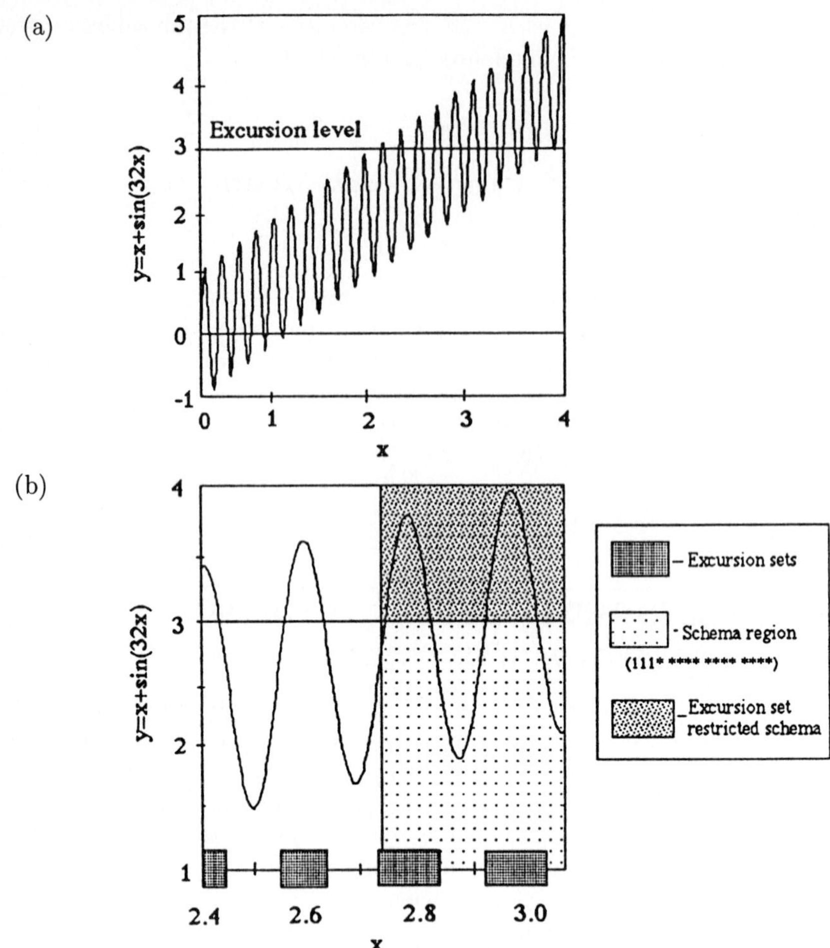

FIGURE 1 Basics of excursion sets and schemata. (a) The plot of the trial function $y = x + \sin 32x$ along with an excursion level at $y = 3.0$. Note the portions of the function lying in the excursion set. (b) The same trial function as in (a) but amplified and restricted to the excursion set. The 16-bit sample schema 111 ***** **** **** is indicated by the sparsely filled dots. The densely filled boxes lying on the x-axis show the disconnected components of the excursion set (4 in number). The combined effects of the excursion set and schema results in the area marked by double-dot density. This region corresponds to the subspace of this schema cut by the excursion set.

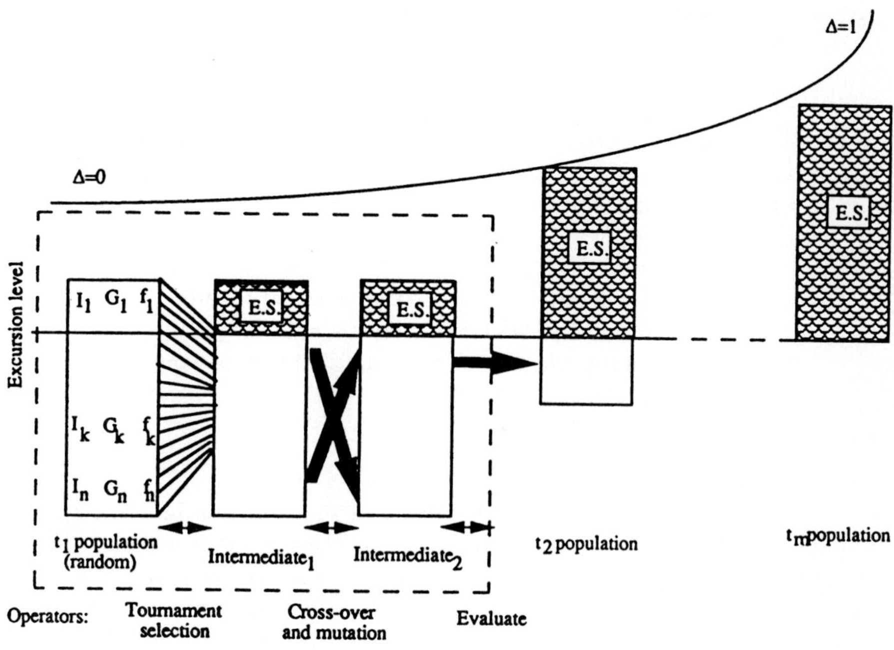

FIGURE 2 Algorithm schematic of the modified GA (ESMGA). The diagram shows the various operational stages of the algorithm. At each intermediate stage, the excursion sets are indicated by the hatched box. The crossed arrows indicate the operation of crossover. On the top, the functional variation of the stochastic promotion rate, Δ, is sketched. The indices I_k, G_k, f_k represent the index, genome string and the fitness of the $k_{\bar{h}}$ individual.

The three stages of selection, modification, and evaluation complete one cycle of operations and in practice correspond to a single generation. The same cycle is repeated iteratively until all the members in the population get pushed up above the excursion level. Subsequent experiments repeat the internal GA dynamics for higher and higher excursion levels. The entire modified protocol is represented within the generational model of the ESMGA by the following pseudo-code:

```
procedure ESMGA
  begin
     t=0
     initialize n(t);
     evaluate structures in n(t) and identify fitnesses
            greater than the excursion level, nε(t);
     while termination condition not satisfied (Δ<1) do
     begin
        t=t+1;
        select n(t) from n(t − 1) using excursion set
             mediation and tournament selection;
        modify structures in n(t)-nε(t) by applying genetic operators;
        evaluate structures in n(t)
     end
  end.
```

Qualitatively the modified GA with excursion sets can be understood to act similarly to variable elite selection within a given generation, but differs in its multigenerational behavior. Any individuals scoring fitnesses above the excursion level get passed to the next generation without alterations. For those individuals scoring fitnesses below the excursion level, however, the modified GA applies the stochastic operators of crossover and mutation. In this way, an increasing number of less-fit individuals are improved, then promoted above the excursion level during subsequent generations.

By not only favoring the best individual, generation after generation (as in the conventional GA), we favor solutions within the excursion set and thus maintain the population's diversity as members in the excursion set distribute themselves over various solutions (i.e., disconnected components of the search space). At the same time, this procedure drives the less-fit population towards successively higher excursion levels.

CONVERGENCE AND A STOCHASTIC Δ PARAMETER

Let some stochastic parameter Δ represent the probability that an unfit individual (one whose fitness rates below the excursion level) will improve enough through crossover and mutation to get carried above the excursion level. Further split the total population, n, into two subsets: the number n_ε of fit individuals rating above the excursion level, ε, and the number $n_{1-\varepsilon}(= n - n_\varepsilon)$ of unfit individuals rating below the excursion level. For any generation, a population balance follows the stochastic recursive form

$$n_\varepsilon(t+1) = n_\varepsilon(t) + \Delta[n - n_\varepsilon(t)]\,, \tag{3}$$

where Δ gives a measure of the convergent attraction toward fitnesses that exceed the excursion level, ε. Solving for Δ yields

$$\Delta = \frac{n_\varepsilon(t+1) - n_\varepsilon(t)}{n - n_\varepsilon(t)} . \tag{4}$$

The Δ parameter can be understood to act in three ways: (1) to give the stochastic rate of initially unfit individuals that receive promotion above the excursion level; (2) to measure the normalized converging difference between intergenerational steps in the gene bank above the excursion level (as an increasing function of $[n_\varepsilon(t+1) - n_\varepsilon(t)]$; and (3) to provide the probabilistic variable of an AR(1) process which follows the recursion

$$n_\varepsilon(t+1) = \rho n_\varepsilon(t) + \beta = (1-\Delta)n_\varepsilon(t) + n\Delta \tag{5}$$

for $\Delta = \beta/n$.

In these ways, the excursion set parameter Δ acts as *a general measure of convergence into suboptimal solutions.* Thus at higher and higher excursion levels, the modified GA will reach successively better suboptimal solutions. It is this fundamental stepping-up which serves as a signature for the excursion set modifications.

SCHEMA THEOREM IN LIGHT OF EXCURSION SETS

The physical meaning of schemas and excursions sets is apparent: schemas subdivide the solutions in the genome space, while the excursions sets subdivide the solutions in the objective function space. These two entities act to partition the search space orthogonally using independent rules.

Holland's schema theorem[5] arises from the observation that the fitness evaluation of a given bit string also provides implicit knowledge about the more general *schemata* that describe that string (genome). Naturally, the amount of knowledge available depends on the specificity (order) of the given schemata; lower order schemas carry less specific information about the fitness of a best individual. Thus while the GA's microlevel search takes place within the space of strings, the essence of the schema theorem maintains that GA can likewise view the changing populations as a search through the set of schematas. Holland calls this implicit parallelism, and this feature of the GA is thought to account for much of its search and optimization power.

Excursion sets introduce further necessary conditions for using implicit parallelism within the schema theorem. In order to give the schemata processing its maximum leverage, the GA should minimize the disruption of above-average fitness schemata (the terms shown within square brackets of Eq. (6)), by allocating trials which satisfy the inequality:

$$n(H, t+1) \geq n(H,t)\frac{f(h)}{\bar{f}}\left[1 - P_c\frac{\delta H}{(L-1)} - O(H)P_m\right], \tag{6}$$

where H represents the schemata, L is the genome length, δH is the crossover length, f is the fitness, and the P_i's correspond to the crossover probability ($i = c$) and mutation probability ($i = m$). In general four arguments affect the scale of disruption: (1) the probability for crossover, P_c, (2) the defining length of a schema, δH; (3) the probability for mutation, P_m, and (4) the order of the schema, $O(H)$. Among these, mutation and crossover represent the most disruptive components. In this view, excursion-set-mediated selection can be seen to greatly reduce disruptive effects, because the ESMGA necessarily *protects and preserves those schemas that are already above the excursion level.* In other words, the excursion level maintains a stronger condition of implicit parallelism. The ESMGA additionally favors and preserves the higher order building blocks that score above the excursion level, thus maintaining a more robust variant for applying the building block hypothesis.

In any genetic algorithm, the excursion sets at higher levels grow at least as fast as the less dominant excursion sets at lower level. Using the language originated in multiobjective decision making,[7] Baker and Grefenstette have proven a more restricted version of the preceding generalization. As Grefenstette noted, the effect of changing the selection algorithm (e.g., from proportional to a rank-based selection) is not only to alter the relative magnitudes of their growth rates, but, more importantly, to leave their relative order invariant. Thus by introducing excursion sets, the modified GA method defines the salient features required for implicit parallelism, namely that trials should be allocated in an exponentially differentiated way into a large number of subsets which implicitly compete. This conclusion follows because the process now represents a superposition of two events: (1) the strictly increasing selection of fitter schemas and (2) the strictly increasing selection of excursion components having these fitter schemas. In this way, excursion sets prepare the GA for a stronger version of what constitutes implicit parallelism within the schema theorem.

THEOREM 1. If A_{u_1} and A_{u_2} are excursion sets such that $u_l < u_2$, then trials are allocated in an exponentially differentiated way to a larger number of subsets of A_{u_2} than A_{u_1}.

A simple mathematical proof for this theorem can be obtained directly from Eq. (4). Solving formally for $\Delta(t)$ in the limit when the difference equation can be written as a differential (valid for late convergence or many generations), one obtains the variation in the number of individuals above the excursion level:

$$n_\varepsilon(t) = n - \exp\left[-\int_1^t \Delta(t')dt'\right] . \tag{7}$$

The power of relation (7) is that for a given Δ (which characterizes the stochastic rate of promotion), the total number of individuals rising into the excursion set increases exponentially. The computational process acts to select increasingly more preferred building blocks from the pool of favored schemata. Thus for the GA-modified version with excursion sets, a simple formalism captures the essential feature of the schema theorem in a particularly transparent way.

RESULTS AND DISCUSSION

1. TRIAL FUNCTION AND ITS OPTIMIZATION

As a demonstration of the performance measures of the new GA, experiments were conducted to optimize the trial function, $y = x + \sin 32x$, on the interval $[0, \pi]$. The function is steeply graded with many local maximum and a global maximum of $y_{\max} = 4.09299342$ for $x = 3.09346401$. Figure 1 shows the trial function and excursion level, $y = 3.0$. In the lower exploded view, the same function appears but, in the region of interest from $x = 2.4$ to $x = 3.1$, the excursion set components are marked by the hatched boxes. The ESMGA was written using modifications to the code supplied by Rialo.[6] Figures 3, 4, and 5 show the results of the experimental optimizations.

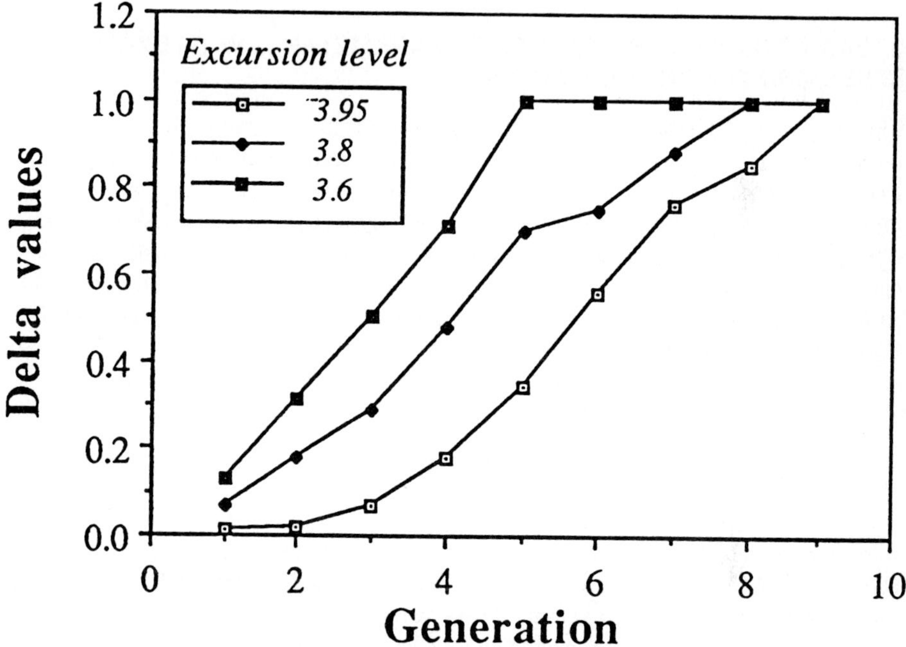

FIGURE 3 Evolution of the stochastic promotion rate, Δ. The figure depicts the way the promotion rate changes with generation and with excursion level as a parameter.

2. EVOLUTION OF FITNESS ABOVE AND BELOW THE EXCURSION LEVEL

Figure 3 tracks the generational evolution of fitnesses above and below the excursion level and compares the two subsets with the fitness evolution of both the average population and the best individual. For a split population consisting of individuals whose fitnesses rate either above or below the excursion level, the fitness dynamics change initially, then saturate towards convergence. As seen in the GA simulations, both the overall population fitness and the average fitness of individuals above the excursion level increases according to an "S-curve" which flattens after several generations. In contrast, the average fitness of individuals below the excursion level decreases and tends towards zero when all individuals get promoted. Less marked changes are seen in the best individual's fitness as these schemas get well-conserved across successive generations. The modified GA dynamics preserves this highly fit genetic heritage without interbreeding or disruption, thus working like a genetic bank.

3. VARIATION OF THE STOCHASTIC PARAMETER Δ WITH EXCURSION LEVEL AND GENERATION NUMBER

Figure 4 shows the variation of the stochastic promotion rate with the generation number for three given excursion levels. Higher excursion levels converge with a less steep ascent; this slow increase corresponds physically to the larger diversity maintained by higher excursion levels. Generally as the modified GA proceeds, the stochastic parameter, Δ, increases until it equals unity near convergence; in this case, all individuals carry a fitness which exceeds the excursion level, $n_\varepsilon(t+1) = n$. In the first generation, no promotion above the excursion level occurs and $\Delta = 0$. Figure 5 shows the variation of the the rate of increase in Δ is large if crossover and mutation rapidly modify and promote previously unfit individuals.

4. INTELLIGENT SLICING OF THE HYPERCUBE

Standard GAs represent their search space on the hypercube. The usual hyperplane partitions the hypercube into closely related bit segments or schemas; in contrast, the excursion level can be understood to partition the hyperplane into disconnected niches which share an equivalent fitness value. The general shape of these niches cannot be predicted. At higher and higher levels, the excursion set represents more and more fit schemas. This induced disconnectedness on the hypercube contrasts markedly with the behavior of hyperplanes. The underlying principle can be understood as follows. While the excursion sets are defined in relation to the objective function, the hyperplanes are defined in the entire space of all possible genomes. Thus for a given excursion level, the dynamic distribution of samples is related to the number of hyperplanes being processed by the GA at that same level. Hence

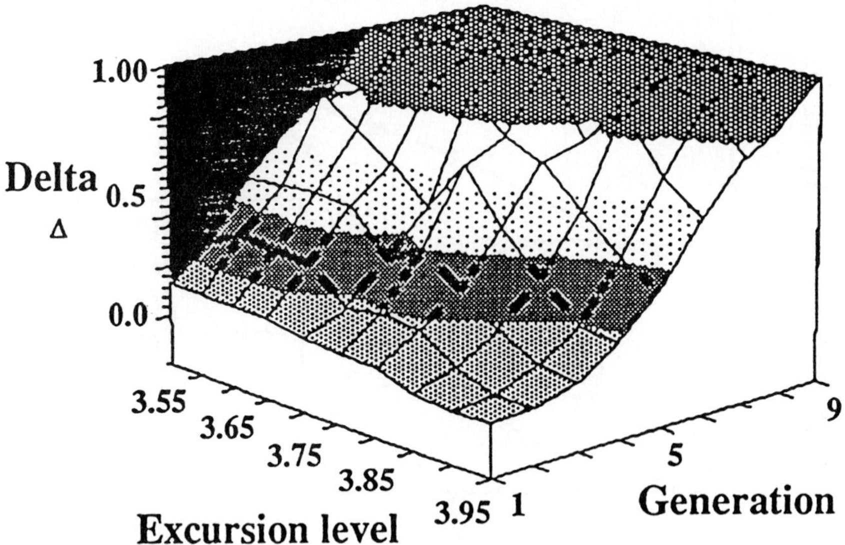

FIGURE 4 The functional variation of the stochastic promotion rate, Δ, as a function of excursion level and generation.

while the hyperplane may belong to several decreasingly fit groups of schemas, the excursion set cuts the hypercube in a quite unique way, such that disconnected regions of the search space contain only those increasingly fit groups. It is these fit groups of schemas which both represent the higher order building blocks and whose objective function values rate them above the prescribed excursion level. This ESMGA behavior contrasts markedly with the computational path followed by a standard GA. Because of its biassed sampling within the hyperplane, the traditional GA's view of competing hyperplanes *may bear little relation to the underlying mean values of their objective function.* In the search space, excursion sets thus represent a second level of shifting or sorting of schemas which intelligently cuts the hypercube and henceforth describes a more robust selection criterion to couple successive generations. In general, excursion sets introduce a fitness-based subset into the search space, rather than merely restricting the search to the particular hyperplanes induced by a given binary representation. Hence, excursion sets mandate that the two closely related features of implicit parallelism and multiple sampling act forcefully throughout a GA simulation.

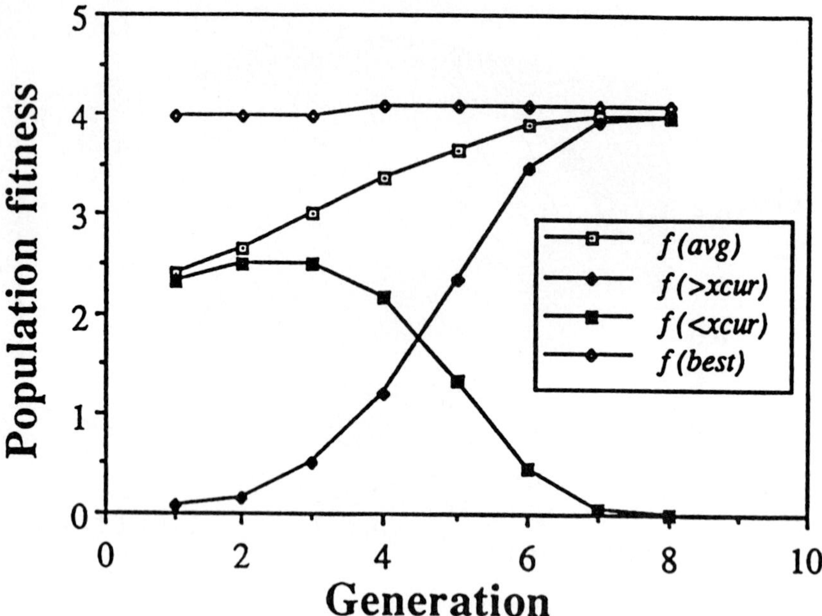

FIGURE 5 Evolution of various fitness measure.[5] The plot shows the evolution of fitness above and below the excursion level, then compares the two subsets to the evolution of total population fitness and the best individual's fitness.

CONCLUSIONS AND FUTURE WORK

In summary, a comparison between GA dynamics and excursion sets can:

1. transparently contain complex stochastic processes (such as crossover, mutation, etc.) in a single parameter, Δ, which corresponds to the difference in proportion of individuals above and below the excursion level for each generation;
2. connect convergence of GA simulations directly to the generational evolution of this stochastic parameter, Δ;
3. identify GA dynamics with an approximate AR (1) process for a given excursion level, such that a recursion relation holds for the number of above-threshold individuals, $n_{t+1} = \rho n_t + \beta$, where β varies stochastically and corresponds physically to the rate of promotion of substandard individuals above the excursion level, ρ introduces the Markov chain character; and

4. link the standard GA version of variable elite selection to a broader interpretation of implicit parallelism within the excursion set.

Future work will introduce the excursion parameter as a variable parameter which can be adjusted to pressure the best individual's fitness continuously. Taking advantage of this cycle of initial relaxation, followed by additional pressure, initial experiments performed on sample functions have shown rapid convergence to near global optima. The final aim is to push GA dynamics to its limits, thus guaranteeing a strong version of implicit parallelism while also preserving higher-order building blocks through excursion-set-mediated selection and relaxation.

ACKNOWLEDGMENTS

This material is based upon work supported in part by the 1992 Summer School on Complex Systems and the Santa Fe Institute. Support was also provided (for S.B.) by the Austrian fellowship agency, Oestereicher Auslandsstudentendienst, and (for D.N.) by the Microgravity Sciences and Application Division, NASA. S.B. gratefully thanks Prof. P. L. Arasappan, Prof. S. Karpaga Vinayagam, and Prof. Peter Schuster for their instruction, unceasing support, and interest.

REFERENCES

1. Adler, D. *The Geometry of Random Fields.* New York: Wiley & Sons, 1981.
2. Davis, L., ed. *Handbook of Genetic Algorithms.* New York: Van Nostrand Reinhold, 1991.
3. Goldberg, D. E. *Genetic Algorithms in Search, Optimization and Machine Learning.* Reading, MA: Addison-Wesley, 1989.
4. Grefenstette, J. J. "Conditions for Implicit Parallelism." In *Foundations of Genetic Algorithms*, edited by G. J. E. Rawlins. San Mateo, CA: Morgan Kaufmann, 1991.
5. Holland, J. H. *Adaptation in Natural and Artificial Systems.* Ann Arbor, MI: University of Michigan Press, 1975.
6. Rialo, R. GA Code Supplied in the Fourth Annual Summer School on Complex Systems, Santa Fe Institute, May 31–June 28, 1992.
7. Schaffer, J. D. "Some Experiments in Machine Learning Using Vector Evaluated Genetic Algorithms." Ph.D. Thesis, Department of Electrical Engineering, Vanderbilt University, Nashville, 1984.

Cathleen Barczys,† Laura Bloom,‡ and Leslie Kay†
†Biophysics Graduate Group, University of California, Berkeley, CA; and ‡Computer Science Department, University of California, San Diego, CA

Applying Genetic Algorithms to Improve EEG Classification and to Explore GA Parametrization

A genetic algorithm (GA) is used to identify outliers in a sorting task where human electroencephalographic (EEG) data are classified by stimulus type using a Euclidean distance measure. To assist in selecting the GA parameter values and to probe the relationships among parameters, a GA is also applied to the simpler theoretical task of maximizing the number of 1s in a bit string. Regarding the human EEG data, the GA with the Euclidean distance fitness function dramatically improves classification, but it does so by excluding records other than outliers; we devise a quantity by which these unexpected exclusions can be explained. Using tuning curves from the simpler problem, we demonstrate qualitative ways to assist in parametrization of mutation rate, number of generations, and population size, and to provide insight into how these parameters change with the complexity of the solution space.

1992 Lectures in Complex Systems, Eds. L. Nadel & D. Stein, SFI Studies in the Sciences of Complexity, Lect. Vol. V, Addison-Wesley, 1993

1. INTRODUCTION

The purposes of this research are: (1) to test whether a genetic algorithm (GA) can improve classification of human electroencephalographic (EEG) data, by applying the GA to identify outliers; and (2) to explore qualitatively the relationship among several GA parameters, by running a simplified version of the above problem (the maximization of 1's in a bit string, called "max_ones"). The biological research to which we apply the GA is a study of human somatosensory (touch) perception, to determine how stimulus patterns are manifested in the brain and how the brain recognizes sensory stimuli. The hypothesis being tested, based on previous experiments with rabbits in olfactory (smell) perception,[4] is that stimulus patterns are manifested as a spatial pattern of amplitudes of a global waveform that extends across the somatosensory area of the brain and lasts about 0.1 sec. Subjects are trained to respond differently to two different stimuli, and we then try to classify their EEG recordings by stimulus type. If the classification is successful, then we hypothesize that perhaps the features of the brain's electrical activity that we used are similar to those used by the organism in its stimulus recognition activity.

There are three parts to the data analysis for the somatosensory project: (1) preprocessing, which includes editing and filtering; (2) application of the Fourier transform to extract the global waveform by decomposing the data into cosines; and (3) classification using a Euclidean distance measure. The purpose of the GA is to improve part 3 of the analysis—the classification—by identifying outliers.

2. EEG APPLICATION

2.1 METHODS

The EEG data are time series derived from an 8×8 array of voltage-recording electrodes (with 1-cm spacing), located directly on the cortex of an epileptic subject and straddling the somatosensory, motor, premotor, and temporal lobe areas of the brain. Data are digitized at 256 Hz and band-pass filtered at 20–56 Hz. During data acquisition the subject performs a somatosensory discrimination task in which she is trained to respond to three different intensities of gentle electrical stimulation (low, medium, and high) with three different responses (press softly, press hard, and no press). Two data sets are formed from one day's experiments with one subject, with each data set containing 20 low-intensity records and 20 high-intensity records.

The data are decomposed into Fourier components to yield matrices of gain coefficients and, for each EEG record, the matrix for either the first dominant component (the 64×1 Gain1 matrix) or the second component (the 64×1 Gain2 matrix) is classified. Both the standard and GA-optimized Euclidean distance procedures attempt to classify into two clusters of points in 64-space—one cluster for the low-stimulus condition and one for the high-stimulus condition. In the GA-optimized procedure, the GA is used to locate records within each cluster ("outliers") whose exclusion improves overall classification. The standard procedure does not remove any outliers. Both procedures classify points in 64-space by first calculating the centroid (average) for each cluster and then calculating the two point-to-centroid distances for each point. A record is "correctly classified" if that record's point in 64-space is closer to its own centroid than to the other.

The "chromosomes" or "individuals" for the GA are 40-character strings composed of 1's and 0's—one character for each EEG record (20 low-stimulus records and 20 high-stimulus records). A "1" in position i of a string signifies that the ith record is included in the calculation of the centroid and in the classification, whereas a "0" signifies that it is excluded. The number of correctly classified records is totalled to give the fitness value for that string, with the maximum possible fitness being 40. The GA parameter settings were selected empirically with insight derived from the qualitative explorations with max_ones. The parameter values used were: # generations = 200, population size = 100, mutation rate = 1/100, crossover method = 2-point, with probability = .75, and selection method = tournament, with probability = .75.

The EEG data were tested under several conditions: (1) using the Gain1 versus Gain2 matrices, (2) using all 64 channels of data versus using a subset of 31 channels believed to be most relevant, and (3) including records that were known to be "bad" (due to preprocessing artifacts) versus excluding them, in order to determine if the GA was picking out the known bad records. For each combination of these data conditions a set of ten GA runs was conducted.

2.2 RESULTS AND DISCUSSION

The GA-optimized classification results are compared with those from the standard procedure in Table 1. The GA dramatically improves classification accuracy for the human EEG data for both data sets under all experimental conditions (see "% Correct" columns, Table 1). The average accuracy with and without GA-optimized classification is 93% and 78%, respectively, yielding an average improvement of 15 percentage points due to the GA.

We performed cross-validations in which the centroids from one data set (the "training set") are used to classify the other data set (the "test set"). The cross-validation results (not shown here) are mixed and, even where there is improvement, the classification accuracies are still low (63% or less). Thus, although the GA dramatically improves classification accuracy under the training set conditions, the improvement does not carry over to the test set.

Six records were commonly excluded by the GA as outliers—three from data set 1 and three from data set 2. They were analyzed to determine whether they

TABLE 1 EEG classification results, for the standard and GA-optimized classification experiments. The GA improves classification accuracy (% Correct) for both data sets under all experimental conditions.

			Standard				GA-Optimized					
Input matrix	With known bad recs?	Chans	Max fitness	Max possible	% Correct	# Incorrect recs	Max fitness; Range	Max possible; Range	Avg % Correct	# Incorrect recs	# Excluded recs	# Identical strings
Data Set 1:												
Gain1	With	All	31	(40)	**78**	9	34–35	(35–37)	**96**	1–2	3–5	3
Gain1	With	Subset	32	(40)	**80**	8	32–35	(34–37)	**96**	1–2	3–6	2;2
Gain1	W/out	All	31	(36)	**86**	5	30–32	(32–33)	**96**	1–2	3–4	7
Gain1	W/out	Subset	30	(36)	**83**	6	29–32	(31–35)	**96**	1–4	1–6	7
Gain2	With	All	31	(40)	**78**	9	33	(34–37)	**92**	1–4	3–7	2
Gain2	With	Subset	29	(40)	**73**	11	29–31	(31–38)	**89**	1–7	2–9	0
Gain2	W/out	All	27	(36)	**75**	9	27–29	(29–33)	**90**	1–6	3–7	0
Gain2	W/out	Subset	26	(36)	**72**	10	27–29	(31–32)	**91**	1–4	4–7	0
Data Set 2:												
Gain1	W/out	All	26	(33)	**79**	7	25–27	(27–32)	**90**	2–5	1–6	3;2

shared a common characteristic that would: (1) provide insight into how the GA was determining which records to exclude and (2) provide a biologically relevant justification for their removal as outliers, and for outlier detection in future data sets. The GA-excluded records are neither outliers in a distance sense with respect to the cluster, nor are they unusual or distinguishable in any way from the rest of the records using standard criteria (such as frequency or power of the dominant component, or total power in the Fourier components).

To explain how the GA may be selecting the records it excludes we derived a quantity, the Z value,[1] which quantifies the conditions under which Euclidean distance classification accuracy is maximized (in the one-dimensional case with two distributions, for example, large inter-mean distance and small mean-to-inner-edge distance). Thus it was predicted that the Z value will be greater for the GA-optimized results than for the Standard results; in all but one of the 40 cases tested this prediction was true. The GA, therefore, as an optimizer of one's fitness function provides an excellent magnifying glass for examining that fitness function and its subtleties, in addition to providing a tool for improving one's results.

3. SETTING THE MUTATION RATE AND POPULATION SIZE PARAMETERS

The fitness function for the EEG application is the most computationally expensive part of the GA, so we wanted to determine what parameter settings would allow us to use a relatively small population size and still obtain reliable results. Since a thorough investigation of all parameters was beyond the scope of this project, we focused our explorations on the interplay among mutation rate, population size, and number of generations. Mutation can be useful for generating new, and possibly more fit, individuals in a population with low variability, which occurs when the population size is small relative to the problem. We looked for a mutation rate that would optimize the relatively small population to which we were restricted by our hardware (SUN SPARC 4/330) and to give us confidence in our solution to the EEG problem. For our measures of performance we use the highest fitness attained and the number of generations required. We found the number of generations more useful than the number of evaluations, because it explicitly gives the amount of crossover that the population experienced, whereas the number of evaluations gives only the raw measure of machine use. As population size increases the number of generations needed decreases, giving an empirical representation of the solution capacity of the original population that can be realized mostly through crossover.

Max_ones is a toy application which maximizes the number of 1's in a fixed-length string, over a binary alphabet. The EEG application also maximizes the number of 1's in the string, but subject to an additional condition (classification accuracy). We suspect that the fitness landscape for the EEG problem is very complex as compared with that for the max_ones problem, because of the presence of seemingly multiple solutions for the EEG application and its requiring a larger population and more generations to reach a satisfactory solution than does max_ones. As for the EEG application, we ran max_ones ten times for each set of parameter values studied.

The relationship between the number of generations needed to reach maximum fitness and the population size is described by Figure 1 for string length 40, to parallel the EEG application, and for other string lengths. Note that for populations larger than a certain critical value—the point where the slope of the curve flattens out—a large increase in population size gives only a marginal decrease in the number of generations required to reach the solution. Thus, increasing the population size past the critical value to decrease the number of generations may not decrease the computational expense.

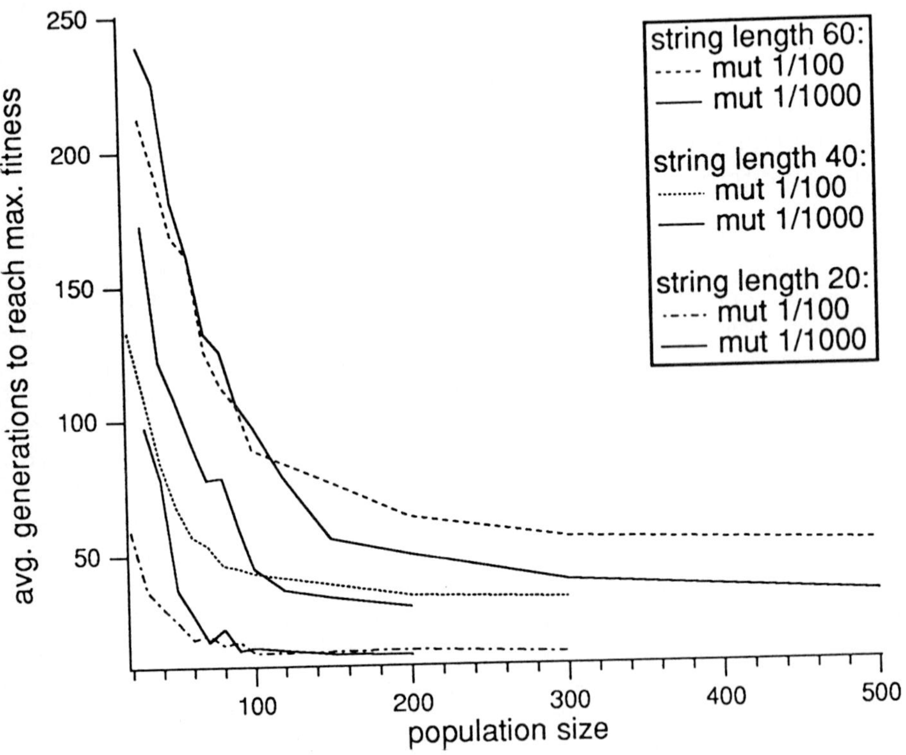

FIGURE 1 The average number of generations to reach maximum fitness, for mutation rates of 1/100 and 1/1000.

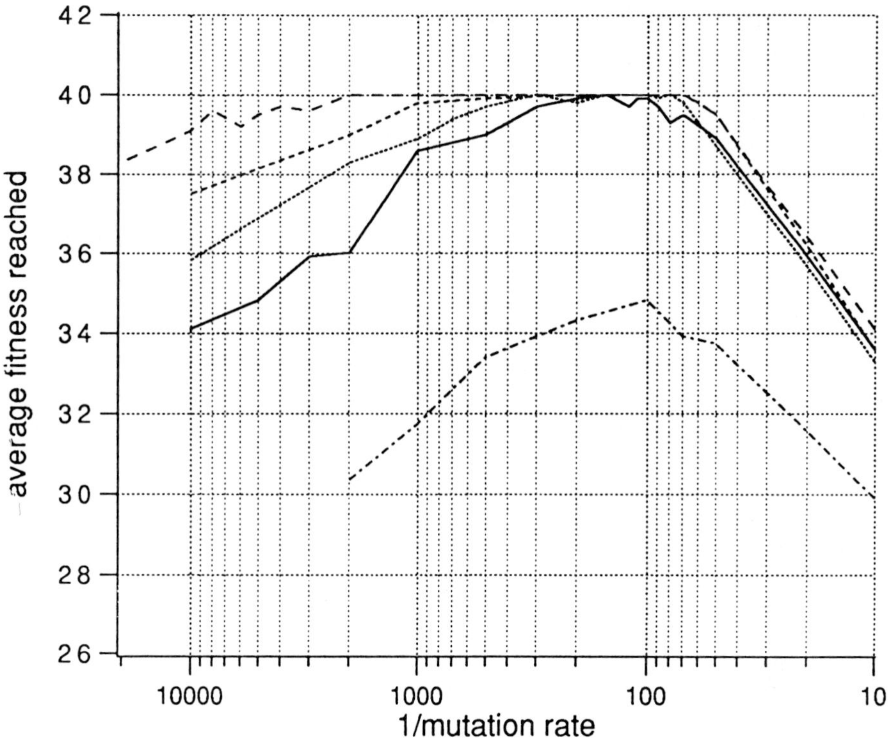

FIGURE 2 Average fitness reached in a set number of generations for varying mutation rates. The top four plots, top to bottom, are max_ones results for population sizes 100, 70, 50, and 40. Each was run 100 generations and the average maximum fitness reached is plotted. The bottom plot is for the EEG application results for population size 100, run for 200 generations. Note the similarity of shape to the max_ones population size 40 plot.

However, one can decrease expense by using a smaller population size from the steeper part of the curve and then adjusting the mutation rate. Four of the curves in Figure 2 exhibit the relationship for max_ones between mutation rate and average fitness reached in 100 generations. These curves suggest that for each population size there may be a range of mutation rates that lead to good performance of the genetic algorithm, where good performance here is defined as attaining fitness 40 in 9 out of 10 runs. The performance of max_ones for mutation rates larger than for those in this range (critical value near 1/70) decreases dramatically. As the population size increases, the range of good mutation rates grows to include lower mutation rates.

In Figure 1 we also compare the number of generations needed by max_ones to reach maximum fitness for two mutation rates as population size varies—the commonly used mutation rate of 1/1000 and the 1/100 rate that lies in the range of good performance for all population sizes in Figure 2. For smaller populations max_ones achieves the optimal solution in fewer generations with a mutation rate of 1/100. What this suggests is that a higher mutation rate on a smaller population decreases the threat of premature convergence, allowing the application to discover a good solution less expensively for a given population size. Notice that with larger populations a higher mutation rate leads to worse performance.

These conclusions may be generalizable to problems having a more complex fitness landscape. We ran the EEG application with varying population sizes and mutation rates and found that the best mutation rate for the EEG application (1/100) is in the same range as for the smaller populations (40–70) in the max_ones experiments (Figure 2). We also found that the performance for the EEG application (the number of generations needed to reach maximum fitness) put the population size of 100 in the steeper part of a curve comparable to those in Figure 1. We infer that a mutation rate of 1/100 allows a smaller than critically sized population to reach a satisfactory solution in fewer generations, and that the critical population size and the number of generations needed increase with the complexity of the fitness landscape.

4. CONCLUSION

The genetic algorithm improved EEG classification accuracy from 83% to 96% for the training set by removing outliers, but did not improve test set classification. The excluded records are not necessarily outliers in the distance sense; a quantity was derived from the fitness function that may describe how the GA selects the records it excludes.

For a complex problem we cannot know *a priori* the minimum population size and the number of generations that will allow the genetic algorithm to effectively search the solution space and to avoid premature convergence. We show that a mutation rate higher than the commonly used 1/1000 can lower the minimum population size needed in order to lessen the effects of hardware constraints.

ACKNOWLEDGMENTS

We thank the Santa Fe Institute and its 1992 Complex Systems Summer School for initial support, Melanie Mitchell for inspiration for this project, and Terry Jones for his superb GA software ("GASSY"). The EEG data were obtained from Alan Gevins, Director of EEG Systems Laboratory, San Francisco, in cooperation with Kenneth Laxer, Northern California Epilepsy Center, University of California, San

Francisco. For computer time we thank the Mathematics Department, University of California, San Diego, and the Graduate Group in Biophysics, University of California, Berkeley. L.K. acknowledges partial support for this project from USPHS grant GM07379. C.B. and L.K. thank Walter J. Freeman, University of California, Berkeley, for support through NIMH grant MH06686.

REFERENCES

1. Barczys, C. "Spatio-Temporal Dynamics of Human Cortical EEG During Somatosensory Perception." Ph.D. Dissertation, Ch. 4, 15–18, University of California, Berkeley, CA 1993.
2. Callahan, K. J., and G. E. Weeks. "Optimum Design of Composite Laminates Using Genetic Algorithms." *Composites Engr.* **2(3)** (1992): 149–160.
3. De Jong, K. A., and W. M. Spears. "An Analysis of the Interacting Roles of Population Size and Crossover in Genetic Algorithms." In *Parallel Problem Solving from Nature, 1st Workshop Proceedings*, edited by H.-P. Schwefel and R. Manner, 38–47. Berlin: Springer-Verlag, 1991.
4. Freeman, W. J., and G. Viana Di Prisco. "EEG Spatial Pattern Differences with Discriminated Odors Manifest Chaotic and Limit-Cycle Attractors in the Olfactory Bulb of Rabbits." In *Brain Theory*, edited by G. Palm and A. Aertsen, 97–119. Berlin: Springer-Verlag, 1986.
5. Goldberg, D. E. "Sizing Populations for Serial and Parallel Genetic Algorithms." In *Proceedings of the 3rd International Conference on Genetic Algorithms*, edited by J. D. Schaffer, 70–79. Arlington, VA: Morgan-Kaufman, 1989.
6. Greffenstette, J. J. "Optimization of Control Parameters for Genetic Algorithms." *IEEE Trans. Sys., Man, & Cyber.* **SMC-16(1)** (1986): 122–128.
7. Manderick, B., M. de Weger, and P. Spiessens. "The Genetic Algorithm and the Structure of the Fitness Landscape." In *Proceedings of the 4th International Conference on Genetic Algorithms*, edited by R. Belew and L. Booker, 143–149. Arlington, VA: Morgan-Kaufman, 1991.
8. Schaffer, J. D., R. A. Caruna, L. J. Eshelman and R. Das. "A Study of Control Parameters Affecting Online Performance of Genetic Algorithms." In *Proceedings of the 3rd International Conference on Genetic Algorithms*, edited by J. D. Schaffer, 51–60. Arlington, VA: Morgan-Kaufman, 1989.

T. David Burns
Department of Economics, George Mason University, Fairfax, VA 22030

Symbiosis in Society and Monopoly in Nature: Mixed Metaphors from Biology and Economics

In a recent talk, Stuart Kauffman[4] used an Edgeworth box to point out the similarities between two traders and two symbiotic species. Mixing these economic and biological metaphors highlights their similar emphases on rivalry and cooperation. Barter leads to symbiosis, then on to competition, cooperation, and monopoly. Cultural strategies give human beings a monopoly position regarding other species. Through culture, human beings influence nature and society both consciously and unconsciously. The most important species to control (humanity itself) presents the most difficulties.

1. SYMBIOSIS AS EXCHANGE

Imagine two traders arranging a barter transaction, perhaps two farmers, Smith and Jones. For this trade, each has given preferences and endowments. Say Smith grows wheat and Jones is a dairy farmer. To construct Edgeworth's model,[6] draw a box whose dimensions each represent the amount of a single farm product. One dimension is milk, the other wheat. Each point in this rectangle represents a division of goods between the traders (see Figure 1), with their original position at upper left-hand corner. If they exchanged all the milk for all the wheat, they would end up in the lower right-hand corner. If Smith simply gave all the wheat to Jones, the upper right hand corner would describe their situation.

Imagine each farmer drawing curves that represent allocations among which she is indifferent (her indifference curves). Typically, there will be a wedge of allocations that both farmers prefer to their original situation. According to the model, bargaining continues until this potential vanishes. There is a locus (the contract curve) where no gains from exchange exist. These consist of the points of tangency between the two farmers' indifference curves. The final trading position of the farmers will be one of these points. The location of the final agreement depends upon the relative bargaining skill of the participants.

What does this model tell us? It depicts an artificial exchange, a thought experiment illustrating the potential for mutual gain among traders. It does not allow prediction but facilitates understanding.

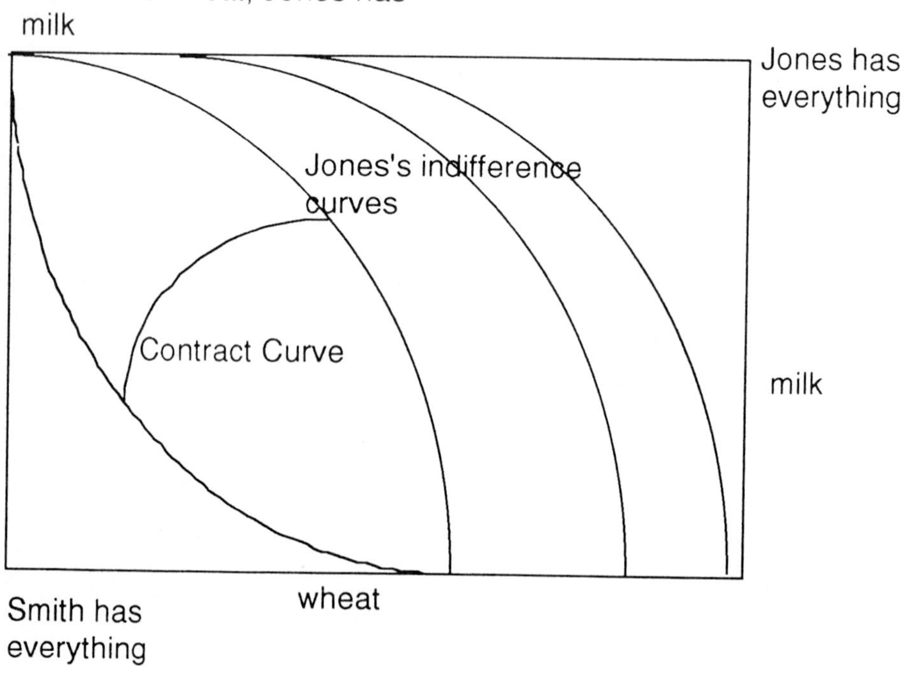

FIGURE 1 An Edgeworth box of a wheat farmers and a dairy farmer's exchange.

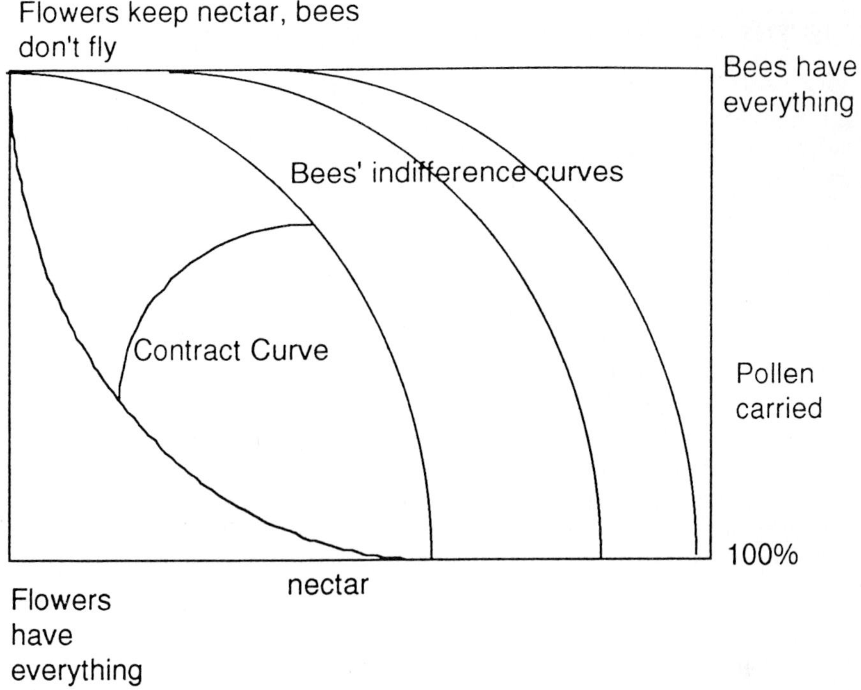

FIGURE 2 An Edgeworth box of the bees' and flowers' symbiosis.

We can draw an Edgeworth box for bees and flowers (Figure 2). Bees carry pollen and flowers produce nectar. Isopopulation or isoenergy lines suggest themselves as analogs of indifference curves. Rather than maximizing as individuals, bees and flowers come pre-equipped with a bargaining strategy that determines their behavior. Symbiosis alters these species' chances to flourish and reproduce. While the barter model said nothing about what would happen after trade took place, its analog demands iteration. Exploitation, where one species gains disproportionately, no longer appears to be an equilibrium. The gains lead to relative population increases, tending to reverse the disproportionality. If flowers over- or underproduce nectar, the bees will die off, yielding the flowers only a temporary gain. Evolution mediates their bargain, giving neither an advantage.

Perhaps bees learn to recognize flowers that tend to have more nectar. If so, we may be using the wrong economic analogy. The absence of exploitation and the feedback effect cause the situation just described to resemble the economists' model of perfect competition.

2. IS THERE NATURAL MONOPOLY IN NATURE?

Under perfect competition, entry or exit of firms from an industry adjusts the level of profit. The resulting feedback guides the industry to the proper price and quantity. As in the biological model, exploitation defeats itself. The assumptions underlying this model are just as restrictive as those of the Edgeworth box. While economists disagree on perfect competition's virtues as a first approximation of reality, none would insist that it provides a literal representation of our world.

The standard model of monopoly shows that monopolistic markets achieve higher prices and profits by restricting output. Assuming that market demand forms a constant relation between quantity and price, the monopolist chooses the price that gives her the most profit. By choosing price, the monopolist effectively chooses quantity. The monopolist will produce fewer goods than they would under competition (assuming the monopolist charges a single price). To restrict output, the monopolist must restrict entry. Monopolists cannot make large profits if those profits tempt new competitors into the market, increasing output and lowering price. They must prevent the arrival of newcomers.

Monopolists may restrict entry by collusion, by government grant, or by exploiting the characteristics of certain markets (natural monopoly). Natural monopoly exists if some technological quirk of the market allows a single firm to exist and prosper in a market where two firms could not coexist profitably. Market demand cannot support both. Cutting each in half either would raise the price the customer must pay or would simply make no sense. Local power utilities, phone companies, and cable television outlets provide familiar examples. Could this occur in our biological analogy?

The economic monopolist maximizes profit by restricting output. How would a monopolistic species act? Above, I speculated that a population might form the feedback link. Break this population link and a species may act as a monopolist.

When factors other than the symbiotic relationship form the binding constraint on the population size of one or both symbiotes, then feedback will fail to occur. If pesticides keep the bee population artificially low, the flowers will be disadvantaged in their exchange. Taken to extremes, this arrives at parasitism. Bears steal the bees' honey, providing nothing in return. Yet this factor does not determine population size and could go on as long as the two species exist. So, monopoly can occur in nature.

Can *natural* monopoly exist in nature? Natural monopoly requires a lumpiness or discontinuity in the production or exchange of symbiotic products. While such discontinuity might exist between symbiotes of an extreme degree of coevolution, such a degree of interdependence seems to imply strong population feedback between the two species, precluding monopoly. For example, the bear's bargaining strength lies in his independence from the existence of bees or, neither of a lichen's component species can live without the other. The commitment of coevolved species is mutually enforcing. Selfish genes[1] must cooperate or die.

While monopolistic interaction may take place between symbiotic species, such interaction depends on temporary circumstances that provide no lasting advantage. Monopolistic species cannot exclude competitors; they can only take advantage of their absence. The more each species depends upon symbiosis, the less they can afford to act as a monopolist. Monopoly depends upon an asymmetric dependence. Selfish genes that attempt to arrange this asymmetry are vulnerable to competition and risk their investment in coevolution. Where genes fail, however, culture may succeed.

Cultural strategies have a time advantage over genetic strategies. They can adapt to a change more quickly. Human beings have exploited this advantage, forming a symbiotic network of domesticated plants and animals. Agricultural societies have shaped species through selective breeding and culling, forcing a biased kind of coevolution. Our culture coevolves with the domesticated species. Human practices, not human lives, depend upon their existence.

Humankind does not depend directly upon any single species. If a disease destroyed every chicken on Earth, individual humans might lose fortunes, but humanity would continue. We depend upon the whole, but can live without any specific part. Humans determine the bargain's terms and, so, act as monopolists.

3. ARM WRESTLING THE INVISIBLE HAND

Human influence takes an even more central role in society. Through science, engineering, law, government, and business, human beings seek to articulate knowledge of the world and to manipulate the world by using that knowledge. Humanity's dilemma lies in our need to do more than we can know how.[2] We must guide nature and culture, although we cannot fully understand them. In both cases, humanity must influence a global phenomenon through local action.

We cannot choose directly between emergent, global aggregates and outcomes. Individual acts and intentions connect to their aggregate outcome indirectly. We can only influence them indirectly, through choice of rules. But even rules cannot be chosen directly. They emerge from a political process, transformed by social reality.

Economic aggregates and the wealth of peoples emerge from the interactions, customs, and practices of individuals, "as if guided by an invisible hand."[9] The invisible hand selects social strategies as natural selection chooses genes. The visible hand of organization and legislation influence social outcomes in the same way that selective breeding influences nature. Each limits the action of the other. The two hands sometimes work in harmony; other times, they wrestle. Evolution and rationality must cooperate in the development of culture. We must breed institutions as we bred domestic species. In doing so, we can never be confident of infallibility.

ACKNOWLEDGMENTS

I would like to thank the Santa Fe Institute for inviting me to their 1992 Complex Systems Summer School, the Institute's funders for supporting them, and the Center for the Study of Market Processes for their assistance with my expenses.

REFERENCES

1. Dawkins, R. *The Selfish Gene.* Oxford: Oxford University Press, 1976.
2. Hayek, F. A. *The Fatal Conceit: The Errors of Socialism.* Chicago: University of Chicago Press, 1988.
3. Hayek, F. A. *Law, Legislation, and Liberty*, Vol. 1, 22–24. Chicago: University of Chicago Press, 1973.
4. Kaufman, S. "[need title]." This volume.
5. Maynard Smith, J. "Evolution and the Theory of Games." *Amer. Sci.* **64** (1976): 41–45.
6. McCloskey, D. *The Applied Theory of Price*, 2nd ed. New York: Macmillan, 1985.
7. Mirowski, P. *More Heat then Light: Economics as Social Physics.* Cambridge: Cambridge University Press, 1991 .
8. Schelling, T. C. *Micromotives and Macrobehavior.* New York: W. W. Norton, 1978.
9. Smith, A. *An Inquiry into the Nature and Causes of the Wealth of Nations*, edited by Edwin Cannan. Chicago: University of Chicago Press, 1976.

Igor Fedchenia and Villy Sundström
Department of Physical Chemistry, University of Umeå, S-901 87 Umeå, Sweden

Disordered Models of Chemical Reactions in Complex Molecules

1. INTRODUCTION

The time dynamics of chemical reactions in complex molecules possessing a large number of atoms ($> 10^3$) and complex internal structure such as proteins and amino nucleic acids is much richer in types of behavior than the dynamics of smaller molecules (number of atoms $\cong 10^2$). The most challenging feature of the dynamics of large molecules for theoreticians is the observation in experiments of nonexponential reaction kinetics.[3,6,8]

It was found, apart from this,[3,6] that the solvent can qualitatively change the shape of $N(t)$, leading to plateaus or decay rate variations. In its simplest form, solvent effect is taken into account in the Smoluchowski equation by assuming that the solvent action on the reaction coordinate is broadband noise. It has been shown[9] that in some cases it is possible to reduce the two-dimensional Smoluchowski equation for protein and ligand coordinates y, x

$$\frac{\partial}{\partial t} P = D \nabla_{xy} (\nabla_{xy} + h \nabla_{xy} V(x, y)) P \tag{1}$$

1992 Lectures in Complex Systems, Eds. L. Nadel & D. Stein, SFI Studies in the Sciences of Complexity, Lect. Vol. V, Addison-Wesley, 1993

(here D is the diffusion constant and h is the height of the potential barrier in units of $k_B T$) to a one-dimensional equation

$$\frac{\partial}{\partial t} P = D \frac{\partial}{\partial x} \left(\frac{\partial}{\partial x} + h \frac{\partial}{\partial x} \hat{V}(x) \right) P + ks(x)P \tag{2}$$

for the ligand coordinate only. Here $s(x)$ is the coordinate-dependent rate of escape from free-ligand state to bound-ligand state.

The same Eq. (2) naturally appears in a quasi-classical description of photochemical reactions in which case $\hat{V}(x)$ is the mean potential surface of an exited state and $s(x)$ (called a "sink") is the transition rate to the ground state surface. The Landau-Zener formula is used to describe the shape of the sink

$$s(x - x_0) = \exp\left\{ -\frac{(x - x_0)^2}{a^2} \right\}. \tag{3}$$

Equation (2) has been extensively studied in the two above-mentioned contexts: by Agmon et al.[2,9] in the case of ligand binding and by Bagchi et al.[4,5] in the case of barrierless photochemical reactions. Numerical simulation and analytical results show that nonexponential behavior of the survival probability or population of the exited state

$$N(t) = \int dx P(x, t) \tag{4}$$

can be obtained in a transient region between the initial state and the long-time exponential tail. Extensive numerical simulations of Eq. (2) made in Åberg et al.[1] show that, although the slope of exponential part and the length of nonexponential region depend on D and h, the shape of $N(t)$ remains qualitatively the same and becomes only biexponential when $\hat{V}(x)$ has a multi-well configuration with a sink in each well, such that jumps between wells and escapes through the sinks become competitive.

We therefore study Eq. (2), choosing several models of disorder for the potential $\hat{V}(x)$ and sink $s(x)$, and trying to find the most sensitive parameters in the disorder model to describe a variety of time dependences of $N(t)$. We have found that, although relaxation rates are dependent quantitatively on D and h in this model (temperature and viscosity of solvents in real experiments), the most drastic qualitative changes are caused by a variation of distribution width of the sink position and width and position of spectrum of the random potential $\hat{V}(x)$. First, however, we describe briefly the numerical method we used in our simulations.

2. BROWNIAN DYNAMICS SIMULATION OF THE FOKKER-PLANCK EQUATION WITH A SINK FUNCTION

Equation (2) was studied in the past by eigenfunction expansion methods[4,5] and by using special hybrid finite differences—the Galerkin technique.[2] We use here the Brownian dynamics approach combined with the Feynmann-Kac formula as a far more efficient method in application to disordered and multidimensional systems.

The Feynmann-Kac formula states that the solution of Eq. (2) is

$$P(x,t) = \langle \exp\{k \int_{t_0}^{t} dt^* s(x(t^*))\}\rangle_{x(t^*)} \tag{5}$$

where $x(t)$ is a trajectory satisfying the Langevin equation

$$dx = -Dh\frac{\partial}{\partial x}\hat{V}(x)dt + D^{1/2}\Delta W \tag{6}$$

where ΔW is the standard Wiener process. The average in Eq. (3) is taken over all possible realizations of Eq. (6) at all times except the last. The survival probability is obtained by additional averaging at the last time t.

In the case of disorder of any type—either of the sink function or the potential $\hat{V}(x)$—we add one more averaging operation with respect to realizations of the disorder. So, in this case the quantity of main interest—the survival probability—becomes

$$N(t) = \langle\langle \exp k \int_{t_0}^{t} dt^* s(x(t^*))\rangle_{x(t^*)}\rangle_{\text{disorder}}. \tag{7}$$

Eq. (7) gives a very flexible method to study Eq. (2) in the sense that it allows one to carry out calculations in a unified way for different kinds of sinks and potential functions as well as various models of disorder, and it admits straightforward multidimensional generalizations.

3. THE MODELS OF DISORDER

Since the model (1) has not been studied yet in a "disordered" setting, we start our calculations from three basic cases separately: random sink position in a potential well; spatial disorder (i.e., random potential curvature in the vicinity of sink); and time disorder (motion in a potential with random variation of instant potential slope).

To use these models for interpretation of, for example, ligand binding experiments, one probably has to use a combination of both. We choose the potential $\hat{V}(x)$ as

$$\hat{V}(x) = b\cos(\omega x). \tag{8}$$

This gives a possibility of simple interpretation of the curvature fluctuations (ω in this case) in terms of the spectral properties of the potential $\hat{V}(x)$. It is known that a spectral density of $\cos(\omega x)$ coincides with the distribution function of $\omega - p(\omega - \omega_0)$. So, we can easily manipulate the type and position of the $\hat{V}(x)$ spectrum by changing that of $p(\omega - \omega_0)$.

To single out the net effect of disorder, we choose the constant b in Eq. (8) large (-5 in all calculations) and a high magnitude of sink k (equal to 10^2). This makes the contribution of jumps between different minima of $\cos(\omega x)$ to the survival probability $N(t)$ negligible, since the initial conditions for trajectories of x (Eq. (6)) have been chosen such that they start in the middle between maxima and minima of $\hat{V}(x)$.

By "spatial disorder" we understand a random value of ω for every single trajectory of Eq. (6). We take $p(\omega - \omega_0)$ as a Gaussian distribution with average value ω_0 and dispersion $\Delta\omega$ which are, as mentioned, the position and the width of the spectrum of $\hat{V}(x)$. By "time disorder" we understand random variations of ω at each time step for each trajectory (6). The same Gaussian function is chosen for this model. Finally, we understand by random sink position the situation where the position of the sink x_0 is chosen randomly in the period of $\cos(\omega x)$ for each trajectory of Eq. (6). We assume this distribution to be Gaussian with zero mean. So there is only one parameter of disorder in this case. In all results presented below a value $D = 0.1$ has been used. Another value of D gives rise to only quantitative difference in the time dependences of $P(t)$.

4. RESULTS FOR THE SPATIAL DISORDER MODEL

Figure 1 represents typical results for the spatial disorder model. All curves have been calculated for $\Delta\omega = 2.5$. The curves 1–7 correspond to different positions of spectrum $-\omega_0 = 0$, 1, 2.5, 4, 5, 5, and 10, respectively. The closest elementary function to the long-time asymptotic decay for this model is

$$\exp(-\beta e^{t}) \quad \text{for } \omega_0 \lesssim 5 \quad \text{and} \tag{9}$$

$$\exp(-\beta e^{t^\alpha}) \quad \text{for } \omega_0 \gtrsim 5. \tag{10}$$

The most remarkable feature of the curves is the "phase transition" with respect to ω_0 between the type of asymptotic behavior of Eqs. (9) and (10). Indeed, the curves 1–4 and 7 have been obtained by using 10^4 trajectories and do not depend in a noticeable way on the time step and seed number of the random number generator. The curves 5 and 6 correspond to the same ω_0 and differ only in their seed numbers. This indicates the presence of long-lived time fluctuations, which are typical for the near-threshold region, and requires very good statistics to be eliminated.

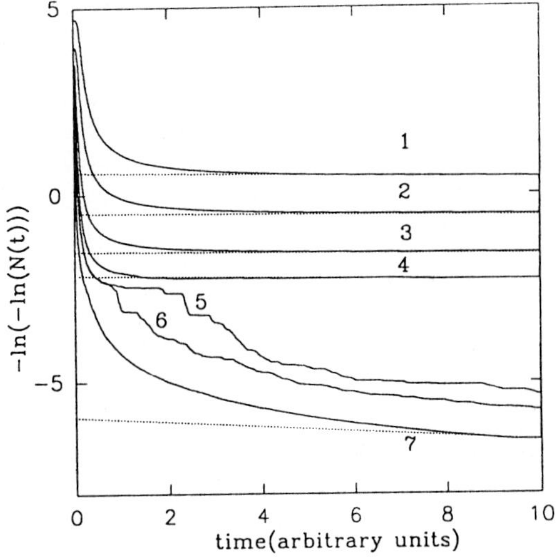

FIGURE 1 Long-time asymptotic for spatial disorder.

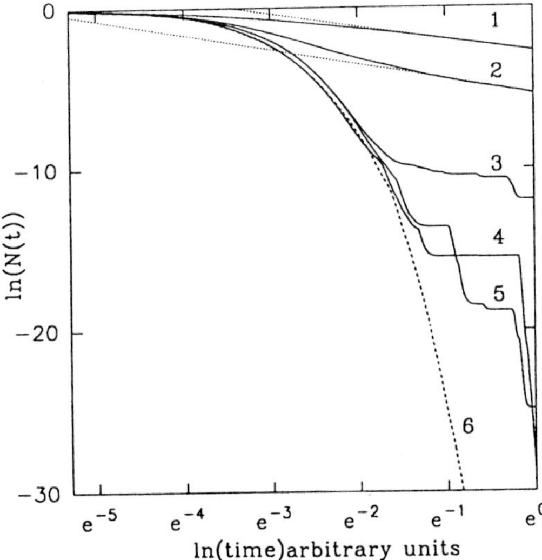

FIGURE 2 Short-time asymptotic for time disorder.

5. RESULTS FOR TIME DISORDER

The curves 6–1 in Figure 2 correspond to fixed spectral position of $\hat{V}(x)$ ($\omega = 10$) and $\Delta\omega = 1, 2.5, 2.5, 2.5, 5$, and 10, respectively. The same kind of phase transition in $\Delta\omega$ can be observed. In this case it takes place between long-time asymptotics

$$e^{-kt} \quad \text{for } \Delta\,\omega \lesssim 2.5 \quad \text{and} \tag{11}$$

$$t^{-\alpha} \quad \text{for } \Delta\omega \gtrsim 2.5. \tag{12}$$

The curves 3, 4, and 5 in Figure 2 correspond to $\Delta\omega = 2.5$. They illustrate the size of fluctuations near the threshold. The curve 3 was obtained with 10^5 trajectories, and curves 4 and 5 with 10^4 trajectories; they correspond to different seed numbers of the random number generator. We note that for a narrow spectrum of disorder (curve 6) this result approaches the well-studied case of the nondisordered system.[1,4,6]

6. RESULTS FOR RANDOM SINK POSITION

It is very difficult to propose an elementary formula to describe the long-time asymptotic behavior of $N(t)$ for this case. Results in Figures 3 and 4 show that the most probable equation is

$$t^{-\alpha}e^{-\beta t}, \tag{13}$$

α and β being smoothly dependent on Δx_0—the width of the distribution of the sink position. In the family of curves 1–6 of Figures 3 and 4, we can observe pure exponential decay for $\Delta x_0 = 0$ (curve 1) (this is the nondisordered case studied in Aberg et al.[1] Bagchi et al.,[4] and Beece et al.[6]), decay which is very close to a pure power law (the curve 2, $\Delta x_0 = 0.1$) and a sequence of mixed cases (of the type given by Eq. (13))—curves 3, 4, 5, and 6 for $\Delta x_0 = 0.15, 0.25, 0.5$, and 1, respectively.

CONCLUSION

The results obtained for three simple models of disorder show a very diversified picture of the time behavior of survival probability $N(t)$ for Eq. (1). This suggests an alternative approach to fitting of experimental data from experiments discussed in the introduction. It is possible to describe complex disordered systems, like proteins,

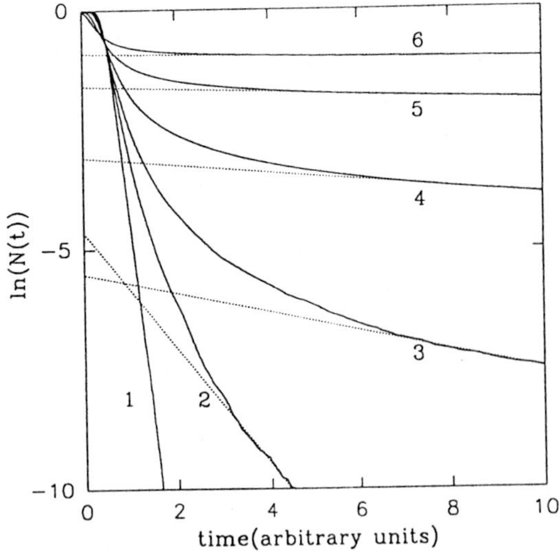

FIGURE 3 Long-time asymptotic for random sink position.

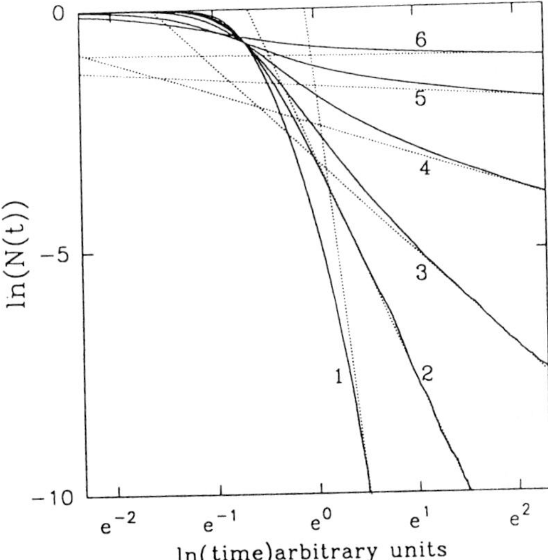

FIGURE 4 Intermediate-time asymptotic for random sink position.

in terms of widths and positions of disorder spectra which can be determined from molecular dynamics simulations along the line outlined by Elber and Karplus.[7]

ACKNOWLEDGMENTS

This work was motivated by lectures of R. Austin and D. Stein during the 5th Complex System Summer School, held in Santa Fe, New Mexico.

REFERENCES

1. Åberg, U., E. Åkesson, I. Fedchenia, and V. Sundström. "Femtosecond Laser Spectroscopy and Computer Simulations of Barrierless Isomerization in Solution." *Israel Chem. J.* (1992): submitted.
2. Agmon, N., and R. Kosloff. "Dynamics of Two-Dimensional Diffusional Barrier Crossing." *J. Phys. Chem.* **91** (1987): 1988–1996.
3. Austin, R., K. Beeson, L. Eisenstein, H. Frauenfelder, and I. Gunsalus. "Dynamics of Ligand Binding to Myoglobin." *Biochemistry* **14** (1975): 5355–5373.
4. Bagchi, B., G. Fleming, and D. Oxtoby. "Theory of Electronic Relaxation in Solution in the Absence of an Activation Barrier." *J. Chem. Phys.* **78** (1983): 7375–7385.
5. Bagchi, B., and G. Fleming. "Dynamics of Activationless Reactions in Solutions." *J. Phys. Chem.* **94** (1990): 9–20.
6. Beece, D., L. Eisenstein, H. Frauenfelder, D. Good, M. Marden, L. Reinisch, A. Reynolds, L. Sorernsen, and K. Yue. "Solvent Viscosity and Protein Dynamics." *Biochemistry* **19** (1980): 5147–5157.
7. Elber, R., and M. Karplus. "Multiple Conformational States of Proteins: A Molecular Dynamics Analysis of Myoglobin." *Science* **235** (1987): 318–321.
8. Frauenfelder, H., G. Nienhaus, and J. Johnson. "Rate Processes in Proteins." *Ber. Bunsenges. Phys. Chem.* **95** (1991): 272–277.
9. Rabinovich, S., and N. Agmon. "Adiabatic Elimination in Multidimensional Diffusive Barrier Crossing." *Chem. Phys. Lett.* **182** (1991): 336–342.

Barry Feldman
Department of Economics, State University of New York at Stony Brook, Stony Brook, New York 11794 USA; e-mail: bfeldman@ccvm.sunysb.edu

A Game Theoretic Interpretation of the Spin Glass

INTRODUCTION

Game theory studies the strategic interaction of groups and individuals, and is increasingly central to modern economic theory. This paper demonstrates some game theoretic properties of the spin glass Hamiltonian and develops an approach to the formal study of coalition formation. It is an addition to the small literature that describes statistical mechanical models used to study economic processes.[3,4,5,6]

Coalition formation is a relatively undeveloped topic in game theory. Typically, coalitions are not thought of as actually forming. Instead, they implicitly bargain over the division among all players of the worth of the all-player coalition. The principal coalition formation model involves the sequential formation of bilateral cooperation agreements.[1,2,11,12]

The spin glass game, as presented here, is a very restricted subset of the space of cooperative games. However, because of its particular representation of the value of association, and its explicit representation of the intrinsic interests of a player, it adds new dimensions to the cooperative game which may be useful in the study of coalition formation. This approach both greatly simplifies the representation and study of coalitional dynamics and also opens up the possibility of applying contemporary statistical mechanical methods to the study of large games.

1. THE SPIN GLASS

The spin glass[14,15] model was originally developed to model complex magnetic phenomena in alloys of atoms with and without magnetic moments, and elaborates the simpler Ising model of ferromagnetism.[8] *Spin* refers to the magnetic moment of the atom, *glass* to the irregular pattern of interactions between spins.

In the infinite range version, all spins interact with all other spins. A spin S^i can only orient itself up ($S^i = +1$) or down ($S^i = -1$). A spin "tries" to orient itself so that it minimizes its *frustration* with respect to all other spins and an outside magnetic field B, if present. The nature of the interaction is represented by a symmetric random matrix J of *quenched* variables, where J_{ij} is the interaction between spins i and j. $J_{ij} > 0$ means that the particles tend to align with each other. Summing over all spin interactions yields the Hamiltonian for a spin glass system:

$$H = -\sum_{i,j} J_{ij} S^i S^j + \sum_i B S^i \,.$$

Note that $S^i S^j$ is positive whenever spins i and j have the same orientation.

In what follows, the interaction matrix will not be required to be symmetric or random and, for notational convenience, will be denoted as Y. The external field will be taken to be a vector with independent values for every player and will be denoted as Z.

2. COOPERATIVE SPIN GLASS GAMES

A cooperative game[10,13] is represented by a characteristic function that assigns a worth to all coalitions formable by a set of players N. The interpretation of the spin glass as a cooperative game results first from viewing the quenched variables as pairwise interaction effects between players which sum to the worth of a coalition. In this section we develop the idea and relate it to some of the central concepts in cooperative game theory. In Section 3 we introduce coalitional structures. Then, in Section 4 we interpret the spin glass external field as the intrinsic preferences of the players and introduce the idea of *orientation*.

DEFINITION 1. A cooperative spin glass game $v = (N, Y)$ is a set of players $N = \{1, 2, \ldots, n\}$ together with an interaction matrix $Y \in \mathbf{R}^{n \times n}$. The interaction matrix defines the characteristic function $v : 2^N \to \mathbf{R}, v(\emptyset) = 0$, which assigns a worth to every coalition of players $\emptyset \neq S \subset N$ as follows:

$$v(S) = \sum_{\substack{i \in S \\ j \in S}} Y_{ij} - \sum_{\substack{i \in S \\ k \notin S}} Y_{ik} \,. \tag{1}$$

We can think of Y_{ij} as being the *worth* of player j to player i. In a spin glass game the worth of the interactions with players not in a coalition must be subtracted in order to determine its worth.

DEFINITION 2. The marginal contribution of a player $i \in S$ to a coalition S, $v^i(S)$ is $v(S) - v(S \setminus \{i\})$, or:

$$v^i(S) = \sum_{j \in S} Y_{ij} - \sum_{k \notin S} Y_{ik} + 2 \sum_{\substack{j \in S \\ j \neq i}} Y_{ji} \,. \tag{2}$$

The Shapley value[10] of a cooperative game assigns to every player her average marginal contribution to all coalitions she is a part of, giving equal weight to the average contribution to all coalitions of a given size. Alternatively, the Shapley value can be calculated as the average marginal contribution of a player over all $n!$ possible orderings of formation of the whole. The Shapley value can be axiomatically derived from the assumptions of efficiency, symmetry, linearity, and zero value to a player who contributes no value. It may be best thought of as the prospective value to a player in playing a game given that all sequences of the formation of the all-player coalition are equally likely.

CLAIM 1. The Shapley value for a player i in a spin glass game is

$$\varphi_i(v) = \sum_{j \in N} Y_{ji} \,. \tag{3}$$

PROOF Think of the Shapley value as the average marginal contribution over all orderings of entry of players into the formation of the all-player coalition. Over all random orderings the likelihood that player j will enter before player i is .5. Thus the first two terms on the right-hand side of Eq. (2) will cancel out except for Y_{ii} and we are left with Eq. (3). ■

The value of a player is the sum of her contributions to other players (and herself).

DEFINITION 3. A game has net positive (non-negative) interaction effects if

$$Y_{ij} + Y_{ji} \geq 0 \, \forall \, i \neq j \in N \, .$$

CLAIM 2. A necessary and sufficient condition for superadditivity in a spin glass game is that it have net positive effects.

PROOF Superadditivity requires $S \cap T = \emptyset \Rightarrow v(S \cup T) \geq v(S) + v(T)$.

$$\begin{aligned} v(S \cup T) - \{v(S) + v(T)\} = & \left(\sum_{i,j \in S \cup T} Y_{ij} - \sum_{\substack{i \in S \cup T \\ k \notin S \cup T}} Y_{ik} \right) \\ & - \left(\sum_{i,j \in S} Y_{ij} - \sum_{\substack{i \in S \\ k \notin S}} Y_{ij} \right) \\ & - \left(\sum_{i,j \in T} Y_{ij} - \sum_{\substack{i \in T \\ k \notin T}} Y_{ij} \right) \\ = & \, 2 \sum_{\substack{i \in S \\ j \in T}} (Y_{ij} + Y_{ji}) \, . \end{aligned}$$

Thus, net positive interaction effects guarantee superadditivity. The property is necessary because, if there exists an i and j for which this condition is not met, then there will be a at least one coalition, $S = \{i, j\}$, that violates the conditions for superadditivity. ■

CLAIM 3. A necessary and sufficient condition for convexity in a spin glass game is that it have net positive interaction effects.

PROOF A cooperative game is convex if $v(S \cup T) + v(S \cap T) \geq v(S) + v(T)$. The following demonstrates sufficiency, necessity follows from the same argument used in the last claim.

$$\begin{aligned} v(S \cup T) + v(S \cap T) - v(S) - v(T) &= \sum_{i,j \in S \cup T} Y_{ij} + \sum_{i,j \in S \cap T} Y_{ji} - \sum_{i,j \in S} Y_{ij} \\ &= -\sum_{i,j \in t} Y_{ji} - \sum_{\substack{i \in S \cup T \\ k \notin S \cup T}} Y_{ik} - \sum_{\substack{i \in S \cap T \\ k \notin S \cap T}} Y_{ik} \\ &\quad + \sum_{\substack{i \in S \\ k \notin S}} Y_{ik} + \sum_{\substack{i \in T \\ k \notin T}} Y_{ik} \\ &= 2 \sum_{\substack{i \in S \cap T^c \\ j \in T \cap S^c}} (Y_{ij} + Y_{ji}) \geq 0. \end{aligned}$$

■

The *core* of a cooperative game is the set of all allocations to players that sum to the worth of the all-player coalition and have the property that the sum of allocations to players in any coalition S is at least as great as the worth of the coalition. It is well known that every convex game has a nonempty core but for a spin glass game with net positive interaction the Shapley value is in the core as well.

CLAIM 4. Every spin glass game with net positive effects has a nonempty core. In particular, $\varphi(v) \in \text{core}(v)$.

PROOF For any $S \subset N$ and any $i \in S$,

$$\begin{aligned} \sum_{i \in S} \varphi_i(v) - v(S) &= \sum_{i \in S} \sum_{j \in N} Y_{ji} - \left(\sum_{\substack{i \in S \\ j \in S}} Y_{ij} - \sum_{\substack{i \in S \\ k \notin S}} Y_{ik} \right) \\ &= \sum_{\substack{j \in S \\ i \notin S}} (Y_{ij} + Y_{ji}) \geq 0. \end{aligned}$$

■

Since net positive effects $\Leftrightarrow$ superadditivity $\Leftrightarrow$ convexity, any one of these properties guarantees a nonempty core and that the Shapley value will be a core allocation. The spin glass formalism thus defines an exceptionally well-behaved cooperative game with well-understood properties under the regime of net positive effects.

3. COALITIONAL STRUCTURES

This section presents a simple approach to coalition formation where we imagine that if the all-player coalition fails to form, then two coalitions will form in its place. Net positive interaction effects are no longer assumed.

Example 1. Consider the following Y matrix for a game of four players:

$$\begin{bmatrix} 4 & 4 & -1 & -1 \\ 4 & 4 & -1 & -1 \\ -1 & -1 & 3 & 3 \\ -1 & -1 & 3 & 3 \end{bmatrix}.$$

$v(N) = 20$ and $\varphi(N) = (6,6,4,4)$ for this game. But if $S = \{1,2\}$ and $T = \{3,4\}$, then we have $v(S) = 20$ and $v(T) = 16$; hence, $v(S) + v(T) = 36 > v(N) = 20$. The breakup of N into S and T is coalitionally rational in as much as there are no other coalitions that can form and do as well or better for all players. ∎

What then is the prospective value of playing this game for each player?

DEFINITION 4. A coalition structure is an element $\omega \in \Omega \equiv \{(S : S^c) | S \subseteq N\}$, the set of unordered complementary subsets of N.

CLAIM 5. The value of a player relative to a coalition structure ω is

$$\varphi_i^\omega(v) = \sum_{j \in S:\, i \in S} Y_{ji} - \sum_{k \in S^c} Y_{ik}\,. \tag{4}$$

PROOF Since the value of a player relative to the all-player coalition is the average marginal contribution over all orderings representing formation of the all-player coalition, it is logical to extend the notion of value relative to a coalition structure to be a player's average marginal contribution to a coalition over all possible orderings of entry into that coalition. Since the players outside player i's coalition never enter, their interaction value must always be subtracted. These values fulfill the essential efficiency condition, $\sum_{i \in S} \varphi_i^\omega(v) = v(S)$. ∎

Note that if $\omega = \{N, \emptyset\}$, $\varphi^\omega(v) = \varphi(v)$. For the example described above, where $\omega = (\{1,2\},\{3,4\})$, $\varphi^\omega(N) = (10,10,8,8)$.

4. ORIENTATED SPIN GLASS GAMES

Orientation expresses the idea that coalitions are forming around a choice between alternatives. In addition to the interaction matrix Y, we now consider an intrinsic preference vector Z. Orientation can be used to study *network externalities* [9] in the choice between two technological standards. Here, Z represents the intrinsic values of one standard versus another to players and Y represents the externalities. Orientation can also describe the legislative process, where Z can be taken to represent the preferences of the representative's constituents while Y represents the personal, ideological or political influences that legislators may bring to bear on each other.

DEFINITION 5. A spin glass game with orientation $v = (N, Y, Z)$ is a set of players $N = \{1, 2, \ldots, n\}$, an interaction matrix $Y \in \mathbf{R}^{n \times n}$, and a preference vector $Z \in \mathbf{R}^n$. Y and Z define a pair of characteristic functions $v_+ : 2^N \to \mathbf{R}, v_+(\emptyset) = 0$, and $v_- : 2^N \to \mathbf{R}, v_-(\emptyset) = 0$, such that:

$$v_+(S) = \sum_{\substack{i \in S \\ j \in S}} Y_{ij} - \sum_{\substack{i \in S \\ k \notin S}} Y_{ik} + \sum_{i \in S} Z_i \,, \tag{5}$$

$$v_-(S) = \sum_{\substack{i \in S \\ j \in S}} Y_{ij} - \sum_{\substack{i \in S \\ k \notin S}} Y_{ik} - \sum_{i \in S} Z_i \,. \tag{6}$$

Clearly, if $Z = 0$, $v_+(S) = v_-(S) = v(S)$ for all $S \subset N$. The Z vector can be thought of as the net intrinsic preference of player i for "+," the positive orientation or up state. We can call "−" the negative orientation or down state, and think of $-Z_i$ as the net intrinsic preference or utility to player i of negative orientation. Thus, $Z_i = u_i(+) - u_i(-)$ where the utility function reflects intrinsic preferences prior to the consideration of the interpersonal externalities represented in the interaction matrix. Usually, the worth of a coalition is simply imagined as the best outcome that the coalition can guarantee itself. Orientation introduces a representation of what this coalition actually does.

Example 2. Consider the following game, a variation on Example 1.

$$(Y, Z) = \left(\begin{bmatrix} 4 & 4 & -1 & -1 \\ 4 & 4 & -1 & -1 \\ -1 & -1 & 3 & 3 \\ -1 & -1 & 3 & 3 \end{bmatrix}, \begin{bmatrix} 5 \\ 5 \\ 5 \\ 5 \end{bmatrix} \right).$$

Again, let $S = \{1, 2\}$ and $T = \{3, 4\}$. We can quickly determine the following results: $v_+(N) = 40$, $v_-(N) = 0$, $v_+(S) = 30$, $v_+(T) = 26$, $v_-(S) = 10$, $v_-(T) = 6$. In this game we should expect that all players will join the same coalition, which will adopt the positive orientation. Orientation allows that intrinsic individual preferences may overcome centrifugal forces inherent in individual interactions. ■

DEFINITION 6. An oriented coalition structure ω is an element of the set of ordered pairs of complementary subsets of N, $\omega \in \Omega \equiv \{(S, S^c) | S \subseteq N\}$, where the first coalition in the pair has positive orientation and the second coalition has negative orientation.

CLAIM 6. The value of a player i in an oriented coalition structure ω in an oriented game is

$$\varphi_i^{\omega}(v) = \sum_{j \in S: i \in S} Y_{ji} - \sum_{k \in S^c} Y_{ik} + O(S\colon i \in S) Z_i\,, \tag{7}$$

where $O(S\colon i \in S)$ is the orientation of the coalition $S \in \omega$ that i belongs to.

PROOF This is the result if we consider the average marginal contribution of a player given Eqs. (5) and (6). It is a simple extension of the argument in Claim 5 to encompass orientation. Note again that the sum of the values of the players in a coalition is equal to the worth of the coalition. ■

Example 3. Consider now an example where orientation leads to coalition formation.

$$(Y, Z) = \left(\begin{bmatrix} 4 & 4 & 1 & 1 \\ 4 & 4 & 1 & 1 \\ 1 & 1 & 3 & 3 \\ 1 & 1 & 3 & 3 \end{bmatrix}, \begin{bmatrix} 5 \\ 5 \\ -5 \\ -5 \end{bmatrix} \right).$$

The unoriented complete game based only on the interaction matrix Y clearly has a core and there are strong benefits to formation of the all-player coalition. With the preference vector Z, however, and with $S = \{1,2\}$ and $T = \{3,4\}$, it can be seen that $v_+(S) = 22$, $v_-(T) = 18$, and $v_+(N) = v_-(N) = 36$. Given the symmetries of the example, thus we should expect the oriented coalitional structure $\omega = (S, T)$ to arise. It is easy to see that $\varphi^{\omega}(v) = (11, 11, 9, 9)$. ■

There may be many possible individually rational coalitional structures in a given game. Several approaches to determining which coalitions should arise in small games based on the structure so far presented are possible depending on the situation being represented. In the context of the formation of a parliamentary governing coalition, the largest party is generally allowed to assemble the coalition of its choice. This would be represented by allowing one player to choose the coalitional structure most favorable to it. In general, we might imagine a variety of noncooperative sequential choice processes (similar to Aumann and Myerson[1] and Bloch[2]) based on an exogenously given ordering of players. For example, the first player selects her most preferred set of coalitions, then the next player selects from these his most preferred, and so on.

5. DISCUSSION

The cooperative spin glass game offers a simple method of both distinguishing interaction effects from intrinsic preferences and studying coalition formation. The concepts presented here can be developed in a number of directions.

Coalitional structures and orientation are presented here as a binary concepts, as in the physical spin glass model, but it should be clear that they can be generalized.

With an appropriate interpretation of temperature,[3] statistical mechanical techniques can be used to evaluate the probability of a particular coalitional structure arising when many are possible. Under these circumstances the path dependent and ultrametric [14] aspects of the spin glass phase space become relevant. The prospective value of a game to a player then becomes a function of the initial conditions. One perspective on the player's expected value of playing the game could be derived from the partition function, [8,14,15] and would have the interpretation as the sum over all coalitional structures of the probability of the coalition arising times the player's value in it.

Finally, with strong assumptions, including constraints on the distribution of the Y_{ij}, the machinery of the replica method[15] might be applied to examine aspects of coalition formation in large games in a manner that extends Föllmer's [6] scope and results.

ACKNOWLEDGEMENTS

The 1992 Complex Systems Summer School was the ideal environment to start work on this subject. I would like to particularly acknowledge conversations with Ernest Chan, Kai Nagel, Richard Palmer, Dan Stein, and Jonathan Yedidia and the meetings of the Economics Working Group. The Santa Fe Institute, and its wonderful staff and other sponsoring organizations, deserve special thanks for making possible this opportunity for cross-disciplinary study, encounter, and collaboration. I would like also to acknowledge helpful comments from Bob Aumann, Per Bak, J. F. Mertens, Kali Rath, and Yair Tauman.

REFERENCES

1. Aumann, Robert, and Roger Myerson. "Endogenous Formation of Links Between Players and of Coalitions: An Application of the Shapley Value." In *The Shapley Value: Essays in Honor of Lloyd S. Shapley*, edited by Alvin Roth, 175–191. Cambridge, MA: Cambridge University Press, 1988.
2. Bloch, Francis. "Sequential Formation of Coalitions with Fixed Payoff Division." Working Paper 92-5, Department of Economics, Brown University, 1992.
3. Blume, Lawrence. "The Statistical Mechanics of Strategic Interaction." *Games & Econ. Behav.* (1993): forthcoming.
4. Durlauf, Steven N. "Path Dependence in Aggregate Output." Working Paper, Department of Economics, Stanford University, 1991.
5. Durlauf, Steven N. "Non-Ergodic Economic Growth." Technical Report no. 6, Department of Economics, Stanford University, 1991.
6. Föllmer, Hans. "Random Economies with Many Interacting Agents." *J. Math. Econ.* **1** (1974): 51–62.
7. Kandori, Michihiro, George Malaith, and Rafael Robb. "Learning, Mutation and Long Run Equilibria in Games." *Econometrica* (1993): forthcoming.
8. Huang, Kerson. *Statistical Mechanics,* 2nd ed. New York: Wiley, 1987.
9. Katz, M. L., and C. Shapiro. "Network Externalities, Competition, and Compatibility." *Amer. Econ. Rev.* **75** (1985): 424–440.
10. Myerson, Roger. *Game Theory: Analysis of Conflict.* Cambridge, MA: Harvard University Press, 1991.
11. Myerson, Roger. "Cooperation on Graphs." *Math. Oper. Resh.* **2** (1977): 225–229.
12. Okada, Akira. "Noncooperative Bargaining and the Core of an n-Person Characteristic Function Game." Discussion Paper 336, Kyoto Institute of Economic Research, 1991.
13. Owen, Guillermo. *Game Theory.* New York: Academic Press, 1982.
14. Palmer, Richard. "Broken Ergodicity." In *Lectures in the Sciences of Complexity*, edited by D. L. Stein. Santa Fe Institute Studies in the Sciences of Complexity, Lect. Vol. I, 275–300. Reading, MA: Addison-Wesley, 1989.
15. Yedidia, Jonathan, "Quenched Disorder: Understanding Classes Using a Variational Principle and the Replica Method." This volume.

Barry Feldman† and Kai Nagel‡
†Department of Economics, State University of New York at Stony Brook, Stony Brook, New York, 11794, USA; e-mail: bfeldman@ccvm.sunysb.edu and ‡Mathematisches Institut and Zentrum für Paralleles Rechnen, Universität zu Köln, Weyertal 86–90, W–5000 Köln 41, Germany; e-mail: kai@mi.uni-koeln.de

Lattice Games with Strategic Takeover

1. INTRODUCTION

This contribution explores a topic of interest in a surprising number of physical and social sciences, the iterated Prisoner's Dilemma game. We use this game to construct a simple model of strategic interaction on a lattice.

The basic game describes two prisoners, accused of having committed a crime together, who are unable to communicate. Each is told that, if he confesses (*defection*), he will get a lighter sentence, but that he will receive a very heavy sentence if he does not confess and the other prisoner does. However, if neither confesses (*cooperation*), each receives a medium sentence.

Both prisoners defecting is the only equilibrium in the game because they cannot make a binding agreement to cooperate. In the basic game theoretic analysis, cooperation can be sustained only by the indefinite repetition (iteration) of the game. The expected future benefits of cooperation must be greater than defection, and cooperation is difficult to sustain.

Modifications of the basic two-player iterated game such as bounded rationality and noisy communication channels have been studied.[8,10,12,15] Axelrod[1,2] studied repeated play among different players in a well-known computational experiment where researchers were invited to submit arbitrarily complex strategies (computer programs) that played each other in a "round-robin" tournament. Rapoport[14] contributed the "Tit-for-Tat" strategy which started off by cooperating and then simply repeated the opponent's last move and became famous by beating all other strategies.

The principal variation in our work is that we arrange agents on a lattice and have them play the same strategy *simultaneously*, but only against their *immediate* neighbors. The secondary variation in our work is that in most of our runs we allow payoffs to accumulate and, if an agent goes bankrupt, he is "taken over" by his most successful neighbor and adopts her strategy. Thus successful strategies propagate spatially, in a simple representation of diffusion through economic and social networks.

In our study there are four important factors to consider: (1) how many iterations an agent can remember and what he can remember; (2) the relative advantage to noncooperation; (3) the degree of "selection pressure"; and (4) the geometry of the lattice. In some of our runs we introduce a low rate of mutation in strategies which gives our work some of the quality of genetic algorithm methods. The work reported here focuses primarily on the effects of selection pressure and variations in the incentive to defect.

It is rare that two players would play only against each other or that all agents would play all other agents in realistic economic situations. Typically, we expect a network of connections between agents. One approach to studying such networks is to model them as spatial behavior on a d-dimensional lattice.[2,4,13] Axelrod[2] already reports on experiments similar to ours on a lattice, but, in his work, agents play their neighbors separately. Most studies of evolutionary processes assume random or uniform matching. Here we allow the diffusion of strategies to take place, but do not make prior assumptions as to how complete "mixing" will be.

Local interaction is becoming a significant dimension of economic processes to study. Blume[3] uses the Ising model to study simultaneous play on a two-dimensional lattice without the possibility of "takeover." Durlauf[5] studies production with local externalities. Ellison[6] finds that local interaction results in much quicker transitions to dominant strategy equilibria than conventional assumptions.

2. OUR GAME

We start with a standard payoff matrix with positive payoffs. Each player has two strategies, *cooperate* and *defect*. The payoffs are: 3 points each when two agents cooperate; 0 points to an agent when she cooperates but her opponent defects; 1

point each to agents when they both defect; and 5 points to an agent which defects against another agent that is cooperating (our b parameter, which is varied). For our simulations, the score of one round of the game for each player is the sum of the payoffs of each individual encounter with each neighbor minus a constant r. Typically, we set $r = 8$. It is easier to think of varying r than to change entries in the payoff matrix. Thus, a cooperative agent with three cooperative and one defecting neighbors would get $3 \times 3 + 0 - r = 9 - 8 = 1$ point according to the above rules. In order to allow evolution of behavior, we designed a specific form of takeover: Sites start out with a certain number of points. Unsuccessful sites which go bankrupt can be taken over by the more successful strategy of one of its neighbors.

Our model can be seen as a variation on Kaufman's genetic model.[11] For clarity, our notation will be based on the one-dimensional case. N agents are positioned on the N sites of a quadratic lattice of size $N = L^d$. We denote the state of an agent at site i and time step t with $s(i,t)$. We take $s = 1$ for cooperate and $s = 0$ for defect. The state of any agent at time $t+1$ is determined by the states of his neighbors at time t (cellular automata[16]):

$$s(i,t+1) = f_i(s(i-1,t), s(i+1,t)) \ .$$

There are $2^2 = 4$ different configurations of $\vec{s} := (s(i-1), s(i+1))$ at time t, and two different ways to react (i.e., $f(\vec{s}) = 0$ (cooperate) or $f(\vec{s}) = 1$ (defect)) and thus $4^2 = 16$ different strategies. Strategies can be numbered by their binary representation: Strategy number 5 is 0101 in binary notation. The left-most bit codes the answer (here "0" = D) towards a configuration 11 (i.e., both neighbors cooperate), the next bit codes the answer towards a configuration 10 (i.e., left neighbor C and right neighbor D), and so on. As we want to allow that different agents follow different strategies, each agent may have its own strategy f_i.

Following cellular automata methodology, lattice sites are updated in parallel; i.e., every agent bases its time t action on the time $t-1$ actions of other players. This is computationally advantageous, but also seems realistic in some cases (see below). Another numerically motivated decision was to take (in two dimensions) helical boundary conditions in one direction, i.e., connecting agent $(L,1)$ to the right with agent $(1,2)$ and so on. The other direction was periodic.

To understand this and the following in the context of economics, a simple illustrative example in two dimensions (its idea taken from Cowan & Miller[4]) might be helpful. Imagine a large city with streets on a square grid. At each crossing a shop is situated, which may only be reached by people living in the four neighboring street sections (a section of a street reaching from one crossing to the next). Thus, when all the shops are open, they have the clients from $4 \times 1/2 = 2$ sections. But if a shop closes, its clients can turn to the shop at the other node of their section, but they cannot go beyond their two nodes.

Now imagine that the shops want to decide if to open on Sundays. If they all stay closed, the clients have to do their shopping at another time. If all shops open, there are not enough clients to cover the costs of the additional opening time. So, obviously, keeping the shop closed corresponds to the cooperate option and opening to the defect option. And the fact that shop managers have to make their decision in advance (organizing personnel, etc.) corresponds to the parallel update. Choosing only one action in response to four neighbors adds an element of spatial frustration to the game, reminiscent of spin glass theory: How should an agent react if he has three usually cooperative neighbors and one that defects?

3. A GAME WITH ACCUMULATED PAYOFFS AND TAKEOVER (GAME A)

On the two-dimensional lattice there are $2^{16} = 65536$ different strategies. To study characteristics of successful strategies, we performed the following computer experiments on a 40×40-lattice: Initially, each site obtained a randomly chosen strategy, a randomly selected state (C or D), and 50 points. Then the game started, each agent playing according to his strategy accumulating points according to the payoffs. But, during each round of the game, the score of each player is at the same time reduced by r points, r usually being 8 or 9 in our simulations. Thus, cumulative negative scores are possible. Each time an agent crosses the threshold of 0 points, he is taken over by the wealthiest of his neighbors. (No takeover takes place if all neighbors are negative as well.) If, say, site i is taken over by site $i+1$, site i now gets the strategy from site $i+1$. The debts of site i are subtracted from the current score of $i+1$, and the result (which might be negative) is equally distributed between i and $i+1$. In order to keep this scheme consistent, it is performed by going through the lattice sequentially, thus introducing some kind of anisotropy. An increasing r leads to a higher number of "dying" agents. r is therefore reminiscent to the biological notion of a higher selection pressure. It should be noted that—after the initialization—the whole game is still totally deterministic. The idea of a takeover is the same as in Nowak and May,[13] but there only two strategies are allowed.

Figure 1 is a visualization of a run of such a game in one dimension (i.e., the players are placed on a ring, like around a lake). For further details, the reader is referred to the figure caption.

In the following, we describe some statistical properties of the strategy space of the system after the simulation of 2,560,000 time steps, for $r = 8$ and $r = 9$. Averages over about 100 runs are used for each of the parameters. We performed most of the simulations on a net of five loosely coupled IBM RS/6000 workstations, using PVM for the distribution of different Monte Carlo runs on different machines and for recollecting the results. In some cases, to obtain results in shorter time, we used in addition a 32-node INTEL iPSC/860 parallel computer with the same technique. The perhaps most interesting result of our simulations is that there is a clear difference between success and efficiency of a strategy. Figure 2 compares our measure for success (i.e., the frequency of a strategy at the end of the runs) to the other measure, not less important in normal life: We define the effectiveness of a strategy as the (normalized) average number of points a strategy scores per time step. We observe that in order to have higher *efficiency* it is, contrary to intuition, useful to have many C reactions to a mainly defecting neighborhood (at least for lower r). The reason behind this is that a relatively good way to survive under the circumstances of the game is to have a larger cluster of a chessboardlike pattern of D's and C's. Then the D's get all the points from the scoring and, each time step again, take over their C neighbors. For our example with the shops in a latticelike town, it would simply mean that opening every second shop and paying the closed ones out would be a good strategy.

Other results are that higher "selection pressure" (higher r) enforces a higher degree of cooperation, and that directionality does not, contrary to our intuition, play a significant role.[7] (The idea was that strategies exploiting the possibility to, say, answer a C offer from the right differently than a C offer from the left, might be more or less successful than others, which behave in a more normal way like "cooperate, when more than two neighbors cooperate, else defect.")

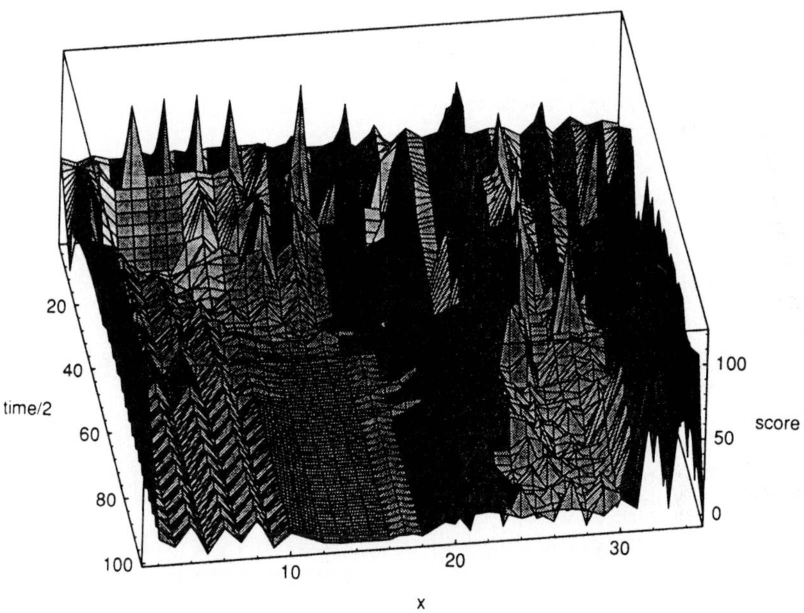

FIGURE 1 This gives one example of the evolution of a one-dimensional lattice game (game A, see text) over 200 time steps. It is a game of 35 players, placed on a one-dimensional "x" array with periodic boundary conditions (i.e., on a ring). The height of the surface illustrates the development of the score of each individual player during time. Different gray shadings stand for different strategies. For $t \rightarrow \infty$, only three strategies will survive.

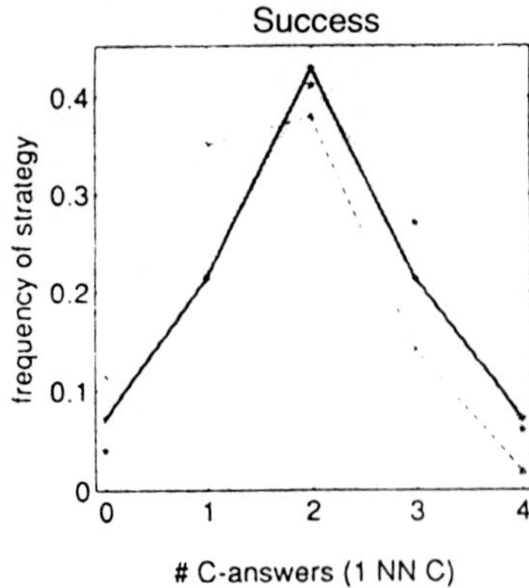

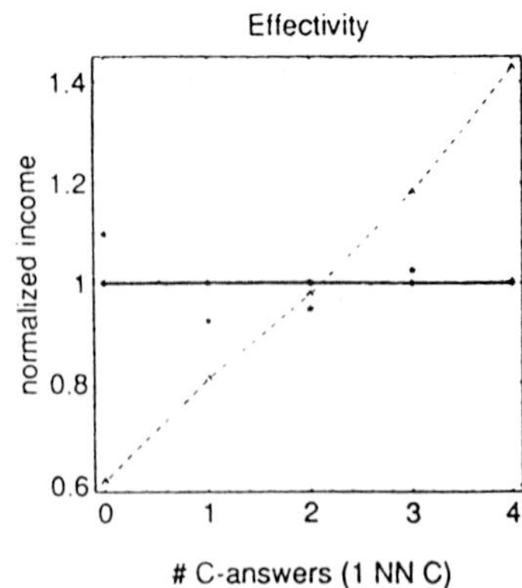

FIGURE 2 Success and effectivity as functions of the number of C answers to configurations where only one neighbor cooperates. Game A (see text), 40×40 lattice after 2,560,000 time steps, averaged over about 300 runs each). Dashed line: $r = 8$ (low selection pressure, see text). Dotted line: $r = 9$ (higher selection pressure). *Left:* Frequency of strategies. Lower selection pressure (dashed line) favors defecting strategies, whereas higher selection pressure does not change much here. The broader gray line gives the initial distribution. *Right:* Effectivity of strategies, defined as normalized "income" per agent. We see that in order to have higher *effectivity* it is, contrary to intuition, useful to have many C reactions to a mainly defecting neighborhood (at least for lower r).

4. A STRICTER GAME: SAVINGS FOR BAD TIMES NO LONGER ALLOWED (GAME B)

In order to relate our game to the experiments of Nowak and May,[13] we did some more simulations with a simplified version of the above game: This time, the scores are not accumulated, but, after each round, an agent is taken over by the strategy of the richest one between his four nearest neighbors and himself. In addition, as a consequence of our findings concerning directionality and our desire to make the strategy space more manageable, we reduced the possible strategies to choosing separate responses to 4, 3, 2, 1, or 0 cooperating neighbors. Therefore, a string of 5 bits is now sufficient, and only 32 different strategies remain.

Simulations with this model turned out to settle down relatively fast to only one or two surviving strategies (on a 40×40 lattice), depending strongly on the initial conditions. We therefore added a gentle *mutation* to the game: In each time step, an agent is randomly selected, then a bit is randomly selected and inverted. For this reason, game B is no longer totally deterministic. This rule led to configurations with many different strategies living together, independent of the initial conditions. Even under a change of the environment during a simulation run, the system is able to react: New strategies are formed by mutation, and they can diffuse by takeover. This corresponds to the ability of an agent in an economy or social setting to find new answers to emerging problems.

Following Nowak and May,[13] we tried games with varying payoffs for the defecting agent (to be exact, we varied the score a D agent gets against a C agent, which was 5 in game A). We, too, will call this the b-parameter, although there are some differences because we wanted to stay with the set of parameters given above (i.e., a payoff of 1 point for each of two defecting agents and a payoff of 3 for each of two cooperative agents).

An overview over the long-term behavior is given in Figure 3. The figure shows the frequency of C agents in the game after 2,560,000 time steps. We find three clearly marked different regimes: One for $b < 2.25$, a second for $2.25 \leq b \leq 3$, and a third one for $b > 3$. Except for the last regime, the systems equilibrate rapidly.

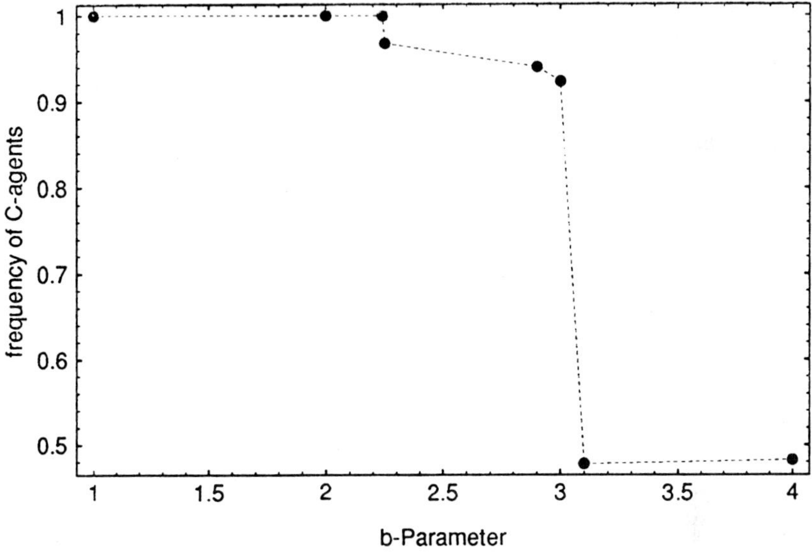

FIGURE 3 Frequency of cooperating agents in game B (see text) on a 40×40-lattice after 2,560,000 time steps for different values of the b-value, which gives the number of points a defecting agent scores against a cooperative agent. For $b < 2.25$, all agents always cooperate; for $2.25 \leq b \leq 3.0$, there is a small portion of defecting agents; for $b > 3.0$, more than one half of the agents defect.

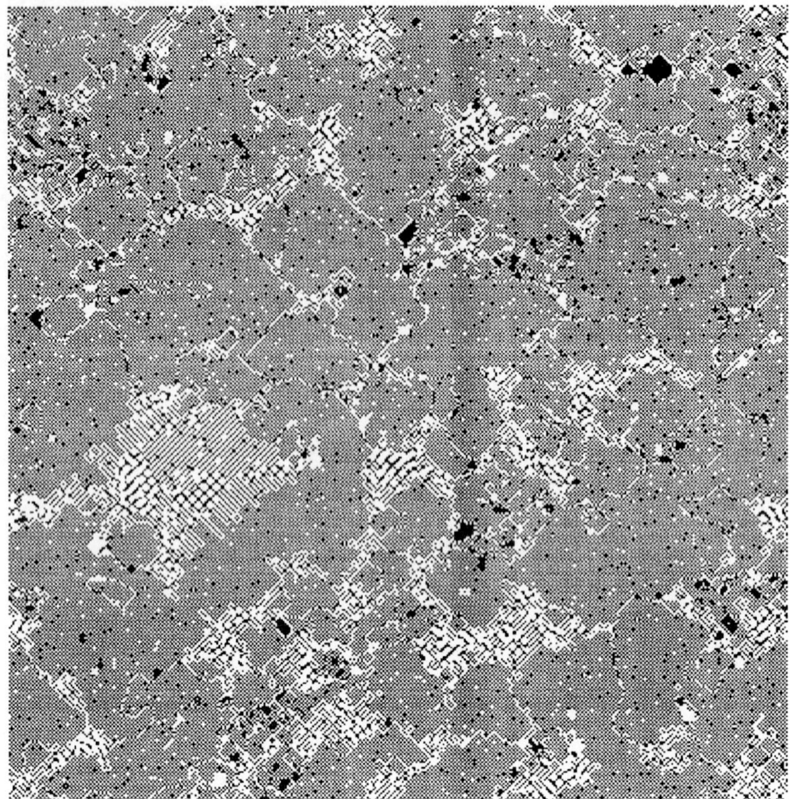

FIGURE 4 A snapshot at game B ($b = 3.001$, lattice size 400×400: much larger than for the statistical results; and after 12,800 iterations: much earlier than for the statistical results). White pixels are defecting agents, black pixels are cooperative agents. Note the large areas of the chessboard pattern (which look gray at this resolution) and the areas, where C and D agents are arranged in diagonal rows. Both are versions of the anti-coordination ("chessboard") solution (see text). "Tentacles" fill the space between incompatible chessboard regions (see Figure 5). The configuration is relatively stationary; during the previous 10,000 time steps, there were only minor changes of the overall configuration. Note the visual resemblance to the results of Nowak and May[13], i.e., a uniform background, and patches on this background which seem to be connected by a network.

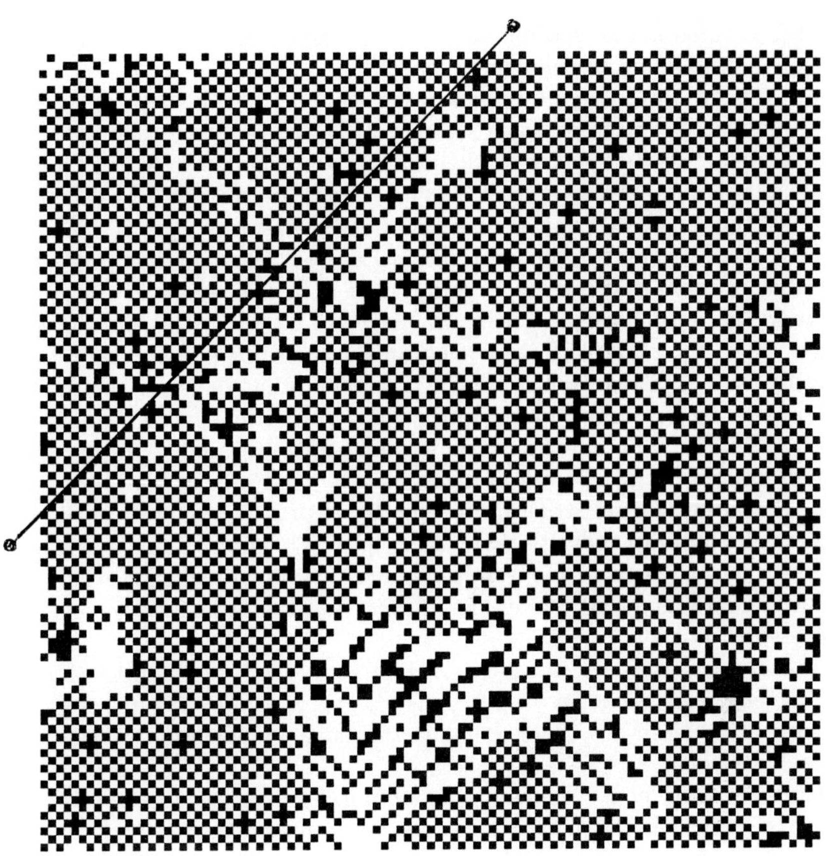

FIGURE 5 Magnification of the center region of Figure 4. It is clearly seen that chessboard patterns on different sides of the dividing tentacles are incompatible.

The results clearly show that large-scale cooperation in this game emerges only for $b \leq 3$, where it is trivial. In fact, simulation results show that the difference between $b \leq 3$ and $b > 3$ is even more marked: The long-term average behavior is independent of the initial conditions, which means that, for $b > 3$, cooperation breaks down even if imposed initially: Starting from a configuration where all agents cooperate, and where all agents have the same "CCCCC" strategy (i.e., whatever you find, cooperate in the next move), we find that this state does not last forever. After a certain time, one finds an "outbreak of defection" (due to one or more successful mutation steps).

In order to further understand the system, we looked in detail at the strategy distributions. The overall result is that the three different regimes of Figure 3 correspond to three markedly different distributions of strategies; a more detailed description can be found in Feldman & Nagel. [7]

The states for $b > 3$ (Figure 4) look rather complex. In this regime, even after 2,560,000 simulated time steps, the frequency of C agents still increases further with simulation time. This, together with the fact that a configuration like in Figure 4 is relatively stable in its overall structure, seems to indicate that we have encountered a slow relaxation process in this regime, like in spin glasses.

Nevertheless, combining the information from pictures and from the strategy distribution,[7] we find that the preferred mechanism in this regime is a chessboard-like pattern of C and D agents. The mechanism behind this is the same as already explained above for game A. So this mechanism proves to be very important for this specific class of lattice games. It should be noted that the exact version of the chessboard pattern strongly depends on boundary conditions, at least for the very small lattices of 40×40 and the boundary conditions we used.[7] Our experiences with larger systems (e.g., Figure 4) suggest that larger systems should be studied.

The central organizing structure in regimes of practical interest is the chessboard. Our observations indicate that, when "profits" can accumulate, high selection pressure will favor more cooperative regimes (game A) and that, when profits cannot accumulate (game B), there is an abrupt transition from mostly cooperation to the chessboard at $b = 3$.

A typical agent inside a chessboard region gets $(4 \times b)/2 = 2b$ per time step. A typical agent inside a cooperating regions gets $4 \times 3 = 12$ every time step. Thus, strict cooperation is always the *Pareto-superior* strategy when $b < 6$ (everyone is as well or better off).

The discontinuity at $b = 3$ when accumulation is not allowed (game B) reflects the situation in which chessboard defectors bordering a cooperative region gain more than their cooperating neighbors. When $b > 3$: $4 \times b > 4 \times 3$. The inherent limitation on the ability of cooperative regions to survive has to do with agent's ability to distinguish between "deviant" or noncooperative agents and agents of the cooperative region responding to these challengers. For example, a single "defector" in a cooperative region can be defeated if its neighbors defect against it. But then the neighbors of these agents must not defect against them. This means to defect against one defection, but not against two. On the other hand, when a cooperative region borders on a chessboard region, the bordering agents must often defect against two agents defecting in order to stop the chessboard.

5. SOME CONCLUDING REMARKS

Simultaneous play of the Prisoner's Dilemma with neighbors on a a lattice did not lead to dominance of the cooperation strategy. Large-scale defection, however, did not occur either. Instead of strict cooperation, we observed widespread local coordination. This is what one often observes from the operation of decentralized economic processes.

The issue of bounded rationality on a lattice is more complex than in the two-person repeated game. The need to respond at the same time to cooperators and defectors changes the nature of the problem. We argue that this is a more realistic model of many economic environments.

Because agents play each other simultaneously, our results seem to contradict Axelrod's[2] proposition that "it is no harder for a strategy to be territorially stable than it is to be collectively stable" (Axelrod,[2] p. 160). However, to our knowledge the "non-territorial stability" of simultaneous play against several randomly matched agents has not been analyzed.

An important aim of our experiments was to display some aspects of the evolution which is inherent in economic behavior. Especially game B seems to be quite successful in this regard because of its capability to react to changes in the environment due to the embedded mutation mechanism.

Finally, we believe that our local interaction approach has yielded useful insights that would be difficult to develop otherwise and is a direction of work worthy of further development.

ACKNOWLEDGEMENTS

We acknowledge computer time on the workstation net of the Mathematisches Institut of the University of Cologne and on the iPSC/860 of the ZAM (Jülich Research Center).

This project began during the 1992 Complex Systems Summer School. We would like to explicitly acknowledge conversations with L. E. Cederman, E. Chan, and D. Burns; the "economics working group;" a discussion of this group with J. H. Miller and R. Palmer, where we were motivated to go on with this work; and the lecturers on "bit-coded dynamics," including M. Mitchell and S. Rasmussen. Finally, we would like to thank the Santa Fe Institute and its wonderful staff, the U.S. National Science Foundation and all others who helped make possible this opportunity for cross-disciplinary study, encounter, and collaboration.

REFERENCES

1. Axelrod R., and W. D. Hamilton. "The Evolution of Cooperation." *Science* **211** (1981): 1390.
2. Axelrod, R. *The Evolution of Cooperation.* NY: Basic Books, 1984.

3. Blume, L. E. "The Statistical Mechanics of Strategic Interaction." *Games & Econ. Behav.* (1993): In press.
4. Cowan, R. A., and J. H. Miller. "Economic Life on a Lattice: Some Game Theoretic Results." Working Paper #90-010, Santa Fe Institute, Santa Fe, NM, 1989.
5. Durlauf, S. "Non-Ergodic Economic Growth." Technical Report No. 6, Department of Economics, Stanford University, 1991.
6. Ellison, G. "Learning, Local Interaction, and Coordination." Department of Economics, MIT, Cambridge, MA, 1991.
7. Feldman, B., and K. Nagel. "Lattice Games with Strategic Takeover." Report No. 92.120, Angewandte Mathematik und Informatik, Universität zu Köln, 1992.
8. Fudenberg, D., and E. Maskin. "Evolution and Cooperation in Noisy Repeated Games." *Amer. Econ. Rev.* **80** (1990): 274–79.
9. Glance, N. S., and B. A. Huberman. "Dynamics of Social Dilemmas." *Sci. Am.* (1992): preprint.
10. Kalai, E. "Bounded Rationality and Strategic Complexity in Repeated Games." In *Game Theory and Applications*, edited by T. Ichiishi, A. Neyman, and Y. Tauman, 131–157. New York: Academic Press, 1987.
11. Kauffman, S. A. "Metabolic Stability and Epigenesis in Randomly Constructed Genetic Nets." *J. Theor. Biol.* **22** (1969): 437–467.
12. Neyman, A. "Bounded Complexity Justifies Cooperation in the Finitely Repeated Prisoner's Dilemma." *Econ. Lett.* **19** (1985): 227–229.
13. Nowak, M. A., and R. M. May, "Evolutionary Games and Spatial Chaos." *Nature* **359** (1992): 826.
14. Rapoport, A., and A. M. Chammah. *Prisoner's Dilemma: A Study in Conflict and Cooperation.* Ann Arbor: University of Michigan Press, 1965.
15. Samuelson, L., and K. Binmore. "Evolutionary Stability in Repeated Games Played by Finite Automata." Social Systems Research Institute 9029, University of Wisconsin, Madison, WI, October 1991.
16. Wolfram, S. *Theory and Applications of Cellular Automata.* Singapore: World Scientific, 1986.

Brian L. Keeley† and E. Bonabeau‡
†Experimental Philosophy Laboratory, Department of Philosophy (0302), University of California at San Diego, 9500 Gilman Drive, La Jolla, CA 92093-0302, USA, e-mail: bkeeley@ucsd.edu; and ‡CNET Lannion B–OCM/TEP, Route de Tregastel, 22301 Lannion Cedex, France, e-mail: bonabeau@lannion.cnet.fr

Is There Room for Philosophy in the Science(s) of Complexity?

INTRODUCTION

A striking feature—if not the most striking—of the 1992 Complex Systems Summer School was the presence of at least 1.5 philosophers among the students. This was sufficient to reach a critical mass and generate many discussions on philosophical aspects of the science(s) of complex systems (hereafter, "SoC"). In this short manifesto, we make a summary of some very naive (experimental) observations we made during one month in the midst of senior, prospective, irresolute, or repentant complex systems scientists.

"What the hell is a complex system?" is a question often asked both by people who sincerely believe that a satisfactory answer can be had and by those who believe not.[1,2,3,5,9,13] Unfortunately, there seem to be as many possible answers as there are researchers in the new field of complex systems. The nonobviousness of any answer to this question begins to open a little space within which philosophy can dwell. At the very least, philosophy can perform the useful function of conceptual clarification.

To begin with, an emerging discipline ought to be able to *point*; it ought to be able to say: "We are interested in studying *these* phenomena, because they seem to share some interesting properties." Otherwise, why should we need to gather numerous heterogeneous disciplines within one unified-looking field? Actually, the epistemological aspects of this new science can be dealt with, to some extent, using mostly sociological concepts, seemingly close to Kuhn's[7] ideas, but far in spirit, since we don't apply them to the acceptance or rejection of a new paradigm (cf. Kuhn's notion of "paradigm shifts"), but rather to the deep motivations that helped constitute SoC. References to other major epistemological works can only be marginal in this context, because one of the key questions we should ask relates to how new concepts can emerge from the interactions between very different disciplines, and we believe these interactions—though they rely on very scientific foundations—take place at the level of social relationships. In this way, this new discipline reflects the changes in post-Kuhnian philosophy and sociology of science, which has seen a shift from viewing science as a primarily linguistic phenomena to seeing it as a *social construction*.[4,8] This latter approach is the relevant level of description of these interactions, and thus the right level for looking for clues. Note that this approach makes SoC itself part of its own object of study. (If SoC were to take this self-referentiality seriously, would that make it the first *postmodern* science?[11])

We will elaborate on this idea very briefly and will try to evaluate the future impact that SoC may have on science and philosophy at large. But first, we should attempt to say something about what, in the sense above, SoC points at, and what those phenomena are said to have in common. We will do this by presenting a series of points, which might be considered candidate criteria, whose relationship to one another is not altogether clear.

COMPLEX SYSTEMS?

1. A NETWORK OF INTERACTING ENTITIES

The most commonly shared definition of a complex system states that it is a network of (relatively simple and similar) interacting entities, agents, elements, or processes that exhibit dynamic aggregate behavior. The action of an object affects subsequent actions of other objects in the network, so that "the action of the whole is greater than the simple sum of the actions of its parts." In other words, a system is complex if it is not *reducible*, in some sense. In exactly *what* sense complex systems resist reduction is another question for philosophy. This resistance to reduction is usually couched in terms of "emergence"; complex systems are those that exhibit some property, or properties, that *emerges* from the actions of its components.[6] The presence of emergent properties indicates that the level of explanation containing these properties is in some way *autonomous*; even a complete account of a lower level of description (which mentions only the actions of individual components) will fail to fully explain the behavior of the system. For example, if the phenomena we call "life" is an emergent property of the action of interacting chemical molecules, then biology is autonomous of chemistry, while at the same time being compatible with chemistry. But at the same time, it should be made clear that complexity (as studied by SoC) *is* reductionist in the sense that what scientists try to do is describe highly complicated phenomena in terms of simple(r) mathematical equations.

2. MANY DEGREES OF FREEDOM

One way of cashing out this resistance to reduction is to notice that since the beginning of time, science has been very busy dealing either with systems whose behaviors are reducible to a few degrees of freedom and thus can be characterized by low-dimensional deterministic equations, or with systems that have many degrees of freedom but whose behavior is reducible to a statistical description. A complex system has many degrees of freedom that strongly interact with each other, preventing either of the two classical reductions. In a nutshell, it exhibits what W. Weaver called *organized complexity* (as opposed to *organized simplicity* and *disorganized complexity*).[14] In this sense, complexity resists the reduction of phenomena to low-dimensional descriptions. In particular, many complex systems are not amenable to "pairwise" reduction—the traditional methodology of physics which seeks to explain the behavior of large aggregates of interacting entities by a simple linear extrapolation from the behavior of just *two* such interacting entities.

3. COMPLEX SYSTEMS VS. COMPLEXITY IN GENERAL

Yet, the idea of chaos has taught us that complex behavior can arise from low-dimensional systems. Even systems that can be explained by simple equations can show "sensitivity to initial conditions," such that their *behavior* is complex. Perhaps then, the connection between complex *systems* and complexity *in general* is not as obvious as it might at first appear. Some apparently simple systems can generate complex behavior, while some complex systems exhibit simple behavior.

What, then, is the scope of SoC? What elements of complexity will it eventually seek to explain and what will fall outside its ken? These important questions will need to become clearer if this new endeavor is to develop and grow.

HOW TO MAKE A SCIENCE

The question now is: why should the previous characterizations of complexity become a common basis for the constitution of an interdisciplinary field? Actually, the answer may be very simple: systems such as those described as complex are by far the most numerous in nature. The breadth of phenomena we saw described at the summer school is a testament to this realization—spin glasses, computational ecologies, protein folding, disease epidemiology, nervous systems, automobile traffic, and on and on. Since science has not yet been able to investigate (quantitatively or qualitatively) many of these systems in a satisfactory manner, it seems a good idea to look for laws, tools, and methods originating from *any* field that has to deal with such systems, with no *a priori* restrictions on the scope of research. Disciplines adopting this type of approach can only benefit from others' experience. Why SoC did not emerge earlier also seems to have a simple reason: complex systems, be they low-dimensional and chaotic, or high-dimensional with nonlinearly coupled degrees of freedom, could not be studied before the last decades because they require high computational power—far beyond the human brain's unaided capabilities. Computer studies gave birth not only to quantitative results but also to theories. The best example is the scientific activity that has developed around dynamical systems, 10 to 15 years after the discovery (on a computer) of the notion of sensitivity to initial conditions in low-dimensional systems.[11]

Another essential aspect of the creation of SoC is related to the people themselves (yes, human beings). Of course, it is a good thing to discover the quark and/or to be a Nobel Prize winner to have a chance to influence the course of science—not only in one's own field. But, as far as money is concerned, even if you are a renowned scientist, you still have to spend a lot of energy to influence people. (Consider the impact on the recent history of artificial intelligence of John Hopfield's publicity campaign for neural nets in the early 1980s and of Hubert Dreyfus' numerous forays to M.I.T. on the behalf of Heideggerianism.) SoC, we believe, owes its current success to the people who participate in its development. This is not as trivial a statement as it may seem at first glance: the scientific community can be seen as a web, which can be locally densely connected, but mostly with a very sparse connectivity (though this is less and less true, due to the recent explosive growth of new media of communication—electronic mail, mailing lists, newsgroups, satellite conferences, etc.). Thus, the construction of a cross-disciplinary field necessitates a highly developed social life (from cocktails to parties without a rest). One must not hesitate to attend conferences that (apparently) have nothing to do with one's own field. Some other classical prejudices also have to be fought: most universities in the United States and the world have well-established departments, with little communication across these "artificial" barriers. Transversal motion of concepts implies removing, at least partially, these barriers: this can be done only by strong social interactions. While it is natural that irreducible systems have to be studied, it is not at all obvious that it should be done in a unified manner. With respect to this particular aspect, the people from the Santa Fe Institute (SFI) have contributed greatly to the unification of SoC. This Summer School is another example of the social foundations of the new science—it is part of a huge "marketing" effort—though we won't complain for being the victims. It must be emphasized that the social aspect is a highly characteristic feature of the development of this science.

Finally, it should be noted that SoC is not alone in embracing such transdisciplinarity. The interdisciplinary field of cognitive science was born out of the realization that academics in a variety of fields (artificial intelligence, cognitive psychology, linguistics, philosophy, and the neurosciences) were dealing with many of the same issues, and that a general "science of the mind" was possible and potentially fruitful. The emerging paradigms of cognitive and computational neuroscience have grown out of the interactions between neuroscientists, psychologists, and computer scientists. In this new age of diversity and inclusion, we are even beginning to see attempts to bridge the gap between science and the arts: *Avant garde* artists like Survival Research Laboratories and Stelarc are invited to an artificial life conference or a cognitive science summer school; and a philosopher, a computer scientist, a sociologist of science, and a post-modern painter join forces to understand the concept of representation.[13] Of the fruitfulness of such intellectual liberalness, only time will tell; but, if it be a delusion, SoC will not be counted alone in suffering it.

THE IMPACT OF THE NEW SCIENCE

There are many examples in SoC of successful interactions between disciplines, ranging from the application of statistical mechanics to the study of neural networks, to the generic use of concepts originating from theoretical physics to the study of evolution (e.g., spin glasses, Boolean nets), economics, biology (e.g., "self-organized criticality," a concept originating from condensed matter physics, has been applied to economics, biology, cloud formation, earthquakes, etc.). It is worth noticing that SoC (so far) seems to be somewhat dominated by theoretical physicists (do they have a more intense social life; i.e., do they like to dance more than other ordinary people?). It is true that physics carried along with it a bunch of new theoretical concepts. Besides new opportunities offered in analytical treatment thanks to theoretical physics and mathematics, there is also a lot being done with computers: Artificial Life is a good example of a field with no strong theoretical or empirical basis—other than the computer—which nevertheless gives great insights into the dynamical principles of life.

Complex systems are certainly changing the way scientists look at science, but it also modifies the way non-scientists look at reality. The notion of sensitivity to initial conditions has become popular—maybe too popular; emergence is more and more accepted as something that happens in everyday life, that helps in understanding sociopolitical concepts. On the other hand, sociopolitical projects can also lead to unified scientific programs, as is illustrated by the SFI's Sustainable World project. And, even as we write this, complexity is poised on the brink of the same kind of popularity that has recently been showered on chaos and nonlinearity.[9,13]

Finally, SoC should ultimately have an impact back on philosophy itself. The intellectual exchange between these two fields will not be entirely one way. We have already mentioned three possible important impacts. First, if by serving as a "case study" this new field can shed light on the notions of emergence and reduction, then this alone will be of enormous help to philosophy, which has been struggling more or less successfully with these topics for centuries. SoC's promise is to finally provide us with a *principled* story of a nonreductionist scientific explanation; principled in the sense that such an account does not make recourse to occult, metaphysical, or "magical" essences. Such an account would hold out the possibility of similar accounts of sociology, psychology, ethology, neuroscience, etc., whereby these classes of explanation won't necessarily *reduce to physics* in some Laplacian image of the world.

Second, SoC adds fuel to current work within Science Studies (a mass term referring to the joint study of science carried out by philosophy, sociology, and history—yet another contemporary interdisciplinary endeavor) focusing on the social aspects of sciences. As an epistemological endeavor, our new discipline seems to be more adequately described in terms of who is conducting the work, rather than in terms of any commonly accepted set of documents and theories.

Finally, the issue with which we began this paper, the proper analysis of the concept of complexity itself, has potentially far-reaching effects through an understanding of the complementary concept of *simplicity*. Simplicity is typically considered a desirable pragmatic virtue of scientific theory; in the sense that all things being equal (predictive and explanatory power, coherence with other accepted theories, etc.), the "simpler" of two candidate theories is considered to be the preferential one. However, cashing out what this pragmatic principle means in practice is no mean feat. How does one measure simplicity? Occam's razor, a celebrated version of this principle, opts for simplicity in terms of a theory's posited ontology. An interesting open question then is: How does the Occam's razor notions of simplicity and complexity relate to the SoC notions?

CONCLUSION

In these modest "philosophical investigations," we have tried to clarify the notion of complex system, not from the scientific but from the *epistemological* point of view, so as to suggest a real underlying unity of SoC. Once again, this unity does not necessarily lie in the very resemblance of all the disciplines that share the field, but rather on phenomenological relationships that allow the application of methods and tools used in a particular field to another one. This is the idea of transdisciplinary concepts, which do not imply the transdisciplinarity of meanings: one must always be cautious with phenomenological resemblances.

The prospects are not at all gloomy for this type of approach: on the contrary, the notion of transversal flows of ideas between disciplines should be generalized, given the successes encountered. We know that cross-fertilization is not new, but it never occurred on such a scale as observed in the study of complex systems.

We argued that this is so because (1) complex systems, as characterized above, are everywhere in nature, (2) computers have allowed for the quantitative and qualitative—leading to unsuspected advances in theory—study of systems which otherwise could not be approached, and (3) social factors have greatly influenced the constitution of the field (was it unavoidable?).

These were only a few philosophical and sociological observations about the creation of a new field. We have admittedly affected an upbeat and relatively uncritical attitude here, in the spirit of charity to a field still in its infancy. Of the eventual success of the science(s) of complexity, only time will tell. This being the case, we reserve deeper comment on the consequences of this field for a future communication.

REFERENCES

1. Anderson, P. W. "Is Complexity Physics? Is It Science? What is It?" *Physics Today* **44(7)** (July 1991): 9–10.
2. Anderson, P. W. "Complexity II: The Santa Fe Institute." *Physics Today* **45(6)** (June 1992): 9–10.
3. Bak, P., C. Tang, and K. Wiesenfeld. "Self-Organized Criticality." *Phys. Rev. A* **38** (1988): 364–374.
4. Bloor, David. *Knowledge and Social Imagery*, 2nd ed. Chicago: University of Chicago Press, 1991.
5. Casti, J. "The Simply Complex: Trendy Buzzword or Emerging New Science?" *The Bulletin of the Santa Fe Institute* **7(1)** (1992): 10–13.
6. Forrest, S. "Emergent Computation." *Physica D* **42(1-3)** (1990): 1–11.
7. Kuhn, T. *The Structure of Scientific Revolutions*, 2nd ed. Chicago: University of Chicago Press, 1970.
8. Latour, B. *Science in Action.* Cambridge: Harvard University Press, 1987.
9. Lewin, R. *Complexity: Life at the Edge of Chaos.* New York: Macmillan, 1992.
10. Lorenz, E. N. "Deterministic Non-Periodic Flow." *J. Atmosph. Sci.* **20** (1963).
11. Lyotard, J.-F. "Answering the Question: What is Postmodernism?" *The Postmodern Condition: A Report on Knowledge.* Minneapolis: University of Minnesota Press, 1984.
12. *Registration Marks: Metaphors for Subobjectivity.* Published to accompany an exhibition of paintings by Adam Lowe, with essays by Adrian Cussins, Bruno Latour, and Brian Smith. London: Pomeroy Purdy Gallery, October 1992.
13. Waldrop, M. Mitchell. *Complexity: The Emerging Science at the Edge of Order and Chaos.* New York: Simon & Schuster, 1992.
14. Weaver, W. "Science and Complexity." *Am. Scientist* **36** (1968): 536–544.

1. *Registration Marks: Metaphors for Subobjectivity*. Published to accompany an exhibition of paintings by Adam Lowe, with essays by Adrian Cussins, Bruno Latour, and Brian Smith. London: Pomeroy Purdy Gallery, October 1992.

Garry D. Peterson
Arthur R. Marshall Jr. Ecological Sciences Laboratory, University of Florida, Gainesville, FL 32611

Animal Aggregation: Experimental Simulation by Using Vision-Based Behavioral Rules

This paper describes a computer program that simulates the behavior of a population of generic animals whose movement is controlled by vision-based behavioral rules. This program was used to evaluate a set of hypothetical mechanisms of animal aggregation. A number of the examined rules produce simulated animal aggregates that behave naturalistically.

INTRODUCTION

Many animals aggregate into organized groups such as flocks, herds, and schools.[3,4,6] These aggregates move as cohesive units and persist over time. This persistence implies that the individual animals comprising the aggregation each possess efficient means of both monitoring the movements of other animals, and behaviorally responding to change in a manner that maintains the cohesion of the aggregate. This paper describes a simulation program that allows a variety of theorized mechanisms of animal aggregation to be tested.

1992 Lectures in Complex Systems, Eds. L. Nadel & D. Stein, SFI Studies in the Sciences of Complexity, Lect. Vol. V, Addison-Wesley, 1993

The simulation program models a simple artificial world, containing a population of animals who move according to specified vision-based behavioral rules. Each simulation of a behavioral rule is an experiment, testing the simulated behavior against actual behavior, and subsequently suggesting elimination, modification, or further testing of hypothetical behavioral rules.

This paper includes a description of the structure of the simulation environment, a description of simulated aggregate behavior, and a discussion of the directions this research will take in the future.

SIMULATION ENVIRONMENT

The simulation environment described in this paper was developed to experiment with hypothesized vision-based mechanisms of animal aggregation and consequently, it is designed to explicitly model animal vision. This design results in a simulation quite different from other investigations of animal aggregation.[5]

Reynold's ground-breaking simulation of animal aggregation was developed from a biologically inspired animation perspective, rather than a theoretical ethological perspective. Consequently, Reynolds focused on generating realistic aggregate behavior, rather than discovering which realistic theoretical aggregation mechanisms can generate realistic aggregate behavior. This alternative approach is a more concrete form of Braitenburg's[1] thought experiments.

The simulation environment contains a population of simulated animals. These animals have been named Zooids following the precedent set by Reynold's Boids. The following sections will describe the physics of the model world, zooid environmental perception, and behavioral rules.

SIMULATED PHYSICS

Currently, the simulated world of the zooids is two-dimensional, and time passes in discrete increments. The world is governed by Aristotelian physics, so that a zooid must continually exert force to remain in motion. Zooids are circle shaped, a form that simplifies the simulation of vision. At every tick of the simulation clock, each zooid views its environment, and then uses a behavioral rule to act upon its perceptions.

Zooid motion is constrained by physical limits to their abilities. Zooids cannot exceed a maximum velocity, nor drop below a minimum velocity. Zooid acceleration and deceleration each cannot exceed fixed values. The rate of change of zooid orientation is similarly constrained. This assortment of limits corresponds to abilities of the zooids under the physical laws of the artificial world that they inhabit. The modeling of specific animal aggregations, such as actual herring schools or flocks of

Canada geese, would likely require a more accurate model of the specific physics of their respective situations.

PERCEPTION

A zooid perceives its environment visually. Each zooid has a 180° field of vision and cannot recognize any object occupying less than 1/2° of arc. The field of vision for actual birds, fish, or herd animals is larger than 180°, but Dill et al.[2] suggest that the central 120° of an aggregating animal's field of vision is the most important for maintaining flock cohesion and, therefore, any field of view larger than that should be adequate for aggregation. A 180° field of view was chosen for its simplicity.

If a zooid can see other zooids, then that zooid will fix upon the zooid that occupies the greatest amount of its visual field. It will consider this zooid its "leader" and will attempt to follow it (see Figure 1). If after a time another zooid occupies a greater proportion of a zooid's field of view than its leader does, that zooid will become the following zooid's new leader.

A zooid monitors two characteristics of its leader, the angle its leader occupies within its visual field (α), and the leader's heading (β). These measurements are shown in Figure 2.

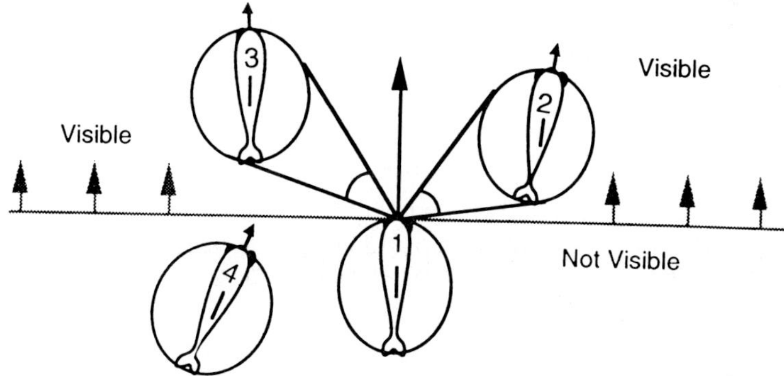

FIGURE 1 A zooid perceives a 180° zone. Therefore, Zooid 1 can see Zooid 2 and Zooid 3, but it is unaware of Zooid 4. In this case, Zooid 1 would follow Zooid 2 rather than Zooid 3, because Zooid 2 occupies more of Zooid 1's visual field.

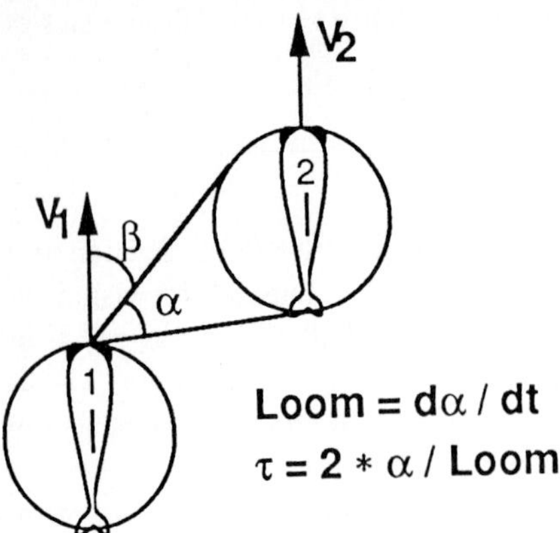

FIGURE 2 Zooid 2 is Zooid 1's leader. Zooid 1 monitors two characteristics of Zooid 2, (b) its bearing β and (a) the amount of Zooid 1's visual field it occupies, α. Loom equals the rate of change of α, and τ is an optical estimate of the time to collision.

A zooid's mind stores one parameter, the previous α angle. This angle and the current α angle are used to calculate visual measurements of a zooid's movement relative to its leader. Two measures are calculated, Loom and τ. Loom is proportional to the rate of change of α, which corresponds to the rate at which the zooid's leader is growing, or shrinking, in a zooid's visual field. In the simulation environment, Loom is approximated by calculating the difference between the current and previous α. τ is proportional to the α angle divided by Loom. τ provides the predicted time to collision between a zooid and its leader.[2]

BEHAVIORAL RULES

The visual cues α, β, Loom, and τ are used by a zooid's behavioral rule to produce a behavior. The visual cues used in the simulated environment can vary, but currently they are fixed and only the behavioral rules have been varied.

The exploration of possible behavioral rules has not been conducted in an arbitrary manner; rather, we have attempted to construct biologically realistic rules that contain a hierarchical structure. Hierarchical structure simplifies the decision-making process by only requiring the necessary information at any level in the process. An example of a rule and its hierarchical structure is shown in Figure 3.

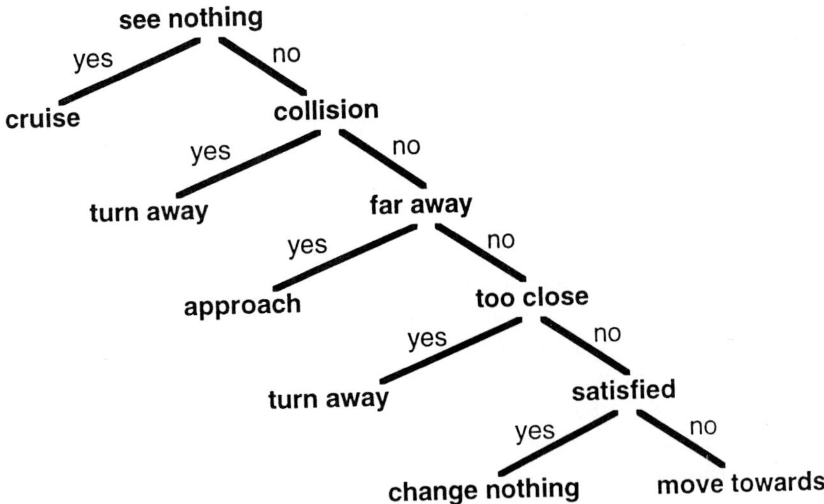

FIGURE 3 An example of a behavioral rule. The behaviors specified by the decision tree vary based upon the values of α, β, Loom, and τ.

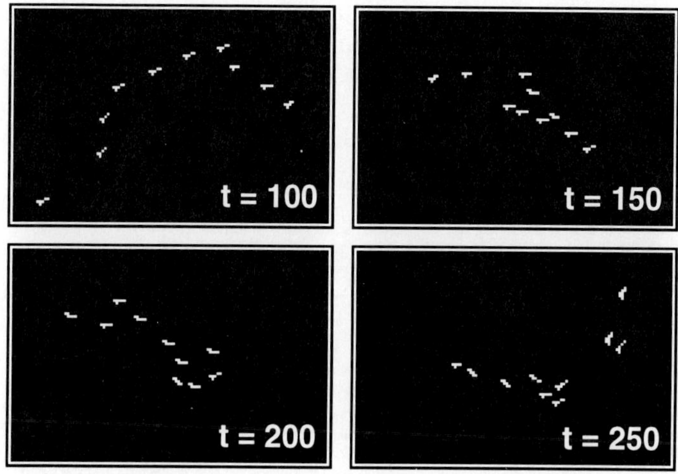

FIGURE 4 The changing form of a zooid aggregate over time (units are arbitrary).

DISCUSSION

The behavioral rules examined in this paper produced aggregations. Thus, when a number of zooids are initially placed at random in the environment, they quickly aggregate into several groups, generally consisting of four to five individuals. These groups have a polarized structure, because all the individuals in the groups face and move in approximately the same direction. The individuals in these groups generally, but not always, avoid collisions with one another. Over time these groups will themselves aggregate, if they are confined within a bounded area. Examples of aggregate structure and its change over time are shown in Figure 4.

A group is lead by a leader that cannot see any other zooids and, therefore, is following nothing. The other zooids in the aggregation follow each other and the leader. Over time leadership of the group will switch from one zooid to another. The most common leadership change occurs when a zooid approaching from behind overshoots the group's leader, so that its former leader begins to follow it. This type of change can be seen in Figure 4. Another common leadership change occurs when a group leader sees another group and begins to follow one of that group's members, bringing the two groups together.

Current aggregation structure ranges between two extreme types: a staggered line and a clump. Over time, as aggregations move and encounter one another, they will often change their forms. Clumpy and linear aggregations are opposing ends of a continuum: linear flocks can become more clumpy as Y structures form (see Figure 5), and clumps may stretch out into lines as is partially seen in Figure 4. The linear form is generally stable, in that its members maintain fixed positions, while the lump experiences a lot of jostling between its members. This jostling will

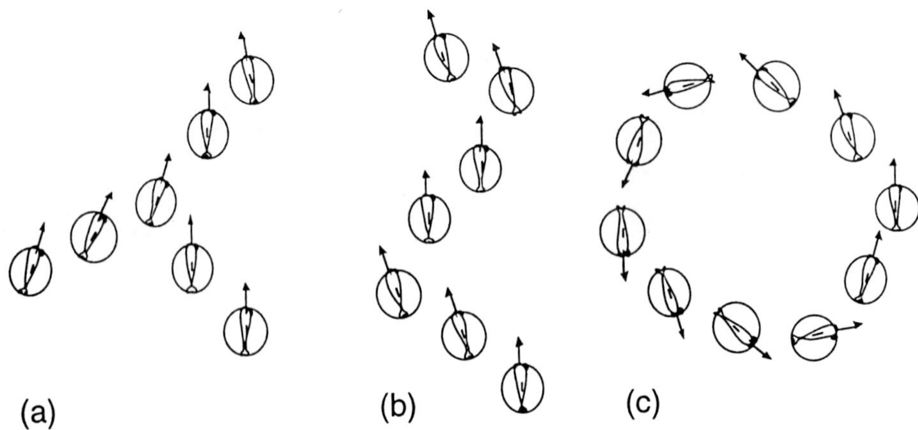

FIGURE 5 (a) An example of a "Y" formation. (b) An example of a "Z" formation. (c) An example of a carousel structure.

occasionally become intense enough to split an aggregate into two or more parts. When two groups meet, they will either join, shatter into smaller groups, or, most commonly, avoid one another and continue intact.

Many types of bird flocks from "V" formations; however, none of the current simulation rules has generated a V aggregate structure. The rules generate Y and Z formations (see Figure 5). One interesting phenomenon that does occur is the formation of mills. In this situation a group of zooids form a circle, where each zooid follows the zooid in front of it. Mill structures are stable and require an external disturbance, such as a passing zooid, to be broken apart. Mills also form in many types of natural aggregations including flocks, schools, and herds.

FUTURE DIRECTIONS

To improve the breadth and realism of the testing environment the simulation environment will be modified in several ways. These changes include expanding the environment from two dimensions to three dimensions, experimenting with alternative animal body forms, developing a better understanding of the effects of physics upon aggregation dynamics, and adding predators and obstacles to the simulation environment.

One further modification to the program, which is currently about half complete, is the construction of a statistics collection module that will allow the detailed comparison of simulated and observed aggregates. Aggregate characteristics, such as the frequency distribution of an aggregate's nearest neighbor distances and its bearings of nearest neighbors, will be calculated.

CONCLUSIONS

Simulated animals following simple behavioral rules can form aggregates. To test a more complete repertoire of aggregate behavior will require an expansion of the present simulation environment. Experimental computing provides a method for testing the emergent properties of behavioral theories, allowing the set of plausible theories to be reduced in number and to be refined in detail.

ACKNOWLEDGMENTS

The development of the ideas associated with visual cues is due to my collaborators Dr. C. S. Holling of the University of Florida and Dr. Larry Dill of Simon Fraser University. I also benefited from discussions with Craig Reynolds, Maarten Boerlijst, and Larry Yaeger at the Third Artificial Life Workshop, which was held during the 1992 Complex Systems Summer School.

The simulation environment is written in C and runs on a Macintosh Quadra 950.

REFERENCES

1. Braitenburg, V. *Vehicles: Experiments in Synthetic Psychology.* Cambridge, MA: MIT Press, 1984.
2. Dill, L. M., C. S. Holling, and L. H. Palmer. "Predicting the Three-Dimensional Structure of Animal Aggregations From Functional Considerations: The Role of Information." In *The Proceedings of the Animal Aggregation Workshop*, edited by J. Parish and W. Hammer. In press.
3. Partridge, B. L. "The Structure and Function of Fish Schools." *Sci. Am.* **(June 82)** (1982): 114–123.
4. Radakov, D. V. *Schooling in the Ecology of Fish.* New York: J. Wiley, 1973.
5. Reynolds, C. W. "Flocks, Herds, and Schools: A Distributed Behavioral Model." *Computer Graphics: Proceedings of SIGGRAPH '87* **21(4)** (1987): 25–34.
6. Wilson, E. O. *Sociobiology: The New Synthesis.* Cambridge, MA: The Belknap Press of Harvard University Press, 1974.

Stefan Schaal† and Dagmar Sternad‡
†Department of Brain and Cognitive Sciences and the Artificial Intelligence Laboratory, MIT, 545 Technology Square, Cambridge, MA 02139 (e-mail: sschaal@ai.mit.edu); and ‡Center for the Ecological Study of Perception and Action, University of Connecticut, 406 Babbidge Road, Storrs, CT 06268

Learning of Passive Motor Control Strategies with Genetic Algorithms

This study investigates learning passive motor control strategies. Passive control is understood as control without active error correction; the movement is stabilized by particular properties of the controlling dynamics. We analyze the task of juggling a ball on a racket. An approximation to the optimal solution of the task is derived by means of optimization theory. To model the learning process, we code the problem for a genetic algorithm in representations without sensory or with sensory information. For all representations the genetic algorithm is able to find passive control strategies, but the learning speed and the quality of the outcome are significantly different. A comparison with data from human subjects shows that humans seem to apply yet different movement strategies from the ones proposed. Regarding feedback representation, some implications arise for learning from demonstration.

1. INTRODUCTION

Despite research advances regarding human motor coordination and motor learning, little understanding has been gained so far as to how these skills are accomplished. From the perspective of control theory,[5] two major control approaches are distinguished. *Closed loop control* requires continuous sensing of the current state of the system: if the planned state differs from the actually achieved one, a modification of the next actuator command compensates for this error. In *open loop control*, on the other hand, the spatio-temporal sequence of actuator commands is determined before the movement starts and then is executed according to plan. Since no feedback is provided, there is no possibility of error correction. While closed loop control can be considered (re-)active control, open loop control is essentially passive.

The central nervous system (CNS) possesses two control circuits with resembling properties. On the spinal level, a short feedback loop takes care of fast-movement regulation. Although feedback is involved, this control receives no input of cortical areas and, in a figurative sense, can be considered passive. In addition to this low-level regulation, higher brain areas may influence the low-level circuitry at any time via long feedback loops through the cortical motor centers.

Passive control is appealing because of its low computational load during movement. One could imagine that for movement initiation a "control package" could be delivered to the spinal level which could trigger an autonomous control circuit to sustain the movement afterwards. The higher brain functions would be free for other tasks and only check for correctness and stability of the movement at discrete events. Some evidence for the biological plausibility of such control procedures has been shown in the work about central pattern generators.[6]

Here we investigate the cyclic movement of juggling a ball on a paddle to find out whether some form of passive control can be learned. An approximation of the optimal solution of the task in Section 2 provides an evaluation criterion for a series of learning experiments with genetic algorithms, presented in Section 3. Section 4 discusses the results of the experiments and compares them to empirical data from human subjects.

2. ANALYSIS OF PADDLE JUGGLING

Figure 1 displays the setup and notation of paddle juggling. The ball bounces on the paddle due to gravity, and the movement of the paddle tries to sustain a regular bouncing motion. A coefficient of restitution $\alpha \in [0,1]$ models the elastic ball-paddle impact. The system can be discretized with a Poincaré section $\sum = \{(\mathbf{x_B}, \mathbf{x_P}) \in \Re^4 | x_B - x_P = 0\}$. From the notation of Figure 1(b), the discrete system equations are:

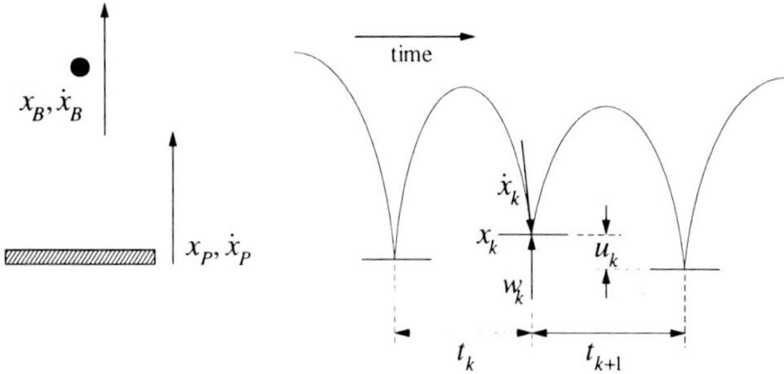

FIGURE 1 (a) Sketch of the paddle juggling setup; (b) notation for the discretization.

$$
\begin{aligned}
\dot{x}_{k+1} &= -\sqrt{((1+\alpha)w_k - \alpha\dot{x}_k)^2 - 2gu_k}, \\
x_{k+1} &= x_k + u_k, \\
t_{k+1} &= \frac{1}{g}\Big(((1_\alpha)w_k - \alpha\dot{x}_k)\dot{x}_{k+1}\Big),
\end{aligned}
\tag{1}
$$

where $(\dot{x}_k)$ denotes ball velocity and (x_k) the ball position immediately before the impact at paddle velocity (w_k), and (u_k) the vertical position shift between consecutive impacts.

Paddle juggling was investigated by several groups in recent years. For the vibrating paddle (high oscillation frequency with small amplitude), it could be shown analytically and experimentally that the system exhibits period bifurcations, strange attractors, and chaos-like motion.[7] Systems that had to control and learn this task (at a moderate juggling frequency) were examined in robotics.[1,2,8,9] Depending on the control algorithm, the execution of this task does not necessarily need feedback: driving the paddle with a sinusoidal motion results in a dynamic system that has a trapping region and that exhibits stable bouncing patterns under certain parameter settings.

2.1 PADDLE JUGGLING AS AN OPTIMIZATION PROBLEM

Paddle juggling can be formulated as an optimization problem. For these calculations, assume that the setpoint, at which the ball shall be juggled, is a given impact state $\mathbf{x}_s = (\dot{x}_s, x_s)^T$, whose setpoint controls $\mathbf{u}_s = (w_s, u_s)$ result implicitly from a periodic paddle trajectory $(\mathbf{x}_P(t))$. The setpoint is entirely determined by one parameter of the juggling motion, which can be the period (τ), the maximal ball height (h), or the impact velocity $(\dot{x}_s)$. In this regulator problem, the task of the controller is to keep the system at the setpoint. If a perturbation displaces the ball from its setpoint, it has to be guided back. By modeling the paddle motion as an rth-order Fourier series:

$$x_P(t) = \frac{a_0}{2} + \sum_{i=1}^{r} a_r \cos(r\omega t) + b_r \sin(r\omega t), \tag{2}$$

a multistage optimization problem[3] is formed, subject to minimizing the cost function:

$$\begin{aligned} J &= \phi(\mathbf{x}_n, \mathbf{p}) + \sum_{k=0}^{n-1} L(\mathbf{x}_k), \quad \text{where} \\ \phi(\mathbf{x}_n, \mathbf{p}) &= (\mathbf{x}_n - \mathbf{x_s})^T \Phi (\mathbf{x}_n - \mathbf{x_s}) + c\frac{1}{2}\sum_{i=1}^{r}(i\omega)^6(a_i^2 + b_i^2), \\ L(\mathbf{x}_k) &= (\mathbf{x}_k - \mathbf{x}_s)^T \mathbf{Q} (\mathbf{x}_k - \mathbf{x}_s)^T, \quad \omega = \frac{2\pi}{\tau}. \end{aligned} \tag{3}$$

$(\mathbf{x}_k)$ denotes the ball state vector at stage (k) and the matrices $(\Phi, \mathbf{Q}, \mathbf{R})$ are weight matrices. The last term of the equation for terminal cost (ϕ) in Eq. (3) represents a so-called jerk term and is weighted by the factor (c). Jerk denotes the third derivative $(\dddot{x}_P)$ of paddle position with respect to time, imposing a biologically motivated smoothness constraint on the paddle acceleration.[4] Without this term in the cost function, any optimization would minimize deviations from the setpoint with unrealistically sharp movements. In sum, the formalism of Eq. (3) tries to guide the perturbed ball smoothly back to the setpoint in an n-stage sequence: given the initial conditions of the perturbed ball $\mathbf{x}_{k=0} = (\dot{x}_0, x_0)^T$, the Fourier coefficients $\mathbf{p} = (a_0, a_1, \ldots, a_r, b_2, \ldots, b_r)^T$ are to be calculated such that Eq. (3) is minimized. Note that (u_k) and (w_k) are not present in Eq. (3), meaning that the paddle trajectory is independent of the ball motion; stability can only come from an appropriate choice of the Fourier coefficients. The system is thus passively controlled.

The numerical solution of Eq. (3) was obtained with the gradient method of dynamic programming.[3] The optimization was done for a ten-stage task, a fifth-order Fourier series, and a coefficient of restitution $\alpha = 0.7$. Initial conditions yielded:

$$\begin{aligned} &\dot{x}_0 \in [-0.12\dot{x}_S, +0.12\dot{x}_S] \\ &x_0 \quad \text{such that } t_{\text{init}} \in [-0.12\tau, +0.12\tau], \quad \text{where } t_{\text{init}} = \tau - t_0 \end{aligned} \tag{4}$$

and the weight matrices $(\Phi, \mathbf{Q})$ were chosen according to a common heuristic:

$$\mathbf{Q} = \frac{1}{100}\Phi = \begin{pmatrix} 1/\dot{x}_{\text{max}}^2 & 0 \\ 0 & 1/x_{\text{max}}^2 \end{pmatrix}. \tag{5}$$

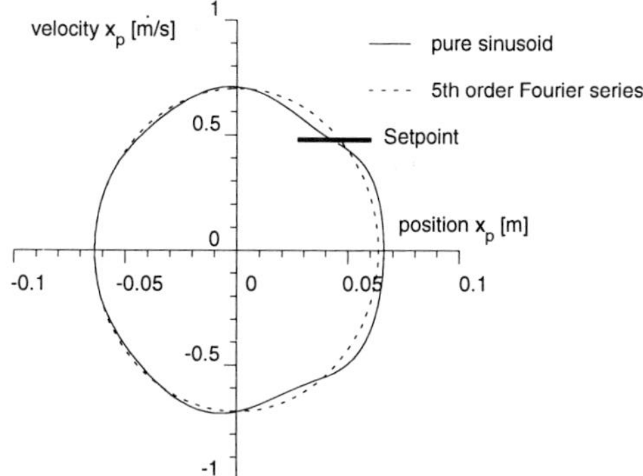

FIGURE 2 State space plot of an optimal paddle trajectory (period $\tau = 0.57$ sec).

The values $(\dot{x}_{\text{max}}, x_{\text{max}})$ correspond to the maximal ranges of the intervals in Eq. (4). A relatively large weight of (Φ) makes sure that the ball comes back to the setpoint. A moderate choice of (c) in Eq. (3), in order to make the penalty of jerk terms to be roughly one forth of the total cost, resulted in a paddle motion that was close to a sinusoid but still had some power in the higher harmonics. Figure 2 shows one result in comparison with a sine function in state space. Linear stability analysis of Eq. (1) modified by Eq. (2) holds that for stable juggling the ball must be hit in the first quadrant of Figure 2.

3. LEARNING EXPERIMENTS WITH GENETIC ALGORITHMS

The following numerical experiments will explore genetic algorithms (GA) (cf. contribution of M. Mitchell in this book) to simulate a reinforcement learning process. Similar to optimization analysis, reinforcement learning requires a performance index to evaluate the quality of the outcome. To apply genetic algorithms to paddle juggling, the task must be encoded as a gene string, and an appropriate fitness function has to be found.

In a first approach to the problem, the periodic paddle movement is divided into a set of 20 real-valued position values (Figure 3), represented by the first 20 genes in the gene string. The 21st gene codes the period (τ) and the 22nd gene codes a scaling factor that multiplies each of the 20 position genes. Given a population of randomized initial genomes, the task of the GA is to find a sequence of position values to perform a paddle movement by which the ball is juggled in a stable fashion. The scaling factor allows the movement in the spatial dimension to stretch; the period value is the corresponding temporal stretch factor.

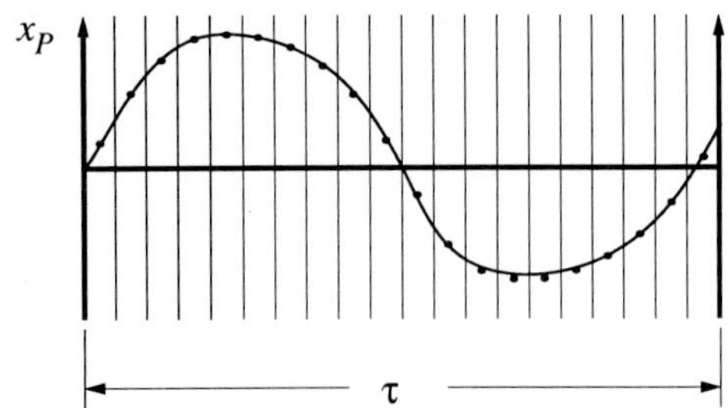

FIGURE 3 Discretized representation of periodic paddle movement for genetic algorithm.

The performance (or fitness) of a paddle trajectory is determined by:

$$J = \frac{10.0}{\bar{j} + 5(\sigma_h/\bar{h}) + 5(\sigma_p/\bar{\tau}_b) + 0.8(\tau/\bar{\tau}_b)} \tag{6}$$

where $(\bar{j})$denotes the mean jerk per period; (σ_h, σ_p) the standard deviation of maximal height of the ball and the standard deviation of the bounce period, respectively; and $(\bar{h}, \bar{\tau}_b)$ the mean maximal height and mean bounce period of the ball, respectively. The distinction between bounce period $(\bar{\tau}_b)$ and period (τ) is necessary because the ball can bounce several times on the paddle during one paddle period (τ). The cost function J penalizes jerk, irregular juggling height, irregular bouncing period, and, with the last term in the denominator, a large number of bounces during one period (τ). By taking the inverse of the cost function, the minimization problem of Section 2 becomes a maximization task in accordance with the usual GA formulation.

The statistical parameters of Eq. (6) were derived by submitting the paddle trajectory of each genome to a paddle-juggling simulation. The discrete trajectory values were treated as the desired position values $x_{P(\text{desired})}$ of a PD controller.[5] By differentiating this trajectory with respect to time, the desired velocities $\dot{x}_{P(\text{desired})}$ for PD control at each discrete time event were derived. With this information and the specifications of the paddle mass, the PD controller is able to generate a smooth pursuit of the encoded trajectory if the trajectory is smooth enough. The paddle movement had a limited workspace of ± 0.5 m, the maximal acceleration of the paddle was restricted to 60 m/s^2, the paddle mass was 0.5 kg, and the ball's coefficient of restitution was $\alpha = 0.7$. The position and velocity gains were set to constant values of $K_P = 200$ and $K_V = 20$. At the start of the simulation, the ball was dropped from 0.4 m above the paddle. After a transient time of roughly 5 periods, the statistical values of Eq. (6) were derived from 20 subsequent periods. All GA experiments had a population size of 100 genomes, mutation probability $P_{\text{mut}} = 0.01$ per gene, crossover probability $P_{\text{cross}} = 0.8$ using double-crossover, and a proportional offspring reproduction mechanism allowing at most five offspring for the best genome.

3.1 REPRESENTATION WITHOUT PERCEPTION

In the setup of the first experiment all genomes where randomized within a reasonable range of the individual genes. After about 400 to 800 generations, the GA solution converged to a steady value. Figure 4(a) shows the phase portrait of the best result that the algorithm developed.

Each chart in Figure 4 contains a segment of a paddle trajectory over three successive periods after the transient time had elapsed. To enable an assessment of the solutions, the mean period ($\overline{\tau}_b$) was used to calculate an approximately optimal paddle trajectory for each GA outcome. To make the setpoint of optimal solution and GA solution align as well as possible, different position coordinates of the setpoints were adjusted for; a shift in position coordinate corresponds to a redefinition of the reference coordinate system and does not change the results. The ideal paddle trajectory in phase space should be a smooth cycle. If the trajectories of successive periods in the graph traverse each other, the PD controller was not able to follow the discrete trajectory plan; i.e., the encoded trajectory was not smooth enough. This is rather pronounced in the solution of the absolute coding (Figure 4(a)). Such a juggling pattern would be vulnerable to perturbations.

The first coding of the paddle trajectory was based on the assumption that a movement plan is based on a discrete position representation of one period. An alternative representation would be in relative coordinates and the movement could be described as an "up-up-up-up-down-down-down-down-up-up-up-..." plan. To find out whether such relative position representation is more suitable, we have

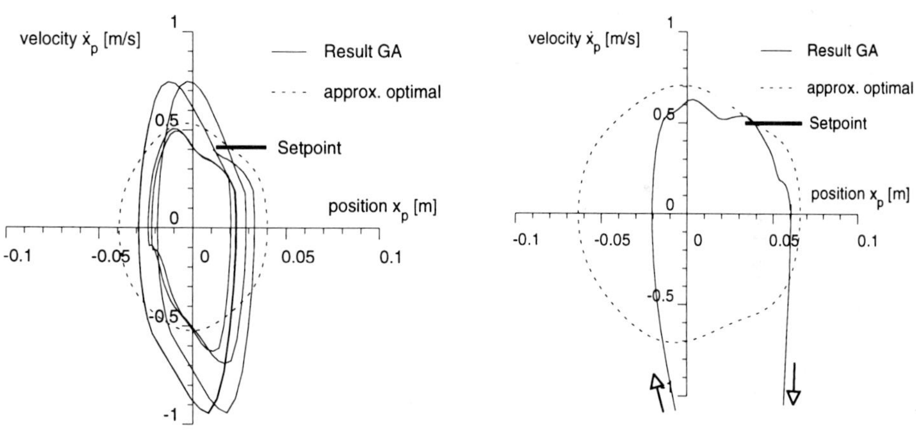

FIGURE 4 Outcome of learning without feedback: (a) best result of absolute coding ($J = 3.8$); (b) best result of relative coding ($J = 4.2$).

the second simulation encode the relative change of position from step to step. The genes were still continuously valued, but the permissible range of their values was decreased appropriately. The spatial and temporal scaling as described above remained unchanged. An exemplary result is displayed in Figure 4(b). This trajectory applied a "one-leave-out" strategy which can be inferred from the high negative velocity of the trajectory in Figure 4(b) (which is partly clipped). Instead of hitting the ball in every cycle, it hit the ball every other cycle. Although this might be considered cheating, it is a valid solution to the given problem and is particularly rewarded by the last term in the denominator of the cost function. Therefore, the seemingly high fitness of $J = 4.6$ does not reflect the real quality of this trajectory and has to be corrected to $J = 3.7$. On the whole, this representation achieved a significantly faster speed of learning (200 to 300 generations) as well as a higher maximum fitness.

3.2 REPRESENTATION WITH VISUAL PERCEPTION

The third experiment addressed the questions: to what extent does perception improve the speed of learning and the quality of the outcome and, in particular, with perception can the passively stable control strategy of the other experiments still be found?

To address this question, the representation of the problem had to be changed again. It was assumed that in visual perception the absolute position (x_B) of the ball, its velocity ($\dot{x}_B$), and the ball's relative distance to the paddle (x_R) can be perceived. These terms are multiplied by appropriate coefficients and summed up to specify the next desired paddle position and the next desired paddle velocity:

$$\begin{aligned} \dot{x}_{P_{k+1}} &= (c_{11}\dot{x}_{B_k} + c_{12}x_{B_k} + c_{13}x_{R_k}) \cdot s, \\ x_{P_{k+1}} &= (c_{21}\dot{x}_{B_k} + c_{22}x_{B_k} + c_{23}x_{R_k}) \cdot s. \end{aligned} \tag{7}$$

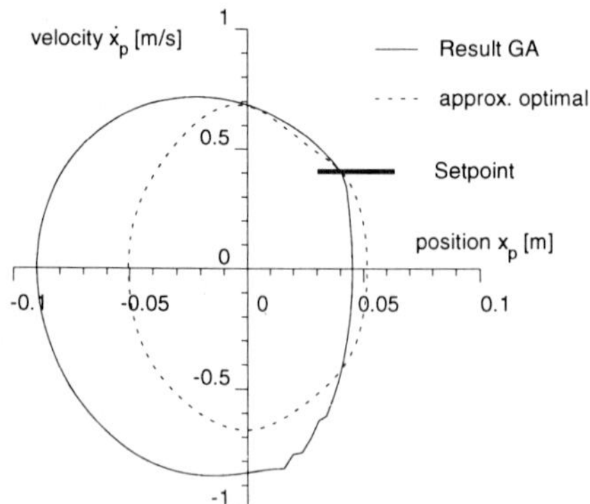

FIGURE 5 Best result of feedback GA ($J = 4.6$) (for the optimal trajectory $J > 4.8$).

The task of the GA was to optimize the "ball-paddle coupling factors" (c_{ij}) and the scaling factor (s). This type of coding only requires seven continuously valued genes. In contrast to the previous representations, the ordering of the genes in the genomes was no longer essential.

Figure 5 depicts the best result out of five trials; the other trials were qualitatively the same. In the crucial impact region (1st quadrant) the GA solution comes very close to what was calculated as the optimal solution, and its entire fitness evaluation $J = 4.6$ differs only marginally from the score of the optimal ($J > 4.8$). The optimal trajectory would score better if its initial phase was perfectly adjusted to minimize the transient time to reach the setpoint. Learning speed was significantly improved by using this representation. Within 10 to 30 generations very good trajectories were accomplished.

4. DISCUSSION

The goal of this paper was to study learning of passive motor control strategies by using genetic algorithms. Passive control strategies do not need continuous replanning of the movement to compensate for perturbations during the movement task, but accomplish the task by relying on a self-organized stabilization due to special properties of the control method or other parameters.[9] Three different representations of the task "to juggle a ball on a paddle" were compared with respect to their learning speed and their quality of solutions, in particular, whether a passive control strategy could be found.

4.1 SIMULATION RESULTS

The results of all three setups were positive in that they converged onto a passive control regime. However, learning speed and the quality of the outcome differed substantially between the three conditions. Figure 6 illustrates this by showing the mean population fitness of representative runs as a function of the number of generations. The relatively poor results of the absolute position coding are not surprising. This kind of coding had a large permissible range of values for each gene and was thus able to jump from one extreme position to another in every single time step. The initial trajectory coded by the randomized genomes looks like a zigzag line. Since GAs do not make use of local information as given by gradients, it takes many iterations until the zigzag line is smoothed to a trajectory fit to perform juggling. The likelihood that the algorithm converges to a local minimum is large. On the other hand, relative position encoding does not tend to have the same jaggedness as absolute coding, because the jump from time step to time step is confined to a rather small range. The crossover operator in GAs also will do less harm to a relative

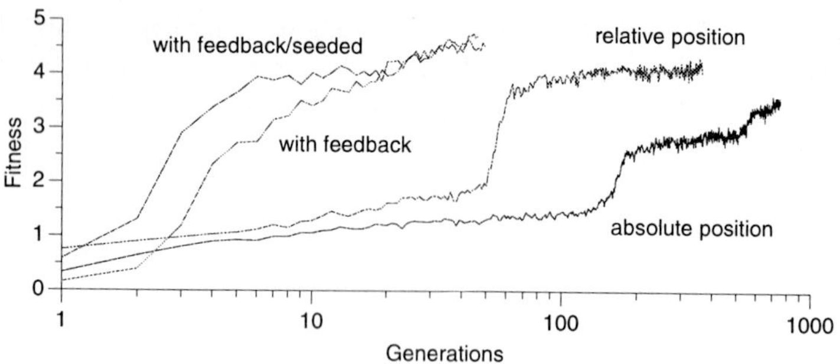

FIGURE 6 Comparison of learning speed of the three different representations.

position coding. Swapping an absolute position gene with another genome will always be detrimental if this gene codes anything but the position actually needed. In contrast, exchanging a relative position gene will not destroy too much as long as the gene retains an appropriate "up" or "down" information. The relative position genes also facilitate the formation of building blocks. The essence of successful juggling is an "up-up-up" sequence in the impact area of the ball. Via crossover, this generally valid building block can easily be tested in several places on the paddle trajectory. On the other hand, a building block containing absolute position information is unlikely to be useful in other genomes: it is difficult to smoothly integrate it into the already existing genes. In sum, if a genome represents a smooth function, relative coding seems to be advantageous.

The quality of the results and of the learning speed was unexpectedly good for the representation with perception. Whereas learning speed must necessarily improve due to the comparatively short genomes, the almost optimal outcome was by no means self-evident. A particularly interesting property of the resultant paddle trajectory is that it still possesses the major characteristics of a passive juggling strategy; i.e., the impact takes place while the paddle position still increases and the velocity decreases. This passive stability property even allows sustainment of stable juggling when perception is cut off (this requires a change to an open-loop-control algorithm). Apparently, movement learning can profit from active control. If the learned control scheme also allowed passive control, the system could gradually switch from active to passive control at an advanced level; mechanisms for that shall not be considered here. This seems to be plausible from monitoring the attention humans devote to a task in the learning and skilled stages.

4.2 COMPARISON WITH HUMAN JUGGLING

To investigate the biological plausibility of the simulation results, the trajectory of the feedback GA was compared with data collected from an experiment in which a human subject juggled a tennis ball on a tennis racket. The phase plot in Figure 7(a) shows an average phase plot over a 30-second run. Since the coefficient of restitution was different from that used in the GA simulations, the phase plots in Figure 7(a) can only serve for qualitative comparison.

As the most noticeable difference in human juggling, the balls impacts with the racket shortly after the positive paddle velocity peak while the optimal result and the GA feedback result have the setpoint farther along the declining part of the trajectory. Thus, the human juggling data stays very close to the limit of passive stability, and it cannot be resolved whether this juggling strategy is more on the passive or active control side. So far, not enough data has been collected from human subjects to allow any generalization. One reason why the simulations did not produce a juggling strategy more similar to humans' strategies may be that the dynamic and kinematic properties of the human arm where not taken into account. Another reason may be that the chosen performance criterion was inadequate.

An intriguing advantage of human motor learning over machine learning is that humans do not start out absolutely "uninitiated." First of all, past experience seems to play a non-negligible role. Secondly, humans often seem to extract an idea of how to approach the task by watching somebody else's performance, reading a "how-to" book, etc. These sources can partially specify the initial strategy and avoid long and fruitless experimentation. An important question, therefore, is what the learner extracts from a demonstration. The successful performance of the feedback GA may suggest that picking up something like coupling coefficients could be advantageous. Such a notion of coupling between perception and action also could be a promising route to transfer knowledge between different tasks. To test this hypothesis,

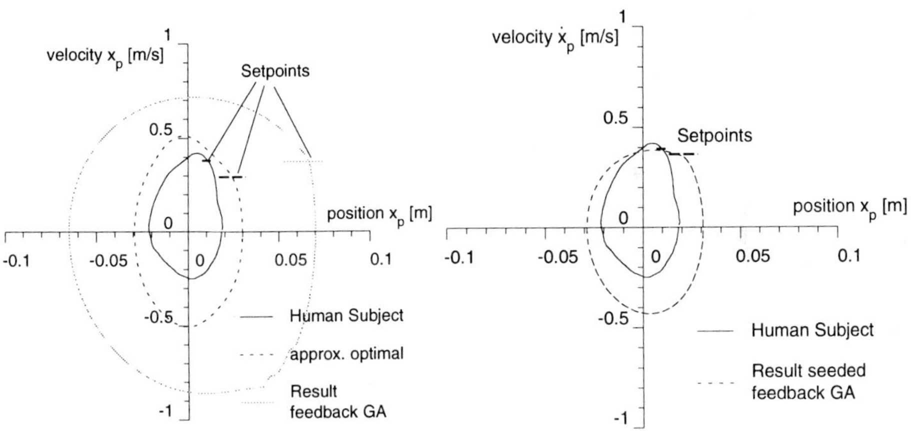

FIGURE 7 Comparison of results with data from human subject.

the coupling coefficients of Eq. (7) were regressed from the human data, and the feedback GA was seeded with random variations of these coefficients. The result is shown in Figure 6 and Figure 7(b). As can be seen, the GA solutions converged faster, although, as discussed before, human juggling produced different results to the simulations and although the properties of tennis ball and tennis racket were different to the ones in the simulation. Now, the juggling setpoint lies closer to the peak velocity which is more similar to the human data. However, note that the unperturbed regression data could not be used for the juggling simulation right away. It resulted in a paddle movement that was too large and continuously increasing.

In summary, this study showed once again the importance of how a motor task is represented in terms of its structure and variables. Different representations change the learning speed and the quality of results dramatically. So far we have little knowledge of how appropriate representations can be developed to obtain more insight into what should be learned.

ACKNOWLEDGMENTS

The work of this paper was supported by grants of the German National Scholarship Foundation to each of the authors; a fellowship of the Alexander von Humboldt Foundation to the first author; an NSF Grant BNS-9109880 awarded to M. T. Turvey, University of Connecticut; an Air Force Office of Scientific Research grant AFOSR-89-05000; and a National Science Foundation Presidential Young Investigator Award awarded to Christopher G. Atkeson, MIT.

REFERENCES

1. Aboaf, E. W., S. M. Drucker, and C. G. Atkeson. "Task-Level Robot Learning: Juggling a Tennis Ball More Accurately." In *Proceedings of IEEE International Conference on Robotics and Automation*, held May 14–19, in Scottsdale, Arizona, 1989, pp. 123–128. Los Alamitos, CA: IEEE Computer Society Press, 1989.
2. Bühler, M. "Robotic Tasks with Intermittent Dynamics." Ph.D. Dissertation, Department of Electrical Engineering, Yale University, 1990.
3. Dyer, P., and S. R. McReynolds. *The Computation and Theory of Optimal Control.* New York: Academic Press, 1970.
4. Flash, T., and N. Hogan. "The Coordination of Arm Movements: An Experimentally Confirmed Mathematical Model." *J. Neurosci.* **5(7)** (1985): 1688–1703.
5. Friedland, B. *Control System Design: An Introduction to State-Space Methods.* New York: McGraw-Hill, 1986.
6. Getting, P. A. "Comparative Analysis of Invertebrate Central Pattern Generators." In *Neural Control of Rhythmic Movements in Vertebrates*, edited by H. Cohon, S. Rossignol, and S. T. Grillner, 101–121. New York: Wiley, 1988.
7. Guggenheimer, J., and P. Holmes. *Nonlinear Oscillations, Dynamical Systems, and Bifurcations of Vector Fields.* New York: Springer, 1983.
8. Rizzi, A. A., and D. E. Koditschek. "Progress in Spatial Robot Juggling." In *Proceedings of IEEE International Conference on Robotics and Automation*, held in Nice, France, 1992, pp. 775–780. Los Alamitos, CA: IEEE Computer Society Press, 1992.
9. Schaal, S., C. G. Atkeson, and S. Botros. "What Should Be Learned?" In *Proceedings of Seventh Yale Workshop on Adaptive and Learning Systems*, 199–204. New Haven, CT: Yale University, 1992.

U. R. Smith
Department of Biology, Yale University, New Haven, CT 06511
e-mail: una@peaplant.biology.yale.edu

How to Get "A Biologist's Guide to Internet Resources"

To answer some of the most frequently asked questions heard among biologists who use the Internet, I have written an Internet "FAQ" document titled "A Biologist's Guide to Internet Resources" that is updated and distributed over the Internet on a monthly basis. The guide contains an overview and lists of free Internet resources that are of specific interest to biologists, such as

- scientific discussion groups and mailing lists;
- research newsletters, directories, and bibliographic databases;
- huge data and software archives;
- tools for finding and retrieving information; and.
- a bibliography of useful books and Internet documents.

The current version of this free 30-page guide can be obtained via the Internet.

- In Usenet, look in sci.bio or sci.answers.

- Gopher to sunsite.unc.edu, and choose this sequence of menu items:
  Sunsite Archives
  ecology and evolution
  Or, from any gopher offering other biology gophers by topic, look for the menu item "Ecology and Evolution at UNC and Yale." There you will find "A Biologist's Guide to Internet Resources" stored two ways: as a single file for easy retrieval and nicely broken up in a meanu for browsing.
- FTP to rtfm.mit.edu. Give the username "anonymous" and your e-mail address as the password. Use the "cd" command to go to the

  pub/usenet/news.answers/biology/

  directory and use "get guide" to copy the file to your computer. The file is actually stored as "guide.Z," which is a compressed binary file, but, if you specify "guide," it will be uncompressed and translated to readable ASCII before it is transferred to your computer. You can also use anonymous FTP to sunsite.unc.edu, where this guide is stored as

  pub/academic/biology/ecology+evolution/FAQ.

- Send e-mail to mail-server@rtfm.mit.edu with the message "send usenet/news.answers/biology/guide." Because the guide is long, you will probably receive it in several parts: save each part separately, delete the e-mail headers, and merge them.

Dagmar Sternad† and Stefan Schaal‡
†Center for the Ecological Study of Perception and Action, University of Connecticut, 406 Babbidge Road, Storrs, CT 06268 (e-mail: sternad@uconnvm.bitnet); and ‡Department of Brain and Cognitive Sciences and the Artificial Intelligence Laboratory, MIT, 545 Technology Square, Cambridge, MA 02139 (e-mail: sschaal@ai.mit.edu)

A Genetic Algorithm for Evolution from an Ecological Perspective

In the population model presented, we explore an evolutionary dynamic based on the operator characteristics of genetic algorithms. Essential modification of genetic algorithms are dynamic boundary conditions and the inclusion of a constraint in the mixing of the gene pool. The pairing for crossover is governed by a selection principle based on a complementarity criterion derived from the theoretical tenet of perception-action mutuality of ecological psychology. According to Swenson and Turvey this mutuality principle is a consequence when evolution is viewed from a thermodynamical perspective. The second law of thermodynamics becomes a physical selection principle by which increasing complexity produces an increase in the rate of dissipation. The present simulation tested the contribution of selective recombination on the rate of energy dissipation as well as three operationalized aspects of complexity. The results support the predicted increase in the rate of energy dissipation, paralleled by an increase in the average complexity of the population. The spatio-temporal evolution of this system, i.e., its frequency distribution of changes in population size, displays the characteristic power-law relations of a nonlinear system poised in a critical state.

1992 Lectures in Complex Systems, Eds. L. Nadel & D. Stein, SFI Studies in the Sciences of Complexity, Lect. Vol. V, Addison-Wesley, 1993

1. INTRODUCTION

In Darwin's account of evolution the central principle for a species' successful adaptation and development is natural selection, a purely *a posteriori* fitness evaluation for randomly created individuals. Our search for an *a priori* account to the prototypical question of how a complex dynamical system evolves toward functional efficiency has led to the theory of complexity which has provided a new perspective, largely based on results from molecular biology. Recently Swenson and Turvey[9] adopted a somewhat unorthodox stance, which could be seen as a reaction to neo-Darwinistic approaches in proposing thermodynamic principles for selection.[9,10] Their argument combines two assumptions. Firstly, in accordance with irreversible thermodynamics, the basic unit is conceived as an open system, embedded in a global system that obeys the second law of thermodynamics. Secondly, in allegiance to the central tenet of ecological psychology, animal and environment are believed to form a cyclically related system, in which perceiving and acting mutually condition each other.[2,3,7,10] The relation between perceiving and acting is lawful, and behavior is goal-directed. In contrast to neo-Darwinist theories that focus on the genetic code as the analytic level of choice, this conceptual framework chooses interactive behavior as the focus of analysis.

Swenson and Turvey's argument comprises four major points: (1) Thermodynamic principles, as foremost expressed by the second law, are the fundamental laws that govern the evolution of matter in the universe. (2) Highly ordered states, which at first sight seem to defy the development toward final maximum disorder, are factually in concordance with classical thermodynamics, because complex states increase the *rate* of energy dissipation and, hence, entropy production. (3) Matter increases the rate of energy dissipation, by the assembly of living matter into higher-order states via an active, goal-directed behavior. (4) The key unit in this self-organizing process is the dual relation between organism and environment via Gibson's notion of information and the mutual conditioning of perception and action. Information from the environment is determined by the organism's action which, in turn, creates the information.

The following population model applies mainly the operators from genetic algorithms, but introduces the notions of open systems and intentional behavior. We start with a brief outline of some assumptions of ecological psychology, followed by the specific goals and the required modifications of genetic algorithms. The numerical experiment presented here tests the effects of a goal-directed behavior in the evolution of a population.

2. SOME TENETS FROM ECOLOGICAL PSYCHOLOGY

In line with irreversible thermodynamics,[11,12] the ecological perspective on adaptive behavior, in particular its focus on the perceptual control of movement coordination, emphasizes that biological systems are open and constantly absorb energy from, and dissipate energy to, the environment.[5] The central tenet of ecological psychology is that an animal together with its environment constitutes such a dynamical system and is defined over their mutual relation, but is not reducible to its two components.[9] The principle of the *mutuality of perceiving and acting* emphasizes that an animal's perception of the environment provides control constraints for the animal's actions; in turn, the animal's actions provide constraints on the perceptual information from the environment. This is referred to as the *perception-action cycle.* In this theoretical framework, the concept of information is reformulated. In contrast to Shannon's concept of information as a quantity, neutral with respect to its subject matter, ecological information is specific as to its subject and is symmetrically defined over perceiver and environment. Observables are defined for the dual pair of perceiver and environment. The detection of information in the environment controls the behavior, which in turn enables the detection of relevant information. Two points will be picked up in the model: (1) the inexorability of goal-directed behavior which is lawfully specified by information and (2) this information as mutually defined over animal and environment. As a result, organisms evolve toward states of higher complexity, the rate of energy dissipation increases and entropy production in the whole system is increased.

3. GOALS OF THE MODEL AND DEFINITION OF OPERATIONS AND THEIR DYNAMICS

In the following simulation, some modifications of the original idea of genetic algorithms are introduced, which, in fact, transform the original optimization strategy into an artificial population that displays complex behavior. Foremost amongst these changes is the construction of individual units as open systems, which are characterized by an inflow and outflow of energy. In total, the modifications can be viewed as an emphasis of the deterministic aspect of the model. The major modifications are:

1. The modeling of bit strings as open systems with energy flow, where inflow and outflow are determined independently.
2. The notion of random selection of bit strings—where, in each generation, every bit string can cross with every other string—is replaced by a goal-directed selection of the crossing partner.

3. The predetermined and static fitness function is replaced by dynamic boundary conditions for the population's development.
4. The interactions between these open systems and their dynamic environment are governed by nonlinear functions.

In a numerical experiment we want to show that these modifications lead to a behavior in which:

1. Evolution towards a higher degree of complexity is achieved.
2. The rate of energy dissipation on the *global* scale increases with the rising complexity in the *individual* strings.
3. This state of higher complexity displays the properties of a *critical state* with a power-law distribution in the fluctuations of the population size.
4. The fractal characteristics, as predicted by the hypothesis of *self-organized criticality*, are more pronounced when perception-action is included.

To aid the intuitive understanding of the algorithmic operators and parameters, illustrative language is frequently chosen to describe the structural ingredients of the model.

POPULATION. The individuals of the population are strings of binary units. These units can be either M (for meat units) or V (for vegetable units). The initial population consists of a random selection of such strings of equal initial length. Each string is attributed a fitness value that is uniformly set for all individuals in the beginning and that will become a function of energy flow and its complexity.

ENERGY INFLOW. Energy inflow, or "food uptake" by the individuals, is determined by their present configuration. The minimal requirement for a string to absorb energy is a sequence of adjacent M's or V's. Beyond a minimal length of this "eating block," food of the same type as the block can be eaten. The amount of energy inflow is a weak power function of the actual length of this eating block. Simultaneously, the same quantity of food is subtracted from the respective energy resource. M's are subtracted from the "meat basket" and the stock in the "vegetable basket" is reduced when V's are eaten. To account for the typical dynamic of supply and demand, as known from market economy, an additional nonlinearity governs the increase of the string's fitness value: the greater the supply, the easier the increase in fitness; the lesser the supply, the lesser the addition of energy to the fitness balance. Subsequently, the fitness value of each bit string is updated. The length of the strings is not affected by this energy uptake. In the case when one bit string has more than one eating block, of the same or a different type, its fitness can be increased through all of them. Food depletion, on the other hand, is a linear function of the available supply.

ENERGY SUPPLY. An initial energy supply is provided by the two "baskets" of V's (vegetables) and M's (meat), which are refilled at each iteration. To obtain the nonconstant replenishment, observed in real market situations, the replenishment is scaled by a Gaussian distribution function: If the food supplies are high or low, filling up is relatively low, while in the midrange the replenishment is optimal. This nonlinearity amplifies the situation when food is scarce, yet prevents unbounded growth when food is available in abundance.

ENERGY DISSIPATION. The energy inflow is counterbalanced by an energy outflow. This dissipation of energy is determined by the "effort" required for the string to find the best mating partner. Thus, energy dissipation is closely linked to reproduction and the goal of improving its offspring's fitness. This is the point where the principle of the perceiving-acting cycle comes into play.

ENERGY FLOW ON A MICRO- AND MACRO-SCALE. For the individual, food intake and energy dissipation determine a flow through the system, which is monitored by the fitness index. Each flow is governed by different nonlinearities, but there is a balance between energy uptake and consumption. Likewise, on the macro-level of the ecosystem, which comprises the total population and the food resources, an energy flow is set up through the replenishment of food and the summative dissipation of energy of all the individuals.

MATING AND PERCEPTION-ACTION CYCLE. As the fitness index is a function of the ability to eat, which itself is determined by the length of the eating block, a viable search principle must lead to an increase in the length of this homogeneous block. To implement a degree of directedness in the selection of a mating partner, a complementary measure is introduced that attempts to capture information as a dually defined concept. The degree of complementarity is quantified by the Hamming distance, defined as the sum of the differences between the genes at corresponding locations in the sequence (identical genes yield zero; nonidentical genes count as 1). The larger the H, the higher the probability that the (double) crossover will replace the units of less favored type. In other words, given its own particular sequence, the most complementary string is selected for crossover. However, this search is not conducted over the whole population, but only over a subset. The size of this subset is calculated as a linear function of the string's length and its fitness. This takes into account the fact that the longer and fitter the bit string, the larger the subset must be to provide an adequate choice. It also pays tribute to the fact that, as the degree of heterogeneity increases, more choice is necessary to find a matching partner that provides potential improvement. This advantage of a large subset is counterbalanced by the disadvantage that, as the subset becomes larger, the search lengthens and more energy is dissipated. Additionally, the dissipation of energy is proportional to the string's length. As a result, the subset together with dissipation is a nonlinear function of string length. Consequently, energy outflow is the price the organism has to pay for a higher probability of increased energy inflow in the next generation.

MATING AND CROSSOVER. Unlike the Mendelian view, the pairing for the crossover is not random, but rather is directed by the individual's intention to increase the chance of survival. Since fitness is a function of the ability to eat, i.e., the length of the eating block, the search must lead to an increase in the length of this homogeneous block. In a first step, a partner for crossover is selected according to maximal Hamming distance. Then double crossover, instead of single crossover, is used to provide an operation that optimizes the possible gain from the selected partner. To strike a balance between chance and self-directed improvement, the two points of crossover remain random. Each string can partake in crossover only once per iteration.

MUTATION. Stochasticity is incorporated at two instances. Firstly, the biological principle of mutation is instantiated as the flipping of a single bit. In the situation when the population of individuals has settled on an equilibrium with a predominantly meat-eating or vegetable-eating population, random mutation becomes the source for change when the respective food resources are depleted. It also acts as a disturbing factor to goal-directed development. Secondly, mutation can change the length of a genome. Longer genomes have a higher chance to assemble a sizable eating block, but also dissipate more energy in the pairing process. Hence, the advantages of eating ability and dissipation are counterbalanced.

REPRODUCTION. Reproduction is not linked to crossover, but is independently regulated by the fitness index. When the fitness index reaches a critical threshold, duplication of the genome occurs. The two identical offspring then start with half of the parent's fitness value.

EXTINCTION. Reproduction is counterbalanced by extinction. When the fitness value decreases to zero, the string dies. As reproduction and extinction are defined individually, this leads to overlapping generations.

4. A NUMERICAL EXPERIMENT

The pivotal point of the present endeavor is an evaluation of the role of a purposive search principle in the restructuring of components, in particular its effect on the system's entropy production and degree of complexity obtained by the system's individuals. According to Swenson and Turvey's hypothesis, the perception-action principle is a necessary constituent to obtain an increase in the rate of entropy production and an increase in complexity. Extending these suppositions we also expect that the system's spatial and temporal variables satisfy the characteristic features of a system at its critical state.[1] In particular we anticipate power-law relations with an exponent between -1 and -2.

4.1 METHOD

The population experiment was run with two conditions. The first condition includes the perception-action principle, referred to as the "Perception-Action Run" (P-A run). In the second condition the selection of mating partners purely follows chance, and will be called the "Random-Search Run" (R-S run). All other parameters and relationships are kept the same. For each condition, 25 runs over 10,000 generations were performed.

4.2 OPERATIONALIZATION OF COMPLEXITY (C) AND ENTROPY PRODUCTION (EP)

COMPLEXITY. To express complexity in a single quantity poses practical and theoretical problems. To set apart our approach, based on combinatorial and probability considerations, from more philosophically grounded definitions, the measure will be called heterogeneity. Possible candidates are absolute length, the proportion of M's and V's in the sequence, and the number of alternations from M to V within one string. The complicating factor is the theoretical requirement that complexity, viz., heterogeneity, should correspond to fitness and constitute a fairly improbable state. For instance, when looking at the number of alternations alone, the maximal number does express maximal heterogeneity, but it is not the most improbable state. The most probable state lies between an uninterrupted sequence of alternations and absolute homogeneity, and the probability follows the binomial distribution. A subsequent comparison with the binomial distribution therefore can give an estimate of its probability. Absolute length also has to be considered, because energy dissipation, as defined, is directly proportional to absolute length. Rather than forcing these aspects into one quantity, we opted to leave these interrelations transparent and use three separate measures to describe heterogeneity:

1. Absolute Length: L = number of bits in the string.
2. Relative number of alternations: $A_{\text{rel}} = A/A_{\max}$ (with $A_{\max} = L - 1$).
3. Relative content of M's or V's : $M_{\text{rel}} = M/L$, or $V_{\text{rel}} = V/L$

RATE OF ENTROPY PRODUCTION. Within the confines of the model it seems viable to operationalize entropy production with the total energy dissipation of the individuals. The rate of change in the entropy production ΔEP is calculated from the average change in energy dissipation per generation ΔE_{diss} over the number of individuals N_{pop}:

$$\Delta EP = \frac{\Delta E_{\text{diss}}}{N_{\text{pop}}}.$$

SELF-ORGANIZED CRITICALITY. To test for the predictions of self-organized criticality, we chose the summative measure of population size N_{pop} and its change to capture the dynamic of the different effects. For the operationalization, N_{pop} was registered every ten generations and the consecutive measures were subtracted. The data points were binned and displayed in a histogram. The exponent of the power function was obtained from the double-logarithmic plot, in which the slope of the linear regression yields the exponent.

PARAMETERS. The experiment was conducted with the following parameters: initial population (100), initial string length (10), initial fitness index (6.0), initial food supply (300/300), food replenishment (50 per iteration, multiplied by a Gaussian as a function of the current food supply), minimal length for an eating block (5), reproduction threshold (15), crossover rate (50%), mutation rate for bit flip (.005), and mutation rate for length change (.00002).

5. RESULTS

Two exemplary time series in Figure 1 and 2 illustrate the behavior of the population over 2,000 iterations.

HETEROGENEITY. Figure 3 shows the time evolution of absolute length L of the strings. In a P-A run the length increases markedly after about 2,000 generations and rises by 3 bits to approximately 13 bits. Further increases, however, happen very slowly. In a R-S run, length L increases after 1,000 generations, but then clearly stays below the value obtained in the P-A run. The second measure of heterogeneity is the average number of alternations in the sequence A_{rel}, as displayed in Figure 4. It starts from the value 0.5, the highest probable number according to the binomial distribution, and decreases throughout the evolution. In contrast to length L, it does not show a significant difference to the runs with random search and, therefore, is not separately shown in the figure. This decrease in both conditions indicates the trend toward more homogeneity.

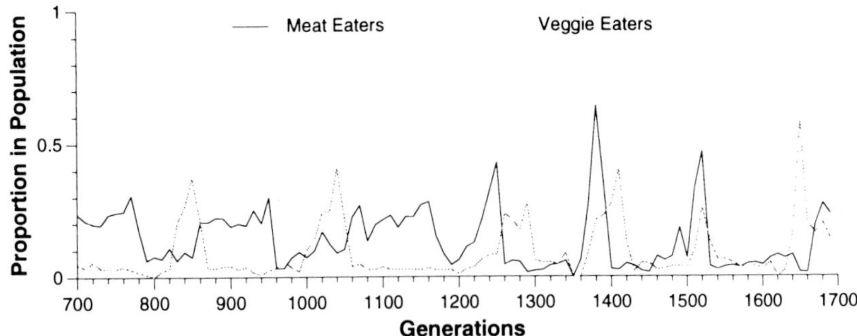

FIGURE 1 Example of temporal evolution of meat and vegetable eaters in the population.

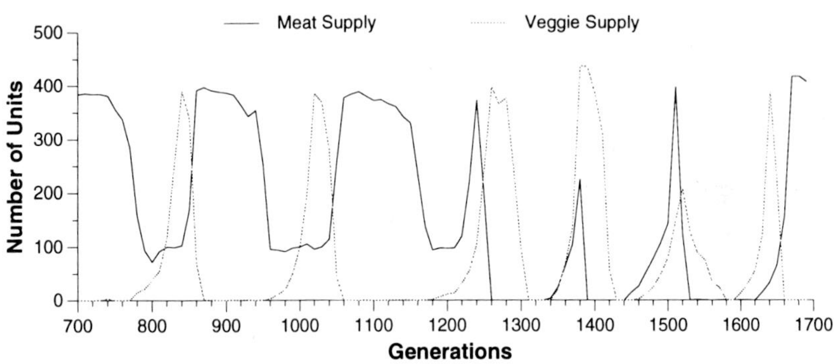

FIGURE 2 Example of temporal dynamics of meat and vegetable supplies.

The relative proportion of meat and vegetable units, $M_{\rm rel}$ and $V_{\rm rel}$—the third measure for heterogeneity—fluctuates around the value 0.5 in both conditions and shows no significant trend.

RATE OF ENTROPY PRODUCTION. In Figure 5 the rate of energy dissipation is plotted for both conditions. The dissipation rate visibly rises with increasing generations and the difference between the two search principles is evident for these exemplary runs. The average slopes (not shown in the graph) of the linear regression over 10,000 generations is 3.46×10^{-6} in the P-A run, whereas in the R-S run it is 1.84×10^{-6}. The confidence intervals only slightly overlap.

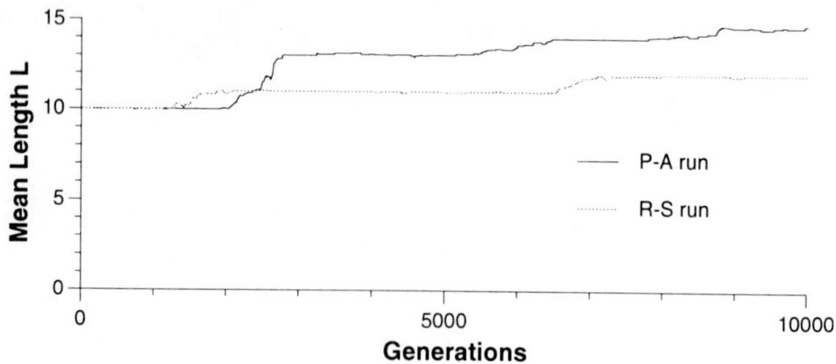

FIGURE 3 String growth over time for P-A and R-S runs.

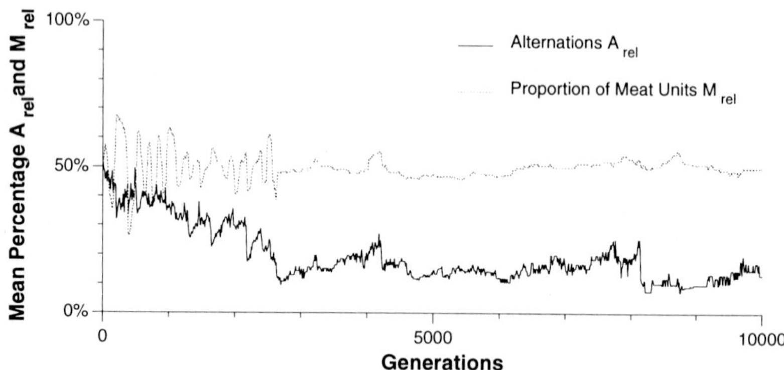

FIGURE 4 Change of alternations A_{rel} and proportion of meat units M_{rel} over time.

FITNESS. The fitness index is graphed over 10,000 generations for the two runs in Figure 6. In both conditions the average fitness value starts from 4.5 points and reaches approximately 5.0 points, which is almost constant. The value shows fluctuations between its 15.0 ceiling (the parameter value where reproduction sets in) and a lower value (after reproduction). It is noticeable that in the P-A run this oscillating dynamic spans about 100 generations, whereas in the control condition it only stretches over about only 50 iterations.

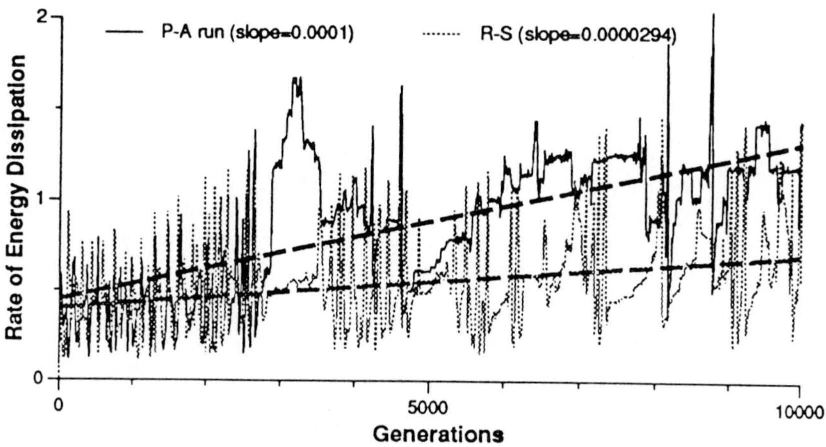

FIGURE 5 Energy dissipation over time for the P-A and the R-S runs.

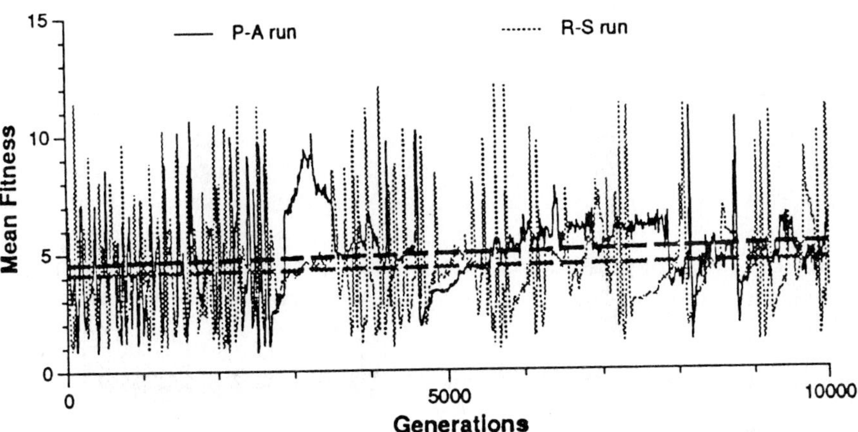

FIGURE 6 Comparison of mean fitness over time between P-A and R-S run.

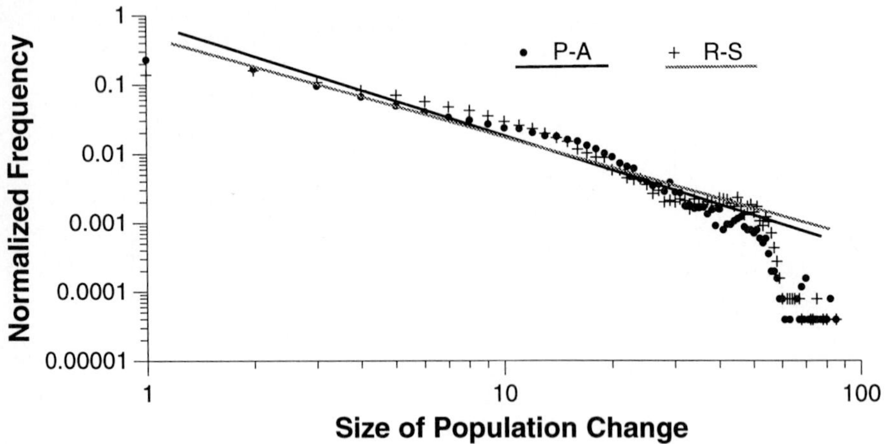

FIGURE 7 Power-law fit for the size of population change over all P-A and R-S runs.

SELF-ORGANIZED CRITICALITY. For the determination of the power-law function, the data of the 25 runs of one condition were pooled to provide an adequate number of data points. The change in population size between at every tenth generation was recorded and grouped into integer bins. Their normalized frequency is plotted in Figure 7. The logarithmic plot is approximated by a linear regression. The slope of −1.76 in the P-A run conforms to the hypothesized slope between 1 and 2 at the critical state ($r^2 = .93$, $p < .0001$). For the R-S trials the plot is slightly more convex, but the linear approximation is still significant ($r^2 = .93$). The range in which the linear approximation is satisfactory is also larger for the P-A runs than for R-S runs.

6. DISCUSSION

To date, genetic algorithms have followed the Mendelian view on genetic transmission and rested on random crossover; i.e., individuals contact one another by mass action alone. The present model was inspired by an advance made by Swenson and Turvey and includes the element of information-constrained behavior, thereby cutting short pure probabilistic considerations. Such informational constraints were implemented for the "choice" of crossover by using a complementarity criterion defined over two individuals. The simulation experiment compared the evolutionary dynamic of a population where an active search principle was implemented against conventional random search. The results yield a clear evolutionary advantage for the P-A runs. The major phenomenon is the increase in the rate of entropy production as hypothesized by Swenson and Turvey. Further support and some explanation for this effect comes from the simultaneous increase in heterogeneity of the individuals. Looking at the three operationalized measures, the results showed that, while the relative proportion of alternations and the relative proportion of M or V units show no difference, string length grows significantly more with the intentional component than without. These findings together suggests that the P-A condition favors a chunking into eating blocks, which then leads to a higher inflow of energy and, consequently, to an increase in fitness. Fitter and "more complex" individuals have a better chance of adaptation, but this also entails a longer search and more energy dissipation. Here the link between energy dissipation and complexity is established. It is still noteworthy, however, that it is the rate measure which is sensitive to the active search principle.

The constant fitness average in both conditions is no real surprise. According to the reproduction rule, fitness rises to the threshold of 15 points and then lapses back to half of this value. As the average is also influenced by weaker strings, the value fluctuates around 5 points. More interesting is the oscillation pattern: In the P-A condition the cycles stretch over approximately 50 generations, compared to 100 generations in R-S runs. This can be interpreted as a more vehement dynamic in the experimental condition, in the sense that there is a stronger tendency and readiness to change, or adapt. Hence, the dynamic in the fitness value can be interpreted as an indicator for the "fluidity" of the state.[6] A quantitative analysis of the fluctuation pattern is in progress.

The results of the heterogeneity measurement also point to another aspect of this dynamical system: the stratification into a micro- and a macro-level and the competition between the "goals" on the micro- and the macro-scale. On the level of the individual, the P-A target is greater homogeneity in composition, because homogeneity ensures the ability to eat and increase fitness. The long-term disadvantage is that the species cannot change between meat- and vegetable-preference as readily and become prone to extinction when supply of their respective food is low. On the other hand, heterogeneity ensures greater adaptability and exploitation of the available food resources, but keeps the energy inflow at a lower level. This discrepancy between the local, short-term goals, and global, long-term advantages is captured in the size of population, which expresses the balance between extinction and reproduction. In order to evaluate the balance between short-term advantage and long-term adaptability, the data were tested for their $1/f$ properties predicted by the theory of self-organized criticality. When adaptability is the balance between the readiness to change and more conservative properties, or between short-term profits and long-term precautions, the present population reflects this as the trade-off between the increase in eating blocks and overreliance on one food resource. Both aspects are captured in the size of the population. In both conditions the data of the change in population size could be significantly approximated by a power-law relationships. When the complementarity criterion guided the reshuffling of the "genes," the exponent was slightly more negative than in the control condition; however, the regression fit provided no basis for a differentiation between the two conditions, although qualitative inspection shows more curvilinearity in the R-S runs. In the P-A trials, the linear fit stretches over a wider range. This latter result allows our tentative conclusion that the active search favors the organization to a critical state. More variables will be tested to differentiate this conclusion.

ACKNOWLEDGMENTS

The work of this paper was supported by grants of the German National Scholarship Foundation to each of the authors; a fellowship of the Alexander von Humboldt Foundation to the second author; an NSF Grant BNS-9109880 awarded to M. T. Turvey, University of Connecticut; an Air Force Office of Scientific Research grant AFOSR-89-05000; and a National Science Foundation Presidential Young Investigator Award awarded to Christopher G. Atkeson, MIT.

REFERENCES

1. Bak, P., C. Tang, and K. Wiesenfeld. "Self-Organized Criticality." *Phys. Rev.* **38** (1988): 364–374.
2. Gibson, J. J. *The Ecological Approach to Visual Perception.* Boston: Houghton Mifflin, 1979.
3. Gibson, J. J. *From Aristotle to Darwin and Back Again*, translated by J. Lyon. Notre Dame, IN: University of Notre Dame Press, 1984.
4. Iberall, A., and H. Soodak. "A Physics for Complex Systems." In *Self Organizing Systems. The Emergence of Order*, edited by F. E. Yates. New York: Plenum, 1987.
5. Kugler, P., J. A. S. Kelso, and M. T. Turvey. "On the Control and Coordination of Naturally Developing Systems." In *The Development of Movement Control and Coordination*, edited by J. A. S. Kelso and J. E. Clark. New York: Wiley, 1982.
6. Langton, C. "Life at the Edge of Chaos." In *Artificial Life II*, edited by C. G. Langton, C. Taylor, J. D. Farmer, and S. Rasmussen. SFI Studies in the Sciences of Complexity, Proc. Vol. X, 41–92. Redwood City, CA: Addison-Wesley, 1991.
7. Shaw, R. E., and M. T. Turvey. "Coalitions as Models for Ecosystems: A Realistic Perspective on Perceptual Organization." In *Perceptual Organization*, edited by M. Kubovu and J. R. Pomerantz, 343–408. Hillsdale, NJ: Lawrence Erlbaum, 1981.
8. Swenson, R. "Emergent Attractors and the Law of Maximum Entropy Production: Foundations of a Theory of General Evolution." *Sys. Res.* **6** (1989): 187–197.
9. Swenson, R., and M. T. Turvey. "Thermodynamic Reasons for Perception-Action Cycles." *Ecol. Psych.* **3** (1991): 317–348.
10. Turvey, M. T., R. Shaw, E. Reed, and W. Mace. "Ecological Laws of Perceiving and Acting: In Reply to Fodor & Pylyshyn." *Cognition* **9** (1981): 245–251.
11. von Bertalanffy, L. *Problems of Life.* London: Watts, 1952.
12. von Bertalanffy, L. *General Systems Theory.* New York: Braziller, 1968.

Kay-Pong Yip† and Henggui Zhang‡
†Department of Physiology and Biophysics, University of Southern California, School of Medicine, Los Angeles, CA 90033, and ‡Centre for Nonlinear Studies and the Department of Physiology, The University, Leeds LS2 9JT, England

Bifurcation of Kidney Hemodynamics in Hypertension

The tubular hydraulic pressure in rat kidney oscillates at around 35 mHz because of the operation of an intrarenal negative feedback system—tubuloglomerular feedback. In a strain of rats (spontaneously hypertensive rats) that develops hypertension at the age of 10–12 weeks, the periodic oscillations of tubular pressure are replaced with irregular, random-appearing fluctuations. Similar patterns of fluctuations were also found in normal rats made hypertensive by clipping one of the renal arteries. Since two different models of experimental hypertension have similar changes in the tubular dynamics. It was speculated that the development of hypertension resets the operating parameters of tubuloglomerular feedback, and drives the system to operate in the chaotic domain. Correlation dimension and Lyapunov exponent spectrum estimated from these random-appearing time series suggest that the measured time series were derived from deterministic chaos. Surrogate data analysis substantiates this conclusion.

INTRODUCTION

The primary function of mammalian kidney is to maintain the volume and composition of body fluids within narrow bounds through the formation and excretion of urine. The process of urine formation begins with the filtration of blood plasma through the glomerular capillaries in each nephron, the functional unit of kidney. The driving force for the filtration is the arterial blood pressure, which is known to exhibit the $1/f$ properties in their power spectra.[6] The fluctuations in the arterial blood pressure could easily cause the rate of glomerular filtration to vary over a range so large that regulation of excretion becomes impossible. The primary responsibility for limiting the variation in the glomerular capillary pressure rests with a negative feedback system that senses flow rate-dependent changes in the composition of the tubular fluid and adjusts the diameter of the arteriole feeding the nephron. The structures comprising the feedback system are shown in Figure 1. The incoming blood in each nephron is carried by the afferent arteriole to the glomerular capillaries and then leaves by the efferent arteriole. The glomerulus acts as a filter through which low-molecular-weight blood constituents drain into the tubule. The tubule forms a loop and makes contact with the afferent arteriole.

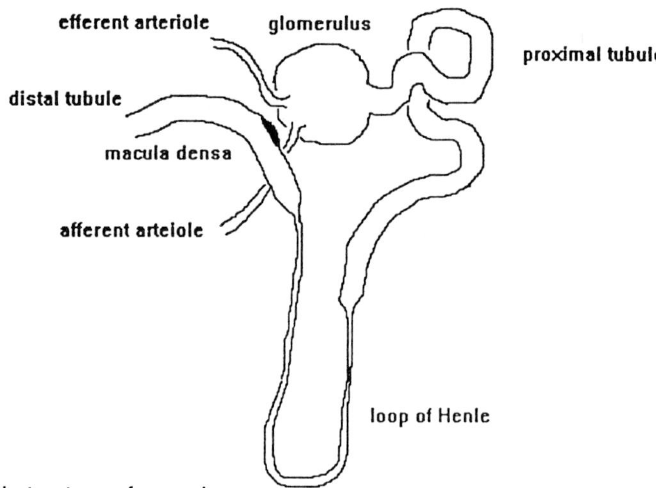

FIGURE 1 Anatomical structure of a nephron.

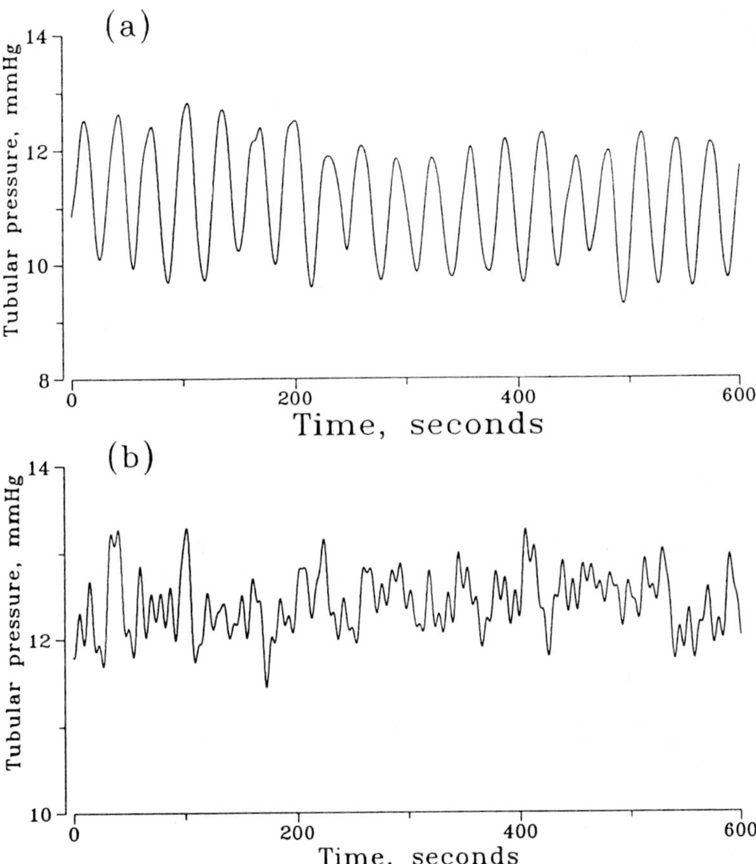

FIGURE 2 Temporal variations of tubular pressure in (a) a normal rat and (b) a spontaneously hypertensive rat.

A specialized collection of tubular cells, the macula densa, is found at the returning point of the tubule to the afferent arteriole. These cells monitor the concentration of NaCl in the tubular fluid, which depends on tubular flow rate, and signal to the afferent arteriole to adjust the diameter. This negative feedback loop is known as tubuloglomerular feedback (TGF).

BIFURCATION OF TGF DYNAMICS IN HYPERTENSION

Because of time delays for the signal propagation along the feedback loop, the hydrostatic pressure, flow, and NaCl concentration in the tubule, as well as the blood flow in the afferent arteriole of the same nephron, are found to oscillate spontaneously at around 35 mHz in anesthetized rats.[3,10] The dependency of this oscillation on TGF has been confirmed with pharmacological intervention and mathematical simulation.[4,5] In a strain of rats (Spontaneously Hypertensive Rat or SHR) that develops hypertension spontaneously at the age of 10–12 weeks, the periodic oscillations of tubular pressure are replaced with irregular, random-appearing fluctuations[9] (Figure 2). The power spectra of tubular pressure time series from SHR are broadband with most of the power localized between 10–200 mHz, instead of unimodal as in normal rats. Similar patterns of tubular pressure fluctuations are also found in rats with renovascular hypertension,[9] in which the high blood pressure is induced by partial obstruction of one renal artery in normal rats for two weeks or longer. These observations suggest that the changes of tubular dynamics from periodic oscillations to irregular random-appearing fluctuations are not specific to SHR, but are common among different models of experimental hypertension.

CHARACTERIZATION OF MEASURED TIME SERIES IN PHASE SPACE

It is well known that in a dissipative physical system, nonlinearities might give rise to deterministic chaotic behaviors. Since TGF is a feedback system with several well-characterized nonlinearities,[4] it is possible that hypertension changes some of the operating parameters in the TGF and thus drives the system to operate in chaotic domain. In this study, two nonlinear measures were employed to quantify the attractor in phase space reconstructed from tubular pressure time series to determine whether they are due to stochastic processes or deterministic chaos. The tubular pressure time series were recorded from the kidney surface of anesthetized rats with a micropipette attached to a servo-nulling hydraulic pressure system. The sampling rate was 12.5 Hz. The recorded tubular pressure time series were filtered by a symmetric digital low-pass filter with a cutoff at 0.1 Hz to remove the oscillations due to the respirator.[9] Ten time series of 15–20 minutes from SHR were used for the analysis.

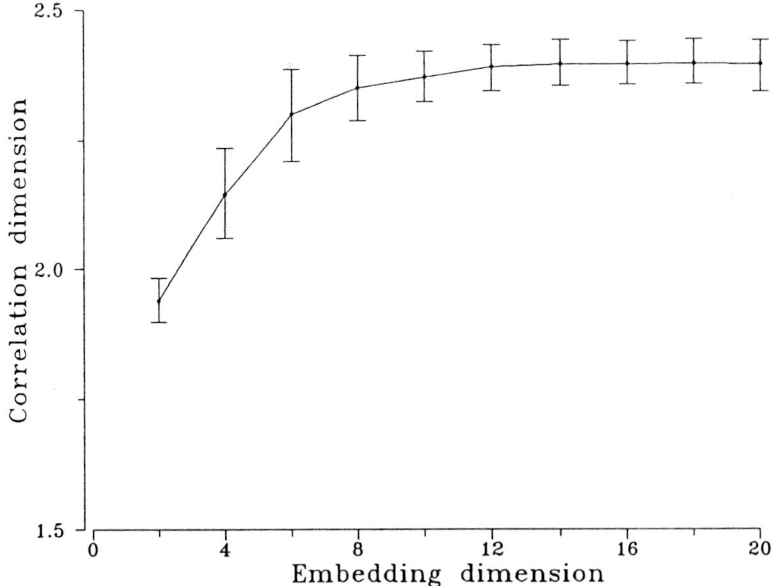

FIGURE 3 Estimated correlation dimension as a function of embedding dimension. Error bar is S.E. ($n = 10$).

CORRELATION DIMENSION OF THE ATTRACTOR

The attractor for the measured time series was reconstructed in phase space using the lagging method introduced by Takens.[7] The algorithm of Grassberger and Procaccia[2] was employed to estimate the correlation dimension from the reconstructed phase-space vectors. The phase-space construction was performed with a time lag of nine time steps, and only retaining those vectors whose first coordinates were five time steps apart. Well-defined scaling regions were found routinely in all measured time series. The estimated correlation dimensions are at the range of 2.3–2.4, which are stable in the embedding dimension from 8–20 (Figure 3). The calculated correlation dimensions are noninteger values, which indicates the presence of fractal scaling in the attractor, and, hence, the attractor is a strange attractor.

Correlated noise, which is induced by filtering on random noise, is also known to have scaling phenomenon when correlation dimension is estimated. Scaling due to nonlinear structure or correlated noise could be discriminated by surrogate data analysis as suggested by Theiler et al.[8] A surrogate data set with the same power spectrum and autocorrelation function as the original experimental data set is generated by taking the Fourier transform of the original data set. The set of calculated phase values is then randomized, and the inverse transformed is taken. No scaling region was detected from the surrogate data set of the original time series in embedding dimension 8–20. However, a scaling region is predicted to be conserved in surrogate data set if the original time series is correlated noise.

LYAPUNOV EXPONENT SPECTRUM

Lyapunov exponent is a standard measure of divergence or convergence in phase space. Chaotic systems have at least one positive Lyapunov exponent in their Lyapunov spectrum. The algorithm derived by Eckmann and Ruelle[1] was used to estimate the Lyapunov exponent spectra for the ten measured time series. One difficulty in applying this algorithm to experimental time series was to build up the local map for a specific point in the trajectory of the attractor. The solution is to take the advantage of singular value decomposition to extract the local orthogonal basis at the specified point for the trajectory of the attractor, and then project the vectors from the specified point to its neighborhoods to this orthogonal basis.[11] An embedding dimension of 3 and a time lag of 9 time steps were used to estimate the Lyapunov exponent spectrum from the measured time series. The results are shown in Table 1. In all ten measured time series, the first exponent is absolutely positive, the second is very close to zero, and the third is negative. The average for the first Lyapunov exponent is $0.204 \pm .015$ ($n = 10$, $p < .001$), which is significantly greater than zero. These are consistent with the notion that the measured time series were derived from deterministic chaos.

CONCLUSION

Periodic oscillations are found in the tubular pressure in normal rats because of the operation of tubuloglomerular feedback. These regular oscillations are replaced by irregular, random-appearing fluctuations in rats with hypertension. Scaling regions were found and well conserved in different embedding dimensions during the estimation of the correlation dimensions from the tubular pressure time series. The possibility that the scaling phenomenon is due to correlated noise induced by filtering was excluded by the surrogate data analysis. Positive Lyapunov exponents were found in all Lyapunov spectra estimated from the measure time series. All

TABLE 1 Lyapunov Exponent Spectra of the Measured Time Series

Record No.	1st Exponent	2nd Exponent	3rd Exponent
1	0.259	0.06574	−1.08
2	0.191	0.02264	−1.11
3	0.262	0.05094	−1.23
4	0.205	0.00344	−1.18
5	0.235	0.00164	−1.13
6	0.236	0.00494	−1.22
7	0.181	−0.00164	−1.22
8	0.164	0.01574	−1.26
9	0.182	0.01394	−1.20
10	0.125	0.00424	−1.21

these analyses suggest that the development of hypertension causes a bifurcation of tubular dynamics in the rat kidney from limit cycle oscillation to chaos. This is an integrated physiological system that shows bifurcation in dynamics associated with pathological conditions, while most of the claims for biological chaos are in isolated systems or the results of pharmacological interventions. The exact parameters that are altered by hypertension in TGF, and the effects of bifurcated tubular dynamics on the whole kidney autoregulation dynamics, are not known and are still under investigation.

ACKNOWLEDGMENTS

Supported by NIH Grants DK 15968 and HL 45623.

REFERENCES

1. Eckmann, J. P., S. O. Kamphorst, D. Ruelle, and S. Ciliberto. "Liapunov Exponent from Time Series." *Phys. Rev. A* **34** (1986): 4971–4979.
2. Grassberger, P., and I. Procaccia. "Characterization of Strange Attractors." *Phys. Rev. Lett.* **50** (1983): 346–349.
3. Holstein-Rathlou, N.-H., and D. J. Marsh. "Oscillations of Tubular Pressure, Flow, and Distal Chloride Concentration in Rats." *Am. J. Physiol.* **256** (1989): F1007–F1014.
4. Holstein-Rathlou, N.-H., and D. J. Marsh. "A Dynamic Model of the Tubuloglomerular Feedback Mechanism." *Am. J. Physiol.* **258** (1990): F1448–F1459.
5. Leyssac, P. P., and N.-H. Holstein-Rathlou. "Effects of Various Transport Inhibitors on Oscillating TGF Pressure Responses in the Rat." *Pflugers Arch.* **407** (1986): 285–291.
6. Marsh, D. J., J. L. Osborn, and A. W. Cowley, Jr. "1/f Fluctuations in Arterial Pressure and the Regulation of Renal Blood Flow in Dogs." *Am. J. Physiol.* **258** (1990): F1394–F1400.
7. Takens, F. "Detecting Strange Attractors in Turbulence." In *Dynamical System and Turbulence (Warwick)*, edited by D. A. Rand, and L.-S. Young. Lecture Notes in Mathematics, Vol. 898, 366–376. Berlin: Springer-Verlag, 1981.
8. Theiler, J., B. Galdrikian, A. Longtin, S. Eubank, and J. D. Farmer. "Using Surrogate Data to Detect Nonlinearity in Time Series." In *Nonlinear Modeling and Forecasting*, edited by M. Casdagli and S. Eubank. Santa Fe Institute Studies in Sciences of Complexity, Proc. Vol. XII, 163–188. Reading, MA: Addison-Wesley, 1992.
9. Yip, K.-P., N.-H. Holstein-Rathlou, and D. J. Marsh. "Chaos in Blood Flow Control in Genetic and Renovascular Hypertensive Rats." *Am. J. Physiol.* **261** (1991): F400–F408.
10. Yip, K. P., G. T. Smedley, and D. J. Marsh. "Blood Flow in a Single Renal Arteriole Measured *in vivo* with Laser-Doppler Velocimetry." *FASEB J.* **6(5)** (1992): A1812.
11. Zhang, H., A. V. Holden, M. Lab, and M. Moutoussis. "Estimating the Persistence of Strain from Time Series Recording from Cardiac Tissue." *Physica D* **58(1-4)** (1992): 489–492.

H. Zhang and A. V. Holden
Center for Nonlinear Studies and Physiology Department, The University, Leeds, L52 9JT, England

Measuring the Complexity of Attractors from Single and Multichannel EEG Signals

We estimate the generalized fractal dimensions and the Lyapunov exponent spectrum of EEG attractors reconstructed from single and multichannel time series recorded from normal and epileptic subjects.

1. INTRODUCTION

In 1985, Rapp and Babloyantz's groups showed that neuronal activity in the monkey[14] and the human EEG during sleep[2] are produced by deterministic chaos rather than stochastic processes. Similar results were found for recordings from cat and rabbit brain.[7] In recent years, experimental and clinical EEG recordings have been studied to quantify brain functions.[5]

In this paper, we quantify the EEG attractors reconstructed from both single and multichannel time series by estimating the generalized dimensions and Lyapunov exponent spectrum. The EEG time series analyzed here are from a normal and an epileptic subject.

1992 Lectures in Complex Systems, Eds. L. Nadel & D. Stein, SFI Studies in the Sciences of Complexity, Lect. Vol. V, Addison-Wesley, 1993

2. RECONSTRUCTION OF ATTRACTOR FROM TIME SERIES

The measured signal from a physical system is a time series, e.g., $x(t)$, $x(t+\tau),\ldots,x(t+(n-1)\tau)$, where τ is the sampling time interval. The reconstructed state vectors in embedding space by the time delay method are[13,15]:

$$\mathbf{x}(t) = [x(t), x(t+\tau), \ldots, x(t+(m-1)\tau]^T\,, \tag{2.1}$$

where m is the embedding dimension. m and τ need to be chosen.[1]

An alternative method is the multichannel method proposed by Eckmann and Ruelle.[6] For m channels the time series becomes $x_i(t), x_i(t+\tau), \ldots, x_i(t+n\tau)$, $i = 1, 2, \ldots, m$, which are recorded concurrently at m different sites. The multichannel method consists in taking each channel as one component of the vector of $\mathbf{x}(t)$, that is

$$\mathbf{x}(t) = [x_1(t), x_2(t), \ldots, x_m(t)]\,. \tag{2.2}$$

The distance between recording sites influences the reconstruction.[4]

3. GENERALIZED DIMENSIONS

For an m-dimensional attractor in a phase space divided into a lattice of hypercubes with size l^m, each of the hypercubes is indexed by $\mathbf{x}_i$, $i = 1, 2, \ldots, N$. The generalized dimensions D_q are defined as[10,9]:

$$D_q = \lim_{l\to 0}\lim_{N\to\infty}\left(\frac{1}{\ln l}\right)\ln C^q(l)\,, \tag{3.1}$$

where C^q is the generalized correlation integral, and

$$C^q = \left[\left(\frac{1}{N}\right)\sum_{i=1}^{N}\left[\left(\frac{1}{N}\right)\sum_{j=1}^{N} H(l - |\mathbf{x}_i - \mathbf{x}_j|)\right]^{(q-1)}\right]^{1/(q-1)}\,. \tag{3.2}$$

$H(x)$ is the Heaviside function. With different order q, the D_q has different physical meanings: D_0 is the Hausdorff dimension, D_1 is the information dimension, and D_2 is the correlation dimension. q can also be negative. For an uniform attractor, the D_q are equal for all q. For a nonuniform attractor, the D_q are ordered with $D_q \leq D_q$ if $q' > q$. The difference between D_q measures the nonuniformity of an attractor.

We cannot calculate the D_1 directly from the above equations. After applying Rolle's theorem, we have

$$D_1 = \lim_{l \to 0} \lim_{N \to \infty} \left(\frac{1}{\ln l}\right) \left(\frac{1}{N}\right) \sum_{i=1}^{N} \ln \left[\left(\frac{1}{N}\right) \sum_{j=1}^{N} H(l - |\mathbf{x}_i - \mathbf{x}_j|) \right] . \tag{3.3}$$

In terms of an experimental time series, $x(1), x(2), \ldots, x(N)$, from the time delay method, a point on the reconstructed attractor is defined by Eqs. (2.1) or (2.2). In numerical computation, we use the box-counting method.[10] By using an efficient algorithm, both the computation time and data space required can be reduced.[11,12]

4. THE LYAPUNOV EXPONENT SPECTRUM

The divergence of two nearby trajectories on the attractor for a chaotic system is quantified by the Lyapunov exponent spectrum.[3,6,16] For a dynamical system with an n-dimensional phase space, an infinitesimal n-sphere on the attractor will evolve into an n-ellipsoid. The average growth rate of the norm of the ith principal axis $a_i(t)$ of this n-ellipsoid gives the ith Lyapunov exponent

$$\lambda = \lim_{t \to \infty} \frac{1}{t} \log_2 \left(\frac{|a_i(t)|}{|a_0(t)|} \right) \text{bits}/s\,. \tag{4.1}$$

In numerical experiments, it is hard to trace the evolution of an infinitesimal n-sphere on the attractor. We avoid this problem by working on the tangent space of the trajectory on the attractor.

In the tangent space, an initial difference vector $\delta\mathbf{x}(0)$ for a given point $\mathbf{x}(0)$ on the trajectory can be mapped into $\delta\mathbf{x}(t)$ at the point $\mathbf{x}(t)$ after a time interval t by the tangent map T^t:

$$\delta\mathbf{x}(t) = T^t(\mathbf{x}(0))\delta\mathbf{x}(0)\,. \tag{4.2}$$

If we divide the time t into k intervals, that is, $t = k\Delta t$, then according to the chain rule,

$$T^t(\mathbf{x}(0)) = T^{k\Delta t}(\mathbf{x}(0)) = T^{\Delta t}(\mathbf{x}(k-1))T^{\Delta t}(\mathbf{x}(k-2)) \ldots T^{\Delta t}(\mathbf{x}(0))\,. \tag{4.3}$$

If $\mathbf{e}_i$ is the ith base vector of the tangent space, then

$$\lambda_i = \lim_{t \to \infty} \left(\frac{1}{t}\right) \log_2 ||T^t(\mathbf{x}(0))\mathbf{e}_i|| \tag{4.4}$$

and λ_i is called the ith Lyapunov exponent of the system.

For a given time series, $x(1), x(2), \ldots, x(n)$, where $x(i) = x(i\tau)$ (τ is the time interval), we start with the reconstruction of an attractor in an n-dimensional embedding space. Mathematically we can think that the evolution of the state on the attractor is produced by the tangent map matrix $T^{\Delta}(\mathbf{x}(i))$ mapping the state $\mathbf{x}(i)$ to $\mathbf{x}(i+1)$. To calculate the Lyapunov exponent spectrum is to find the local tangent map matrix $T^{\Delta t}(\mathbf{x}(i))$. The method to find the local tangent map $T^{\Delta t}(\mathbf{x}(i))$ is discussed in Eckmann et al.[6] and Zhang and Holden.[17]

5. CASE STUDIES

We have estimated the generalized fractal dimensions and the Lyapunov exponent spectrum of EEG attractors reconstructed from single and multichannel time series. The EEG signals were recorded from a normal man and an epileptic woman, both in the resting state.

The multichannel EEG time series are formed by grouping these 19 channel recordings. To study the local and global behavior of the brain, we use multichannel time series from both local area and the whole area. The electrode sites and groups are given in Holden et al.[12]

The correlation dimension corresponds to $q = 2$ in Eq. (3.1). We calculate the correlation dimension for 19 channel epileptic EEG time series using 50 s of data sampled at 200 Hz. The results are compared with the estimated correlation dimension for 19 channels of normal EEG recordings. We also give the first five Lyapunov exponents for the EEG attractors in epileptic case. In all calculations, we assume embedding dimension $m = 10$. The results are shown in Table 1.

In Table 1, we can see that in both cases the correlation dimension takes non-integer values. The Lyapunov exponent spectrum for the epileptic case has two positive definite Lyapunov exponents. The estimated correlation dimension in the epileptic case is significantly lower than in normal case, which means the correlation dimension could be used as a diagnostic index. For single time series the estimated correlation dimension varies (standard deviation = 0.11 for epileptic subject, standard deviation = 0.08 for normal subject) from each other. When we use multichannel time series, the variability of the correlation dimension is smaller (standard deviation = 0.02 for epileptic subject, standard deviation = 0.02 for normal subject). See Table 2.

TABLE 1 The correlation dimension D_2 for 19 channels of EEG in both normal and epileptic state. The first five Lyapunov exponents for $\lambda_1 - \lambda_5$ 19 channels of EEG in epileptic state.

	epileptic case						normal case
channel	D_2	λ_1	λ_2	λ_3	λ_4	λ_5	D_2
1	3.32	18.6	7.5	0.2	-21.4	-62.7	3.99
2	3.21	19.4	10.1	0.5	-26.4	-59.9	3.90
3	3.05	17.8	6.5	0.2	-25.4	-63.4	4.32
4	2.95	14.8	5.6	-0.1	-28.9	-63.1	5.16
5	3.15	17.5	7.6	-0.8	-24.7	-62.1	4.67
6	3.18	17.2	4.8	0.1	-22.4	-60.1	4.74
7	3.61	14.8	5.1	-0.5	-25.1	-55.6	4.32
8	3.26	15.1	4.6	-0.2	-28.4	-60.5	4.53
9	3.25	15.5	5.4	-0.8	-25.7	-57.7	4.63
10	3.17	15.0	5.1	0.3	-31.4	-62.9	4.82
11	3.25	17.8	8.2	0.5	-25.6	-59.7	4.68
12	3.54	19.6	9.2	0.3	-22.4	-62.8	4.51
13	3.73	14.9	5.1	-0.6	-20.9	-58.7	4.18
14	3.71	17.6	5.5	-1.2	-22.7	-57.9	4.58
15	3.62	20.1	8.7	-0.8	-30.5	-68.4	4.27
16	3.78	21.7	8.1	0.6	-24.9	-65.2	4.77
17	3.71	20.3	6.1	-2.1	-27.6	-67.8	3.90
18	3.84	21.1	10.1	0.7	-22.4	-57.6	4.49
19	3.72	14.9	5.3	-0.8	-20.1	-63.4	4.44

Since the brain is a complicated system composed of a number of local functional subsystems, the EEG signals have many coexisting subattractors and the EEG attractor is nonuniform.[8] The nonuniformity of an attractor is measured by the difference between the generalized dimensions D_q. We calculate the D_q with q changing from -2 to 2 for EEG attractors reconstructed from a single time series and multichannel time series in epileptic case. The results are shown in Table 3.

The nonuniformity of EEG attractors is illustrated by the difference in D_q. In the single channel case, the difference between D_q is more significant than the difference of D_q in multichannel case. The attractor reconstructed from multichannel time series is more uniform than the one reconstructed from a single time series.

TABLE 2 The correlation dimension for EEG attractors from multichannel time series in epileptic and normal states.

	epileptic case	normal case
group	D_2	D_2
1	3.42	4.19
2	3.53	4.27
3	3.40	4.14
4	3.34	4.18
5	3.31	4.11
6	3.46	4.30
7	3.36	4.20

TABLE 3 The generalized dimensions for multisite recording.

	multisite	single site
D_{-2}	3.8	4.6
D_{-1}	3.6	4.5
D_0	3.5	3.4
D_1	3.4	3.4
D_2	3 3	3.2

7. CONCLUSIONS

Ideas from nonlinear dynamics have permeated experimental and clinical neurophysiology. Before we apply nonlinear dynamics to study a biological system, we need to note that most biological systems are spatially extended systems with open boundaries through which a biological system exchanges information with its environment. Further, biological systems are nonstationary systems. When we use the methods derived from nonlinear dynamics to quantify the attractor reconstructed from experimental time series, we usually assume that the system is closed and stationary.

Strange attractors can be reconstructed and quantified; however, the significance (biological or clinical) of these measures requires extensive empirical evaluation.

REFERENCES

1. Albano, A. M., A. I. Mees, G. C. de Guzman, and P. E. Rapp. "Data Requirements for Reliable Estimation of Correlation Dimension." In *Chaos in Biological Systems*, edited by H. Degn, A. V. Holden, and L. F. Olsen. New York: Plenum Press, 1987.
2. Babloyantz, A., C. Nicolis, and J. M. Salazar. "Evidence of Chaotic Dynamics of Brain Activity During the Sleep Cycle." *Phys. Lett. A* **111** (1985): 152–156.
3. Bennetin, G., L. Galgani, A. Giorgilli, and J.-M. Strelcyn. "Lyapunov Characteristic Exponents from Smooth Dynamic System and for Hamiltonian System; Method for Calculation of All of Them." *Meccanica* **15** (1980): 9–20.
4. Destexhe, A., J. A. Sepulchre, and A. Babloyantz. "A Comparative Study of the Experimental Quantification of Deterministic Chaos." *Phys. lett. A* **132** (1988): 101–106.
5. Dvorak, I., and A. V. Holden, eds. *Mathematics Approaches to Brain Function Diagnostics.* Manchester, UK: Manchester University Press, 1991.
6. Eckmann, J.-P., and D. Ruelle. "Ergodic Theory of Chaos." *Rev. Mod. Phys.* **57** (1985): 617–656.
7. Freeman, W. J. "Simulation of Chaotic EEG Patterns with a Dynamical Model of the Olfactory System." *Biol. Cybern.* **56** (1987): 139–150.
8. Gallez, D., and A. Babloyantz. "Lyapunov Exponents for Nonuniform Attractors." *Phys. Lett. A* **161** (1991): 247–254.
9. Grassberger, P. "Generalized Dimensions of Strange Attractors." *Phys. Lett. A* **97** (1983): 227.
10. Grassberger, P., and I. Procaccia. "Characterization of Strange Attractors." *Phys. Rev. Lett.* **50** (1983): 346–349.
11. Grassberger, P. "An Optimized Box-Assisted Algorithm for Fractal Dimensions." *Phys. Lett A* **148** (1990): 63.
12. Holden, A. V., J. Hyde, and H. Zhang. "Computing with the Unpredictable; Chaotic Dynamics and Fractal Structures in the Brain." In *Application of Fractals and Chaos*, edited by A. Crilly, R. A. Eanshaw, and H. Jones. Berlin: Springer-Verlag, 1993.
13. Mane, R. "On the Dimension of the Compact Invariant Sets of Certain Nonlinear Maps." In *Dynamical Systems and Turbulence (Warwick)*, edited by D. A. Rand and L.-S. Young. Berlin: Springer-Verlag, 1981.

14. Rapp, P. E., T. R. Bashore, J. M. Martinerie, A. M. Albano, I. D. Zimmerman, and A. I. Mees. "Dynamics of Brain Electrical Activity." *Brain Topography* **2** (1989): 99–118.
15. Takens, F. "Detecting Strange Attractors in Turbulence." In *Dynamical Systems and Turbulence (Warwick)*, edited by D. A. Rand and L.-S. Young. Berlin: Springer-Verlag, 1981.
16. Wolf, A., J. B. Swift, H. L. Swinney, and J. A. Vastano. "Determining Lyapunov Exponent from Time Series." *Physica D* **16** (1985): 285–317.
17. Zhang, H., A. V. Holden, M. Lab, and M. Moutoussis. "Estimating the Persistence of Strain from Time Series Recording from Cardiac Tissue." *Physica D* **58(1-4)** (1992): 489–492.

Index

A

B

C

Y

Z